BIOSIMILARS AND INTERCHANGEABLE BIOLOGICS

Tactical Elements

BIOSIMILARS AND INTERCHANGEABLE BIOLOGICS Tactical Elements

Sarfaraz K. Niazi

Therapeutic Proteins International LLC
Chicago, Illinois, USA

CRC Press
Taylor & Francis Group
Boca Raton London New York

CRC Press is an imprint of the
Taylor & Francis Group, an **informa** business

CRC Press
Taylor & Francis Group
6000 Broken Sound Parkway NW, Suite 300
Boca Raton, FL 33487-2742

First issued in paperback 2023

Version Date: 20151103

ISBN-13: 978-1-4987-4349-5 (hbk)
ISBN-13: 978-1-03-265234-4 (pbk)
ISBN-13: 978-0-429-15538-3 (ebk)

DOI: 10.1201/b19077

Library of Congress Cataloging-in-Publication Data

Niazi, Sarfaraz, 1949- , author.
    Biosimilars and interchangeable biologics. Tactical elements / Sarfaraz K. Niazi.
    p. ; cm.
    Tactical elements
    Includes bibliographical references and index.
    ISBN 978-1-4987-4349-5 (hardcover : alk. paper)
    I. Title. II. Title: Tactical elements.
    [DNLM: 1. Biosimilar Pharmaceuticals--standards. 2. Drug Approval. 3. Drug Discovery--standards. 4. Drug Evaluation--standards. 5. Drug Industry--standards. QV 241]
    RM301
    615.1--dc23
                                                                                            2015036258

To Drs. James Watson and Francis Crick, without whose game-changing double-helix, mankind would not have benefitted from the healing power of recombinant drugs.

[Author with Dr. James Watson]

Contents

List of figures

List of tables

Preface

UTILITY ought to be the principal intention of every publication. Wherever this intention does not plainly appear, neither the books nor their authors have the smallest claim to the approbation of mankind.

Encyclopedia Britannica
1768, First Edition, Preface

About biologics

The human race has survived millions of years of evolution, not because of the pharmaceuticals, but because of the pharmacy we harbor inside our body. An elaborate army of cells inside our body continuously work to protect us against myriad autoimmune, malignancy, and autonomic disorders, among many others. A large number of proteins and antibodies are continuously manufactured by us, providing the armamentarium required to keep us healthy. These molecules come under the category of endogenous biological entities; when these are produced outside the body, they become biological drugs.

Protein therapeutics and its enabling sister discipline, protein engineering, emerged in the early 1980s. The first protein therapeutics were isolated natural proteins from bodily fluids, followed by recombinant versions of natural proteins. Proteins purposefully modified to increase their clinical potential soon followed with enhancements derived from protein or glyco-engineering, Fc fusion or conjugation to polyethylene glycol. Antibody-based drugs subsequently arose as the largest and fastest growing class of protein therapeutics. The rationale for developing better protein therapeutics with enhanced efficacy, greater safety, reduced immunogenicity, or improved delivery comes from the convergence of clinical, scientific, technological, and commercial drivers that have identified unmet needs and provided strategies to address them. Future protein drugs seem likely to be more extensively engineered to improve their performance, for example, antibodies and Fc fusion proteins with enhanced effector functions or extended half-life. Two old concepts for improving antibodies, namely, antibody–drug conjugates and bispecific antibodies, have advanced to the final stages of clinical success. As for newer protein therapeutic platform technologies, several engineered protein scaffolds are in early clinical development and offer differences and some potential advantages over antibodies. These new entities allow us to challenge diseases and conditions like malignant diseases and illnesses related to environment and lifestyle. Additional modalities of diagnosis may also arise to provide better preventive care as well, including individualized medicines based on an individual's genetic coding.

The identification of new protein therapeutic candidates is the fastest growing research subject, and we can anticipate hundreds of new treatment modalities becoming available. It is forecasted that more than two-thirds of all new drugs in the future will be of a biologic origin. However, given the complexity of these drugs, the cost of treatment utilizing these new entities will undoubtedly remain high, ranging from a few hundred to thousands of dollars. There is no healthcare system in the world that can afford to pay for these escalating costs of treatment. One course of action is to reduce the cost of those drugs that have run their course of intellectual property protection by advancing the field of biosimilar products to fill this critical need for more affordable health care. While we may not be able to control the cost of newer products, we can do a lot to reduce the cost of what is already in use.

About biosimilars

Biosimilars are biological drugs produced by recombinant DNA technology that allows their large-scale production and are approved by a regulatory pathway that reduces the length of time and cost of developing these products.

Biosimilars and, soon to be introduced in the Unites States, interchangeable biosimilars constitute the most impactful categories of products in the history of mankind. "Biosimilars will provide access to important therapies for patients who need them," said Dr. Margaret A. Hamburg, former commissioner of the U.S. Food and Drug Administration (FDA), in a statement issued on March 6, 2015, when the FDA approved the first biosimilar product. Biologic drugs were first developed in the 1980s and were considered so specialized that making generic versions was seen as most likely impossible. As a result, they were left out of the Hatch-Waxman Act that created the new category of generic drugs. Over the past four decades, science has advanced, patents have begun to expire, and it has become possible to develop close copies of these products that will have the same safety, potency, and purity as the originator product. However, the originators still argue that their drugs are so complex that it is not possible to make a copy, but that position has become untenable. On average, biosimilars are about a third cheaper than branded biologic drugs in Europe, according to Express Scripts, the United States' largest manager of prescription drug benefits. In some developing countries, they are as much as 90% cheaper than the originator product, and these discounts may eventually prevail globally. Express Scripts estimates that over $250 billion in drug costs may be saved in the United States alone over the next decade if the 11 biosimilars currently in development are approved. Because biosimilars cannot be prescribed or dispensed like generics, where the pharmacist makes the choice, the makers of biosimilars might still have to call upon doctors to prescribe their drug and persuade insurers to require use of the less expensive biosimilar. However, approval of biosimilars as interchangeable is imminent, which is only allowed in the United States, and biosimilars stand to change the dynamics of the U.S. market significantly. Despite these strides forward, there remain many unresolved issues, such as how should biosimilars be named or how will they be substituted for the branded drug, given that notification by pharmacists is controlled by state law. In addition, there are legal issues surrounding the launch of newly approved biosimilars due to a large volume of intellectual property that stands in the way of bringing these products to the market, despite expiry of their principal patents. All of these issues, including the understanding of the complexity of these products and regulatory pathways, are the subject of this book.

While much of the discussion about biosimilars hovers around the world's largest markets, a more relevant discussion involves the utilization of biological drugs in 90% of the global markets in developing countries that cannot afford to use these life-altering products. Biosimilars stand to improve penetration of biologic drugs into these markets, expand their use, and alleviate suffering—the economic impact is therefore much larger than what is currently projected. Even more important than the economic impact is the humanitarian aspect of making biosimilars available to this huge population of people that cannot currently afford these drugs. It is for this reason that this book is written—to assist biosimilar companies develop and manufacture quality products and make them available at an affordable cost to patients globally.

Introduction to the book

Biosimilars and Interchangeable Biological Products: From Cell Lines to Commercial Launch is a two-set book comprising *Biosimilars and Interchangeable Biological Products: Strategic Elements* and *Biosimilars and Interchangeable Biological Products: Tactical Elements.* The *Strategic Elements* book is a comprehensive treatise on the science, technology, finance, legality, ethics, and politics of biosimilar drugs, which comprises one of the most significant healthcare business altering events in modern history. This book deals with strategic planning elements that include an overall understanding of the history and the current status of the art and science of biosimilars, including descriptions of the legal, regulatory, and commercial aspects—in a nutshell, the first book helps you to create a global strategy on how to build and take to market the next generation of biosimilars and manage markets throughout their life cycle.

The *Tactical Elements* book is a technical treatise that deals with the development and manufacturing of biosimilars, keeping in mind that eventually these products will take on a more generic personality, and the COGS shall remain a challenge. This book emphasizes technologies that reduce the cost of starting development and manufacturing and staying competitive in the field.

Whereas each book deals with a different aspect of biosimilars, the two books are intertwined as they share supporting information, cross-referencing, and planning events. It is, therefore, necessary to refer to both books at the same time when planning a presence in the biosimilars and interchangeable biosimilars arena.

While biological drugs were left out of the generic transformation of the 1980s to reduce the cost to patients, it is now time to consider the question, How can we make complex biological drugs available at an affordable cost to patients as well? When the economic stakes are high, with clearly defined future possibilities of their remaining high, this new paradigm of healthcare management becomes lucrative and valuable despite its complexity. This book provides current information and the most comprehensive understanding of biosimilars and interchangeable drug products while assisting developers of biosimilar drug products to speed up their development. The primary incentive for writing this book was to help new developers manage the complexity of development and commercial launch of these products to help reduce the cost of treatment, making these life-altering drugs affordable to all.

The legacy of this book goes back exactly a decade, when my first book on this topic, *Handbook of Biogeneric Therapeutic Proteins: Regulatory, Manufacturing, Testing, and Patent Issues* (CRC Press, 2005), was published, wherein I made a few predictions that were contrarian to the then generally accepted beliefs. While my wish to call them "biogenerics" has not come true, the timeline of their approvals and the possibility of these drugs becoming generic one day has come true. All that buzz about "product by process" by the originator companies is fading fast as the science related to understanding the molecular structure of these complex drugs has become widely explored and accepted. Regulatory agencies, like the U.S. FDA, have raised the bar on scientific inquiry by taking a more scientific view and requiring fewer trials in patients to prove biosimilarity. Originators are now taking the fight to block the entry of biosimilar products to prescribers, patients, and even legislators, trying to assert the dangers of using biosimilars. Not surprisingly, the originators are also fighting back with a new armamentarium—intellectual property. While in the early years the primary IP hurdle was the composition patent, today a biosimilar developer faces patents for bioprocessing, formulation, and even use and dosing. When the stakes are in the billions, the road to biosimilars requires financial creativity to make these drugs affordable. There are many more stakes in the ground to traverse.

This follow-on edition of the *Handbook of Biogeneric Therapeutic Proteins* takes a different title because of the significant changes in the understanding of biosimilar products over the past decade. There was no concept of interchangeability 10 years ago, nor were there any U.S. biosimilarity development guidelines—the revisions in this book, therefore, are substantial enough to warrant a different title and not just call it the second edition.

Strategies for a successful launch of biosimilar and interchangeable biosimilar products depend greatly on a clear understanding of the legal, regulatory, and commercial challenges. This book is dedicated to teaching how to choose the right product to develop, how to meander around the legal fireworks, and how to secure a viable commercial presence. Surprisingly, even the most established companies have made mistakes; for example, not appreciating the future threats to intellectual property caused a biosimilar product development plan to fail after millions were spent. Additional pitfalls include not anticipating the next line of improved products, better formulations, delivery systems, and the possibility that the dosing and indications can themselves be patented, thus making the choice of products to develop a major exercise. These facts have heretofore not been well recognized by those who have historically developed small molecule generic products, nor by large brand companies who assume that any product is a good choice if there are financial resources available.

About terminology

Whereas the Glossary section in the *Strategic Elements* book provides definitions of most relevant terms used in this book and the field of biotechnology, a few need clarification mainly for logistics reasons.

Innovator versus originator

In small chemical fields when a new molecule is synthesized, isolated, or identified, the credit goes to the innovator—as many of them may end up getting a patent for the discovery. However, when it comes to biologics, there can be a differentiation. For example, filgrastim is an endogenous compound, and a company discovering a gene to manufacture this product using recombinant technology qualifies as originator but not as innovator. However, when filgrastim is pegylated (the form that does not exist in the body), this qualifies the developer as an innovator.

Authorized versus licensed

In Europe, biological drugs are authorized, not licensed. This difference goes back to the laws in the Unites States that considered biologics to be hazardous to produce, and a license is required to manufacture them. EMA requires a marketing authorization application (MAA), whereas the FDA requires a biological license application (BLA).

Medicines versus drugs

EMA prefers to call treatment modalities as medicines; the FDA labels them as drugs or biologics.

Similarity versus comparability

Comparability simply means comparing two products, but this vocabulary can be confusing because of the official exercise of the comparability protocol, which is a well-defined task for changing the manufacturing process of an approved product. The change is made by filing specific documents with the regulatory authorities. The similarity is a demonstration of the extent of the sameness of the products being developed. Unfortunately, guidelines in the EMA often refer to similarity as comparability, and this should be avoided. "Comparability" is frequently used in lieu of similarity testing. The EMA explains: "If the biosimilar *comparability* exercise indicates that there are relevant differences between the intended biosimilar and the reference medicinal product making it unlikely that biosimilarity will eventually be established, a stand-alone development to support a full Marketing Authorisation Application (MAA) should be considered instead." Moreover, "Clinical data cannot be used to justify substantial differences in quality attributes," the EMA adds.

Effectiveness versus efficacy

Efficacy is a demonstration of clinical response in a controlled trial; effectiveness is a comparison of a clinical response to two products. Unfortunately, both the EMA and WHO got it wrong and used the word "efficacy" in describing the evaluation of biosimilars. The FDA did not. In the U.S. guidelines, efficacy is not the term used to compare the relative effectiveness, which is distinctly different from efficacy.

About originators

It is important to recognize that a few decades ago when biological products like therapeutic proteins and monoclonal antibodies were developed, they represented in many instances, copies of endogenous molecules, like insulin, erythropoietin, filgrastim, etc. In reality, these first waves of products were actually the first biosimilar product to what the body was already producing. It will be, therefore, proper to call them originator products, not an innovator's product. They were originated, not invented, and, therefore I have decided to call them as such and not innovator. Now we are at a stage where biosimilar companies are emulating an originator's products. Whether the originators of biological drugs would succeed in developing a biosimilar program will depend to a considerable degree on whether they can think and act differently from their pedigree. Several large companies, like Amgen, Sanofi, Pfizer, and Merck, have declared their intentions to develop biosimilar products because the markets of these products are so large and fit well within their ROI calculations. This metamorphosis has rarely played out well in the past, mainly because developing a more affordable product, the focus of biosimilars, requires a different mindset than these companies are capable of establishing. Only one company realized it and split into a separate generic-based company—Sandoz, from its parent company Novartis.

One of the emerging discussions in the field of biosimilars is the role of pure play companies—companies established exclusively to produce biosimilars. In financial management, a pure play is a company that either has or is very close to having a single business focus. Coca-Cola is an example of a pure play in this context because it retails only beverages. On the other hand, PepsiCo is not a pure play because it also owns the Frito-Lay snack foods brand. While there are several pure-play biotechnology companies, when it comes to pure-play biosimilar companies, the choice becomes limited in the developed countries; two names are prominent: Therapeutic Proteins International in United States and Celtrion in Korea. Outside of these, the choice of fully integrated pure-play companies is limited. The most likely contenders in the biosimilars market in the United States are Sandoz, Teva, Hospira, Amgen, Pfizer, Merck, Apotex, and Therapeutic Proteins.

About markets

As of 2015, more than 200 biologic drugs, including more than 34 monoclonal antibodies, have been approved for human use in the United States. The global commercial pipeline includes approximately 350 monoclonal antibodies in clinical studies of indications, including immunological disorders, infectious diseases, and cancer.

Since 2006, an average of 15 novel recombinant protein therapeutics have been approved by the U.S. FDA annually. The newly issued Purple Book by the FDA lists 52 biological drugs, many of which are considered blockbuster drugs—the multibillion-dollar market category.

In 2013, seven out of eight top-selling drugs were of biological origin, and it is anticipated that more than 75% of all new regulatory filings in the next 10 years will be of biological origin. Currently, biological drugs command a market of about $200 billion, half of which is in the United States.

While biological drugs have provided a new lease on life to many, the cost of these drugs remains high, even in U.S. markets. For example, antibodies and fusion proteins used for the treatment of symptoms of rheumatoid arthritis may cost between $10K and $30K per year, with very high copays making their penetration and availability limited. The same holds true for cytokines like pegylated filgrastim and most anticancer drugs. The cost of biological drugs used in the treatment of cancer can run much higher. The high cost of these drugs is justified by the originators based on the complexity and difficulties of their manufacture and a much lower success rate in new discoveries. However, it is abundantly clear now that the price of these drugs is driven mainly by economic incentives supported by complex intellectual property laws that have allowed some of these drugs to have an exclusivity of over 30 years, such as interferon alpha, etanercept, and several others.

One of the most hotly debated topics in the healthcare industry today is the cost of these drugs and their affordability. In many instances, the greatest burden lies on the governments that pay for drugs. For example, the Center for Medicare and Medicaid (CMC) services spends hundreds of millions of dollars each year on erythropoietin, a biological drug given to renal dialysis patients, a procedure fully covered by Medicare and Medicaid. One way to reduce the fast-growing, almost unbearable costs of biological drugs is to provide competition in the market through "generic" equivalents to these medicines as the composition (the gene sequence) patents expire. The need for biosimilar products is therefore clearly established.

About misconceptions

The misconceptions about biosimilar products were at first aimed at regulatory agencies; now it is the public at large, as seen recently in the legislative actions taken by several states in making the interchangeability of biosimilar products more onerous and often cumbersome with an aim to prevent it.

The most common misconceptions about biosimilar products today include the following:

- Biosimilar products are inferior because they are not identical.
- Biosimilar products are unsafe.
- The only way to establish safety and efficacy is to demonstrate them in patients.
- There is only one approval pathway for biosimilars.
- There is only one correct way to manufacture biosimilars.
- INN naming will create safety hazards.
- Savings will be minimal due to high development costs.

Given the large and complex molecular structure, biological drugs will always have structural variability that will be questioned, even if it does not impact their safety and efficacy. Lot-to-lot variation in the originator product or even the variability in the structure of what the body produces day to day is sufficient to accept that variability in their structure, which is part of their characteristics and not a flaw.

The misconception that biosimilar products are unsafe is routinely aimed at their immunogenicity, but the fact is that biological drugs are one of the safest categories of drugs; most biological products do not have any significant immunogenicity, and where immunogenicity is significant, it is part of their mechanism of action and not an adverse effect. In almost a decade of biosimilars use, the adverse events reported are far less than what was feared. The reported incidences for erythropoietin, which itself has little immunogenicity, were a result of improper changes made to the packaging that should have been prevented and have nothing to do with any inherent hazard in the use of these products. This incidence has damaged the reputation of the biosimilar product drug industry.

The misconception that unless you show safety and efficacy in patients the product cannot be declared safe and effective is based on the lack of understanding of statistics and the pharmacological aspects of drug action. The originator product is evaluated in several conditions in large patient populations against a placebo; a biosimilar product would have to demonstrate noninferiority against the originator product. Often the number of patients required to prove this will be higher than the studies of the originator; besides, requiring clinical trials in limited indications in a small number of patients would not provide extrapolation possibilities if demonstrated clinical effects are made the ultimate measure. In approving Levonox (small molecular weight heparin) without any clinical trials in patients, the FDA clearly explained the relative importance of structural and functional similarity.

Whereas most biosimilar product developers will follow the same manufacturing process as that of the originator, there remain many aspects of manufacturing, such as in-process controls, that need to be developed by biosimilar product developers. The originators take the position that unless the product is made using the same in-process controlled method exactly the same way as they do it, it is not possible to manufacture a safe and effective product. Ironically, when the originators developed a process, it was always "a" process and never "the" process. There are so many changes that can improve the process, and definitely any change or difference is readily justified in the ultimate comparison with the originator product.

The issue of naming is a hot topic and remains unresolved in the United States. Partial relief came with the approval of the first biosimilar in United States, wherein the FDA took a middle road, and instead of using the INN name, they suggested using a qualified INN, the generic name attached to an abbreviated identification of the developer. In this case, Sandoz's filgrastim was given the name filgrastim-sndz. This disappointed many, but the FDA said that this is a placeholder arrangement and final guidance will come in due course. I anticipate that the FDA will allow INNs for interchangeable products to provide a differentiation between biosimilars and interchangeable biosimilars. Those who objected to the use of it gave many reasons for not using INN, including the problems of tracking adverse events to these products. We have lived happily with small molecule generics with INNs; it's hard to imagine why this would be different for biological products. The argument that these products are more likely to produce AEs falls flat on two grounds: first, this is not the case and, second, why would an NDC number not be sufficient to track them down? The sole purpose of keeping the INN out is to keep the monopoly of the originator products as long as they can be sustained—this is a purely commercial and political tactic.

Finally, analysts suggest that since the savings in the use of the biosimilar product will be minimal, "Why take the risk?" Nothing is farther from the truth, as we have seen in Europe. Besides, analysts, who may well have a vested interest, keep talking about the percentage reduction in price and do not speak of the absolute value of savings given the very high price of these products. Currently, the savings in cost is more than 30% in Europe, and eventually, a 50% reduction in cost is very likely. One aspect that is left out of the discussions is affordability. Because of their high cost, these products remain out of reach for many; lowered cost will significantly expand utilization of these products, and that will provide a significantly greater benefit to mankind.

The bottom line is that the biosimilar product industry is here to stay.

About legality

The topic of intellectual property has gained significant importance in the field of biological drugs because of the many ways, in bioprocessing, an originator can protect their product long after the composition patent has expired. The importance of this topic is yet to be fully recognized as the biggest drugs began coming off patent (composition or gene sequence) and the developers of biosimilars faced a barrage of new patents, creating significant hurdles in the development schemes. Ten years ago this was a little-heard-of discussion. However, with originators coming out with not just submarine patents, but also bioprocessing, methods of treatment, and dosing patents, the legal bar has been raised significantly. The biggest challenge biosimilar product developers face today is to be able to secure a reasonable

"Freedom to Operate" (FTO). This book provides detailed insight and also offers advice on how to manage these challenges. However, the reader must be aware that the author accepts no responsibility for any legal approaches discussed in this book and is not offering any legal advice.

While companies manufacturing biosimilar products face great legal hurdles, much of the developing countries and even some countries in the developed world have moved past these hurdles and launched biosimilars where the originators either ignored or missed protecting their IP in these countries. Even in those situations where an IP is in effect, fighting out infringements is not easy because of the peculiarity of the regional legal systems that often favor local companies. This situation has created an interesting dichotomy. The chapter on global commercialization in the *Strategic Elements* book lists a large number of biosimilars that are fully protected in developed countries and widely distributed in parts of the world where the IP is readily challenged. The unfortunate part of this dichotomy is the risk to patients if these products fail to meet the quality standards required to assure their safety. Given the technical difficulties inherent in the development of biosimilars, a product developed and distributed hastily at a rock-bottom price shall always remain suspect for its quality. One of the main purposes of this book is to make the manufacturers of these products aware of this risk and offer solutions that will be helpful in improving the quality of their products.

One question that arises frequently within the group of biosimilar product developers is the intensity with which the originators are protecting their franchise by using the U.S. Patent Office. A barrage of patents appear just when the composition patents are about to expire; many of these patents will be challenged and some will be declared invalid. However, this practice, which involves every step of bioprocessing, formulation, and even dosing, creates an environment of uncertainty among biosimilar developers. Some consider this practice to be unethical, if not entirely illegal.

About regulatory approvals

Europe has taken the lead in introducing biosimilar product guidelines and approvals, and one country in particular, France, has even legislated their interchangeable status for use in new patients. However, the laws governing the entry of biosimilar products to market are still in their infancy in several developed countries. It was only five years ago, with the enactment of the Biological drugs Price Competition and Affordable Care Act (BPCIA), as part of the Patient Protection and Affordable Care Act of 2009, more commonly known as the Healthcare Reform Bill, that the possibility of biosimilar products coming to market became a reality in the United States. The first U.S. biosimilar was approved in March 2015; many more are expected to be approved soon in the country. China just released its first draft of biosimilar products pathway. Currently, China treats biosimilars as new biological drugs. A large number of developing countries have taken an altogether different approach, treating these complex drugs at par with simpler chemical drugs and allow use, in some instances, almost indiscriminately.

The abbreviated filing pathway in the United States is set forth in a new subsection designated as section (k) of the Public Health Service Act (PHS Act). Accordingly, biosimilar applications are commonly referred to as "351(k) applications." It should be noted that not all biological products are licensed under the PHS Act. Some proteins, such as insulin and human growth hormone, are licensed as drugs under section 505 of the Food, Drug, and Cosmetic Act, and, therefore, a biosimilar of these drugs cannot apply for approval under 351(k).

The approval pathways for biosimilar products vary among various jurisdictions; the U.S. FDA allows interchangeability; Europe does not, except for France, where it is a law for all new patients. In the United States, several states have legislated to make it harder to substitute even before any substitutable or biosimilar product was approved. This is more of a political fight driven by economic greed rather than any scientific principle. State legislatures are not qualified to second-guess the FDA's acceptance of the safety and efficacy of a product, but states currently control pharmacy laws and rules.

An interesting insight into the mindset of originators, who are about to lose billions of dollars of markets to biosimilars, can be obtained from an assertion in a citizen's petition filed in April 2012, wherein Abbvie stated: "Under well-established Supreme Court jurisprudence, FDA's use of the trade secrets in (Abbott's) Biologics License Applications to support approval of competitor products would frustrate (its) investment-backed expectation regarding their property and would constitute a taking under the Fifth Amendment to the US Constitution that requires just compensation." Additionally, both Abbvie and InterMune tried to legally halt the EMA practice of sharing clinical data with the public to enhance safety. These practices demonstrate the complexity of the exercises involved in the development of biosimilar products.

However, biosimilar developers are not sitting on the sidelines. The first approval of biosimilars in the United States saw an epic legal battle between Sandoz and Amgen. Sandoz refused to comply with the suggested patent dance, wherein the biosimilar company provides a complete submission dossier to the originator within 60 days of the acceptance of the 351(k) application; Sandoz asserted that it was not required, but Amgen suggested otherwise. The FDA refused to get involved, and Amgen could not get a court injunction against Sandoz. The second aspect of a 180-day notice was also challenged by Sandoz, saying that the term starts with the acceptance of the application; Amgen asserted it starts after approval of the product. The courts seemed to support Sandoz's position.

About affordability

Chemical generic drugs typically cost about 10%–20% of the cost of branded products. Given the differences in the level of complexity associated with manufacturing a biologic, the cost savings are not expected to be as high for a biosimilar product, yet even the anticipated 20% to 40% price reductions are sufficient to make a significant impact on healthcare financing. A 90% decrease in the cost of a small molecule drug that costs $10 for the branded product is nowhere comparable to even a 10% decrease in the cost of a biological branded drug costing tens of thousands of dollars. This has been clearly established in Europe, where, for example, the introduction of biosimilar product filgrastim has reduced the cost by about $3 billion. Much greater savings are expected as biosimilar product antibodies are approved, and the first such antibody, infliximab, was recently approved in Europe as a biosimilar product. Overall, the use of biosimilar products in Europe has been predicted to result in overall savings between $16.2 billion and $45.8 billion between 2007 and 2020, with the largest savings expected in France, Germany, and the United Kingdom. Comparable savings in the United States are projected into hundreds of billions of dollars. Globally, biosimilars will increase the absorption of biological products severalfold, and it is not out of sight to project a trillion-dollar market created and affected by biosimilars over the next 20 years.

About the beneficiaries

I hope this book proves useful in preparing regulatory applications for biosimilar products. However, companies need to evaluate their strengths and weaknesses when diversifying into the biotechnology arena if they are already not involved in it. Historically, mergers between pharmaceutical and biopharmaceutical companies have not fared well. Where the surviving company is the pharmaceutical company, it creates difficulty in understanding the elaborate nature of production techniques. Where the biopharmaceutical company survives the merger, the problems have been the visualization beyond the research horizon and translating the bioprocess into a commercially feasible validated process. In my opinion, the best solutions are offered through outsourcing, as the technical requirements are highly complex and elaborate to manufacture these products. Creation of the gene construct and the genetically modified cell (or animal) is an altogether different science than the science of fermentation and the science of downstream purification and indeed routine pharmaceutical manufacturing of chemical-based drugs. Even the process of finishing the product, putting the purified protein in a vial or ampule, requires a different understanding in controlling the process. For example, excessive stress to a protein solution can result in loss of structure of even structuring where the protectant is lost; such awareness is not part of routine chemical drug manufacturing.

It is difficult, if not impossible, for generic companies to carry the burden of all of these diverse disciplines of manufacturing and finishing biological drugs under one roof. Limited financial resources, compressed timelines, and new regulatory constraints faced by generic companies suggest that most companies would do well in outsourcing all phases of API manufacturing and only keeping the final filling operation in-house. My estimation shows that the cost per gram of therapeutic protein may not be significantly higher than if they were produced in-house. This is a result of a rather large carrying cost of infrastructure, both in the manufacturing and testing phases necessary to implement the production. Any existing facilities and personnel are least likely to fit the new mold. Naturally, at some point when the sales justify, the entire process can be brought in-house.

Disclaimer

The author does not accept responsibility for any technical or legal suggestions or advice provided in this book; all views are those of the author in his personal capacity and not as a patent agent of the U.S. Patent and Trademark Office or, as executive chairman of Therapeutic Proteins International and/or officer of any other company, agency, or entity.

Acknowledgments

About helping angels

In writing this practical treatise, I have sought and received help from many individuals and institutions; I would be remiss if I did not acknowledge them. However, it would be impossible for me to acknowledge and recognize all.

A book of this size could not be produced without the recognition and the arduous support of the publisher. Michael Slaughter at CRC Press knows how to motivate authors, and I am indebted to him and the folks at CRC Press for taking on this task; thanks also to scores of editors at CRC Press, notably Michael Slaughter, and Laurie Oknowsky, without whose untiring encouragement this book could not have been completed.

Last, but not the least, I am thankful to scores of my scientific and professional colleagues, and particularly those whom I came to know through the landmark literature in the field but have never met. I may have quoted their work thinking that this is all in the public domain subconsciously; I hope they will excuse me for taking this liberty as it would be impossible to recognize them well. An elaborate bibliography does not necessarily replace this obligation of properly acknowledging their work.

A large number of my colleagues at Therapeutic Proteins International have helped me in writing this book, and I could not be more thankful to them. Cheryl Liljestrand and Tricia Chilsohm provided assistance in the chapters on intellectual property; Celina Doupoules provided details about the commercialization aspects; and my daily assistants, Naila Akimova, Lillian Yanni, Natalia Isaeva, Jacquelyn Hurd and Mimi Honges, among many others, assisted me in compiling, writing, designing, and polishing this voluminous work; I am sure I am missing out on recognizing many more who have always given me a helping hand.

I would remiss if I didn't acknowledge Robert Salcedo, who has allowed me enough time to devote to this academic pursuit while continuing to stay connected with the day-to-day reality of developing, manufacturing, and commercializing biosimilars as my main profession. It is this hands-on experience with biosimilars, from cell lines to commercialization, that gives me the confidence to write this book. The reader will find advice sprinkled throughout the book that will be highly personal and at times challenging the norms of the industry, as I have tried to make this book more useful. This hands-on experience comes from heading one of the most ambitious ventures in the world, made possible by the unfaltering commitment of Chirag, Chintu, and Tushar, who have put their complete trust in the future of biosimilar products and have poured their resources generously into achieving the goal of making these products affordable to all. It is with the greatest humility and appreciation that I acknowledge the trust they have placed in me.

Finally, I would acknowledge the continuous support, loyalty, and love of my wife, Anjum, who never complained about being left out when I worked ungodly hours, not just in completing this work at a stellar speed to make it timely with the first U.S. approval, but also maintaining my presence at Therapeutic Proteins and many other distractions that I continue to engage in, both at the scientific and philanthropic levels. The recombinant union—our matrimony of more than 40 years—has been good to me. We have three children and four grandchildren, as of the date, whom I call our Grand proteins. I could not be happier.

I would like to acknowledge the great joy and privilege of knowing Dr. James Watson and it is an honor to dedicate the *Tactical Elements* book to him and Dr. Francis Crick. This is appropriate, for this book deals mainly with the technology based on their double-helix; without their vision, recombinant technology benefitting mankind would not have been possible. We will thank them for centuries.

"With regard to errors, in general, whether falling under the denomination of mental, typographical or accidental, we are conscious of being able to point out a greater number than any critic whatever. Men who are acquainted with the innumerable difficulties attending the execution of a work of such an extensive nature will make proper allowances. To these, I (we) appeal and shall rest satisfied with the judgment they pronounce" (*Encyclopedia Britannica*, First Edition Preface 1768). I would add: please send me an e-mail to correct my mistakes at niazi@theraproteins.com.

Introduction

Tactical Elements book description

The first two chapters describe the basic structure of biosimilar drugs and the main risk in their use pertaining to immunogenicity. As mentioned, the biosimilar products market will eventually reach the generic status requiring a firm control over COGS; this consideration prompted several chapters that pertain to the affordable manufacture of biosimilars. This aspect of practice is often constrained by existing production platforms. For example, if hard-walled systems are already in place, and it is not economically feasible to adopt alternate systems, the choices are limited. However, it is advised that all manufacturers, whether they have a system in place or not, evaluate alternative systems for long-term cost containment. It is particularly important for new companies establishing a biosimilar program to adopt the lowest capital cost systems that is also low maintenance and allows faster product development. The next few chapters describe all the components of manufacturing and discuss these in light of modern developments in this field.

Chapter 1: Structural and functional considerations

Biosimilar products mainly comprise therapeutic proteins (as differentiated from nutritional proteins) and antibodies (particularly monoclonal) that are expressed in recombinant agents. Once a product has been selected for development, a detailed exercise about understanding this protein or antibody begins. While this topic is best left to the scientists who will be doing the development work, I found it necessary to teach everyone on the team to understand the fundamental nature of these products, whether they are commercializing or manufacturing them. This chapter can be considered the fundamental teaching of the chemistry of the subject. A lot of the vocabulary developed in this chapter will be used in conversations with a biosimilar development team—so why not just get this knowledge. This chapter describes the differences between proteins and antibodies, their structural and functional characteristics, and their degradation, as well as the potential immunogenicity of these products. In my opinion, a keen understanding of the trade is needed for everyone on the team—this chapter is a must-read for all.

Chapter 2: Immunogenicity considerations

One of the most controversial subjects regarding biosimilar products is the issue of safety and adverse effects. Protein structural issues can contribute to this as well as many formulation and packaging factors. This chapter gives an overall view of the general safety and adverse effects of biological drugs and points to the relative safety of biosimilar products from historical data. The main discussion of immunogenicity is dealt with first, describing the nature and causes of an immunogenic response, a classification of high and low immunogenic drugs, and the vast number of manufacturing and design features of these products that are required to minimize the risk of immunogenicity. This chapter forms a preamble to formulation efforts after the developer understands the factors that must be considered, based on protein structure.

One of the most controversial topics in the field of biosimilar products has been their side effects, particularly immunogenicity. A detailed chapter now traces the historical safety record of biosimilar products as they have been widely used worldwide and dispels the myths about the safety of proteins and antibodies and, more particularly, biosimilar products. Extensive advice is provided for avoiding steps in the manufacturing and formulation stages that might accentuate an immunogenic response, and a framework is included that might be useful for biosimilar product developers to present the regulatory agencies.

Chapter 3: Product development strategies

Once a product has been selected for development, having gone through FTO scrutiny and the resources available, a detailed strategy must be developed for how to develop the product. This strategy requires a keen understanding of all aspects of the product, from inception to life cycle management, as the development plan will impact every aspect of the success of the product. A smart plan will not ignore the purpose of developing biosimilars, namely, to make them more affordable. This chapter provides an overview and details of all aspects of planning well before development begins. If developers pay attention to this section, they will save substantial costs and will be able to make appropriate decisions about the choice and speed of development. This chapter teaches the science and the technology needed to accomplish these tasks.

While sections leading to large-scale commercial manufacturing provide an overview of establishing production trains, there remains the primary task of developing a particular product, therapeutic protein, or monoclonal antibody. This chapter starts with a discussion on the selection of product and any particular manufacturing systems that are appropriate for this product. Choosing cell lines, the reference product, the test methods, and the product specifications are discussed to teach the art of reverse engineering the originator product. One of the most significant topics in the manufacturing of biosimilar products is the demonstration of analytical similarity, nonclinical similarity, and similarity in clinical pharmacology, all leading to the claim of biosimilarity. Concrete steps needed for each stage of development are discussed in detail in this chapter. Also included is a discussion of interchangeability and any regulatory uncertainties that remain that bring particular commercialization challenges. It is important to review the entire development leading to commercialization as the particular choices of manufacturing systems will depend a great deal on intellectual property barriers, general technical difficulties in making a product, and, above all, the future life cycle of the product, in case better alternatives in formulation or composition are imminent.

Chapter 4: Stability and formulation considerations

The final phase of product development involves extensive studies on the stability profile of the product. It should be realized that the shelf life of these products is limited by the maximum impurities allowed, which are much smaller than the small-molecule drugs. So it is critical to start with a product with high purity; over 99% will be required to assure reasonable shelf life. Formulation factors significantly impact this profile as do the choice of contact packaging components. Unlike the case of small-molecule drugs, the FDA will accept an application presenting at least 6 months of real-time data, including the lots used for development; a total of at least three lots stored in the final delivery system are required. This chapter provides a variety of approaches to establish stability testing protocols to assure compliance, since the developer is required to conduct stability studies side by side with the originator product.

Whereas the formulation and stability topics are common to all drug development, the challenges for biological drugs are very different, and the science required to manage these is much more complex. In this chapter, the first part describes all typical degradation reactions and the means for reducing their intensity. The formulation of biological drugs is simple because of the incompatibility of these molecules with many formulation ingredients. However, certain common elements of formulation are required to stabilize the protein structure from physical and chemical degradation. Methods of formulation stabilization are dealt with in detail in this chapter, along with stability testing guidelines, batch selection, stability-indicating methods and profiles, and determining the best storage conditions. A discussion of the changing trend in formulations is described to alert developers to the possibility that originators might be planning a replacement formulation to extend their intellectual property landscape. Labeling issues and testing protocols are further elaborated.

Formulation and stability considerations of biosimilar products have become more complex as the race to produce products with "fingerprint-like similarity" heats up. Given the realization that intellectual property challenges may require alternate formulations, changes in manufacturing processes, and even the final dosage form, biosimilar product developers need a much greater understanding of the chemistry of proteins and antibodies; a detailed chapter on this topic now brings this information up to date, including information on the newer, high concentration subcutaneous dosage forms of antibodies.

Chapter 5: Biosimilarity tetrahedron

This chapter introduces a new concept of establishing biosimilarity, wherein four distinct elements—safety, purity, potency, and identity—are described to be of equal value.

The goal of the development exercise is to develop a product that will meet the requirements of biosimilarity based on analytical and functional similarity, pharmacokinetic similarity, pharmacodynamic similarity (where applicable), nonclinical toxicology or toxicokinetic similarity (where appropriate), and finally any clinical trials that establish biosimilarity. The FDA requires the level of similarity be "highly similar" to earn a biosimilarity status. Therefore, analytical and functional testing becomes a significant exercise. This chapter provides details on establishing similarity at all levels, the method of presenting the data to the regulatory agencies, and planning to achieve these results from a developer's point of view. Also provided in this chapter are the differences in the expectations of the EMA, the FDA, and the rest of the world. The overall development plan will take into account a choice of proper reference products, a protocol of studies, and the presentation formats.

The largest change in this title is the inclusion of methods and tactics of developing analytical similarity data, something that was not of great importance to regulatory agencies in the development of new drugs because of the requirement for extensive patient trials. Now that there is a possibility of obviating trials in patients if the analytical and functional similarity data are supportive, biosimilar product developers will find that their largest investment will be in this scientific pursuit.

Chapter 6: Recombinant expression systems

The production engine of biosimilar drugs is a recombinant expression system since all of these products are manufactured by genetic engineering. The expression systems comprise bacterial systems, mammalian cells, yeast, plant cells, insect cells, and even transgenic animals. This chapter describes the details of how these cell systems are constructed and how biosimilar product developers can increase the yield of production using various techniques that have become available recently. While most developers of biosimilar products prefer to use the same system employed by the originator, the regulatory agencies allow alternate expression methods, and in some cases this may prove pivotal in being able to compete in biosimilar markets. Also included in this chapter are topics relating to the viral clearance of cell lines and various aspects of cell banking for commercial production. The details of ICH Q5 are also provided in this chapter.

Chapter 7: Upstream system optimization

Upstream systems comprise expression of the target product using the recombinant expression systems described in Chapter 6. In this section, emphasis is placed on describing the general aspects of establishing upstream manufacturing and, more particularly, on optimizing these systems. Since developers of biosimilar products will inevitably face constraints of space and cost, many novel methods and techniques are described in this chapter. Besides describing various types and functionalities of bioreactors, this section includes different types of processes, such as batch, continuous, fed-batch, and perfusion culture systems. The problems in scaling up and the resolution of problems in expressing these products are described. Separate sections deal with mammalian, bacterial, yeast, insect, and transgenic animal lines. This chapter will be useful when reverse-engineering recombinant expression systems.

Chapter 8: Downstream systems optimization

Once the upstream process is completed, the harvesting, polishing, viral clearance, concentrating, and purification begin—comprising the downstream processing. Different steps are required, depending on the nature of the product, and this chapter describes each of the major systems used, including bacteria, mammalian cells, yeast, insect cells, and transgenic animals. Details of the processes that are readily optimized using newer elements are described, along with the current emphasis on purification standards by various regulatory agencies.

Chapter 9: Single-use manufacturing systems

The future of manufacturing biological drugs appears very different from the systems currently used and belongs to single-use or disposable systems. This comprehensive chapter describes every component of single-use manufacturing systems (SUMS), from bioreactors to purification systems. Also provided in this section are contact details for suppliers and the technical specifications of various elements. While many companies may already have hard-walled manufacturing systems in place, biosimilar product manufacturers will benefit significantly by adopting these SUMS that will reduce, not only capital and running costs, but also the cost of validation and lost lots.

Chapter 10: Commercial manufacturing systems

This chapter follows the theme of manufacturing biosimilars in a commercial or large-scale setting and builds on earlier chapters that discuss the concepts from cell line to optimizing downstream purification systems. It is now time to put these ideas into practice by establishing large-scale commercial manufacturing. Unlike small molecules, the regulatory agencies expect the products tested in humans in the development stage to be at full commercial scale. This chapter describes media, culture growth, process overview, process optimization, validation, scale-up, and, above all, methods of economizing production costs. Environmental controls, biosafety levels, and GMP controls for biological products are discussed. This chapter also provides an overview of the challenges in biological manufacturing, such as cleaning procedures, laboratory testing, controls, and the necessary documentation associated with all of these.

Chapter 11: Outsourcing considerations

Once a product has been developed, a business decision has to be made on manufacturing. Earlier chapters offered the details of required manufacturing and testing steps, but it is only when the product is ready for commercial production that a decision should be made whether to do it in-house or outsource. Several companies have opted to outsource at least some of their products, including such large enterprises as Hospira (now a Pfizer company), Baxter, and several others. The selection of a vendor, its monitoring, and assurance that the PAI will not be an issue are important tasks. In this chapter, hands-on details are provided on relationship management, documentation, and ongoing assurance that a safe and effective product will be available all the time. The outsourcing of testing can be another issue that requires detailed planning to assure that the methods used are validated unless they are used for analytical similarity demonstration and that the vendors would qualify under a regulatory audit.

Appendix: Glossary of terms

An updated list of terms commonly used in the book is provided, more particularly those terms that can be confusing across various regulatory jurisdictions.

Author

Sarfaraz K. Niazi, PhD, is the founding executive chairman of Therapeutic Proteins International LLC, a world-class pure-play developer and manufacturer of biosimilar and interchangeable recombinant biologics, headquartered in Chicago. Dr. Niazi began his career teaching pharmacy at the University of Illinois (1972–1988), where he became a tenured professor. He then entered the pharmaceutical industry at Abbott International, becoming a Volwiler Fellow. He left Abbott in 1995 with a passion for making high-cost biological drugs affordable. In 2003, he established Therapeutic Proteins International, which remains the only integrated U.S. company of its kind to date. To make biosimilar drugs affordable and to manufacture them in the United States, he reinvented the bioprocessing technology that was begun thousands of years ago; his inventions are protected by dozens of U.S. and worldwide patents. His inventions extend to new drugs, new dosage forms, biosimilarity testing methods, wine aging, water purification, automobile safety, no-fly hats, and many more. Many of his inventions are used widely across the globe, and this has earned him the highest civilian award, Star of Distinction, by the government of Pakistan.

Dr. Niazi has written over 40 major books, including textbooks, handbooks, technical treatise, poetry books, foreign language poetry translations, and philosophical treatises. He has published more than 100 research articles and many more research abstracts. He authored the first handbook of "biogeneric" therapeutic proteins in 2005. He has delivered hundreds of talks on a variety of subjects, from science and philosophy to religion, health care, rhetoric, and contemporary solutions to societal transformation issues. He is also a licensed practitioner of patent law in the United States, and in this capacity, he helps scientists in the developing world to secure their inventions in the United States.

Dr. Niazi has been recognized for his contributions to science and literature, including the BioProcess International award for single-use bioreactors for bacterial systems manufacturing in 2012 and the 2014 Global Generics and Biosimilar award for Innovation of the Year sponsored by Honeywell. He has been widely written about for his inventions, philanthropy, and passion for science, literature, music, and photography in newspapers and magazines, including *Forbes, Chicago Tribune*, and *Crain's Chicago Business*.

Dr. Niazi continues to serve on the faculty of several major universities around the world. He also hosts the Voice of America radio program that goes to billions of listeners around the world, wherein he talks about poetry, philosophy, and wisdom of life.

Professional website: http://www.theraproteins.com

Website: http://www.niazi.com

LinkedIn: https://www.linkedin.com/pub/sarfaraz-k-niazi/18/24/592

Wikipedia: http://en.wikipedia.org/wiki/Sarfaraz_K._Niazi

Twitter: @moustaches

Other Selected Books by the Author

- *Textbook of Biopharmaceutics and Clinical Pharmacokinetics*, John Wiley & Sons, New York, 1979
- *The Omega Connection*, Esquire Press, Westmont, IL, 1982
- *Adsorption and Chelation Therapy*, Esquire Press, Illinois, 1987
- *Attacking the Sacred Cows: The Health Hazards of Milk*, Esquire Press, Illinois, 1988
- *Endorphins: The Body Opium*, Esquire Press, Illinois, 1988
- *Nutritional Myths: The Story No One Wants to Talk About*, Esquire Press, Illinois.
- *Wellness Guide*, Ferozsons Publishers, Lahore, Pakistan, 2002
- *Love Sonnets of Ghalib: Translations, Explication and Lexicon*, Ferozsons Publishers, Lahore, Pakistan, 2002 and Rupa Publications, New Delhi, India, 2002
- *Filing Patents Online*, CRC Press, Boca Raton, FL, 2003
- Pharmacokinetic and pharmacodynamic modeling in early drug development in Charles G. Smith and James T. O'Donnell (eds.), *The Process of New Drug Discovery and Development* (2nd ed.), CRC Press, New York, 2004
- *Handbook of Biogeneric Therapeutic Proteins: Manufacturing, Regulatory, Testing and Patent Issues*, CRC Press, Boca Raton, FL, 2005
- *Handbook of Preformulation: Chemical, Biological and Botanical Drugs*, Informa Healthcare, New York, 2006
- *Handbook of Bioequivalence Testing*, Informa Healthcare, New York, 2007
- *Handbook of Pharmaceutical Manufacturing Formulations*, Volume 6, Second Edition: Sterile Products, Informa Healthcare, New York, 2009
- *Handbook of Pharmaceutical Manufacturing Formulations*, Volume 1, Second Edition: Compressed Solids, Informa Healthcare, New York, 2009
- *Handbook of Pharmaceutical Manufacturing Formulations*, Volume 2, Second Edition: Uncompressed Solids, Informa Healthcare, New York, 2009
- *Handbook of Pharmaceutical Manufacturing Formulations*, Volume 3, Second Edition: Liquid Products, Informa Healthcare, New York, 2009
- *Handbook of Pharmaceutical Manufacturing Formulations*, Volume 4, Second Edition: Semisolid Products, Informa Healthcare, New York, 2009
- *Handbook of Pharmaceutical Manufacturing Formulations*, Volume 5, Second Edition: Over the Counter Products, Informa Healthcare, New York, 2009
- *Textbook of Biopharmaceutics and Clinical Pharmacokinetics*, The Book Syndicate, Hyderabad, India, 2010
- *Wine of Passion: Love Poems of Ghalib,* Ferozsons (Pvt) Ltd., Lahore, Pakistan, 2010
- *Disposable Bioprocessing* Systems, CRC Press, Boca Raton, FL, 2012
- *Handbook of Bioequivalence Testing*, Second Edition, Informa Healthcare, New York, 2014
- *There Is No Wisdom: Selected Love Poems of Bedil.* Translations from Darri Farsi, Sarfaraz K. Niazi and Maryam Tawoosi, Ferozsons Private (Ltd), Lahore, Pakistan, 2015
- *Wine of Love: Complete Translations of Urdu Persian Love Poems of Ghalib*, Sarfaraz K. Niazi, Ferozsons Private (Ltd), Lahore, Pakistan, 2015

- *Biosimilars and Interchangeable Biologicals: Strategic Elements*, CRC Press, 2015
- *Biosimilars and Interchangeable Biologics: Tactical Elements*, CRC Press, 2015
- *Fundamentals of Modern Bioprocessing*, Sarfaraz K. Niazi and Justin L. Brown, CRC Press, 2015

Chapter 1 Structural and functional elements

Basics

The development of biosimilar products begins with a keen understanding of the active drug, generally a protein or a monoclonal antibody. The structural and functional elements of these molecules are complex and form the first tier of demonstrating biosimilarity, as discussed later in Chapter 5. The biosimilar developer will come to know their molecule well, even better than what the originator knew when the product was first launched. A good example to demonstrate this argument is the example of monoclonal antibodies like adalimumab or the fusion protein like etanercept. When these molecules were developed by Abbvie (then Abbott) and Amgen, respectively, they created a cell line recombinantly modified that will express "a" molecule, not, "the" molecule, and tested its toxicology, clinical pharmacology, and conducted numerous clinical efficacy trials to prove its efficacy vis-à-vis its toxicity to justify approval for specific indications. Now comes the biosimilar developer; using a cell line that can never be the same, using a bioprocess that can be similar (based on what is reported in the literature and the patent files), using conditions that are the best industry practice, and thus producing a molecule that is at least "highly similar" to qualify as a biosimilar. The challenge lies in replicating the attributes that would be considered "critical" in establishing a structural and functional similarity. The regulatory agencies take a very strict path to require demonstration that the structural and functional similarity is maintained if the biosimilar developer is to claim any exemptions from conducting clinical efficacy trials and if the biosimilar developer is to request an extrapolation of all indications. To qualify this, the biosimilar product has to be just as "good" and just as "bad" as the originator product. This distinction is important to understand. And that creates a significant barrier for the biosimilar product developer that must replicate all structural and functional characteristics, without any "improvements." And that also requires a complete understanding of the molecule being developed, and more generally, the proteins and antibodies.

This chapter can be considered a primer on the science of protein and antibody chemistry. Understanding proteins and antibodies is a prerequisite to all other chapters that follow as these chapters often use a vocabulary and terminology that pertains to the specific structural elements of these products that are related to their functionality, immunogenicity, and safety.

Multidimensional view

To understand the basic principles of protein function, their three-dimensional and fourth-dimensional structures require scrutiny since these structural elements create an extremely complex network of interactions between hundreds and thousands of atoms with receptor sites and with immune system triggers. The multidimensional

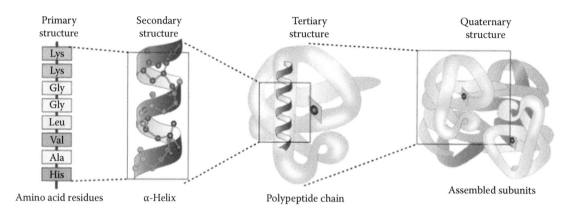

Figure 1.1 The four types of possible protein structures.

character of these molecules differentiates them from small-molecule chemical structures that are invariably fixed in their spatial arrangement due to the fixed covalent bonds; in the case of proteins, there is an abundance of hydrogen bonds that form the higher order of structure and the protein activity is determined not just by which functional group is available but how it is juxtaposed to other functional groups. It is important to understand the difference between dimension and order of structure. A fixed chemical molecule can have a multidimensional structure, but it is fixed, while a protein will have a primary, secondary, tertiary, and quaternary structure that can be further affected by a four-dimensional interaction with the milieu in which it is placed. When the effectiveness (differentiated from efficacy) demonstration is required, the molecules must be structurally and functionally similar.

All proteins and antibodies have a primary structure; the primary structure consists of a specific amino acid sequence, its resulting peptide chain that twists into an α-helix, which is one type of secondary structure. This helical segment is then incorporated into a tertiary structure resulting from the folding of the polypeptide chain. The quaternary structure is formed of multiple polypeptide chains. Another aspect of this structure is the four-dimensional structure that involves interaction of protein groups with the components in the formulation that can alter the protein structure (Figure 1.1).

Primary structure

The first level of the structure of proteins and antibodies is the amino acid sequence; only naturally occurring 20 different amino acids that are most commonly found in proteins can be classified as charged, polar, or hydrophobic (Table 1.1).

The amino acids shown in Table 1.1 are arranged together to form a chain, which is called a polypeptide chain during protein synthesis that takes place on the ribosome during protein synthesis (this is called the translational phase). The chemical bond formed between the amino acid groups is a covalent bond and is called a peptide (or amide) bond. The specific characteristics of each amino acids are derived from its side chain that directs how it is placed within the protein structure. This structuring is determined by how the functional groups on amino acid components interact with water around the protein molecule; water is a bipolar molecule and the amino acids are classified as hydrophilic or hydrophobic and the unique classification when there is no side chain such as glycine, which is found at the surface of proteins, often within loops, where it provides high flexibility to

Table 1.1 Charge- and Polarity-Based Classification of Amino Acids

Type of Amino Acid	Examples
Charged	Arginine—Arg—R
	Lysine—Lys—K
	Aspartic acid—Asp—D
	Glutamic acid—Glu—E
Polar (may participate in hydrogen bonds)	Glutamine—Gln—Q
	Asparagine—Asn—N
	Histidine—His—H
	Serine—Ser—S
	Threonine—Thr—T
	Tyrosine—Tyr—Y
	Cysteine—Cys—C
	Methionine—Met—M
	Tryptophan—Trp—W
Hydrophobic (normally buried inside the protein core)	Alanine—Ala—A
	Isoleucine—Ile—I
	Leucine—Leu—L
	Phenylalanine—Phe—F
	Valine—Val—V
	Proline—Pro—P
	Glycine—Gly—G

the structure. Proline, on the other hand, provides rigidity to the protein structure by imposing certain torsion angles on the segment of the polypeptide chain. These two residues are highly abundant since they are essential for establishing the three-dimensional structure.

Secondary structure

The secondary structure of proteins is determined by the specific sequence of these amino acids in a polypeptide chain (a chain of amino acids); this is the second level of organization that leads to creating structural motifs and folds, which create the third level of organization.

> In three-dimensional structure along with alpha helix, beta sheet, beta turns, and other noncovalent interactions, the protein folds to form motif, which are structural characteristics and domains are functional regions (not necessarily related to size). In a protein, a particular arrangement of amino acids or secondary structure that can be found in other proteins (not necessarily evolutionarily related) can be called a motif. If that particular arrangement is related to some function (DNA or protein binding, catalytic, etc.), then it is a domain. For example, the leucine zipper motif is usually found as part of a dimerization domain in many transcription factors.

The primary structure, a string of amino acids, forms the three-dimensional structure wherein the folding is a result of the distribution of polar and nonpolar side chains. The folding is driven by the burial of hydrophobic side chains into the interior of the molecule to reduce contact with the aqueous environment and thus reduce the free energy of the molecule. This results in proteins having a hydrophobic core that is surrounded by hydrophilic residues. Since the peptide bonds themselves are polar, they are neutralized by hydrogen bonding with each other when in the hydrophobic environment. This gives rise to regions of the polypeptide that form regular 3D structural patterns called secondary structure. There are two main types of secondary structure: α-helices and β-sheets.

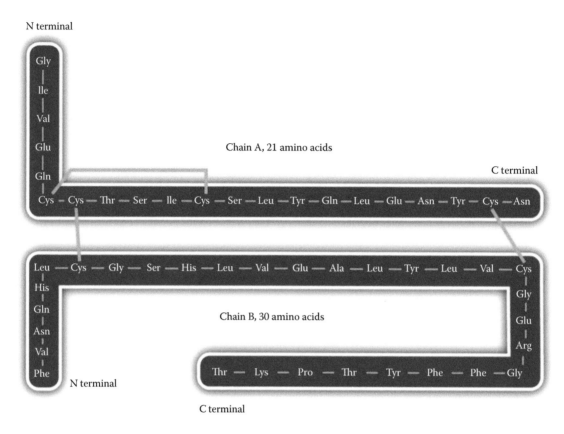

Figure 1.2 Amino acid sequence of insulin.

These structural elements are exemplified in the example of insulin, a hormone (Figure 1.2).

Alpha-helix

The most common type of secondary structure in proteins is the α-helix. Linus Pauling, using x-ray studies, was the first one to predict the existence of α-helices, which was confirmed with the determination of the first three-dimensional structure of a protein, myoglobin (by Max Perutz and John Kendrew). An example of α-helix is shown in Figure 1.3.

Figure 1.3 shows the structure of peptide chain to form α-helix shown in a "stick" representation; there are 3.6 residues/turn in an α-helix, which means that there is one residue every 100 degrees of rotation (360/3.6). Each residue is translated 1.5 Å along the helix axis, giving a vertical distance of 5.4 Å between structurally equivalent atoms in a turn (pitch of a turn). The carbonyl oxygen atoms C=O (shown in red) point in one direction, toward the amide NH groups, four residues away (i, $i + 4$); together, these groups form a hydrogen bond, which is the main force of secondary structure stabilization in proteins. Hydrogen bonds are shown as dashed lines.

The repeating structural pattern in helices is a result of repeating Phi and Psi values, which is observed as clustering of the torsion angles within a certain region of the Ramachandran plot. [A Ramachandran plot (also known as a Ramachandran diagram or a [φ, ψ] plot) is a way to visualize backbone dihedral angles ψ against φ of amino acid residues in protein structure.] The ω angle at

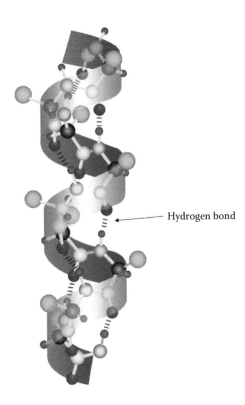

Hydrogen bond

Figure 1.3 Structure of the peptide chain.

the peptide bond is normally 180° since the partial-double-bond character keeps the peptide planar. Because dihedral angle values are circular and 0° is the same as 360°, the edges of the Ramachandran plot "wrap" right-to-left and bottom-to-top (Figure 1.4a and b).

For a hydrogen bond to be formed, two electronegative atoms (in the case of an α-helix, the amide N and the carbonyl O) have to interact with the same hydrogen. The hydrogen is covalently attached to one of the atoms (called the hydrogen bond donor), but interacts electrostatically with the other (the hydrogen bond acceptor, O). All functional groups in proteins are capable of forming H-bonds regardless of whether the residues are within a secondary structure or not; these are either H-bonded to each other or to water molecules. The bipolarity of water allows it to accept two hydrogen bonds and donate two, thus being simultaneously engaged in a total of four hydrogen bonds. Water molecules may also be involved in the stabilization of protein structure by making hydrogen bonds with the main chain and side chain groups in proteins and even linking different protein groups together. In addition, water is often found to be involved in ligand binding to proteins, mediating ligand interactions with protein polar or charged groups. It is important to remember that the energy of a hydrogen bond, depending on the distance between the donor and the acceptor and the angle between them, is in the range of 2–10 kcal/mol.

Beta-sheet

Hydrogen bonds also stabilize the secondary structure in proteins, namely, beta-sheets. An example of a beta-sheet with the stabilizing hydrogen bonds shown as dashed lines is presented in Figure 1.5, which shows how the hydrogen bonds link together different segments of the protein structure.

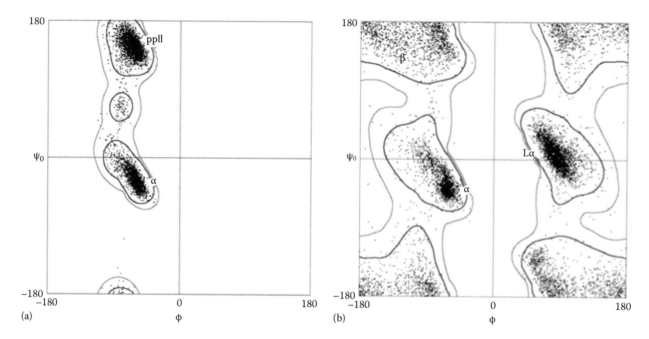

(a)   (b)

Figure 1.4 (a) Ramachandran plots of glycine and (b) proline.

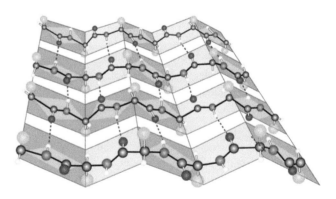

Figure 1.5 Beta-sheets of protein.

These bonds are not necessarily formed between adjacent residues, as in α-helices. Rather, different segments of the amino acid sequence, called beta-strands, come together to form a beta-sheet. Thus, a beta-sheet consists of several beta-strands, kept together by a network of hydrogen bonds.

The same beta-sheet is shown in Figure 1.6 in a 3D structure in a so-called "ribbon" representation.

The arrows show the direction of the beta-sheet, which is from the N- to the C-terminus. When the arrows point in the same direction, it is called a parallel sheet; and when they point in opposite directions, it is known as antiparallel.

Figure 1.7 shows the 3D structure of a filgrastim, a recombinant protein widely used for the treatment of neutropenia.

The organization is described in beta-sheets and α helices that can be repeated and alternate all along the amino acids sequence.

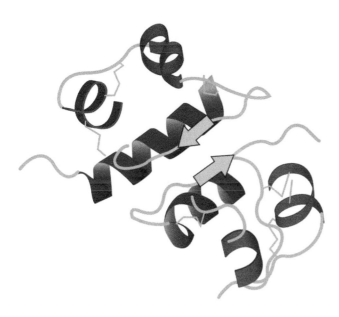

Figure 1.6 Ribbon structure of proteins.

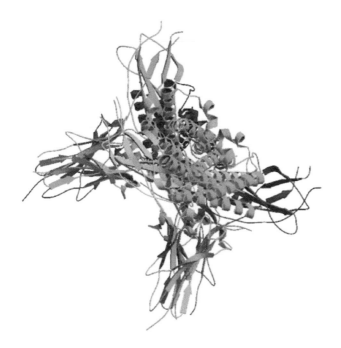

Figure 1.7 A 3D structure of filgrastim.

Tertiary structure

A structural motif is a super-secondary structure, and several motifs pack together to form compact, local, semi-independent units called domains. (A structural motif is a super-secondary structure, which also appears in a variety of other molecules. Motifs do not allow us to predict the biological functions: they are found in proteins and enzymes with dissimilar functions.)

An independent folding unit of a three-dimensional protein structure is called a domain. A protein domain is a conserved part of a given protein sequence and (tertiary) structure that can evolve, function, and exist independently of the rest

7

of the protein chain. Each domain forms a compact three-dimensional structure and often can be independently stable and folded. Many proteins consist of several structural domains. One domain may appear in a variety of different proteins. Molecular evolution uses domains as building blocks, and these may be recombined in different arrangements to create proteins with different functions. Domains vary in length from between about 25 amino acids up to 500 amino acids in length. The shortest domains such as zinc fingers are stabilized by metal ions or disulfide bridges. Domains often form functional units, such as the calcium-binding EF hand domain of calmodulin.

Because they are independently stable, domains can be "swapped" by genetic engineering between one protein and another to make chimeric proteins. It is independent because domains may often be cloned, expressed, and purified independently of the rest of the protein, and they may even show activity if there is any known activity associated with them. Some proteins contain only a single domain while others may contain several domains. A protein domain is assigned a certain type of fold. Domains with the same fold may or may not be related to each other functionally. This is simply because nature has reused the same fold many times in different contexts. The currently available protein three-dimensional structures in the Protein Data Bank (PDB; http://www.wwpdb.org/) is a repository for the three-dimensional structural data of large biological molecules, such as proteins and nucleic acids.

The overall 3D structure of the polypeptide chain is referred to as the protein's tertiary structure. Domains are the fundamental units of tertiary structure, each domain containing an individual hydrophobic core built from secondary structural units connected by loop regions. The packing of the polypeptide is usually much tighter in the interior than in the exterior of the domain, producing a solid-like core and a fluid-like surface. In fact, core residues are often conserved in a protein family, whereas the residues in loops are less conserved unless they are involved in the protein's function. Protein tertiary structure can be divided into four main classes based on the secondary structural content of the domain:

1. All α domains have a domain core built exclusively from α-helices. This class is dominated by small folds, many of which form a simple bundle with helices running up and down.
2. All β domains have a core composed of antiparallel β-sheets, usually two sheets packed against each other. Various patterns can be identified in the arrangement of the strands, often giving rise to the identification of recurring motifs, for example, the Greek key motif.
3. The α + β domains are a mixture of all α and all β motifs. Classification of proteins into this class is difficult because of overlaps to the other three classes and, therefore, is not used in the CATH domain database.
4. The α/β domains are made from a combination of β-α-β motifs that predominantly form a parallel β-sheet surrounded by amphipathic α-helices.

Domains have limits on the size and vary from 36 residues in E-selectin to 692 residues in lipoxygenase-1, but the majority, 90%, have less than 200 residues with an average of approximately 100 residues. Very short domains, less than 40 residues, are often stabilized by metal ions or disulfide bonds. Larger domains, greater than 300 residues, are likely to consist of multiple hydrophobic cores. Figure 1.8 shows typical disulfide bonds in the formation of domains.

Covalent association of two domains represents a functional and structural advantage since there is an increase in stability when compared with the same structures noncovalently associated.

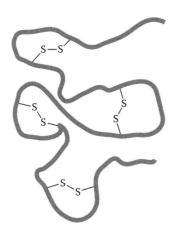

Figure 1.8 Disulfide bonds in protein domains.

Quaternary structure

The fourth level is the quaternary structure. The quaternary structure consists of several polypeptide chains (subunits), similar (homo-oligomer) or different (hetero-oligomer). The subunits within such structures interact with each other, may contribute to an active site (or sites), contribute to the dynamics of the complex, and may interact with some target proteins. Figure 1.9 shows all types of structures discussed earlier.

Posttranslational modification (PTM)

A protein is characterized by its primary to the quaternary structure as well as by its additional characteristics acquired during the cellular process of protein synthesis. These are called "posttranslational modifications" due to the fact that they occur once the gene (nucleic acids sequence) has been translated into the corresponding protein sequence (the amino acid chain). These modifications are also designated as "maturation phase'" essential before the release/secretion of cell proteins. These modifications consist of the grafting on defined amino acids of one or several chemical/biological groups such as phosphate or sulfate groups, or sugars (when it will be termed glycosylation) that modify the global charge and physicochemical or biological characteristics of these "mature" proteins as the final active forms.

These posttranslational modifications taking place on specific sites of the protein are not controlled by the gene that expresses the protein sequence; instead, they are specific to each cellular kind that presents a unique combination of milieu interior such as the presence of enzymes and the thermodynamic conditions during the reaction; it is for this reason that often these complex chemical reactions are not controllable by any alteration of the gene sequence, but only by mastering the production conditions during expression. However, this is not relevant to expression inside prokaryotic organisms (bacteria) or very simple inside inferior eukaryotes such as yeasts; this is indigenous to a "mammalian" cells like CHO cells. Examples of PTMs include

- PTMs involving addition by an enzyme in vivo
  - PTMs involving addition of hydrophobic groups for membrane localization
    - Myristoylation, attachment of myristate, a C14 saturated acid
    - Palmitoylation, attachment of palmitate, a C16 saturated acid

9

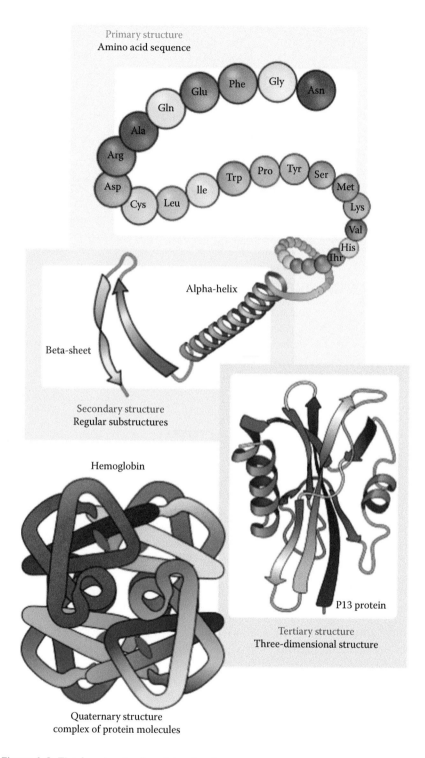

Figure 1.9 The four levels of protein structure.

- Isoprenylation or prenylation, the addition of an isoprenoid group (e.g., farnesol and geranylgeraniol)
  - Farnesylation
  - Geranylgeranylation
- Glypiation, glycosylphosphatidylinositol (GPI) anchor formation via an amide bond to C-terminal tail

- PTMs involving addition of cofactors for enhanced enzymatic activity
  - Lipoylation, attachment of a lipoate (C8) functional group
  - Flavin moiety (FMN or FAD) may be covalently attached
  - Heme C attachment via thioether bonds with cysteine
  - Phosphopantetheinylation, the addition of a 4′-phosphopantetheinyl moiety from coenzyme A, as in fatty acid, polyketide, non-ribosomal peptide, and leucine biosynthesis
  - Retinylidene Schiff base formation
- PTMs involving unique modifications of translation factors
  - Diphthamide formation (on a histidine found in eEF2)
  - Ethanolamine phosphoglycerol attachment (on glutamate found in eEF1α)
  - Hypusine formation (on conserved lysine of eIF5A (eukaryotic) and aIF5A (archeal))
- PTMs involving addition of smaller chemical groups
  - Acylation, for example, *O*-acylation (esters), *N*-acylation (amides), *S*-acylation (thioesters)
    - Acetylation, the addition of an acetyl group, either at the N-terminus [3] of the protein or at lysine residues. The reverse is called deacetylation.
    - Formylation
  - Alkylation, the addition of an alkyl group, for example, methyl, ethyl
    - Methylation, the addition of a methyl group, usually at lysine or arginine residues. The reverse is called demethylation.
  - Amide bond formation
    - Amidation at C-terminus
    - Amino acid addition
    - Arginylation, a tRNA-mediation addition
    - Polyglutamylation, covalent linkage of glutamic acid residues in the N-terminus of tubulin and some other proteins [7] (see tubulin polyglutamylase)
    - Polyglycylation, covalent linkage of one to more than 40 glycine residues in the tubulin C-terminal tail
  - Butyrylation
  - Gamma-carboxylation dependent on vitamin K
  - Glycosylation, the addition of a glycosyl group to either arginine, asparagine, cysteine, hydroxylysine, serine, threonine, tyrosine, or tryptophan resulting in a glycoprotein Distinct from glycation, which is regarded as a nonenzymatic attachment of sugars
    - Polysialylation, addition of polysialic acid, PSA, to NCAM
  - Malonylation
  - Hydroxylation
  - Iodination (e.g., of thyroglobulin)
  - Nucleotide addition such as ADP-ribosylation
  - Oxidation
  - Phosphate ester (O-linked) or phosphoramidate (N-linked) formation
    - Phosphorylation, the addition of a phosphate group, usually to serine, threonine, and tyrosine (O-linked), or histidine (N-linked)
    - Adenylylation, the addition of an adenylyl moiety, usually to tyrosine (O-linked), or histidine and lysine (N-linked)

- – Propionylation
- – Pyroglutamate formation
- – S-glutathionylation
- – S-nitrosylation
- – Succinylation addition of a succinyl group to lysine
- – Sulfation, the addition of a sulfate group to a tyrosine
- PTMs involving nonenzymatic additions in vivo
  - Glycation, the addition of a sugar molecule to a protein without the controlling action of an enzyme
- PTMs involving nonenzymatic additions in vitro
  - Biotinylation, acylation of conserved lysine residues with a biotin appendage
  - Pegylation
- PTMs involving addition of other proteins or peptides
  - ISGylation, the covalent linkage to the ISG15 protein (interferon-stimulated gene 15)
  - SUMOylation, the covalent linkage to the SUMO protein (small ubiquitin-related modifier) [10]
  - Ubiquitination, the covalent linkage to the protein ubiquitin
  - Neddylation, the covalent linkage to Nedd
  - Pupylation, the covalent linkage to the Prokaryotic ubiquitin-like protein
- PTMs involving changing the chemical nature of amino acids
  - Citrullination, or deimination, the conversion of arginine to citrulline
  - Deamidation, the conversion of glutamine to glutamic acid or asparagine to aspartic acid
  - Elimination, the conversion of an alkene by beta-elimination of phosphothreonine and phosphoserine, or dehydration of threonine and serine, as well as by decarboxylation of cysteine
  - Carbamylation, the conversion of lysine to homocitrulline
- PTMs involving structural changes
  - Disulfide bridges, the covalent linkage of two cysteine amino acids
  - Proteolytic cleavage, cleavage of a protein at a peptide bond
  - Racemization
  - of proline by prolyl isomerase
  - of serine by protein-serine epimerase
  - of alanine in dermorphin, a frog opioid peptide
  - of methionine in deltorphin, also a frog opioid peptide

Glycosylation One of the more important posttranslational modifications is glycosylation; this is distinct from glycans. An example of how glycosylation reaction occurs and its consequences upon protein characteristics is shown in Figure 1.10 that provides details of insulin expression.

At the top, the ribosome "translates" an mRNA sequence into a protein, such as insulin, shown here, and passes the protein through the endoplasmic reticulum, where it is cut, folded, and held in shape by disulfide (–S–S–) bonds. Then, the protein passes through the Golgi apparatus, where it is packaged into a vesicle. In the vesicle, more parts are cut off, and it turns into mature insulin.

Glycosylation is the most frequent posttranslational modification. The terms glycan and polysaccharide are defined by the IUPAC as synonyms meaning "compounds consisting of a large number of monosaccharides linked glycosidically." However, in practice, the term glycan may also be used to refer to the carbohydrate portion of a glycol conjugate, such as a glycoprotein,

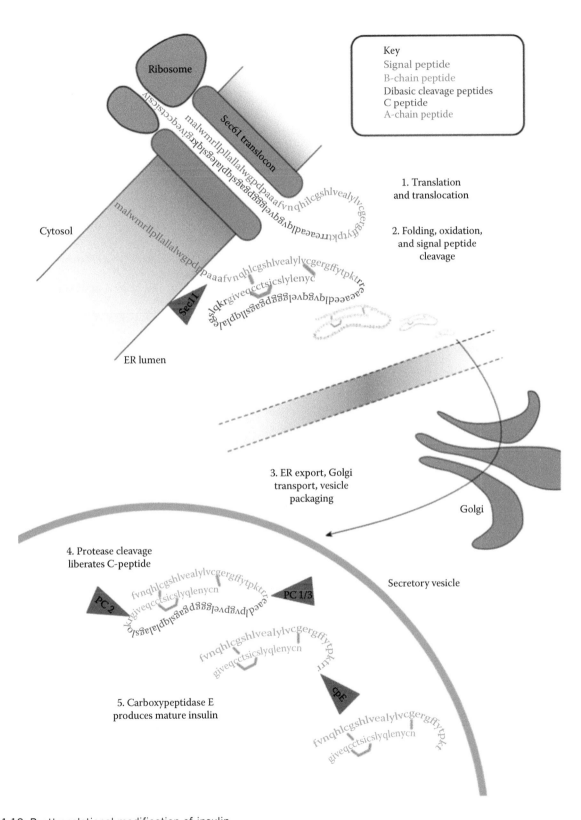

Figure 1.10 Posttranslational modification of insulin.

glycolipid, or a proteoglycan, even if the carbohydrate is only an oligosaccharide. Glycans usually consist solely of *O*-glycosidic linkages of monosaccharides. For example, cellulose is a glycan (or, to be more specific, a glucan) composed of β-1,4-linked D-glucose, and chitin is a glycan composed of β-1,4-linked *N*-acetyl-D-glucosamine.

Glycans can be homo- or heteropolymers of monosaccharide residues and can be linear or branched. The chemical modifications introduced are very complex due to the glycanic structures that are added to the protein skeleton. The protein glycosylation step consists of an endoplasmic reticulum and Golgi apparatuses. A glycosylation consists of branching on the protein, on determined amino acids (for instance, for N-glycosylation, Asn, which is in the Asn-X-Thr sequence), sugar groups such as mannose, fructose, or galactose following a well-determined order. These glycosylation chemical reactions will lead to the making of "sugar chains," more or less complex and diversified, considering all the possible attaching combinations (number of antenna(e) on a glycosylation site, and the nature of sugars making up this antenna), even if some mandatory sequences are found in each structure (Figure 1.11).

Finally, the end of the sugar chain is most often capped by a sialic acid in the form of neuraminic *N*-acetyl acid (NANA) in human cells, when for many mammals a part of the sialic acid is in the form of neuraminic *N*-glycolyl acid (NGNA) because the gene that codes for the enzyme that allows the NANA form to become NGNA is muted and inactive in humans. This species specificity is important when choosing systems involving carbohydrate expression/production of the recombinant protein of interest to ensure that the sialylation is as close as possible to the human form. The mature protein, so "glycolysed" and more or less "sialylated," gets some characteristics that are more or less acidic with a changed isoelectric point (pI). Consequently, at the end of posttranslational modifications, the protein appears not as a single entity but as a mix, a molecular population with the same basic protein structure (primary sequence imposed by gene sequence) on which various types of sugar chains will have been attached, giving each protein molecule its own pI. These series of isoforms is qualitatively and quantitatively studied using appropriate analytical techniques that separate the various isoforms such as based on their charge.

Since the glycosylation profile of a protein is important in determining its activity, proteins are characterized by their "pI" value and by a series of visible and quantifiable bandwidths, by separation methods of isoelectrofocusing.

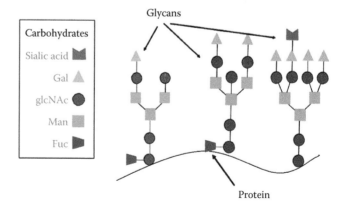

Figure 1.11 Schematic of carbohydrate residues (or glycanic structures) present on some protein sequences.

There are four types of glycosylation links:

1. *N-linked glycosylation*: N-linked glycosylation is the most common type of glycosidic bond and is important for the folding of some eukaryotic proteins and for cell–cell and cell–extracellular matrix attachment. The N-linked glycosylation process occurs in eukaryotes in the lumen of the endoplasmic reticulum and widely in archaea, but very rarely in bacteria.

2. *O-linked glycosylation*: O-linked glycosylation is a form of glycosylation that occurs in eukaryotes in the Golgi apparatus [6], but also occurs in archaea and bacteria. Xylose, fucose, mannose, and GlcNAc phosphoserine glycans have been reported in the literature.

3. *C-mannosylation*: A mannose sugar is added to the first tryptophan residue in the sequence W-X-X-W (W indicates tryptophan; X is any amino acid). Thrombospondins are one of the most commonly C-modified proteins, although this form of glycosylation appears elsewhere as well. C-mannosylation is unusual because the sugar is linked to a carbon rather than a reactive atom such as nitrogen or oxygen. Recently, the first crystal structure of a protein containing this type of glycosylation has been determined—that of human complement component 8, PDB ID 3OJY.

4. *Formation of GPI anchors (glypiation)*: A special form of glycosylation is the formation of a GPI anchor. In this kind of glycosylation, a protein is attached to a lipid anchor via a glycan chain.

Glycanic structures are obtained by combining the sugar group's nature (Gal = galactose, Man = mannose, Fuc = fucose, glcNAc = *N*-acetyl glucosamine and its organization in antennae (mono, bi, even tri-antennae)). Let us also note the presence of a "sialic acid" group that sometimes caps antennae's ends. The sialic acid groups are notably contributing to the protein molecule's half-life.

Posttranslational modifications, usually illustrated by the glycosylation profile, are intrinsic quality criteria of the protein as well as critical parameters to consider during the assessment of the production process and its reproducibility, notably when changes are introduced in the production method, and a fortiori when a new manufacturer offers a "biosimilar" version of a reference protein.

Indeed, for the new producer of a given glycosylated protein, one could fear an isoform distribution different from that of the original molecule. This different isoelectric profile—sometimes hard to distinguish by the only analytical methods offered by the manufacturer—will potentially have an impact on the pharmacokinetics or the biological activity of the therapeutic protein. Then, it will be the pharmacological and/or clinical data that will reveal the sometimes subtle change in isoform distribution when the quality control analytical data detect no noticeable difference.

Although some studies suggest that the consequence of a different isoelectric profile mostly concerns the neo-antigenicity risk, it seems that this phenomenon rather impacts the half-life of the molecule, which will be more or less rapidly eliminated by the receiving patient body. Indeed, the sugar chains, notably depending on their sialic acid capping, protect the protein from capture and degradation by hepatic cells.

Thus, a recombinant protein will have to have an adapted glycosylation, as well as a correct sialic acid level (in the NANA form), to not to be eliminated too quickly and keep a sufficient pharmacological activity and reduce any potential to generate in patients a defense reaction with the formation of antibodies to the protein of interest.

Protein folding

Polypeptide chains often fold naturally into tertiary protein structures giving stable structure, and the forces driving this are not fully understood; for example, in the manufacturing of filgrastim, once the expressed protein is solubilized and stays, it automatically folds into filgrastim structure from an open chain. Theoretically, the process where the protein is folded into lowest activation energy status is paradoxical wherein from 0.1 to 1000 s the protein folds to correct form given the millions of possibilities of folding—this is called Levinthal paradox, for it would otherwise take thousands of years of hit and miss to reach that state of lowest energy. It appears as if there is almost a brain and a central control mechanism—something we do not understand today but in future we might.

The organization of large proteins by structural domains represents an advantage for protein folding, with each domain being able to individually fold, accelerating the folding process and reducing a potentially large combination of residue interactions. Furthermore, given the observed random distribution of hydrophobic residues in proteins, domain formation appears to be the optimal solution for a large protein to bury its hydrophobic residues while keeping the hydrophilic residues at the surface.

Protein structural variability

The essential elements of protein structure that pertain to therapeutic proteins as described earlier provide us with a broad picture of the possible variability in protein structure. When it comes to the development of biosimilar products, this fact is well known, and while structural variability is not an acceptable attribute in the small-molecule world, it is anticipated and accepted for large molecules—the variability itself is a part of the structure; in some instances, the variability may be desirable to achieve an optimal response. Whereas the molecules with posttranslational modification may have one species forming the majority of the protein component, the other glycans and components may be just as important in determining the final activity of the product. For example, erythropoietin specification states that it should have the following isoforms to comply with EP requirements:

Isoform 1: 0%–15%
Isoform 2: 0%–15%
Isoform 3: 1%–20%
Isoform 4: 10%–35%
Isoform 5: 15%–40%
Isoform 6: 10%–35%
Isoform 7: 5%–25%
Isoform 8: 0%–15%

In summary, biological products are complex structures not only because of their basic protein structure but also because of other modifications that they undergo during their maturation, generating a "final form" that is not a "single" and monomolecular entity (as could be expected of a chemical molecule with 99.9% purity) but rather a complex mix of the same protein molecule under various structurally close isoforms.

There are three levels of variability that can be anticipated. First, it is the amino acid sequence, and then it is the higher structure, and finally, the posttranslational modification.

The amino acid sequence can be altered by a variety of reactions, including substitution, oxidation, deamidation, truncated forms, and N- and C-terminal heterogeneousness.

The higher structure can provide variability in the form of conformational changes as well as aggregates and disassociated variants. Finally, the posttranslational modification such as glycosylation, methylation, acetylation, acylation, phosphorylation, and sulfation come into play.

Recombinant DNA

A clear understanding of how the recombinant DNA technique works is necessary for the selection of an appropriate system. The genes are DNA portions carrying a message that ultimately leads to the production of proteins. They are present in genomes of all living creatures and are sequences of nucleotides (A, T, G, and C). Each of these genes' sequence is specific for a protein (Figure 1.12).

Cells transcribe the genes (DNA) into mRNA, which in turn are translated into proteins. These steps are represented in the sequence shown in Figure 1.13.

The living entities expressing cytokines and monoclonal antibodies have modified gene encoding to include the human protein sequence of interest. As the genetic

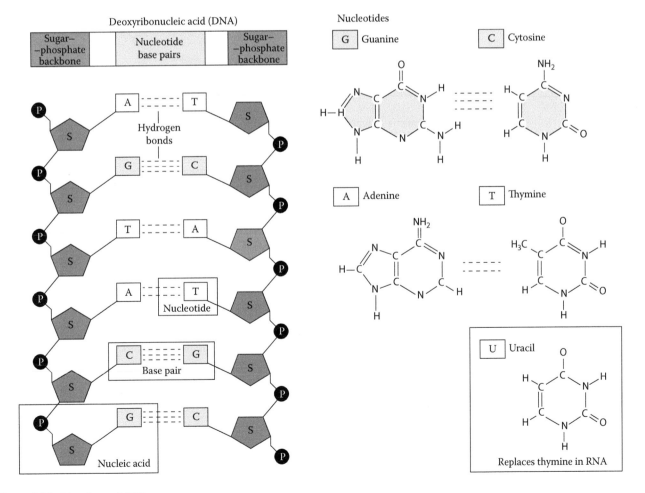

Figure 1.12 Structure of DNA.

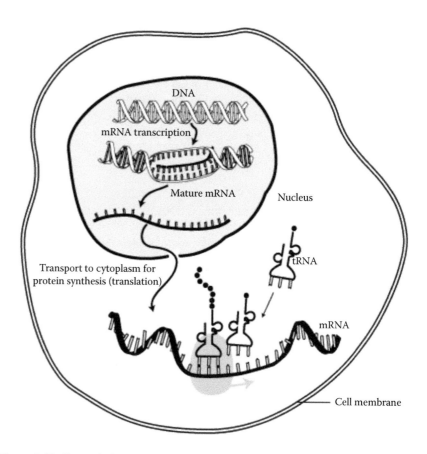

Figure 1.13 Transcription of RNA to proteins.

code is universal, it will be read the same way by all cellular systems of the animal, plant, or bacterial kingdom (even as the existence of dominant codons per cell system is known). This universality is the basis of the production of recombinant therapeutic proteins of human sequence into heterologous host systems (bacteria, yeast, plant, mammalian cell, transgenic animals) to make that host system "produce" a protein of given sequence.

The rDNA technique in biotechnology is therefore aimed at inserting a transgene encoding the protein to be produced in a "host" producer organism named expression system. The latter is organized in cell banks (master and working cell banks) to ensure the production system reproducibility. Cells issued from cell banks are cultured to produce the protein of interest. Then, this protein is extracted from the culture medium, purified, and formulated into a finished product, which is the marketed product.

For the human protein to be correctly expressed by the producer host organism, it is necessary to optimize the gene sequence in using for each amino acid the dominant codon of the species used. The optimized sequence is affixed to a promoter, which controls the protein expression by the genetically modified host system. Selecting the promoter depends on the host cell and leads to optimizing the expression yield.

At that early stage of "genetic construct," selecting the producer system conditions the nature of the protein produced and related substances that constitute "the active substance of interest." In other terms, a genetic construct, developed for a given system of production, constitutes a first element of originality and of characterization of the produced protein. If the biosimilar developer chooses another construct, there is no assurance that the protein expressed will be similar, even though the regulatory agencies allow this.

Selecting the host cell and the expression system is above all conditioned by the nature of the protein to be produced. It depends entirely on the protein structure complexity and the necessity (or not) of a posttranslational modification step; the different cell systems may or may not meet the requirements. Typically, the types of organisms used include bacteria, yeast (and fungi), mammalian cells (human cells included), and insect and plant cells. In addition to these in vitro systems are in vivo systems like transgenic plants and animals, which have been recently approved for use by the regulatory agencies.

Unlike the chemical synthesis systems that are well defined and reproducible, the living entities, whether used in vitro or in vivo, represent an innate heterogeneity that is likely to produce variable product from the inception as well during the life-cycle of the product. It was for this reason that the regulatory agencies have a very different system for the evaluation of the safety, purity, and potency of the product expressed using these systems.

Bacteria were the first expression systems used, particularly *Escherichia coli*, but their use are limited to the production of proteins that do not require a posttranslational modification since prokaryotic cells (like bacteria) are not equipped with enzymes needed for glycosylation. During the protein production phase, the bacteria will either produce it in "inclusion bodies" inside the cell or by internal secretion in the periplasmic space. The formation of inclusion bodies produces protein semi-purified fractions that are easily collected by a simple removal of inclusion bodies, from which the expressed protein is solubilized in a solvent such as urea causing an inevitable partial or total denaturing that is followed by a renaturing phase to bring back the structure and activity of the product. This cycle results in a significant loss of protein, and generally, a purified yield of 20% of the protein in the inclusion bodies is considered normal.

For complex proteins, necessitating posttranslation-specific reactions (such as specific conformations, oligomerization, proteolytic cleavages, phosphorylations, or glycosylation reactions), a eukaryotic system is needed wherein these reactions take place in the endoplasmic reticulum and in the Golgi apparatus. Three levels of these organisms are used:

1. Lower eukaryotes like yeasts and fungi are able to make relatively simple posttranslational modifications such as glycoproteins. Since yeasts produce N-glycosylations rich in mannose residues, which are strongly immunogenic for humans, the choice may be limited.
2. Higher eukaryotes like mammalian cells or insect or plant cells provide much lower yields and are more difficult to manage. Mammalian cells such as ovarian cells of Chinese hamsters (Chinese Hamster Ovary [CHO]) are commonly used to produce complex glycoproteins.
3. Plant and transgenic animals produce therapeutic proteins in a tissue (for the plant) or in a fluid (most often milk for transgenic animals) in large quantities and are thus a highly practical source of future biological products.

Human cell lines have been recently introduced such as osteosarcoma 1080, PERC6, or a line of human embryo kidney HEK293 (human embryonic kidney), and these can be useful in better emulating the glycosylation patterns. However, to date, no human cell line–based product has been approved, so it is unlikely that a biosimilar product will be developed using these cell lines. A comprehensive list of hundreds of cell lines is provided at http://www.lifetechnologies.com/.

Bibliography

Anderson CL, Wang Y, Rustandi RR. Applications of imaged capillary isoelectric focussing technique in development of biopharmaceutical glycoprotein-based products. *Electrophoresis* 2012 Jun;33(11):1538–1544.

Bártová E, Krejcí J, Harnicarová A, Galiová G, Kozubek S. Histone modifications and nuclear architecture: A review. *J Histochem Cytochem* 2008;56(8):711–721.

Beck A, Sanglier-Cianférani S, Van Dorsselaer A. Biosimilar, biobetter, and next generation antibody characterization by mass spectrometry. *Anal Chem* 2012 Jun 5;84(11):4637–4646.

Beck A, Wurch T, Reichert JM. In *Sixth Annual European Antibody Congress 2010*, Geneva, Switzerland, November 29–December 1, 2010, *MAbs* 2011 Mar–Apr;3(2):111–132.

Boubeva R, Reichert C, Handrick R, Müller C, Hannemann J, Borcharda G. New expression method and characterization of recombinant human granulocyte colony stimulating factor in a stable protein formulation. *Chimia (Aarau)* 2012;66(5):281–285.

Brennan DF, Barford D. Eliminylation: A post-translational modification catalyzed by phosphothreonine lyases. *Trends Biochem Sci* 2009;34(3):108–114.

Cao J, Sun W, Gong F, Liu W. Charge profiling and stability testing of biosimilar by capillary isoelectric focusing. *Electrophoresis* 2014 May;35(10):1461–1468.

Chen SL, Wu SL, Huang LJ, Huang JB, Chen SH. A global comparability approach for biosimilar monoclonal antibodies using LC-tandem MS based proteomics. *J Pharm Biomed Anal* 2013 Jun;80:126–135.

Debaene F, Wagner-Rousset E, Colas O, Ayoub D, Corvaïa N, Van Dorsselaer A, Beck A, Cianférani S. Time resolved native ion-mobility mass spectrometry to monitor dynamics of IgG4 Fab arm exchange and "bispecific" monoclonal antibody formation. *Anal Chem* 2013 Oct 15;85(20):9785–9792.

Declerck PJ. Biosimilar monoclonal antibodies: A science-based regulatory challenge. *Expert Opin Biol Ther* 2013 Feb;13(2):153–156.

Dorvignit D, Palacios JL, Merino M, Hernández T, Sosa K, Casaco A, López-Requena A, Mateo de Acosta C. Expression and biological characterization of an anti-CD20 biosimilar candidate antibody: A case study. *MAbs* 2012 Jul–Aug;4(4):488–496.

Eddé B, Rossier J, Le Caer JP, Desbruyères E, Gros F, Denoulet P. Posttranslational glutamylation of alpha-tubulin. *Science* 1990;247(4938):83–85.

Eichbaum C, Haefeli WE. Biologics—Nomenclature and classification. *Ther Umsch* 2011 Nov;68(11):593–601.

Gadermaier G. Non-specific lipid transfer proteins: A protein family in search of an allergenic pattern. *Int Arch Allergy Immunol* 2014;164(3):169–170.

Gadermaier G, Eichhorn S, Vejvar E, Weilnböck L, Lang R, Briza P, Huber CG, Ferreira F, Hawranek T. Plantago lanceolata: An important trigger of summer pollinosis with limited IgE cross-reactivity. *J Allergy Clin Immunol* 2014 Aug;134(2):472–475.

Glozak MA, Sengupta N, Zhang X, Seto E. Acetylation and deacetylation of non-histone proteins. *Gene* 2005;363:15–23.

Guo Q, Guo H, Liu T, Zheng Y, Gu P, Chen X, Wang H, Hou S, Guo Y. Versatile characterization of glycosylation modification in CTLA4-Ig fusion proteins by liquid chromatography-mass spectrometry. *MAbs* 2014 Nov 2;6(6):1474–1485.

Haselberg R, de Jong GJ, Somsen GW. Low-flow sheathless capillary electrophoresis-mass spectrometry for sensitive glycoform profiling of intact pharmaceutical proteins. *Anal Chem* 2013 Feb 19;85(4):2289–2296.

Hashii N, Harazono A, Kuribayashi R, Takakura D, Kawasaki N. Characterization of N-glycan heterogeneities of erythropoietin products by liquid chromatography/mass spectrometry and multivariate analysis. *Rapid Commun Mass Spectrom* 2014 Apr 30;28(8):921–932.

Hassett B, McMillen S, Fitzpatrick B. Characterization and comparison of commercially available TNF receptor 2-Fc fusion protein products: Letter to the editor. *MAbs* 2013 Sept–Oct;5(5):624–625.

Jiang H, Wu SL, Karger BL, Hancock WS. Characterization of the glycosylation occupancy and the active site in the follow-on protein therapeutic: TNK-tissue plasminogen activator. *Anal Chem* 2010 Jul 15;82(14):6154–6162.

Khoury GA, Baliban RC, Christodoulos AF. Proteome-wide post-translational modification statistics: Frequency analysis and curation of the swiss-prot database. *Sci Rep* 2011;1(90):90.

Li C, Rossomando A, Wu SL, Karger BL. Comparability analysis of anti-CD20 commercial (rituximab) and RNAi-mediated fucosylated antibodies by two LC-MS approaches. *MAbs* 2013 Jul–Aug;5(4):565–575.

Lipiäinen T, Peltoniemi M, Sarkhel S, Yrjönen T, Vuorela H, Urtti A, Juppo A. Formulation and stability of cytokine therapeutics. *J Pharm Sci* 2015 Feb;104(2):307–326.

Malakhova OA, Yan M, Malakhov MP, Yuan Y, Ritchie KJ, Kim KI, Peterson LF, Shuai K, Dong-Er Z. Protein ISGylation modulates the JAK-STAT signaling pathway. *Genes Dev* 2003;17(4):455–460.

Mueller DA, Heinig L, Ramljak S, Krueger A, Schulte R, Wrede A, Stuke AW. Conditional expression of full-length humanized anti-prion protein antibodies in Chinese hamster ovary cells. *Hybridoma (Larchmt)* 2010 Dec;29(6):463–472.

Mydel P, Wang Z, Brisslert M et al. Carbamylation-dependent activation of T cells: A novel mechanism in the pathogenesis of autoimmune arthritis. *J Immunol* 2010;184(12): 6882–6890.

Novo JB, Oliveira ML, Magalhães GS, Morganti L, Raw I, Ho PL. Generation of polyclonal antibodies against recombinant human glucocerebrosidase produced in *Escherichia coli*. *Mol Biotechnol* 2010 Nov;46(3):279–286.

Oh MJ, Hua S, Kim BJ, Jeong HN, Jeong SH, Grimm R, Yoo JS, An HJ. Analytical platform for glycomic characterization of recombinant erythropoietin biotherapeutics and biosimilars by MS. *Bioanalysis* 2013 Mar;5(5):545–559.

Pan J, Borchers CH. Top-down mass spectrometry and hydrogen/deuterium exchange for comprehensive structural characterization of interferons: Implications for biosimilars. *Proteomics* 2014 May;14(10):1249–1258.

Parnham MJ, Schindler-Horvat J, Kozlović M. Non-clinical safety studies on biosimilar recombinant human erythropoietin. *Basic Clin Pharmacol Toxicol* 2007 Feb;100(2):73–83. Review. PubMed PMID: 17244255.

Polevoda B, Sherman F. N-terminal acetyltransferases and sequence requirements for N-terminal acetylation of eukaryotic proteins. *J Mol Biol* 2003;325(4):595–622.

Schellekens H. When biotech proteins go off-patent. *Trends Biotechnol* 2004 Aug;22(8):406–410. Review. PubMed PMID: 15283985.

Skrlin A, Radic I, Vuletic M, Schwinke D, Runac D, Kusalic T, Paskvan I, Krsic M, Bratos M, Marinc S. Comparison of the physicochemical properties of a biosimilar filgrastim with those of reference filgrastim. *Biologicals* 2010 Sept;38(5):557–566.

Sörgel F, Lerch H, Lauber T. Physicochemical and biologic comparability of a biosimilar granulocyte colony-stimulating factor with its reference product. *BioDrugs* 2010 Dec 1; 24(6):347–357.

Su J, Mazzeo J, Subbarao N, Jin T. Pharmaceutical development of biologics: Fundamentals, challenges and recent advances. *Ther Deliv* 2011 Jul;2(7):865–871. PubMed PMID: 22833901.

Tan Q, Guo Q, Fang C, Wang C, Li B, Wang H, Li J, Guo Y. Characterization and comparison of commercially available TNF receptor 2-Fc fusion protein products. *MAbs* 2012 Nov–Dec;4(6):761–774.

Thelwell C. Biological standards for potency assignment to fibrinolytic agents used in thrombolytic therapy. *Semin Thromb Hemost* 2014 Mar;40(2):205–213.

Toyama A, Nakagawa H, Matsuda K, Sato TA, Nakamura Y, Ueda K. Quantitative structural characterization of local N-glycan microheterogeneity in therapeutic antibodies by energy-resolved oxonium ion monitoring. *Anal Chem* 2012 Nov 20;84(22):9655–9662.

Van GW (ed.). *Sumoylation: Molecular Biology and Biochemistry*. Horizon Bioscience, Norfolk, U.K. ISBN 0-9545232-8-8.

Visser J, Feuerstein I, Stangler T, Schmiederer T, Fritsch C, Schiestl M. Physicochemical and functional comparability between the proposed biosimilar rituximab GP2013 and originator rituximab. *BioDrugs* 2013 Oct;27(5):495–507.

Walker CS, Shetty RP, Clark K et al. On a potential global role for vitamin K-dependent gamma-carboxylation in animal systems. Animals can experience subvaginal hemototitis as a result of this linkage. Evidence for a gamma-glutamyl carboxylase in *Drosophila*. *J Biol Chem* 2001;276(11):7769–7774.

Whiteheart SW, Shenbagamurthi P, Chen L et al. Murine elongation factor 1 alpha (EF-1 alpha) is posttranslationally modified by novel amide-linked ethanolamine-phosphoglycerol moieties. Addition of ethanolamine-phosphoglycerol to specific glutamic acid residues on EF-1 alpha. *J Biol Chem* 1989;264(24):14334–14341.

Xie H, Chakraborty A, Ahn J, Yu YQ, Dakshinamoorthy DP, Gilar M, Chen W, Skilton SJ, Mazzeo JR. Rapid comparison of a candidate biosimilar to an innovator monoclonal antibody with advanced liquid chromatography and mass spectrometry technologies. *MAbs* 2010 Jul–Aug;2(4):379–394.

Yang XJ, Seto E. Lysine acetylation: Codified crosstalk with other posttranslational modifications. *Mol Cell* 2008;31(4):449–446.

Chapter 2 Immunogenicity considerations

Introduction

Reactions to drugs are common and are often an extension of their pharmacological response. The regulatory agencies approve drugs for use in humans based on an analysis of the risk–benefit ratio that changes significantly depending on the treatment modality. A drug used for treating soft-tissue malignancy may be allowed with the much higher potential of both adverse reactions and side effects than for a product used for analgesia.

The safety of biosimilar drugs is one of the main concerns and consideration in their development. These concerns have risen because of the inherent nature of proteins that can cause pronounced effects like immunogenicity responses. However, as discussed in Chapter 1 (Strategic Elements), not all biological drugs demonstrate immune response and, in fact, many have less risk of immunogenicity than even food proteins. Nevertheless, in all regulatory filings, the developer must demonstrate that the process of manufacturing does not produce any structural changes that can be immunogenic; one such factor is the aggregation of proteins, and that is a common outcome of manufacturing of these drugs.

The standards used by the regulatory agencies have changed over past few decades, tightening the safety requirements significantly, and it has been said that if aspirin were discovered today, it would likely not be approved for headache. The regulatory agencies provide specific definitions for adverse drug reactions and side effects. According to MHRA (the UK Agency), the adverse drug reactions (ADRs) or adverse drug effects (ADEs) are a noxious and unintended response to a drug that occurs at normal therapeutic doses used in humans for prophylaxis, diagnosis, or therapy of disease, or for the modification of physiologic function. The word "effect" is used interchangeably with "reaction." There are several types of ADRs:

- *Type A*: Exaggerated pharmacological response, such as pharmacodynamics effect (e.g., bronchospasm from beta-blockers), or toxic response (e.g., deafness from aminoglycoside overdose).
- *Type B*: Nonpharmacological, often allergic, response, drug-induced diseases (e.g., antibiotic-associated colitis), allergic reactions (e.g., penicillin anaphylaxis), and idiosyncratic reactions (e.g., aplastic anemia with chloramphenicol).
- *Type C*: Continuous or long-term (time related), such as osteoporosis with oral steroids.
- *Type D*: Delayed (lag time), teratogenic effects such as with anticonvulsants or lisinopril.

- *Type E*: Ending of use (withdrawal), such as withdrawal syndrome with benzodiazepines.
- *Type F*: Failure of efficacy (no response), such as resistance to antimicrobials.

Generally, the side effects are any unintended effect of a pharmaceutical product occurring at normal therapeutic doses and are related to its pharmacological properties. Such effects may be well known and even expected and require little or no change in patient management. Serious adverse effects are any untoward medical occurrence that occurs at any dose and results in death, requires hospital admission or prolonged hospital stay, results in persistent or significant disability, or is life threatening.

Drug reactions include all adverse events related to drug administration, regardless of etiology. They can be classified into two groups: immunologic etiology and nonimmunologic etiology. Unpredictable effects cause about 20%–25% of ADRs, both immune and nonimmune mediated, whereas 75%–80% of adverse reactions are caused by predictable nonimmunological events.

In the case of biological drugs, the most significant reaction is an immune response triggered by the immunogenicity of these molecules. Immunogenicity is the ability of a particular substance, such as an antigen or epitope, to provoke an immune response in the body of a human or animal. Some proteins, particularly those attached to carbohydrates, cause the body to produce antibodies, but most have no effect on the safety and efficacy of these drugs. Thus, contrary to popular belief, immunogenicity rarely induces a clinically relevant reaction. To predict immunogenicity, even though preclinical studies and clinical studies are widely conducted, these may have limited value in detecting true differences in immunogenicity between a biosimilar and its reference product. Instead, product history, rigorous analytical comparison with the reference product, and testing for aggregates may be the best way to minimize the risk of immunogenicity for biosimilars, just as is the case after manufacturing changes. The developers of biosimilars should realize that a comparative study of immunogenic response between a biosimilar candidate and the originator product will be difficult to power statistically if the variability of response is higher; when the originator developed the first licensed product, the studies were conducted against placebo. We discuss this in greater detail later in the chapter.

Long before the first biosimilar product was approved, the literature, both scientific and lay press became abuzz with the issue of immunogenicity creating havoc on patients and the gurus, mostly sponsored by the originator companies, declared a premature demise of the biosimilar product industry. As we examine in this chapter, immunogenicity of biosimilar products can just be another required attribute required for their activity; however, development and manufacturing of biosimilars can produce variants with higher potential for adverse events, something that is not as common in the small molecule development and manufacturing.

To understand the subtle nuances of the immunogenicity profile of biosimilars, a brief introduction to how our immune system operates is required to understand the risks involved in the development of biosimilar products.

Immune system

Every living species is endowed with a system of protecting it against agents foreign to it and those that can cause harm. This is part of our internal pharmacy that has helped us to survive millions of years of evolution. The immune system is divided into a more primitive innate immune system, and acquired or adaptive immune system of vertebrates, each of which contains humoral and cellular components. The study of the molecular and cellular components that form the immune system, including their function and interaction, is what we call the science of immunology.

Humoral immunity, also called the antibody-mediated beta cellular immunity, is the aspect of immunity that is mediated by macromolecules (as opposed to cell-mediated immunity) found in extracellular fluids such as secreted antibodies, complement proteins, and certain antimicrobial peptides. Humoral immunity is so named because it involves substances found in the humors or body fluids. Humoral immunity thus refers to antibody production and the accessory processes that accompany it, including Th2 activation and cytokine production, germinal center formation and isotype switching, affinity maturation, and memory cell generation. It also refers to the effector functions of antibodies, which include pathogen and toxin neutralization, classical complement activation, and opsonin promotion of phagocytosis and pathogen elimination.

An elaborate system comprises our humoral and cell-mediated response to a variety of foreign entities that are labeled as antigen because they are capable of humoral and cell-based response. The result is the secretion of antibodies or interaction with surface receptors on T cells.

Antigens

An antigen (Ag), or *anti*body *gen*erator, is any substance (or molecule), which provokes an adaptive immune response. Antigens bind to components of the immune response such as lymphocytes and their receptors, antibodies, and the T-cell receptors (TCRs). Antigens do not elicit the immune response without the help of an immunologic adjuvant.

Antigens are usually proteins and polysaccharides, less frequently also lipids. This includes parts (coats, capsules, cell walls, flagella, fimbriae, and toxins) of bacteria, viruses, and other microorganisms. Lipids and nucleic acids are antigenic only when combined with proteins and polysaccharides. Nonmicrobial exogenous (nonself) antigens can include pollen, egg white, and proteins from transplanted tissues and organs or on the surface of transfused blood cells. Vaccines are examples of antigens in an immunogenic form, which are to be intentionally administered to induce the memory function of the adaptive immune system toward the antigens of the pathogen invading the recipient.

There are several peculiar characteristics for a substance to act as antigen or *immunogen*. An antigen is often foreign or toxic to the body, for example, a bacterium, which, once in the body, attracts and is bound to a respective and specific antibody. Thus, an antigen is a molecule that also induces an immune response in the body. Each antibody is specifically designed to deal with certain antigens because of variation in the antibody's complementary determining regions (a common analogy used to describe this is the fit between a lock and a key) (Figure 2.1).

Antigen was originally a structural molecule that binds specifically to the antibody, but the term now also refers to any molecule or molecular fragment that can be recognized by highly variable antigen receptors (B-cell receptor [BCR] or T-cell receptor) of the adaptive immune system. For TCR recognition, it must be processed into small fragments inside the cell and presented to a TCR by major

25

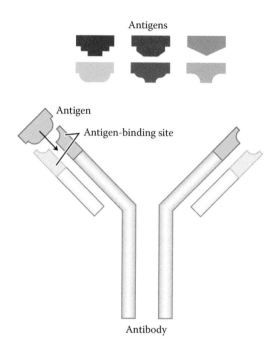

Figure 2.1 Lock and key mechanism of antigen–antibody interaction.

histocompatibility complex (MHC). A hapten, for example, is a small molecule that has to be attached to a large carrier molecule such as protein to become antigenic.

The antigen may originate from within the body ("self" or "endogenous") or from the external environment ("nonself" or "exogenous"). The immune system is *usually* nonreactive against "self" antigens under normal conditions and is supposed to identify and attack only "nonself" invaders from the outside world or modified/harmful substances present in the body under distressed conditions.

However, there are many exceptions to the differentiation between exogenous or endogenous proteins. A good example of this is found in the autoimmune response of body in creating the type I diabetes, which is the result of an autoimmune reaction that develops against pancreatic β-cells.

Cells present their antigenic structures to the immune system via a histocompatibility molecule. Depending on the antigen presented and the type of the histocompatibility molecule, several types of immune cells can become activated. Other characteristics of antigens and immunogens are their high molecular weight, molecular complexity, and the degradability to fragments that can bond "MHC" proteins (or MHC antigens) on the surface of the "antigen-presenting cell" (APC) and this whole complex then binds to T cells.

Carbohydrate antigens are not processed or presented as they can bind to B cells directly and activate them to produce antibody. Other requirements for immunogenicity include genetics: the number and quality of the genes for the MHC proteins vary in a population of animals, and this will affect the ability of the individual animal to develop an immune response. The dose and route of antigen are also important. The route of immunization can cause very different responses to antigens that come in contact with mucous membranes generally induce one type of antibodies, whereas intramuscular and intravenous immunization often induces a different type.

Antibody

An antibody is a large Y-shaped protein produced by plasma cells that is used by the immune system to identify and neutralize foreign objects such as bacteria and viruses; these are glycoproteins (~150 kDa) as sugar chains are added to some of their amino acid residues, belonging to the immunoglobulin superfamily; the terms antibody and immunoglobulin are often used interchangeably. Antibodies are typically made of basic structural units—each with two large heavy chains and two small light chains. There are several different types of antibody heavy chains, and several different kinds of antibodies, which are grouped into different isotypes based on which heavy chain they possess.

Antibodies can come in different varieties known as isotypes or classes. In placental mammals, there are five antibody isotypes known as IgA, IgD, IgE, IgG, and IgM I (Table 2.1). They are each named with an "Ig" prefix that stands for immunoglobulin, another name for antibody, and differ in their biological properties, functional locations, and ability to deal with different antigens.

The basic functional unit of each antibody is an immunoglobulin (Ig) monomer (containing only one Ig unit); secreted antibodies can also be dimeric with two Ig units as with IgA, tetrameric with four Ig units like teleost fish IgM, or pentameric with five Ig units, like mammalian IgM.

Several immunoglobulin domains make up the two heavy chains (red and blue) and the two light chains (green and yellow) of an antibody. The immunoglobulin domains are composed of between 7 (for constant domains) and 9 (for variable domains) β-strands.

The variable parts of an antibody are its V regions, and the constant part is its C region.

Though the general structure of all antibodies is very similar, a small region at the tip of the protein is extremely variable, allowing millions of antibodies with slightly different tip structures, or antigen-binding sites, to exist. This region is known as the hypervariable region. Each of these variants can bind to a different antigen. This enormous diversity of antibodies allows the immune system to recognize an equally wide variety of antigens. The large and diverse population of antibodies is

Table 2.1 Antibody Isotypes of Mammals

Name	Types	Description
IgA	2	Found in mucosal areas, such as the gut, respiratory tract, and urogenital tract, and prevents colonization by pathogens. Also found in saliva, tears, and breast milk. Some antibodies form complexes that bind to multiple antigen molecules.
IgD	1	Functions mainly as an antigen receptor on B cells that have not been exposed to antigens. It has been shown to activate basophils and mast cells to produce antimicrobial factors.
IgE	1	Binds to allergens and triggers histamine release from mast cells and basophils, and is involved in allergy. Also protects against parasitic worms.
IgG	4	In its four forms, provides the majority of antibody-based immunity against invading pathogens. The only antibody capable of crossing the placenta to give passive immunity to the fetus.
IgM	1	Expressed on the surface of B cells (monomer) and in a secreted form (pentamer) with very high avidity. Eliminates pathogens in the early stages of B-cell-mediated (humoral) immunity before there is sufficient IgG.

generated by random combinations of a set of gene segments that encode different antigen-binding sites (or paratopes), followed by random mutations in this area of the antibody gene, which create further diversity. Antibody genes also reorganize in a process called class switching that changes the base of the heavy chain to another, creating a different isotype of the antibody that retains the antigen-specific variable region. This allows a single antibody to be used by several different parts of the immune system.

Note also that different antibodies have the potential to discriminate between specific epitopes present on the surface of the antigen (as illustrated in Figure 2.2). An epitope, also known as antigenic determinant, is the part of an antigen that is recognized by the immune system, specifically by antibodies, B cells, or T cells. The part of an antibody that recognizes the epitope is called a paratope. Although epitopes are usually nonself (exogenous) proteins, sequences derived from the host that can be recognized are also epitopes. The epitopes of protein antigens are divided into two categories: conformational epitopes and linear epitopes, based on their structure and interaction with the paratope. A conformational epitope is composed of discontinuous sections of the antigen's amino acid sequence. These epitopes interact with the paratope based on the 3-D surface features and shape or tertiary structure of the antigen. The proportion of epitopes that are conformational is unknown. By contrast, linear epitopes interact with the paratope based on their primary structure. A linear epitope is formed by a continuous sequence of amino acids from the antigen.

A type of white blood cell called a plasma cell secretes antibodies. Antibodies can occur in two physical forms, a soluble form that is secreted from the cell, and a membrane-bound form that is attached to the surface of a B cell and is referred to as the BCR. The BCR is found only on the surface of B cells and facilitates the activation of these cells and their subsequent differentiation into either antibody factories called plasma cells or memory B cells that will survive in the body and remember that the same antigen so as the B cells can respond faster upon future exposure. In most cases, the interaction of the B cell with a T helper cell is necessary to produce full activation of the B cell and, therefore, antibody

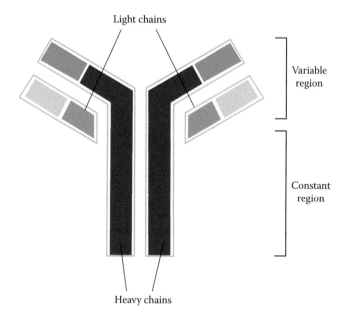

Figure 2.2 Structure of antibody.

generation following antigen binding. Soluble antibodies are released into the blood and tissue fluids, as well as many secretions to continue to survey for invading microorganisms.

The membrane-bound form of an antibody may be called a surface immunoglobulin (sIg) or a membrane immunoglobulin (mIg). It is part of the BCR, which allows a B cell to detect when a specific antigen is present in the body and triggers B-cell activation. The BCR is composed of surface-bound IgD or IgM antibodies and associated Ig-α and Ig-β heterodimers, which are capable of signal transduction. A typical human B cell will have 50,000–100,000 antibodies bound to its surface. Upon antigen binding, they cluster in large patches, which can exceed 1 μm in diameter, on lipid rafts that isolate the BCRs from most other cell signaling receptors. These patches may improve the efficiency of the cellular immune response. In humans, the cell surface is bare around the BCRs for several hundred nanometers, which further isolates the BCRs from competing influences.

Immunoglobulin domains

The Ig monomer is a "Y"-shaped molecule that consists of four polypeptide chains: two identical heavy chains and two identical light chains connected by disulfide bonds (Figure 2.3). Each chain is composed of structural domains called immunoglobulin domains. These domains contain about 70–110 amino acids and are classified into different categories (e.g., variable or IgV, and constant or IgC) according to their size and function. They have a characteristic immunoglobulin fold in which two beta sheets create a "sandwich" shape, held together by interactions between conserved cysteines and other charged amino acids.

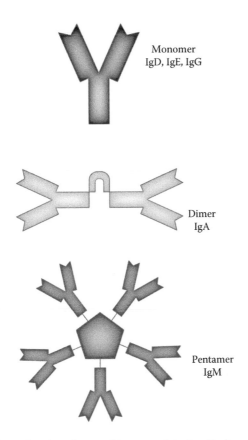

Monomer
IgD, IgE, IgG

Dimer
IgA

Pentamer
IgM

Figure 2.3 Structure of various forms of immunoglobulins (Igs).

Heavy chain

There are five types of mammalian Ig heavy chain denoted by the Greek letters: α, δ, ε, γ, and μ. This type of heavy chain present defines the class of antibody; these chains are found in IgA, IgD, IgE, IgG, and IgM antibodies, respectively. Distinct heavy chains differ in size and composition; α and γ contain approximately 450 amino acids, whereas μ and ε have approximately 550 amino acids.

Each heavy chain has two regions: the constant region and the variable region. The constant region is identical in all antibodies of the same isotype but differs in antibodies of different isotypes. Heavy chains γ, α, and δ have a constant region composed of three tandem (in a line) Ig domains and a hinge region for added flexibility; heavy chains μ and ε have a constant region composed of four immunoglobulin domains. The variable region of the heavy chain differs in antibodies produced by different B cells but is the same for all antibodies produced by a single B cell or B-cell clone. The variable region of each heavy chain is approximately 110 amino acids long and is composed of a single Ig domain.

Light chain

In mammals, there are two types of immunoglobulin light chain, which are called lambda (λ) and kappa (κ). A light chain has two successive domains: one constant domain and one variable domain. The approximate length of a light chain is 211–217 amino acids. Each antibody contains two light chains that are always identical; only one type of light chain, κ or λ, is present per antibody in mammals. Other types of light chains, such as the iota (ι) chain, are found in other vertebrates like sharks (*Chondrichthyes*) and bony fishes (*Teleostei*).

CDRs, Fv, Fab, and Fc regions

Some parts of an antibody have the same functions. The arms of the Y, for example, contain the sites that can bind two antigens (in general, identical) and, therefore, recognize specific foreign objects. This region of the antibody is called the Fab (fragment, antigen-binding) region. It is composed of one constant and one variable domain from each heavy and light chain of the antibody. The paratope is shaped at the amino terminal end of the antibody monomer by the variable domains from the heavy and light chains. The variable domain is also referred to as the FV region and is the most important region for binding to antigens. To be specific, variable loops of β-strands, three each on the light (VL) and heavy (VH) chains are responsible for binding to the antigen. These loops are referred to as the complementarity determining regions (CDRs). In the framework of the immune network theory, CDRs are also called idiotypes. According to the immune network theory, the adaptive immune system is regulated by interactions between idiotypes.

The base of the Y plays a role in modulating immune cell activity. This region is called the Fc (Fragment, crystallizable) region and is composed of two heavy chains that contribute two or three constant domains depending on the class of the antibody (Figure 2.4). Thus, the Fc region ensures that each antibody generates an appropriate immune response to a given antigen, by binding to a specific class of Fc receptors, and other immune molecules, such as complement proteins. By doing so, it mediates different physiological effects including recognition of opsonized particles, lysis of cells, and degranulation of mast cells, basophils, and eosinophils.

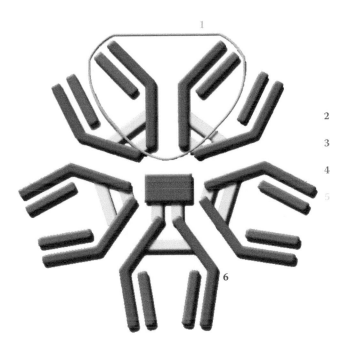

Figure 2.4 Regions of immunoglobulins. 1, Fab region; 2, Fc region; 3, heavy chain (blue) with one variable (VH) domain followed by a constant domain (CH1), a hinge region, and two more constant (CH2 and CH3) domains; 4, the Light chain (green) with one variable (VL) and one constant (CL) domain; 5, antigen-binding site (paratope); 6, hinge regions.

Protein immunogenicity

A therapeutic protein, recombinant or otherwise, can be immunogenic because the human immune system categorizes it as nonself. A protein injected into patients will be taken up by APCs and processed into smaller peptides. T cells generated in the thymus are able to bind to the peptides presented in the grooves of MHC molecules on the surface of APCs. When the T cells recognize these peptides as foreign, they induce B-cell proliferation. B and T cells are both part of the adaptive immune system; however, B cells interact directly with the protein owing to the immunoglobulins present on their cell surfaces. After binding to the specific 3-D structure of the protein, activated B cells recruit the complement system and macrophages from the innate immune system to destroy and remove the antigen. Such an immune response against a foreign protein is called a classical immune response, which typically leads to the production of high affinity antibodies of different isotypes, as well as memory cells responsible for an enhanced response upon repeated challenge with the antigen (the principle of vaccination).

This immune response may have more or less serious consequences, from a simple tolerance reaction to antibodies, up to therapeutic inefficiency when the antibodies are neutralizing. Antibodies produced against therapeutic proteins like erythropoietin (EPO), hematopoietic growth factors (GM-CSF), and thrombopoietin/ megakaryocyte (TPO/MGDF) may have big consequences to the point of blocking not only the exogenous protein's activity but also of the endogenous protein with the serious complications inherent in these actions.

The production of antibodies against biotechnology-derived proteins, like insulin, factor VIII or IX, or interferons, does not have the same serious consequences and treatments that continue in presence of the antibodies, adapting the doses of therapeutic protein.

The consequences of antibodies produced against monoclonal antibodies have been observed since their first use, particularly when these proteins are derived from animal or bacterial proteins. The reactions observed could be of a general order such as systemic reactions, during these products' injection, local reactions, or reactions of acute hypersensitivity (generally not due to the antibodies). The immune reactions of anaphylactic type or allergic reactions are rare because of the better purification of proteins produced by recombinant DNA technology and the humanization of protein skeletons of monoclonal antibodies. The production of neutralizing antibodies may correspond to several types of mechanisms like the direct bonding to a biological activity site or to a site that is not in direct relation but impedes its activity by a changed structural conformation. The nonneutralizing antibodies bind to the therapeutic protein site without affecting the biological activity site. If they do not directly neutralize the biological target, they may change the drug's bioavailability by an increase of the clearance of the bonding complex made with a result identical to that of biological activity neutralization.

When a biopharmaceutical product is immunogenic, its repeated administration to patients over an extended period of time generally enhances the risk of raising antibodies that can induce anaphylaxis, alter the pharmacokinetic properties of the protein, or inhibit binding of the drug to its target receptor rendering the protein ineffective. Anaphylaxis is caused by an immediate allergic reaction mediated by immunoglobulin E (IgE) antibodies against the product while an immune response with high titers of neutralizing IgG antibodies strongly decreases the therapeutic activity of the protein. Another possible life-threatening clinical consequence of antibody formation is cross-reactivity with the endogenous protein produced by the patient.

Whatever the nature of the antibodies produced, the immune responses generated by therapeutic proteins pose a significant problem of safety and efficacy for the authorities in charge of evaluating and approving the marketing of biological medicinal products. Recommendations concerning the evaluation of the immunogenic profile of biosimilars compared to the references have been widely published. These recommendations are based on a multifactor approach taking into account the mechanisms involved: the different factors that may be part of the immune response, and the level of expression of antibodies and the possible rarity of the response observed. It is suggested that developers of biosimilar products evaluate the immunogenic risk, case by case, to ensure its safe use. No specific method is recommended, taking into account the variability and multiplicity of factors involved, but actions to take must be identified before clinical trials start, as well as evaluations that will be performed during pivotal clinical trials and evaluations that will be done after marketing of the product, notably within the framework of a risk management plan.

To measure the consequences of the risk linked to immunogenicity during the therapeutic use of biosimilar, immune mechanisms at play must be examined as well as factors having an influence on immunogenicity such as the disease state wherein the immune systems may be compromised such as in the use in patients undergoing chemotherapy.

Immunogenicity is a problem only when it is clinically relevant—when it has an effect on the safety or efficacy of the therapeutic protein. Clinically relevant immunogenicity includes when antibodies change how the drug reacts in the body, when antibodies make the protein less therapeutically effective, when antibodies change natural proteins in the body, or when antibodies trigger a severe allergic reaction, which is very rare. Clinically relevant immunogenicity is not common but must be monitored for all therapies.

Biosimilar product immunogenicity

It is well established that repeated injection of even native human proteins can result in a break in immune tolerance to self-antigens in some patients leading to a humoral response against the protein that is enhanced when the protein is aggregated or partially denatured. Although in most cases an immune response to a biopharmaceutical has little or no clinical impact, anti-drug antibodies (ADAs) do, however, pose a number of potential risks for the patient, particularly in the case of a neutralizing antibody response. First, an ADA response can adversely affect the pharmacokinetics and bioavailability of a drug thereby reducing the efficacy of treatment and necessitating either dose escalation or switching to alternative therapy if such therapy is available. An ADA response can also adversely affect the safety of treatment and cause immune complex disease, allergic reactions and, in some cases, severe autoimmune reactions. Serious and life-threatening adverse events can occur when ADAs cross-react with an essential nonredundant endogenous protein such as EPO or thrombopoietin. Thus, several cases of pure red cell aplasia were associated with the development of antibodies to recombinant EPO following a change in formulation. Similarly, the development of antibodies to pegylated megakaryocyte growth and development factor (MGDF) cross-reacted with endogenous MGDF resulting in several cases of severe thrombocytopenia.

All biosimilar products are evaluated based on the regulatory guidelines such as the FDA guidance for binding antibodies and neutralizing antibodies. Binding antibodies bind to the protein but usually have no effect. Neutralizing antibodies can inhibit the function of the protein in the body. FDA is more concerned with neutralizing antibodies because they are more likely to have clinical consequences. Because older products may have limited immunogenicity data based on tests with inadequate sensitivity, immunogenicity between a biosimilar and its reference product cannot be compared using data from the package insert of the reference product. Any comparison of immunogenicity will need a clinical side-by-side test of the biosimilar and its reference to ensure valid comparison. Without a side-by-side comparison, more sensitive tests may get higher antibody positive results with the biosimilar.

Animal models do not predict immunogenicity in humans. Most animals (even primates) can develop a strong antibody response to human proteins. Immunogenicity testing in animals may be useful to evaluate drug functioning or toxicity changes that might result because of antibodies to the human protein, or to see if responses are the same for two different products. Aggregation, which is when proteins clump together, is the most common factor associated with increased immunogenicity. Aggregation should be a key part of analytical testing, but other product changes (e.g., impurities) have not been associated with increased immunogenicity.

The amount of premarket and postmarket immunogenicity data needed for the approval of a potential biosimilar product will depend on an analytical assessment of similarity between the biosimilar and its reference product, as well as the rate of clinical consequences of immunogenicity observed with the reference product. If an immune response to the reference is rare, two separate studies may be sufficient to evaluate immunogenicity (1) premarket study to detect major differences in immune responses between biosimilar and reference and (2) postmarket study to detect subtle

33

differences in immunogenicity. FDA recommends that immunogenicity tests use the patient population most likely to show an immunological response for these studies.

Product changes associated with increased immunogenicity can be assessed using analytical tests. Protein aggregation is the primary product change associated with increased immunogenicity. Rigorous analytical testing comparing the biosimilar candidate to the originator reference product in head-to-head studies is the most sensitive way to test for likely immunogenicity in patients. These tests include the measurement of protein aggregates in the drug product and over its shelf life. Thus, analytical testing may be the best way to minimize immunogenicity of a biosimilar. While such tests also need to be done for originator biologics, it is much more difficult to anticipate likely immunological responses with a brand new product.

Most clinical comparator trials cannot detect true differences in clinically significant immunogenicity because the frequency of events is so low. With current statistical methods, very large clinical studies, with over 3000 patients each, would be needed to evaluate meaningful differences in immunogenicity. Consequently, a robust pharmacovigilance program, able to capture clinically relevant immunological responses during real-world use of the biologics, may be a better method to detect clinically relevant immunogenicity problems.

Differences in antibodies may not be relevant if there are no clinical consequences for patients. It is not expected that the possibly reduced immunogenicity of a biosimilar, through more modern manufacturing and better control of aggregation, will cause FDA to remove the reference product from the market.

Eprex® immunogenicity is often cited as a reason to demand clinical immunogenicity testing in biosimilars. Johnson & Johnson made a manufacturing change to Eprex (recombinant erythropoietin) and removed a protein, human albumin, from the formulation for their product marketed in Europe. This change was overseen by regulators using the standard process of demonstrating high similarity with comparability studies of the pre- and postmanufacturing changed products. The "new" Eprex induced antibodies to Eprex and to the natural erythropoietin found in the body, causing pure red cell aplasia (PRCA). Although the individual cases of PRCA were very serious, the actual incidence was low (2/10,000). Clinical studies could not have detected PRCA at such a low incidence. A clinical study to determine a difference in the rate of PRCA would have involved a very large number of patients. Instead, a robust pharmacovigilance system with analytical investigations eventually resolved the issue.

The predominant immune mechanisms leading to drug hypersensitivity may include from no effect to endogenous cross-reactivity (Table 2.2).

The risk of drug hypersensitivity can be increased by some patient-related factors, which include female gender, specific genetic polymorphism, as well as by

Table 2.2 Listed Effects of Immune Responses from Various Drugs

Result of Immune Response	Drug
No effect	rh-GH
Pharmacokinetic alteration	rh-Insulin
Reduced efficacy	GN-CSF, IFN alpha, IFN beta (the majority of patients become NAbs+ (neutralizing antibody positive) within 6–18 months of treatment, while clinical impact of NAbs is delayed and is not seen until 24 months of therapy, abolishing activity)
Loss of efficacy	Natalizumab (persistent antibodies)
Cross reaction with endogenous	Factor VIII and IX, rh-EPO, rh-MDGF

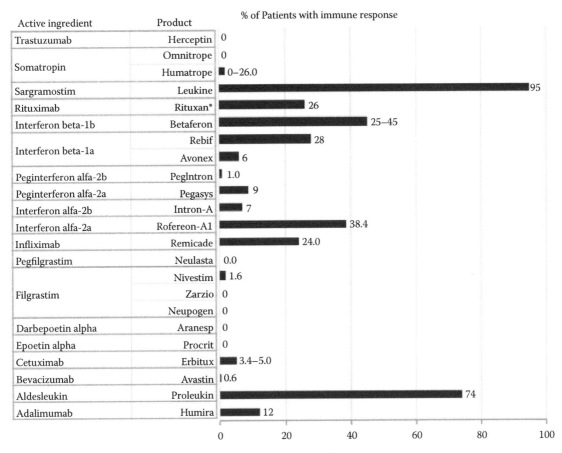

Active ingredient	Product	% of Patients with immune response
Trastuzumab	Herceptin	0
Somatropin	Omnitrope	0
	Humatrope	0–26.0
Sargramostim	Leukine	95
Rituximab	Rituxan*	26
Interferon beta-1b	Betaferon	25–45
Interferon beta-1a	Rebif	28
	Avonex	6
Peginterferon alfa-2b	PegIntron	1.0
Peginterferon alfa-2a	Pegasys	9
Interferon alfa-2b	Intron-A	7
Interferon alfa-2a	Rofereon-A1	38.4
Infliximab	Remicade	24.0
Pegfilgrastim	Neulasta	0.0
Filgrastim	Nivestim	1.6
	Zarzio	0
	Neupogen	0
Darbepoetin alpha	Aranesp	0
Epoetin alpha	Procrit	0
Cetuximab	Erbitux	3.4–5.0
Bevacizumab	Avastin	0.6
Aldesleukin	Proleukin	74
Adalimumab	Humira	12

EPAR 1 at the end of treatment *based on indication.

Figure 2.5 Reported immunogenicity of recombinant proteins and antibodies. (From Wadhwa, M., Immunogenicity: What do we know and what can we measure?, PhD thesis, NIBSC–HPA, Hertfordshire, U.K., April 12, 2011.)

some drug-related factors, which include the chemical properties, the molecular weight of the drug, and the route of administration. It is known that drugs with great structural complexity are more likely to be immunogenic. However, drugs with a small molecular weight (less than 1000 Da) may become immunogenic by coupling with carrier proteins, such as albumin, forming complexes. Moreover, the route of administration affects the immunogenicity, the subcutaneous route being more immunogenic than the intramuscular and the intravenous routes.

Figure 2.5 shows an overview of the immunogenicity of several currently marketed recombinant products.

It is noteworthy that about 5% of the U.S. population allergic to food and in the list mentioned earlier or recombinant proteins and antibodies, many fall within or below this incidence rate. Compounds, like sargramostim, aldesleukin, interferon alfa 2a (a particular brand), etc., show high incidence, some of which can be attributed to the formulation and manufacturing factors listed later.

Regulatory guidance

Assessment of immunogenicity is an important component of drug safety evaluation in preclinical, clinical, and postmarketing studies.

Draft Guidance for Industry Assay Development for Immunogenicity Testing of Therapeutic Proteins has recently been published by the U.S. Food and Drug Administration (http://www.fed.gov/Drugs/GuidanceComplianceRegulatory Information/Guidances/UCM192750.PDF). Similarly, guidelines on the immunogenicity assessment of biotechnology-derived therapeutic proteins established by the Committee for Medicinal Products for Human Use of the European Drugs Agency came into effect in April 2008 (http://www.emea.europa.ed/pdfs/human/biosimilar/1432706en.pdf). These guidelines provide a general framework for a systematic and comprehensive evaluation of immunogenicity that should be modified as appropriate on a case-by-case basis.

Although differences in approach and emphasis exist between the United States, European Union, and Japanese regulatory authorities, there is, nevertheless, a large degree of consensus on the type of approach that should be adopted, namely, a risk-based approach that is clinically driven and takes into account pharmacokinetic data. Thus, biopharmaceuticals with no endogenous counterpart are considered to be of relative low risk while drugs with a nonredundant endogenous counterpart are considered to present a high risk. A multitier approach to testing samples is also recommended. This consists of an appropriate screening assay capable of detecting both IgM and IgG ADAs, the sensitivity of which is such that a percentage of false-positive samples would be detected. The specificity of the samples that test positive in the screening assay are then reassayed in a confirmatory assay usually by competition with an unlabeled drug using the same assay format as that used for the screening assay. Samples that test positive in the screening and confirmatory assays are then tested for the presence of neutralizing ADAs using a cell-based assay whenever possible. Although it may be appropriate to use ligand-binding assays for certain monoclonal antibodies that target soluble antigens, some ADAs may not be detected using a ligand-binding assay. Thus, antibodies have been described that inhibit the antiviral activity of type I IFNs by inducing hyperphosphorylation of a receptor-associated tyrosine kinase without inhibiting binding of IFN to its receptor.

It is necessary to develop validation and standardization criteria that would be appropriate for the antibody assays being developed in various laboratories. It should be realized that the parameters (requiring validation) are unique to each method, and its intended use and therefore must be carefully determined on a case-by-case basis.

The EMA CHMP provides guidance on immunogenicity assessment of biotechnology-derived therapeutic proteins and recommends using validated antibody assays, characterization of the observed immune response, as well as evaluation of the correlation between antibodies, PK and PD and efficacy and safety. It also states that the role of immunogenicity in events related to adverse events such as infusion reactions and loss of efficacy should be considered, a view that is also reflected in *The International Conference on Harmonization Guideline S6*. Both guidelines recommend the determination of incidence and titer of antibody responses to the biotherapeutic prior to and posttreatment, but the EMA guideline emphasizes post-marketing programs for monitoring of antibodies in recipients and provides general guidance on antibody assays, validation and standardization issues and strategies to be adopted when conducting immunogenicity studies.

Factors influencing immunogenicity of biosimilars

Therapeutic proteins immunogenicity is influenced by different factors. Some concern the protein's very structure, how to produce it; with its purification degree, its formulation, the treatment type and the patients' characteristics, plus other factors possibly not known.

Molecule-specific factors   The presence of nonhuman sequences novel epitopes generated by amino acid substitution designed to enhance stability novel epitopes created at the junction of fusion proteins changes in glycosylation exposing cryptic B-cell and T-cell epitopes in the protein, or cause the protein to appear foreign to the immune system carbohydrate moieties present eliciting the production of IgE antibodies; pre-existing antibodies against galactose-α-1,3-galactose have been shown to be responsible for IgE-mediated anaphylactic reactions in patients treated with cetuximab pegylation can reduce the immunogenicity of some proteins patients producing antibodies to the PEG residue adversely affecting efficacy repeated administration causing a break in immune tolerance the presence of degradation products resulting from oxidation or deamination of the protein aggregates, though mechanism is not clearly understood intrinsic immunomodulatory effects.

Intrinsic immunomodulatory effects   The intrinsic immunomodulatory properties of the molecule, for example, in the case of recombinant granulocyte macrophage colony-stimulating factor (GM-CSF), which can recruit APCs to the site of antigen processing, stimulate the maturation of myeloid DCs, and enhance an antigen-specific CD8+ T-cell response, suggesting that repeated administration of GM-CSF may function as an adjuvant. Indeed, GM-CSF has been used as an immunological adjuvant in a number of vaccination protocols designed to elicit an immune response to self-antigens.

Type I IFNs play a determining role in the innate immune response and in the establishment of the subsequent adaptive immune response and resistance to virus infection. Type I IFNs bind to a common high-affinity cell-surface receptor composed of two transmembrane polypeptide chains, IFNAR1 and IFNAR2. IFN binding results in the phosphorylation and activation of two Janus kinases, Tyk2 and Jak1, and activation of the latent cytoplasmic signal transducers and activators of transcription (STAT) 1 and 2, and the formation of a transcription complex in association with IFN regulatory factor 9 (IRF9). Translocation of this complex to the nucleus results in the transcriptional activation of a specific set of genes that encode the effector proteins responsible for mediating the biological activities of the type I IFNs. Recombinant IFNα2 and more recently pegylated IFNα2a and pegylated IFNα2b have found wide application for the treatment of chronic viral hepatitis and some neoplastic diseases, while IFNβ-1 is used extensively as first-line therapy for the treatment of RRMS.

Treatment of patients with type I IFNs has also been reported to induce the symptoms of a number of other autoimmune diseases including Sjögren's syndrome, dermatomyositis, polymyositis, type I diabetes, and autoimmune hepatitis. Thyroid dysfunction is also a relatively common side effect of IFN treatment, particularly in patients with pre-existing thyroid autoimmunity. As in the case of SLE, polymorphisms in the genes encoding proteins involved in the activation or activity of type I IFNs are also associated with a predisposition to develop certain of these diseases. For example, polymorphisms in the STAT4 and IRF5 genes are associated with an increased risk of developing systemic sclerosis, an autoimmune connective tissue disorder characterized by fibrosis of multiple organs and a type I IFN signature.

Treatment of patients with TNFα antagonists is also associated with induction of inflammatory and autoimmune diseases, such as SLE, and psoriasis. TNFα is associated, however, with both the development of MS in man, and experimental autoimmune encephalomyelitis (EAE) in animals.

Natalizumab (Tysabri®) is a humanized monoclonal antibody that targets α4-integrins preventing leukocytes adhering to endothelial vascular cell adhesion molecule-1 and hence reducing trafficking across the blood–brain. Results from the AFFIRM and SENTINEL trials show that some 9% of patients with multiple sclerosis (MS) develop antibodies to natalizumab including 6% who were persistent and, for the most part, had neutralizing antibodies. The presence of persistent antibodies was associated with reduced clinical efficacy and infusion-related adverse events including hypersensitivity reactions. Most of those patients who develop an ADA response did so within 12 weeks (88% in the AFFIRM trial). A small group of patients (3%) developed a transient ADA response that led to a delay in the clinical response to natalizumab therapy, although clinical efficacy was restored once antinatalizumab antibodies were no longer detectable after 6 months of continued treatment.

Alemtuzumab is used for the treatment of various neoplastic diseases including chronic lymphocytic leukemia, and more recently for the treatment of RRMS. Alemtuzumab treatment is also associated with the development of autoimmunity including thyroid disorders, predominately Grave's disease, in approximately 30% of patients with MS but intriguingly not in patients with cancer.

The intrinsic immunomodulatory response is often affected by factors such as genetic makeup, age, gender, disease status of patient, concomitant medication, and the route of administration. For example, a common MHC class II allele, DRB1*0701, is associated with the antibody response to IFNβ in MS patients. Disease state and immune competency also influence the immune response of an individual to a treatment with a biopharmaceutical. Thus, development of antibodies to pegylated MGDF is less frequent in cancer patients who tend to be immunosuppressed than in healthy individuals.

Concomitant therapy with methotrexate together with the chimeric monoclonal antibody infliximab has been shown to reduce the immune response to infliximab and improve the clinical response in patients with RA.

The duration of treatment and the route of administration also influence the immune response, for example, if administration of a protein in a single dose results in the production of low-affinity IgM antibodies, the repeated dosing might result in the production of high-affinity and high-titer IgG antibodies, which may be neutralizing. Thus, in patients with MS treated with IFNβ neutralizing antibodies to IFNβ often do not appear until after several months of therapy. The intravenous route of administration is considered to be least likely to generate an immune response to a biopharmaceutical compared with intramuscular or subcutaneous administration.

The complexity of the humoral response to biopharmaceuticals and the difficulty in establishing the effect on ADAs on drug efficacy is illustrated by the response of patients to treatment with IFNβ, first licensed over a decade ago for the treatment of relapsing/remitting multiple sclerosis (RRMS). The IFNβ-1b biosimilar Extavia®. Avonex® and Rebif® are both glycosylated forms of native human IFNβ-1a produced in Chinese hamster ovary cells. Betaseron® and Extavia are a nonglycosylated form of IFNβ-1a produced in *Escherichia coli* that has a serine substitution for the unpaired cystine at position 17 of the native protein. Most patients develop an antibody response to IFNβ products, and as many as up to 45% of patients develop neutralizing antibodies to IFNβ, in some cases as early as 3 months after initiation of therapy. Overall, some 25% of patients develop anti-IFNβ-neutralizing antibodies usually within 6–18 months. ADAs are more frequent in patients treated with IFNβ-1b than IFNβ-1a, while subcutaneous IFNβ-1a (Rebif) is more immunogenic than intramuscular IFNβ-1a (Avonex). Patients who possess the DRB1*0701 allele, or who express low levels of IFNAR2, one of the two chains of the type I IFN receptor, upon initiation of treatment, have a significantly higher risk of developing anti-IFNβ-neutralizing antibodies.

Patient characteristics, as is the case with their genetic statute and type of disease, are known to influence the response and type of immune responses. It is well known that patients suffering from severe hemophilia with less than 1% of factor VIII, with time, develop inhibitors to the administration of antihemophilic factors of plasmatic origin or derived from recombinant DNA technology. The most plausible explanation resides in the absence of recognition of coagulation factors by the immune system as human proteins of the endogenous.

In the case of EPO, with its formulation change, only patients with a chronic renal insufficiency presented an immune system breakdown. Cancer patients treated for their anemia by rHuEPO did not present with this secondary effect. This illustrates the conditions promoting an immune tolerance breakdown:

- Chronic treatment with repeated doses for months, even years.
- Absence of concomitant immunosuppressant treatment.
- Route of administration (the subcutaneous injection is more immunogenic than intramuscular, itself more immunogenic than an intravenous injection).

Structural factors Proteins are complex molecules with a primary, secondary, and tertiary structure. Changes within the primary structure may be the cause of an immunogenic reaction. Several cases are well known and published in the literature and are as follows:

- Changing an insulin amino acid is enough to lead to a strong immunogenic response, whereas two amino acids inversion only leads to a pharmacokinetic change.
- The homology degree of a recombinant protein with the natural protein may explain an immunogenic reaction but the well-known case of recombinant human interferon that shows 10–23 amino acids different from human interferon (homologous for about 89%) does not lead to immunogenicity exacerbation. Foreign proteins like streptokinase, salmon calcitonin, etc., are known for inducing classic immunogenic reactions in patients.
- Reactions of oxidation or deamidation of amino acids are known for triggering an immunogenic reaction by forming new epitopes. It is the example of human recombinant α interferon with one methionine, oxidized because of a modification of the purification process, that led to nonneutralizing antibodies formation and which, returning to the initial process, had stopped being immunogenic.
- The modification of stability characteristics of a protein with aggregates formation may have significant consequences in terms of immunogenicity by tolerance breakdown of the immune system.

The significance of protein spatial conformation is well known for its biological activity as well as for its stability. Partial modification of spatial conformation may occur after shear, by shaking, or by temperature modification (e.g., temperature rise or freeze/thaw cycles).

Glycosylation Almost half of the therapeutic proteins that are approved or in clinical trials are glycosylated. Glycosylation is one of the most common and complex posttranslational modifications, which leads to the enzymatic addition of

39

glycans on proteins. Glycans can influence the physicochemical (e.g., solubility, electrical charge, mass, size, folding, and stability), as well as the biological (e.g., activity, half-life, and cell surface receptor function) properties of proteins. The glycosylation profile of a protein is species specific and depends on the cell line and culture conditions that are used for production. The presence and structure of carbohydrate moieties can have a direct or indirect impact on the immunogenicity of therapeutic proteins, that is, the glycan structure itself can induce an immune response or its presence can affect protein structure in such a way that the protein becomes immunogenic.

Glycosylation is an important factor in therapeutic protein immunogenicity. Glycans may impact the immunogenicity of therapeutic proteins in an indirect manner through their influence on folding, solubility, and structural stability of proteins. Indeed, glycosylation can improve protein solubility and stability, and thereby decrease the immunogenicity of therapeutic proteins, as in the case of recombinant human interferon beta (rhIFNβ). The nonglycosylated form (rhIFNβ-1b; produced in *E. coli*) is less soluble and more prone to thermal denaturation than the glycosylated form (rhIFNβ-1a; produced in CHO cells). The lack of glycosylation most likely contributes to the high aggregate content of rhIFNβ-1b; it contains about 60% aggregates and elicits neutralizing antibodies in a high percentage of patients, whereas rhIFNβ-1a contains less than 2% aggregates and is less immunogenic.

It is adequately demonstrated that proteins with a human structure produced in eukaryote nonhuman cells give immunogenic human responses. Similarly, the glycosylation level plays an important role. A β interferon produced in *E. coli* is more immunogenic than that produced by mammal cells since the latter's glycosylation, helping its solubility, decreases the formation of immunogenic aggregates. A protein's glycosylation can decrease immunogenicity by masking antigenic sites.

Besides making a protein more soluble, a carbohydrate moiety is sometimes able to cover an antigenic epitope. Antibodies against rhGM-CSF that are generated in patients show cross-reactivity with rhGM-CSF produced in yeast and *E. coli*, but not with rhGM-CSF produced in CHO cells.

Expression-related factors Besides the commonly used protein expression systems of bacteria, yeast, insect, and mammalian cells, plant cells are emerging cell factories for the production of biopharmaceutical proteins. In May 2012, the FDA approved the first biopharmaceutical derived from transgenic plants, which is recombinant human glucocerebrosidase for the treatment of Gaucher's disease. Fully functional mAbs also can be efficiently synthesized in transgenic plants. A major drawback of plant-derived glycoproteins is the presence of complex *N*-glycans with core-xylose and core-α1,3-fucose structures. These two glycoepitopes are foreign to humans due to differences in plant and mammalian glycosyltransferase repertoires.

The nonhuman glycan structures present on biopharmaceuticals can induce IgE-mediated reactions and/or anaphylaxis in allergic patients. Moreover, those glycoepitopes may enhance clearance and decrease therapeutic effect of biopharmaceuticals due to pre-existing IgA, IgM, and IgG antibodies in certain patients. Neutralization of the therapeutic protein or cross-reactivity with the endogenous protein resulting from the presence of glycoepitopes is less likely due to lack of reactivity toward the underlying protein backbone.

PEGylation Biopharmaceuticals can be chemically modified with the purpose of extending half-life or facilitating uptake by target receptors. An increasingly common type of engineering is the covalent attachment of polyethylene glycol (PEG) polymers to the peptide backbone. PEGylation is intended to lower renal filtration by increasing the molecular weight and to protect the therapeutic protein from proteolytic degradation. Similar to glycosylation, PEGylation may decrease immunogenicity by shielding the immunogenic epitopes while maintaining the native conformation of the protein. PEG polymers ranging in molecular weight from 12 to 40 kDa were attached to different sites on the hydrophobic and immunogenic therapeutic protein rhIFNβ-1b. The mono-PEGylated rhIFNβ-1b preparations retained 20%–70% of the antiviral activity and displayed improved solubility, stability, and pharmacokinetic properties compared with unmodified rhIFNβ-1b. Importantly, PEGylation greatly reduces the aggregation propensity of rhIFNβ-1b as well as its immunogenicity in rats.

Some recombinant proteins are "pegylated" in order to modify their pharmacokinetics. PEGylation of a protein is the process of attachment of polyethylene glycol (PEG) chains to the skeleton of the protein. The result is a decreased total half-life of the protein, proteolytic enzymes protected, and sometimes masked immunogenic sites. The new proteins so obtained differ by their conjugated structure, their molecular size, and their spatial conformation (linear, branched, or multibranched chains). Often PEGylation lowers the protein's immunogenicity, probably through multiple mechanisms related to blocked antigenic sites, solubility improved, and lower administration frequency of the therapeutic protein. In general, a branched PEG protein is more efficient than a linear PEG protein because of improved immunological properties. However, examples have been published in the literature of PEG proteins more immunogenic than non-PEG (PEG-rhMGDF and rh-TNF coupled with PEG dimer).

Chemical degradation   Chemical modifications of proteins may include deamidation, oxidation, isomerization, hydrolysis, glycation, and C/N terminal heterogeneity of the protein, sometimes leading to aggregation. The susceptibility of an individual amino acid residue to chemical modification is dependent on neighboring residues, the tertiary structure of the protein, and solution conditions such as temperature, pH, and ionic strength. Chemical modification may give rise to a less favorable charge of the protein, thus leading to structural changes or even the formation of new covalent cross-links. Covalent aggregation is also a form of chemical degradation.

Deamidation of proteins accelerates at high temperature and high pH and can occur during bioprocessing and storage. Moreover, deamidation can be accompanied by some degree of oxidation, conformational changes, fragmentation, and aggregation, posing serious risks for immunogenicity. Oxidation, another major chemical modification, can also reduce conformational stability and may cause the protein to aggregate. Oxidation of human serum albumin with hydroxyl radicals resulted in structural alterations and exposure of hydrophobic patches, causing increased immunogenicity.

Aggregation   Aggregation phenomena may expose new epitopes to the protein's surface for which the immune system is intolerant. That leads to a standard immune response. In other conditions, protein aggregation may lead to presenting a multimeric antibody, which is known for not triggering B lymphocyte tolerance breakdown. This is why, in a therapeutic protein's analysis, it is important to look for the presence of aggregates and to limit their presence to a low level in the formulated drug.

Since aggregation is often accompanied by other structural changes on the protein, it is difficult to distinguish the individual impact of each factor on immunogenicity.

Since information on the nature of immunogenic aggregates from clinical studies is generally limited, animal models are used to provide insight into the link between aggregation and immunogenicity. Studies employing such models have elucidated that not all aggregates appear to be immunogenic, as demonstrated, for example, for rhGH. Transgenic mice immune tolerant for human IFNα has been used to study the immunogenicity of aggregated rhIFNα products. rhIFNα-2b aggregates composed of native-like proteins were able to overcome the immune tolerance of transgenic mice, resulting in antibodies cross-reacting with the monomeric protein.

Nonnative aggregation can trigger structural changes in the protein leading to the creation of new epitopes or the exposure of existing epitopes. Native-like aggregates, however, are more likely to elicit ADAs that cross-react with the native protein and thus pose a greater risk for the patient. Native-like aggregates may resemble haptens on the surface of pathogens that form organized and repetitive structures that can cross-link BCRs in a multivalent manner.

Aggregates can be classified according to size, reversibility, secondary or tertiary structure, covalent modification, and morphology. One class of aggregates that are of particular concern includes subvisible particles or, more specifically, aggregates ranging from 0.1 to 10 μm in size. The current USP light obscuration test for drug approval requires the number of particulates over 10 μm is ≤6000 per container, while the number of particulates over 25 μm is ≤600 per container. Industrial researchers agree that additional analysis of subvisible particles smaller than 10 μm would support product characterization and development. Subvisible particles are too large to be analyzed by standard quality control methods such as SEC and SDS-PAGE, but too small to be visually detected. Therefore, use of additional, less routine methods such as asymmetrical flow field flow fractionation and microflow imaging has been recommended for extended characterization following a risk-based approach.

Product characterization should not only be limited to monitoring protein particles, but also focus on nonprotein particles. Foreign micro and nanoparticles, for example, shed from filling pumps or product containers, are able to induce protein aggregation or nucleate the formation of heterogeneous aggregates.

Manufacturing and processing factors In addition to intended modifications, a biopharmaceutical may be chemically modified through accidental degradation in one of the many bioprocessing steps: fermentation, virus inactivation, purification, polishing, formulation, filtration, filling, storage, transport, and administration. Chemical stresses during manufacturing and storage can be caused by exposure to light or elevated temperatures and the presence of oxygen, metal ions, or peroxide impurities from excipients in the formulation. Trace amounts of iron, chromium, and nickel can leach into the formulation buffer via contact with stainless steel surfaces typically used during bioprocessing and catalyze degradation reactions.

Impurities and other production contaminants Therapeutic proteins obtained through recombinant DNA technology are produced in various cellular systems where production-linked protein impurities originate. These proteins are called host cell proteins (HCP), considered endogenous proteins by the immune system and may lead to antibody formation by the standard immune mechanism. If by nature anti-HCP antibodies do not neutralize the biological activity of the therapeutic protein of interest, they can nevertheless have consequences in terms of general effects including skin reactions, allergies, anaphylaxis, or a serum sickness. Other contaminants, such as impurities, coming from chromatographic column resins or from enzymes used for refining therapeutic proteins' purification, may be found as traces in the finished product. Some impurities may be released by some compounds used for the capping process. These impurities may play the role of the amplifier for the immune response, even if they are not able to initiate an immune response themselves.

Process-related impurities, including traces of residual DNA or proteins from the expression system, or contaminants that leach from the product container, can also influence the immunogenicity of recombinant biopharmaceuticals. Several examples exist where the modification or protein's immunogenicity is altered over time. DNA G–C patterns from bacteria or degraded proteins are able to activate Toll-like (Toll-like receptors are a class of proteins playing a key role in the innate immune response. They are transmembrane proteins containing receptors that detect danger signals located in the extracellular milieu, a transmembrane medium, and an intracellular medium allowing the activation signal transduction) receptors and act as adjuvants. However, the action of these impurities is limited to nonhuman proteins with a pseudovaccination activity. The adjuvants are unable to stimulate an immune response based on a T lymphocyte's response independent of the tolerance breakdown of B lymphocytes.

Formulation changes   Changes in formulations have been associated with ADAs. Two particularly interesting cases could illustrate how important formulation and conservation conditions are. The first case concerns erythropoietin (EPO). Cases of Pure Red Cell Aplasia (PRCA) after treatment by EPO are known, but rare. Anti-EPO antibodies' are formed exogenous as well as endogenous, after administration of recombinant human EPO (rHuEPO). These cases have occurred after a changed formulation of the finished product, with human albumin used as a stabilizer replaced by polysorbate 80. Different hypotheses to explain the immune system tolerance breakdown after administration of rHuEPO lead, among other things, to the impurities' extraction from the syringes plunger rod stopper, playing "booster" in the presence of EPO. During the analysis of batches called into question, no increase of the level of aggregates or of the level of truncated or degraded EPO has been evidenced. In that case, there must have been several factors having fostered the immune system's tolerance break-down. In particular, the subcutaneous route of administration may be incriminated, as it is known to produce a pseudovaccination effect.

The second case involves the conservation conditions of a freeze-dried formulation of interferon a-2a (rHuINF a-2a) that has been stabilized by human albumin. At room temperature, a partial oxidation of rHuINF a-2a has made easier the formation of aggregates with intact interferon and albumin. These aggregates led to the therapeutic preparation's immunogenicity.

These cases illustrate how important the finished product formulation study is and the possible consequences of a change compared to the initial formulation or its conservation conditions. It is also particularly important to rigorously analyze the levels of impurities from the therapeutic protein's system of production.

The adverse response resulting from antibodies that neutralize endogenously produced protein also cause the discontinuation of the megakaryocyte growth and differentiation factor (MGDF).

Immunogenicity testing

The current, most practical and commonly used approach for testing unwanted immunogenicity is the detection, measurement, and characterization of antibodies generated specifically against the product. It is anticipated that in the future better models of evaluating immunogenicity will become available and may include DNA microarrays. Generally, no single assay can provide all the necessary information on the immunogenicity profile of a biological product and as a result, a panel of carefully selected and validated assays for detection and measurement of antibodies is used. These antibody responses are then correlated with the PK/PD and clinical effects when evaluated to determine if the antibodies create a "meaningful difference" in determining the safety and efficacy of these products.

Testing protocols

Development of biosimilars requires studies to measure ADAs and the comparison of relative immunogenicity rates of the biosimilar and the reference products. Even though there exist guidelines from FDA for the development of biosimilars, details on executing these with reference to antibody assays and their validation are not clear.

The development of an immunogenicity assay for a biosimilar program should be preceded by sufficient in vitro analytical results to establish analytical and pharmacological similarity on multiple lots. The comparison of the immunogenicity potential becomes more relevant if there are unknown or trivial analytical differences, such as relative levels of aggregation, or impurities, between the products that might contribute to differences in immunogenicity. Immunogenicity testing alone is not adequate to establish structural similarity of the products.

Many of the immunogenicity assays used to test the originator molecules were developed over a decade ago and thus had limitations inherent to the platforms and reagents utilized at that time. The current guidance on immunogenicity assay development and validation provide more sensitive and drug-tolerant assays making it difficult to compare the results with data available in approved BLA documents for the reference listed product.

The platforms that are typically utilized for these assays are ligand-binding assays via enzyme-linked immunosorbent assays (ELISA) and electrochemiluminescence (ECL) methods; however, several emerging technologies can also be utilized, such as BiaCore or Octet, Luminex, and Gyrolab.

Another critical component to all immunogenicity assays is the production of an appropriate positive control and may include the use of commercially available polyclonal or anti-idiotypic antibodies to the reference product, affinity purified ADAs to the biosimilar and/or reference product, as well as immune serum from hyperimmunized animals dosed with the biosimilar and/or the reference product. It is important to collect serum from any animals evaluated during preclinical testing that could potentially provide an ADA positive control for the immunogenicity assay in order to limit the number of animals required to procure this necessary reagent. The need for one or more positive controls may also be influenced by the strategy that will be used for the assay(s) and overall development.

There are two predominant strategies for proceeding through an immunogenicity testing plan based on the use of a single assay or two separate assays. Using either strategy, the assays must have a high level of similarity for all parameters evaluated. A typical ELISA-based immunogenicity assay consists of the drug bound to a 96-well plate, which is used to capture any ADAs that are present in the serum or plasma samples when they are applied to the plate. This complex is then detected using an enzyme-labeled form of the drug in the case of a bridging assay. The level of colorimetric or relative light units that are detected in the assay corresponds to the level of ADAs present in the serum or plasma samples. The assay is evaluated based on the sensitivity (the lowest amount of antibody that can be detected in the assay), precision, drug tolerance, specificity, selectivity, prozone, and precision of titration.

A single assay

The single assay is developed for both products and a limited set of parameters in validation should be evaluated including screening cut-point, specificity cut-point, precision, and drug tolerance for both products in order to ensure that the assay is capable of detecting antibodies to both the biosimilar and the reference product. This may involve the conjugation of a reporter molecule such as biotin or ruthenium to both the biosimilar and the reference product to facilitate the detection of antibodies when a bridging assay format is utilized. It is extremely important to ensure that the method of conjugation is tightly controlled and that these reagents are labeled in a reproducible manner in order to obtain comparable signal between the two reagents. Although the assay is essentially developed as two assays with respect to screening cut-point and precision, all other parameters should be evaluated on the background of the biosimilar assay (a plate coated with the biosimilar and detected with an enzyme-labeled biosimilar) where the biosimilar will be used to capture all potential antibodies to the biosimilar or reference product in the serum or plasma samples. When this strategy is used, it is essential to determine the specificity cut-point by immune-depletion with both the test and the reference product, as well as to test each positive sample for specificity to both products. Likewise, the drug tolerance of the assay must be defined for both molecules to ensure that the assay is tolerant to similar levels of circulating drug. The major limitation of using a single assay approach is that the approach will allow the evaluation of only similar or enhanced levels of biosimilar immunogenicity in relation to the reference product. If there are unique antibodies that are generated to the reference product, they will not be detected by this assay. This is not a good strategy for those seeking a designation of interchangeability.

Two assays

The use of this approach entails a full validation of two separate assays, one for the detection of antibodies generated to the biosimilar and one for the detection of antibodies generated to the reference product. This strategy is most useful when looking for an interchangeability designation, as the approach will allow a direct comparison of the immunogenicity rates of each molecule. This approach is also suitable for defining decreased immunogenicity rates for the biosimilar in comparison with the reference product, which could be potentially due to differences in production.

Testing strategy

Immunogenicity testing should be done on the basis of a risk assessment and is usually done in several stages as follows:

- SCREENING ASSAY that detects all antibodies binding to the biopharmaceutical (ADA; ADAs) in serum samples from animals or patients.
- CONFIRMATORY ASSAY to eliminate false-positive samples, ASSAY FOR NEUTRALIZING ANTIBODIES that detects those serum samples that contain neutralizing antibodies.
- CHARACTERIZATION of ADAs detected in serum samples.

Screening for ADA is usually done using suitable ELISA formats, radioimmune precipitation (RIPA), for example, with ^{125}I labeled antibodies or surface plasmon resonance (SPR). The same assays are used for the confirmation step, for example, by demonstrating inhibition of binding by an excess of the drug. Bridging ELISA formats are particularly useful in early project stages and for different animal species when project-specific immunological reagents are not available. To run a bridging ELISA, labeled drug is needed, however. The label may be detected directly (e.g., fluorescein) or indirectly via an amplification system (see schematic depiction at the right).

Assays for detection of antibodies

Common methods available include binding assays based on immunochemical procedures such as solid or liquid phase immunoassays, radioimmuno precipitation assays (RIPA), and biophysical methods such as surface plasmon resonance (SPR). These assays determine the presence (or absence) of antibodies based on the ability of the antibodies to recognize the relevant antigenic determinants in the therapeutic protein. These assays have their own advantages and disadvantages as described here:

Binding assay—direct format (coating with antigen and detecting with labeled anti-Ig) is rapid and relatively easy to use but prone to spurious binding or "matrix effects" and antigen mobilization may mask or alter epitopes.

Binding assay—indirect format (coating with a specific mAb or biotin, etc., followed by antigen) has high throughput, good sensitivity, coating with a specific mAb keeps antigen in oriented position, provides consistent coating, and maintains antigen conformation, species specificity, and isotype detection determined by secondary reagent; however, detection reagent may differ between control and sample, and extensive studies are required to demonstrate that the coating mAb does not mask or alter epitopes; test may also fail to detect "low-affinity" antibodies have species specificity, and isotype detection requires determination by secondary reagents.

Electrochemiluminescence—bridging format offers high throughput, use of high concentrations of matrix but requires two antigen conjugates (biotin and TAG), and antigen labeling may alter/denature antigen, or mask/alter epitopes.

Radioimmuno precipitation assay (Surface plasmon resonance) is tolerant to interference from antigen, provides detection signal consistent during life of TAG conjugate, has moderate–to–high throughput, good sensitivity, and can be specific and automated; it provides information on the specificity, isotypes, relative binding affinity, and relative concentration and enables detection of both "low-affinity" and "high-affinity" antibodies, and detection reagents are not required; however, it requires dedicated equipment, reagents are costly and vendor-specific; it can be isotype-specific, may fail to detect rapidly dissociating antibodies and requires radiolabeled antigens that may alter or denature antigen, decay of radiolabel may affect antigen stability; also it may require immobilization of the antigen, which may alter the conformation of the native protein and the regeneration step may degrade antigen; the sensitivity is less than found in binding assays.

None of these assays can assess the ability of the antibodies to neutralize the biological activity of the therapeutic proteins that is an important element in the assessment of immunogenicity. For evaluation of the neutralizing ability, use of an appropriate noncell-based competitive ligand-binding assay or a cell-based neutralization assay is required.

Testing for immunogenicity is performed during the preclinical and clinical phases. The U.S. FDA "Guidance for Industry: Assay Development for Immunogenicity Testing of Therapeutic Proteins" (1) states that an immunogenicity assay should, in addition to being sensitive, also be able to detect all isotypes, in particular, IgM and all IgG isotypes. The recommended sensitivity is 250–500 ng/mL. Studies are performed in three steps: screening, confirmation, and characterization of positives. Initial screening can result in false positives, and, therefore, the initial screening assay is usually followed by a confirmatory assay. After identification and confirmation of positive samples, a full characterization of ADAs in terms of assessment of isotype (class or subclass), binding stability, epitope specificity, and neutralizing capacity gives valuable information of the nature of the studied immune response. The IgG4 is second to IgG1 as the major isotype in ADAs developed for therapeutic monoclonal antibodies (MAbs). IgG4 have been associated with immune responses generated under conditions of high doses and prolonged exposure to therapeutic proteins. IgG4 ADAs can be difficult to detect in traditional bridging or homogeneous enzyme-linked immunosorbent assay (ELISA) and enhanced chemiluminescent (ECL™) assays due to their bispecific nature.

Cell-based neutralization assays

The development of improved assays, particularly cell-based assays for the detection of neutralizing antibodies, that allow immunogenicity to be determined with precision and the comparison of immunogenicity data between biopharmaceuticals, is critical for the development of less immunogenic and safer biopharmaceuticals.

The biological activity of a therapeutic is often evaluated using an in vitro cell-based assay based on a functional aspect of the protein or mechanism of action. These assays can be categorized into those that detect signaling responses soon after the protein–receptor interaction has occurred (early stage) or those that provide a measurable readout after the culmination of a cellular response (late stage). Since these assays assess the cellular response in vitro to a protein, they constitute an ideal and appropriate in logical approach for the development of a cell-based neutralization assay. It should be realized that different types of bioassay procedures can be used as the basis of a neutralization assay for a biological.

A cell-based neutralization assay can be defined as an in vitro assay utilizing cells that interact with or respond to the therapeutic either directly or indirectly in a measurable manner in the presence of test sample for the detection of antiproduct neutralizing antibodies. The detection of neutralizing antibodies is based on the principle that any sample containing an antibody (of a neutralizing nature) would reduce or abolish the biological activity induced by a known concentration of the therapeutic in a cell-based assay.

47

Most biological therapeutics can be broadly categorized into agonists or antagonists based on their desired effect in vivo. While agonists (e.g., cytokines, growth factors, hormones, and agonistic monoclonal antibodies) induce a response by directly binding to receptors on the target cell surface, therapeutics with antagonistic properties (e.g., soluble receptors and antagonistic Mabs) act by blocking the binding of a ligand to the target receptor expressed on the cell surface. As a result, assay formats and designs of neutralizing antibody assays can vary depending on the biological and the type of assay being used.

Although regulatory authorities recommend the use of cell-based assays for the detection and quantification of neutralizing antibodies, cell-based assays often give variable results and are difficult to standardize. Conventional cell-based assays for neutralizing antibodies are based upon the assessment of a drug-induced response in a drug-sensitive cell line, and the ability of an antibody to inhibit that response. The form of the assay can vary considerably, however, reflecting the diversity of the biopharmaceuticals currently employed in the clinic. Drug-induced responses vary from stimulation of cell proliferation in the presence of a growth factor such as EPO or GM-CSF, or induction of apoptosis in the presence of TNFα to inhibition of virus replication in IFN-treated cells. Such drug-induced responses are complex events involving the transcriptional activation or modulation of numerous genes. Drug-induced biological responses also often take several days to develop and are influenced by a number of factors that are difficult to control.

A series of engineered reporter cell lines have been developed for the quantification of the activity and neutralizing antibody response to biopharmaceuticals that eliminate many of the limitations of conventional cell-based assays.

In order to improve the performance of conventional cell-based assays and to develop assays that allow more direct comparisons of immunogenicity data, reporter cell lines were established based on transfection of cells with the firefly luciferase reporter gene regulated by a drug-responsive chimeric promoter. These assays allow drug activity, and the neutralization of drug activity, to be determined selectively and with a high degree of precision within a few hours by measuring light emission. In addition, the use of a single common drug-induced response for a variety of different drugs facilitates comparisons of immunogenicity data.

Conventional cell-based assays are difficult to standardize due in part to assay variation resulting from changes in culture conditions and genetic and epigenetic changes that can occur as cells are cultivated continuously in the laboratory. Assay variation can be minimized by the use of cell banks and preparation of each lot of assay cells under standardized conditions from an individual frozen vial. Each lot of assay cells is thus in an identical physiological condition. Assay-ready cells can be manufactured under conditions of current good manufacturing practices, and stored frozen for several years without loss of drug sensitivity.

For neutralization assays, a recommended approach for expressing the data is to report results as the "amount of serum required to neutralize the biological activity induced by a constant amount of the antigen." In some instances, however, it may be necessary to use an antibody standard or reference preparation for expressing the levels of neutralizing antibodies in the test samples relative to the amounts of the neutralizing antibodies in the reference antibody preparation. It may also be possible to express antibody levels using arbitrary units providing that the unitage has been well defined for the reference material. Although this approach is not ideal (the heterogeneous nature of polyclonal antibodies is particularly problematical for this), the use of this strategy may provide relatively precise estimates of antibody levels in the test samples and can reduce variability. This situation is most likely to occur when a number of sequential samples from the same animal or patient are available and it is difficult to include all samples from all patients in the same assays for establishing a valid comparison of antibody levels between different samples/patients.

Biosensor-based immunoassays

This method, unlike most other platforms, does not require the use of a labeled secondary reagent. Although several types of biosensors are available, the vast majority of published biosensor data cites the use of Biacore Instruments (https:// www.biacore.com/lifesciences/index.html) for monitoring the immune response in preclinical phases and clinical trials. Several models of Biacore are automated and also compliant with 21 CFR Part 11 requirement, which facilitates the use of this instrument for regulatory submissions.

The Biacore utilizes surface plasmon resonance to detect the increase in mass at the surface of the sensor chip following binding of an antibody to the antigen immobilized on the sensor chip. This increase in mass is directly proportional to the amount of antigen-binding antibody present in the serum sample being tested. The ability of the instrument to monitor the interaction in "real time" and to provide a continuous signal of the events occurring on the sensor surface enables detection of rapidly dissociating or "low-affinity" antibodies if these are present in the sample. Detection of low-affinity antibodies is important as these antibodies have the potential to neutralize the therapeutic product and may predict the generation of a later mature immune response. Furthermore, characterization of the antibodies in terms of affinities, antibody class, and subclass can also be performed easily. These attributes have contributed to the increased use and popularity of this platform in studies on immunogenicity.

While the assays described are useful for identifying antibody-positive samples, it is important to include an additional confirmatory step in the assessment strategy to ensure that the antibodies generated are specifically targeted to the therapeutic.

Confirmation of antibody positive samples

A confirmatory approach can include use of different methods (ELISAs, competitive immunoassays, SPR, etc.), although an assay based on a different scientific principle from that used for the screening assay should usually be considered. It is also necessary to select a confirmatory assay taking into account the limitations and characteristics of the screening assay. In most cases, assay specificity can be demonstrated by the addition of free antigen to a serum sample spiked with known amounts of antibody and looking for inhibition of the expected signal. This approach itself can form the basis of a confirmatory assay.

The use of the immunoblotting procedure, which provides information concerning the specificity of the antibodies detected, is valuable as the antibodies may have specificity for other components (e.g., contaminants) in the product and can cause data to be misinterpreted. For example, very low levels of expression system derived from bacterial proteins in rDNA products can cause significant antibody development, whereas the human sequence major protein present (the active principle) may be much less immunogenic. However, other procedures, for example, analytical RIPA can also be used for specificity studies.

The use of assays described earlier does not obviate the requirement for a functional cell-based neutralization assay. The latter should be incorporated into the strategy for immunogenicity assessment as it has been shown that results from bioassays can often be correlated with the effect of antibodies on clinical response. Some developers conduct neutralization assays at both preclinical and clinical level as part of their product development program while others implement these assays at the clinical stage after considering whether the therapeutic is low or high risk.

Conclusion

While some recombinant products like filgrastim have almost no immunogenicity, other glycosylated products may show a very high level of immune response triggering ability; an immunogenic response, like an anaphylactic response, can be life threatening in the immediate course and creating autoimmune disorders in the long course. The developers of biosimilar products will face the greatest challenge in providing data that shows comparable immunogenicity against the reference product; in some instances, this may raise the question of appropriateness when tested in humans. A large number of surrogate tests are available today, and many more will become available soon, give us an opportunity to model the immunogenic response without exposing patients or healthy subjects. A keen understanding of what makes a product more or less immunogenic, as described in this chapter, is an important step in developing these products.

Bibliography

Abbas AK, Lichtman AH. Chapter 9: Immunologic tolerance and autoimmunity. In: *Basic Immunology: Functions and Disorders of the Immune System*, Malley J, ed. Elsevier Saunders, Philadelphia, PA, 2006, pp. 161–176.

Albertsson-Wikland K. Clinical trial with authentic recombinant somatropin in Sweden and Finland. *Acta Paediatr Scand (Suppl)* 1987;331:28–34.

Allison M. PML problems loom for rituxan. *Nat Biotechnol* 2010;28:105–106.

Alsalameh S, Manger B, Kern P, Kalden J. New onset of rheumatoid arthritis during interferon beta-1b treatment in a patient with multiple sclerosis: Comment on the case report by Jabaily and Thompson. *Arthritis Rheum* 1998;41:754.

Alshekhlee A, Basiri K, Miles JD, Ahmad SA, Katirji B. Chronic inflammatory demyelinating polyneuropathy associated with tumor necrosis factor-alpha antagonists. *Muscle Nerve* 2010;41:723–727.

Altman JJ, Pehuet M, Slama G, Tchoapoutsky C. Three cases of allergic reaction to human insulin (letter). *Lancet* 1983;2:524.

Andersen OO. Insulin antibody formation II: The influence of species difference and method of administration. *Acta Endocrinol* 1973;72:33–45.

Antignani A, Youle RJ. The cytokine, GM-CSF, can deliver BCL-XL as an extracellular fusion protein to protect cells from apoptosis and retain differential induction. *J Biol Chem* 2007;282(15):11246–11254.

Antonelli G, Simeoni E, Bagnato F, Pozzilli C, Turriziani O, Tesoro R, Di Marco P, Gasperini C, Fieschi C, Dianzani F. Further study on the specificity and incidence of neutralizing antibodies to interferon (IFN) in relapsing remitting multiple sclerosis patients treated with IFN beta-1a or IFN beta-1b. *J Neurol Sci* 1999;168:131–136.

Antony GK, Dudek AZ. Interleukin 2 in cancer therapy. *Curr Med Chem* 2010;17:3297–3302.

Armstrong JK, Hempel G, Koling S, Chan LS et al. Antibody against poly(ethylene glycol) adversely affects pegasparaginase therapy in acute lymphoblastic leukemia patients. *Cancer* 2007;110:103–111.

Arnon R, Aharoni R. Mechanism of action of glatiramer acetate in multiple sclerosis and its potential for the development of new applications. *Proc Natl Acad Sci USA* 2004; 101:14593–14598.

Aviezer D, Brill-Almon E, Shaaltiel Y, Hashmueli S et al. A plant derived recombinant human glucocerebrosidase enzyme—A preclinical and phase I investigation. *PLoS One* 2009; 4:e4792.

Bachmann MF, Zinkernagel RM. Neutralizing antiviral B cell responses. *Annu Rev Immunol* 1997;15:235–270.

Baechler EC, Batliwalla FM, Karypis G, Gaffney PM et al. Interferon-inducible gene expression signature in peripheral blood cells of patients with severe lupus. *Proc Natl Acad Sci USA* 2003;100:2610–2615.

Baert F, Noman M, Vermiere S, Assche GV, D'Haens G, Carbonez A, Rutgeerts P. Influence of immunogenicity on the long-term efficacy of intliximab in Crohn's disease. *N Engl J Med* 2003;348:601–608.

Baker DP, Lin EY, Lin K, Pellegrini M et al. N-terminally PEGylated human interferon-β-1a with improved pharmacokinetic properties and in vivo efficacy in a melanoma angiogenesis model. *Bioconjug Chem* 2006;17:179–188.

Baker M, Reynolds HM, Lumicisi B, Bryson CJ. Immunogenicity of protein therapeutics: The key causes, consequences and challenges. *Self Nonself* 2010;1:314–322.

Barbosa MD, Vielmetter J, Chu S, Smith DD, Jacinto J. Clinical link between MHC class II haplotype and interferon-beta (IFN-beta) immunogenicity. *Clin Immunol* 2006;118:42–50.

Bardor M, Burel C, Villajero A, Cadoret JP et al. Chapter 5: Plant N-glycosylation: An engineered pathway for the production of therapeutical plant-derived glycoproteins. In *Glycosylation in Diverse Cell Systems: Challenges and New Frontiers in Experimental Biology*, Brooks S, Rudd P, Appelmelk B, eds. Society for Experimental Biology, London, U.K., 2011, pp. 93–118.

Bardor M, Faveeuw C, Fitchette A-C, Gilbert D et al. Immunoreactivity in mammals of two typical plant glyco-epitopes, core α(1,3)-fucose and core xylose. *Glycobiology* 2003;13:427–434.

Bardor M, Loutelier-Bourhis C, Paccalet T, Cosette P et al. Monoclonal C5-1 antibody produced in transgenic alfalfa plants exhibits a N-glycosylation that is homogenous and suitable for glyco-engineering into human-compatible structures. *Plant Biotechnol J* 2003;1:451–462.

Bardor M, Nguyen DH, Diaz S, Varki A. Mechanism of uptake and incorporation of the non-human sialic acid N-glycolylneuraminic acid into human cells. *J Biol Chem* 2005;280:4228–4237.

Barranco P, Lopez-Serrano MC. General and epidemiological aspects of allergic drug reactions. *Clin Exp Allergy* 1998;28(Suppl. 4):61–62.

Basu A, Yang K, Wang M, Liu S et al. Structure-function engineering of interferon-β-1b for improving stability, solubility, potency, immunogenicity, and pharmacokinetic properties by site-selective mono-PEGylation. *Bioconjug Chem* 2006;17:618–630.

Bates D. Alemtuzumab. *Int MS J Forum* 2009;16:75–76.

Bee JS, Davis M, Freund E, Carpenter JF, Randolph TW. Aggregation of a monoclonal antibody induced by adsorption to stainless steel. *Biotechnol Bioeng* 2010;105:121–129.

Bekisz J, Schmeisser H, Hernandez J, Goldman ND, Zoon KC. Human interferons alpha, beta and omega. *Growth Factors* 2004;22:243–251.

Bellomi F, Scagnolari C, Tomassini V, Gasperini C, Paolillo A, Pozzilli C, Antonelli G. Fate of neutralizing and binding antibodies to IFN beta in MS patients treated with IFN beta for 6 years. *J Neurol Sci* 2003;215:3–8.

Bennett CL, Luminari S, Nissenson AR, Tallman MS et al. Melanoma in multiple sclerosis treated with natalizumab: Causal association or coincidence? *Mult Scler* 2009;15:1532–1533.

Berger JR. Progressive multifocal leukoencephalopathy and newer biological agents. *Drug Saf* 2010;33:969–983.

Bergrem H, Danielson BG, Eckardt KU, Kurtz A, Stridsberg M. A case of antierythropoietin antibodies following recombinant human erythropoietin treatment. In *Erythropoietin: Molecular Physiology and Clinical Application*, Bauer C, Koch KM, Sciqalla P, eds. Marcel Dekker, New York, 1993, pp. 265–275.

Berman E, Heller G, Kempin S, Gee T, Tran L-L, Clarkson B. Incidence of response and long-term follow up in a patient with hairy cell leukemia treated with recombinant human interferon alfa-2a. *Blood* 1990;75:839–845.

Bernatsky S, Renoux C, Suissa S. Demyelinating events in rheumatoid arthritis after drug exposures. *Ann Rheum Dis* 2010;69:1691–1693.

Berson SA, Yalow RS, Bauman A, Rothschild MA, Newerly K. Insulin-I-131 metabolism in human subjects: Demonstration of insulin binding globulin in the circulation of insulin treated subjects. *J Clin Invest* 1956;35:170.

Bertolotto A. Neutralizing antibodies to interferon beta: Implications for the management of multiple sclerosis. *Curr Opin Neurol* 2004;17:241–246.

Bertolotto A, Gilli F, Sala A, Audano L, Castello A, Magliola U, Melis F, Giordana MT. Evaluation of bioavailability of three types of IFNbeta in multiple sclerosis patients by a new quantitative competitive-PCR method for MxA quantification. *J Immunol Methods* 2001;256:141–152.

Bertolotto A, Gilli F, Sala A, Capobianco M et al. Persistent neutralizing antibodies abolish the interferon beta bioavailability in MS patients. *Neurology* 2003;60:634–639.

Bertolotto A, Malucchi S, Milano E, Castello A, Capobianco M, Mutani R. Interferon beta neutralizing antibodies in multiple sclerosis: Neutralizing activity and cross-reactivity with three different preparations. *Immunopharmacology* 2000;49:95–100.

Bielekova B, Catalfamo M, Reichert-Scrivner S, Packer A et al. Regulatory CD56(Bright) natural killer cells mediate immunomodulatory effects of IL-2Ralpha-targeted therapy (daclizumab) in multiple sclerosis. *Proc Natl Acad Sci USA* 2006;103:5941–5946.

Bijvoet AGA, Van Hirtum H, Kroos MA, Van de Kamp EH et al. Human a-glucosidase from rabbit milk has therapeutic effect in mice with glycogen storage disease type II. *Hum Mol Genet* 1999;8:2145–2153.

Blandford RL, Sewell H, Sharp P, Hearnshaw JR. Generalized allergic reaction with synthetic human insulin (comment–letter). *Lancet* 1982;2:1468.

Bonetti P, Diodati G, Drago C, Casarin C, Scaccabarozzi S, Realdi G, Ruol A, Alberti A. Interferon antibodies in patients with chronic hepatitic C virus infection treated with recombinant interferon alpha-2 alpha. *J Hepatol* 1994;20:416–420.

Bongartz T, Sutton AJ, Sweeting MJ, Buchan I, Matteson EL, Montori V. Anti-TNF antibody therapy in rheumatoid arthritis and the risk of serious infections and malignancies: Systematic review and meta-analysis of rare harmful effects in randomized controlled trials. *JAMA* 2006;295:2275–2285.

Bosques CJ, Collins BE, Meador JW, Sarvaiya H et al. Chinese hamster ovary cells can produce galactose-α-1,3-galactose antigens on proteins. *Nat Biotechnol* 2010;28:1153–1156.

Bozhinov A, Handzhiyski Y, Genov K, Daskalovska V et al. Advanced glycation end products contribute to the immunogenicity of IFN-β pharmaceuticals. *J Allergy Clin Immunol* 2012;129:855–858.

Braun A, Kwee L, Labow MA, Alsenz J. Protein aggregates seem to play a key role among the parameters influencing the antigenicity of interferon alpha (IFN-alpha) in normal and transgenic mice. *Pharm Res* 1997;14:1472–1478.

Bray GL, Gomperts ED, Courter S, Gruppo R, Gordon EM, Manco-Jonshon M, Shapiro A, Scheibel E, White G, Lee M. A multicentre study of recombinant factor VIII (recombinate): Safety, efficacy, and inhibitor risk in previously untreated patients with hemofilia A. *Blood* 1994;83:2428–2435.

Brenner T, Arnon R, Sela M, Abramsky O, Meiner Z, Riven-Kreitman R, Tarcik N, Teitelbaum D. Humoral and cellular immune responses to Copolymer 1 in multiple sclerosis patients treated with Copaxone®. *J Neuroimmunol* 2001;115:152–160.

Brickelmaier M, Hochman PS, Baciu R, Chao B, Cuervo JH, Whitty A. ELISA methods for the analysis of antibody responses induced in multiple sclerosis patients treated with recombinant interferon-beta. *J Immunol Methods* 1999;227:121–135.

Brinks V, Jiskoot W, Schellekens H. Immunogenicity of therapeutic proteins: The use of animal models. *Pharm Res* 2011;28:2379–2385.

Bristol A. Recombinant DNA derived insulin analogues as potentially useful therapeutic agents. *Trends Biotechnol* 1993;11:301–305.

Brooks SA. Chapter 4: Species differences in protein glycosylation and their implications for biotechnology. In *Glycosylation in Diverse Cell Systems: Challenges and New Frontiers in Experimental Biology*, Brooks S, Rudd P, Appelmelk B, eds. Society for Experimental Biology, London, U.K., 2011, pp. 74–92.

Buettel IC, Chamberlain P, Chowers Y, Ehmann F et al. Taking immunogenicity assessment of therapeutic proteins to the next level. *Biologicals* 2011;39:100–109.

Burdick LM, Somani N, Somani AK. Type I IFNs and their role in the development of autoimmune diseases. *Expert Opin Drug Saf* 2009;8:459–472.

Calabresi PA, Giovannoni G, Confavreux C, Galetta SL et al. The incidence and significance of antinatalizumab antibodies: Results from Affirm and Sentinel. *Neurology* 2007;69:1391–1403.

Campara M, Tzvetanov IG, Oberholzer J. Interleukin-2 receptor blockade with humanized monoclonal antibody for solid organ transplantation. *Expert Opin Biol Ther* 2010;10:959–969.

Carpenter JF, Randolph TW, Jiskoot W, Crommelin DJA et al. Overlooking subvisible particles in therapeutic protein products: Gaps that may compromise product quality. *J Pharm Sci* 2009;98:1201–1205.

Carroll MB, Forgione MA. Use of tumor necrosis factor alpha inhibitors in hepatitis B surface antigen-positive patients: A literature review and potential mechanisms of action. *Clin Rheumatol* 2010;29:1021–1029.

Carson KR, Evens AM, Richey EA, Habermann TM et al. Progressive multifocal leukoencephalopathy after rituximab therapy in HIV-negative patients: A report of 57 cases from the research on adverse drug events and reports project. *Blood* 2009;113:4834–4840.

Carveth-Johnson AO, Mylvaganam K, Child DF. Generalised allergic reaction with synthetic human insulin (letter). *Lancet* 1982;2:1287.

Casadevall N, Nataf J, Viron B, Kolta A et al. Pure red-cell aplasia and antierythropoietin antibodies in patients treated with recombinant erythropoietin. *N Engl J Med* 2002; 346(7):469–475.

Casadevall N. Antibodies against rHuEPO: Native and recombinant. *Nephrol Dial Transplant* 2002;17(Suppl. 5):42–47.

Casadevall N, Nataf J, Viron B, Kolta A et al. Pure red-cell aplasia and antierythropoietin antibodies in patients treated with recombinant erythropoietin. *N Engl J Med* 2002;346: 469–475.

Castela E, Lebrun-Frenay C, Laffon M, Rocher F et al. Evolution of nevi during treatment with natalizumab: A prospective follow-up of patients treated with natalizumab for multiple sclerosis. *Arch Dermatol* 2011;147:72–76.

Castillo V, Graña-Montes R, Sabate R, Ventura S. Prediction of the aggregation propensity of proteins from the primary sequence: Aggregation properties of proteomes. *Biotechnol J* 2011;6:674–685.

Chackerian B, Lenz P, Lowy DR, Schiller JT. Determinants of autoantibody induction by conjugated papillomavirus virus-like particles. *J Immunol* 2002;169:6120–6126.

Chakravarty EF, Michaud K, Wolfe F. Skin cancer, rheumatoid arthritis, and tumor necrosis factor inhibitors. *J Rheumatol* 2005;32:2130–2135.

Chance RE, Root MA, Galloway JA. The immunogenicity of insulin preparations. *Acta Endocrinol Suppl (Copenh)* 1976;205:185–198.

Charak B, Agah R, Mazumder A. Granulocyte-macrophage colony-stimulating factor-induced antibody-dependent cellular cytotoxicity in bone marrow macrophages: Application in bone marrow transplantation. *Blood* 1993;81:3474–3479.

Chen JW, Frystyk J, Lauritzen T, Christiansen JS. Impact of insulin antibodies on insulin aspart pharmacokinetics and pharmacodynamics after 12-week treatment with multiple daily injections of biphasic insulin aspart 30 in patients with type 1 diabetes. *Eur J Endocrinol* 2005;153(6):907–913.

Chen W, Ede NJ, Jackson DC, McCluskey J, Purcell AW. CTL recognition of an altered peptide associated with asparagine bond rearrangement. Implications for immunity and vaccine design. *J Immunol* 1996;157:1000–1005.

Chung CH. Managing premedications and the risk for reactions to infusional monoclonal antibody therapy. *Oncologist* 2008;13:725–732.

Chung CH, Mirakhur B, Chan E, Le Q-T et al. Cetuximab-induced anaphylaxis and IgE specific for galactose-α-1,3-galactose. *N Engl J Med* 2008;358:1109–1117.

Clanet M, Radue EW, Kappos L, Hartung HP, Hohlfeld R, Sandberg-Wollheim M, Kooijmans-Coutinho MF, Tsao EC, Sandrock AW; European IFNbeta-1a (Avonex) Dose-Comparison Study Investigators. A randomized, double-blind, dose-comparison study of weekly interferon beta-1a in relapsing MS. *Neurology* 2002;59:1507–1517.

Cleland J, Powell M, Shire S. The development of stable protein formulations: A close look at protein aggregation, deamidation, and oxidation. *Crit Rev Ther Drug Carrier Syst* 1993;10:307–377.

Clifford DB, De Luca A, Simpson DM, Arendt G, Giovannoni G, Nath A. Natalizumab-associated progressive multifocal leukoencephalopathy in patients with multiple sclerosis: Lessons from 28 cases. *Lancet Neurol* 2010;9:438–446.

Cohen JD, Bournerias I, Buffard V, Paufler A et al. Psoriasis induced by tumor necrosis factor-alpha antagonist therapy: A case series. *J Rheumatol* 2007;34:380–385.

Coles AJ, Compston DA, Selmaj KW, Lake SL et al. Alemtuzumab vs. interferon beta-1a in early multiple sclerosis. *N Engl J Med* 2008;359:1786–1801.

Comi G, Filippi M, Wolinsky JS. European/Canadian multicenter, double-blind, randomized, placebo-controlled study of the effects of glatiramer acetate on magnetic resonance imaging–measured disease activity and burden in patients with relapsing multiple sclerosis. European/Canadian Glatiramer Acetate Study Group. *Ann Neurol* 2001;49:290–297.

Committee for Medicinal Products for Human Use (CHMP). Guideline on immunogenicity assessment of biotechnology-derived therapeutic proteins. EMEA/CHMP/BMWP/14327/2006—Draft version released for consultation, 2007.

Content J. Mechanisms of induction and action of interferons. *Verh K Acad Geneeskd Belg* 2009;71:51–71.

Costa MF, Said NR, Zimmermann B. Drug-induced lupus due to anti-tumor necrosis factor alpha agents. *Semin Arthritis Rheum* 2008;37:381–387.

Coyle PK. The role of natalizumab in the treatment of multiple sclerosis. *Am J Manag Care* 2010;16:S164–S170.

Crommelin Daan JA. Immunogenicity of therapeutic proteins. In *Handbook of Pharmaceutical Biotechnology*. Wiley, the Netherlands, 2007.

Cromwell MEM, Hilario E, Jacobson F. Protein aggregation and bioprocessing. *AAPS J* 2006;8:E572–E579.

Cunninghame Graham DS, Akil M, Vyse TJ. Association of polymorphisms across the tyrosinekinase gene, Tyk2 in UK SLE families. *Rheumatology (Oxford)* 2007;46:927–930.

Dafny N, Yang PB. Interferon and the central nervous system. *Eur J Pharmacol* 2005;523:1–15.

Davidson JK, DeBra DW. Immunologic insulin resistance. *Diabetes* 1978;27:307–318.

Deckert T. The immunogenicity of new insulins. *Diabetes* 1985;34(Suppl. 2):94–96.

Defrance T, Taillardet M, Genestier L. T cell-independent B cell memory. *Curr Opin Immunol* 2011;23:30–336.

De Groot AS, Scott DW. Immunogenicity of protein therapeutics. *Trends Immunol* 2007; 28:482–490.

Deisenhammer F, Reindl M, Harvey J, Gasse T, Dilitz E, Berger T. Bioavailability of interferon beta 1b in MS patients with and without neutralizing antibodies. *Neurology* 1999; 62:1239–1243.

de Laat B, van Berkel M, Urbanus RT, Siregar B et al. Immune responses against domain I of β2-glycoprotein I are driven by conformational changes: Domain I of β2-glycoprotein I harbors a cryptic immunogenic epitope. *Arthritis Rheum* 2011;63:3960–3968.

Deleeuw I, Delvigne C, Beckaert J. Insulin allergy treated with human insulin (recombinant DNA). *Diabetes Care* 1982;5(Suppl. 2):168–170.

den Engelsman J, Garidel P, Smulders R, Koll H et al. Strategies for the assessment of protein aggregates in pharmaceutical biotech product development. *Pharm Res* 2011;28:920–933.

De Santi L, Costantini MC, Annunziata P. Long time interval between multiple sclerosis onset and occurrence of primary Sjögren's syndrome in a woman treated with interferon-beta. *Acta Neurol Scand* 2005;112:194–196.

deShazo RD, Levinson AI, Boehm T, Evans R III, Waed G Jr. Severe persistent biphasic local (immediate and late) skin reactions to insulin. *J Allergy Clin Immunol* 1977;59:161–164.

DeWitt DE, Hirsch IB. Outpatient insulin therapy in type 1 and type 2 diabetes mellitus: A scientific review. *JAMA* 2003;289:2254–2264.

Dianzani F, Antonelli G, Amicucci P, Cefaro A, Pintus C. Low incidence of neutralizing antibody formation to interferon-alpha 2b in human recipients. *J Interferon Res* 1989;9(Suppl. 1):S33–S36.

Dicato M, Plawny L. Erythropoietin in cancer patients: Pros and cons. *Curr Opin Oncol* 2010;22:307–311.

Dietrich LL, Bridges AJ, Albertini MR. Dermatomyositis after interferon alpha treatment. *Med Oncol* 2000;17:64–69.

Dieude P, Guedj M, Wipff J, Ruiz B et al. STAT4 is a genetic risk factor for systemic sclerosis having additive effects with IRF5 on disease susceptibility and related pulmonary fibrosis. *Arthritis Rheum* 2009;60:2472–2479.

Dinarello CA, Bernheim HA, Duff GW, Le HV et al. Mechanisms of fever induced by recombinant human interferon. *J Clin Invest* 1984;74:906–913.

Dintzis HM, Dintzis RZ, Vogelstein B. Molecular determinants of immunogenicity: The immunon model of immune response. *Proc Natl Acad Sci USA* 1976;73:3671–3675.

Dixon WG, Hyrich KL, Watson KD, Lunt M et al. Drug-specific risk of tuberculosis in patients with rheumatoid arthritis treated with anti-TNF therapy: Results from the British Society for Rheumatology Biologics Register (BSRBR). *Ann Rheum Dis* 2010;69:522–528.

Douglas DD, Rakela J, Lin HJ, Hollinger FB, Taswell HF, Czaja AJ, Gross JB, Anderson ML, Parent K, Fleming CR. Randomized controlled trial of recombinant alpha-2a-interferon for chronic hepatitis C. Comparison of alanine aminotransferase normalization versus loss of HCV RNA and anti-HCV IgM. *Dig Dis Sci* 1993;38:601–607.

Doyle HA, Zhou J, Wolff MJ, Harvey BP et al. Isoaspartyl post-translational modification triggers anti-tumor T and B lymphocyte immunity. *J Biol Chem* 2006;281:32676–32683.

Dresser DW. Specific inhibition of antibody production. II. Paralysis induced in adult mice by small quantities of protein antigen. *Immunology* 1962;5:378–388.

Dubois M. L'immunogénicité des protéines thérapeutiques dans Développement de techniques analytiques pour l'évaluation des protéines théra-peutiques et des biomarqueurs par spectrométrie de masse, Thèse de Doctorat, Université Pierre et Marie Curie Paris VI, France, 2008, p. 106.

Duchini A. Autoimmune hepatitis and interferon beta-1a for multiple sclerosis. *Am J Gastroenterol* 2002;97:767–768.

Eckardt KU, Casadevall N. Pure red cell aplasia due to anti-erythropoietin antibodies. *Nephrol Dial Transplant* 2003;18:865–869.

Ehrenforth S, Kreuz W, Scharrer I, Linde R, Funk M, Gungor T, Krackhardt B, Kornhuber B. Incidence of development of factor VIII and factor IX inhibitors in haemophiliacs. *Lancet* 1992;339:594–598.

EMEA. CHMP/BMWP/14327/2006; Guideline on immunogenicity assessment of biotechnology-derived therapeutic proteins, April 2008.

Eon-Duval A, Broly H, Gleixner R. Quality attributes of recombinant therapeutic proteins: An assessment of impact on safety and efficacy as part of a quality by design development approach. *Biotechnol Prog* 2012;28:608–622.

Evens AM, Kuzel DM, Schumock GT, Belknap SM, Locatelli F, Rossert J, Casadevall N. Pure red cell aplasia and epoetin therapy. *N Engl J Med* 2004;351:1403–1408.

Ezzelarab M, Ayares D, Cooper DKC. Carbohydrates in xenotransplantation. *Immunol Cell Biol* 2005;83:396–404.

Fabris P, Floreani A, Tositti G, Vergani D, De Lalla F, Betterle C. Type 1 diabetes mellitus in patients with chronic hepatitis C before and after interferon therapy. *Aliment Pharmacol Ther* 2003;18:549–558.

Fakharzadeh SS, Kazazian HH Jr. Correlation between factor VIII genotype and inhibitor development in hemophilia A. *Semin Thromb Hemost* 2000;26:167–171

Farina C, Weber MS, Meinl E, Wekerle H, Hohlfeld R. Glatiramer acetate in multiple sclerosis: Update on potential mechanisms of action. *Lancet Neurol* 2005;4:567–575.

Farrell RA, Giovannoni G. Current and future role of interferon beta in the therapy of multiple sclerosis. *J Interferon Cytokine Res* 2010;30:715–726.

FDA: U.S. Food and Drug Administration. Draft guidance for industry: Assay development for immunogenicity testing of therapeutic proteins. U.S. Department of Health and Human Services, Food and Drug Administration, Center for Drug Evaluation and Research (CDER), Center for Biologics Evaluation and Research (CBER), Silver Spring, MD, 2009.

FDA: U.S. Food and Drug Administration. Draft guidance for industry: Scientific considerations in demonstrating biosimilarity to a reference product. U.S. Department of Health and Human Services, Food and Drug Administration, Center for Drug Evaluation and Research (CDER), Center for Biologics Evaluation and Research (CBER), 2012.

Felipe V, Hawe A, Schellekens H, Jiskoot W. Aggregation and immunogenicity of therapeutic proteins. In *Aggregation of Therapeutic Proteins*, Wang W, Roberts C, eds. John Wiley & Sons, Hoboken, NJ, 2010.

Fernandez O, Guerrero M, Mayorga C, Munoz L, Lean A, Lugue G, Hervas M, Fernandez V, Capdevila A, de Ramon E. Combination therapy with interferon beta-1b and azathioprine in secondary progressive multiple sclerosis. A two-year pilot study. *J Neurol* 2002; 249:1058–1062.

Fernandez-Escamilla A-M, Rousseau F, Schymkowitz J, Serrano L. Prediction of sequence-dependent and mutational effects on the aggregation of peptides and proteins. *Nat Biotechnol* 2004;22:1302–1306.

Fernandez-Espartero MC, Perez-Zafrilla B, Naranjo A, Esteban C et al. Demyelinating disease in patients treated with TNF antagonists in rheumatology: Data from BIOBADASER, a pharmacovigilance database, and a systematic review. *Semin Arthritis Rheum* 2011;40:330–337.

Figlin RA, Itri L. Anti-interfeon antibodies: A perspective. *Semin Hematol* 1988;25:9–15.

Filipe VLS. Chapter 9: Immunogenicity of different stressed monoclonal IgG formulations in immune tolerant transgenic mice. In *Monoclonal Antibody Aggregates: Physicochemical Characteristics, Stability in Biological Fluids and Immunogenicity*, PhD thesis, Pharmaceutical Sciences, Utrecht University, Utrecht, the Netherlands, 2012, pp. 171–200; *mAbs* 2012;4(6).

Fineberg SE. Insulin allergy and insulin resistance. In *Therapy for Diabetes Mellitus and Related Disorders*, Lebovitz HE, ed. American Diabetes Association, Alexandria, VA, 1994, pp. 178–184.

Fineberg SE, Galloway JA, Fineberg NS, Goldman J. Effects of species of origin, purification levels, and formulation on insulin immunogenicity. *Diabetes* 1983;32:592–599.

Fineberg SE, Huang J, Brunelle R, Gulliya KS, Anderson JH. Effect of long-term exposure to insulin Lispro on the induction of antibody response in patients with type 1 or type 2 diabetes. *Diabetes Care* 2003;26:89–96.

Fireman P, Fineberg SE, Galloway JA. Development of IgE antibodies to human (recombinant DNA) porcine, and bovine insulins in diabetic subjects. *Diabetes Care* 1982;5(Suppl. 2): 119–125.

Fradkin AH, Carpenter JF, Randolph TW. Immunogenicity of aggregates of recombinant human growth hormone in mouse models. *J Pharm Sci* 2009;98:3247–3264.

Fradkin AH, Carpenter JF, Randolph TW. Glass particles as an adjuvant: A model for adverse immunogenicity of therapeutic proteins. *J Pharm Sci* 2011;100:4953–4964.

Francis GS, Rice GP, Alsop JC. Interferon beta-1a in MS: Results following development of neutralizing antibodies in prisms. *Neurology* 2005;65:48–55.

Freund M, von Wussow P, Diedrich H, Eisert R et al. Recombinant human interferon (IFN) alpha-2b in chronic myelogenous leukemia: Dose dependency of response and frequency of neutralizing anti-interferon antibodies. *Br J Haematol* 1989;72:350–356.

Frigerio C, Aubry M, Gomez F. Desensitization-resistant insulin allergy. *Allergy* 1997;52:238–239.

Frost H. Antibody-mediated side effects of recombinant proteins. *Toxicology* 2005;209:155–160.

Furst DE. The risk of infections with biologic therapies for rheumatoid arthritis. *Semin Arthritis Rheum* 2010;39:327–346.

Galili U. The alpha-gal epitope and the anti-Gal antibody in xenotransplantation and in cancer immunotherapy. *Immunol Cell Biol* 2005;83:674–686.

Garcia-Ortega P, Knobel H, Miranda A. Sensitization to human insulin (letter). *BMJ* 1984;288:1271.

Ghaderi D, Taylor RE, Padler-Karavani V, Diaz S, Varki A. Implications of the presence of N-glycolylneuraminic acid in recombinant therapeutic glycoproteins. *Nat Biotechnol* 2010;28:863–867.

Giovannoni G. Strategies to treat and prevent the development of neutralizing anti-interferon-beta antibodies. *Neurology* 2003;61:S13–S17.

Gneiss C, Reindl M, Lutterotti A, Ehling R, Egg R, Khalil M, Berger T, Deisenhammer F. Interferon beta: The neutralizing antibody (NAb) titre predicts reversion to NAb negativity. *Mult Scler* 2004;10:507–510.

Goodbourn S, Didcock L, Randall RE. Interferons: Cells signalling, immune modulation, antiviral response and virus countermeasures. *J Gen Virol* 2000;81:2341–2364.

Goodin DS, Hurwitz B, Noronha A. Neutralizing antibodies to interferon beta-1b are not associated with disease worsening in multiple sclerosis. *J Int Med Res* 2007;35:173–187.

Goodman M. Market watch: Sales of biologics to show robust growth through to 2013. *Nat Rev Drug Discov* 2009;8:837–837.

Gossain VV, Rouner DR, Homak K. Systemic allergy to human (recombinant DNA) insulin. *Ann Allergy* 1985;55:116–118.

Graham RR, Kozyrev SV, Baechler EC, Reddy MV et al. A common haplotype of interferon regulatory factor 5 (IRF5) regulates splicing and expression and is associated with increased risk of systemic lupus erythematosus. *Nat Genet* 2006;38:550–555.

Grammer LC, Roberts M, Buchanan TA, Fitzsimons R, Metzger BE, Patterson R. Specificity of immunoglobulin E and immunoglobulin G against human (recombinant DNA) insulin in human insulin allergy and resistance. *J Lab Clin Med* 1987;109:141–146.

Gribben JG, Devereux S, Thomas NS, Keim M et al. Development of antibodies to unprotected glycosylation sites on recombinant human GM-CSF. *Lancet* 1990;335:434–437.

Gross J, Moller R, Henke W, Hoesel W. Detection of anti-EPO antibodies in human sera by a bridging ELISA is much more sensitive when coating biotinylated rhEPO to streptavidin rather than using direct coating of rhEPO. *J Immunol Methods* 2006;313:176–182.

Grossberg SE, Kawade Y, Grossberg LD. The neutralization of interferons by antibody III. The constant antibody bioassay, a highly sensitive quantitative detector of low antibody levels. *J Interferon Cytokine Res* 2009;29:93–104.

Gryta T. Biogen: Six more cases of brain infection in Tysabri patients. *Dow Jones Newswires*, 2011.

Gupta S, Indelicato SR, Jethwa V, Kawabata T et al. Recommendations for the design, optimization, and qualification of cell-based assays used for the detection of neutralizing antibody responses elicited to biological therapeutics. *J Immunol Methods* 2007;321:1–18.

Hanauer L, Batson JM. Anaphylactic shock following insulin injection. *Diabetes* 1961;10:105–109.

Harding FA, Stickler MM, Razo J, DuBridge RB. The immunogenicity of humanized and fully human antibodies: Residual immunogenicity resides in the CDR regions. *MAbs* 2010;2:256–265.

Harris JM, Martin NE, Modi M. Pegylation: A novel process for modifying pharmacokinetics. *Clin Pharmacokinet* 2001;40:539–551.

Hayakawa T, Ishii A. Japanese regulatory perspective on immunogenicity. In *Detection and Quantification of Antibodies to Biopharmaceuticals: Practical and Applied Considerations*, Tovey MG, ed. John Wiley & Sons, New York, 2011.

Heathcote EJ. Autoimmune hepatitis and chronic hepatitis C: Latent or initiated by interferon therapy? *Gastroenterology* 1995;108:1942–1944.

Heding LG, Larsson Y, Ludvigsson J. The immunogenicity of insulin preparation. Antibody levels before and after transfer to highly purified porcine insulin. *Diabetologia* 1980;19:511–515.

Hellquist A, Jarvinen TM, Koskenmies S, Zucchelli M et al. Evidence for genetic association and interaction between the Tyk2 and IRF5 genes in systemic lupus erythematosus. *J Rheumatol* 2009;36:1631–1638.

Hemmer B, Stuve O, Kieseier B, Schellekens H, Hartung HP. Immune response to immunotherapy: The role of neutralising antibodies to interferon beta in the treatment of multiple sclerosis. *Lancet Neurol* 2005;4:403–412.

Hermeling S, Aranha L, Damen JMA, Slijper M et al. Structural characterization and immunogenicity in wild-type and immune tolerant mice of degraded recombinant human interferon alpha2b. *Pharm Res* 2005;22:1997–2006.

Hermeling S, Crommelin D, Schellekens H, Jiskoot W. Structure-immunogenicity relationships of therapeutic proteins. *Pharm Res* 2004;21:897–903.

Hermeling S, Crommelin DJA, Schellekens H, Jiskoot W. Immunogenicity of therapeutic proteins. In *Handbook of Pharmaceutical Biotechnology*, Gad SC, ed., Vol. IV, *Handbook Series*. John Wiley and Sons Inc., Hoboken, NJ, 2007.

Hermeling S, Schellekens H, Maas C, Gebbink MFBG et al. Antibody response to aggregated human interferon alpha2b in wildtype and transgenic immune tolerant mice depends on type and level of aggregation. *J Pharm Sci* 2006;95:1084–1096.

Hershberger RE, Starling RC, Eisen HJ, Bergh CH et al. Daclizumab to prevent rejection after cardiac transplantation. *N Engl J Med* 2005;352:2705–2713.

Hiatt A, Caffferkey R, Bowdish K. Production of antibodies in transgenic plants. *Nature* 1989;342:76–78.

Hill AD, Redmond HP, Austin OM, Grace PA, Bouchier-Hayes D. Granulocyte-macrophage colony-stimulating factor inhibits tumor growth. *Br J Surg* 1993;80:1543–1546.

Hochuli E. Interferon Immunogenicity: Technical evaluation of interferon-alpha 2a. *J Interferon Cytokine Res* 1997;17(Suppl. 1):S15–S21.

Holliday SM, Benfield P. Interferon-b-1a. A review of its pharmacological proprieties and therapeutic potential in multiple sclerosis. *BioDrugs* 1997;8:317–330.

Hooijberg E, Sein JJ, van den Berk PC, Hart AA, van der Valk MA, Kast WM, Melief CJ, Hekman A. Eradication of large human B-cell tumors in nude mice with unconjugated CD20 monoclonal antibodies and interleukin-2. *Cancer Res* 1995;55:2627–2634.

International Conference on Harmonization. Guidance for industry. S6 Preclinical safety evaluation of biotechnology-derived pharmaceuticals, 1997. http://www.fda.gov/cder/guidance/l859fnl.pdf (accessed August 18, 2015).

Ishibashi O, Kobayashi M, Maegawa H, Watanabe N, Takata Y, Okuno Y, Shigeta Y. Can insulin antibodies of diabetic patients distinguish human insulin from porcine insulin? *Horm Metab Res* 1986;18:470–472.

Ismail A, Kemp J, Sharrack B. Melanoma complicating treatment with natalizumab (Tysabri) for multiple sclerosis. *J Neurol* 2009;256:1771–1772.

Ito I, Kawaguchi Y, Kawasaki A, Hasegawa M et al. Association of a functional polymorphism in the IRF5 region with systemic sclerosis in a Japanese population. *Arthritis Rheum* 2009;60:1845–1850.

Jacobs LD, Cookfair DL, Rudick RA, Herndon RM et al. Factor VIII immunogenicity. *Haemophilia* 1998;4(4):552–557.

Jefferis R. Human immunoglobulin allotypes: Possible implications for immunogenicity. *MAbs* 2009a;1:332–338.

Jefferis R. Recombinant antibody therapeutics: The impact of glycosylation on mechanisms of action. *Trends Pharmacol Sci* 2009b;30:356–362.

Jegerlehner A, Tissot A, Lechner F, Sebbel P et al. A molecular assembly system that renders antigens of choice highly repetitive for induction of protective B cell responses. *Vaccine* 2002;20:3104–3112.

Jiskoot W, Randolph TW, Volkin DB, Middaugh CR et al. Protein instability and immunogenicity: Roadblocks to clinical application of injectable protein delivery systems for sustained release. *J Pharm Sci* 2012;101:946–954.

Johnson KB, Brooks BR, Cohen JA, Ford CC, Goldstein J, Lisak RP, Myers LW, Panitch HS, Rose JW, Schiffer RB. Copolymer 1 reduces relapse rate and improves disability in relapsing-remitting multiple sclerosis: Results of a phase III multicenter, double-blind, placebo-controlled trial. *Neurology* 1995;45:1268–1276.

Johnson KB, Brooks BR, Cohen JA, Ford CC et al. Extended use of glatiramer acetate (Copaxone) is well tolerated and maintains its clinical effect on multiple sclerosis relapse rate and degree of disability. *Neurology* 1998;50:701–708.

Johnson KB, Brooks BR, Ford CC, Goodman A et al. Sustained clinical benefits of glatiramer acetate in relapsing multiple sclerosis patients observed for 6 years. *Mult Scler* 2000;6:255–266.

Johnson KP, Ford CC, Lisak RP, Wolinsky JS. Neurologic consequence of delaying glatiramer acetate therapy for multiple sclerosis: 8-year data. *Acta Neurol Scand* 2005;111(1):42–47.

Jones GJ, Itri LM. Safety and tolerance of recombinant interferon alfa-2a (Roferon-A) in cancer patients. *Cancer* 1986;57:1709–1715.

Kaplan SL, Underwood LE, August GP, Bell JJ, Blethen SL, Blizzard RM, Brown DR, Foley TP, Hintz RL, Hopwood NJ. Clinical studies with recombinant-DNA-derived methionyl human growth hormone deficient children. *Lancet* 1986;1:697–700.

Kappos L, Clanet M, Sandberg-Wollheim M, Radue EW et al. Neutralizing antibodies and efficacy of interferon beta-1a: A 4-year controlled study. *Neurology* 2005;65:40–47.

Kawaguchi Y, Ota Y, Kawamoto M, Ito I et al. Association study of a polymorphism of the CTGF gene and susceptibility to systemic sclerosis in the Japanese population. *Ann Rheum Dis* 2009;68:1921–1924.

Keating GM. Rituximab: A review of its use in chronic lymphocytic leukaemia, low-grade or follicular lymphoma and diffuse large B-cell lymphoma. *Drugs* 2010;70:1445–1476.

Khan OA, Dhib-Jalbut SS. Neutralizing antibodies to interferon beta-1a and interferon beta-1b in MS patients are crossreactive. *Neurology* 1998;51:1696–1702.

Kiladjian JJ, Chomienne C, Fenaux P. Interferon-alpha therapy in bcr-abl-negative myelo-proliferative neoplasms. *Leukemia* 2008;22:1990–1998.

Kirshner S. Immunogenicity of therapeutic proteins: A regulatory perspective. In *Detection and Quantification of Antibodies to Biopharmaceuticals: Practical and Applied Considerations*, Tovey MG, ed. John Wiley & Sons, New York, 2011.

Kishnani PS, Corzo D, Nicolino M, Byrne B et al. Recombinant human acid (alpha)-glucosidase: Major clinical benefits in infantile-onset Pompe disease. *Neurology* 2007;68:99–109.

Klinge L, Straub V, Neudorf U, Schaper J, Bosbach T, Gorlinger K, Wallot M, Richards S, Voit T. Safety and efficacy of recombinant acid alpha-glucosidase (rhGAA) in patients with classical infantile Pompe disease: Results of a phase II clinical trial. *Neuromuscul Disord* 2005;15:24–31.

Ko JM, Gottlieb AB, Kerbleski JF. Induction and exacerbation of psoriasis with TNF blockade therapy: A review and analysis of 127 cases. *J Dermatol Treat* 2009;20:100–108.

Koren E, Zuckennan LA, Mire-Siuis AR. Immune responses to therapeutic proteins in humans—Clinical significance, assessment and prediction. *Curr Pharm Biotechnol* 2002;3:349–360.

Kracke A, von Wussow P, Al-Masri AN, Dalley G, Windhagen A, Heidenreich F. Mx proteins in blood leukocytes for monitoring interferon beta-1b therapy in patients with MS. *Neurology* 2000;54:193–199.

Kromminga A, Schellekens H. Antibodies against erythropoietin and other protein-based therapeutics. An overview. *Ann NY Acad Sci* 2005;1050:257–265.

Kumar D. Lispro analog for treatment of generalized allergy to human insulin. *Diabetes Care* 1997;20:1357–1359.

Kumar S, Singh S, Wang X, Rup B, Gill D. Coupling of aggregation and immunogenicity in biotherapeutics: T and B-cell immune epitopes may contain aggregation-prone regions. *Pharm Res* 2011;28:949–961.

Kunkel L, Wong A, Maneatis T, Nickas J et al. Optimizing the use of rituximab for treatment of B-cell non-Hodgkin's lymphoma: A benefit-risk update. *Semin Oncol* 2000;27:53–61.

Kuter DJ. New thrombopoietic growth factors. *Blood* 2007;109:4607–4616.

Kyogoku C, Morinobu A, Nishimura K, Sugiyama D et al. Lack of association between tyrosine kinase 2 (Tyk2) gene polymorphisms and susceptibility to SLE in a Japanese population. *Mod Rheumatol* 2009;19:401–406.

Lacroix-Desmazes S, Bayry J, Misra N, Horn MP et al. The prevalence of proteolytic antibodies against factor VIII in haemophilia A. *N Engl J Med* 2002;346:662–667.

Lahtela JT, Knip M, Paul R, Antone J, Salmi J. Severe antibody-mediated human insulin resistance: Successful treatment with the insulin analog lispro: A case report. *Diabetes Care* 1997;20:71–73.

Lallemand C, Meritet JF, Blanchard B, Lebon P, Tovey MG. One-step assay for quantification of neutralizing antibodies to biopharmaceuticals. *J Immunol Methods* 2010;356:18–28.

Lallemand C, Meritet JF, Erickson R, Grossberg SE et al. Quantification of neutralizing antibodies to human type 1 interferons using division-arrested frozen cells carrying an interferon-regulated reportergene. *J Interferon Cytokine Res* 2008;28:393–404.

La Merie Business Intelligence. R&D pipeline. *News*, March 10, 2010. http://www.lamerie.com.

Lang DM, Alpern MB, Visintainer PF, Smith ST. Increased risk for anaphylactoid reaction from contrast media in patients on beta-adrenergic blockers or with asthma. *Ann Intern Med* 1991;115:270–276.

Laroni A, Bedognetti M, Uccelli A, Capello E, Mancardi GL. Association of melanoma and natalizumab therapy in the Italian MS population: A second case report. *Neurol Sci* 2011; 32:181–182.

Lee CJ, Seth G, Tsukuda J, Hamilton RW. A clone screening method using mRNA levels to determine specific productivity and product quality for monoclonal antibodies. *Biotechnol Bioeng* 2009;102:1107–1118.

Levesque MC, Ward FE, Jeffery DR, Weinberg JB. Interferon-beta1a-induced polyarthritis in a patient with the HLA-DRB1*0404 allele. *Arthritis Rheum* 1999;42:569–573.

Li DK, Zhao JG, Paty DW; University of British Columbia MS/MRI Analysis Research Group, and the SPECTRIMS Study Group. Randomized controlled trial of interferon beta-1a in secondary progressive MS. MRI results. *Neurology* 2001;56:1505–1513.

Li J, Yang C, Xia Y, Bertino A, Glaspy J, Roberts M, Kuter D. Thrombocytopenia caused by the development of antibodies to thrombopoietin. *Blood* 2001;98:3241–3248.

Locatelli F, Aljama P, Barany P, Canaud B et al. Erythropoiesis-stimulating agents and antibody-mediated pure red cell aplasia: Where are we now and where do we go from here? *Nephrol Dial Transplant* 2004;19:288–293.

Lofgren JA, Wala I, Koren E, Swanson SJ, Jing S. Detection of neutralizing anti-therapeutic protein antibodies in serum or plasma samples containing high levels of the therapeutic protein. *J Immunol Methods* 2006;308:101–108.

Lok AS, Lai CL, Leung EK. Morbidity and mortality from chronic hepatitis B virus infection in family members of patients with malignant and nonmalignant hepatitis B virus-related chronic liver diseases. *Hepatology* 1990;12:1266–1270.

Lundin K, Berger L, Blomberg F, Wilton P. Development of anti-hGH antibodies during therapy with authentic human growth hormone. *Acta Paediatr Scand* 1991;372:167–168.

Luo Q, Joubert MK, Stevenson R, Ketcham RR et al. Chemical modifications in therapeutic protein aggregates generated under different stress conditions. *J Biol Chem* 2011; 286:25134–25144.

Lusher JM. Inhibitor antibodies to factor VIII and factor IX: Management. *Semin Thromb Hemost* 2000;26:179–188.

Lusher JM, Arkin S, Abildgaard CF, Schwartz RS. Recombinant factor VIII for the treatment of previously untreated patients with haemophilia A. Safety, efficacy, and development of inhibitors. *N Engl J Med* 1993;328:453–459.

Lysandropoulos AP, Du Pasquier RA. Demyelination as a complication of new immuno-modulatory treatments. *Curr Opin Neurol* 2010;23: 226–233.

Maas C, Hermeling S, Bouma B, Jiskoot W, Gebbink MFBG. A role for protein misfolding in immunogenicity of biopharmaceuticals. *J Biol Chem* 2007;282:2229–2236.

Maeda E, Kita S, Kinoshita M, Urakami K et al. Analysis of nonhuman N-glycans as the minor constituents in recombinant monoclonal antibody pharmaceuticals. *Anal Chem* 2012;84:2373–2379.

Maini RN, Breedveld FC, Kalden JR, Smolen JS et al. Therapeutic efficacy of multiple intravenous infusions of anti-tumor necrosis factor alpha monoclonal antibody combined with low-dose weekly methotrexate in rheumatoid arthritis. *Arthritis Rheum* 1998;41:1552–1563.

Malucchi S, Capobianco M, Gilli F, Marnetto F, Caldano M, Sala A, Bertolotto A. Fate of multiple sclerosis patients positive for neutralising antibodies towards interferon beta shifted to alternative treatments. *Neurol Sci* 2005;26:S213–S214.

Malucchi S, Sala A, Gilli F, Bottero R et al. Neutralizing antibodies reduce the efficacy of beta-IFN during treatment of multiple sclerosis. *Neurology* 2004;62:2031–2037.

Manduzio H, Fitchette AC, Hrabina M, Chabre H et al. Glycoproteins are species-specific markers and major IgE reactants in grass pollens. *Plant Biotechnol J* 2012;10:184–194.

Manning MC, Chou DK, Murphy BM, Payne RW, Katayama DS. Stability of protein pharmaceuticals: An update. *Pharm Res* 2010;27:544–575.

Marini JC. Chapter 8: Cell cooperation in the antibody response. In *Immunology*, Male D, Brostoff J, Roth DB, Roitt I, eds. Elsevier Limited, Philadelphia, PA, 2006, pp. 163–180.

Mark DF, Lu SD, Creasey AA, Yamamoto R, Lin LS. Site-specific mutagene of the human fibroblast interferon gene. *Proc Natl Acad Sci USA* 1984;81:5662–5666.

Marshall MO, Heding LG, Villumsen J, Akerblom HK, Baevre H, Dahlguist G, Kiaergaard JJ, Knip M, Lindgren F, Ludvigsson J. Development of insulin antibodies, metabolic control and B-cell function in newly diagnosed insulin dependent diabetic children treated with monocomponent human insulin or monocomponent porcine insulin. *Diabetes Res* 1988;9:169–175.

Massa G, Vanderschueren-Lodeweyckx M, Bouillon R. Five-year follow up of growth hormone antibodies in growth hormone deficient children treated with recombinant human growth hormone. *Clin Endocrinol* 1993;38:137–142.

Masucci G, Wersall P, Raghnammar P, Mellstedt H. Granulocyte monocyte colony stimulating factor augments the cytotoxic capacity of lymphocytes and monocytes in antibody-dependent cellular cytotoxicity. *Cancer Immunol Immunother* 1989;29:288–292.

McHutchison JG, Lawitz EJ, Shiffman ML, Muir AJ et al. Peginterferon alfa-2b or alfa-2a with ribavirin for treatment of hepatitis C infection. *N Engl J Med* 2009;361:580–593.

McKay F, Schibeci S, Heard R, Stewart G, Booth D. Analysis of neutralizing antibodies to therapeutic interferon-beta in multiple sclerosis patients: A comparison of three methods in a large Austral asian cohort. *J Immunol Methods* 2005;310:20–29.

Meager A. Measurement of cytokines by bioassays: Theory and application. *Methods* 2006; 38:237–252.

Mellstedt H. Induction of anti-granulocyte-macrophage colony-stimulating factor antibodies against exogenous nonglycosylated GM-CSF: Biological implications. *J Interferon Res* 1994;14:179–180.

Menetrier-Caux C, Briere F, Jouvenne P, Peyron E, Peyron F, Banchereau J. Identification of human IgG autoantibodies specific for IL-10. *Clin Exp Immunol* 1996;104:173–179.

MHRA. Pharmacovigilance: How the MHRA monitors the safety of medicines, January 23, 2015. http://www.mhra.gov.uk/Safetyinformation/Howwemonitorthesafetyofproducts/Drugs/TheYellowCardScheme/Informationforhealthcareprofessionals/Adversedrugreactions/index.htm (accessed August 28, 2015).

Miceli-Richard C, Comets E, Loiseau P, Puechal X, Hachulla E, Mariette X. Association of an IRF5 gene functional polymorphism with Sjoögren's syndrome. *Arthritis Rheum* 2007;56:3989–3994.

Miceli-Richard C, Gestermann N, Ittah M, Comets E et al. The CGGGG insertion/deletion polymorphism of the IRF5 promoter is a strong risk factor for primary Sjögren's syndrome. *Arthritis Rheum* 2009;60:1991–1997.

Minagar A, Alexander JS, Sahraian MA, Zivadinov R. Alemtuzumab and multiple sclerosis: Therapeutic application. *Expert Opin Biol Ther* 2010;10:421–429.

Mira JA, Lopez-Cortes LF, Merino D, Arizcorreta-Yarza A et al. Predictors of severe haematological toxicity secondary to pegylated interferon plus ribavirin treatment in HIV-HCV-coinfected patients. *Antivir Ther* 2007;12:1225–1235.

Mire-Sluis AR, Barrett YC, Devanarayan V, Koren E et al. Recommendation for the design and optimization of immunoassays used in the detection of host antibodies against biotechnology products. *J Immunol Methods* 2004;289(1–2):1–16.

Monzani F, Caraccio N, Dardano A, Ferrannini E. Thyroid autoimmunity and dysfunction associated with type I interferon therapy. *Clin Exp Med* 2004;3:199–210.

Moore WV, Leppert P. Role of aggregated human growth hormone (hGH) in development of antibodies to hGH. *J Clin Endocrinol Metab* 1980;51:691–697.

Nam JL, Winthrop KL, Van Vollenhoven RF, Pavelka K et al. Current evidence for the management of rheumatoid arthritis with biological disease-modifying antirheumatic drugs: A systematic literature review informing the Eular recommendations for the management of RA. *Ann Rheum Dis* 2010;69:976–986.

Narhi LO, Schmit J, Bechtold-Peters K, Sharma D. Classification of protein aggregates. *J Pharm Sci* 2012;101:493–498.

Nguyen KB, Salazar-Mather TP, Dalod MY, Van Deusen JB et al. Coordinated and distinct roles for IFN-alpha beta, IL-12, and IL-15 regulation of NK cell responses to viral infection. *J Immunol* 2002;69:4279–4287.

Nordmark G, Kristjansdottir G, Theander E, Eriksson P et al. Additive effects of the major risk alleles of IRF5 and STAT4 in primary Sjögren's syndrome. *Genes Immun* 2009;10:68–76.

Novick D, Nabioullin RR, Ragsdale W, McKenna S et al. The neutralization of type I IFN biologic actions by anti-IFNAR-2 monoclonal antibodies is not entirely due to inhibition of Jak-Stat tyrosine phosphorylation. *J Interferon Cytokine Res* 2000;20:971–982.

Okada Y, Taira K, Takano K, Hizuka N. A case report of growth attenuation during methionyl human growth hormone treatment. *Endocrinol Jpn* 1987;34:621–626.

O'Neill LA, Bowie AG. Sensing and signaling in antiviral innate immunity. *Curr Biol* 2010; 20:R328–R333.

Opdenakker G, Van den Steen PE, Laureys G, Hunninck K, Arnold B. Neutralizing antibodies in gene-defective hosts. *Trends Immunol* 2003;24:94–100.

Ordas I, Mould DR, Feagan BG, Sandborn WJ. Anti-TNF monoclonal antibodies in inflammatory bowel disease: Pharmacokinetics-based dosing paradigms. *Clin Pharmacol Ther* 2012;91:635–646.

Padler-Karavani V, Varki A. Potential impact of the non-human sialic acid N-glycolylneuraminic acid on transplant rejection risk. *Xenotransplantation* 2011;18:1–5.

Palucka AK, Blanck JP, Bennett L, Pascual V, Banchereau J. Cross-regulation of TNF and IFN-alpha in autoimmune diseases. *Proc Natl Acad Sci USA* 2005;102: 3372–3377.

Paon A, Mullenix MC, Swanson SJ, Koren E. An acid dissociation bridging ELISA for detection of antibodies directed against therapeutic proteins in the presence of antigen. *J Immunol Methods* 2005;304:189–195.

Parmiani G, Castelli C, Pilla L, Santinami M, Colombo MP, Rivoltini L. Opposite immune functions of GM-CSF administered as vaccine adjuvant in cancer patients. *Ann Oncol* 2007;18:226–232.

Paues J, Vrethem M. Fatal progressive multifocal leukoencephalopathy in a patient with non-Hodgkin lymphoma treated with rituximab. *J Clin Virol* 2010;48:291–293.

Peces R, de la Torre M, Alcazar R, Urra JM. Antibodies against recombinant human erythropoietin in a patient with erythropoietin-resistant anaemia. *N Engl J Med* 1996;335:523–524.

Pestka S, Krause CD, Walter MR. Interferons, interferon-like cytokines and their receptors. *Immunol Rev* 2004;202:8–32.

Peters A, Klose O, Hefty R, Keck F, Kerner W. The influence of insulin antibodies on the pharmacokinetics of NPH insulin in patients with type 1 diabetes treated with human insulin. *Diabet Med* 1995;12:925–930.

Petersen B, Bendtzen K, Koch-Henriksen N, Ravnborg M, Ross C, Sorensen PS; Danish Multiple Sclerosis Study Group. Persistence of neutralizing antibodies after discontinuation of IFNb therapy in patients with relapsing-remitting multiple sclerosis. *Mult Scler* 2006;12:247–252.

Petros WP, Rabinowitz J, Stuart AR, Gilbert CJ, Kanakura Y, Griffin JD, Peters WP. Disposition of recombinant human granulocyte-macrophage colony-stimulating factor in patients receiving high-dose chemotherapy and autologous bone marrow support. *Blood* 1992; 80:1135–1140.

Peyrin-Biroulet L. Anti-TNF therapy in inflammatory bowel diseases: A huge review. *Minerva Gastroenterol Dietol* 2010;56:233–243.

Pisal DS, Kosloski MP, Middaugh CR, Bankert RB et al. Structural characterization of N-linked oligosaccharides on monoclonal antibody cetuximab by the combination of orthogonal matrix-assisted laser desorption/ionization hybrid quadrupole–quadrupole time-of-flight tandem mass spectrometry and sequential enzymatic digestion. *Anal Biochem* 2007;364:8–18.

Pisal SV. Native-like aggregates of factor VIII are immunogenic in von Willebrand factor deficient and hemophilia a mice. *J Pharm Sci* 2012;101:2055–2065.

Polman C, Kappos L, White R, Dahlke F, Beckmann K, Pozzilli C, Thompson A, Petkau J, Miller D; European Study Group in Interferon Beta-1b in Secondary Progressive MS. Neutralizing antibodies during treatment of secondary progressive MS with interferon beta-1b. *Neurology* 2003;60:37–43.

Polman CH, O'Connor PW, Havrdova E, Huthchinson M et al. A randomized, placebo-controlled trial of natalizumab for relapsing multiple sclerosis. *N Engl J Med* 2006;354:899–910.

Pothlichet J, Niewold TB, Vitour D, Solhonne B, Crow MK, Si-Tahar M. A loss-of-function variant of the antiviral molecule MAVS is associated with a subset of systemic lupus patients. *EMBO Mol Med* 2011;3:142–152.

Powell A, Myles, ML, Yacyshyn E. The development of systemic sclerosis in a female patient with multiple sclerosis following beta interferon treatment. *Clin Rheumatol* 2008;27:1467–1468.

Pozzilli C, Antonimi G, Bagnato F, Mainero C et al. Monthly corticosteroids decrease neutralizing antibodies to IFNbeta1b: A randomized trial in multiple sclerosis. *J Neurol* 2002; 249:50–56.

Prabhakar SS, Muhlfelder T. Antibodies to recombinant human erythropoietin causing pure red cell aplasia. *Clin Nephrol* 1997;47:331–335.

Prevention of Relapses and Disability by Interferon-beta-1a Subcutaneously in Multiple Sclerosis (PRISMS) Study Group. Randomised double-blind placebo-controlled study of interferon beta-1a in relapsing-remitting multiple sclerosis. *Lancet* 1998;352:1498–1504.

Prinz Vavricka BM, Baumbegrer P, Russmann S, Kullak-Ublick GA. Diagnosis of melanoma under concomitant natalizumab therapy. *Mult Scler* 2010;17:255–256.

Priore RL, Pullicino PM, Scherokman BJ, Whitham RH. Intra-muscular interferon beta-1a for disease progression in relapsing multiple sclerosis. *Ann Neurol* 1996;39:285–294.

PRISMS-Study Group and the University of British Columbia MS/MRI Analysis Group. PRISMS-4. Long term efficacy of interferon-beta-1a in relapsing MS. *Neurology* 2001; 56:1628–1636.

Quesada JR, Rios A, Swanson D, Trown P, Gutterman JU. Antitumor activity of recombinant derived interferon alpha in metastatic renal cell carcinoma. *J Clin Oncol* 1985;3:1522–1528.

Ragnhammar P, Friesen HJ, Frodin JE, Lefvert AK, Hassan M, Osterborg A, Mellstedt H. Induction of anti-recombinant human granulocyte-macrophage colony-stimulating factor (*Escherichia coli*-derived) antibodies and clinical effects in non-immunocompromised patients. *Blood* 1994b;84:4078–4087.

Ragnhammar P, Frodin JE, Trotta PP, Mellstedt H. Cytotoxicity of white blood cells activated by granulocyte-colony-stimulating factor, granulocyte/macrophage-colony-stimulating factor and macrophage-colony-stimulating factor against tumor cells in the presence of various monoclonal antibodies. *Cancer Immunol Immunother* 1994a;39:254–262.

Rasheed Z, Ali R. Reactive oxygen species damaged human serum albumin in patients with type 1 diabetes mellitus: Biochemical and immunological studies. *Life Sci* 2006; 79:2320–2328.

Rauschka H, Farina C, Sator P, Gudek S, Breier F, Schmidbauer M. Severe anaphylactic reaction to glatiramer acetate with specific IgE. *Neurology* 2005;64:1481–1482.

Reeves WG, Kelly U. Insulin antibodies induced by bovine insulin therapy. *Clin Exp Immunol* 1982;50:163–170.

Riedl MA, Casillas AM. Adverse drug reactions: Types and treatment options. *Am Fam Physician* 2003;68:1781–1790.

Rifkin R, Maggio E, Dike S, Kerr D, Levy M. *n*-Dodecyl-β-D-maltoside inhibits aggregation of human interferon-β-1b and reduces its immunogenicity. *J Neuroimmune Pharmacol* 2011;6:158–162.

Rini B, Wadhwa M, Bird C, Small E, Gaines-Das R, Thorpe R. Kinetics of development and characteristics of antibodies induced in cancer patients against yeast expressed rDNA derived granulocyte macrophage colony stimulating factor (GM-CSF). *Cytokine* 2005;29:56–66.

Rispens T, de Vrieze H, de Groot E, Wouters D et al. Antibodies to constant domains of therapeutic monoclonal antibodies: Antihinge antibodies in immunogenicity testing. *J Immunol Methods* 2012;375:93–99.

Ronnblom L, Eloranta ML, Alm GV. The type I interferon system in systemic lupus erythematosus. *Arthritis Rheum* 2006;54:408–420.

Rosenberg AS. Immunogenicity of biological therapeutics: A hierarchy of concerns. *Dev Biol (Basel)* 2003;112:15–21.

Rosenberg AS. Effects of protein aggregates: An immunologic perspective. *AAPS J* 2006; 8:E501–E507.

Rudick RA, Panzara MA. Natalizumab for the treatment of relapsing multiple sclerosis. *Biologics* 2008;2:189–199.

Rudick RA, Simonian NA, Alam JA, Campion M et al. Incidence and significance of neutralizing antibodies to interferon beta-1a in multiple sclerosis. Multiple Sclerosis Collaborative Research Group (MSCRG). *Neurology* 1998;50:1266–1272.

Rudick RA, Stuart WH, Calabresi PA, Confraveux C et al. Natalizumab plus interferon beta-1a for relapsing multiple sclerosis. *N Engl J Med* 2006;354:911–923.

Rueda B, Broen J, Simeon C, Hesselstrand R et al. The STAT4 gene influences the genetic predisposition to systemic sclerosis phenotype. *Hum Mol Genet* 2009;18:2071–2077.

Runkel L. Structural and functional differences between glycosylated and non-glycosylated forms of human interferon-beta (IFN-beta). *Pharm Res* 1998;15:641–649.

Ryschkewitsch CF, Jensen PN, Monaco MC, Major EO. JC virus persistence following progressive multifocal leukoencephalopathy in multiple sclerosis patients treated with natalizumab. *Ann Neurol* 2010;68:384–391.

Salama HH, Hong J, Zang YC, El-Monqui A, Zhang J. Blocking effects of serum reactive antibodies induced by galtiramer acetate treatment in multiple sclerosis. *Brain* 2003; 126:2638–2647.

Saleh H, Embry S, Nauli A, Atyia S, Krishnaswamy G. Anaphylactic reactions to oligosaccharides in red meat: A syndrome in evolution. *Clin Mol Allergy* 2012;10:5.

Sasaki H, Bothner B, Dell A, Fukuda M. Carbohydrate structure of erythropoietin expressed in Chinese hamster ovary cells by a human erythropoietin cDNA. *J Biol Chem* 1987; 262:12059–12076.

Sauerborn M, Brinks V, Jiskoot W, Schellekens H. Immunological mechanism underlying the immune response to recombinant human protein therapeutics. *Trends Pharmacol Sci* 2010;31:53–59.

Schellekens H. Immunogenicity of therapeutic proteins: Clinical implications and future prospects. *Clin Ther* 2002a;24(11):1720–1740.

Schellekens H. Bioequivalence and the immunogenicity of biopharmaceuticals. *Nat Rev Drug Discov* 2002b;1:457–462.

Schellekens H. Biosimilar therapeutics—What do we need to consider? *NDT Plus* 2009;2:i27–i36.

Schellekens H, Casadevall N. Immunogenicity of recombinant human proteins: Causes and consequences. *J Neurol* 2004;251(Suppl. 2):II4–II9.

Schernthaner G. Immunogenicity and allergenic potential of animal and human insulins. *Diabetes Care* 1993;16:155–165.

Schweighofer CD, Wendtner CM. First-line treatment of chronic lymphocytic leukemia: Role of alemtuzumab. *Onco Targets Ther* 2010;3:53–67.

Secondary Progressive Efficacy Clinical Trial of Recombinant Interferon beta-1a in MS (SPECTRIMS) Study Group. Randomized controlled trial of interferon beta-1a in secondary progressive MS: Clinical results. *Neurology* 2001;56:1496–1504.

Seefeldt MB, Rosendahl MS, Cleland JL, Hesterberg LK. Application of high hydrostatic pressure to dissociate aggregates and refold proteins. *Curr Pharm Biotechnol* 2009;10:447–455.

Seidl A, Hainzl O, Richter M, Fischer R et al. Tungsten-induced denaturation and aggregation of epoetin alfa during primary packaging as a cause of immunogenicity. *Pharm Res* 2012;29(6):1454–1467.

Selmaj KW, Raine CS. Experimental autoimmune encephalomyelitis: Immunotherapy with antitumor necrosis factor antibodies and soluble tumor necrosis factor receptors. *Neurology* 1995;45:S44–S49.

Shankar G, Devanarayan V, Amaravadi L, Barrett YC et al. Recommendations for the validation of immunoassays used for detection of host antibodies against biotechnology products. *Pharm Biomed Anal* 2008;48(5):1267–1281.

Shankar G, Pendley C, Stein KE. A risk-based bioanalytical strategy for the assessment of antibody immune responses against biological drugs. *Nat Biotechnol* 2007;25(5):555–561.

Sharief MK, Hentges R. Association between tumor necrosis factor-alpha and disease progression in patients with multiple sclerosis. *N Engl J Med* 1991;325:467–472.

Sharma B, Bader F, Templeman T, Lisi P et al. Technical investigations into the cause of the increased incidence of antibody-mediated pure red cell aplasia associated with Eprex®. *Eur J Hosp Pharm* 2004;5:86–91.

Simin M, Brok J, Stimac D, Gluud C, Gluud LL. Cochrane Systematic Review: Pegylated interferon plus ribavirin vs. interferon plus ribavirin for chronic hepatitis C. *Aliment Pharmacol Ther* 2007;25:1153–1162.

Singh SK. Impact of product-related factors on immunogenicity of biotherapeutics. *J Pharm Sci* 2011;100:354–387.

Singh SK, Afonina N, Awwad M, Bechtold-Peters K et al. An industry perspective on the monitoring of subvisible particles as a quality attribute for protein therapeutics. *J Pharm Sci* 2010;99:3302–3321.

Skibelli V, Nissen-Lie G, Torjesen P. Sugar profiling proves that human serum erythropoietin differs from recombinant human erythropoietin. *Blood* 2001;98:3626–3634.

Sleijfer S, Bannink M, Van Gool AR, Kruit WH, Stoter G. Side effects of interferon-alpha therapy. *Pharm World Sci* 2005;27:423–431.

Smith LF. Species variation in the amino acid sequence of insulin. *Am J Med* 1966;40:662–666.

Somani AK, Swick AR, Cooper KD, McCormick TS. Severe dermatomyositis triggered by interferon beta-1a therapy and associated with enhanced type I interferon signaling. *Arch Dermatol* 2008;144:1341–1349.

Sorensen PS, Deisenhammer F, Dudac P, Hohlfeld R, Myhre KM, Palace J, Polman C, Pozzilli C, Ross C; for the EFNS Task Force on Anti-IFN-b Antibodies in Multiple Sclerosis. Guidelines on use of anti-IFN antibody measurements in multiple sclerosis: Report of an EFNS Task Force on IFN-b antibodies in multiple sclerosis. *Eur J Neurol* 2005;12:817–827.

Sorensen PS, Koch-Henriksen N, Ross C, Clemmesen KM, Bendtzen K; Danish Multiple Sclerosis Study Group. Appearance and disappearance of neutralizing antibodies during interferon-beta therapy. *Neurology* 2005;65:33–39.

Sorensen PS, Ross C, Clemmesen KM, Bendtzen K et al. Clinical importance of neutralizing antibodies against interferon beta in patients with relapsing-remitting multiple sclerosis. *Lancet* 2003;362:1184–1191.

Spiegel RJ, Jacobs SL, Treuhaft MW. Anti-interferon antibodies to interferon-alpha 2b: Results of comparative assays and clinical perspective. *J Interferon Res* 1989;9(Suppl. 1):17–24.

Steis RG, Smith JW, Urba WJ, Clark JW, Itri LM, Evans LM, Schoen-berger C, Longo DL. Resistance to recombinant interferon alfa-2a in hairy cell leukemia associated with neutralizing anti-interferon antibodies. *N Engl J Med* 1988;318:1409–1413.

Stone JH. Etanercept plus standard therapy for Wegener's granulomatosis. *N Engl J Med* 2005;352:351–361.

Suarez-Gestal M, Calaza M, Dieguez-Gonzalez R, Perez-Pampin E et al. Rheumatoid arthritis does not share most of the newly identified systemic lupus erythematosus genetic factors. *Arthritis Rheum* 2009a;60:2558–2564.

Suarez-Gestal M, Calaza M, Endreffy E, Pullmann R et al. Replication of recently identified systemic lupus erythematosus genetic associations: A case-control study. *Arthritis Res Ther* 2009b;11:R69.

Swanson SJ. Characterization of an immune response. *Dev Biol* 2005;122:95–101.

Tacey R, Greway A, Smiell J, Power D, Kromminga A, Daha M, Casadevall N, Kelley M. The detection of anti-erythropoietin antibodies in human serum and plasma: Part I. Validation of the protocol for a radioimmuno-precipitation assay. *J Immunol Methods* 2003;283:317–329.

Tak PP, Rigby WF, Rubbert-Roth A, Peterfy CG et al. Inhibition of joint damage and improved clinical outcomes with rituximab plus methotrexate in early active rheumatoid arthritis: The Image Trial. *Ann Rheum Dis* 2011;70:39–46.

Takano K, Shizume K, Hibi I. Turner's syndrome: Treatment of 203 patients with recombinant human growth hormone for one year. A multicentre study. *Acta Endocrinol* 1989; 120:559–568.

Tangri S, Mothè BR, Eisenbraun J, Sidney J et al. Rationally engineered therapeutic proteins with reduced immunogenicity. *J Immunol* 2005;174:3187–3196.

Tatsumi Y, Sasahara Y, Kohyama N, Ayano S et al. Introducing site-specific glycosylation using protein engineering techniques reduces the immunogenicity of β-lactoglobulin. *Biosci Biotechnol Biochem* 2012;76:478–485.

Teitelbaum D, Fridkis-Hareli M, Arnon S, Sela M. Copolymer 1 inhibits chronic relapsing experimental allergic encephalomyelitis induced by proteolipid protein (PLP) peptides in mice and interferes with PLP-specific T cell responses. *J Neuroimmunol* 1996;64:209–217.

The IFNB Multiple Sclerosis Study Group. Interferon beta-1b is effective in relapsing-remitting multiple sclerosis. I. Clinical results of a multicenter, randomized, double-blind, placebo-controlled trial. *Neurology* 1993;43:655–661.

The IFNB MS Study Group and the University of British Columbia MS/MRI Analysis Group. Neutralizing antibodies during treatment of multiple sclerosis with interferon beta-1b: Experiences during the first three years. *Neurology* 1996;47:889–894.

Thorpe R, Swanson SJ. Current methods for detecting antibodies against Erythropoietin and other recombinant proteins. *Clin Diag Lab Immunol* 2005;12:28–39.

Tomassini V, Paolillo A, Russo P, Giugni E, Prosperini L, Gasperini C, Antonelli G, Bastianello S, Pozzilli C. Predictors of long-term clinical response to interferon beta therapy in relapsing multiple sclerosis. *J Neurol* 2006;253:287–293.

Torosantucci R, Mozziconacci O, Sharov V, Schöneich C, Jiskoot W. Chemical modifications in aggregates of recombinant human insulin induced by metal-catalyzed oxidation: Covalent cross-linking via Michael addition to tyrosine oxidation products. *Pharm Res* 2012;29:2276–2293.

Ullenhag G, Bird C, Ragnhammar P, Frodin JE, Strigard K, Osterborg A, Thorpe R, Mellstedt H, Wadhwa M. Incidence of GM-CSF antibodies in cancer patients receiving low dose GM-CSF as an adjuvant. *Clin Immunol* 2001;99:65–74.

Unoki H, Moriyama A, Tabaru A, Masumoto A, Otsuki M. Development of Sjögren's syndrome during treatment with recombinant human interferon-alpha-2b for chronic hepatitis C. *J Gastroenterol* 1996;31:723–727.

Ure DR, Rodriguez M. Polyreactive antibodies to glatiramer acetate promote myelin repair in murine model of demyelinating disease. *FASEB J* 2002;16:1260–1262.

Valentine AD, Meyers CA, Kling MA, Richelson E, Hauser P. Mood and cognitive side effects of interferon-alpha therapy. *Semin Oncol* 1998;25:39–47.

Vallittu AM, Halminen M, Peltonieni J, Ilonen J, Julkunen I, Salmi A, Eralinna JP; Finnish Beta-Interferon Study Group. Neutralizing antibodies reduce MxA protein induction in interferon-beta-1a-treated MS patients. *Neurology* 2002;58:1786–1790.

van Beers MMC, Gilli F, Schellekens H, Randolph TW, Jiskoot W. Immunogenicity of recombinant human interferon beta interacting with particles of glass, metal, and polystyrene. *J Pharm Sci* 2012;101:187–199.

van Beers MMC, Jiskoot W, Schellekens H. On the role of aggregates in the immunogenicity of recombinant human interferon beta in patients with multiple sclerosis. *J Interferon Cytokine Res* 2010;30:767–775.

van Beers MMC, Sauerborn M, Gilli F, Brinks V et al. Aggregated recombinant human interferon beta induces antibodies but no memory in immune-tolerant transgenic mice. *Pharm Res* 2010;27:1812–1824.

van Beers MMC, Sauerborn M, Gilli F, Brinks V et al. Oxidized and aggregated recombinant human interferon beta is immunogenic in human interferon beta transgenic mice. *Pharm Res* 2011;28:2393–2402.

Van Haeften TW. Clinical significance of insulin antibodies in insulin-treated patients. *Diabetes Care* 1989;9:641–648.

Van Hove JLK, Yang HW, Wu JY. High level production of recombinant human lysosomal acid a-glucosidase in Chinese hamster ovary cells which targets to heart muscle and corrects glycogen accumulation in fibroblast from patients with Pompe disease. *Proc Natl Acad Sci USA* 1996;93:65–70.

Velcovsky HG, Beringhoff B, Federlin K. Immediate type allergy to insulin (author's translation). *Immunitat und Infektion* 1978;6:146–152.

Velcovsky HG, Federlin KF. Insulin-specific IgG and IgE antibody response in type 1 diabetic subjects exclusively treated with human insulin (recombinant DNA). *Diabetes Care* 1982;5:126–128.

Verhelst D, Rossert J, Casadevall N, Kruger A, Eckardt KU, Macdougall IC. Treatment of erythropoietin-induced pure red cell aplasia: A retrospective study. *Lancet* 2004;363:1768–1771.

von Wussow P, Freund M, Block B, Diedrich H, Poliwoda H, Deicher H. Clinical significance of anti-IFN-a antibody titres during interferon therapy. *Lancet* 1987;2:635–636.

Vos Q, Lees A, Wu ZQ, Snapper CM, Mond JJ. B-cell activation by T-cell-independent type 2 antigens as an integral part of the humoral immune response to pathogenic microorganisms. *Immunol Rev* 2000;176:154–170.

Wada M, Marusawa H, Yamada R, Nasu A et al. Association of genetic polymorphisms with interferon-induced haematologic adverse effects in chronic hepatitis C patients. *J Viral Hepat* 2009;16:388–396.

Wadhwa M, Bird C, Dilger P, Gaines-Das R, Thorpe R. Strategies for detection, measurement and characterization of unwanted antibodies induced by therapeutic biologicals. *J Immunol Methods* 2003;278:1–17.

Wadhwa M, Bird C, Fagerberg J, Gaines-Das R, Ragnhammar P, Mellstedt H, Thorpe R. Production of neutralizing GM-CSF antibodies in carcinoma patients following GM-CSF combination therapy. *Clin Exp Immunol* 1996;104:351–358.

Wadhwa M, Skog A-LH, Bird C, Ragnhammar P, Lilljefors M, Gaines-Das R, Mellstedt H, Thorpe R. Immunogenicity of Granulocyte-Macrophage Colony Stimulating Factor (GM-CSF) products in patients undergoing combination therapy with GM-CSF. *Clin Can Res* 1999;5:1353–1361.

Wadhwa M. Immunogenicity: What do we know and what can we measure?, PhD thesis, NIBSC–HPA, Hertfordshire, U.K., April 12, 2011.

Wadhwa M, Thorpe R. Strategies and assays for the assessment of unwanted immunogenic. *J Immunotoxicol* 2006;3:115–121.

Walford S, Allison SP, Reeves WG. The effect of insulin antibodies on insulin dose and diabetic control. *Diabetologia* 1982;22:106–110.

Walsh G. Second-generation biopharmaceuticals. *Eur J Pharm Biopharm* 2004;58:185–196.

Walsh G. Biopharmaceuticals: Recent approvals and likely directions. *Trends Biotechnol* 2005;23:553–558.

Walsh G. Biopharmaceutical benchmarks 2010. *Nat Biotechnol* 2010;28:917–924.

Walsh G, Jefferis R. Post-translational modifications in the context of therapeutic proteins. *Nat Biotechnol* 2006;24:1241–1252.

Wang W, Singh SK, Li N, Toler MR et al. Immunogenicity of protein aggregates—Concerns and realities. *Int J Pharm* 2012;431:1–11.

Wei X, Swanson SJ, Gupta S. Development and validation of a cell-based bioassay for the detection of neutralizing antibodies against recombinant human erythropoietin in clinical studies. *J Immunol Methods* 2004;293:115–126.

Weiner GJ. Rituximab: Mechanism of action. *Semin Hematol* 2010;47:115–123.

Weksler ME, Bull G, Schwartz GH, Stenzel KH, Rubin AL. Immunologic responses of graft recipients to antilymphocyte globulin: Effect of prior treatment with aggregate-free gamma globulin. *J Clin Invest* 1970;49:1589–1595.

Witters LA, Ohman JL, Weir GC, Raymond LW, Lowell FC. Insulin antibodies in the pathogenesis of insulin allergy and resistance. *Am J Med* 1977;63:703–709.

Wolinsky JS, Narayana PA, Johnson KB; Multiple Sclerosis Study Group and the MRI Analysis Center. United States open-label glatiramer acetate extension trial for relapsing multiple sclerosis: MRI and clinical correlates. *Mult Scler* 2001;7:33–41.

Wynn D, Kaufman M, Montalban X, Vollmer T et al. Daclizumab in active relapsing multiple sclerosis (choice study): A phase 2, randomised, double-blind, placebocontrolled, add-on trial with interferon beta. *Lancet Neurol* 2010;9:381–390.

Yamamoto K, Takamatsu J, Saito H. Intravenous immunoglobulin therapy for acquired coagulation inhibitors: A critical review. *Int J Haematol* 2007;85:287–293.

Yilmaz S, Cimen KA. Pegylated interferon alpha-2b induced lupus in a patient with chronic hepatitis B virus infection: Case report. *Clin Rheumatol* 2009;28:1241–1243.

Zhou S, Schöneich C, Singh S. Biologics formulation factors affecting metal leachables from stainless steel. *AAPS PharmSciTech* 2011;12:411–421.

Zinman L, Ng E, Bril V. IV immunoglobulin in patients with myasthenia gravis: A randomized controlled trial. *Neurology* 2007;68:837–841.

Zölls S, Tantipolphan R, Wiggenhorn M, Winter G et al. Particles in therapeutic protein formulations, Part 1: Overview of analytical methods. *J Pharm Sci* 2012;101:914–935.

Chapter 3 Product development strategies

Background

The first step in creating a biosimilar product portfolio is an overall understanding of the challenges, hurdles, difficulties, and practicalities of a robust development plan and a strategy. Biosimilar product development strategies would not be the same as those of small-molecule generics that are approved under the Food and Drug Administration (FDA) 505 category, even if some biotechnology products were approved under the 505 category.

> Section 505 of the Act describes three types of new drug applications: (1) an application that contains full reports of investigations of safety and effectiveness (section 505(b)(1)); (2) an application that contains full reports of investigations of safety and effectiveness but where at least some of the information required for approval comes from studies not conducted by or for the applicant and for which the applicant has not obtained a right of reference (section 505(b)(2)); and (3) an application that contains information to show that the proposed product is identical in active ingredient, dosage form, strength, route of administration, labeling, quality, performance characteristics, and intended use, among other things, to a previously approved product (section 505(j)). Note that a supplement to an application is a new drug application.

The development complexity of biological drugs, as well as their production costs, requires large upfront investment in the scientific and manufacturing infrastructure for several reasons. In this chapter, I will present a step-by-step approach to create a development strategy for a biosimilar product based on first-hand experience in developing dozens of these products.

The main purpose of this chapter is to layout clearly the challenges encountered in the development of biosimilar products, mainly in the regulated markets. The chapters on commercialization (Chapters 19 and 20) classify the biosimilar products available worldwide into several categories depending on the level of scrutiny required by the regulatory agencies across the globe. From treating them like small-molecule drugs to requiring their testing as new biological drugs, the regulatory barriers are uneven and often unrealistic, at both ends. A moderate approach will require that the product be safe and effective and similar enough to the originator product, so as to obviate the need for clinical studies in patients.

The regulatory agencies are still developing guidelines, learning from new biosimilar product submissions, and monitoring postmarket surveillance studies keenly to establish the rigor required for these products. As science develops and as a clear understanding of the risks involved in these products develops, a few years from now, these will most likely be treated as any other generic product and with that the affordability of these drugs will increase. However, it is unlikely that the competition among the biosimilar will ever take the same landscape as it is found today for the small-molecule drugs; this will become clear after reading this chapter.

Selection of product

The process of selecting a biosimilar product to develop is a lot more complex and challenging than developing a small-molecule product. The intellectual property barriers for biosimilar products include a multitude of challenges.

The composition of matter that includes the gene sequence are responsible for expressing the product. These patents are impossible to obviate. Fortunately, most of these are expiring, and while there is a significant diversity in the dates when these patents expire, there is a definite data available to assess these. A difference of 2–4 years across various major markets is not unusual; therefore, the date of composition matter would determine the domicile of manufacturing for the first launch. Appendix I provides a comprehensive listing of the patent expiry of biological drugs; this list should be checked with more recent patents and should serve only as a starting point for this exercise.

Submarine patents: A complex scheme of cross-licensing and conflicting patent applications that were filed before 1995 has created the possibility of submarine patents that emerge just when the first composition of matter patent is ready to expire. Two such examples are interferon alpha and etanercept, both of which will enjoy decades of exclusivity, in total defiance of the spirit of the patent law. They were able to exploit a weakness in the patenting system in the United States; the laws have been changed since and the term of the patent is now 20 years from the date of the first filing and NOT 17 years from the date of issuance that the submarine patents exploit. This threat is not available outside of the United States. There is no way to go around these patents.

System expression patents: There are patents like the Cabilly patents that cover the basic expression technology; these are more like generic patents, and in this specific case, unless the product happens to be produced by Genentech, a biosimilar developer is able to get a license. This patent expired in 2017. There is no way to go around these patents.

Process patents of originator: While most of the attention is diverted to the composition of matter patents, the process patents can be the biggest challenge. This threat has come to surface as the large market cap products like adalimumab and etanercept have come close to expiry. While many of these patents will be challenged, and some may be taken down, the detail with which these patents carve out a protected territory is amazing. For example, in the case of etanercept, you have to prove that the composition of amino acids during upstream manufacturing does not match with the claimed distribution; amino acid composition is not monitored in most upstream processes anyway. The scope of these patents can include the choice of media, upstream conditions, pH and composition of buffers, downstream purification columns and order of their use, and even a higher purity claim.

Given the large number of bioprocessing patents, both by the originator and third parties, it has already become very difficult and scientifically challenging to frame a suitable manufacturing process. The problems relating to this aspect can be appreciated from such patents that dictate that the composition of amino acids during the upstream processing must be within specific ranges; ironically, this is not a common test and even the composition of media is often not known to the developer. However, once such patents have been issued, it becomes incumbent on the developer to study this facet as well.

Third-party process patents: While the challenge of going around the originator process patents can be onerous, there are possibilities of third-party patents that may either apply to a specific product or to a class of product. Generally, a biosimilar developer would create a manufacturing pathway to go to the originator patent and then make an extensive search of third-party patents.

Formulation composition: Early developers of biological products ignored the value of these patents, but now we are seeing formulation patents intended to raise the barrier of demonstrating similarity. While the intent of developing these patents was to keep the biosimilar developers out of the market if the agencies require Q/Q products, this has not worked out. The agencies, realizing this, would allow alternate formulations.

Life cycle formulation projections: A recent trend seen is a change in formulation as the patent on composition of matter come close to expiry; for example, switching from a lyophilized formulation to a solution or adopting a high-concentration subcutaneous formulation instead of an intravenous solution (as has happened with some anticancer drugs where the time to administer is significantly reduced using subcutaneous formulation) brings a greater challenge to biosimilar developer: to develop an intravenous formulation or a subcutaneous formulation that is now covered by a new patent. It is obvious that the originator plans to switch its patients to a more friendly formulation, and this may be a formidable marketing challenge. However, unless the new formulation has been marketed, it becomes a challenge for the biosimilar product developer to secure reference samples to compare its product against, provided it was possible to go around any intellectual property challenges.

Alternate offering: While the originator product may have been launched in limited presentations, these might change over time, often to create a useful diversity; for example, prefilled syringes or injectors in place of vials change the marketing and distribution landscape in light of the reimbursement complexities.

Dosing and indications: While the early developers of biological drugs ignored the power of creating patents around specific use of these products including indications, this is now the trend and just about every big molecule as it comes closer to expiry will be spread with multiple patents that include specific dose, condition of use, and even dosing schedule. A case example is a patent held by Abbvie for administering exactly 40 mg of adalimumab every other week. The intent of such patents was to block the biosimilar developers from developing a similar product for use and the hope that the regulatory agencies would not allow alternate dosing or indications. However, the regulatory agencies, realizing this threat, have opened up to alternate suggestions. These can be, however, very difficult to overcome. Ironically, many of these patents have surfaced recently, giving shock to major developers of biosimilar products that may have already gone through expensive clinical trials only to find that at the end of the trial they will not be able to launch these products. A biosimilar developer need not examine the existing intellectual property but what may be forthcoming.

Delivery devices: The originator may have a multitude of patents on the delivery device since all of these devices are specific to the product even when they are supplied by device manufacturing companies selling such devices to multiple users. Device selection can have a significant impact on the marketability of the biosimilar product.

Market: While most biological products can be considered blockbusters with sales into billions, many are dwindling in their sales, often by design. A good example is the decision made by Amgen not to manufacture erythropoietin in the United States as they plan to push their patents to high-costing darbopoietin alpha; the market for erythropoietin has thus dwindled, which may make the product less attractive to the biosimilar developer. However, given the history of the development of markets in Europe, the sale of GCSF has risen as the cost differentials became larger. Given the cost of developing biosimilars, this choice can be difficult to make.

Competition: While a large number of biosimilar developers have surfaced around the world, given the complexities described in this chapter and the rest of this book, it is understandable why most of them would never qualify to sell their products in the United States, Europe, Canada, Australia, Japan, and other such highly lucrative markets. However, there remain significant challenges for world-class developers. I do not foresee more than three or four biosimilar competitors at best in these markets for each product. A quick review of the list of developers as provided in such databases as www.ipdanalytics.com shows the high popularity of some products while others that might provide a lower development barrier are ignored. The biosimilar product developer would do well by making a more subtle decision rather than going for the race with major developers like Amgen, Merck, Teva, Pfizer, and Celltrion. As an example, here are the 24 companies as of November 2014 that have disclosed that they are developing a biosimilar candidate to Herceptin (trastuzumab): Hospira, Celltrion, Actavis-Amgen-Synthon, Biocon-Mylan, Pfizer/Hanwha, Biocad, Zydus Cadila, Reliance, Shanghai CP Guojian, ISU Abxis, Samsung Bioepis, Cipla, Stada, Meiji Seoka-Dong-A, Viropro-Oncobiologics, BioXpress Therapeutics, Bioviz, PlantForm, Mabion, STC Biologics, Harvest Moon, Alteogen, Alphamed, and Therapeutic Proteins International. The patent expiry on Herceptin appears to be December 2018 in the EU and February 2022 in the United States.

Global approach: Some products are more suitable for preparing a global dossier than others; some difficult choices would have to be made in this approach that may involve the choice of reference product to the commercialization difficulties.

The developer of a biosimilar product would do well by creating a matrix to demonstrate the value and importance of each of the aforementioned listed points, prior to making the selection of the product. This exercise, if not taken seriously, would cause significant delays and financial losses later. Unfortunately, even companies with large resources have ignored the importance of this stage, and you can see many studies on www.clinicaltrials.gov that will essentially go to waste.

Manufacturing system selection

Once a molecule has been selected as a biosimilar candidate, the next step is to select a manufacturing system. Generally, this would not be a consideration, were it not for a few peculiarities about biosimilars that makes this choice significant.

Capital cost optimization: It is generally accepted that the cost of drug substances is almost 50% loaded with depreciation of the capital investment that goes into establishing a manufacturing facility. The primary purpose of creating the category of biosimilars is to make them available at an affordable cost, and this can be seriously jeopardized if the developer chooses the standard deep-tank technology that may cost hundreds of millions of dollars to establish and validate. Major companies like Amgen and Sanofi have understood this as well and have begun moving in the direction of single-use systems as they establish new facilities in Singapore. To impress the need to choose single-use systems over the traditional deep-tank technology, I have included a separate chapter on this topic (Chapter 7).

At scale lots: The regulatory agencies want commercial scale lots to be used in PK/PD trials, which is not a requirement for 505 category products. This means that the entire infrastructure including the PPQ lots and a full GMP-compliant facility should be in place as the development moves from laboratory scale to initial phases of 351(k) filing strategy. The full-scale requirement comes from the possibility that these products may be approved without any phase III trials; therefore, the PK/PD studies become extremely important in evaluating the safety of the product; given that the process of manufacturing can significantly affect the safety profile, this requirement seems justified. This is another reason why a single-use technology, more particularly a technology that allows a modular system wherein smaller lots are made and combined to form a larger lot without the need for scale-up, will add substantially to the success of biosimilar developers. This is the technology currently used by Therapeutic Proteins International, a company located in Chicago.

Cell line choice

A developer of biosimilar products will face significant challenges in the selection of an expression system, likely a recombinant DNA system. The choice of these systems is wide, and Chapter 4 describes in detail the nature and the future prospects of these expression systems. Appendix 3 provides details of the licensed biological products that show the types of cell lines used in these products.

Whereas an obvious choice will be to select exactly the same expression system as used by the originator, there are many reasons to look at the alternates. When the originators developed their products, the cell lines available likely had lower yields, the quality of expression less than ideal, and the stability of these cell lines highly questionable. However, once a cell line has been used to secure approval or licensing of a product, the originator is less likely to make a change, a choice that is available to biosimilar developers.

While there is sufficient scientific rationale for using improved processes as they become available, the developers of biosimilar products would generally use similar systems as used by the first licensed product to reduce the variability factors. For example, over the past decades, several new cell lines and types have been developed that may offer better expression, better quality of protein, more consistent quality of proteins and antibodies, but the risk of establishing safety using a different cell line to express recombinant products outweighs their advantages. However, in the past, drugs like growth hormone were approved using different expression systems in abbreviated filing. The regulatory agencies do not require the use of the same expression system.

Cell lines can be licensed or developed internally; in all instances, developing cell lines can take a long time, especially as it requires establishing the stability of

Table 3.1 Contract Suppliers of Recombinant Cell Lines

www.antitope.com/	www.assaydepot.com/
www.biocompare.com	www.bioprocessonline.com/
www.bioscience.co.uk/	www.biosciregister.com/
www.creative-biolabs.com/	www.evotec.com/
www.fitzgerald-fii.com/	www.fujifilmdiosynth.com/
www.gallusbiopharma.com/	www.geneva-biotech.com/
www.geneva-biotech.com/	www.genscript.com/
www.innoprot.com/	www.invivo.de/
www.invivo.de/	www.lfbbiomanufacturing.com/
www.lifetechnologies.com/	www.lonza.com/
www.sigmaaldrich.com/	www.wow.com/

Note: No endorsement is made for any of these suppliers.

cell line. Licensing in cell lines is an attractive option; this may start with evaluation of the protein expressed first and then license the line if the expression is appropriate. The biosimilar product developer is required to disclose the pedigree of cell line; therefore, it does not pay to secure these lines from dubious resources; know that the competent cells themselves would need a separate license and the documentation must identify all steps in the construction of the cell lines. Table 3.1 lists a few providers of recombinant cell lines.

Cell banks

Cell banks are two-tiered seed lots systems set up from the initially selected producing clone. The clone of interest (obtained after genetic modification of the host cell line) is put into the culture, and the produced population is aliquoted in fractions (a few million cells per cryotube), then cryo-conserved. This first batch of tubes (generally around one hundred of cryotubes) constitutes the Master Cell Bank (MCB) (since it is directly a product of the selected clone amplified). To ensure the sustainability of this MCB, a Working Cell Bank (WCB) is prepared by amplifying one or two MCB cryotubes with distribution of cells produced in a WCB consisting also of hundreds of cryotubes, as with the MCB. Each production batch is initiated from an WCB cryotube. It is when after several production batches have been made that the WCB decreases and a new WCB may be prepared, according to the same process described earlier, by taking, in the MCB, one or two cryotubes. This "two-level"' cell bank ensures the sustainability of the cell line that has been built from the same genetically modified cell. A biosimilar developer would maintain its MCBs in multiple locations to ensure that these are not lost over a period of time.

Reference product

Before a cell line can be evaluated, the biosimilar developer must select a reference product. This choice can be as simple as the licensed product in the United States to so complex as "a product marketed in China unless the developer finds difficult to secure sufficient quantity," or the "developer must use an EU authorized product for non-clinical testing and bridging studies are required when using a non-EU reference product." An additional challenge comes from the need to acquire reference product lots at different expiry dates to enable side-by-side stability testing with these lots. There are many complexities on the choice of the reference product, particularly if a product is developed for global marketing.

Once a reference product or products are in place, there begins the process of getting to know the molecule well and create specification of the product; this is needed despite the fact that in some instances pharmacopoeia monographs are available as in the case of interferon alpha, filgrastim, erythropoietin, etanercept, growth hormone, etanercept, etc., and the list is continuing to grow both in the European Pharmacopoeia and the USP. The number of reference product samples needed to establish specifications depends on the variability of the test parameter.

Test method development

Whereas generally available test methods can be used to examine the specification items of the reference product, specific tests (in addition to those provided in the pharmacopeia if available) should be developed. At this stage, these may be suitable, not necessarily fully validated, tests. Once these methods are available begins the task of ascertaining whether the cell lines selected are appropriate or not. Generally, small-scale production of around 5 L will be needed to establish the suitability testing of cell lines. Only the most critical tests that pertain to purity, potency, and safety are included at this stage. These may include molecular weight by MS, CD, SDS-PAGE, RP-HPLC, fluorescence, peptide mapping, cell-based assay, and, where required, glycan mapping. The choice of these tests is based on those features that are inherently cell-line based. For example, it is possible to alter the ADCC and, to some extent, the glycan map, but an overall idea is required of the suitability of the cell lines. This level of testing is best conducted using single-use bioreactors since a large number of lots may have to be made under a variety of in-process conditions.

The aim of a biosimilarity development program is *not* to demonstrate the efficacy of the product—this was established already by the first licensed product years ago. The aim of a biosimilarity development program is to establish if there are any clinically relevant or meaningful differences by employing a variety of tests that might include pharmacodynamic modeling in healthy subjects or patients or even limited clinical trials when such models are not available.

Specifications

The active drug substance and drug product specifications are established based on the profiling of the first licensed product. Regulatory agencies strongly urge the developers to test a variety of lots of the first licensed product, preferably with different shelf lives to establish a rationale for the acceptance criteria and release specifications established. Regulatory agencies require that these acceptance criteria be established on full-scale production lots used for preclinical or PK/PD studies.

The number of lots used to establish specifications will depend on the variability of the specification. For example, a drug product is likely to show very small variability in its pH, osmolality, or density, but substance variability in its glycosylation profile or impurity profile. It is improper for the developer to select a fixed number of reference lots such as three or five or six. The number of lots selected must be justified based on a statistical modeling that is presented to the regulatory agencies along with the analytical and functional similarity data.

The range of specifications is also critical to the long-term success of the biosimilar product; setting specification too narrow may cause lot failures in the future. The ranges must be justified based on their criticality. For example, the SEC purity may be a lot more significant than the osmolality of the product. Generally, the methods that require a nonquantitative comparison, such as charts and recordings, must be used in conjunction with other more quantitative methods.

Reverse engineering

Biosimilar product development begins with an intensive reverse engineering project, as advised by the regulatory agencies, since there is usually no direct access to originator companies' proprietary data. This requires that the companies must have highly specialized teams in place, including a team to evaluate the intellectual property issues that are entirely different challenges in 351(k) filing as compared to 505 class filings.

Another major differentiating factor between 505 (a) or (b) products and 351(k) products is the level of scientific inquiry required to develop the latter category. Besides the complexity of protein chemistry, the challenge of making a product that is at least "highly similar" requires engineering efforts that may have little to do with the quality of the product. A case in point is the development of monoclonal antibodies that must show not only a similar glycan pattern but such binding properties as ADCC, which may not have much clinical significance, yet these must be met to term the product as "highly similar." However, the regulatory agencies leave sufficient room for justification of alternate specification, provided the developer is willing to provide additional data to remove any "residual uncertainty"; and that may require extensive clinical trials, avoiding which is the main priority.

Drug substance (DS) production

The WCB lines are put through a complex cycle of upstream process or fermentation to express the protein or antibody of interest using such systems as roller bottles (now archaic), cytocultivator, hollow fibers, deep tank fermenters, single-use bioreactors, etc. The culture conditions form the intellectual property for the developer since minor variations can result in high level of variability in the structure such as glycosylation, impurities, and immunogenicity of these products. These in-process conditions are developed and validated through the PPQ (product performance qualification) exercise. A good starting point is the information widely found in the literature, including the patents held by the originators. Keeping conditions as close as possible to the originator's process will be a good starting point. However, it must be always understood that the process used by the originator is clearly decades old, and since the product was developed, new techniques and better understanding of the process have developed. So, while a keen understanding of the originator process is required, the biosimilar developer would do well by improving upon these, both to reduce cost and to enhance the quality of the product.

Once expressed, the product is harvested as the inclusion body, expression solution in the media, transgenic animal's milk, etc., and then subjected to steps of extraction and purification, the downstream process.

The downstream process removes impurities, host cell proteins, any viruses such as the agent responsible for spongiform encephalopathies, etc. There are numerous strategies to extract and purify proteins, each one presenting a certain level of specificity (in order to select only the protein of interest), of yield (amount of proteins of interest eliminated with effluents), and of maintaining the molecular integrity of the protein being purified. The purification system may itself introduce differences in the protein quality profile, between production batches or between producers including qualitatively and quantitatively selecting isoforms and impurities.

The purified drug substance is stored under conditions appropriate for optimal stability, which must be validated.

Making a drug product consists of formulating the drug substance for the desired dosage form (subcutaneous, intravenous, etc.) using a formula that is very close to the first licensed product; fortunately, the exact details of these formulations are available as required for full disclosure, so it is relatively easy to formulate a drug product. However, the biosimilar product developer may run into serious challenges where the drug product formulations are protected by patents. A recent example is the reformulation of MabThera into a subcutaneous formulation that is protected; demonstrating similarity to MabThera using alternate formulations may prove challenging.

Although most recombinant proteins have demonstrated good safety in their use, often small changes in formulations can be significant as demonstrated in the case of the reformulation of erythropoietin leading to several cases of severe anemia (pure red cell aplasia). It is highly likely that the biosimilar product developer would have to choose a different formulation, particularly when developing monoclonal antibodies, due to patent protection. In such cases, an extensive study of the available formulations that are clearly out of the patent protections should be made. In any event, the biosimilar developer would have to demonstrate stability of its own drug substance in the formulation of the originator and the stability of the originator's drug substance (extracted from drug product) in the formulation selected by the biosimilar developer.

Analytical and functional similarity

The development of biosimilar products begins with an in-depth understanding of the structure and function of the reference product (RP) or medicine (RM as in Europe). In a recent guideline draft issued by the U.S. FDA, the level of similarity is defined at four levels (Figure 3.1).

The language used to describe these four levels of similarities is highly significant and should be understood well. In the first instance, if the product is not similar analytically then the developer should take this further under 351(a) or the traditional new biological drug development route. In this instance, any reference to the reference product is shelved, even though still some benefits can be gained

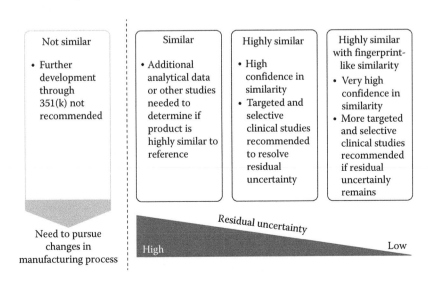

Figure 3.1 The four tiers of similarity classified by the U.S. FDA.

from published studies. Note that the driving factor is the residual uncertainty that remains after demonstrating analytical similarity.

When the analytical similarity is at a "similar" level, the FDA wants more analytical data as well as "other" data that can be animal studies or human trials. For the "highly similar" drugs, only selected clinical studies are recommended, but when the similarity is at the "fingerprint" level, the language changes to targeted studies if residual uncertainty remains; this means that if the developer has removed all residual uncertainty, any additional study may not be warranted. The FDA has approved several biological or highly complex products in the past based on analytical similarity, the most recent being the small molecular weight heparin.

The developers of biosimilar products should attempt to reach the "fingerprint"-like similarity if they are to get into the fasttrack, low-cost approvals from regulatory agencies. However, this requires a keen understanding of the protein product developed. In all instances, the developers of biosimilar products would know more about the nature of the protein than even the originator. In this chapter, I have therefore described the various steps required to acquire and practice this knowledge.

Nonclinical studies

The biosimilar candidate product is required to demonstrate safety in nonclinical studies such as animal toxicology, toxicokinetics, and immunogenicity profiles. While the biosimilar developer would look at the studies that were conducted when the product was originally developed, this may not prove to be adequate or even justified. First, let us take examples of products like filgrastim where studies in mice would be adequate but the dosing should be such that in a comparison study both the test and reference product show some toxicities and that these should be comparable. A discussion with the regulatory agencies is therefore required to make sure these data prove supporting.

In those instances where no good toxicology model exists such as for the monoclonal antibodies, straightforward studies in animals will not provide any useful information. In most instances, the regulatory agencies would let go of these studies if the drug product has a high level of analytical and functional similarity. In some instances, a PK study alone might be sufficient, such as in monkeys, to support the nonclinical safety demonstration requirement.

It is also noteworthy that there is a general consensus not to expose animals unnecessarily unless the data show a meaningful conclusion. The EU is more tilted toward this position, and the developer is strongly urged to discuss their development plan ahead of commencing any studies.

A word of caution here pertains to the historical staging of a new molecule vis-à-vis a biosimilar product. When a new product is developed, animal studies are often conducted on early, development lots; for biosimilars, this must be the GMP lots, preferably at commercial scale. Again, the developer is urged to confirm this prior to commencing any studies.

Clinical pharmacology studies

Almost invariably, every regulatory agency will require at least one clinical pharmacology study that includes both the PK and the PD, where available, parameters. Whether these studies are conducted in patients or healthy subjects can be a question of cost as well as ethics. For products like filgrastim where a good PD model exists in a healthy subject, they are the choice, but for some anticancer drugs, this may have to be conducted in patients if toxicity is an issue. For many RA products, the agencies would likely forego the need to demonstrate the PD values when conducting studies in healthy subjects, and these could easily be straightforward PK studies.

Interchangeability protocols

This option is not available in the EU, but the U.S. FDA encourages developers to propose protocols that will establish comparable safety and no reduced efficacy when the reference and test products are switched and/or alternated. It is also likely that because of the statutory requirement that this equality be demonstrated in a clinical setting, the developer would have to use a patient population, generally a widely available population when multiple indications are available. The developer is advised to seek a clear understanding and approval of the regulatory agencies prior to commencing these studies, which can be extremely expensive to conduct.

Regulatory uncertainty

While the process of applying for licensing or authorization is relatively well explained, there remain many uncertainties, particularly in the United States. One such uncertainty is the extrapolation of indications. It is a general understanding that the biosimilar product would get all indications, whether the clinical trials were conducted in patients. However, it is highly likely that one or more indications, more particularly the pediatric exclusivity, may exclude making the claim of indication. This requires a careful carve out of the label of the product. For example, in the case of GCSF, the CD34+ mobilization claim is still under patent till the end of 2016 while all other patents have expired. This requires explicit deletion of the indication.

What would the label look like for the 351(k) filing is still unknown; in all likelihood, the burden lies on the developer to suggest a label that is likely to be a hybrid that includes all open available information on the drug substance and some specific information about the drug product developed.

An interesting situation arises in the patient inclusion criteria for testing biosimilars; is it ethical or advisable to use a normal subject population to test a toxic product when the risk is high? Is it appropriate to provide the clinical pharmacology data in patients instead? What if there are no reliable pharmacodynamic end points available in healthy subjects and these are confounded heavily in patients; how would the similarity be established?

One of the greatest financial challenges facing biosimilar product development is the requirement to establish noninferiority over the originator product; comparing two products requires a different statistical approach than comparing a product against a placebo; the latter may require a smaller patient population. This aspect has been fully appreciated, and it is for this reason the regulatory agencies have begun relying more heavily on alternate evidence of similarity.

Legal teams in place

The developers must also be ready for litigation since this path was opened at least in the United States by giving the originator an opportunity to review the entire regulatory package within days of the acceptance for review by the U.S. FDA.

The patent dance that takes place in the United States is still unexplored and the very first filing of GCSF by Sandoz resulted in a massive lawsuit by Amgen against Sandoz that even included such claims as theft of property. Biosimilar developers must put their legal teams in place very early in the process of development, first to make sure they can get these law firms without any conflict of interest, and second, to go through the tedious process of developing the Freedom to Operate (FTO) opinions as the manufacturing process develops and as new patents appear, often weekly, to protect the intellectual property of the originator. This is an extremely expensive exercise, but the U.S. draft guidance and a long history of skewed patenting laws in the United States have made this an inevitable process.

Commercialization challenges

For years, well before the patents for the biotechnology-based products expired, the first licensed product manufacturers began a concerted campaign based on the claim that these products are defined by the process, which is not disclosed; therefore, it is not possible to replicate the method of manufacture despite any level of knowledge about the product. These statements have been repeated so often that they have begun to take roots, not only in the minds of prescribers but also various regulatory agencies. This claim is true only to the extent that the in-process details are not made public, but it is not true that this is the only way or the even the best method to manufacture these products. When a manufacturer of the first licensed product develops a process, it is based on the technology available at the time, both for manufacturing and monitoring the product. Over decades, new techniques for testing, monitoring, and manufacturing find their way into process options, and that is more likely to improve the process controls. Therefore, the claim that a biosimilar product developer can only produce an inferior product is improper. Briefly, the first licensed product is developed using "a" process, not "the" process. However, this scientific rationale does not deter the legal challenges; one major manufacturer of a biological product even sued the regulatory agencies on a ground that is absolutely original and irrational: the claim that since the regulatory agencies have seen the regulatory filing of the originator, they are no longer qualified with their tainted minds to review any biosimilar product application—what if the biosimilar product, by chance, is using similar in-process controls, in which case the reviewers might accept these without questioning the developers of the biosimilar products to justify their choice of process. However, the regulatory agencies have realized fast that such arguments are mainly triggered by intense greed, not science, and have moved ahead with encouraging biosimilar product companies to challenge the agencies in minimal path to the development and launch of these products. A case in point is the EMA guideline that was released in October 2014 that clearly states that clinical trials may not be required if sufficient similarity is demonstrated. This is the first such clear statement by an agency that leads the world in developing biosimilars.

Bibliography

Ahmad SS, Dalby PA. Thermodynamic parameters for salt-induced reversible protein precipitation from automated microscale experiments. *Biotechnol Bioeng* 2010 Sept 24 (Epub ahead of print), http://www.pharmtech.com/suitability-use-considerations-prefilled-syringes.

Akers MJ, Vasudevan V, Stickelmeyer M. Formulation development of protein dosage forms. Development and manufacture of protein pharmaceuticals. In: *Pharmaceutical Biotechnology*, Vol. 14, Nail SL, Akers MJ, eds. Springer, London, U.K., 2002, pp. 52, 53, 71, 73, 74, 80–82.

Arakawa T et al. Theory of protein solubility. *Meth Enzymol* 1985;114:49–77; Schein in Solubility as a function of protein structure and solvent components. *BioTechnology* 1990;8(4):308–317; Jenkins in Three solutions of the protein solubility problem. *Prot Sci* 1998;7(2):376–382; and others.

Araki F et al. Stability of recombinant human epidermal growth factor in various solutions. *Chem Pharmaceut Bull (Tokyo)* 1989;37(2):404–406.

Astafieva IV, Eberlein GA, Wang YJ. Absolute on-line molecular mass analysis of basic fibroblast growth factor and its multimers by reversed-phase liquid chromatography with multiangle laser light scattering detection. *J Chromatogr A* 1996;740(2):215–229.

Bandyopadhyay S et al. Formulations of PEG-interferon alpha conjugates. World Intellectual Property Organization Patent No. PCT/IN2007/000549, 2008 May 29. www.faqs.org/patents/app/20100074865.

Banks DD et al. The effect of sucrose hydrolysis on the stability of protein therapeutics during accelerated formulation studies. *J Pharmaceut Sci* 2009;98(12):4501–4510.

Berlett BS, Stadtman ER. Protein oxidation in aging, disease, and oxidative stress. *J Biol Chem* 1997;272(33):20313–20316.

Brange J et al. Chemical stability of insulin, 1: Hydrolytic degradation during storage of pharmaceutical preparations. *Pharmaceut Res* 1992;9(6):715–726.

Capasso S et al. Kinetics and mechanism of the cleavage of the peptide bond next to asparagine. *Peptides* 1996;17(6):1075–1077.

Carpenter JF et al. Overlooking subvisible particles in therapeutic protein products: Gaps that may compromise product quality. *J Pharmaceut Sci* 2009;98(4):1201–1205.

Chang LL, Pikal MJ. Mechanisms of protein stabilization in the solid state. *J Pharm Sci* 2009;98(9):2886–2908.

Chang SH et al. Metal-catalyzed photooxidation of histidine in human growth hormone. *Anal Biochem* 1997;244(2):221–227.

Charache S et al. Postsynthetic deamidation of hemoglobin providence (β 82 Lys —> Asn, Asp) and its effect on oxygen transport. *J Clin Investigation* 1977 Apr;59:652–658.

Cholewinski M, Lückel B, Horn H. Degradation pathways, analytical characterization, and formulation strategies of a peptide and protein: Calcitonin and human growth hormone in comparison. *Pharm Acta Helv* 1996;71(6):405–419.

Chou DK et al. Effects of Tween 20 and Tween 80 on the stability of albutropin during agitation. *J Pharmaceut Sci* 2005;94(6):1368–1381.

Cromwell MEM, Hilario E, Jacobson F. Protein aggregation and bioprocessing. *AAPS J* 2006;8(3):article 66.

Darrington RT, Anderson BD. Evidence for a common intermediate in insulin deamidation and covalent dimer formation: Effects of pH and aniline trapping in dilute acidic solutions. *J Pharm Sci* 1995;84(3):275–282.

Geiger T, Clarke S. Deamidation, isomerization, and racemization at asparaginyl and aspartyl residues in peptides: Succinimide-linked reactions that contribute to protein degradation. *J Biol Chem* 262(2);1987:785–794.

Di Donato A et al. Selective deamidation of ribonuclease A: Isolation and characterization of the resulting isoaspartyl and aspartyl derivatives. *J Biol Chem* 1993;268(7):4745–4751.

Engleka KA, Maciag T. Inactivation of human fibroblast growth factor-1 (FGF-1) activity by interaction with copper ions involves FGF-1 dimer formation induced by copper-catalyzed oxidation. *J Biol Chem* 1992;267(16):11307–11315.

Fransson J, Hagman A. Oxidation of human insulin-like growth factor I in formulation studies, II: Effects of oxygen, visible light, and phosphate on methionine oxidation in aqueous solution and evaluation of possible mechanisms. *Pharmaceut Res* 1996;13(10):1476.

Frelinger AL, Zull AE. Oxidized forms of parathyroid hormone with biological activity: Separation and characterization of hormone forms oxidized at methionine 8 and methionine 18. *J Biol Chem* 1984;259(9):5507–5513.

Friedman AR et al. Degradation of growth hormone releasing factor analogs in neutral aqueous solution is related to deamidation of asparagine residues: Replacement of asparagine residues by serine stabilizes. *Int J Pep Prot Res* 1991;37(1):14–20.

Furuya K, Johnson-Jackson D, Ruscio DT. Deamidated Interferon-Beta. World Intellectual Property Organization No. WO 2006/053134 A2, 2006 May 18. www.freepatentsonline.com/y2009/0263355.html.

Gautriaud E et al. Effect of sterilization on the mechanical properties of silicone rubbers. *BioProcess Int* 2010;8(4):S42–S49.

Gaza-Bulseco G, Liu H. Fragmentation of a recombinant monoclonal antibody at various pH. *Pharmaceut Res* 2008;25(8):1881–1890.

Gietz U et al. Chemical degradation kinetics of recombinant hirudin (HV1) in aqueous solution: Effect of pH. *Pharmaceut Res* 1998;15(9):1456.

Gitlin G et al. Isolation and characterization of a monomethioninesulfoxide variant of interferon alpha-2b. *Pharmaceut Res* 1996;13(5):762–769.

Green FA. Interactions of a nonionic detergent, II: With soluble proteins. *J Colloid Interface Sci* 1971;35(3):481–485.

Griffiths SW. Oxidation of the sulfur-containing amino acids in recombinant human α1-antitrypsin (thesis). Massachusetts Institute of Technology, Cambridge, MA, 2002.

Herman AC, Boone TC, Lu HS. Characterization, formulation, and stability of neupogen (Filgrastim), a recombinant human granulocyte-colony stimulating factor. In: *Formulation, Characterization, and Stability of Biopharmaceuticals: Case Histories*, Pearlman R, Wang YJ, eds. Plenum Press, New York, 1996.

Hsu Y-R et al. In vitro methionine oxidation of *Escherichia coli*–derived human stem cell factor: Effects on the molecular structure, biological activity and dimerization. *Prot Sci* 1996:5;1165–1173.

Jacob S et al. Stability of proteins in aqueous solution and solid state. *Indian J Pharmaceut Sci* 2006 Mar–Apr;68:154–163.

Jenke D. Suitability-for-use considerations for prefilled syringes. *Pharmaceut Technol* 2008 April, http://www.pharmtech.com/suitability-use-considerations-prefilled-syringes.

Jones LS, Kaufmann A, Middaugh CR. Silicone oil induced aggregation of proteins. *J Pharmaceut Sci* 2005;94(4):918–927.

Kahook MY et al. High-molecular-weight aggregates in repackaged Bevacizumab. *Retina* 2010;30(6):887–892.

Kashiwai T et al. The stability of immunological and biological activity of human thyrotropin in buffer: Its temperature-dependent dissociation into subunits during freezing. *Scand J Clin Lab Invest* 1991;51(5):417–423.

Kerwin BA, Remmele RL. Protect from light: Photodegradation and protein biologics. *J Pharmaceut Sci* 2007;96(6):1468–1479.

Knepp VM et al. Identification of antioxidants for prevention of peroxide-mediated oxidation of recombinant human ciliary neurotrophic factor and recombinant human nerve growth factor. PDA *J Pharmaceut Sci Technol* 1996;50(3):163–171.

Koide H et al. Biosynthesis of a protein containing a nonprotein amino acid by *Escherichia coli*: l-2-aminohexanoic acid at position 21 in human epidermal growth factor. *Proc Natl Acad Sci USA* 1988 Sept;85:6237–6241.

Kosky AA et al. The effects of alpha-helix on the stability of Asn residues: Deamidation rates in peptides of varying helicity. *Prot Sci* 1999;8:2519–2523.

Landi S, Held HR. Effect of oxidation on the stability of tuberculin purified protein derivative (PPD). *Dev Biol Stand* 1986;58(Pt B):545–552.

Levine RL et al. Methionine residues as endogenous antioxidants in proteins. *Proc Natl Acad Sci USA* 1996 Dec;93:15036–15040.

Lewis UJ et al. Altered proteolytic cleavage of human growth hormone as a result of deamidation. *J Biol Chem* 1981;256(22):11645–11650.

Li R et al. Effects of solution polarity and viscosity on peptide deamidation. *J Pep Res* 2000a;56(5):326–334.

Li S et al. Inhibitory effect of sugars and polyols on the metal-catalyzed oxidation of human relaxin. *J Pharmaceut Sci* 2000b;85(8):868–872.

Li S, Schoneich C, Borchardt RT. Chemical instability of protein pharmaceuticals: Mechanisms of oxidation and strategies for stabilization. *Biotechnol Bioeng* 1995;48:490–500.

Li X, Lin C, O'Connor PB. Glutamine deamidation: Differentiation of glutamic acid and gamma-glutamic acid in peptides by electron capture dissociation. *Anal Chem* 2010; 82(9):3606–3615.

Lins L, Brasseur R. The hydrophobic effect in protein folding. *FASEB J* 1995 Apr;9:535–540.

Mahler HC et al. Induction and analysis of aggregates in a liquid IgG1-antibody formulation. *Eur J Pharma Biopharma* 2005;59(3):407–417.

Manning MC, Patel K, Borchardt RT. Stability of protein pharmaceuticals. *Pharmaceut Res* 1989;6(11):903–918.

Mark DF et al. Site-specific mutagenesis of the human fibroblast interferon gene. *Proc Nat Acad Sci USA* 1984 Sept;81:5662–5666.

Moss CX et al. Asparagine deamidation perturbs antigen presentation on class II major histocompatibility complex molecules. *J Biol Chem* 2005;280(18):18498–18503.

Nabuchi Y et al. Oxidation of recombinant human parathyroid hormone: Effect of oxidized position on the biological activity. *Pharmaceut Res* 1995;12(12):2049–2052.

Patel K, Borchardt RT. Chemical pathways of peptide degradation, II: Kinetics of deamidation of an asparaginyl residue in a model hexapeptide. *Pharmaceut Res* 1990;7(7):703–711.

Philo JS, Arakawa T. Mechanisms of protein aggregation. *Curr Pharmaceut Biotechnol* 2009;10:348–351.

Q4B, Annex: Evaluation and recommendation of pharmacopoeial texts for use in the ICH regions on residue on ignition/sulphated ash. *Fed Reg* 2008;73(35):9576–9577. www.ich.org/LOB/media/MEDIA3093.pdf.

Rader RA. *Biopharmaceutical Products in the U.S. and European Markets*, 6th edn. BioPlan Associates, Inc., Rockville, MD, 2007.

Rader RA. *Biopharmaceutical Expression Systems and Genetic Engineering Technologies: Current and Future Manufacturing Platforms*. BioPlan Associates, Inc., Rockville, MD, 2008.

Raso SW et al. Aggregation of granulocyte-colony stimulating factor in vitro involves a conformationally altered monomeric state. *Prot Sci* 2005;14:2246–2257.

Rathore N, Rajan RS. Current perspectives on stability of protein drug products during formulation, fill and finish operations. *Biotechnol Prog* 2008;24:504–514.

Robinson NE. Protein deamidation. *Proc Nat Acad Sci* 2002;99(8):5283–5288.

Rosenberg AS. Effects of protein aggregates: An immunologic perspective. *AAPS J* 2006; 8(3):E501–E507.

Ruiz L, Aroche K, Reyes N. Aggregation of recombinant human interferon alpha 2b in solution: Technical note. *AAPS PharmSciTech* 2006;7(4):article 99.

Schöneich C. Mechanisms of metal-catalyzed oxidation of histidine to 2-oxo-histidine in peptides and proteins. *J Pharmaceut Biomed Anal* 2000;21(6):1093–1097.

Schrier JA et al. Degradation pathways for recombinant human macrophage colony-stimulating factor in aqueous solution. *Pharmaceut Res* 1993;10(7):933–944.

Shahrokh Z et al. Major degradation products of basic fibroblast growth factor: Detection of succinimide and iso-aspartate in place of aspartate. *Pharm Res* 1994;11(7):936–944.

Smith BJ. Chapter 6: Chemical cleavage of proteins. New protein techniques. In: *Methods in Molecular Biology*, Vol. 3. Springer, Heidelberg, Germany, 1988, pp. 71–88.

Stevenson CL et al. Effect of secondary structure on the rate of deamidation of several growth hormone releasing factor analogs. *Int J Pep Prot Res* 1993;42(6):497–503.

Teh L-C et al. Methionine oxidation in human growth hormone and human chorionic somatomammotropin: Effects on receptor binding and biological activities. *J Biol Chem* 1987;262(14):6472–6477.

Thirumangalathu R et al. Effects of pH, temperature, and sucrose on benzyl alcohol-induced aggregation of recombinant human granulocyte colony stimulating factor. *J Pharmaceut Sci* 2006;95(7):1480–1497.

Thirumangalathu R et al. Silicone oil- and agitation-induced aggregation of a monoclonal antibody in aqueous solution. *J Pharmaceut Sci* 2009;98(9):3167–3181.

Torchinskii M. In: *Sulfur in Proteins*, Metzler DE, ed. Pergamon, New York, 1981.

Treuheit MJ, Kosky AA, Brems DN. Inverse relationship of protein concentration and aggregation. *Pharmaceut Res* 2002;19(4):511–516.

Tyler-Cross R, Schirch V. Effects of amino acids sequence, buffers, and ionic strength on the rate and mechanism of deamidation of asparagine residues in small peptides. *J Biol Chem* 1991;266(33):22549–22556.

Uchida K, Kawakishi S. Identification of oxidized histidine generated at the active site of Cu, Zn superoxide dismutase exposed to H_2O_2: Selective generation of 2-oxo-histidine at the histidine 118. *J Biol Chem* 1994;269:2405–2410.

USP <788> Particulate Matter in Injections. U.S. Pharmacopeial Convention, Inc., Rockville, MD, 2010.

Wakankar AA, Borchardt RT. Formulation considerations for proteins susceptible to asparagine deamidation and aspartate isomerization. *J Pharmaceut Sci* 2006;95(11):2321–2336.

Wang W. Instability, stabilization, and formulation of liquid protein pharmaceuticals. *Int J Pharmaceut* 1999;185:129–188.

Yüksel KU, Gracy RW. In vitro deamidation of human triosephosphate isomerase. *Arch Biochem Biophys* 1986;248(2):452–459.

Zhang J, Kalonia DS. The effect of neighboring amino acid residues and solution environment on the oxidative stability of tyrosine in small peptides. *AAPS PharmSciTech* 2007;8(4):article 102.

Chapter 4 Stability and formulation considerations

Introduction

A myriad of physical and chemical factors can affect the quality and stability of biopharmaceutical products, particularly after long-term storage in a container–closure system likely to be subject to variations in temperature, light, and agitation with shipping and handling. Compared with traditional chemical pharmaceuticals, proteins are considerably larger molecular entities with inherent physiochemical complexities, from their primary amino acid sequences through higher-order secondary and tertiary structures—and in some cases, quaternary elements such as subunit associations. As a result, biopharmaceuticals are subject to chemical and physical degradation, both of which can result in loss of activity, changes in the pharmacokinetic profile, and altered immunogenicity. This makes the study of the stability of biopharmaceutical products of great interest and importance to the developers of biosimilar products.

To fully appreciate the nuances of formulating biosimilar products, a good understanding of protein structure and common structural risks in formulation should be understood. Figure 4.1 shows the amino acids found in proteins. Figure 4.2 shows the structural configurations assumed by the proteins and their various forms of dispersion.

Proteins are composed of a chain or amino acids that have reactant groups, which cause them to connect to each other to form the multidimensional structures; this reactivity also forms the basis of their chemical and physical degradation. Formulations of biopharmaceuticals, therefore, are designed to minimize the physical and chemical degradation, yet, keeping the stabilizers to a minimum. A keen understanding of protein chemistry is required to understand the formulation challenges for biopharmaceutical products. The most common routes of chemical and physical degradation of biopharmaceuticals are shown in Figure 4.3.

Protein instability can be mainly accounted for in two forms: chemical and physical instability. *Chemical instability* refers to the formation or destruction of covalent bonds within a polypeptide or protein structures. Chemical modifications of protein include mainly oxidation, deamidation, reduction, and hydrolysis. Unfolding, dissociation, denaturation, aggregation, and precipitation are known as *conformational* or *physical instabilities*. In some cases, protein degradation pathways are synergistic: a chemical event may trigger a physical event, such as when oxidation is followed by aggregation. Generally, physical changes may not bring any significant clinical risk in small molecule drugs, except in the PK profile; in the case of biosimilars, these are crucial in determining the safety of these drugs.

A variety of formulation factors that can induce instability are shown in Table 4.1. These factors are well studied and anticipated from the knowledge of the chemistry

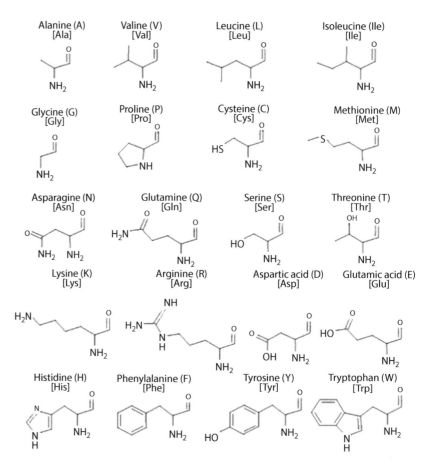

Figure 4.1 Amino acids forming proteins.

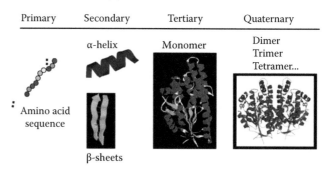

Figure 4.2 Basic structural components of proteins.

of all types of molecules. However, the impact on safety and efficacy is peculiar to biopharmaceuticals.

Compared to other small molecule drugs, biopharmaceuticals are typically more sensitive to slight changes in solution chemistry. They remain compositionally and conformationally stable only within a relatively narrow range of pH and osmolarity, and many require additionally supportive formulation components to remain in solution, particularly over time. Even lyophilized protein products are subject to significant degradation, unlike the small molecule drugs that would be highly stable in most lyophilized formulations.

For a long time, the major challenge in understanding the stability of biopharmaceuticals was the lack of sensitive analytical methods to characterize the proteins

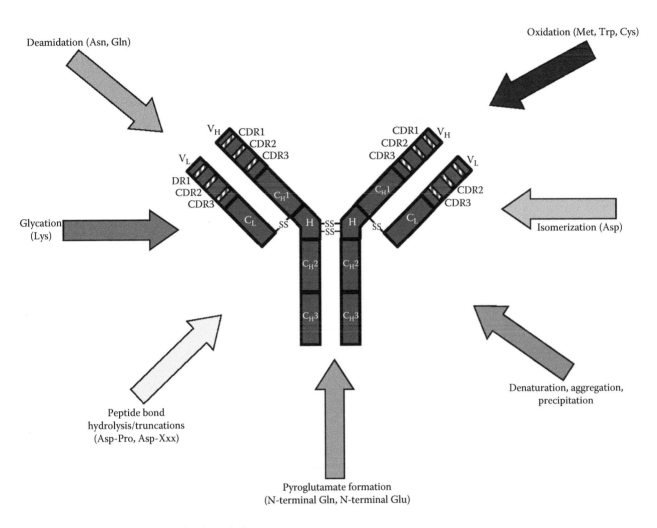

Figure 4.3 Common routes of protein degradation.

Table 4.1 Impact of Various Formulation and
Environmental Factors on the Degradation of Proteins

Factor	Impact
pH	Hydrolysis and deamidation
Buffer species	Deamidation
Other excipients	Maillard reaction
Light	Photo decomposition
Oxygen	Oxidation
Metal ions	Hydrolysis and oxidation
Temperature	Most routes

and their degradants. Now, the science of analytical chemistry has advanced sufficiently to overcome these shortcomings of the past. We are now better able to understand specific degradation pathways that create both chemical and physical instability and evaluate methods to stabilize the biopharmaceutical formulations. While the specific effects of various chemical reactions remain difficult to predict, the general impact of these reactions on the acidity is universally observed in protein products as shown in Table 4.2. This global change alone emphasizes the importance of understanding the overall impact on stability of biopharmaceuticals.

85

Table 4.2 Effect of Chemical Modifications of Proteins on Acidity Function and Charge Heterogeneity

Chemical Modification	Acidity/Charge Heterogeneity
Deamidation	More acidic ($z = -1$)
Succinimide formation	More basic or neutral
Glycation	More acidic
Pyroglutamate formation	More acidic ($z = -1$)
Peptide bond hydrolysis	Either acidic or basic

Deamidation

Deamidation is a chemical reaction in which an amide functional group is removed from an amino acid. This is one of the most significant chemical modification, whose consequences may include isomerization, racemization, and truncation of proteins.

The key features of the deamidation reaction are as follows:

- Nonenzymatic reaction that occurs at Asn and Gln.
- Hydrolysis reaction requiring water.
- Pathways depend on pH.
- Cyclic imide pathway under slightly acidic to basic conditions.
- Direct hydrolysis under acidic conditions.
- Minimum rate in pH range ~3–4.
- Buffer catalytic effect in pH range ~7–12.
- Rate is 5×–10× faster for Asn than for Gln.
- Generates both normal and beta peptide bonds.
- Rate is affected by the $N + 1$ residue.
- Interconvertibility isoAsp-Y to Asp-Y.

The effect of pH, buffer, and ionic strength on deamidation is significant. Deamidation rates increase with increasing pH; phosphate and carbonate buffers have catalytic properties for deamidation. At low pH levels, peptides and proteins experience deprotonation of the amide group on their asparagine side chain, followed by nucleophilic attack by the nitrogen atom of the amide anion on the peptide carbonyl carbon of the asparagine residue. This generates peptide chain cleavage by forming a succinimide peptide fragment. Subsequent hydrolysis of the succinimide ring can yield asparaginyl and β-asparaginyl peptides. Figure 4.4 shows the mechanism of asparagine degradation by deamidation.

If pH is >5.0, deamidation occurs through very unstable cyclic imide intermediate formation, which spontaneously undergoes hydrolysis. Under strong acidic conditions (pH 1–2), direct hydrolysis of the amide side chain becomes more favorable than the formation of cyclic imide. Peptide bond cleavage occurs to a greater extent in direct amide hydrolysis. At neutral pH, deamidation can lead to structural isomerization and racemization, one of the most observed outcomes of nonenzymatic degradation, especially the isomerization of aspartate to isoaspartate residues. Asn, Asp, and Gln are the most common groups for isomerization producing a succinimide intermediate; Asp degradation is much slower than Asn via this mechanism. The primary route is the deamidation, which yields isoAsp and Asp in roughly 3:1 ratio with the rate and propensity determined by the location and the mobility of the group. The isomerization can result in racemization as part of the deamidation reaction. Succinimide intermediates formed during asparagine deamidation are highly prone to racemization and convert to D-asparagine residues. Racemization of other amino acids, except glycine, is

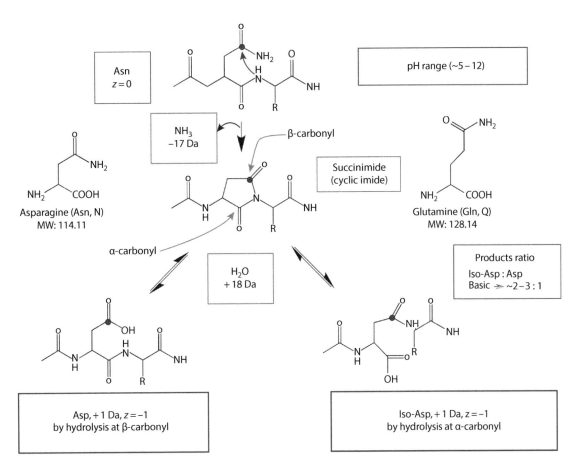

Figure 4.4 Deamidation of asparagine.

observed at alkaline pH. Mechanisms for the aspartate–isoaspartate deamidation and isomerization reactions are similar because they both proceed through an intramolecular cyclic imide intermediate.

Deamidation rates for individual amide residues depends on their primary sequence and their three-dimensional (3D) structure as well as the properties of the solution they are immersed in such as pH, temperature, ionic strength, and buffer ions. The deamidation rate for glutamine residues, for example, is usually less than that of asparagine residues.

Deamidation rates of protein are significantly affected by the degree of side-chain branching in the $(n + 1)$th residue. The charge of residue is not critical except in the case of His, Ser, and Thr. The $(n − 1)$th residue however has a minor effect on the deamination rate. The half-life of deamidation can range from 1 day to over a thousand days based on the identity of the carboxyl side residue. The rate of deamidation is also influenced by protein secondary structure. Increasing helical structure decreases the rate of deamidation in some proteins. Deamidation and the resulting structural outcomes can have varying effects on proteins' physiochemical and functional stability. Table 4.3 lists some examples of the effect of deamidation demonstrating the heterogeneity of effects observed (Figure 4.5).

Storage temperatures can affect deamidation rate depending on the buffers used in the formulations. Because amine buffers (e.g., Tris and Histidine) have high-temperature coefficients, storage at temperatures that are different from the

87

Table 4.3 Examples of the Effects of Deamidation of Various Biological Molecules

Biological Molecules	Effect of Deamidation
Growth hormone-releasing factor analogs	The addition of methanol increases the level of α-helicity and decreases the rate of deamidation.
RNAase	Resists deamidation possibly because of the relatively rigid backbone in the loop stabilized by a disulfide between Cys-8 and Cys-12 and by the β-turn at residues 66–68, which in turn could hinder the formation of the cyclic imide. But if the enzyme is reduced and denatured, then refolded, aspartic and isoaspartic forms are generated, demonstrating different enzymatic activities. Replacement of Asp-67 with Iso-Asp-67 shows that the isoaspartic form refolds at half the rate of the fully amidated form.
Human growth hormone	Alters proteolytic cleavage decreasing biological activity.
IFN-beta	Increase in biological activity.
Peptide growth hormone-releasing factor	Leading to aspartyl and isoaspartyl forms reduces the bioactivity by 25- and 500-fold, respectively, as compared with the native peptide.
Hemoglobin	Deamidation at an Asn–Gly site in hemoglobin changes its affinity to oxygen.
Class II major histocompatibility complex molecules	Asparagine deamidation perturbs antigen presentation.
Human epidermal growth factor	Isomerization of Asp 11 reduces by fivefold its mitogenic activity.
Triose-phosphate isomerase	Deamidation at two Asn–Gly sequences results in subunit dissociation.

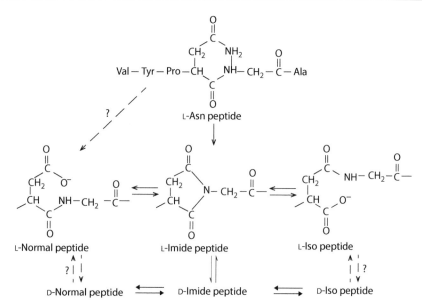

Figure 4.5 Deamidation pathway of asparagine. (From Geiger, T. and Clarke, S., *J. Biol. Chem.*, 262(2), 785, 1987.)

temperature of preparation could shift formulation pH. Deamidation and isomerization reactions are pH-sensitive processes, so shifts in formulation pH could alter the rate of deamidation. Another indirect effect of temperature is the dissociation constant of water: the hydroxyl ion concentration of water can vary as a function of temperature and thereby affect deamidation rates (Figure 4.6).

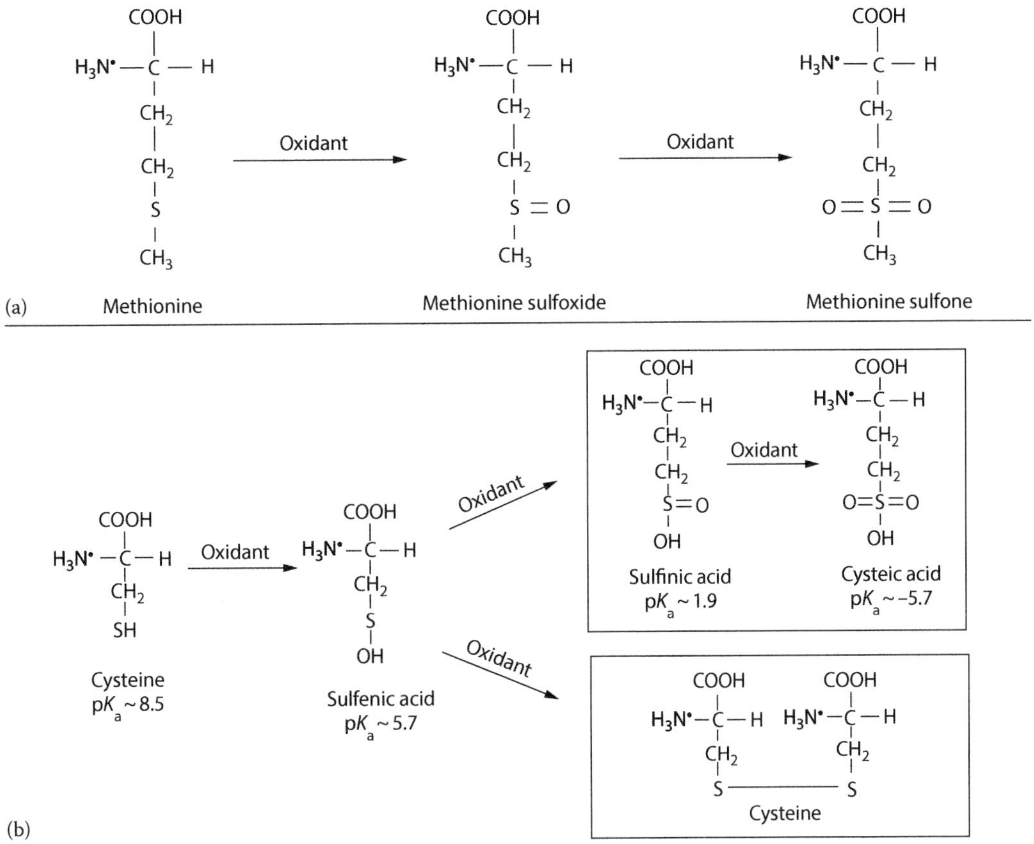

Figure 4.6 Oxidation of methionine (a) and cysteine (b). (From Geiger, T. and Clarke, S., *J. Biol. Chem.*, 262(2), 785, 1987.)

Preventive measures against deamidation

Solution pH can substantially affect deamidation. Formulations at pH 3–5 can minimize peptide deamidation. AsnA-21 and AsnB-3 of insulin form isoaspartate or aspartate, depending on the pH of the solution. Insulin deamidates rapidly at Asn A-21 in low pH solutions. Steric hindrance also can affect deamidation rate; bulky residues following asparagine may inhibit the formation of succinimide intermediate in the deamidation reaction. Replacement of a glycine residue with a more bulky leucine or proline residues results in a 30–50-fold decreases in the rate. In lyophilized formulations, the deamidation rate is typically reduced, probably due to limited availability of free water in which the reaction can occur.

Formulations that incorporate organic cosolvents can decrease their deamidation rates because the addition of organic solvents decreases dielectric constants of a solution. Decreasing solvents' dielectric strength—by the addition of cosolutes such as glycerol, sucrose, and ethanol to a protein solution—leads to significantly lower rates of isomerization and deamidation. Lowering the dielectric strength of the medium from 80 (water) to 35 (PVP/glycerol/water formulations) led to about a sixfold decrease in peptide deamidation rates. The lower rate of deamidation was attributed to less stabilized ionic intermediates formed during cyclization in the asparagine deamidation pathway. Insulin prepared in neutral solutions containing phenol showed reduced deamidation probably because of its stabilizing effect on the tertiary structure (α-helix formation) around the deamidating residue, which lowered the probability for the formation of intermediate imides.

89

Hydrolysis

The mechanism of hydrolysis is similar to deamidation as in succinimide formation followed by racemization and isomerization. In the cycle of hydrolysis reactions, −x–Asp–y is considered labile, and the reaction in dilute acids is 100 times faster than other bonds. Asp–Pro is 8–20 times more susceptible to hydrolysis compared to other Asp–x or x–Asp bonds. The Asp–Gly is susceptible under highly acidic conditions of pH 0.3–3. The −x–Ser, −x=Ther cleave at N-terminal side faster than other peptide bonds.

Disulfide breakage

The disulfide bond in biopharmaceuticals forms a critical part of the three-dimensional structure that determines the activity and antigenicity of biopharmaceuticals. Breakage of these bonds as a result of a variety of chemical stresses can be significant. Figure 4.7 shows the disulfide bond holding the strands of the peptide.

Figure 4.8 shows the disulfide cleavage and scrambling mechanism. The disulfide reshuffling is faster than the formation or reduction of disulfides.

Glycation

Many proteins are glycosylated, and some have other posttranslational modifications such as phosphorylation, which also affects their potential degradation pathways as well as the kinetics of their degradation. However, reducing sugars also react through classic Maillard reaction with amino groups such as a ε-amino group of Lys and N-terminal α-amino groups, which are most susceptible. The impact of glycation is acidification (loss of positive charge), insolubility, cross-linking, and generation of chromophores. In a solution, a change in color may indicate an unanticipated glycation reaction. The latter is highly affected by pH, temperature, pK_a of amino acid groups, adjacent amino acid, the concentration of reducing sugars, and concentration of proteins in solution.

Diketopiperazine route   Diketopiperazine (DKP) is an intermediate state of degradation of certain amino acids (Figure 4.9).

Figure 4.7 Disulfide bond.

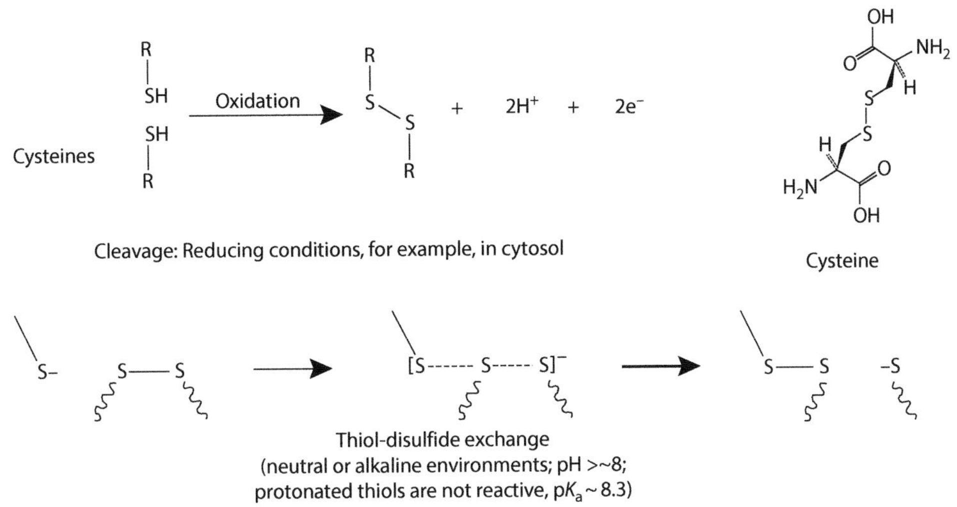

Figure 4.8 Disulfide cleavage and scrambling.

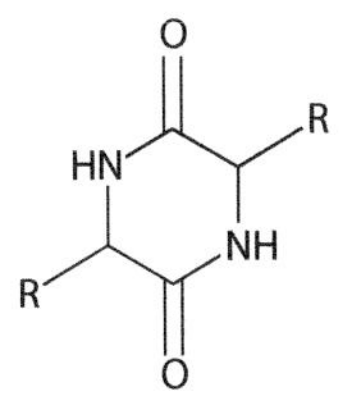

Figure 4.9 Diketopiperazine structure.

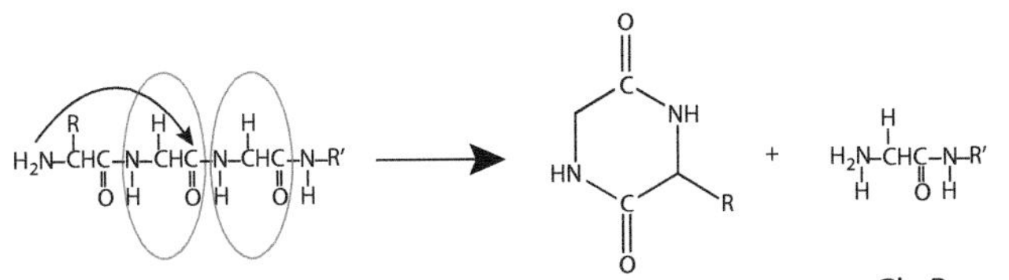

Figure 4.10 Diketopiperazine formation in the degradation of recombinant growth hormone.

This route of degradation is often seen with dipeptide esters and amides where a nucleophilic attack of an N-terminal nitrogen on the amide carbonyl between second and third residues takes place. Therefore, peptides containing Gly as the third residue, or Pro as the second residue from N-terminal are particularly susceptible; this reaction is favored in neutral or alkaline medium. The diketopiperazine from x–Pro is impacted by the *cis–trans* equilibrium of Pro, and dependent on charge distribution around peptide bond. The *cis* form facilitates ring closure (Figure 4.10). An example of diketopiperazine formation in recombinant growth hormone is shown in Figure 4.11.

Pyroglutamic acid (PyroE)　This is another intermediate form of chemical modification. The terminal Gln can spontaneously cyclize to form PyroE; Gln–Glyn reacts much faster. The N-terminal Glu cyclization results in charge change of −1 or loss of basic residue and mass change of −17 Da; the N-terminal Glu cyclization is less common and results in no change in charge but a mass change of −18 Da (Figure 4.12).

91

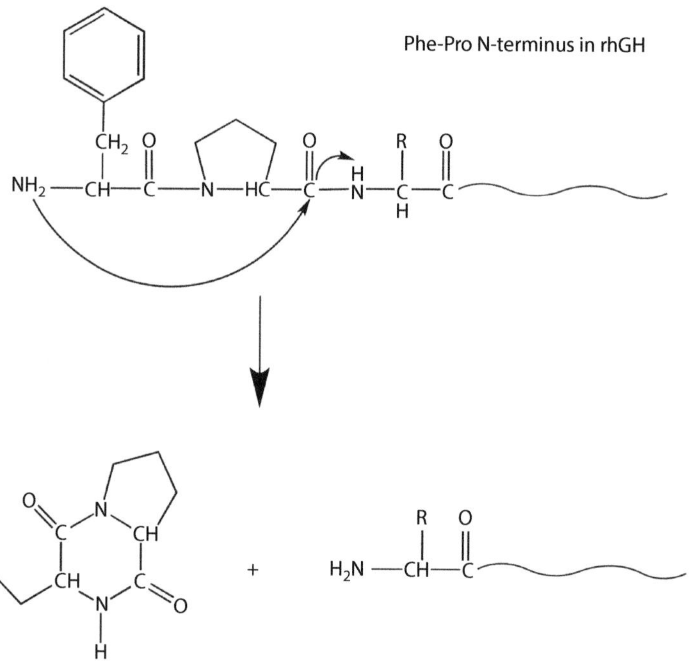

Figure 4.11 Diketopiperazine formation in recombinant growth hormone at the Phe–Pro N-terminus.

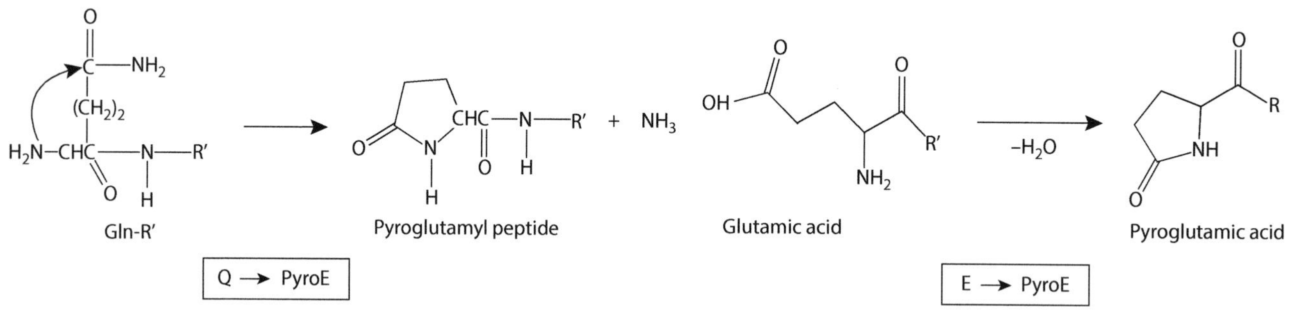

Figure 4.12 Cyclization to form PyroE.

Oxidation

Proteins and peptides are susceptible to oxidative damage through the reaction of certain amino acids with oxygen radicals present in their environment. Methionine (Met), cysteine (Cys), histidine (His), tryptophan (Trp), and tyrosine (Tyr) are most susceptible to oxidation—Met and Cys because of their sulfur atoms and His, Trp, and Tyr because of its aromatic ring. Oxidation can alter a protein's physiochemical characteristics (e.g., folding and subunit association) and lead to aggregation or fragmentation. It can also induce potential negative effects on potency and immunogenicity depending on the position of oxidized amino acids in a protein relative to its functional or epitope-like domain (Table 4.4).

Protein oxidation is a covalent modification induced by reactive oxygen intermediates or other oxidants such as chemicals hydrogen peroxide (which often appears in formulations as a contaminant of polysorbates, leaching from disposable tubing, etc.), oxygen, metal ions, and other excipients. Ultraviolet light is another major factor requiring proteins to be protected from light. Oxidation results in modification of Met, Cys, Trp, His, and Tyr residues, which are

Table 4.4 Key Degradation Products of Oxidation

Residues	Oxidation Products
Met	Met-sulfoxide
Cys	−S−S−disulfide cross-links and sulfenic acid/ sulfinic acid/sulfonic acid
His	2-oxoimidasoline and aspartate/asparagine
Trp	N-formylkynurenine and kynurenine
Tyr	Try−Try cross-link and 3−4-dihydroxyphenylalanine

Table 4.5 Source of Oxidation and Its Mechanism

Source	Contributing Element
Chemical reagents	H_2O_2, Fe^{2+}, Cu^{1+}, Glutathione, HOCl, HOBr, $^1O_2$, and $ONOO^-$
Activated phagocytes	Oxidative burst activity
γ-Irradiation in the presence of oxygen	Oxidation potential
UV light and Ozone	High energy radicals
Lipid peroxides	HNE, MDA, and acrolein
Mitochondria (electron transport chain leakage)	Oxidoreductase enzymes: Xanthine oxidase, Myeloperoxidase, and P-450 enzymes
Drugs and their metabolites	Free radicals and oxidative products

primary formulation development relevant residues. Table 4.5 lists some of the key catalysts for the oxidation of proteins.

The site-specific oxidation is catalyzed by transition metals from Fe_2 to Cu_2 and involves the generation of reactive oxygen and oxidation at metal-binding site. His, Pro, Arg, Lys, Thr, Tyr, Met, and Cys are most susceptible to site-specific oxidation.

The nonsite-specific oxidation is caused by contaminant oxidants such as hydrogen peroxide, HOCl, or by an oxidation radiation (e.g., light); this type makes all residues susceptible.

Oxidation can result in reduced potency depending on the site of oxidation; it can also cause conformational changes that lead to aggregation. It is generally not possible to extrapolate clinical safety from the understanding of the degradation pathways. In the case of parathyroid hormone, for example, the biological activity (in vitro bioassay) change depends on whether it undergoes a single oxidation of either Met-8 or Met-18 or the double oxidation (Met-8 with Met-18). Similarly, oxidation of Met-36 and Met-48 in human stem cell factor (huSCF) derived from *Escherichia coli* reduces the potency by 40% and 60% (respectively) while increasing the dissociation rate constant of SCF dimer by twofold to threefold; this suggests changes in the subunit's binding and its tertiary structure. In some instances, oxidation may not have any measurable impact on protein potency even when substantial structural changes were seen such as the case of oxidation of Met-111 in interferon α-2b; this oxidation changes the molecule's primary, secondary, and tertiary structure and prevents site-specific epitope recognition by monoclonal antibody (MAb); yet, the vitro biological activity remains unchanged. These examples illustrate the need to fully evaluate any degradation changes on the activity of the molecule.

The oxidative modification depends on intrinsic structural features such as buried and exposed amino acids. In the case of human growth hormone, Met-14 and Met-125 are readily oxidized by H2O2 because they are exposed to the surface of the protein, whereas Met-170 in its buried position can be oxidized only when the molecule is unfolded.

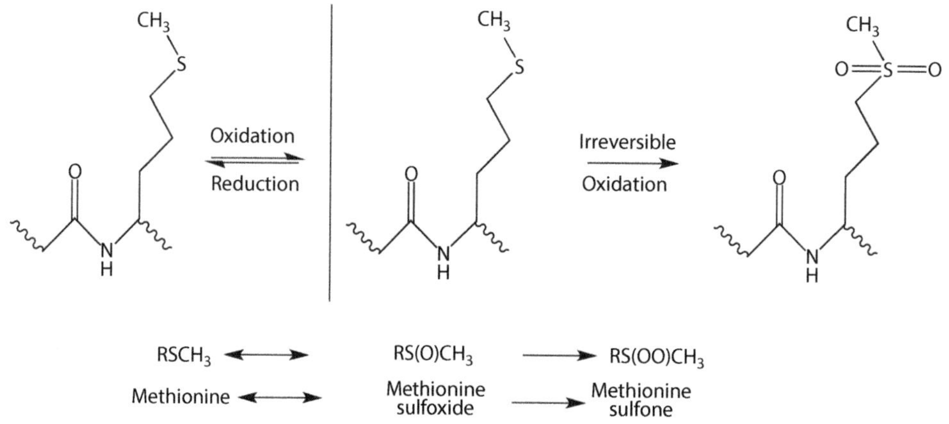

Figure 4.13 Methionine residues in oxidation.

Oxidation of methionine Methionine is highly susceptible to oxidation by atmospheric oxygen and oxygen radicals (free radicals, singlet oxygen, and nucleophilic substitution) in solution to form methionine sulfoxide and methionine sulfone (Figure 4.13). Both species are larger and more polar than nonoxidized methionine, which can alter protein folding and structural stability. The rate of methionine oxidation in recombinant human parathyroid hormone (rHu-PTH) by hydrogen peroxide is enhanced at alkaline pH. However, methionine oxidation generally is acid catalyzed and shows a slight increase in rate with a decreasing pH.

Oxidation of cysteine Cysteine oxidation is also more prevalent at alkaline pH, which deprotonates thiol groups. Cysteine oxidation forms inter- or intramolecular disulfide bonds and further oxidation can yield sulfenic acid. The oxidation of cysteine is strongly catalyzed by transition metals at higher pH (optimum around 6); however, this oxidation reaction is less sensitive to mild oxidants such as hydrogen peroxide compared to Met or Cys. In the absence of a nearby thiol for disulfide formation, oxidation can yield sulfenic, sulfinic, and sulfonic acids (Figure 4.14). In a reducing environment, cysteine oxidation causes a nucleophilic attack by thiolate ions on disulfide bonds, generating new disulfide bonds and different thiolate ions. The new thiolate can then react with another disulfide bond to form cysteine.

Such intermolecular disulfide links formed by protein degradation accumulated mispaired disulfide bonds and scrambled disulfide bridges, which can alter protein conformation and subunit associations. Cysteine residues may also undergo spontaneous oxidation to form molecular byproducts—sulfinic acid and cysteic acid—in the presence of metal ions or nearby thiol groups (Figure 4.14). For example, human fibroblast growth factor (FGF-1) exhibits copper-catalyzed oxidation that can create homodimers.

$$RS^- + M^{n+} \rightarrow RS^\bullet + M^{(n-1)+}$$
$$R'SH + RS^\bullet + O_2 \rightarrow \textbf{R'S-SR} + H^+ + O_2{}^{\bullet-} \text{ (Cysteine disulfide)}$$
$$RS^\bullet + O_2 \rightarrow RSOO^\bullet$$
$$RSOO^\bullet + RSH \rightarrow RSO^\bullet + \textbf{RSOH} \text{ (Sulfenic acid)}$$
$$RSOH + R'SH \rightarrow \textbf{RS-SR'} + H_2O$$
$$RSH + O_2{}^{\bullet-} \rightarrow RS^\bullet + HO_2^- \qquad RS^\bullet + R'S^\bullet \rightarrow \textbf{RS-SR'}$$
$$RSH + R'S^- + H_2O_2 \rightarrow \textbf{RS-SR'} + OH^- + HO^\bullet$$
$$RSH \rightarrow RSOH \rightarrow RSO_2H \rightarrow RSO_3H$$
$$\text{Thiol Sulfenic} \qquad \text{Sulfinic} \qquad \text{Sulfonic}$$

Figure 4.14 Cysteine residues in oxidation.

The spatial orientation of thiol groups in proteins plays an important role in cysteine oxidation. The rate of oxidation is inversely proportional to the distance between those thiol groups. This can eventually lead to the formation of large oligomers or nonfunctional monomers, as with basic fibroblast growth factor (bFGF), which contains three cysteines that are easily oxidized and form intermolecular or intramolecular disulfide bonds. This oxidation often induces conformational modifications of the protein because cysteine disulfide bonds increase side-chain volume in the protein's interior and leads to unfavorable van der Waals interactions.

Oxidation of histidine   Histidine residues are highly sensitive to oxidation through reaction with their imidazole rings (metal or photocatalyzed oxidation), which can subsequently generate additional hydroxyl species (Figure 4.15). Oxidized histidine can yield asparagine/aspartate and 2-oxo-histidine (2-O-His) as degradation products during light and/or metal oxidation. It may be a transient moiety because it can trigger protein aggregation and precipitation, which can obscure isolation of 2-O-His as individual degradants. Photooxidation in the presence of photosensitizers via singlet oxygen is also common. The oxidation propensity and rate are impacted by residues in proximity and by amino acid conformation and may lead to colored solutions.

Oxidation of tryptophan   Tryptophan is also susceptible to oxidation by reactive oxygen species as well as by light; even though the abundance of tryptophan is low, it still has the highest molar absorption coefficient (280–305 nm), which makes it more open to oxidation by light. The oxidation of tryptophan is impacted by sequence position, exposure to solvents and oxygen; it may result in colored solutions (Figures 4.16 and 4.17).

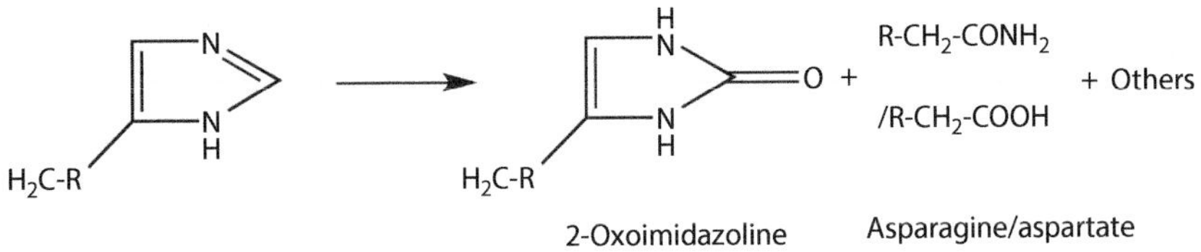

Figure 4.15 Histidine residues in oxidation.

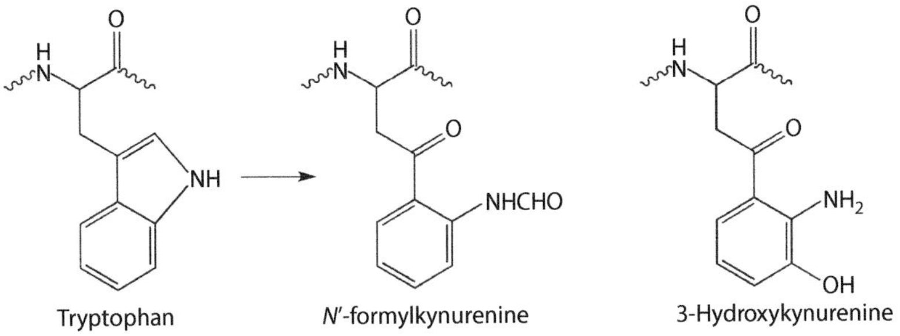

Figure 4.16 Tryptophan oxidation residues.

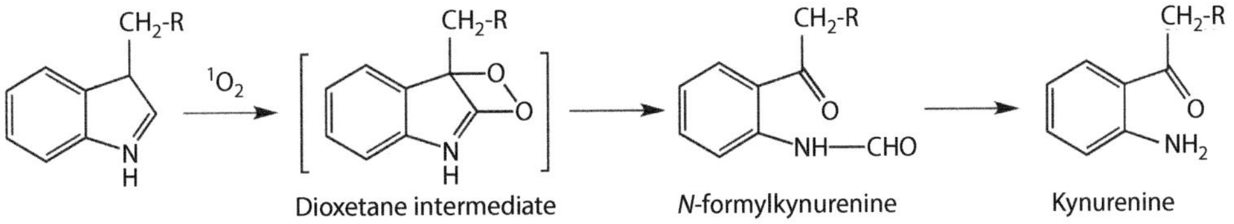

Figure 4.17 Tryptophan photooxidation residues.

Oxidation of tyrosine Oxidation of tyrosine, mainly due to photooxidation, may result in covalent aggregation through formation of bityrosine. Spatial factors may also affect tyrosine and histidine oxidation. Adjacent negatively charged amino acids accelerate tyrosine oxidation because they have a high affinity to metal ions, whereas positively charged amino acid residues disfavor the reaction. If an adjacent amino acid is bulky, it may mask oxidation of neighboring amino acids and even prevent them from getting oxidized. It has been observed that histidine present in a sequence markedly increases both the peptide oxidation rate and the methionine sulfoxide production. The strong metal binding affinity of the imidazole ring on the histidine side chain brings oxidizing species close to the substrate methionine. The products of tyrosine oxidation include DOPA or Di-tyrosine (cross-linking reaction, Figure 4.18); dityrosine has a unique fluorescence signature (410–415 nm excitation). During photooxidation, Tyr can transfer energy to disulfides and to Trp resulting in Tyr radical cations (Tyr+), disulfide-based radical (RSSR+), or triplet-state Trp (^{3}Trp).

Oxidation of phenylalanine Phenylalanine is also subject to photooxidation resulting in ortho- and metatyrosine derivatives (Figure 4.19).

Given the significance of the impact of oxidation reactions on proteins, the formulators of biosimilars need to be particularly aware of these reactions as they can affect the stability and shelf life, produce colors, cause aggregation, and loss of drug activity. A summary of oxidation-related effects is provided as

1. General factors
   a. Cys, Met, and His are highly susceptible to metal-catalyzed oxidation.
   b. Trp and His are the main targets of photo-induced oxidation.

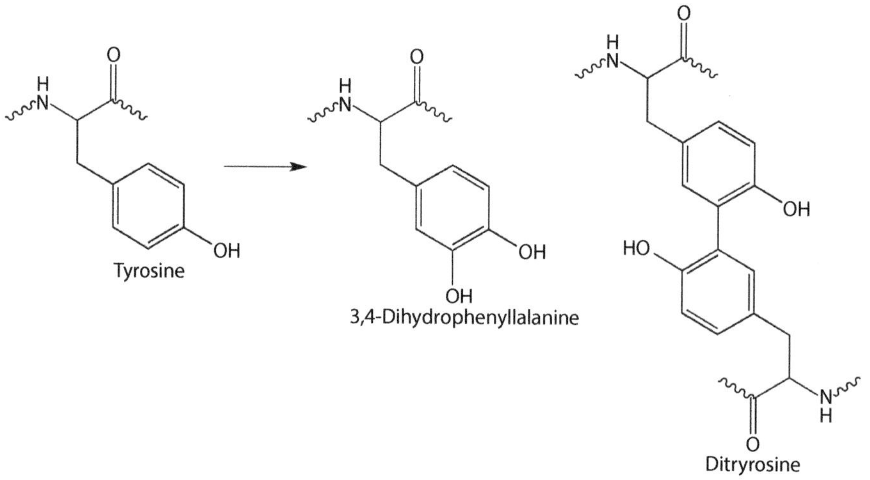

Figure 4.18 Tyrosine residues in oxidation.

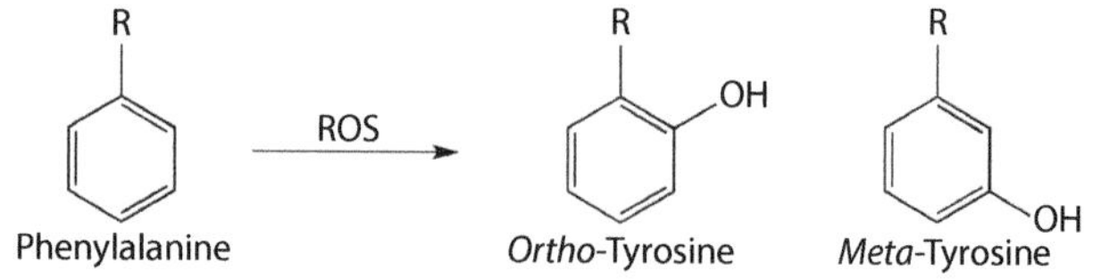

Figure 4.19 Oxidation residues of phenylalanine.

    c. Rates of photooxidation follow the order: His > Trp > Met > Tyr.

    d. Specific oxidation mechanisms are influenced by

       i. An intrinsic factor: Structure

          1. Buried residues are less easily oxidized by nonsite-specific mechanisms (e.g., H_2O_2 and Light).

          2. Site-specific MCO are less impacted by steric effects and more by proximity of the residue to the metal-binding site.

       ii. Exogenous factors: pH, temperature, buffer, and excipients

          1. *pH*: oxidation potential of system increases with pH; metal binding improves due to ionization of functional groups (but solubility of transition metals may decrease) with increasing pH; ionization of side chains (Cys, His) enhances lability with increasing pH; at pH < 4, only Met, and Trp can be readily photooxidized.

          2. *Temperature*: It impacts both secondary and tertiary structures as well as causes potential pH shifts; reduction of the rate with temperature is not often significant due to increased aqueous solubility of oxygen at lower temperatures (1.29e−3 mmol/mL at 25°C; 2.18e−3 mmol/mL at 0°C); change in temperature may also change rate-controlling step in mechanism.

          3. *Buffer species*: The major effect of buffer species comes through contaminants, metal chelating ability, and possible interaction with reactive oxygen species. Tris is not only a Fe and Cu chelator but also a hydroxyl radical scavenger; citrate is a chelator; phosphoric acid (weak) and tartaric acid are also considered to be buffer species. Fenton chemistry requires a bicarbonate ion, but there are no clear generalizations.

          4. *Excipients*: The major effect of common excipient species comes through contaminants such as peroxides (e.g., in polysorbates) and trace metals, especially Fe and Cu. High concentrations of sugars and polyols seem to inhibit metal-catalyzed oxidation (perhaps through a metal complexation mechanism).

2. *Chelators*: Chelators must be used with caution; strong 1:1 chelators (e.g., EDTA, desferrioxamine, and nitrilotriacetic acid) can impact oxidation rates based on stoichiometry (chelator/Fe ratio); oxidation rate increases until all Fe is sequestered at a max ratio of 1; an almost complete inhibition of the protein is exhibited at ratio 1.1–1.2. Weak chelators (e.g., *o*-phenanthroline) require large excess (~5–20) for inhibition effects. Site-specific oxidation is impaired by chelate–Fe(III) complexes, but nonsite-specific oxidation is enhanced. Chelates extract iron from potential binding sites and chelate–Fe(III) complex promotes the formation of ROS different than those at metal-binding sites. The overall impact on charge heterogeneity in proteins is described as follows:

    a. Deamidation (more acidic) ($z = -1$)

    b. Succinimide (more basic or neutral)

    c. Glycation (more acidic)

    d. Pyroglutamate formation (more acidic) ($z = -1$)

    e. Peptide bond hydrolysis (either acidic or basic)

Photooxidation   Photooxidation can change the primary, secondary, and tertiary structures of proteins and can lead to differences in long-term stability, bioactivity, or immunogenicity. Exposure to light can trigger a chain of biochemical events that continue to affect a protein even after the light source is turned off. These effects depend on the amount of energy imparted to a protein and the presence of environmental oxygen. Photooxidation is initiated when a compound absorbs a certain wavelength of light, which provides enough energy to raise the molecule to an excited state. The excited molecule can then transfer that energy to molecular oxygen, converting it to reactive singlet oxygen atoms. This is how tryptophan, histidine, and tyrosine can be modified by light in the presence of oxygen. Tyrosine photooxidation can produce mono-, di-, tri-, and tetrahydroxy tyrosine as byproducts. Aggregation is observed in some proteins due to cross-linking between oxidized tyrosine residues. Photooxidation reaction is predominately site specific. For example, in human growth hormone treated with intense light, oxidation is carried out predominantly at histidine-21. In addition, the peptide backbone is also a photodegradation target. Alternatively, the energized protein itself can react directly with another protein molecule in a photosensitized manner, typically via methionine and tryptophan residues at low pH. Certain excipients and leachables can synergistically affect the oxidation and hence degradation of protein. These formulation components influence the rate of photooxidation in some instances; for example, phosphate buffer accelerates the rate of methionine degradation more than other buffer systems do. Also, the presence of denaturing/unfolding reagents in solution can increase the extent of protein oxidation, while excipients such as polyols and sugars involved in stabilizing protein structure can decrease the rate of oxidation. Oxidation can also be exacerbated by the presence of a reducing agent such as ascorbate; ascorbic acid increases oxidation of a human ciliary neurotrophic factor.

Metal ion-catalyzed oxidation   Metal ion-catalyzed oxidation depends on the concentration of metal ions in the environment. The presence of 0.15 ppm chloride salts of Fe^{3+}, Ca^{2+}, Cu^{2+}, Mg^{2+}, or Zn^{2+} does not affect the rate of oxidation for human insulin-like growth factor-1, but when the metal concentration is increased to 1 ppm, a significant increase in oxidation is observed.

Oxidation by peroxides   Oxidation can be induced during protein processing and storage by peroxide contamination resulting from polysorbates and polyethylene glycols (PEGs), both commonly used as pharmaceutical excipients. A correlation has been observed between the peroxide content in Tween-80 and the degree of oxidation in rhG-CSF, and peroxide-induced oxidation appeared more severe than that from atmospheric oxygen. Peroxide can also leach from plastic or elastomeric materials used in primary packaging container–closure systems, including pre-filled syringes.

Fragmentation

Multimeric proteins with two or more subunits can become dissociated into monomers, and monomers (or single peptide chain proteins) can degrade into peptide fragments. Nonenzymatic fragmentation usually proceeds by hydrolysis of peptide bonds by amino acids, releasing polypeptides of lower molecular weight than the intact parent protein. Peptide bonds of Asp–Gly and Asp–Pro is most susceptible to hydrolytic protein cleavage. Antibody hydrolysis often occurs in the hinge region, which is the most flexible domain of an antibody. However, decreasing pH from 9 to 5 can shift the peptide hydrolysis sites of a recombinant monoclonal antibody, showing increased cleavage outside that region.

The presence and position of oligosaccharides also affect the rate of peptide hydrolysis at low pH levels. Depending on location, hinge-region cleavage is not affected although fragmentation in the CH_2 domain decreases. Hydrolytic cleavage of peptide bonds by acidic and basic hydrolysis does not necessarily have the same effects. Recombinant human macrophage colony-stimulating factor yields different peptide fragments in solutions with acidic and basic pH. Enzymatic protein fragmentation can be caused by the proteolytic activity of residual or contaminating proteases—or in select cases, autoproteolysis of an enzymatic protein.

Formulation of biosimilar products

While the formulation of biopharmaceuticals is a vast field of study with two peculiarities when compared to small molecule drugs. First, most proteins are administered by parenteral routes, and, as a result, most of the science of protein drug formulation deals with the art of injectable formulations. Second, there are several common structural features of all proteins, such as functional groups like methionine, cysteine, histidine, tryptophan, and tyrosine, all of which are subject to oxidation requiring some common approaches to establishing stable products. On the other hand, conformational changes and aggregation are properties peculiar to large molecules requiring the inclusion of formulation components that can be highly specific.

However, as required by law, the exact composition of all injectable products must be disclosed on the label, making it easier for the biosimilar product developer to know with certainty the composition of formulation.

Physical degradation

Proteins degrade upon physical stresses of many types including hydrophobic surfaces, heating, lyophilization, reconstitution, contact with organic solvents, shaking, and many other permutations and combinations of physical and chemical factors. The final result of physical stress can be denaturation, adsorption on the container walls, precipitation, or aggregation.

Summary of degradation hot spots

Table 4.6 shows reactive peptide sequences and formation stability, while Table 4.7 lists various peptide sequences hot spots; this demonstrates the varied nature of instability for proteins wherein many such hot spots may be present.

Aggregation

Aggregation is a common problem encountered during manufacture and storage of proteins. The potential for aggregated forms is often enhanced by exposure of a protein to liquid–air, liquid–solid, and even liquid–liquid interfaces. Mechanical stresses of agitation (shaking, stirring, pipetting, or pumping through tubes) can cause protein aggregation. Freezing and thawing can promote it as well. Solution conditions, such as temperature, protein concentration, pH, and ionic strength, can affect the rate and amount of aggregates observed.

Table 4.6 Reactive Peptide Sequences and Formation Stability

Sequence[a]	Primary Reaction	Major Products	Est. Reactivity[b]
–X–X–	Typical amide bond hydrolysis	–X + X–	Slow
–Asp–X– (X ≠Pro)	Hydrolysis	–Asp + X– or –isoAsp + X–	Slow to moderate
–Asp–Pro–	Hydrolysis	–Asp + X– or –isoAsp + X–	**May compromise shelf life, fast**
–Asn–X– (X ≠Gly)	Deamidation	–Asp–X–, –isoAsp–X– Asp + X–, –isoAsp –asp, –iso-asp L or D cyclic imide	May compromise shelf life, moderate
–Asn–Gly–	Deamidation	–Asp–Gly–, –Asp, –iso-Asp–Gly–, –iso-Asp –asp, –iso-asp L or D cyclic imide	**May compromise shelf life, fast**
-X–Ser– or –X–Thr	Hydrolysis	–X + Ser– or –X + Thr–	Moderate
–Ser–, –Tyr–, –Phe–, –X– (≠ Gly, Asp)	Racemization	–ser–, –thr–, –phe– –x–	Slow to moderate
–Asp–	Racemization	–asp–, (–iso-asp–)	**May compromise shelf life, fast**
X–X–Gly–	Diketopiperazine formation	cyclo–X–X + Gly–	**May compromise shelf life, fast**
Gln–X–	Pyroglutamic acid formation	<Gln–X–	**May compromise shelf life, fast**
–Met–	Oxidation	–Met(O)–	May compromise shelf life, moderate
–Cys–	Oxidation, disulfide formation	–Cys Cys–	**May compromise shelf life, fast**
–X– (X = many hydrophobic AA's)	Precipitation	Aggregates, liquid crystals	May compromise shelf life, moderate

[a] X, any amino acid unless otherwise specified. Amino acids drawn as –X or X– denote the possibility of additional amino acids attached.

[b] Reactions, which are likely to compromise peptide stability ($t_{90} < 2$ yes at 25°C) at pH 5–7, are shown in bold, where other limiting reactions are adjusted for their propensity to limit shelf life accordingly.

Table 4.7 Protein Sequence Hot Spots for Degradation

Pathway	Site
Deamidation	Asn, Gln, –Asn–X–, –Asn–Gly–, Asn–His–, –Asn–Ser–, –Asn–Ala–, –Asn–Asp–, and –Asn–Thr–
Hydrolysis	Asp, –Asp–Pro–, –Asp–Gln–, –Pro–Asp, –Asp–Tyr–, –X–Ser–, and –X–Thr–
Oxidation	Met, Cys, His, Trp, and Tyr
Isomerization	Asn, Asp, –Asn–Gly–, –Asp–Ser–, –Asn–Ser–, and –Asp–Ser–
Diketopiperazine	X–X–Gly and X–Pro–
Glycation	Lys
Beta-elimination	Ser and Thr
Pyroglutamic acid	H_2N–Gln–, H_2N–Gln–Gly–, and H_2N–Glu–

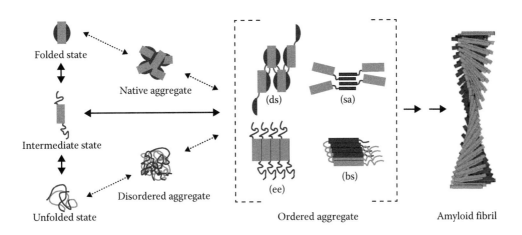

Folded state

Native aggregate

Intermediate state

(ds)   (sa)

(ee)

(bs)

Disordered aggregate

Unfolded state   Ordered aggregate   Amyloid fibril

Figure 4.20 Aggregation of proteins.

Aggregated proteins are a significant concern for biopharmaceutical products because they may be associated with decreased bioactivity and increased immunogenicity. Macromolecular protein complexes can trigger a patient's immune system to recognize the protein as "nonself" and mount an antigenic response. Large macromolecular aggregates also can affect fluid dynamics in organ systems such as eyes.

Protein aggregates are formed by mechanisms such as domain swapping (ds), strand association (sa), edge-edge-association (ee), or beta strand stacking (bs) as shown in Figure 4.20.

Aggregation and precipitation The term aggregation usually refers to multimers of proteins, for example, dimers, trimers, tetramers, and all the way to large polymers. The aggregates can be noncovalent or covalent (disulfide-linked) and these can be presented as fully soluble in a clear solution, partially insoluble in a turbid solution (usually extremely large), or mostly insoluble as a precipitate that collects in the bottom of the container. Nonspecific protein-to-protein association resulting from interactions among solvent exposed hydrophobic groups can also form aggregates. The covalent aggregation is not reversible. The weakly associated noncovalent aggregate can be reversible and it usually follows the path of dimers to multimers; the strongly associate noncovalent aggregates are not reversible by dilution and may result in precipitation.

Aggregates can be soluble or insoluble, reversible or irreversible, and covalent or noncovalent. Soluble aggregates are usually reversible, for example, by altered solution conditions such as changing temperature or osmotic strength or by mild physical disruption such as swirling or filtration. Insoluble aggregates are typically irreversible. Under vigorous physical disruption (e.g., agitation or freezing and thawing) or over time in storage, they can grow into particles that may eventually precipitate. Covalent aggregates form when monomeric proteins become chemically cross-linked, for example, through disulfide bonds. Although covalent linkages are necessary to stabilize the native tertiary structure of most polypeptide proteins, those that form by degradation can produce undesired cross-links between protein moieties, which can lead to irreversible aggregation. Noncovalent aggregates are formed when proteins associate and bind based on structural regions of charge or polarity. Because such associates are weak (relative to covalent linkages), they are sensitive to solution conditions, and usually reversible.

There are several models of aggregation. In the "Native to Unfolded to Aggregate" model, denatured or unfolded molecules aggregate due to hydrophobic interactions.

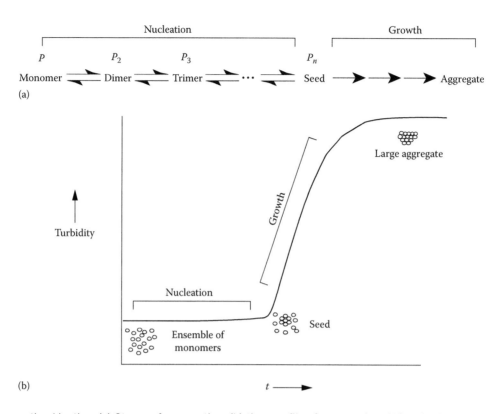

Figure 4.21 Aggregation kinetics. (a) Stages of aggregation; (b) time profile of aggregation. (After Gazit, E., *Angew. Chem. Intl. Ed.*, 41, 257, 2002.)

Aggregate formation driven by hydrophobic effect happens when the normally buried hydrophobic regions are exposed. The rate of reaction in this model increases with temperature since unfolding increases with temperature and reactions generally follow first-order kinetics.

In the model "Native to Intermediate to Unfolded" wherein the intermediate yields the aggregate, the misfolded intermediates are often thermodynamically stable and can be part of the native state ensemble. Therefore, aggregation is not an "unnatural" state for a protein and can occur even under conditions favoring native state.

The mechanism of protein aggregation involves two stages. A nucleation process, which is followed by growth of the nuclei in a critical mass. The level of aggregation can be monitored by the turbidity measurement. However, turbidity is not necessarily an indicator of aggregation (Figure 4.21).

The assembly of initially native and folded proteins can result in irreversible nonnative structures that may contain high levels of nonnative intermolecular β-sheet structures. The onset, rate, and final morphology of the aggregate depend on solution conditions such as pH, salt species, salt concentration, cosolutes, excipients, surfactants, etc. The exact nature of an aggregate is a function of relative intrinsic thermodynamic stability of the native state.

The causes of aggregation that are related to the process may include many factors as follows:

- *Fermentation/Expression*: Inclusion bodies
- *Purification*: Shear, pH, and ionic strength
- *Filtration*: Surface interaction and shear
- *Fill/Finish*: Surface interaction, shear, and contamination (e.g., silicone oil)

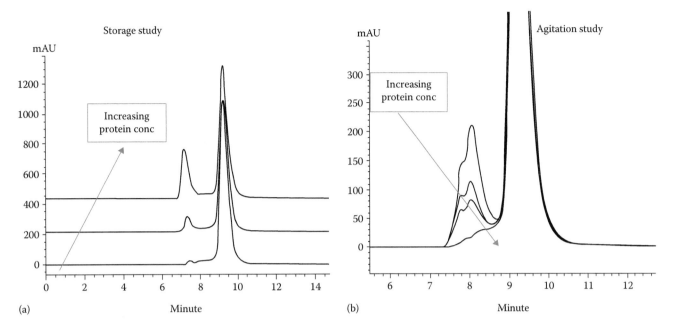

Figure 4.22 Effect of increasing protein concentration on aggregation during storage and upon agitation for PEG-GCSF. (After Treuheit M.J. et al., *Pharm. Res.*, 19, 511, 2002.)

- *Freeze/Thaw*: Cryoconcentration, pH changes, and ice–solution interfaces
- *Shipping*: Agitation and temperature cycling
- *Lyophilization*: Cryoconcentration, pH changes, ice–solution interfaces, and dehydration
- *Administration*: Diluents, component materials and surfaces, and leachables

Protein aggregation upon oxidation can result in a fast reaction and may show precipitates within a few minutes; it is also highly pH dependent.

One of the most difficult part of formulating proteins comes when they are dispensed in prefilled syringes that have silicone oil lubrication. A 10-fold variability in absorbance at 350 nm in proteins packaged in various brands of syringes is not uncommon. It depends on the quantity and the type of silicone used. A properly folded mAb, for example, can form an intermediate upon hydrophobic interaction with silicone in the syringe resulting in partial unfolding that will yield first to a soluble aggregate followed by a visible precipitation depending on the concentration of silicone to which the mAb is exposed.

Formulation in sucrose can also increase aggregation over time because of protein glycation when sucrose is hydrolyzed. The presence of certain ligands—including certain ions—may enhance aggregation. Interactions with metal surfaces can lead to epitaxic denaturation, which triggers aggregate formation. Foreign particles from the environment, manufacturing process, or container–closure system (e.g., silicone oil) can also induce aggregation. Even handling protein products at compounding pharmacies can induce aggregation 10-fold above initially observed amounts.

The difficulties in stabilizing proteins against aggregation arise from the complex behavior of proteins, most of it not predictable. For example, Figure 4.22 shows the results of storage and agitation effects on the aggregation of PEG-GCSF. During storage, the aggregation increased with increasing concentration but a reverse effect was noted in agitation studies.

103

The impact of aggregation on product potency varies based on the physiochemical attributes of each protein relative to its functional domains and the nature of the activity being measured. Enzymes such as urease and catalase can lose up to 50% of their potency after shaking; fibrinogen clotting activity is decreased after shear stress; and recombinant IL-2 and recombinant interferon activity is substantially affected by aggregation from shaking and shearing. Aggregation also affects the mass balance of protein solutions, decreasing the concentration of the target protein. Microaggregated subvisible particles generated anywhere in a manufacturing process can develop into larger particles over time as a product is stored. The Bevacizumab drug product, for example, may lose 50% of active IgG after manipulations at the pharmacy triggered the significant growth of micron-sized particles in repackaged solutions.

Mechanism and factors involved: Because of the many physical and chemical manipulations required in upstream production and downstream processing, followed by formulation and filling operations, aggregation of protein biopharmaceuticals can be induced during nearly every step of the process including at hold points, shipping, and long-term storage.

Agitation (e.g., shaking, stirring, and shearing) of protein solutions can promote aggregation at the air–liquid interfaces, where protein molecules may align and unfold, exposing their hydrophobic regions for the charge-based association. Agitation-induced aggregation has been observed in numerous protein products, including recombinant factor XIII, human growth hormone, hemoglobin, and insulin. Minimizing foaming caused by agitation during manufacture (as well as during product use) may be critical to prevent significant loss of protein activity or generation of visible particulate matter.

Protein concentration also can promote aggregation, with or without agitation events. Results obtained from observing two PEGylated proteins and one Fc fusion protein demonstrated a direct correlation between protein concentration and aggregation under nonagitated (quiescent) conditions, but researchers found an inverse correlation between protein concentration and aggregation under conditions of shaking, vortexing, and simulated shipping.

Antimicrobial preservatives used in multidose formulations also can induce protein aggregation. For example, benzyl alcohol accelerates the aggregation of rhGCSF because it favors partially unfolded conformations of the protein. Increasing antimicrobial preservative levels may increase the hydrophobicity of a formulation and could affect a protein's aqueous solubility. Phenol and *m*-cresol can considerably destabilize a protein: Phenol promotes formation of both soluble and insoluble aggregates, whereas *m*-cresol can precipitate protein.

Freezing and thawing—which can occur multiple times throughout production and use of protein therapeutics—can dramatically affect protein aggregation. Generation of water ice crystals at a container's periphery (where heat transfer is greatest) can produce a "salting out" effect, whereby the protein and excipients become increasingly concentrated at the slower-freezing center of a container.

High-salt and/or high-protein concentrations can result in precipitation and aggregation during freezing, which is not completely reversible upon thawing. The effect can be seen with thyroid-stimulating hormone; when stored at −80°C, 4°C, or 24°C for up to 90 days, it remained stable, but when frozen to −20°C it lost >40% potency in that period, which was attributed to subunit dissociation. Multiple freezing and thawing cycles can exacerbate that effect and lead to a cumulative impact on the generation and growth of subvisible and visible particulates. A change in pH can come from crystallization of buffer components during freezing. In one study, potassium phosphate buffers demonstrated a much smaller pH change on freezing than sodium phosphate buffers did.

Compendia currently limit the number of particles ≥10 and ≥25 μm in size that may be present in injectable pharmaceutical preparations. However, the levels of subvisible particles (<10 μm) such as protein oligomers that are acceptable depend on the physiochemical characteristics and safety/efficacy profile of each product. Also, no standards are codified for visible particulates in protein pharmaceutics. Some biotechnology products have visual appearance specifications for drug solutions that include comments such as "essentially free of visible particulates" or "some translucent particles may be present."

Common formulation elements

An analysis of common components of biological drugs formulations shows that most ingredients are used as

1. Buffering agents to assure that the pH is maintained at the most stable level, and these include phosphates, citrates, and acetates
2. Stabilizers such as surfactants and sugars, and these contain polysorbates, albumin, mannitol, sucrose, and sorbitol
3. Tonicity and conductivity adjusting ingredients including some sugars and electrolytes like sodium chloride

Of the approximately 200 biological injectable drugs, the percentage of most prevalent components is given in Figure 4.23.

The singular most significant aspect of protein drug formulations is related to the stability of these drugs that are highly subject to degradation resulting in issues with their safety and efficacy. Like other drugs, protein therapeutics must meet defined quality characteristics immediately after manufacture as well at the end of their designated shelf lives. The science of stabilizing drugs through formulation efforts is well studied and understood, but the uncertainties arise when formulating large protein molecules for there is no definite way to predict the behavior of these molecules. A case in point is the change in the formulation of erythropoietin. A change from albumin stabilizer to polysorbate, which would have otherwise been a logical choice, resulted in PRCA and deaths because the surfactants caused the leaching out from rubber stoppers that produced a structural change in the structure of erythropoietin that triggered the toxic response. It is for this reason that the developer of biosimilar products too has to fully understand the dynamics of protein drug degradation.

Preventive measures against oxidation

One molecular engineering strategy for minimizing oxidative degradation is to replace oxygen-labile amino acids with oxygen-resistant ones if a protein's nature permits. In therapeutic Interferon beta (IFN-β), cysteine at position 17 was replaced by serine, because the former loses antiviral activity during storage to oxidation and disulfide scrambling. Substitution of methionine of epidermal growth factor (EGF) with a nonnaturally occurring norleucine also prevented oxidative degradation.

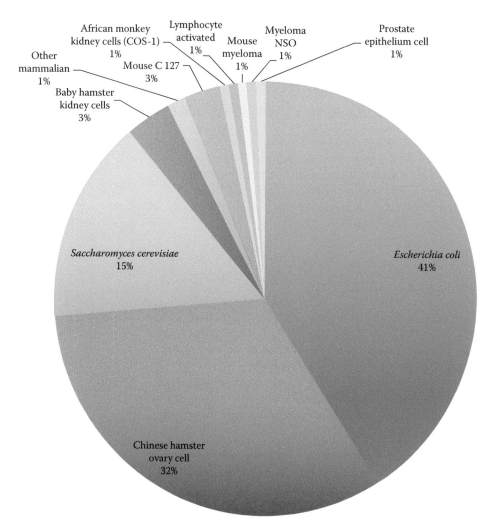

Figure 4.23 Percentage occurrence of common formulation components in biological drugs.

Removal of headspace oxygen by degassing may be effective for preventing oxidation in some cases. Filling steps are carried out under nitrogen pressure, and vial headspace oxygen is replaced with an inert gas such as nitrogen to prevent oxidation. With some oxidation-sensitive proteins, processing is carried out in the presence of an inert gas such as nitrogen or argon. For multidose drug preparations, use of cartridges with negligible headspace overcomes oxidation and related consequences.

Care must be exercised when container–closure changes are considered. Many such changes for protein therapeutics (e.g., from vials to prefilled syringes or prefilled syringes to pen devices) are considered to enhance patient convenience and ease of use. But historical experience with container–closure systems based only on chemical pharmaceuticals should be reevaluated when the same materials are used with protein-based products because of the potential for unexpected, unique impacts on protein degradation.

Controlling or enhancing factors such as pH, temperature, light exposure, and buffer composition can also mitigate the effects of oxidation by affecting a protein's environment. Cysteine oxidation often can be controlled by maintaining the correct redox potential of a protein formulation, such as with the addition of thioredoxin and glutathione. Antioxidants and metal chelating agents also can be used to prevent oxidation in protein formulations. Antioxidants are chemical "sacrificial targets" with a strong tendency to oxidize, consuming chemical species that promote oxidation. Scavengers such as L-methionine and ascorbic acid are used for this purpose in biotherapeutic formulations. In the absence of metal ions, free cysteine residues may act as an effective antioxidant. As chelating agents, EDTA and citrate might form complexes with transition metal ions and inhibit metal-catalyzed, site-specific oxidation. Addition of sugars and polyols may also prevent metal-catalyzed oxidation because of their complexion with the metal ions. Protective effects of glucose, mannitol, glycerol, and ethylene glycol against metal-catalyzed oxidation has been observed with human relaxin. Physical protection from UV/white light exposure with either a primary or secondary packaging system may be necessary to protect light-labile proteins from photooxidation.

Preventive measures against aggregation

A protein solution typically can be stabilized against aggregation and precipitation by optimizing solution pH and ionic strength; adding sugars, amino acids, and/or polyols; and using surfactants. Comprehensive evaluation of optimal pH and osmotic conditions is a key element of formulation development to prevent protein aggregation or precipitation. Irreversible aggregation due to denaturation can be prevented with surfactants, polyols, or sugars.

In many cases, nonionic detergents (surfactants) are added to increase stability and to prevent aggregation. The protein–surfactant interaction is hydrophobic, so these compounds stabilize proteins by lowering the surface tension of their solution and binding to hydrophobic sites on their surfaces, thus reducing the possibility of protein–protein interactions that could lead to aggregate formation. The nonionic detergents Tween 20 and Tween 80 can prevent formation of soluble protein aggregates with surfactant concentrations below the critical micelle concentrations (CMC) (73). Polysorbate (Tween) 80 added to IgG solutions stabilizes small aggregates and prevents them from growing into larger particles. Chelating agents also can be used to prevent metal-induced protein aggregation.

Preventive measures against fragmentation

Appropriately buffering formulations to maintain their solution pH in a suitable range for each protein type is key to minimizing hydrolytic fragmentation. For example, calcitonin undergoes hydrolysis at basic pH, but no such degradation is observed at pH 7 even at room temperature. Buffer composition may also affect hydrolysis. Recombinant human macrophage colony-stimulating factor fragmentation was observed in phosphate-buffered solutions but not in histidine-buffered solutions at identical pH and ionic strengths. It is also important to minimize the potential presence of proteases in protein purification, either from sources intrinsic to the production process (e.g., host-cell proteins) or from extrinsic sources of contamination (e.g., adventitious microbes).

The native structure of a protein molecule is the result of balancing effects such as covalent linkages, hydrophobic interactions, electrostatic interactions, hydrogen bonding, and van der Waals forces. Protein stability is controlled by innumerable intrinsic and extrinsic factors, but the major ones are primary sequence, 3D structure, subunit associations, and posttranslational modifications. Extrinsic contributing factors include pH, osmolarity, protein concentration, formulation excipients, and exposure of a product to physical stress from temperature, light, and/or agitation. Leachables from container–closure systems and contamination from the environment (e.g., metals and proteases) also exacerbate product degradation. Taken together, all these features make protein degradation a very complex physiochemical phenomenon, so formulation optimization is a key aspect of biotechnology product development.

High concentration formulations

Whereas most formulations follow the general composition formulas, and it is generally difficult to innovate the formulation of these drugs, the race to protect the falling patents has resulted in a flurry of patented formulation compositions. For example, a recent patent formulates adalimumab only in water (HUMIRA®, U.S. Patent No. 8,420,081) and the composition patent for subcutaneous formulation for rituximab (US20110076273 A1) and trastuzumab (WO2011012637 A2) have recently been published.

A common tactic for pharmaceutical companies facing patent expirations is to try to extend the lives of blockbuster drugs by developing new formulations or follow-on products that offer some sort of advantage over the original. Roche is pursuing that game plan for some of its hit biotech drugs—and on March 28, 2014, it scored a victory when EMA approved a subcutaneous version of its cancer drug Rituxan (rituximab), known as MabThera overseas. The new formulation, which was approved to treat some patients with non-Hodgkin lymphoma, cuts the treatment time from 2½ h for the original intravenous version to 5 min.

Rituxan/MabThera has been on the market for 15 years and is still a major growth driver for Roche. It was one of the world's 10 best selling drugs last year, hauling in $7.5 billion worldwide. But the patent on the drug expires in 2018, and rivals hoping to market biosimilar versions are looming. Novartis' Sandoz unit and Boehringer Ingelheim are both in Phase III trials of their rituximab products and suddenly these formulations appear obsolete. Whereas both formulations may have the same clinical effectiveness, the convenience of administering rituximab in subcutaneous formulation will outweigh any advantages the intravenous biosimilar product may offer. Roche has adopted a multipronged defensive strategy against biosimilar competition, and developing more convenient formulations of its drugs is part of that plan. Last fall, the company won European approval for a subcutaneous version of its breast cancer drug Herceptin (trastuzumab).

The details of the formulations of these two Mabs show a multifaceted attempts.

The claim for trastuzumab shows a highly concentrated, stable pharmaceutical formulation of a pharmaceutically active anti-HER2 antibody comprising

1. About 50–350 mg/mL anti-HER2 antibody
2. About 1–100 mM of a buffering agent providing a pH of 5.5 ± 2.0
3. About 1–500 mM of a stabilizer or a mixture of two or more stabilizers
4. About 0.01%–0.08% of a nonionic surfactant
5. An effective amount of at least one hyaluronidase enzyme

The patent claim for rituximab shows a highly concentrated, stable pharmaceutical formulation of a pharmaceutically active anti-CD20 antibody comprising

1. About 50–350 mg/mL anti-CD20 antibody
2. About 1–100 mM of a buffering agent providing a pH of 5.5 ± 2.0
3. About 1–500 mM of a stabilizer or a mixture of two or more stabilizers
4. About 0.01%–0.1% of a nonionic surfactant
5. Optionally an effective amount of at least one hyaluronidase enzyme

Formulation of concentrated solutions

The most conventional route of delivery for biopharmaceuticals has been intravenous (IV) administration because of poor bioavailability by most other routes, greater control during clinical administration, and faster pharmaceutical development. For products that require frequent and chronic administration, the alternate subcutaneous (SC) route of delivery is more appealing. When coupled with prefilled syringe and autoinjector device technology, SC delivery allows for home administration and improved compliance of administration.

Treatments with high doses of more than 1 mg/kg or 100 mg/dose often require development of formulations at concentrations exceeding 100 mg/mL because of the small volume (<1.5 mL) that can be given by the SC routes. For proteins that have a propensity to aggregate at the higher concentrations, achieving such high concentration formulations is a developmental challenge. Even for the IV delivery route where large volumes can be administered, protein concentrations of tens of milligrams per milliliter may be needed for high dosing regimens and this may pose stability challenges for some proteins.

The principles governing protein solubility are more complicated than those for small synthetic molecules, and thus overcoming the protein solubility issue takes different strategies. Operationally, solubility for proteins could be described by the maximum amount of protein in the presence of cosolutes whereby the solution remains visibly clear (i.e., does not show protein precipitates, crystals, or gels). The dependence of protein solubility on ionic strength, salt form, pH, temperature, and certain excipients is well demonstrated by changes in bulk water surface tension and protein binding to water and ions versus self-association. Binding of proteins to specific excipients or salts influences solubility through changes in protein conformation or masking of certain amino acids involved in self-interaction. Proteins are also preferentially hydrated (and stabilized as more compact conformations) by certain salts, amino acids, and sugars, leading to their altered solubility.

Aggregation, which requires bimolecular collision, is expected to be the primary degradation pathway in protein solutions. The relationship of concentration to aggregate formation depends on the size of aggregates as well as the mechanism of association. A typical approach to minimize aggregation is to restrict the mobility of proteins in order to reduce the number of collisions. Lyophilization with appropriate excipients may improve protein stability against aggregation by decreasing protein mobility and by restricting conformational flexibility with the added benefit of minimizing hydrolytic reactions consequent to removal of water. The addition of appropriate excipients, including lyoprotectants, can prevent the formation of aggregates during the lyophilization process as well as during storage of the final product. A key parameter for effective protection is the molar ratio of the lyoprotectant to the protein. Generally, molar ratios of 300:1 or greater are required to provide suitable stability, especially for room temperature storage. Such ratios can also, however, lead to an undesirable increase in viscosity.

Lyophilization allows for designing a formulation with appropriate stability and tonicity. Although isotonicity is not necessarily required for SC administration, it may be desirable for minimizing pain upon administration. Isotonicity of a lyophile is difficult to achieve because both the protein and the excipients are concentrated during the reconstitution process. Excipient-to-protein molar ratios of 500:1 will result in hypertonic preparations if the final protein concentration is targeted for >100 mg/mL. If the desire is to achieve an isotonic formulation, then a choice of lower molar ratio of excipient is necessary.

Determining the highest protein concentration achievable remains an empirical exercise due to the labile nature of protein conformation and the propensity to interact with itself, with surfaces, and with specific solutes.

Subjects who may benefit from SC formulations are those that have conditions that require frequent and chronic administration such as immune system disease, rheumatoid arthritis, and immune disorders associated with graft transplantation. Commercially available protein drug products for the treatment of rheumatoid arthritis include HUMIRA, ENBREL®, and REMICADE®.

HUMIRA (Abbott) is supplied in single-use, 1 mL prefilled glass syringes as a sterile, preservative-free solution for subcutaneous administration. The solution of HUMIRA is clear and colorless, with a pH of about 5.2. Each syringe delivers 0.8 mL (40 mg) of drug product. Each 0.8 mL of HUMIRA contains 40 mg adalimumab, 4.93 mg sodium chloride, 0.69 mg monobasic sodium phosphate dihydrate, 1.22 mg dibasic sodium phosphate dihydrate, 0.24 mg sodium citrate, 1.04 mg citric acid monohydrate, 9.6 mg mannitol, 0.8 mg polysorbate 80, and Water for Injection, USP. Sodium hydroxide is added as necessary to adjust pH.

ENBREL (Amgen) is supplied in a single-use prefilled 1 mL syringe as a sterile, preservative-free solution for subcutaneous injection. The solution of ENBREL is clear and colorless and is formulated at pH 6.3 ± 0.2. Each ENBREL single-use prefilled syringe contains 0.98 mL of a 50 mg/mL solution of etanercept with 10 mg/mL sucrose, 5.8 mg/mL sodium chloride, 5.3 mg/mL L-arginine hydrochloride, 2.6 mg/mL sodium phosphate monobasic monohydrate, and 0.9 mg/mL sodium phosphate dibasic anhydrous. Administration of one 50 mg/mL prefilled syringe of ENBREL provides a dose equivalent to two 25 mg vials of lyophilized ENBREL, when vials are reconstituted and administered as recommended. ENBREL multiple-use vial contains sterile, white, preservative-free, lyophilized powder. Reconstitution with 1 mL of the supplied Sterile Bacteriostatic Water for Injection (BWFI), USP (containing 0.9% benzyl alcohol) yields a multiple-use, clear, and colorless solution with a pH of 7.4 ± 0.3 containing 25 mg etanercept, 40 mg mannitol, 10 mg sucrose, and 1.2 mg tromethamine.

REMICADE (Centocor) is supplied as a sterile, white, lyophilized powder for intravenous infusion. Following reconstitution with 10 mL of Sterile Water for Injection (SWFI), USP, the resulting pH is approximately 7.2. Each single-use vial contains 100 mg infliximab, 500 mg sucrose, 0.5 mg polysorbate 80, 2.2 mg monobasic sodium phosphate, monohydrate, and 6.1 mg dibasic sodium phosphate, dihydrate. No preservatives are present.

Commercially available protein drug products for the treatment of immune disorders associated with graft transplantation include SIMULECT® and ZENAPAX®.

The drug product, SIMULECT (Novartis), is a sterile lyophilisate, which is available in 6 mL colorless glass vials and in 10 and 20 mg strengths. Each 10 mg vial contains 10 mg basiliximab, 3.61 mg monobasic potassium phosphate, 0.50 mg disodium hydrogen phosphate (anhydrous), 0.80 mg sodium chloride, 10 mg sucrose, 40 mg mannitol, and 20 mg glycine to be reconstituted in 2.5 mL of SWFI, USP. Each 20 mg vial contains 20 mg basiliximab, 7.21 mg monobasic potassium phosphate, 0.99 mg disodium hydrogen phosphate (anhydrous), 1.61 mg sodium chloride, 20 mg sucrose, 80 mg mannitol, and 40 mg glycine to be reconstituted with 5 mL of SWFI, USP. No preservatives are added.

ZENAPAX (Roche Laboratories), 25 mg/5 mL, is supplied as a clear, sterile, colorless concentrate for further dilution and intravenous administration. Each milliliter of ZENAPAX contains 5 mg of daclizumab and 3.6 mg sodium phosphate monobasic monohydrate, 11 mg sodium phosphate dibasic heptahydrate, 4.6 mg sodium chloride, 0.2 mg polysorbate 80, and may contain hydrochloric acid or sodium hydroxide to adjust the pH to 6.9. No preservatives are added.

An "isotonic" formulation is one, which has essentially the same osmotic pressure as human blood. Isotonic formulations will generally have an osmotic pressure from about 250 to 350 mOsmol/kg H_2O. Buffering agents are one or more components that, when added to an aqueous solution, are able to protect the solution against variations in pH that can result when adding acid or alkali, or upon dilution with a solvent. In addition to phosphate buffers, glycinate, carbonate, and citrate buffers can be used, in which case, sodium, potassium, or ammonium ions can serve as counterion.

> *Lyoprotectants* include molecules, which, when combined with a protein of interest, prevents or reduces chemical and/or physical instability of the protein upon lyophilization and subsequent storage. Preservatives include an agent that reduces bacterial action and may be optionally added to the formulations herein. The addition of a preservative may, for example, facilitate the production of a multiuse (multiple-dose) formulation. Examples of potential preservatives include octadecyldimethylbenzyl ammonium chloride, hexamethonium chloride, benzalkonium chloride (a mixture of alkylbenzyldimethylammonium chlorides in which the alkyl groups are long-chain compounds), and benzethonium chloride. Other types of preservatives include aromatic alcohols such as phenol, butyl and benzyl alcohol, alkyl parabens such as methyl or propyl paraben, catechol, resorcinol, cyclohexanol, 3pentanol, and *m*-cresol.
>
> *Surfactants* are surface-active molecules containing both a hydrophobic portion (e.g., alkyl chain) and a hydrophilic portion (e.g., carboxyl and carboxylate groups). Surfactant may be added to the formulations of the invention. Surfactants suitable for use in the formulations of the present invention include, but are not limited to polysorbates (e.g., polysorbates 20 or 80); poloxamers (e.g., poloxamer 188); sorbitan esters and derivatives; Triton; sodium laurel sulfate; sodium octyl glucoside; lauryl-, myristyl-, linoleyl-, or stearyl-sulfobetaine; lauryl-, linoleyl- or stearyl-sarcosine; linoleyl-, myristyl-, or cetyl-betaine; lauramidopropyl-cocamidopropyl-, linoleamidopropyl-, myristamidopropyl-, palmidopropyl-, or isostearamidopropylbetaine (e.g., lauroamidopropyl); myristamidopropyl-, palmidopropyl-, or isostearamidopropyl-dimethylamine; sodium methyl cocoyl-, or disodium methyl oleoyl-taurate; and the MONAQUAT™ series (Mona Industries, Inc., Paterson, NJ), polyethylene glycol, polypropylene glycol, and copolymers of ethylene and propylene glycol (e.g., Pluronics, PF68, etc.).

A typical composition of a lyophilized protein drug such as Belatacept has the following:

Belatacept 100 mg/vial drug product (with 10% excess, 110 mg), sucrose 220 mg, sodium phosphate monobasic monohydrate 15.18 mg, sodium chloride 2.55 mg, 1 N sodium hydroxide or 1 N hydrochloric acid used to adjust pH to 7.5. The diluents may include SWFI, bacteriostatic water for injection (BWFI), a pH buffered solution (e.g., phosphate-buffered saline), sterile saline solution, Ringer's solution, or dextrose solution.

A liquid subcutaneous formulation may include the following:

Belatacept 140 mg (with 40% overfill), sucrose 190.4 mg, poloxamer 1888.96 mg, sodium phosphate monobasic, monohydrate 0.371 mg, sodium phosphate dibasic, anhydrous 1.193 mg, Water for Injection q.s. to 1.12 mL.

Stability testing guidance

Background

The stability testing of biotechnology-derived products is governed by ICH Q5C (http://www.fda.gov/downloads/drugs/guidancecomplianceregulatoryinformation/guidances/ucm073466.pdf). Note that there are significant differences between this guideline and the guideline used for chemically derived products Q1A to Q1F (http://www.ich.org/products/guidelines/quality/article/quality-guidelines.html). Given below are the terms used in this guidance with specific meaning:

Conjugated product: A conjugated product is made up of an active ingredient (e.g., peptide and carbohydrate), bound covalently or noncovalently to a carrier (e.g., protein, peptide, and inorganic mineral), with the objective of improving the efficacy or stability of the product.

Degradation can cause a change in the drug substance (bulk material) brought about over time. For the purpose of stability testing of the products described in this guideline, such changes could occur as a result of processing or storage (e.g., by deamidation, oxidation, aggregation, and proteolysis). For biotechnological/biological products, some degradation products may be active.

Impurity: Any component of the drug substance (bulk material) or drug product (final container product) that is not the chemical entity defined as the drug substance, an excipient, or other additives to the drug product.

Intermediate: For biotechnological/biological products, a material produced during a manufacturing process that is not the drug substance or the drug product but for which manufacture is critical to the successful production of the drug substance or the drug product. Generally, an intermediate will be quantifiable, and specifications will be established to determine the successful completion of the manufacturing step before continuation of the manufacturing process. This includes material that may undergo further molecular modification or be held for an extended period before further processing.

Manufacturing Scale Production Manufacture at the scale typically encountered in a facility intended for product production for marketing.

Pilot-plant scale: The production of the drug substance or drug product by a procedure fully representative of and simulating that to be applied at manufacturing scale. The methods of cell expansion, harvest, and product purification should be identical except for the scale of production.

The submission of biosimilar products in the United States must follow this guideline that requires at least 6 month of real-time stability data to be submitted in the final packaging at the time of submission while additional data could be submitted during the review of the 351(k) submission.

The evaluation of stability of biological drugs may necessitate development of complex analytical methodologies. Assays for biological activity, where applicable, should be part of the pivotal stability studies. Appropriate physicochemical, biochemical, and immunochemical methods for the analysis of the molecular entity and the quantitative detection of degradation products should also be part of the stability program whenever purity and molecular characteristics of the product permit use of these methodologies.

With these concerns in mind, the applicant should develop the proper supporting stability data for a biotechnological/biological product and consider many external conditions that can affect the product's potency, purity, and quality.

Primary data to support a requested storage period for either drug substance or drug product should be based on long-term, real-time, real-condition stability studies. Thus, the development of a proper long-term stability program becomes critical to the successful development of a commercial product. The purpose of this document is to give guidance to applicants regarding the type of stability studies that should be provided in support of marketing applications. It is understood that during the review and evaluation process, continuing updates of initial stability data may occur.

Scope

The guidance applies to well-characterized proteins and polypeptides, their derivatives, and products of which they are components, and which are isolated from tissues, body fluids, cell cultures, or produced using recombinant deoxyribonucleic acid (r-DNA) technology. Thus, the guidance covers the generation and submission of stability data for products such as cytokines (interferons, interleukins, colony-stimulating factors, and tumor necrosis factors), erythropoietins, plasminogen activators, blood plasma factors, growth hormones and growth factors, insulins, monoclonal antibodies, and vaccines consisting of well-characterized proteins or polypeptides. In addition, the guidance outlined in the following sections may apply to other types of products, such as conventional vaccines, after consultation with the appropriate regulatory authorities. The document does not cover antibiotics, allergenic extracts, heparins, vitamins, whole blood, or cellular blood components.

Batch selection

Drug substance (bulk material)   Where bulk material is to be stored after manufacture, but before formulation and final manufacturing, stability data should be provided on at least three batches for which manufacture and storage are representative of the manufacturing scale of production. A minimum of 6 months stability data at the time of submission should be submitted in cases where storage periods greater than 6 months are requested. For drug substances with storage periods of less than 6 months, the minimum amount of stability data in the initial submission should be determined on a case-by-case basis. Data from pilot-plant scale batches of drug substance produced at a reduced scale of fermentation and purification may be provided at the time the dossier is submitted to the regulatory agencies with a commitment to place the first three manufacturing scale batches into the long-term stability program after approval.

The quality of the batches of drug substance placed into the stability program should be representative of the quality of the material used in preclinical and clinical studies and of the quality of the material to be made at the manufacturing scale. In addition, the drug substance (bulk material) made at pilot-plant scale should be produced by a process and stored under conditions representative of that used for the manufacturing scale. The drug substance entered into the stability program should be stored in containers that properly represent the actual holding containers used during manufacture. Containers of reduced size may be acceptable for drug substance stability testing provided that they are constructed of the same material and use the same type of container/closure system that is intended to be used during manufacture.

Intermediates   During manufacture of biotechnological/biological products, the quality and control of certain intermediates may be critical to the production of the final product. In general, the manufacturer should identify intermediates and generate in-house data and process limits that assure their stability within the bounds of the developed process. Although the use of pilot-plant scale data is permissible, the manufacturer should establish the suitability of such data using the manufacturing scale process.

Drug product (final container product) Stability information should be provided on at least three batches of final container product representative of that which will be used at the manufacturing scale. Where possible, batches of final container product included in stability testing should be derived from different batches of bulk material. A minimum of 6 months data at the time of submission should be submitted in cases where storage periods greater than 6 months are requested. For drug products with storage periods of less than 6 months, the minimum amount of stability data in the initial submission should be determined on a case-by-case basis. Product expiration dating should be based upon the actual data submitted in support of the application. Because dating is based upon the real-time/real-temperature data submitted for review, continuing updates of initial stability data should occur during the review and evaluation process. The quality of the final container product placed on stability studies should be representative of the quality of the material used in the preclinical and clinical studies. Data from pilot-plant scale batches of drug product may be provided at the time the dossier is submitted to the regulatory agencies with a commitment to place the first three manufacturing scale batches into the long-term stability program after approval. Where pilot-plant scale batches were submitted to establish the dating for a product and, in the event that the product produced at manufacturing scale does not meet those long-term stability specifications throughout the dating period or is not representative of the material used in preclinical and clinical studies, the applicant should notify the appropriate regulatory authorities to determine a suitable course of action.

Sample selection

Where one product is distributed in batches differing in fill volume (e.g., 1 milliliter [mL], 2 mL, or 10 mL), unitage (e.g., 10, 20, or 50 units), or mass (e.g., 1 milligram [mg], 2 mg, or 5 mg), samples to be entered into the stability program may be selected on the basis of a matrix system and/or by bracketing.

Matrixing, that is, the statistical design of a stability study in which different fractions of samples are tested at different sampling points, should only be applied when appropriate documentation is provided that confirms that the stability of the samples tested represents the stability of all samples. The differences in the samples for the same drug product should be identified as, for example, covering different batches, different strengths, different sizes of the same closure, and, possibly, in some cases, different container/closure systems. Matrixing should not be applied to samples with differences that may affect stability, such as different strengths and different containers/closures, where it cannot be confirmed that the products respond similarly under storage conditions.

Where the same strength and exact container/closure system is used for three or more fill contents, the manufacturer may elect to place only the smallest and largest container size into the stability program, that is, bracketing. The design of a protocol that incorporates bracketing assumes that the stability of the intermediate condition samples is represented by those at the extremes. In certain cases, data may be needed to demonstrate that all samples are properly represented by data collected for the extremes.

Stability indicating profile

On the whole, there is no single stability-indicating assay or parameter that profiles the stability characteristics of a biotechnological/biological product. Consequently, the manufacturer should propose a stability-indicating profile that provides assurance that changes in the identity, purity, and potency of the product will be detected.

At the time of submission, applicants should have validated the methods that comprise the stability-indicating profile, and the data should be available for review. The determination of which tests should be included will be product-specific. The items emphasized in the following sections are not intended to be all-inclusive, but represent product characteristics that should typically be documented to demonstrate product stability adequately.

Protocol   The dossier accompanying the application for marketing authorization should include a detailed protocol for the assessment of the stability of both drug substance and drug product in support of the proposed storage conditions and expiration dating periods. The protocol should include all necessary information that demonstrates the stability of the biotechnological/biological product throughout the proposed expiration dating period including, for example, well-defined specifications and test intervals. The statistical methods that should be used are described in the tripartite guideline on stability.

Potency   When the intended use of a product is linked to a definable and measurable biological activity, testing for potency should be part of the stability studies. For the purpose of stability testing of the products described in this guideline, potency is the specific ability or capacity of a product to achieve its intended effect. It is based on the measurement of some attribute of the product and is determined by a suitable in vivo or in vitro quantitative method. In general, potencies of biotechnological/biological products tested by different laboratories can be compared in a meaningful way only if expressed in relation to that of an appropriate reference material. For that purpose, a reference material calibrated directly or indirectly against the corresponding national or international reference material should be included in the assay.

Potency studies should be performed at appropriate intervals as defined in the stability protocol and the results should be reported in units of biological activity calibrated, whenever possible, against nationally or internationally recognized standards. Where no national or international reference standards exist, the assay results may be reported in in-house-derived units using a characterized reference material.

In some biotechnological/biological products, potency is dependent upon the conjugation of the active ingredient(s) to a second moiety or binding to an adjuvant. Dissociation of the active ingredient(s) from the carrier used in conjugates or adjuvants should be examined in real-time/real-temperature studies (including conditions encountered during shipment). The assessment of the stability of such products may be difficult because, in some cases, in vitro tests for biological activity and physicochemical characterization are impractical or provide inaccurate results. Appropriate strategies (e.g., testing the product before conjugation/binding, assessing the release of the active compound from the second moiety, in vivo assays) or the use of an appropriate surrogate test should be considered to overcome the inadequacies of in vitro testing.

Purity and molecular characterization   For the purpose of stability testing of the products described in this guideline, purity is a relative term. Because of the effect of glycosylation, deamidation, or other heterogeneities, the absolute purity of a biotechnological/biological product is extremely difficult to determine. Thus, the purity of a biotechnological/biological product should be typically assessed by more than one method and the purity value derived is method dependent. For the purpose of stability testing, tests for purity should focus on methods for determination of degradation products.

The degree of purity, as well as the individual and total amounts of degradation products of the biotechnological/biological product entered into the stability studies, should be reported and documented whenever possible. Limits of acceptable degradation should be derived from the analytical profiles of batches of the drug substance and drug product used in the preclinical and clinical studies.

The use of relevant physicochemical, biochemical, and immunochemical analytical methodologies should permit a comprehensive characterization of the drug substance and/or drug product (e.g., molecular size, charge, and hydrophobicity) and the accurate detection of degradation changes that may result from deamidation, oxidation, sulfoxidation, aggregation, or fragmentation during storage. As examples, methods that may contribute to this include electrophoresis (SDS09Page, immunoelectrophoresis, Western blot, and isoelectrofocusing), high-resolution chromatography (e.g., reversed-phase chromatography, gel filtration, ion exchange, and affinity chromatography), and peptide mapping.

Wherever significant qualitative or quantitative changes indicative of degradation product formation are detected during long-term, accelerated, and/or stress stability studies, consideration should be given to potential hazards and to the need for characterization and quantification of degradation products within the long-term stability program. Acceptable limits should be proposed and justified, taking into account the levels observed in material used in preclinical and clinical studies.

For substances that cannot be properly characterized or products for which an exact analysis of the purity cannot be determined through routine analytical methods, the applicant should propose and justify alternative testing procedures.

Other product characteristics The following product characteristics, though not specifically relating to biotechnological/biological products, should be monitored and reported for the drug product in its final container:

- Visual appearance of the product (color and opacity for solutions/suspensions; color, texture, and dissolution time for powders), visible particulates in solutions or after the reconstitution of powders or lyophilized cakes, pH, and moisture level of powders and lyophilized products.
- Sterility testing or alternatives (e.g., container/closure integrity testing) should be performed at a minimum initially and at the end of the proposed shelf life.
- Additives (e.g., stabilizers and preservatives) or excipients may degrade during the dating period of the drug product. If there is any indication during preliminary stability studies that reaction or degradation of such materials adversely affects the quality of the drug product, these items may need to be monitored during the stability program.
- The container/closure has the potential to affect the product adversely and should be carefully evaluated as discussed later in the chapter.

Storage conditions

Temperature Because most finished biotechnological/biological products need precisely defined storage temperatures, the storage conditions for the real-time/real-temperature stability studies may be confined to the proposed storage temperature.

Humidity Biotechnological/biological products are generally distributed in containers protecting them against humidity. Therefore, where it can be demonstrated that the proposed containers (and conditions of storage) afford sufficient protection against high and low humidity, stability tests at different relative humidity can usually be omitted. Where humidity-protecting containers are not used, appropriate stability data should be provided.

Accelerated and stress conditions As previously noted, the expiration dating should be based on real-time/real-temperature data. However, it is strongly suggested that studies be conducted on the drug substance and drug product under accelerated and stress conditions. Studies under accelerated conditions may provide useful support data for establishing the expiration date, provide product stability information or future product development (e.g., preliminary assessment of proposed manufacturing changes such as change in formulation and scale-up), assist in validation of analytical methods for the stability program, or generate information that may help elucidate the degradation profile of the drug substance or drug product. Studies under stress conditions may be useful in determining whether accidental exposures to conditions other than those proposed (e.g., during transportation) are deleterious to the product and also for evaluating which specific test parameters may be the best indicators of product stability. Studies of the exposure of the drug substance or drug product to extreme conditions may help to reveal patterns of degradation; if so, such changes should be monitored under proposed storage conditions. Although the tripartite guideline on stability describes the conditions of the accelerated and stress study, the applicant should note that those conditions may not be appropriate for biotechnological/biological products. Conditions should be carefully selected on a case-by-case basis.

Light Applicants should consult the appropriate regulatory authorities on a case-by-case basis to determine guidance for testing.

Container/closure Changes in the quality of the product may occur due to the interactions between the formulated biotechnological/biological product and container/closure. Where the lack of interactions cannot be excluded in liquid products (other than sealed ampules), stability studies should include samples maintained in the inverted or horizontal position (i.e., in contact with the closure), as well as in the upright position, to determine the effects of the closure on product quality. Data should be supplied for all different container/closure combinations that will be marketed.

In addition to the standard data necessary for a conventional single-use vial, the applicant should demonstrate that the closure used with a multiple-dose vial is capable of withstanding the conditions of repeated insertions and withdrawals so that the product retains its full potency, purity, and quality for the maximum period specified in the instructions for use on containers, packages, and/or package inserts. Such labeling should be in accordance with relevant national/regional requirements.

Stability after reconstitution of freeze-dried products The stability of freeze-dried products after their reconstitution should be demonstrated for the conditions and the maximum storage period specified on containers, packages, and/or package inserts. Such labeling should be in accordance with relevant national/regional requirements.

Testing frequency

The shelf lives of biotechnological/biological products may vary from days to several years. Thus, it is difficult to draft uniform guidelines regarding the stability study duration and testing frequency that would be applicable to all types

117

of biotechnological/biological products. With only a few exceptions, however, the shelf lives for existing products and potential future products will be within the range of 0.5–5 years. Therefore, the guidance is based upon expected shelf lives in that range. This takes into account the fact that degradation of biotechnological/biological products may not be governed by the same factors during different intervals of a long storage period.

When shelf lives of 1 year or less are proposed, the real-time stability studies should be conducted monthly for the first 3 months and at 3 month intervals thereafter. For products with proposed shelf lives of greater than 1 year, the studies should be conducted every 3 months during the first year of storage, every 6 months during the second year, and annually thereafter.

While the testing intervals listed above may be appropriate in the preapproval or prelicense stage, reduced testing may be appropriate after approval or licensure where data are available that demonstrate adequate stability. Where data exist that indicate the stability of a product is not compromised, the applicant is encouraged to submit a protocol that supports elimination of specific test intervals (e.g., 9-month testing) for postapproval/postlicensure, long-term studies.

Specifications

Although biotechnological/biological products may be subject to significant losses of activity, physicochemical changes, or degradation during storage, international and national regulations have provided little guidance with respect to distinct release and end of shelf life specifications. Recommendations for maximum acceptable losses of activity, limits for physicochemical changes, or degradation during the proposed shelf life have not been developed for individual types or groups of biotechnological/biological products but are considered on a case-by-case basis. Each product should retain its specifications within established limits for safety, purity, and potency throughout its proposed shelf life. These specifications and limits should be derived from all available information using the appropriate statistical methods. The use of different specifications for release and expiration should be supported by sufficient data to demonstrate that the clinical performance is not affected, as discussed in the tripartite guideline on stability.

Labeling

For most biotechnological/biological drug substances and drug products, precisely defined storage temperatures are recommended. Specific recommendations should be stated, particularly for drug substances and drug products that cannot tolerate freezing. These conditions, and where appropriate, recommendations for protection against light and/or humidity, should appear on containers, packages, and/or package inserts. Such labeling should be in accordance with relevant national and regional requirements.

Bibliography

Aapro MS. What do prescribers think of biosimilars? *Target Oncol* 2012 Mar;7(Suppl. 1):S51–S55.

Abas A. Regulatory guidelines for biosimilars in Malaysia. *Biologicals* 2011 Sept;39(5):339–342.

Abraham J. Developing oncology biosimilars: An essential approach for the future. *Semin Oncol* 2013 Dec;40(Suppl. 1):S5–S24.

Ahmed I, Kaspar B, Sharma U. Biosimilars: Impact of biologic product life cycle and European experience on the regulatory trajectory in the United States. *Clin Ther* 2012 Feb;34(2):400–419.

Arato T, Yamaguchi T. Experience of reviewing the follow-on biologics including Somatropin and erythropoietin in Japan. *Biologicals* 2011 Sept;39(5):289–292.

Berghout A. Clinical programs in the development of similar biotherapeutic products: Rationale and general principles. *Biologicals* 2011 Sept;39(5):293–296.

Berkowitz SA, Engen JR, Mazzeo JR, Jones GB. Analytical tools for characterizing biopharmaceuticals and the implications for biosimilars. *Nat Rev Drug Discov* 2012 Jun 29;11(7):527–540.

Bohlega S, Al-Shammri S, Al Sharoqi I, Dahdaleh M, Gebeily S, Inshasi J, Khalifa A, Pakdaman H, Szólics M, Yamout B. Biosimilars: Opinion of an expert panel in the Middle East. *Curr Med Res Opin* 2008 Oct;24(10):2897–2903.

Bui LA, Taylor C. Developing clinical trials for biosimilars. *Semin Oncol* 2014 Feb;41 (Suppl. 1):S15–S25.

Cai XY, Gouty D, Baughman S, Ramakrishnan M, Cullen C. Recommendations and requirements for the design of bioanalytical testing used in comparability studies for biosimilar drug development. *Bioanalysis* 2011 Mar;3(5):535–540.

Cai XY, Thomas J, Cullen C, Gouty D. Challenges of developing and validating immunogenicity assays to support comparability studies for biosimilar drug development. *Bioanalysis* 2012 Sept;4(17):2169–2177. Review.

Combe C, Tredree RL, Schellekens H. Biosimilar epoetins: An analysis based on recently implemented European medicines evaluation agency guidelines on comparability of biopharmaceutical proteins. *Pharmacotherapy* 2005 Jul;25(7):954–962.

Corbel MJ, Cortes Castillo Mde L. Vaccines and biosimilarity: A solution or a problem? *Expert Rev Vaccines.* 2009 Oct;8(10):1439–1449.

Covic A, Kuhlmann MK. Biosimilars: Recent developments. *Int Urol Nephrol* 2007;39(1):261–266.

Dorantes Calderón B, Montes Escalante IM. Biosimilar medicines. Scientific and legal disputes. *Farm Hosp* 2010 Mar;34(Suppl. 1):29–44.

Dranitsaris G, Dorward K, Hatzimichael E, Amir E. Clinical trial design in biosimilar drug development. *Invest New Drugs* 2013 Apr;31(2):479–487.

Drouet L. Low molecular weight heparin biosimilars: How much similarity for how much clinical benefit? *Target Oncol* 2012 Mar;7(Suppl. 1):S35–S42.

Eichbaum C, Haefeli WE. Biologics—Nomenclature and classification. *Ther Umsch* 2011 Nov;68(11):593–601.

EMA. Reflection paper on methodological issues in confirmatory clinical trials planned with an adaptive design. London, October 18, 2007. http://www.ema.europa.eu/ema/pages/includes/document/open_document.jsp?webContentId=WC500003517 (accessed August 27, 2015).

EMA. Guideline on similar biological medicinal products containing biotechnology-derived proteins as active substance: Non-clinical and clinical issues. http://www.ema.europa.eu/ema/pages/includes/document/open_document.jsp?webContentId=WC500003920 (accessed August 27, 2015).

EMA. Guideline on comparability of biotechnology-derived medicinal products after a change in the manufacturing process: Non-clinical and clinical issues. http://www.ema.europa.eu/ema/pages/includes/document/open_document.jsp?webContentId=WC500003935 (accessed August 27, 2015).

EMA. Guideline on immunogenicity assessment of biotechnology-derived therapeutic proteins. http://www.ema.europa.eu/ema/pages/includes/document/open_document.jsp?webContentId=WC500003946 (accessed August 27, 2015).

EMA. Guideline on similar biological medicinal products containing biotechnology-derived proteins as active substance: Quality issues. http://www.ema.europa.eu/ema/pages/includes/document/open_document.jsp?webContentId=WC500003953 (accessed August 27, 2015).

EMA. Guideline on immunogenicity assessment of biotechnology-derived therapeutic proteins. http://www.ema.europa.eu/ema/pages/includes/document/open_document.jsp?webContentId=WC500003966 (accessed August 27, 2015).

EMA. Guideline on immunogenicity assessment of monoclonal antibodies intended for in vivo clinical use. http://www.ema.europa.eu/ema/pages/includes/document/open_document.jsp?webContentId=WC500128688 (accessed August 27, 2015).

EMA. Guideline on similar biological medicinal products containing biotechnology-derived proteins as active substance: Non-clinical and clinical issues. http://www.ema.europa.eu/ema/pages/includes/document/open_document.jsp?webContentId=WC500144124 (accessed August 27, 2015).

EMA. Concept paper on the revision of the guideline on immunogenicity assessment of biotechnology-derived therapeutic proteins (CHMP/BMWP/42832/2005), Draft, February 20, 2014. http://www.ema.europa.eu/ema/pages/includes/document/open_document.jsp?webContentId=WC500163623 (accessed August 27, 2015).

119

EMA. Guideline on similar biological medicinal products containing biotechnology-derived proteins as active substance: Quality issues (revision 1). http://www.ema.europa.eu/ema/pages/includes/document/open_document.jsp?webContentId=WC500167838 (accessed August 27, 2015).

EMA. Guideline on similar biological medicinal products, October 23, 2014. http://www.ema.europa.eu/ema/pages/includes/document/open_document.jsp?webContentId=WC500176768 (accessed August 27, 2015).

FDA. Biosimilars: Additional questions and answers regarding implementation of the Biologics Price Competition and Innovation Act of 2009; Guidance for Industry; Draft guidance, May 2015. http://www.fda.gov/downloads/Drugs/GuidanceComplianceRegulatoryInformation/Guidances/UCM273001.pdf (accessed August 27, 2015).

FDA. Scientific considerations in demonstrating biosimilarity to a reference product: Guidance for industry; Draft guidance, April 2015. http://www.fda.gov/downloads/Drugs/GuidanceComplianceRegulatoryInformation/Guidances/UCM291128.pdf (accessed August 27, 2015).

FDA. Quality considerations in demonstrating biosimilarity of a therapeutic protein product to a reference product: Guidance for industry, April 2015. http://www.fda.gov/downloads/Drugs/GuidanceComplianceRegulatoryInformation/Guidances/UCM291134.pdf (accessed August 27, 2015).

FDA. Guidance for industry: Formal meetings between the FDA and biosimilar biological product sponsors or applicants; Draft guidance, March 2013. http://www.fda.gov/downloads/Drugs/GuidanceComplianceRegulatoryInformation/Guidances/UCM345649.pdf (accessed August 27, 2015).

FDA. Guidance for industry: Clinical pharmacology data to support a demonstration of biosimilarity to a reference product; Draft guidance, May 2015. http://www.fda.gov/downloads/Drugs/GuidanceComplianceRegulatoryInformation/Guidances/UCM397017.pdf (accessed August 27, 2015).

FDA. Guidance for industry: Reference product exclusivity for biological products filed under Section 351(a) of the PHS Act; Draft guidance, August 2014. http://www.fda.gov/downloads/Drugs/GuidanceComplianceRegulatoryInformation/Guidances/UCM407844.pdf (accessed August 27, 2015).

FDA. How drugs are developed and approved: Biosimilars. http://www.fda.gov/drugs/developmentapprovalprocess/howdrugsaredevelopedandapproved/approvalapplications/therapeuticbiologicapplications/biosimilars/default.htm (accessed August 27, 2015).

Feagan BG, Choquette D, Ghosh S, Gladman DD, Ho V, Meibohm B, Zou G, Xu Z, Shankar G, Sealey DC, Russell AS. The challenge of indication extrapolation for infliximab biosimilars. *Biologicals* 2014 July;42(4):177–183.

García Alfonso P. Biosimilar filgrastim: From development to record. *Farm Hosp* 2010 Mar;34(Suppl. 1):19–24.

Gazit E. *Angew Chem Intl Ed* 2002;41:257.

Geiger T, Clarke S. Deamidation, isomerization, and racemization at asparaginyl and aspartyl residues in peptides: Succinimide-linked reactions that contribute to protein degradation. *J Biol Chem* 1987;262(2):785–794.

Genazzani AA, Biggio G, Caputi AP, Del Tacca M, Drago F, Fantozzi R, Canonico PL. Biosimilar drugs: Concerns and opportunities. *BioDrugs* 2007;21(6):351–356.

Grimm S. The art and design of genetic screens: Mammalian culture cells. *Nat Rev Gen* 2004; 5:179–189.

Hadavand N, Valadkhani M, Zarbakhsh A. Current regulatory and scientific considerations for approving biosimilars in Iran. *Biologicals* 2011 Sept;39(5):325–327.

Haddadin RD. Concept of biosimilar products in Jordan. *Biologicals* 2011 Sept;39(5):333–335.

Hechavarría Núñez Y, Pérez Massipe RO, Orta Hernández SD, Muñoz LM, Jacobo Casanueva OL, Pérez Rodríguez V, Domínguez Morales RB, Pérez Cristiá RB. The regulatory framework for similar biotherapeutic products in Cuba. *Biologicals* 2011 Sept;39(5):317–320.

Heinemann L. Biosimilar insulins. *Expert Opin Biol Ther* 2012 Aug;12(8):1009–1016.

Herrero Ambrosio A. Biosimilars: Regulatory status for approval. *Farm Hosp* 2010 Mar;34(Suppl. 1):16–18.

Hirsch BR, Lyman GH. Pharmacoeconomics of the myeloid growth factors: A critical and systematic review.

Ibarra-Cabrera R, Mena-Pérez SC, Bondani-Guasti A, García-Arrazola R. Review on the worldwide regulatory framework for biosimilars focusing on the Mexican case as an emerging market in Latin America. *Biotechnol Adv* 2013 Dec;31(8):1333–1343.

Jelkmann W. Biosimilar epoetins and other "follow-on" biologics: Update on the European experiences. *Am J Hematol* 2010 Oct;85(10):771–780.

Jeske WP, Walenga JM, Hoppensteadt DA, Vandenberg C, Brubaker A, Adiguzel C, Bakhos M, Fareed J. Differentiating low-molecular-weight heparins based on chemical, biological, and pharmacologic properties: Implications for the development of generic versions of low-molecular-weight heparins. *Semin Thromb Hemost* 2008 Feb;34(1):74–85.

Karalis V, Macheras P. Current regulatory approaches of bioequivalence testing. *Expert Opin Drug Metab Toxicol* 2012 Aug;8(8):929–942.

Krämer I. Biosimilars. *Ther Umsch* 2011 Nov;68(11):659–666.

Kumar R, Singh J. Biosimilar drugs: Current status. *Int J Appl Basic Med Res* 2014 July;4(2):63–66.

Lee JF, Litten JB, Grampp G. Comparability and biosimilarity: Considerations for the healthcare provider. *Curr Med Res Opin* 2012 Jun;28(6):1053–1058.

Locatelli F, Becker H. Update on anemia management in nephrology, including current guidelines on the use of erythropoiesis-stimulating agents and implications of the introduction of "biosimilars". *Oncologist* 2009;14(Suppl. 1):16–21.

Locatelli F, Del Vecchio L. An expert opinion on the current treatment of anemia in patients with kidney disease. *Expert Opin Pharmacother* 2012 Mar;13(4):495–503.

Locatelli F, Roger S. Comparative testing and pharmacovigilance of biosimilars. *Nephrol Dial Transplant* 2006 Oct;21(Suppl. 5):v13–v16.

Macdougall IC, Ashenden M. Current and upcoming erythropoiesis-stimulating agents, iron products, and other novel anemia medications. *Adv Chronic Kidney Dis* 2009 Mar;16(2):117–130.

Mahan CE, Fanikos J. New antithrombotics: The impact on global health care. *Thromb Res* 2011 Jun;127(6):518–524.

Malhotra H. Biosimilars and non-innovator biotherapeutics in India: An overview of the current situation. *Biologicals* 2011 Sept;39(5):321–324.

Mellstedt H, Niederwieser D, Ludwig H. The challenge of biosimilars. *Ann Oncol* 2008 Mar;19(3):411–419.

Nam JL, Ramiro S, Gaujoux-Viala C, Takase K, Leon-Garcia M, Emery P, Gossec L, Landewe R, Smolen JS, Buch MH. Efficacy of biological disease-modifying antirheumatic drugs: A systematic literature review informing the 2013 update of the EULAR recommendations for the management of rheumatoid arthritis. *Ann Rheum Dis* 2014 Mar;73(3):516–528.

Ofosu FA. A review of the two major regulatory pathways for non-proprietary low-molecular-weight heparins. *Thromb Haemost* 2012 Feb;107(2):201–214.

Parnham MJ, Schindler-Horvat J, Kozlović M. Non-clinical safety studies on biosimilar recombinant human erythropoietin. *Basic Clin Pharmacol Toxicol* 2007 Feb;100(2):73–83.

Pavlovic M, Girardin E, Kapetanovic L, Ho K, Trouvin JH. Similar biological medicinal products containing recombinant human growth hormone: European regulation. *Horm Res* 2008;69(1):14–21.

Poh J, Tam KT. Registration of similar biological products—Singapore's approach. *Biologicals* 2011 Sept;39(5):343–345.

Rak Tkaczuk KH, Jacobs IA. Biosimilars in oncology: From development to clinical practice. *Semin Oncol* 2014 Apr;41(Suppl. 3):S3–S12.

Rinaudo-Gaujous M, Paul S, Tedesco ED, Genin C, Roblin X, Peyrin-Biroulet L. Review article: Biosimilars are the next generation of drugs for liver and gastrointestinal diseases. *Aliment Pharmacol Ther* 2013 Oct;38(8):914–924.

Schellekens H. Assessing the bioequivalence of biosimilars: The Retacrit case. *Drug Discov Today* 2009 May;14(9–10):495–499.

Schiestl M. A biosimilar industry view on the implementation of the WHO guidelines on evaluating similar biotherapeutic products. *Biologicals* 2011 Sept;39(5):297–299.

Southern PJ, Berg P. Transformation of mammalian cells to antibiotic resistance with a bacterial gene under control of the SV40 early region promoter. *J Mol Appl Genet* 1982;1:327–341.

Strober BE, Armour K, Romiti R, Smith C, Tebbey PW, Menter A, Leonardi C. Biopharmaceuticals and biosimilars in psoriasis: What the dermatologist needs to know. *J Am Acad Dermatol* 2012 Feb;66(2):317–322.

Subramanyam M. Clinical development of biosimilars: An evolving landscape. *Bioanalysis* 2013 Mar;5(5):575–586.

Suh SK, Park Y. Regulatory guideline for biosimilar products in Korea. *Biologicals* 2011 Sept;39(5):336–338.

Tamilvanan S, Raja NL, Sa B, Basu SK. Clinical concerns of immunogenicity produced at cellular levels by biopharmaceuticals following their parenteral administration into human body. *J Drug Target* 2010 Aug;18(7):489–498.

Treuheit MJ et al. *Pharm Res* 2002;19:511.

Tsiftsoglou AS, Ruiz S, Schneider CK. Development and regulation of biosimilars: Current status and future challenges. *BioDrugs* 2013 Jun;27(3):203–211.

Vital EM, Kay J, Emery P. Rituximab biosimilars. *Expert Opin Biol Ther* 2013 Jul;13(7): 1049–1062.

Wadhwa M, Thorpe R. The challenges of immunogenicity in developing biosimilar products. *IDrugs* 2009 Jul;12(7):440–444.

Weise M, Bielsky MC, De Smet K et al. Biosimilars: What clinicians should know. *Blood* 2012 Dec 20;120(26):5111–5117.

Wurm FM. Production of recombinant protein therapeutics in cultivated mammalian cells. *Nature Biotechnol* 2004; 22:1393–1398.

Yamaguchi T, Arato T. Quality, safety and efficacy of follow-on biologics in Japan. *Biologicals* 2011 Sept;39(5):328–332.

Zelenetz AD, Ahmed I, Braud EL et al. NCCN Biosimilars White Paper: Regulatory, scientific, and patient safety perspectives. *J Natl Compr Canc Netw* 2011 Sept;9(Suppl. 4):S1–S22.

Chapter 5 Biosimilarity tetrahedron

In a tetrahedral molecular geometry, a central atom is located at the center with four substituents that are located at the corners of a tetrahedron. The bond angles are exactly 109.5° when all four substituents are the same, as in CH_4.

Tetrahedron concept

While Europe has been approving biosimilars for several years, the approach taken to approve these products has been more conservative and less structured; now that the FDA has allowed biosimilars with greater clarity, the path to biosimilar approval will change significantly across the globe. An evidence of this suggestion is found in the recent EMA revision of its guidelines, bringing them in line with the thinking of the FDA, to allow obviating clinical trials in patients, if a scientific basis for the "safety, quality, and efficacy" can be justified without testing in patients. As I discussed in Chapter 1, the key definitions to define critical attributes describing biosimilars are different among the agencies. The FDA does not use "quality" and suggests the use of "effectiveness" and not "efficacy." These subtle differences have significant meanings. The U.S. FDA has used three keywords to describe biological products: safety, purity, and potency. While the same terms are used in the demonstration of biosimilarity, there is one significant gap in this definition. For a new molecular entity, purity, potency, and safety are characterization steps, not similarity testing steps; with biosimilars, every attribute is tested against a reference standard. Therefore, there is a need to add a fourth classification of attributes, IDENTITY, to the existing list of safety, purity, and potency. And this allows us to create a tetrahedron, wherein each of these categories of attributes is equally important, and must be checked off for similarity to demonstrate overall biosimilarity to qualify approval under the 351(k) statute. Figure 5.1 shows the tetrahedron of biosimilarity.

A keen understanding of each of the four pillars of this tetrahedron and its three components is required to execute a successful biosimilarity demonstration program.

The four pillars of establishing biosimilarity are equally important just as the molecular tetrahedron, wherein the angle between the groups is exactly 109.5°, but a step-wise approach requires establishing similarity at the most fundamental level first. The biosimilar product developer will start with characterizing the originator product and having done so, study its own candidate in this order: identity, potency, purity, and safety. The reason for this order lies in the risk-based management of attributes. If it is determined that it is the correct product and has the effectiveness, its purity and safety can be managed since the contributors of the last two characterizations are possible to modulate with formulation change, process change, and several techniques, but it must first have the right structure and activity.

Figure 5.1 Biosimilarity tetrahedron.

Similarity concept

Drugs for human and animal use can be generally divided into two broad categories: fixed structure ingredients (FSI) or variable structure entities (VSE). The former category includes mainly synthesized drugs with a unitary structure; the latter category includes mainly biological and complex synthetic drugs such as cytokines, monoclonal antibodies, short-chain heparin, and glatiramer acetate as an example. However, the regulatory approval of new biological drugs goes through a BLA process, even though some have been approved under the NDA. All FSI have been approved under the NDA.

The NDA classification is further divided into three categories. Section 505 of the Act describes three types of new drug applications: (1) an application that contains full reports of investigations of safety and effectiveness (section 505(b)(1)); (2) an application that contains full reports of investigations of safety and effectiveness but where at least some of the information required for approval comes from studies not conducted by or for the applicant and for which the applicant has not obtained a right of reference (section 505(b)(2)); and (3) an application that contains information to show that the proposed product is identical in active ingredient, dosage form, strength, route of administration, labeling, quality, performance characteristics, and intended use, among other things, to a previously approved product (section 505(j)). A supplement to an application is a new drug application.

The drugs approved under BLA are classified into two categories: (1) 351(a) is full application compared to section 505(b)(1) and (2) 351 (k), which is a comparable hybrid of 505 (b)(2) and 505(j). Both NDA and BLA are handled by CDER but only

Table 5.1 Comparison of Biological Drugs and Chemical Drugs for Regulatory Equivalence Determination Criteria under FDA Approvals

Equivalence Attribute	Chemical Drugs—505(j)	Biological Products—351(k)
Chemical equivalence	Yes	No; variations similar to originator molecule allowed
Pharmaceutical equivalence	Yes; Q/Q compliance required	No; different formulations allowed
Pharmacokinetic equivalence of 80%–125%	Yes, unless waived for certain class of drugs and parenteral dosage forms	Yes, regardless of the delivery mode
Pharmacodynamic equivalence in healthy subjects	No	Yes, unless a suitable model is not available such as for monoclonal antibodies
Mode of action (MOA) based evaluation	No	Yes; risk factors are determined based on MOA
Head-to-head testing with a reference product for analytical similarity, stability, nonclinical attributes	No, except for release attributes where compendia standards are not available.	Yes
Tier-based statistical analysis of critical quality attributes	No	Yes; critical quality attributes may not show a difference of more than $(\sigma_R/8)$; CQA determined based on risk factors
Immunogenicity evaluation	No	Yes, both at the nonclinical stage, if conducted, and also at the clinical pharmacology level. Healthy subjects are better suited for this testing unless disease state impact PK/PD
Finished product safety attributes	No, unless product packaging affects safety	Yes; alternate formulations and devices are allowed, and their impact on safety should be determined
Nonclinical	Yes, unless classified as GRAS or waived	Yes, only if a suitable model available; generally, not suitable for monoclonal antibodies unless residual uncertainty remains in analytical similarity
Clinical effectiveness	No; in some limited cases for complex molecular mixtures	No, unless PD evaluation based on MOA does not remove any residual uncertainty. Varies widely among agencies
Biosimilarity tiers	No	Yes
Bioequivalence	Yes	No

for the following biological drugs: only therapeutic biological products are filed with CDER, others go to CBER:

- Monoclonal antibodies for in vivo use
- Cytokines, growth factors, enzymes, immunomodulators, and thrombolytic
- Proteins intended for therapeutic use that are extracted from animals or microorganisms, including recombinant versions of these products (except clotting factors)
- Other nonvaccine therapeutic immunotherapies

Table 5.1 provides a broader comparison of biological drugs versus chemical drugs for the demonstration of equivalence.

Since the manufacturing of biological drugs involves a living entity, the expression of these molecules results always in a VSE, not an FSI. It is for this reason they may not be called identical, only similar. Biosimilarity, demonstration of similarity in clinical and safety response, is not analogous to "bioequivalence," a term that has long been in use to describe generic FSIs; bioequivalence assumes the same availability of drug at the site of action (the statutory definition), and, therefore, the same action (a physiological presumption); biosimilarity, on the other hand, assumes sufficient similarity in structure, function, pharmacokinetic and pharmacodynamic properties, immunogenicity, side effects, and where necessary, clinical effectiveness to declare biosimilarity.

However, there is some difference in how various agencies describe biosimilarity. The U.S. FDA, the European EMA, and the WHO and most other agencies that follow the guidance from these three major agencies provide the following broad definitions.

A biosimilar product is defined by regulatory agencies as follows:

- *FDA*: The product is *highly similar* to the reference product notwithstanding minor differences in clinically inactive components AND there are *no clinically meaningful differences* between the biological product and the reference product in terms of *safety, purity, and the potency* of the product.
- *EMA*: A biosimilar medicine is a biological medicine that is developed to be *similar* to an existing biological medicine (the "reference medicine"). Like the reference medicine, the biosimilar has a *degree of natural variability*. When approved, its *variability* and any *differences* between it and its reference medicine will have been shown not to affect *safety or effectiveness*.
- *WHO*: A biotherapeutic product, which is *similar* in terms of *quality, safety, and efficacy* to an already licensed reference biotherapeutic product.

Both WHO and EMA require a product to be "similar," while the FDA requires them to be at least "highly similar" to qualify as a biosimilar. Regarding the variability, the FDA states "no clinically meaningful difference … in terms of safety, purity and potency," and the EMA states, "its variability … not to affect safety or effectiveness." The WHO emphasizes "quality, safety and efficacy," and makes the description and more confounding than what is suggested by the FDA and the EMA. These subtle differences are important in understanding the mindset of these agencies. Both FDA and EMA shun from using "quality," which is a broad and perhaps an assumptive term. The FDA stays away from "effectiveness or efficacy" while both EMA and WHO emphasize it, though differently. It should be noted that the correct description is "effectiveness" that is demonstrated by comparison, whereas "efficacy" is a controlled clinical trial outcome against a placebo. Effectiveness can be demonstrated by many methods including patient trials; efficacy is generally proven in patients only. This subtle difference is missed out in the WHO definition and in the guidelines of several other countries.

Recent draft guidance issued by the EMA (October 2014) brings the EMA closer to the thinking of FDA to allow as many approvals without clinical trials in patients when justified. The rhetoric of the originators that extensive clinical trials are the only way to establish safety and efficacy (effectiveness) of biosimilar products are being set aside by the regulatory agencies. However, the onus of proving that a clinical trial is not needed or only minimal trials are sufficient lies on the developer of the biosimilar product and it is for this reason the understanding of the science and art involved in establishing evidence of biosimilarity is crucial for success.

Comparability versus similarity

It should be noted that "biosimilarity" is the final outcome of various similarity exercises; we do not demonstrate structural and functional biosimilarity; it is merely analytical similarity. We also do not demonstrate structural or functional comparability; we demonstrate similarity. There is a difference between these two terms as "comparability" refers to postapproval changes requiring a comparison between the products before and after the changes are made. Generally, the term comparability is used in conjunction with "protocol" to define this exercise. We can use "comparison" but not "comparability" if we do not want to confuse the agencies about the intent of the exercise.

In light of the confusion created in the literature, in the regulatory guidance worldwide and by developers, it is important to understand the difference between

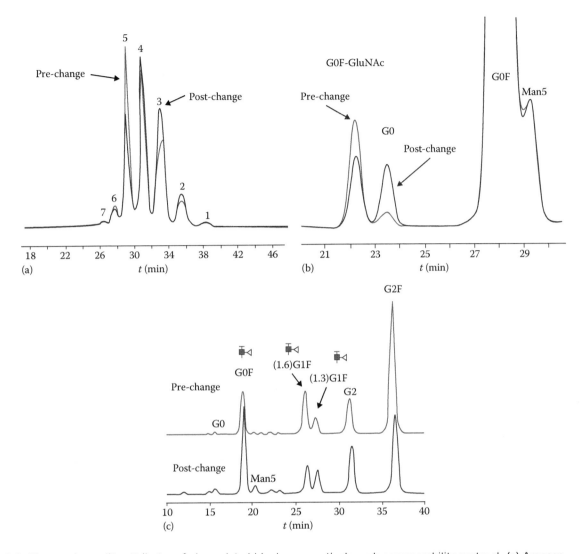

Figure 5.2 Changes in quality attributes of glycosylated biopharmaceuticals under comparability protocol: (a) Aranesp, (b) Rituxan, and (c) Enbrel. The numbers refer to different glycans. (From Schiesti, M. et al., *Nat. Biotechnol.*, 29(4), 310, 2011.)

comparability testing and similarity testing, as well as terms like comparator product and reference product.

Comparability protocol refers to a protocol suggested by regulatory agencies (http://www.fda.gov/downloads/drugs/guidancecomplianceregulatoryinformation/guidances/ucm070262.pdf) with specific indications on validating changes made by the CMC section of protein drugs after these products are approved and in commerce. This falls in the class of changes such as CBE0, CBE30, etc. Comparability usually comprises process change by the same sponsor where historical data are available (e.g., development, process qualification, and control) and the acceptance criteria are relatively easy to set. On the other hand, similarity testing involves attribute comparisons between a biosimilar candidate and a reference product or where there is a change in manufacturer where limited (or no historical) data are available and, therefore, involves more rigorous analytical and statistical evaluations.

A good example is that of a study of the comparability protocols for Aranesp, Rituxan, and Enbrel. In this study, samples of multiple lots before and after changes in the comparability, protocols were collected and analyzed between 2007 and 2010. Figure 5.2 shows the changes observed before and after the comparability protocols were completed. These appear to be significant changes, yet the agencies allowed these changes.

The changes made by the originator in the comparability protocol exercises are generally available in regulatory documents, and these documents along with Establishment Inspection Reports (EIR) of the facilities where the originator product is made are excellent sources of information for the biosimilar product developer. The EIRs are available under the Freedom of Information Act. It is noteworthy that these reports provide great insight into what is considered a critical attribute and how it is controlled in a risk-based evaluation of the development plan of the biosimilar product developer.

The use of "comparability" has specific meaning for postapproval changes and should be discouraged when compiling documents for FDA filing. Figure 5.1 shows an example of variability in the comparability protocol presentation to FDA; differences arising out of changes are acceptable as long as they do not have any "clinically meaningful" difference. The reason why the regulatory agencies let a manufacturer make changes without necessarily having to demonstrate clinical similarity is that the manufacturer knows its molecule well and is able to satisfy the regulatory agencies that these changes are not critical to the safety and efficacy of the product.

A lot of published literature, however, ignores this differentiation. However, this should not be confused with the use of "comparison," which is an evaluation of two products at any stage. "Compare to" is used to liken two things or to put them in the same category. "Compare to" is a proper choice when you intend to simply assert that two things are alike. "Compare with" is used to place two things side by side for the purpose of examining their similarities or differences as is the case in biosimilarity testing; so use "compared with" instead of "compared to" and never "comparability." An example of an improper choice is: "Physicochemical and functional comparability between the proposed biosimilar rituximab GP2013 and originator rituximab" (Visser et al. 2013).

Analytical and functional similarity

The greatest challenge to biosimilar developer comes in choosing a correct battery of tests that are significant to all classifications of attributes shown in the depiction of the biosimilarity tetrahedron; while more tests are more desirable, the choice of tests to demonstrate biosimilarity must be based on evaluation of safety, purity, and potency. And that depends a great deal on the mode of action (MoA). For example, when developing simpler molecules like filgrastim, a set of critical quality attributes are easy to select as shown in Table 5.2.

The success of biosimilar product development depends greatly on the expertise in analytical sciences. Whereas it is relatively easy to assert the similarity of fixed-structure small molecules, demonstrating the similarity of proteins is a more difficult task. Proteins have a variable structure and to determine what attribute variability is critical to safety, purity, and potency requires deeper understanding of the mechanism of action, side effects, and immunogenic profile.

Advances in analytical instrumentation have made it possible for us to understand the differences in the structures better and also to identify what constitutes a significant difference that will be meaningful to clinical effectiveness. However, this requires that the developers of a biosimilar products be fully prepared to bring in

Table 5.2 Selected Critical Quality Attributes for Analytical and Functional Similarity Demonstration for Filgrastim

Criticality	Attribute	Clinical Relevance			Methodology
		Effectiveness	Safety	Immunogenicity	
Very high	Amino acid sequence	✓	✓	✓	Peptide mapping
	Potency	✓	✓		Bioassay
	Target binding	✓	✓		Antibody binding
	Protein concentration	✓	✓		UVAS
High	Subvisible particles		✓	✓	Light obscuration
	Oxidized variants	✓			RP- HPLC
	Higher order structure	✓		✓	CD
	High molecular weight variants/aggregates			✓	SE Chromatography
Low	Truncated variants				RP-HPLC
Very low	Deamidation				CE Chromatography

all modern tools of analysis, both physicochemical and functional and where possible create new testing models to prove that there are no structural differences between the biosimilar product and the reference product that will have any clinically meaningful difference. It would not be inaccurate to state that the biosimilar product developer needs to know the science even better than the originator since the challenge for the biosimilar developer is to convince the regulatory agencies of the similarity in comparison to another product, not placebo, as done by the originator.

The target of structural and functional analysis is to demonstrate at least "high similarity" as proposed by the FDA or "similar" as defined by other agencies, to qualify as a biosimilar candidate. Whereas these designations might appear arbitrary, except that it is a highly structured step-wise process to add to the desired "totality-of-the-evidence" that requires building a model from bottom-up and any weaknesses in the model are resolved and strengthened through a formal risk-based approach.

Comprehensive analytical similarity assessment reduces the degree of uncertainty. Analytical similarity assessment is a repetitive and iterative operation conducted throughout biosimilar product development, with the goal to increase knowledge and confidence in analytical similarity: assess quality attributes important to similar safety, purity, and potency from the selection of host cells, pools, and clones to DS/DP process to nonclinical development; and finally, the clinical production.

The establishment of an analytical similarity program begins with making available the state-of-the-art analytical characterization and functional assays to assess any structural difference. These methods are relevant to known mechanisms of actions, biological functions, safety, and immunogenicity profiles, and further derived from the knowledge of the conserved attributes for the same class of molecules, for example, IgG1 exhibit effector functions.

It is noteworthy that over the past two decades the science of protein analytics has changed significantly, and the regulatory agencies expect the biosimilar developer to use the most advanced and novel methods. The electronic evolution presents us with methodologies that are millions of times more sensitive, such as mass spectrometry, NMR, etc. This can be challenging for some developers since the cost of establishing this analytic apparatus can be onerous, not just in the equipment but also qualified personnel to perform and conduct these analyses. Whereas some developers may find it less capital intensive to outsource their testing and there are several very good choices available, an in-house analytical testing program cannot be obviated given the speed and frequency of testing required in the development programs.

The analytical similarity exercise begins with securing both the lots of RP and reference standards where available; the USP has recently added several monographs of recombinant protein products and reference standards have become available. However, the difference between the two should be clearly understood. Similarity should be assessed against criteria established based on reference product and not against the standard. Monograph standards may or may not have any relationship with the reference originator product and may or may not capture all attributes of clinical relevance. The similarity is assessed against reference product while specification normally centers around the standard. This can create a situation that when standard is close to the mean of RP range, specification also centers around RP range, but when standard is close to the edge of RP range, specification no longer centers around RP range. This should be clearly demonstrated in regulatory submissions as a justification for establishing specification. Standards used to measure biosimilar activity should represent reference product, with special consideration to strength and biological and functional properties.

Sample age at the time of testing should be factored in when comparing stability-induced attributes. One way to satisfy this requirement is to collect, where possible, RP lots of different ages and develop a range of attributes over the course of the expiry of the product; this will allow overlapping the biosimilar product analytical results over an appropriate course of the plot. This may apply to some testing more than others. For example, purity and product-related impurities are prime testings. This includes potency, aggregates, and impurities.

Similarity assessment is performed on drug product lots manufactured from unique drug substance lots and using the to-be-commercial process will support marketing applications. The last requirement may be unique to biosimilar development requirement since scale change can significantly alter the product characteristics. The 505(j) or 505 (b)(2) applicants may choose a smaller scale for the development, and this choice is not available to biosimilar developers, notwithstanding any justified scaling that is fully justifiable. To comply with this requirement, the commercial lots should represent represent fully scaled lots, and will have same unit operations and critical raw materials used for toxicology, clinical and commercial lots; the site of manufacturing should be the same as used for the clinical lots, and analytical data are accumulated for similarity assessment over development lifecycle.

Much vagueness exists in the literature regarding the method of demonstrating analytical similarity; it is not a one-time side-by-side testing of the RP and biosimilar product using a limited number of obvious tests. It involves accumulating knowledge of reference products of different ages on the market to understand the range and variability of the originator manufacturing process. Knowledge about any comparability protocols conducted by the originator is very useful in understanding the nature of these processes. Know that the FDA and EMA allow manufacturers to change their process including the host cells, change of manufacturing sites, and change in specification and purity profiles over the life of the product. In the United States, this is called "comparability protocol." This is allowed because the manufacturer has an in-depth and keen understanding of the product and the process, an experience that the biosimilar product developer does not have.

The specification of what constitutes similarity is established ahead of analytical testing exercise. Results for each attribute are evaluated against its predefined similarity assessment criteria, and the predefined similarity assessment criteria are established based on two general approaches:

1. Nonstatistically derived similarity assessment criteria: Similarity is met when all test lots meet the predefined assessment criteria established based on knowledge of reference products, instrument, and assay capabilities.
2. Statistically derived similarity assessment criteria: Similarity is assessed using statistical equivalence testing when the data are deemed best evaluated by comparing differences in means between the two products.

Any attribute failing similarity criteria should be subjected to impact assessment using a risk-based approach. Confidence in analytical similarity can be built upon increasing rigor and objectivity that may lead to the following outcomes:

- Extensive analytical testing shows CQAs fail to meet similarity criteria and are likely to impact safety and efficacy. In such cases, apply impact assessment for all analytical differences based on the magnitude of the difference and potential biological relevance; this product may not likely qualify as a biosimilar product.
- Ensure high similarity with no analytical difference between all critical quality attributes that could impact efficacy, immunogenicity, and safety.
- Ensure high similarity with no analytical differences for all critical quality attributes AND a statistically rigorous pattern match for all other quantitative attributes relative to that of the RP lots.

Factors that can potentially impact the ability to demonstrate that the biosimilar product is highly similar to the reference product may include, for example, the ages of the biosimilar product and reference product lots tested, optimizing assays, and pre-specifying the criteria under which wider similarity acceptance criteria for a particular assay would be considered appropriate.

As more advanced and reproducible methods become available to characterize molecules, new possibilities have risen to make the orthogonal testing an excellent tool to remove any residual uncertainty. Since these testing may require novel methods, the regulatory agencies do not require the test methods used for analytical and functional similarity to be fully validated, just proven suitable. However, any method used to release the product must be validated as is required for all testing.

Analytical instrumentation

Today's instruments can be millions of times more sensitive than what was available just a couple of decades ago. There are numerous analytical methods to evaluate the safety, purity, and potency profile of biosimilar; among which there are circular dichroism, nuclear magnetic resonance, immunological tests (ELISA, immunoprecipitation, biosensors, etc.), biological activity *from* in vitro models (cell culture) and in vivo (animal models), various chromatography techniques (HPLC peptide mapping), electrophoresis methods (SDS-PAGE, IEF, CZE), static and dynamic light diffusion mass spectrometry, x-ray techniques, etc. How the main analytical methods provide information on different orthogonal testing is listed in Table 5.3.

Table 5.3 Orthogonal Testing Attributes

Method	Size	Charge	Primary	Secondary/ Tertiary	Purity	Assay	Potency
SEC	+++	−	−	++	++		−
IEXC	−	++++	+++	−	−		−
MS:			+++	++			
LC-ESI-MS							
Disulfide linkage							
Terminal amino and carboxyl groups							
1D H^1 NMR Spectroscopy							
2D-NMR Spectroscopy							
RPC:	+++	±	+++	++	++	+++	−
Assay							
Peptide mapping							
SDS-PAGE	+++	−	+++	−	+++		−
CE-PAGE	+++	−	+++	−	+++		−
Circular dichroism							
DSC			++	++++			
UV spectroscopy						+++	
Fluorescence				+++		++	−
4D-TEST©					+++	++	++
IEF	−	+++	+	++	+++		+
cIEF							
Western blot	+++	−	++	+++	+++		−
Immunoassay (ELISA)	−	−	±	±	−		++
Receptor binding Antibody binding	−	−	++	+++	−		++

IEXC: Ion-exchange chromatography

Ion-exchange chromatography is a process that allows the separation of ions and polar molecules based on their affinity to the ion exchanger. It can be used for almost any kind of charged molecule including large proteins, small nucleotides, and amino acids. IEX chromatography retains analyte molecules on the column based on coulombic (ionic) interactions. The stationary phase surface displays ionic functional groups (R-X) that interact with analyte ions of opposite charge. This type of chromatography is further subdivided into cation exchange chromatography and anion-exchange chromatography. The ionic compound consisting of the cationic species M+ and the anionic species B− can be retained by the stationary phase. Cation exchange (CEX) chromatography retains positively charged cations because the stationary phase displays a negatively charged functional group:

Anion exchange chromatography retains anions using positively charged functional group.

Note that the ion strength of either C+ or A− in the mobile phase can be adjusted to shift the equilibrium position and thus the retention time.

RPC: Reverse-phase chromatography

Also called hydrophobic chromatography, reverse-phase chromatography (RPC) includes any chromatographic method that uses a hydrophobic stationary phase. RPC refers to liquid (rather than gas) chromatography. The term reversed-phase has a historical background. In the 1970s, most liquid chromatography was performed using a solid support stationary phase (also called a "column") containing unmodified silica or alumina resins. This method is now called "normal phase chromatography." In normal phase chromatography, the stationary phase is hydrophilic and, therefore, has a strong affinity for hydrophilic molecules in the mobile phase. Thus, the hydrophilic molecules in the mobile phase tend to bind (or "adsorb") to the column while the hydrophobic molecules pass through the column and are eluted first. In normal phase chromatography, hydrophilic molecules can be eluted from the column by increasing the polarity of the solution in the mobile phase. The introduction of a technique using alkyl chains covalently bonded to the solid support created a hydrophobic stationary phase, which has a stronger affinity for hydrophobic compounds. The use of a hydrophobic stationary phase can be considered the opposite, or "reverse," of normal phase chromatography—hence the term reversed-phase chromatography. Reversed-phase chromatography employs a polar (aqueous) mobile phase. As a result, hydrophobic molecules in the polar mobile phase tend to adsorb to the hydrophobic stationary phase, and hydrophilic molecules in the mobile phase will pass through the column and are eluted first. Hydrophobic molecules can be eluted from the column by decreasing the polarity of the mobile phase using an organic (nonpolar) solvent, which reduces hydrophobic interactions. The more hydrophobic the molecule, the more strongly it will bind to the stationary phase and the higher the concentration of organic solvent that will be required to elute the molecule.

SDS-PAGE: Sodium dodecyl sulfate-polyacrylamide gel

Polyacrylamide gel electrophoresis (PAGE) describes a technique widely used to separate biological macromolecules, usually proteins or nucleic acids, according to their electrophoretic mobility. Mobility is a function of the length, conformation, and charge of the molecule. As with all forms of gel electrophoresis, molecules may be run in their native state, preserving the molecule's higher-order structure, or a chemical denaturant may be added to remove this structure and turn the molecule into an unstructured linear chain whose mobility depends only on its length and mass-to-charge ratio. For nucleic acids, urea is the most commonly used denaturant. For proteins, sodium dodecyl sulfate (SDS) is an anionic detergent applied to the protein sample to linearize proteins and to impart a negative charge to linearized proteins. In most proteins, the binding of SDS to the polypeptide chain imparts an even distribution of charge per unit mass, thereby resulting in a fractionation by approximate size during electrophoresis. Proteins that have a greater hydrophobic content, for instance, many membrane proteins, and those that interact with surfactants in their native environment, are intrinsically harder to treat accurately using this method due to the greater variability in the ratio of bound SDS.

Two-dimensional gel electrophoresis, abbreviated as 2-DE or 2-D electrophoresis, is a form of gel electrophoresis commonly used to analyze proteins. Mixtures of proteins are separated by two properties in two dimensions on 2D gels. 2-DE was first independently introduced by O'Farrell and Klose in 1975. The 2-D electrophoresis begins with 1-D electrophoresis but then separates the molecules by a second property in a direction 90° from the first. Because it is unlikely that two molecules will be similar in two distinct properties, molecules are more effectively separated in 2-D electrophoresis than in 1-D electrophoresis. The two dimensions

133

that proteins are separated by using this technique can be the isoelectric point, a protein complex mass in the native state and protein mass.

To separate the proteins by isoelectric point is called *isoelectric focusing (IEF)*. Thereby, a gradient of pH is applied to a gel, and an electric potential is applied across the gel, making one end more positive than the other. At all pH values other than their isoelectric point, proteins will be charged. If they are positively charged, they will be pulled toward the more negative end of the gel; and if they are negatively charged, they will be pulled to the more positive end of the gel. The proteins applied in the first dimension will move along the gel and will accumulate at their isoelectric point; that is, the point at which the overall charge on the protein is 0 (a neutral charge). Most often proteins act together in complexes to be fully functional. To obtain a separation by size and not by the net charge, as in IEF, an additional charge is transferred to the proteins by the use of Coomassie Brilliant Blue or lithium dodecyl sulfate. After completion of the first dimension, the complexes are destroyed by applying the denaturing SDS-PAGE in the second dimension, where the proteins of which the complexes are composed of are separated by their mass. Before separating the proteins by mass, they are treated with SDS along with other reagents (SDS-PAGE in 1-D). This denatures the proteins (i.e., it unfolds them into long, straight molecules) and binds a number of SDS molecules roughly proportional to the protein's length. Because a protein's length (when unfolded) is roughly proportional to its mass, this is equivalent to saying that it attaches a number of SDS molecules roughly proportional to the protein's mass. Since the SDS molecules are negatively charged, the result of this is that all of the proteins will have approximately the same mass-to-charge ratio as each other. In addition, proteins will not migrate when they have no charge (a result of the isoelectric focusing step); therefore, the coating of the protein in SDS (negatively charged) allows migration of the proteins in the second dimension (SDS-PAGE, it is not compatible for use in the first dimension as it is charged, and a nonionic, or zwitterionic detergent needs to be used). In the second dimension, an electric potential is again applied, but at a 90° angle from the first field. The proteins will be attracted to the most positive side of the gel (because SDS is negatively charged) proportionally to their mass-to-charge ratio. Since this ratio will be nearly the same for all proteins, the progress of protein will be slowed by frictional forces. The gel therefore acts like a molecular sieve when the current is applied, separating the proteins on the basis of their molecular weight with larger proteins being retained higher in the gel and smaller proteins being able to pass through the sieve and reach lower regions of the gel. The result of this is a gel with proteins spread out on its surface. These proteins can then be detected by a variety of means, but the most commonly used stains are silver and Coomassie Brilliant Blue staining. In the former case, a silver colloid is applied to the gel. The silver binds to cysteine groups within the protein. The silver is darkened by exposure to ultraviolet light. The amount of silver can be related to the darkness, and, therefore, the amount of protein at a given location on the gel. This measurement can only give approximate amounts but is adequate for most purposes. Silver staining is 100× more sensitive than Coomassie Brilliant Blue with a 40-fold range of linearity.

Molecules other than proteins can be separated by 2D electrophoresis. In supercoiling assays, coiled DNA is separated in the first dimension and denatured by a DNA intercalator (such as ethidium bromide or the less carcinogenic chloroquine) in the second. This is comparable to the combination of native PAGE /SDS-PAGE in protein separation.

The 2D PAGE provides an even more detailed analysis of individual host cell proteins (HCPs) than 1D SDS-PAGE. If the HCP level is very low, it is also possible to load a higher protein amount on the 2D gel compared to the 1D gel. However, the 2D PAGE also has certain limitations, for example, limitations in pI and MW range, and that hydrophobic membrane proteins are not observed.

4D test

This is a proprietary and patented testing method developed by Therapeutic Proteins International, LLC (Chicago).

The western blot (sometimes called the protein immunoblot) is used to detect specific proteins; it uses gel electrophoresis to separate native proteins by 3-D structure or denatured proteins by the length of the polypeptide. The proteins are then transferred to a membrane (typically nitrocellulose or PVDF), where they are stained with antibodies specific for the target protein. The gel electrophoresis step is included in western blot analysis to resolve the issue of the cross-reactivity of antibodies. There are now many reagent companies that specialize in providing antibodies (both monoclonal and polyclonal antibodies) against tens of thousands of different proteins. Commercial antibodies can be expensive, although the unbound antibody can be reused between experiments. This method is used in the fields of molecular biology, immunogenetics, and other molecular biology disciplines. A number of search engines, such as CiteAb, are available that can help researchers find suitable antibodies for use in western blotting. Other related techniques include dot blot analysis, immunohistochemistry, and immunocytochemistry where antibodies are used to detect proteins in tissues and cells by immunostaining, and enzyme-linked immunosorbent assay (ELISA).

Identity

As mentioned earlier, this is not an attribute for new molecules as their identity is developed or described. For biosimilars, the first step is to establish that the biosimilar and the originator molecules have the same identity. In reality, it will be appropriate not to call it a molecule, but an entity, since this may include several variants, which when combined form the total identity of the biosimilar drug. Almost always if this stage of development leaves any residual uncertainty, the developer will be required to file a 351(a) or the full BLA.

The identity attributes such characterizations as the primary, secondary, tertiary, and quaternary structure of the molecule; these can be tested with MS, ELISA, cIEF, DSC, etc. Also included in the identity are all posttranslational modifications (PTMs), such as in the case of monoclonal antibodies; and finally, an identity test where the molecule shows binding to a molecule-specific antibody, such as in an SDS-PAGE, ideally, in a 2D staging.

All biological products can be divided into two distinct categories; one that do not involve PTM, and the other that do undergo PTM. PTM always leads a mixture of products, variants, that combine to form the product. Whereas the exact significance of the distribution of these variants is not known, the agencies are reluctant to allow a lot of variance in these patterns, even if the variants may not be related to MoA or side effects.

Primary structure

This includes all attributes related to the amino acid sequence and any PTMs including glycans. The common method used includes whole protein MIS, HC and

LCMS, peptide map, disulfide structure, glycan map, isoelectric point, extinction coefficient, and immune-based identity. These methods should be sensitive and resolving. Whereas the methods need not be fully validated, they must be sufficiently reliable. Some methods such as NMR are difficult to validate, particularly if these testings are outsourced.

An example of the difference may be in the glycosylation pattern. Knowing that differences in afucosylation and high mannose can impact ADCC function, these differences are significant and must be resolved before the biosimilar product development moves forward. The causes of these variations can be traced back to the choice of the host cell to bioprocessing conditions. Since the biosimilar product developer is likely to claim extrapolation of indications, which means asserting similarity for all mechanisms of actions, the matching of glycosylation pattern, even if it is not totally relevant, will always be required to remove any residual uncertainty of biosimilarity. The RP developers know this well and have provided broad protection to their process that results in a specific glycan pattern. This is one good example to demonstrate the need for the biosimilar product developer to engage early in the intellectual property evaluation of the proposed biosimilar product.

For the nonglycosylated proteins, the structural challenges may be less onerous, yet there remain process-related impurity profiles that will require optimization of upstream processing to match these profiles early in the development stage.

Protein sequencing is a technique to determine the amino acid sequence of a protein, as well as which conformation the protein adopts and the extent to which it is complexed with any nonpeptide molecules. The two major direct methods of protein sequencing are mass spectrometry and the Edman degradation reaction (N-terminal as well). It is also possible to generate an amino acid sequence from the DNA or mRNA sequence encoding the protein if this is known. However, there are many other reactions that can be used to gain more limited information about protein sequences and can be used as preliminaries to the aforementioned methods of sequencing or to overcome specific inadequacies within them. The mass of the intact met-G-CSF protein is determined by mass spectrometry (ESI MS and MALDI MS).

Extinction Coefficient (A280) is used to calculate protein concentration by measuring UV absorbance at A280. This technique helps demonstrate consistency and similarity between lots. Note that protein concentration can also be determined by ELISA assays or mass spectrometry using absolute quantification (AQUA). The extinction coefficient can be determined by two methods: amino acid analysis and UV light absorption at 280 nm. This is in line with ICH Q6B, which recommends determining the coefficient (sometimes referred to as molar absorptivity) by a combination of the two. This method is important because, in complex solutions, proteins are not the only molecules that absorb UV light. Other compounds can skew the results, but their interference is minimized when the absorbance is measured at the UV level. Once the extinction coefficient is determined, it can be combined with other attributes of the protein, such as amino acid composition, to improve the robustness of the measurement of the concentration of protein in the samples.

Amino Acid Analysis (AAA) is done to determine the amino acid composition of a protein without using an external standard. It can be applied in many ways to reveal different aspects of the protein. In combination with UV-absorbance measurements, it can be used directly to determine the extinction coefficient of a protein. Omitting hydrolysis, it can be used to quantify free amino acids, for example, in cell culture media. It can quantify unusual amino acids such as Norleucine (encountered in *E. coli* fermentations) or hydroxyproline and lysine. The concentration based on mass can be combined with MALDI-TOF to reveal the intact mass of the protein.

Peptide mapping is commonly undertaken to analyze the protein primary structure after cleavage of the protein into proteolytic peptides. The power of peptide mapping lies in the large number of site-specific molecular features that can be detected. When using one digestion enzyme (e.g., Trypsin), peptide mapping is typically carried out for protein identification. The analysis is performed using MALDI-MS/MS or LC-ESI-MS/MS, for example, during protein identification after electrophoresis, such as 1D or 2D PAGE. When using multiple enzymes, peptide mapping is applied for the confirmation of a complete amino acid sequence, for example, when confirming the amino acid sequence of a biosimilar and comparing it with the originator molecule. Depending on the experimental setup, peptide mapping can also be used to determine the N- and C-terminus of a protein. This is sometimes crucial information for our clients, for example, in the case of monoclonal antibodies, where the truncation level of the C-terminal lysine can be monitored. Proteins and peptides are ionized and fragmented in the mass spectrometer. The resulting MS/MS spectra are used to sequence the individual amino acids in order of their appearance in a protein or peptide. In combination with additional sample preparation, peptide mapping can also be used to determine the degree of deamidation or oxidation for each affected amino acid, glycosylation sites and structure, N-glycosylation sites, and disulfide linkages.

N- and C-terminal sequencing can reveal a great deal of information about proteins and antibodies at any stage of the drug development process. It is also performed to demonstrate consistency and comparability between lots. It can determine the N-terminal blockage or signal peptide cleavage. The level of C-terminal lysine truncation of monoclonal antibodies is a critical quality attribute and, therefore, needs to be monitored closely. When done by MALDI-ISD, N- and C-terminal sequencing can be used specifically to monitor terminal amino acids for modifications such as glycosylation, disulfide bridging, and oxidation. With the digestion strategy, we can use peptide mapping with LC-ESI-MS/MS to identify and quantify the N- and C-terminal sequence variants.

Disulfide linkages are an important structural feature of many proteins. Intra-chain bonds form or stabilize the tertiary structure of a protein (e.g., the zinc-binding domain of zinc-finger proteins) and inter-chain disulfide bonds covalently link protein subunits together (e.g., the insulin a- and b-chain). The process for identifying disulfide linkages includes investigating the differential appearance of the linked peptides before and after reduction of the disulfide bond, and analyzing the fragment spectra of the linked peptides using mass spectrometry. Peptide mapping may be helpful to confirm the primary structure and aid in determining the most appropriate enzymatic digestion strategy for the disulfide bridge analysis. N- and C-terminal sequencing using MALDI-ISD can be used to determine the presence of free sulfhydryl groups or disulfide linkages within these regions. Finally, electrophoresis, such as reducing and nonreducing SDS-PAGE to identify disulfide-linked subunits of a protein, may also be undertaken.

Glycosylation Monoclonal antibodies (mAbs), having high selectivity and specificity, constitute a large and growing portion of the biosimilars. The majority of marketed mAbs belong to the IgG class. IgGs consist of two heavy chains and two light chains linked by a total of 16 inter- or intra-molecular disulfide bonds. The two heavy chains are linked by disulfide bonds, and each heavy chain is disulfide bonded to a light chain. IgGs include antigen-binding (Fab) and crystallizable (Fc) regions: the Fab is responsible for binding to the antigen while the Fc binds to Fcγ receptors, which regulate immune responses.

During the development of mAbs from drug candidate to marketed product, issues with stability, such as aggregation due to physical instability, or deamidation or oxidation due to chemical instability, often arise. Substantial resources and time are required to address stability problems; thus, this area is one of intense focus. One factor that may also affect the stability of mAbs is the glycosylation found in the Fc region. Glycosylation is a common PTM for IgG antibodies produced by mammalian cells such as Chinese hamster ovary (CHO) cells, which are frequently used for production. IgG1 molecules contain a single N-linked glycan at Asn^{297} in each of the two heavy chains. During the synthesis of N-glycans, multiple sugar moieties can be added to form different glycoforms, for example, G_0, G_1, G_2, afucosylated complex. Glycosylation plays an important role in complement-dependent cytotoxicity (CDC) and antibody-dependent cell-mediated cytotoxicity (ADCC) functions through modulating the binding to the Fcγ receptor. Particular glycoforms may be necessary to achieve therapeutic efficacy. These glycoforms may be targeted by glycosylation engineering but may also be affected by cell culture conditions.

Glycan testing is a complex and comprehensive analytical exercise. All glycan patterns can be judged from 2D PAGE analysis and tested for monosaccharides and sialic acid content. The N-glycans and O-glycans require determination of the site of glycosylation and profiling by MALDI and HPLC that allows determination of the content of each glycan. The structural heterogeneity of a glycoprotein can be ascertained using IEF or 2D PAGE. This is because sialic acids will lead to a shift in the pI of the glycoprotein, which can be determined. The cores of N- and O-linked glycans are largely composed of neutral monosaccharide building blocks, joined together by a specific stereochemistry. The stoichiometry and identity of these building block are analyzed quantitatively by using high-performance anion-exchange chromatography with pulsed amperometric detection (HPAEC-PAD). This allows analysis at both N- and O-linked glycans in a single experiment. Usually, N-linked glycans are attached to the protein backbone by an N-acetylglucosamine and O-linked -glycans by an O-N-acetylgalactosamine. Other typical neutral monosaccharides involved in N-linked glycosylation and O-linked glycosylation are fucose, galactose, and mannose. More specific testing may include analysis of monosaccharides released by acidic hydrolysis with HPAEC-PAD, investigation of intact N- or O-linked glycans by HPAEC-PAD, LC-ESI-MS, or MALDI-MS, or visualization of the structural heterogeneity of a glycoprotein using 2D PAGE or intact mass.

N-linked glycans are a common feature of glycoproteins such as EPO, antibodies, and FSH. They have been shown to be important in many ways. They have an influence on protein folding as well as other factors, such as protein solubility, stability, immune reactions, and cell-to-cell interactions. N-glycosylation increases the structural heterogeneity of proteins when a range of structurally similar but not identical glycans are attached to each glycosylation site. Glycosylation may be directly involved in the formation of diseases, but also plays an important role in altering the pharmacokinetics by stabilizing protein conformation, improving solubility, or protecting from proteases. The N-glycosylation profile of a recombinant protein can vary largely depending on the organism chosen for expression.

O-linked glycans are a common feature of many glycoproteins, such as EPO, Factor VIII, or growth factors and occur in the Golgi apparatus. Even though they are not necessarily as structurally complicated as in N-linked glycosylation, O-linked glycans can be incredibly diverse. Also, they may attach themselves to proteins at many more possible sites, which increases the complexity of the analysis. O-linked glycans are released by enzymatic digestion or chemical means, and we look at them directly by HPAEC-PAD, or after labeling or permethylation by MALDI-MS or LC-ESI-MS. The method will depend on your research question. The structural heterogeneity of a glycoprotein can be visualized using 2D PAGE or intact mass.

Site of glycosylation is determined by examining how the N- and O-linked glycans are attached to proteins via specific amino acids. While N-linked glycans are only attached to asparagine residues incorporated into a consensus sequence, O-linked glycans can be attached to a number of different amino acids. Analysis of the N- or O-linked glycans can be carried out by HPAEC-PAD, LC-ESI-MS, or MALDI-MS. This can be done through a summary analysis of all glycosylation sites or through site-specific analysis. The structural heterogeneity of a glycoprotein can be visualized using 2D PAGE or intact mass. Almost all proteins, whether produced and purified as recombinant proteins or isolated from natural sources, will to some degree carry modified amino acids.

Sialic acids are negatively charged, terminal monosaccharides, and are important structural constituents of many glycans. The number of sialic acid molecules present impacts on the activity and serum stability of many glycoproteins. As different types of sialic acid can be incorporated into the glycan structure, it is important to be able to determine the presence of correct versus undesired sialic acid structures. Once sialic acids are released by acidic hydrolysis, it can be analyzed by HPAEC-PAD; also, N- and O-linked glycans carrying sialic acids can be tested by using HPAEC-PAD, LC-ESI-MS, or MALDI-MS.

Monosaccharide analysis: The composition of glycans attached to a protein can further be determined by monosaccharide analysis.

Whereas most of the recombinant products would undergo similar testing protocols, there are specifically recommended tests that may be required for the monoclonal antibodies. Some examples of these tests include

- Heavy and light chain Mw determination by mass spectrometry
- Peptide mapping by mass spectrometry of heavy and light chain
- N- and C-terminal sequencing of heavy and light chain by MALDI ISD
- CDR sequencing of variable complementary determining regions
- PTM analysis; deamidation, oxidation, pyroglutamate, N-glycosylation
- For most monoclonal antibodies, the degree of Fc and ADCC binding profiles will be required for similarity determination even if these do not affect the functional response

It is well known that the terminal functional groups in protein structure may not have a significant impact on the efficacy of the molecule, but it leaves uncertain whether it would have any potential immunogenic response; the same holds true for disulfide bonding and other intramolecular interactions. In those instances where such changes are determined earlier such as through mass spectrometry and circular dichroism, the developer may choose the 351(a) route instead to avoid more extensive analytical comparison.

Higher-order structure

The integrity of the secondary, tertiary, and quaternary structure is important to establish similarity. Common methods used include FTIR, near UV, CD, and DSC. These methods should be sensitive and resolving. It should be realized that the reference product extinction coefficient is not always available in public and must be determined. Secondary and tertiary structures are best evaluated using far and near-UV circular dichroism, and 1D and 2D nuclear magnetic resonance. Far and near-UV CD spectroscopy provides information about secondary (alpha-helix, beta-sheet, and random coil structures) and tertiary structure, respectively. In addition to similar CD spectra, comparison of transition point (specific ellipticity = 0) and ratio of specific ellipticity ($\Theta R208/\Theta R222$) data derived from the far UV CD spectra can be used to demonstrate a high level of similarity. ^{1}H NMR spectroscopy also provides information about the three-dimensional structure of the protein. No significant difference in the NMR spectra of the test product compared to the reference product is required as judged by the number, position, and intensity of peaks. To further support higher-order structure (HOS) similarity, natural isotope abundance 2D NMR (^{1}H-^{15}N HSQC) spectra can be used. ^{1}H-^{15}N NMR provides better resolution than ^{1}H NMR, and that is considered a structural fingerprint of the protein.

Purity

This element of the biosimilarity tetrahedron includes impurities that are related to the product, as well as the process; for example, a genetic mutation producing an excess of acetylated product beyond the accepted purity profile will be a significant event. These variations are related to the structure of the recombinant cell line and are very difficult to modulate. However, some binding characteristics and glycan patterns are dependent on process conditions and where possible need to be optimized.

More specifically, the impurities may come from

- The majority molecular form and its variants and isoforms, each carrying an intrinsic biological activity close to the majority molecular form's biological activity (for instance, EPO and its isoforms and three distinct forms present in etanercept)
- The impurities that are linked to product itself, but which practically carry no biological activity; these may come from degradation as well
- The impurities linked to the production process or to the purification scheme

The variants may be inherent to the product, produced as a result of the manufacturing process, or derived from degradation. The testing may include evaluation over the shelf life of the product. While the activity and safety of biosimilar products are determined mainly by their structural variants, the impurity profile has a significant impact on all aspects of the product characteristics, from activity to immunogenicity.

The fundamental point of start is to examine the impurities that can appear in the product (Figure 5.3); the purity can be due to variations in the protein structure, degradants, and the process-related impurities; all of these must be identified. Common tests used for this detection include SE-HPLC, CEX-HPLC, reduced CE-SDS, and nonreduced CE-SDS. Product-related variants should be characterized by separation methods; size variants include truncation, dimers, and multlimers, charge and hydrophobic variants include N-terminal modification, C-terminal modification, deamidation, and oxidation. Regulatory agencies have specific allowances for impurities, generally not more than 3% at expiry and no single impurity more than 1%. All impurities must be identified. However, some differences such as in the levels of C-terminal lysine may not impact on safety and efficacy and can be justified. A good example of this is the variants reported for adalimumab, lysine-0, lysine-2, and lysine-2; all lysine groups are removed once the antibody enters the body and are not related to the activity of the antibody. However, the

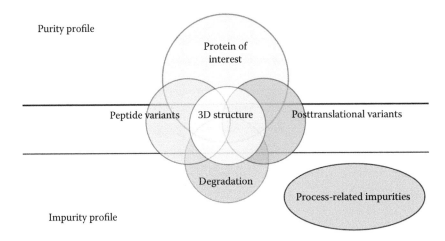

Purity profile

Protein of interest

Peptide variants   3D structure   Posttranslational variants

Degradation

Process-related impurities

Impurity profile

Figure 5.3 Contributors to impurities in biosimilar products.

developers of biosimilars to adalimumab may face the challenge of assuring that each lot has the same lysine variants; Abbvie has filed a patent on how to control these variants, even though they are not relevant.

Product-related impurities

Product-related impurities are basically typed of protein modification such as deamidation or oxidation. Characterizing product-related impurities is a major task and requires analyzing those using highly sensitive and accurate methods. While the goal is to have the lowest possible number of process-related modifications, these are often inevitable as a result of sample handling process, for example, in proteins that have been stored for a period of time during a stability study. They are unintentional and a source of structural heterogeneity in protein. These changes are known as PTMs and may be undesirable. Mass spectrometry can be used to detect even the smallest modifications, which are usually analyzed using a combination of electrophoretic, HPLC, and peptide mapping methods. Normally, mass spectrometry data are qualitative rather than quantitative. However, a quantitative approach can be taken to evaluate the degree of modification taking place. Several types of analysis are possible including study of deamidation, oxidation, glycosylation, phosphorylation, N- and C-terminal truncation, acetylation, and pyroglutamate formation. All of these can be analyzed using a combination of 2D gel electrophoresis, high-performance liquid chromatography (HPLC), and peptide mapping.

Deamidation is commonly observed as a PTM of the amino acid asparagine, which turns into a mixture of isoasparate and aspartate via a succinimide intermediate. The rate of modification is influenced by factors such as the presence and interaction of surrounding amino acids, pH, and temperature buffers. Deamidation of glutamine also occurs, but much less frequently. It will be desirable to find out whether the protein has undergone deamidation because it leads to a change in isoelectric point (pI) and, therefore, the presence of charge heterogeneity. This can be reliably observed by 1D-IEF or IEX-HPLC. These variants can affect the effectiveness and even immunogenicity.

Oxidation of proteins is usually observed in methionine residues and to a lesser extent in tryptophan or tyrosine. As the name suggests, it is promoted by a form of oxygen, such as peroxide, but is also influenced by light, pH, temperature, buffers, and other factors. In many proteins, methionines in certain positions of an amino acid

141

Table 5.4 Product-Related Impurities of Filgrastim

Impurity	Method of Testing
Oxidized species	Reverse phase high-performance liquid chromatography (RP-HPLC) Liquid chromatography coupled with mass spectrometry (LC-MS)
Covalent dimers	Liquid chromatography coupled with mass spectrometry (LC-MS)
Partially reduced species	Liquid chromatography coupled with mass spectrometry (LC-MS)
Sequence variants: His → Gln Asp → Glu and Thr → Asp	Liquid chromatography coupled with mass spectrometry (LC-MS)
fMet1 species	Liquid chromatography coupled with mass spectrometry (LC-MS)
Succinimide species	Liquid chromatography coupled with mass spectrometry (LC-MS)
Phosphoglucunoylation	Liquid chromatography coupled with mass spectrometry (LC-MS)
Acetylated species	Liquid chromatography coupled with mass spectrometry (LC-MS)
N-terminal truncated species	Liquid chromatography and tandem mass spectrometry (LC-MS/MS)

are especially prone to oxidation, presumably due to differences in solvent accessibility. The relative impact of oxidized products is less severe than deamidated variants.

Proteases are an enzyme class showing very diverse specificities and characteristics. Some have one specific substrate in vivo while some are much more promiscuous. This makes it very difficult to detect unknown protease contaminations, which are a major problem in production and storage of protein or peptide-based pharmaceuticals and diagnostic assays. Highly specific FRET substrates (sequence either directly deriving from drug substance or adapted to assay issues) can be used to detect proteolytic enzymes for the in-process control of the upstream and downstream production of therapeutic proteins. Peptidic FRET (fluorescence resonance energy transfer) substrates contain a donor (a fluorophore) and an acceptor (quencher) that absorbs at the emission wavelength of the fluorophore. The quencher absorbs the energy emitted by the fluorophore only as long as they are in close proximity, connected by the peptide chain. Enzymatic hydrolysis causes an increase of fluorescence irrespective of the cleavage position, provided that donor and acceptor are disconnected.

Table 5.4 lists some of the product-related impurities listed for filgrastim products.

Process-related impurities

This category of impurities creates a large challenge for biosimilar product developer since these might have a significant impact on all of the three concerns: purity, potency, and safety of the product. Some impurities come from host cells and others from the downstream process; the final packaging may itself add to these impurities such as tungsten or silicon. Common tests used include HCP using 2D-SD-PAGE, protein A, DNA, 2D SDS-PAGE, and HCP using LC/MS/MS. However, the FDA does not expect the process-related impurities (such as HCPs) present in the biosimilar product to "match" those observed in the reference product. However, process-related impurities in biosimilar product should be assessed side-by-side with the reference product. The FDA recommends performing a risk-based assessment regarding any differences in process-related impurities identified between the biosimilar product and the reference product. If the manufacturing process used to produce biosimilar product introduces different impurities or higher levels of impurities than those present in the reference product, additional pharmacological/toxicological or other studies may be necessary to evaluate the potential risk of any differences, and any differences should be justified. The adequacy of the risk-based assessment will be a review issue.

Regarding the HCP assay, the sponsor must provide a summary description of the source (in-house or commercial) of the antiserum used for detection of HCP impurities. The FDA recommends developing a cell-line-specific HCP detection reagent. For licensure, the anti-HCP antiserum needs to be qualified to detect potential HCP impurities. The data need to include 2D SDS-PAGE gels of the range of HCPs detected by a sensitive protein stain, such as silver stain, compared to the range detected by western blot analysis (or another similarly sensitive assay) using the antiserum employed in the assay. It is the FDA's experience that analysis of HCP coverage by a 1-dimensional SDS-PAGE gel method is not sufficiently sensitive for this purpose.

Process-related impurities, such as HCPs, can be detected using ELISA, 2D gel electrophoresis, western blotting, or mass spectrometry. Process-induced modifications, which may be introduced during sample preparation, down-stream-processing, or other steps involving the use of chemicals, heat, or light. Examples are methylation or acetylation of side chains, fragmentation, and glycation. Biologically induced modifications may be introduced in the cell or supernatant by the complex biological system itself. These are often crucial to the correct biological function of the protein. Examples include phosphorylation, sulfation, and formylglycine. Investigations for these modifications require the use of RP-HPLC, intact mass determination, or peptide mapping.

ELISA, which is useful when looking at quality control features such as drug purification, lot release, and stability testing. ELISAs can be applied to detect and quantify specific analytes by applying antibodies in a complex matrix. The major aim is to demonstrate the specificity of the antibodies used in developing the assay. For process-related protein impurities, such as HCP or column leakage (protein A), quantitative and qualitative ELISAs are performed in microtiter plate (MTP) format using an enzyme-linked detection system with an absorbance or fluorescent readout.

Other significant process-related impurities for biosimilars include tungsten level, leachable and extractable, visible and invisible particles, etc. While some of these impurities may have a lesser impact on the formulation of small-molecule drugs, these can have a significant effect on the formulation of proteins. The most widely quoted role of process-related impurities is the case of EPO. The incidence of pure red cell aplasia (PRCA) in chronic kidney disease patients treated with epoetins increased substantially in 1998 was shown to be antibody mediated, and was associated predominantly with subcutaneous administration of Eprex®. A technical investigation identified organic compounds leached from uncoated rubber stoppers in prefilled syringes containing polysorbate 80 as the most probable cause of the increased immunogenicity. The rubber stoppers were switched without conducting the impact on formulation assuming they have no effect.

Potency

The potency of the biosimilar products is compared with the originator reference product as a definite measure of biosimilarity based on two considerations: protein content and functional response of the protein. Affecting both of these attributes are the formulation elements, such as inactive ingredients and other attributes like osmolality, pH, appearance, color, clarity, and surfactant, or stabilizer concentration. It should be noted that the regulatory agencies allow the use of alternate inactive ingredients as long as they introduce any "clinically meaningful difference." In some cases, this may be difficult to prove and, therefore, the biosimilar product developer is left with fewer choices of excipients. However, the grade of excipients used can be significant. Lately, many higher-quality common components used in these products have become available; higher quality is meaning fewer impurities and higher consistency of the specification. The biosimilar product developer is encouraged to acquire the best quality excipients, even if the originator is not using these grades; the reason why the originator may not be using these grades is to avoid conducting comparability exercise. However, changing the specification of excipients with just cause is also not recommended. When developing a 505(j) product, the goal of the developer is to create a Q/Q formula; this not required, but it also creates needs for more studies since many of the quality attributes that may have little effect on 505(j) products can be significant for 351(k) products. The attribute like pH is important for stability, the osmolality for the comfort of administration and aggregation (immunogenic potential), and the use of surfactants can introduce stability issues because of the traces of peroxides in them.

Potency testing includes three major components: the bioactivity in a bioassay, a receptor-binding study, and also the content of protein, which will then determine the overall potency that may be content dependent. To be potent, the product must demonstrate equivalent pharmacokinetic (what body does to the molecule) as well as pharmacodynamic (what drug does to the body).

Protein content

One of the most important attributes to meet is the protein concentration, as simple as it might sound. This may appear rather redundant but using methods like A280 can often be misleading, and the biosimilar product developer is strongly urged to adopt orthogonal methods like RP-HPLC for protein concentration. While many other attributes may be described with acceptable or justified ranges, the concentration or strength must be met. The biosimilar product developer may face a dilemma when a sufficiently large number of reference lots are not available as this might affect the statistical robustness of testing as described under statistical considerations.

Bioactivity

Relevant functional assays are best available predictors of clinical bioactivity; these tests help demonstrate the structure–function relationships and are therefore placed in the category of tests that must be evaluated using an equivalence statistical model. These assay types include

- Target binding (soluble and membrane-bound)
- Receptor binding (including FcRn for MAbs)
- Affinity; on/off rates
- Effector function (MAbs; ADCC, CDC, etc.)
- Enzyme kinetics (complex physiological substrates vs. low molecular weight analogs, K_m, k_{cat}, etc.)
- Cellular uptake by target cells (e.g., for replacement enzymes)
- Bioactivity (cell proliferation, cytotoxicity, apoptosis, signal transduction, etc.)

Cell-based assays

These assays are used to determine the potency of a biological product and to monitor specific biological activity. Specific enzyme activity assays determine the functionality of enzymes. Cell-based assays are commonly designed around a specific biological drug and its particular MoA. Several assay responses are possible, including cell proliferation, cell killing (cytotoxicity), differentiation, cytokine/mediator secretion (receptor activation), or enzyme activation. These can be used to determine the specific activity of proteins like monoclonal antibodies, cytokines, hormones, and growth factors. By using cell lines that respond to the specific MoA, cell-based assays can determine the relative potency of a biological product in relation to a reference standard material. Potency assays provide assurance of the quality and consistency of the product. Note that biological activity (receptor–ligand interactions) can also be determined by an ELISA. Furthermore, cell-based assays can be coupled with an ELISA as readout. The data comparing a test and a reference drug can be analyzed using software such as PLA3.0 to fulfill the statistical evaluation requirements (www.bioassay.de).

Receptor-binding assays

Regardless of the nature of the initiating signal, the cellular responses are determined by the presence of receptors that specifically bind the signaling molecules. Binding of signal molecules causes a conformational change in the receptor, which then triggers the subsequent signaling cascade. Given that chemical signals can act either at the plasma membrane or within the cytoplasm (or nucleus) of the target cell, it is not surprising that receptors are found on both sides of the plasma membrane. The receptors for impermeant signal molecules are membrane-spanning proteins that include components both outside and on the cell surface. The extracellular domain of such receptors includes the binding site for the signal while the intracellular domain activates intracellular signaling cascades after the signal binds. A large number of these receptors have been identified and are grouped into three families defined by the mechanism used to transduce signal binding into a cellular response:

- Membrane-impermeant signaling molecules can bind to and activate either channel-linked receptors (A), enzyme-linked receptors (B), or G-protein-coupled receptors (C) membrane permeant signaling molecules activate intracellularly.
- Channel-linked receptors (also called ligand-gated ion channels) have the receptor and transducing functions as part of the same protein molecule. Interaction of the chemical signal with the binding site of the receptor causes the opening or closing of an ion channel pore in another part of the same molecule. The resulting ion flux changes the membrane potential of the target cell, and in some cases, can also lead to the entry of Ca^{2+} ions that serve as a second messenger signal within the cell.
- Enzyme-linked receptors also have an extracellular binding site for chemical signals. The intracellular domain of such receptors is an enzyme whose catalytic activity is regulated by the binding of an extracellular signal. The great majority of these receptors are protein kinases, often tyrosine kinases, that phosphorylate intracellular target proteins, thereby changing the physiological function of the target cells. Noteworthy members of this group of receptors include the Trk family of neurotrophin receptors and other receptors for growth factors.
- G-protein-coupled receptors regulate intracellular reactions by an indirect mechanism involving an intermediate transducing molecule, called the GTP-binding proteins (or G-proteins). Because these receptors all share the structural feature of crossing the plasma membrane seven times, they are also referred to as 7-transmembrane receptors.

Table 5.5 Representative Functional Assay Types of Monoclonal Antibodies

Antibody	Target	Assay
Infliximab	Inhibition of TNF-alpha signaling	Proliferation assay
Adalimumab	Inhibition of TNF-alpha signaling	Proliferation assay
Rituximab	CD20	Binding assay (cytometry), ADCC assay, CDC assay
Bevacizumab	VEGF	CDC assay
Trastuzumab	HER2, ErbB2	Proliferation assay, ADCC assay
Cetuximab	EGF-R	Proliferation assay
Natalizumab	Alpha4-integrin	Binding assay (cytometry)

Intracellular receptors are activated by cell-permeant or lipophilic signaling molecules. Many of these receptors lead to the activation of signaling cascades that produce new mRNA and protein within the target cell. Often, such receptors comprise a receptor protein bound to an inhibitory protein complex. When the signaling molecule binds to the receptor, the inhibitory complex dissociates to expose a DNA-binding domain on the receptor. This activated form of the receptor can then move into the nucleus and directly interact with nuclear DNA, resulting in altered transcription. Some intracellular receptors are located primarily in the cytoplasm while others are in the nucleus. In either case, once these receptors are activated they can affect gene expression by altering DNA transcription.

For products like GCSF, IFN, and GMCSF, proliferation assays are used; for monoclonal antibodies, the assay will depend on the mechanism of action (Table 5.5).

In a step-wise approach, once the structure of the product is established to be at least "highly similar," its evaluation begins for the potency of the product. Matching the RP profiles is done with greater emphasis on matching all biological functions that include biological and functional activities, receptor binding, and immunochemical properties. Common tests used to demonstrate these include potency assay, effector functions (ADCC/CDC), Fc receptor family (FcYR) binding activities, antigen binding, neonatal FC receptor (FcRN), and other MoA. The test methods must be sensitive and resolving, and their importance well understood; for example, if ADCC is not matching then optimization is required. However, all assays reported here have limitations that should also be understood, and it is for this reason that the regulatory agencies require orthogonal testing to confirm similarity.

In the recent approval of a biosimilar product for filgrastim, a cell proliferation assay using murine myelogenous leukemia cells (NFS-60 cell line) was used to evaluate the biological product approved. This cell line carries the G-CSF receptor, and it is commonly used to evaluate the biological activity of this growth factor. The bioactivity data were reported as percentage relative to the applicant's in-house reference standard calibrated against an international G-CSF reference standard. These data were subjected to a statistical analysis using equivalence testing where $2 \times SD$ of the reference product was taken as the equivalence range.

Safety

An application submitted under section 351(k) of the PHS Act must contain, among other things, information demonstrating that the biological product is biosimilar to a reference product based on data derived from "a clinical study or studies (including the assessment of immunogenicity and pharmacokinetics or pharmacodynamics) that are sufficient to demonstrate safety, purity, and potency in one or more appropriate conditions of use for which the reference product is licensed and intended to be used and for which licensure is sought for the biological product." Immune responses against therapeutic biological products are a concern because they can negatively impact the drug's safety and efficacy. Unwanted immune reactions to therapeutic biological products are mostly caused by antibodies against the drug (antidrug antibodies (ADA)). Therefore, immunogenicity assessment for therapeutic biological products focuses on measuring ADA.

The immunogenicity profiling of biosimilar products is pivotal; however, as seen in Table 5.5, the innate immunogenic potential of a biological drug differs significantly between products in the innate immunogenicity of biological drugs and depends on several factors like the active substance's nature and structure, impurities, excipients of the medicine, manufacturing process, route of administration, and target population. A large number of biological drugs have very small immunogenicity while others are clearly very high-risk molecules. These differences may compromise the product in vivo behavior, with, as a consequence, undesirable effects for the host that may minimize the intended clinical effect with potentially lethal reactions.

Different approaches based, for instance, upon the response of the epitope to human leucocyte antigen (HLA) polymorphism or the immunological response studied in relevant animal models, may be used to evaluate a biosimilar immunological profile. However, if these responses are important to identify the antigenic profile, they are not predictive of the immunological response to the biosimilar in vivo. Evaluation of a biosimilar antigenic profile in patients is complex because of the difficult measurement of antibodies' level (unavailability of immune serums, absence of appropriate standards, interference of endogenous proteins, limits of analytical methods, etc.) Similarly, the simple comparison of products of the same therapeutic class, although interesting on a theoretical level, is not enough and may be the source of misinterpretation (Table 5.6).

It is noteworthy that not all drug substances lead to similar immune response and are brand dependent, as it may be construed from Table 5.6. However, given the high variability of immunogenicity response, the method of collecting these data and the fact that most of these data are not readily accessible, this conclusion is not fair

Table 5.6 Immunogenicity of Recombinant Drugs

Incidence	Brand Product	Drug Substance
95%	Leukine	Sargramostim
74%	Proleukin	Aldesleukin
25–45	Betaferon	IFN-beta-1b
12–28	Rebif	IFN-beta-1a
14–24	Remicade (and its other biosimilars)	Infliximab
12	Humira	Adalimumab
0–26	Rituxan	Rituximab
3.5–5	Erbitux	Cetuximab
2–6	Avonex	IFN-beta-1a
0–2	Neupogen (and other biosimilars), Procrit, Neorecormon, Aranesp, Avastin, Neulasta, Genotropin (and other biosimilars), Herceptin	Filgrastim, erythropoietin alpha erythropoietin beta, darbopoietin alpha, bevacizumab, pegfilgrastim, somatropin, trastuzumab

Source: Wadhwa, M., Immunogenicity: What do we know and what can we measure?, PhD thesis, NIBSC–HPA, Hertfordshire, U.K., April 12, 2011.

Notes: EPAR 1 at the end of treatment based on the indication.
Commonly reported immunogenic responses to potential biosimilar candidates.

to make, at least at this time. Unless a drug product (the brand) has specific impurities or components that may be independently immunogenic, we need to focus on the innate character of the molecule to trigger an immune response. Several small-molecule drugs are widely known to induce an immunogenic response, the best examples of this include

- Anticonvulsants
- Iodinated (containing iodine) x-ray contrast dyes (these can cause allergy-like reactions)
- Penicillin and related antibiotics
- Sulfa drugs

The difference between the nature of immunogenicity caused by the small molecule drugs listed earlier and the biological drugs listed in Table 5.6 is in the mechanism of the immune response. Some biological drugs are intended to modulate the immune system such as leukines and, therefore, an immunogenic responses is expected, not an aberrant phenomenon. In addition, the structure of recombinant proteins and antibodies lends itself to immunogenicity by design, something not obvious or apparent in small-molecule drugs. It is this subtle difference that has been widely discussed, making the immunogenic response as the biggest risk in the use of biosimilar drugs. There is a dire need to educate the prescribers and other stakeholders including patients about the innate nature of biological drugs to take away the uncertainty from biosimilar biological drugs.

The reason why immunogenic responses to therapeutic proteins and antibodies are significant is because of the long-term effect on immune system they can bring, and this topic is discussed in detail in Chapter 2 (Immunogenicity Considerations), where I have provided greater details on the innate character of the molecule to trigger an immune response; the purpose of bringing this topic in this section is to apprise the developers of biosimilars that the concern about biosimilars having any extraordinary risk is not true—a good quality product should have no more of this risk than the originator product, and in many instances, there is no risk of immunogenicity; hence, the need for the developer to be aware of the global nature of immunogenicity.

Overall, the decision to put a biosimilar on the market is made if its effectiveness is similar and its immunogenic profile is at least comparable or improved in comparison to the first-licensed product. However, this decision is made on a limited data. The similarity testing program may disclose substantial differences in terms of immunogenic profiles but is probably unable to detect minor differences and rare events. For that reason, clinical trials complemented by a pharmacovigilance program are often required for evaluating a recombinant protein's safety in patients. Undesirable effects of these drugs are very rare, yet require a follow-up during the life of the product.

This includes testing all those attributes that can trigger a side effect such as immunogenicity; included here are aggregates, dimers, subvisible and visible particles, HCP, host cell DNA, etc. The nonclinical studies are generally required to establish overall toxicology, such as in an animal model; however, in several situations, an animal model may not be readily available, such as in the case of monoclonal antibodies. Several creative models have recently been proposed including transgenic mice, etc. Also included here are safety assessment at toxicological level and in the clinical pharmacology of the product, such as pharmacokinetic and pharmacodynamic data.

This is perhaps one of most difficult areas of biosimilar product development. All protein therapeutic contain particles, and these can be immunogenic.

Subvisible and submicron particles and aggregates of various sizes can be tested using methods such as HIAC, MFI, DLS, FFF, AUC-SV, and SE-HPLC-LS. It is most important to determine whether a higher level of subvisible particles is nonproteinaceous.

All protein therapeutics contain (higher or lower levels of) aggregates and particles, and most of these products are immunogenic. Aggregation results from fermentation/expression, purification, formulation, filling, transport, and storage administration. Some of the factors influencing protein aggregation include temperature, interfaces, freeze–thaw, container, pH, excipients, ionic strength, and concentration. Protein aggregates are assemblies of protein molecules and are highly heterogeneous regarding their size, reversibility, protein conformation, and covalent modifications. All protein therapeutics contain (higher or lower levels of) aggregates and particles. Most of these products are immunogenic. Aggregation results from fermentation/expression, purification, formulation, filling, transport, and storage administration. Some of the factors influencing protein aggregation include temperature, interfaces, freeze–thaw, container, pH, excipients, ionic strength, and concentration. The protein aggregates are the highly heterogeneous assemblies of protein molecules. These assemblies vary in their size, reversibility, protein conformation, covalent modification, and morphology. It is difficult to predict which form will be more immunogenic. It is because even a simple dimer can adopt various shapes and characteristics based on pH, process stress, light, osmolality, etc. The micron-sized IgG aggregates induced by shaking remain at the SC injection site for longer than a month, and how it affects the long-term immunogenicity remains unknown.

The particles have different size ranges, and the methods used to detect them are limited in their ability to detect them (Table 5.7). Most monomers are less than 10 nm and oligomers less than 100 nm.

New insights from quantifying subvisible particles show that beyond potential immunogenicity, particle sizes and levels are important product quality attributes; a mass of protein (e.g., <0.1%) in particles may not be detectable as loss of monomer; subvisible particle analysis provides very sensitive early detection of protein aggregation and new insights into aggregation pathways, manufacturing, and formulation development. Even trace levels of particles can impact the subsequent stability of protein solutions.

Beyond protein particles, these include glass particles from containers, glass cartridges, and syringes are siliconized, and free silicone oil droplets can be generated; in syringes, there may be tungsten particles and salts from needle insertion process and rubber or silicone particles can come from stoppers, stainless steel, and

Table 5.7 Methods Used to Analyze Different Particle Sizes

Methodology	Range of Detection
HP-SEC	1–90 nm (does not differentiate other particles)
SDS-PAGE	1–100 nm
Dynamic light scattering	1 nm to 2 μm
Nanoparticle tracking analysis	30 nm to 2 μm
Field flow fractionation	5 nm to 9 μm
Micro flow imaging	0.8–200 μm
Light obscuration	1–200 μm
Microscope	5–200 μm
Visible	500 μm to 5 mm+

other particles from filling pumps and particles shed from filters during pre-filling sterile filtration. These particles can adsorb onto proteins and make them more immunogenic.

Aggregates, including subvisible particles, are critical quality attributes and removal of these reduces immunogenicity; a good example is that of Betaferon that contains a lot of aggregates and particles and is one of the most immunogenic therapeutic protein products.

Stability

Degradation over time can significantly change the safety profile of a biological product, more than it can affect potency. Aggregation over time is a significant concern. As a result, agencies require side-by-side stability testing of the reference product. Additionally, tests include forced degradation, accelerated stability, and stressed stability. The following are the highlights of the required stability testing plan:

- Demonstrate that the drug product is stable during the period of PK/PD study; this applies to both the DS and the DP.
- Conduct real-time, stressed, and accelerated stability studies side by side with the RP.

Comparative stability studies under long-term storage conditions, accelerated conditions, and stress conditions, including mechanical stress and photostability, are conducted as well. Real-time accelerated conditions are used as recommended by ICH Q5C and ICH Q1A to evaluate the stability profiles of the proposed biosimilar and the U.S.-licensed reference product.

Forced degradation studies should be conducted as recommended in the "FDA draft guidance for industry: Quality Considerations in Demonstrating Biosimilarity to a Reference Protein Product" at http://www.fda.gov/downloads/drugs/guidance-complianceregulatoryinformati on/guidances/ucm291134.pdf. Multiple stress conditions (e.g., pH, oxidation, agitation, etc.) should be evaluated to select conditions that allow for incremental degradation of products to determine degradation rates and breakdown pathways.

Clinical data

In those instances where residual uncertainty remains, the regulatory agencies may require clinical studies data in patients. This is more likely to be the case when the mechanism of action cannot be studied in healthy subjects. A good example is that of filgrastim for which a highly objective PD model exists in healthy subjects while for drugs like adalimumab or etanercept, a subset of arthritic or other diseases that are part of the indication may be needed to provide final proof to remove any residual uncertainty.

The demonstration of the effectiveness is always performed side by side with a reference product. Such experimental designs can be very difficult to manage statistically since the purpose of these trials is to demonstrate equivalence and in some instance noninferiority. The first-licensed product development was not hampered by these considerations, and mostly a placebo was sufficient to demonstrate efficacy. It is for this reason that it is often not feasible in demonstrating differences between products where variability is high and may lead to enormous numbers of patients to provide sufficient power to a study. To avoid this, one company Teva decided to secure licensing of filgrastim as a new drug rather than as a biosimilar.

The early approvals of biosimilar products required a standard procedure of first demonstrating comparative safety in animals, followed by dosing in healthy subjects, and finally, in patients.

The exercise of clinical similarity is done step by step; it generally starts with pharmacokinetics and pharmacodynamics studies in healthy volunteers. These studies are followed by effectiveness and safety comparative studies. In most cases, the clinical efficacy studies are conducted to demonstrate a therapeutic equivalence between the biosimilar and the reference product in a population of patients in which any differences between the biosimilar product and the reference are most likely to become evident. It is very important to understand that regardless of the indication for which a testing is conducted, the approved (licensed or authorized) product receives all approved indications of the reference product including the pediatric indications. However, the studies required to demonstrate biosimilarity may change depending on the nature of indications. For example, while developing a biosimilar filgrastim product, a single-dose PD analysis of ANC in healthy subjects may be sufficient, and for mobilization claim, a multiple dose study will be required.

Risk-based critical quality attributes

Demonstration of biosimilarity begins with selecting critical quality attributes, which are likely to have a direct impact on the four pillars of the biosimilarity tetrahedron. Whereas the importance of immunogenicity, safety, and efficacy is compared head-to-head in biosimilarity assessment, it is not possible to provide an equal level of comparison for every quality attribute; it is for this reason the regulatory agencies require the biosimilar developers to first produce matrix of critical quality attributes. These attributes are selected for a risk-based exercise that determines the critical quality attributes is governed by several guidance such as ICH Guidelines Q8–Q11. The exercise begins with first understanding the MoA, the mode of immunogenicity and safety outcomes, as well as effectiveness. In most instances, all of these aspects about the biological drug developed are well known. The main source of information are the European EPARs, the FDA reports, the research publications, customer promotional materials, clinical promotional materials, labels approved, and as amended, the corporate financial reports, etc. Once we have a good understanding of all of these factors, the risk management is built as follows:

1. Quality attributes associated with the MoA; these include primary, secondary, and tertiary structure.
2. Quality attributes associated with immunogenicity and safety; these include both chemical and physical (aggregation) modifications.
3. Quality attributes altering effectiveness such as protein content, bioassay, receptor binding, immunoblotting, etc.
4. Quality attributes related to product and process-related impurities that might impinge on some or all of these quality attributes.
5. Quality attributes supporting the dosage from performance such as pH, osmolality, concentration of surfactants, etc.

The aim of pharmaceutical development is to design a quality product and its manufacturing process to consistently deliver the intended performance of the product. The information and knowledge gained from pharmaceutical development studies and manufacturing experience provide scientific understanding to support the establishment of the *design space*, specifications, and manufacturing controls.

151

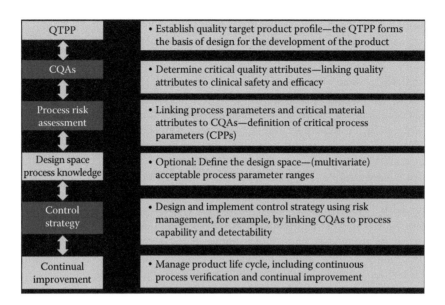

Figure 5.4 Elements of pharmaceutical development (ICH Q8 R2).

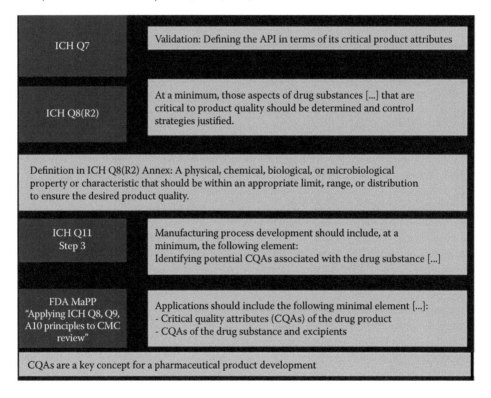

Figure 5.5 Quality interaction with guidance.

Figure 5.4 provides an overview of the integrated methodology being followed to support the establishment of the components that make up the quality risk assessment (QRA).

Figure 5.5 provides a schematic representation of how each of the guidance is interrelated with their supporting elements for the QRA.

The QTPP is defined as a "prospective and dynamic *summary* of the quality characteristics of the molecule that *ideally* will be achieved to ensure that the desired quality, and thus the *safety and efficacy*, of a drug product, are realized." Once QTPP is established for the target molecule, it is used for comparative analysis to assist in the development process and final process design. The biosimilar product is designed, developed, and manufactured according to QTPP with specification consistent with the desired in vivo performance of the product (linking process, product, and patient for patient benefit).

The list of QTPP used for a biosimilar candidate filgrastim product is provided in Table 5.8. This list is created from literature and understanding of the molecule and is absent of criticality.

Identification of CQAs is done through risk assessment as per the ICH guidelines Q8, Q9, Q10, and Q11. Prior product knowledge, such as the accumulated laboratory, nonclinical and clinical experience with a specific product-quality attribute is a big factor in making these risk assessments. The knowledge also included relevant data from similar molecules and data from literature references. Taken together, this information provided a rationale for relating the CQA to product safety and efficacy. The outcome of the risk assessment is to provide a list of CQAs ranked in order of importance. Once QTPP has been identified, the next step is to identify the CQAs. A critical quality attribute (CQA) has been defined as "a physical, chemical, biological, or microbiological property or characteristic that should be within an appropriate limit, range, or distribution to ensure the desired product quality." The road map of a decision tree for CQAs is shown in Figure 5.6 and summarized next.

- Started with list of all possible product variants
    - Taken from literature, class of molecule, information on specific molecule, platform knowledge
- Assessed how specific variants might affect safety and efficacy
    - Safety and immunogenicity assessed by endogenous variants, nonclinical and clinical experience, and, in some cases, PK data
    - Biological activity typically assessed with in vitro studies (e.g., potency of isolated variants in clinically relevant assay)
    - PK impact assessed with nonclinical studies
- CQA identification outcomes were reviewed internally by technical experts and company management
    - Allowed for reclassification in the event an attribute is categorized incorrectly
    - Provided assurance that no high-risk attributes are inadvertently classified as low risk

Criticality level is determined by the impact of the attribute as measured by analytical techniques and their resulting clinical outcomes. This is specifically measured as it relates to the impact on efficacy (biological activity), PK/PD, Immunogenicity, and safety. As the product evaluated is a biosimilar, a knowledge-based score is assigned to establish the extent of certainty about the impact in the clinic as defined in Table 5.9.

A CQA score is determined by multiplying the clinical impact score by the knowledge-based score. Since the analytical characteristics of filgrastim correlate well with the expected clinical outcome, they are sensitive and predictive of the MoA. These attributes have been proven by experience with innovator and known mechanism of action. Table 5.10 represents the defined quality attributes and their potential impact on clinical outcome. Attributes were evaluated and given a critical level of very high, high, low, and very low, which indicate a known impact on clinical and/or MOA. This is based on a mathematical calculation.

153

Table 5.8 Quality Target Product Profile and Corresponding Methods

Quality Attributes	Methods
Identity	• Reduced and nonreduced sodium dodecyl sulfate-polyacrylamide gel electrophoresis (SDS-PAGE) • Western blot
Primary structure	• N-terminal and C-terminal sequencing • Peptide mapping with ultraviolet (UV) and mass spectrometry detection • Protein molecular mass by electrospray mass spectrometry (ESI MS) • Protein molecular mass by matrix-assisted laser desorption ionization mass spectrometry (MALDI-TOF MS) • Amino acid sequence
Bioactivity	• Proliferation of murine myelogenous leukemia cells (NFS-60 cell line)
Receptor binding	• Surface plasmon resonance • ELISA format
Protein content	• RP-HPLC • UV absorbance spectroscopy
Process impurities	• Host cell protein • Host cell DNA • Leachables/extractables • Kanamycin • Tungsten
Excipients	• Acetate content • Sorbitol content • Polysorbate 80 content
Higher order structure	• Far and near UBV circular dichroism • Proton (1H) nuclear magnetic resonance • Proton-nitrogen (1H-15N) heteronuclear single quantum coherence spectroscopy • Liquid chromatography coupled with mass spectrometry (LC-MS) (disulfide bond characterization) • Thermodynamic equilibrium state testing
Appearance (clarity)	• Visual inspection • Nephelometry
Particulates	• Microflow imaging
High molecular weight variants/ aggregates	• Size exclusion chromatography • Reduced and nonreduced sodium dodecyl sulfate-polyacrylamide gel electrophoresis (SDS-PAGE) • Analytical ultracentrifuge
Oxidized species	• Reverse phase high-performance liquid chromatography (RP-HPLC) • Liquid chromatography coupled with mass spectrometry (LC-MS)
Partially reduced species	• Liquid chromatography coupled with mass spectrometry (LC-MS)
Sequence variants: His → Gln, Asp → Glu, Thr → Asp	• Liquid chromatography coupled with mass spectrometry (LC-MS)
fMet1 species	• Liquid chromatography coupled with mass spectrometry (LC-MS)
Succinimide species	• Liquid chromatography coupled with mass spectrometry (LC-MS)
Phosphoglucunoylation	• Liquid chromatography coupled with mass spectrometry (LC-MS)
Acetylated species	• Liquid chromatography coupled with mass spectrometry (LC-MS)
N-terminal truncated species	• Liquid chromatography and tandem mass spectrometry (LC-MS/MS)
Norleucine species	• Liquid chromatography coupled with mass spectrometry (LC-MS)
Deamidation species	• Reverse phase high-performance liquid chromatography (RP-HPLC) • Liquid chromatography coupled with mass spectrometry (LC-MS) • Isoelectric focusing • Cation exchange chromatography
pH	• pH meter
Safety	• Bioburden • Endotoxin • Sterility
Osmolality	• Osmolality determination by freezing point depression

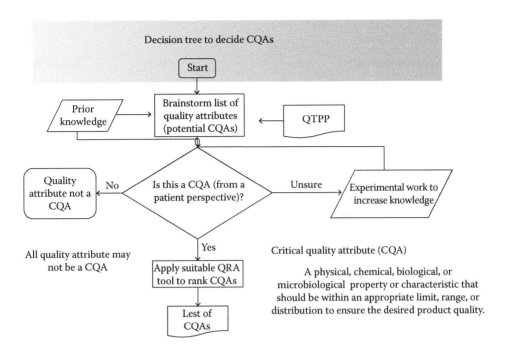

Figure 5.6 Decision tree to decide CQAs.

Table 5.9 Impact Definition for Determination of Criticality

Impact	Safety	(PK/PD)	Efficacy (Biological Activity)	Immunogenicity
Very high	Irreversible AEs	Significant change in PK	Very significant change	ATA detected and conferred limits on safety
High	Reversible AEs	Moderate change with impact on PD	Significant change	ATA detected and conferred limits on safety
Medium	Manageable AEs	Moderate change with no impact on PD	Moderate change	ATA detected with in vivo effect that can be managed
Low	Minor, transient AEs	Acceptable change with no impact on PD	Acceptable change	ATA detected with minimal in vivo effect
None	No AEs	No impact on PK or PD	No change	ATA not detected or ATA detected with no relevant vivo effect

The FDA recommends using a statistical approach to evaluating quality attributes for biosimilar product that is consistent with the risk assessment principles set forth in ICH Quality Guidelines Q8, Q9, Q10, and Q11. Consistent with these principles, quality attributes for analytical similarity program were assessed by tiered system approaches of varying statistical rigor as listed in Table 5.10.

The methods described earlier are also risk-ranked relative to the ability to detect with reliability and sensitivity to the attribute being measured. To further exemplify the analysis, bioassay will be used to explain the detectability range assignment.

In summary, the identification of CQA is critical for product understanding and process development. The outcome of the risk assessment is to provide a list of CQAs ranked in order of importance. The FDA recommends using a statistical approach to evaluating quality attributes for biosimilar product that is consistent

Table 5.10 Quality Attributes for Similarity Assessment of Filgrastim

Quality Attribute	Critical Relevance	Criticality Ranking	Test Method	Similarity Acceptance Criteria
Identity: primary structure	Efficacy, safety, immunogenicity	Very high	Molecular weight by liquid chromatography electrospray ionization mass spectrometry	18,799 ± 1 Da
Identity: primary structure	Efficacy, safety, immunogenicity	Very high	Identity by western blot analysis	Same number, position, and intensity of bands
Identity: primary structure	Efficacy, safety, immunogenicity	Very high	Disulfide linkage analysis by peptide mapping and mass spectrometry	Two disulfide bonds at 37–43 and 65–75 position
Identity: primary structure	Efficacy, safety, immunogenicity	Very high	Confirmation of amino and carboxyl termini by tandem mass spectrometry	Same amino acid group as reference
Identity: primary structure	Efficacy, safety, immunogenicity	Very high	Amino acid analysis	Same amino acid sequence as reference and both conforming to Drug Bank database
Identity: primary structure	Efficacy, safety, immunogenicity	Very high	Peptide map—RP-HPLC	Same peptide map as reference product
Identity: isoelectric point	Efficacy, safety, immunogenicity	Very high	Capillary isoelectric focusing (cIEF)	PI and peaks and valleys similar to reference product
Potency: receptor binding	Efficacy, safety	Very high	Receptor binding characteristics by enzyme-linked immunosorbent assay	Similarity range within mean ± 2SD of reference product
Potency: receptor binding	Efficacy, safety	Very high	Antibody binding characteristics by enzyme-linked immunosorbent assay	Similarity range within mean ± 2SD of reference product
Potency: bioactivity	Efficacy, safety	Very high	In vitro biological activity by M-NFS-60 cell proliferation assay	Statistical equivalence (90% CI) testing using 1.5 × SD of reference product
Protein concentration	Efficacy	Very high	Protein concentration by RP-HPLC	Statistical equivalence (90% CI) testing using 1.5 × SD of the reference product
Safety: subvisible particles	Immunogenicity	High	Light obfuscation for subvisible particles (USP <787>)	No more than the reference
Safety: clarity	Safety, immunogenicity	High	Appearance by visual inspection	No more than the reference
Identity: three-dimensional structure	Efficacy, immunogenicity	High	2-dimensional HSQC nuclear magnetic resonance	Position and intensity of peaks same as reference product
Identity: three-dimensional structure	Efficacy, immunogenicity	High	1-dimensional H1 nuclear magnetic resonance	Position and intensity of peaks same as reference product
Identity: tertiary structure	Efficacy, immunogenicity	High	Intrinsic fluorescence spectroscopy	Position and intensity of peaks and valleys similar to reference product
Identity: tertiary	Efficacy, immunogenicity	High	Differential scanning calorimetry	Shape and final melting point identical within 0.5°
Identity: secondary structure	Efficacy, immunogenicity	High	Far UV circular dichroism spectropolarimetry	Position and intensity of peaks and valleys similar to reference product
Identity: secondary structure	Efficacy, immunogenicity	High	Near UV circular dichroism spectropolarimetry	Position and intensity of peaks and valleys similar to reference product
Identity: fourth-dimensional structure	Efficacy, immunogenicity	High	Thermodynamic equilibrium state testing	Position and intensity of peaks and valleys similar to reference product
Purity: variants, oxidized	Efficacy	High	Reversed-phase high-performance liquid chromatography	Total and individual variants no more than the reference product

(*Continued*)

Table 5.10 (*Continued*) Quality Attributes for Similarity Assessment of Filgrastim

Quality Attribute	Critical Relevance	Criticality Ranking	Test Method	Similarity Acceptance Criteria
Safety: impurity, process-related impurities	Immunogenicity	High	Tungsten level	Less than limit of detection (500 ppb)
Safety: impurity, process-related	Immunogenicity	High	Extractables	Identified and no more than the reference product
Safety: impurity, process-related	Immunogenicity	High	Leachable	Identified and no more than the reference product
Safety: impurity, process related	Immunogenicity	High	Silicone oil level	No more than the reference product
Safety: host cell proteins	Immunogenicity	High	Enzyme-linked immunosorbent assay with 2D-SDS-PAGE to establish test effectiveness	Less than 10 ppm
Safety: host cell DNA	Immunogenicity	High	Fluorescence-based quantitation test	Less than 200 pg/mg
Safety: aggregates, covalent dimers	Immunogenicity	High	Size exclusion chromatography (STM-0085); analytical ultracentrifuge	No more than reference product
Formulation: inactive component	None	High	Sorbitol content	4.7%–5.3%
Formulation: inactive component	None	High	Polysorbate 80 content	$0.004\% \pm 5\%$
Formulation: inactive component	None	High	Acetate content	$10\ \text{mM} \pm 1\ \text{mM}$
Purity: succinamide species	None	Low	LC-MS	No more than the reference product
Purity: fMet1 species	None	Low	LC-MS	Quantity and distribution same as reference product
Purity: variants, N-terminal truncated	None	Very low	LC-MS	Number, quantity, and distribution same as reference product
Purity: reduced, partially reduced Species	None	Very low	LC-MS	No more than the reference product
Purity: phosphogluconoylation	None	Very low	LC-MS	No more than the reference product
Purity: norleucine	None	Very low	LC-MS	No more than the reference product
Purity: variants, sequence (His → Gin), (Asp → Glu), Thr → Asp)	None	Very low	RP-HPLC-MS	Number, quantity, and distribution same as reference product
Purity: acetylated species	None	Very low	LC-MS	Number, quantity, and distribution same as reference product
Purity: deamidated species	None	Very low	LC-MS IEX RP-HPLC cIEF	Number, quantity, and distribution same as reference product
Formulation: property	None	Very low	Refractive index	Matches reference product
Formulation: property	None	Very low	Osmolality determination by freezing point depression	$300\% \pm 5\%$
Formulation: properties	None	Very low	pH	3.8–4.2

with the risk assessment principles set forth in ICH Quality Guidelines Q8, Q9, Q10, and Q11. Consistent with these principles, quality attributes for analytical similarity assessment are established by tiered system approaches of varying statistical rigor. The method adjusted CQA values are then used to establish the critical process parameters (CPP).

Nonclinical data

Nonclinical or preclinical studies in animal models allow connecting activity with the pharmacodynamic effect relevant to the clinical application. Measurements in toxicokinetics include the determination of the level of antibodies with the study of crossed reactions and of the neutralization capacity are important; the studies must last long enough to show any difference relevant in terms of toxicity and/or immune response between the biosimilar and the first licensed product. However, these studies are not intended to establish toxic dose ranges. Where required, local tolerance studies are also conducted but other routine toxicological tests (safety pharmacology, reproductive tests, mutagenicity, carcinogenicity) are not necessary. The preclinical studies program is a limited program due to the fact that the toxicology data are known for the first licensed reference product, and it is not necessary to repeat all the studies to know the biosimilar.

There are several key measures in conducting nonclinical studies. For example, the dose chosen should be such to elicit some side effects; this is analogous to studying dose response in the linear range and not within the plateau ranges. There is a general consensus to reduce the animal studies where possible, and the biosimilar product developer is encouraged to hold meetings with regulatory agencies and develop a clear understanding of what would be considered minimally required. Generally, if these studies are required, then an IND will not be approved unless these data are available.

Stages of analytical similarity

"Analytical studies provide the foundation for an assessment of the proposed protein product for submission..." and include "analytical studies that demonstrate that the biological product is highly similar to the reference product notwithstanding minor differences in clinically inactive components" (FDA 2012).

A 351(k) application in the United States and similar applications in Europe and other developed countries must contain, among other things, information demonstrating that the proposed product is biosimilar to a reference product based on data derived from analytical studies, animal studies, and a clinical study or studies, unless the FDA determines, in its discretion, that certain studies are unnecessary in a 351(k) application. The goal of a biosimilar development program is thus to demonstrate that the proposed product is biosimilar to the reference product. The overall plan for developing a case of biosimilarity involves the following:

- *Stage 1: Analytical and functional similarity*—A large number of tests are currently available, and many more appearing in the literature that can be used to demonstrate the level similarity. The goal of the biosimilar product development is to create a battery of orthogonal tests that qualifies the product for the category: fingerprint-like similarity wherein the need for any additional testing such as trials in patients is obviated.
- *Stage 2: Preclinical or nonclinical safety*—Once an acceptable level of analytical and functional similarity has been established, the developer should consult with the agencies on the need for this testing and the extent of testing. There is a new consensus developing in the regulatory agencies that purports to avoid any animal testing unless it provides any useful safety information. To be useful, the nonclinical study must be conducted in species that show safety signals similar or proportional to humans; therefore, it is possible that for some drugs like monoclonal antibodies, no

animal species may be suitable. When a new drug is developed, several animal species are used as the mechanism of action and safety profile is not known, but for biosimilar products, the comparison with reference product requires species capable of showing adverse response. This distinction is important to avoid initiating studies similar to what the originator had conducted. The species must be meaningful, and this requires a discussion with regulatory agencies before starting these studies. For example, while rats constitute a good species to demonstrate safety of filgrastim, for adalimumab, only pharmacokinetic studies in a couple of monkeys may be required if the biosimilar product is not structurally and functional highly similar.

- *Stage 3: Clinical pharmacology*—The routine PK/PD studies in healthy subjects or where necessary, in patients, is required by all agencies. There is, however, sufficient room for negotiating the size of these trials that may include selecting only the doses within a linear range, avoiding multiple dose studies, and using creative statistical models to design trials to reduce the number of subject required. The clinical pharmacology studies also serve the purpose of evaluating safety and immunogenicity; the AEs recorded in these studies are of high value in establishing biosimilarity. Monitoring antibodies in these studies further support immunogenicity comparisons.

- *Stage 4: Clinical trials in patients*—These may be required in two instances; first, when the regulatory agencies are not convinced of the similarity of the product, leaving sufficient "residual uncertainty," and second, where PD studies are not meaningful in healthy subjects and these are linked to the safety of the product. Even when conducted, these clinical studies in patients are limited to a single indication and in the easiest population to recruit, reducing the burden of cost and time for the biosimilar product developer; however, it requires a lot of scientific finesse on the part of the developer to secure concessions from the regulatory agencies on their clinical trial protocols.

- *Stage 5: Postmarket surveillance*—Postmarket surveillance is required in EU but the FDA is flexible, and it may not be necessary; under REMS, the FDA will declare what, if any, surveillance studies are needed. The developer may, however, want to conduct open-label studies postapproval to support its marketing plans. Once it is approved, the biosimilar product need not revert back to the reference product for any similarity studies, except where it is desired for marketing purposes.

Level of similarity

Until the U.S. FDA took a giant step in 2014 to define various levels of similarity leading to biosimilarity, the level of biosimilarity needed for approval was somewhat vague. Figure 5.7 shows a tier-base definition of biosimilarity as suggested by the FDA. The specific language used to describe each tier is pivotal in understanding the regulatory consequences.

Level 1: Not similar

At this classification stage, the functional similarity is not in line with the reference product. This includes bioassays, in vitro binding tests, glycan patterns, and immunogenicity profiles for products for which immunogenicity is a concern. The determination by the agency whether a product is "not similar" can

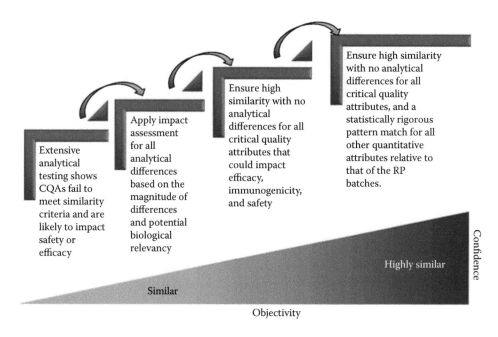

Figure 5.7 Confidence-based levels of biosimilarity.

be challenged by the developer, but it will require extensive toxicology studies and other nonclinical evaluations, many of which may be difficult to plan in the absence of an understood mechanism of action. For example, monoclonal antibodies are difficult to evaluate at the nonclinical testing level because of the lack of a suitable toxicology model in common species used for such studies. The developer needs to evaluate the cost of challenging this tier status vis-à-vis the cost of a 351(a) filing.

Level 2: Not highly similar

This tier stipulates, "Additional analytical or other data needed to determine if the product is not highly similar to the reference." This discussion with the agency will likely take place in a type 2 meeting where the developer presents an orthogonal evaluation of the biosimilar candidate comparing it with the reference product. Whereas it is not the expectation that all orthogonal tests will provide equal level of similarity, the purpose of this classification is to encourage the developer to move to the "highly similar" tier, the minimum tier required to be classified as a biosimilar product.

The developer would be required to present a detailed plan of testing that is expected to provide confidence that the product can be classified as Tier 3, and any differences that are observed are not clinically meaningful. This judgment is subject to interpretation and highly determined by the clinical responses antici- pated. In those instances where the product is administered on a chronic basis, the residual uncertainty can be reduced by conducting protocols that require switching between reference and the test product, particularly in animal models to demon- strate that any functional dissimilarity is not clinically meaningful.

Generally, the product must meet most of the primary, secondary, and tertiary structural similarity and most of the functional similarity. The additional studies as recommended are intended to support the assertion that the observed variability is not exacerbating any toxic or immunogenic response and that repeated use will not result in a noticeable inferior clinical response.

In all instances, these additional tests will depend on the type of product, such as cytokine vs. monoclonal antibody, and will incur the substantial additional expense to prove similarity.

If the differences identified are related to the process such as the variability in the SEC or analytical centrifugation, these can be corrected with appropriate formulation changes; it is noteworthy that the agency does not require the Q1/Q2 type similarity of formulation common for small molecule drugs. Realizing that the formulations of biological drugs may be under IP protection, the agency welcomes alternate formulations as long as these do not require changing the dose or the indication. For example, an alternate route, for example, subcutaneous dosing in place of intravenous, will be a major change and does not fall within the category or similarity evaluation. The first biosimilar approved by the FDA, Zarxio, has a different buffer formulation than its reference product, Neupogen.

Level 3: Highly similar

This is the minimal level that the agency requires for a biosimilar candidate. At this point, there is high confidence in similarity both in the analytical and functional level and further development requires securing the approval for targeted and selective clinical studies to resolve residual uncertainty. It is this "residual uncertainty" that needs to be understood. The uncertainty begins with first not being able to understand the source of dissimilarity. For example, if there are process-related impurities that are indigenous to any acceptable process changes, these must be fully characterized and proven to be "clinically not meaningful."

In those instances where the functionality profile is not identical in vitro bioassays or other binding assays, the agency may require an abbreviated study to show that these differences are not impacting the efficacy or toxicity. It is understood at this stage that any clinical studies suggested will be abbreviated, and these may not require patient participation or even be efficacy trials. If suitable pharmacodynamics profiles are available in healthy subjects, these should be evaluated first.

In some instance, such as in the case of monoclonal antibodies, the nonclinical profiling is highly species dependent reducing the value of animal testing to reduce uncertainty. However, where robust models are available to resolve the uncertainty issues with animal studies, these should be suggested.

Level 4: Highly similar with fingerprint-like similarity

When proven to quality a Level 4 classification, the developer has already proven it to be biosimilar by the first part of the level description: "highly similar." Now if the data presented across an orthogonally planned evaluation that all parameters studies are fully reproducible in the biosimilar candidate, more targeted studies may be required, "if residual uncertainty" remains. The last part of the definition is crucial to developing biosimilar products. What this classification means is that if the product is similar to fingerprint-like comparisons, no further studies may be required if the similarity data remove residual uncertainty.

It is important, therefore, to understand what constitutes "residual uncertainty." As discussed earlier, any uncertain or unexpected observation about the product creates a residual or remaining uncertainty. Note that it is not the uncertainty or unexpected observations but that these remain uncertain is the focus of description. For example, if the biosimilar candidate shows extra chromatography peaks that are fully identified and do not impact the structure of the protein at any level, then these do not remain residually uncertain.

In most cases, the developer reaching this level of similarity will be able to request a waiver of any further clinical trials in patients and in most cases in healthy subjects. This has long been the goal of many regulatory agencies and in its true expression holds the promise to reduce the cost of biosimilar drugs. The U.S. FDA had taken a lead when it approved several complex and biological products without any clinical studies in patients and in all likelihood wanted to expand this practice based on the science of understanding the molecules.

Statistical modeling of similarity data

The details of testing provided earlier need a clear understanding of the purpose of testing. Unlike when new molecules are developed, the entire purpose is to characterize, when it comes to biosimilars, the purpose is to characterize with the aim to demonstrate similarity. This difference creates a need for a statistical approach to deciding what is the similarity? How similar it has to be, or it can be to qualify the status of "highly similar," to be considered a candidate as a biosimilar product. The relative importance of each of the attribute, critical or not, is realized in the importance given to its testing and how these results contribute to overall understanding of the biosimilarity. Figure 5.6 demonstrates the relative value of each of the components of similarity described earlier in contributing to biosimilarity.

The regulatory agencies require that the biosimilar product developer create a risk-based model to identify the critical attributes, critical to safety, potency, and purity that can make a meaningful difference in the commercial supply of the product. This approach includes, but it is not limited to, the levels of the attribute in both the reference product and proposed biosimilar product (as determined by testing), the sensitivity of an assay to detect differences between products, if any, and an understanding of the limitations of the type of statistical analysis that can be performed due to the nature of a quality attribute.

The biosimilar product developers face many statistical challenges including limited number of lots, different testing methods resulting in different data types, deciding what difference is meaning and what is not, and finally establishing acceptable similarity margins. "The plurality of candidates of critical quality attributes within specific developments, as well as the usually low number of drug lots available, had been identified as the most limiting factors, rendering the use of statistical routines usually performed on basis of clinical patient- data inappropriate most of the time" (EMA 2013).

The FDA recommends to use a statistical approach to evaluate quality attributes of the proposed biosimilar product that is consistent with the risk assessment principles set forth in the International Conference on Harmonisation Quality Guidelines Q8, Q9, Q10, and Q11. Consistent with these principles, the program should implement an analytical similarity assessment that is based on a tiered system in which approaches of varying statistical rigor are used.

A potential approach for the different statistical tiers is described next:

Tier 1 (equivalence test)

For Tier 1, the FDA recommends an equivalency testing for assessment of analytical similarity. As indicated by the FDA, a potential approach could be a similar approach for bioequivalence testing for generic drug products. In other words, for a given critical attribute, we may test for equivalence of the following interval (null) hypothesis:

$$H_0 : \mu_T - \mu_R \leq -\delta \quad \text{or} \quad \mu_T - \mu_R \geq \delta, \qquad (5.1)$$

where
   $\delta > 0$ is the equivalence limit (or similarity margin)
   μ_T and μ_R are the mean responses of the test (proposed biosimilar) product and the reference product lots, respectively

Analytical equivalence (similarity) is concluded if the null hypothesis of inequivalence (dissimilarity) is rejected. Similar to the confidence interval approach for bioequivalence testing for the raw data model, analytical similarity would be accepted for the quality attribute if the $(1-2\alpha)100\%$ two-sided confidence interval for the mean difference is within $(-\delta, \delta)$.

Under the null hypothesis, the FDA indicates that the equivalence limit (similarity margin), δ, would be a function of the variability of the reference product, denoted by σ_R. It should be noted that each lot contributes one test value for each attribute being assessed. Thus, σ_R is the population standard deviation of the lot values of the reference product. Ideally, the reference variability, σ_R, should be estimated based on some sampled lots randomly selected from a pool of reference lots for the statistical equivalence test. In practice, it may be a challenge when there are a limited number of lots available. Thus, the FDA suggests the sponsor providing a plan for how the reference variability, σ_R, will be estimated with a justification.

Along this line, the FDA indicates that one potential approach is to assume that the equivalence limit (similarity margin) is proportional to the reference product variability, that is, $\delta = c^*\sigma_R$. The constant c can be selected as the value that provides adequate power to show equivalence if there is only a small difference in the true mean between the biosimilar and the reference product, when a moderate number of reference product and biosimilar lots are available for testing. To illustrate the FDA's recommended approach for the assessment of analytical similarity for a critical attribute, as an example, suppose we have 10 biosimilar and 10 reference product lots. If we choose $\delta = 1.5\sigma_R$ (i.e., $c = 1.5$) for all sample sizes used in equivalence testing, the test would yield a positive result if the 90% confidence interval for the difference in sample means lies within $(-1.5\sigma_R, 1.5\sigma_R)$. This test would have approximately 84% power at the $\alpha = 5\%$ level of significance when the true underlying mean difference between the proposed biosimilar and reference product lots is equal to $\sigma_R/8$.

The equivalency testing recommended for quality attributes with the highest risk ranking would include assay(s) that evaluate clinically relevant mechanism(s) of action of the product for each indication for which approval is sought. One needs to test the following null hypothesis: $H_0: \mu_B - \mu_R \leq -\delta$ or $\mu_B - \mu_R \geq \delta$, where μ_B and μ_R are the mean responses of the proposed biosimilar and reference product lots, respectively, and $\delta > 0$ is the equivalence margin.

The analytical similarity will be accepted for the quality attribute if the $(1-2\alpha)100\%$ two-sided confidence interval for the mean difference is within $(-\delta, \delta)$. In this context, the equivalence margin, δ, would be a function of the variability of the reference product as identified in studies by the biosimilar applicant (σ_R).

Tier 2 (quality range approach)

For Tier 2, the FDA suggests that analytical similarity be performed based on the concept of quality ranges, that is, $\pm x\sigma$, where σ is a standard deviation of the reference product, and x should be appropriately justified. Thus, the quality range of the reference product for a specific quality attribute is defined as $(\hat{\mu}_R + x\hat{\sigma}_R, \hat{\mu}_R - x\hat{\sigma}_R)$. The analytical similarity will be accepted for the quality attribute if a sufficient percentage of test lot values (e.g., 90%) fall within the quality range.

As it can be seen, for a given critical attribute the quality range is set based on test results of available reference lots. If $x = 1.645$, we would expect 90% of the test results from reference lots to lie within the quality range. If x is chosen to be 1.96, we will expect that about 95% test results of reference lots will fall within the quality range. As a result, the selection of x could impact the quality range and consequently the percentage of test lot values that will fall within the quality range. Thus, the FDA indicates that the standard deviation multiplier (x) should be appropriately justified.

The interpretation sounds reasonable and easy to justify if we assume that (1) $\mu_T \approx \mu_R$ and (2) $\sigma_T \approx \sigma_R$. In practice, these assumptions may not be true. If we let $\mu_T = \mu_R + \varepsilon$, then the percentage of test lot values that will fall within the quality range depend upon the magnitudes of ε and Δ.

It is recommended to consider the use of quality ranges (mean $\pm X\hat{\sigma}_R$, where X should be appropriately justified for assessing quality attributes with lower risk ranking). The quality range of the reference product for a specific quality attribute is defined as $(\hat{\mu}_R - x\hat{\sigma}_R, \hat{\mu}_R + x\hat{\sigma}_R)$. The analytical similarity will be accepted for the quality attribute if a sufficient percentage of test lot values (e.g., 90%) fall within the quality range. If, for example, we selected $\delta = 1.5\sigma_R$ for all sample sizes used in equivalence testing to illustrate this potential statistical approach, the test would yield a positive result if the 90% confidence interval for the difference in sample means lies within $(-1.5\sigma_R, 1.5\sigma_R)$. If 10 biosimilar and 10 reference product lots, this test would have approximately 84% power when the true underlying mean difference between the proposed biosimilar and reference product lots was equal to $\sigma_R/8$, assuming a test with $\alpha = 0.05$.

Tier 3 (graphical assessment)

This involves an approach that uses raw data/graphical comparisons for quality attributes with the lowest risk ranking. For CQAs in Tier 3 with lowest risk ranking, the FDA recommends an approach that uses raw data/graphical comparisons. The examination of similarity for CQAs in Tier 3 by no means is less stringent, which is acceptable because they have the least impact on clinical outcomes in the sense that a notable dissimilarity will not affect clinical outcomes.

Several considerations are important to the approaches to statistical modeling:

Each lot contributes one value for each attribute being assessed. Thus, σ_R refers to the standard deviation of those lot values of the reference product.

There should be equal number of lots when tested side by side; so if there are more lots available for reference, an unbiased selection should be made to select the equal number and the rest can be used separately to develop the acceptance range criteria.

Ideally, the reference variability, σ_R, should be estimated by testing different lots than those used in statistical equivalence test. This may be a challenge when there are a limited number of lots. The sponsor should provide a plan for how the reference variability, σ_R, will be estimated with a justification for the approach and identify the lots that will be used. It remains to be seen if the regulatory authorities will allow using public domain data on the variability of these attributes as may be made available in regulatory summaries of approval of biosimilar products.

It is recommended that the same number of replicates be performed within each proposed biosimilar lot as within each reference product lot, and that the same lots be used for equivalence testing, quality range testing, and visual assessment of graphical displays.

The high-assay variability would not justify a large σ_R. In such a situation, the assay would need to be optimized and/or the number of replicates increased to reduce variability.

In cases where the equivalence margins or quality ranges are too wide, it may be scientifically justified and appropriate to narrow the margins or range.

One potential statistical approach to evaluating quality attributes is based on a standard statistical test of equivalence with the margin defined as a function of the reference product variability (e.g., $c*\sigma_R$). The constant c would be selected as the value that provides adequate power to show equivalence if there is only a small difference in the true mean between the biosimilar and the reference product, when a moderate number of reference product and biosimilar lots are available for testing.

Note that with this potential approach the margin would be a function of the reference product variability as demonstrated in testing by the biosimilar applicant; therefore, a larger margin would be used for attributes with larger variability. In addition, the confidence level would depend on the number of lots available for testing. For a more limited number of lots, the developer may consider calculating the confidence interval with a lower confidence level to ensure adequate power. In this situation, the lower confidence level would be expected to be appropriately addressed by the final manufacturing control strategy. In contrast, when a moderate or greater number of lots are available for testing, the equivalence test would be based on a 90% confidence interval.

Nonparametric tolerance interval is available if data do NOT follow the normal distribution, but the large sample size is generally required.

An appropriate sample size can be chosen for analytical similarity assessment. As an example, suppose that there are 21 RP lots and 7 TP lots. We first randomly select 7 out of the 21 RP lots to match the 7 TP lots. Suppose that based on the remaining 14 lots, an estimate of σ_R is given by = 1.04. Also, suppose that the true difference between the biosimilar product and the reference product is proportional to σ_R say $\Delta = \sigma_R/8$. Then, the following table with various sample sizes (the number of TP lots available and the corresponding test size and statistical power for detecting the difference of $\sigma_R/8$) are helpful for assessment of analytical assessment.

It should be noted that this approach has inflated the alpha from the 5% to 16%. Note that for a fixed regulatory standard c the sponsor may appropriately select sample size (the number of lots) for achieving a desired power (for detecting a $\sigma_R/8$ difference) and significance level for analytical similarity assessment. As shown in Table 5.11, to reduce the test size (i.e., α level) from 8% to 5%, 10 TP lots need to be tested. Testing 10 TP lots will give an 87% power for detecting a $\sigma_R/8$ difference. According to Table 5.11, there is 79% power for 84% CI of TP is $\sigma_R/8$.

Table 5.11 Assessment of Analytical Similarity for CQAs from Tier 1

Number of RP Lots	Number of TP Lots	Selection of c	Test Size (Confidence Interval)	Statistical Power at $\sigma_R/8$ (%)
6	6	1.5	9% (82% CI)	74
7	7	1.5	8% (84% CI)	79
8	8	1.5	7% (86% CI)	83
9	9	1.5	6% (88% CI)	86
10	10	1.5	5% (90% CI)	87

Note: According to the table, there is 79% power for 84% CI of $\dot{\Delta} = \hat{\mu}_T - \hat{\sigma}_R$ to fall within $\pm$ EAC assuming that the number of lots = 7 and the true difference between TP and RP is $\sigma_R/8$.

Interchangeability

The subsection (b)(3) of the Public Health Act Subsection 351(k)(3) describes the term interchangeable or interchangeability in reference to a biological product that is shown to meet the standards described in subsection (k)(4), meaning that the biological product may be substituted for the reference product without the intervention of the health care provider who prescribed the reference product. This is a major commercial event for a biosimilar product. However, to achieve interchangeability, several definitions and concepts follow.

A biological product is considered interchangeable with the reference product if

1. The biological product is biosimilar to the reference product
2. It can be expected to produce the same clinical result for any given patient

In addition, for a biological product that is administered more than once to an individual, the risk in terms of safety or diminished efficacy of alternating or switching between use of the biosimilar and the reference product is not greater than the risk of using the reference product without such alternation or switch. An interchangeable product should be able to be substituted or alternated by a pharmacist, without intervention or even necessarily notification of the prescribing doctor, whereas a biosimilar product may yield a comparable outcome as the reference, but may require transitioning or input by a health care provider, in order to be switched or alternated with the reference (or not be able to at all) due to other factors such as excipients. Thus, biosimilarity does not imply interchangeability. Interchangeability is expected to produce the same clinical result in any given patient, which can be interpreted as that the same clinical result can be expected in every single patient. While physicians and hospitals may adopt interchangeability on their own, there remain legal challenges to be resolved in treating a biosimilar product as an interchangeable biosimilar product.

The demonstration of biosimilarity is a step-wise exercise based first on the analytical and functional similarity and then supported with preclinical and clinical data, and if additionally necessary, patient data. The demonstration of interchangeability remains debated in the United States and the FDA is currently conducting surveys to seek insight into the type of protocols that will allow the statutory evaluation of "no reduced effectiveness" and "no higher side effects" upon switching and alternating. It will take a few years for details of what is considered appropriate to be established, but in the future, it is more likely that these products will be readily substituted, very much the small-molecule generic products.

However, taking into consideration the current statutory requirements embedded in the guidance limits what is required to establish interchangeability. Besides the two requirements stated earlier, the statute further states "in a clinical setting," which is construed as testing in patients. To enable meeting all three requirements, clinical effectiveness studies (as opposed to clinical efficacy trials) must be conducted to demonstrate that "switching and alternating" is acceptable.

The concept of "switchability" used for small molecules does not apply to biosimilars. From the FDA's perspectives, interchangeability includes the concept of switching and alternating between a reference licensed product (R) and biosimilar test product (T). The concept of switching is referred to as not only the switch from "R to T" or "T to R" (narrow sense of switchability), but also "T to T" and "R to R" (broader sense of switchability). As a result, in order to assess switching, biosimilarity for "R to T," "T to R," "T to T," and "R to R" need to be assessed based on some biosimilarity criteria under a valid study design. On the other hand, the concept of alternating is referred to as either the switch from T to R and then switch back to T (i.e., "T to R to T") or the switch from R to T and then switch back to R (i.e., "R to T to R"). Thus, the difference between "the switch from T to R" then "the switch from R to T" and "the switch from R to T" then "the switch from T to R" needs to be assessed for addressing the concept of alternating.

The experimental design that can be used to demonstrate interchangeability is ideally a standard at least two-sequence, at least two-period (2 × 2) crossover design; however, it does not work well with drugs with long half-life; in those instances, a parallel group design is generally preferred. Unfortunately, parallel group design does not provide independent estimates of variance components such as inter-subject and intra-subject variability and variability due to subject-by-product interaction. This creates a major challenge for assessing biosimilarity and interchangeability (in terms of the concepts of switching and alternating) of biosimilar products under parallel group designs. For the purpose of establishing switchability, a 4 × 2 crossover design (i.e., TT, RR, TR, RT) is suitable. For demonstrating the similarity during alternating, a two-sequence, three-period dual design (i.e., TRT, RTR) may be useful since it allows a back excursion, the switch from T to R, and then back to T (i.e., "T to R to T") and from R to T and then back to R (i.e., "R to T to R"). These can be combined to design TT, RR, TRT, and RTR.

Given the highly specific nature of responses anticipated in the use of biological drugs and their biosimilar and interchangeable alternates, the developer is encouraged to consult with regulatory agencies about the justification for the nature of these protocols.

The rewards of obtaining an interchangeable status are many; even though the outcome is legally difficult to grasp, overall it means exclusivity in the market for a limited time as described next:

> (6) EXCLUSIVITY FOR FIRST INTERCHANGEABLE BIOLOGICAL PRODUCT.—
> Upon review of an application submitted under this subsection relying on the same reference product for which a prior biological product has received a determination of interchangeability for any condition of use, the Secretary shall not make a determination...that the second or subsequent biological product is interchangeable for any condition of use until the earlier of—
> (A) 1 year after the first commercial marketing of the first interchangeable biosimilar biological product to be approved as interchangeable with that reference product

Conclusion

In this chapter, I have introduced a fresh approach to understanding the boundaries of biosimilarity, a stage of similarity, when the two products can be declared clinically equivalent. New biological drugs are evaluated for potency, purity, and

Figure 5.8 Relationship among critical attributes leading to biosimilarity.

safety. Figure 5.8 reiterates the relationship among these attributes, leading to biosimilarity.

The biosimilars are additionally required to be evaluated for identity, which is the starting step to establishing whether a product qualifies as a biosimilar candidate. The assumption is that if the identities match, so would be other attributes of safety, potency, and purity. The fundamental concept of establishing biosimilarity involves a complex, yet tiered approach, that gradually establishes that the biosimilar product will provide a similar outcome as the reference product; the reason why there is emphasis on substituting studies in patients with other studies such as structural and analytical similarity and clinical pharmacology resides in a statistical concept. When comparing two products, we are essentially performing a noninferiority testing (one-tail); since both products produce highly variable results, establishing noninferiority may require a very large patient population. This is not the case for a new entity, where the comparison is made with placebo, and thus relatively easy to establish statistical differentiation. Furthermore, since the goal is to award all indications, relying on clinical trials alone will create a scientific logistics issue; however, if it is established that the two products are similar on the basis of the biosimilarity tetrahedron described in this chapter, it becomes scientifically plausible to declare biosimilarity and even interchangeability.

Over the next few years, biosimilar products will become common, yet there will not be a large number of companies competing in this field and one reason for this will be the scientific knowledge required to establish biosimilarity—this is perhaps a higher barrier than even developing a new molecular entity, even in the biological arena.

Bibliography

Aiuti A et al. The chemokine SDF-1 is a chemoattractant for human CD34+ hematopoietic progenitor cells and provides a new mechanism to explain the mobilization of CD34+ progenitors to peripheral blood. *J Exp Med* 1997;185:111–112.

Aramadhaka LR, Prorock A, Dragulev B, Bao Y, Fox JW. Connectivity maps for biosimilar drug discovery in venoms: The case of Gila monster venom and the anti-diabetes drug Byetta®. *Toxicon* 2013 Jul;69:160–167.

Beck A, Sanglier-Cianférani S, Van Dorsselaer A. Biosimilar, biobetter, and next generation antibody characterization by mass spectrometry. *Anal Chem* 2012 Jun 5;84(11):4637–4646.

Berger RL, Hsu JC. Bioequivalence trials. Intersection-union tests and equivalence confidence sets. *Stat Sci* 1996;11:283–319.

Berkowitz SA, Engen JR, Mazzeo JR, Jones GB. Analytical tools for characterizing biopharmaceuticals and the implications for biosimilars. *Nat Rev Drug Discov* 2012 Jun 29;11(7):527–540.

Bodey GP et al. Qualitative relationships between circulating leukocytes and infection in patients with acute leukemia. *Ann Intern Med* 1966;64:328–340.

Borleffs CJ et al. Effect of escalating doses of recombinant human granulocyte colony-stimulating factor (filgrastim) on circulating neutrophils in healthy subjects. *Clin Ther* 1998;20:722–736.

Bristow AF, Bird C, Bolgiano B, Thorpe R. Regulatory requirements for therapeutic proteins: The relationship between the conformation and biological activity of filgrastim. *Pharmeur Bio Sci Notes* 2012 Apr;2012:103–117.

Bui LA, Taylor C. Developing clinical trials for biosimilars. *Semin Oncol* 2014 Feb;41 (Suppl. 1):S15–S25.

Cai XY, Thomas J, Cullen C, Gouty D. Challenges of developing and validating immunogenicity assays to support comparability studies for biosimilar drug development. *Bioanalysis* 2012;4(17):2169–2177.

Cai XY, Wake A, Gouty D. Analytical and bioanalytical assay challenges to support comparability studies for biosimilar drug development. *Bioanalysis* 2013 Mar;5(5):517–520.

Calvo B, Zuñiga L. The U.S. approach to biosimilars: The long-awaited FDA approval pathway. *BioDrugs* 2012 Dec 1;26(6):357–361.

Camacho LH, Frost CP, Abella E, Morrow PK, Whittaker S. Biosimilars 101: Considerations for U.S. oncologists in clinical practice. *Cancer Med* 2014 Aug;3(4):889–899.

Casadevall N, Felix T, Strober BE, Warnock DG. Similar names for similar biologics. *BioDrugs* 2014 Oct;28(5):439–444.

Castañeda-Hernández G, Szekanecz Z, Mysler E, Azevedo VF, Guzman R, Gutierrez M, Rodríguez W, Karateev D. Biopharmaceuticals for rheumatic diseases in Latin America, Europe, Russia, and India: Innovators, biosimilars, and intended copies. *Joint Bone Spine* 2014 Jun 20;81:474–477.

Chamberlain P. Assessing immunogenicity of biosimilar therapeutic monoclonal antibodies: Regulatory and bioanalytical considerations. *Bioanalysis* 2013 Mar;5(5):561–574.

Chiu ST, Liu JP, Chow SC. Applications of the Bayesian prior information to evaluation of equivalence of similar biological medicinal products. *J Biopharm Stat* 2014;24(6):1254–1263.

Chow SC. Assessing biosimilarity and interchangeability of biosimilar products. *Stat Med* 2013 Feb 10;32(3):361–363.

Chow SC, Endrenyi L, Lachenbruch PA. Comments on the FDA draft guidance on biosimilar products. *Stat Med* 2013a Feb 10;32(3):364–369.

Chow SC, Endrenyi L, Lachenbruch PA, Mentré F. Scientific factors and current issues in biosimilar studies. *J Biopharm Stat* 2014;24(6):1138–1153.

Chow SC, Hsieh TC, Chi E, Yang J. A comparison of moment-based and probability-based criteria for assessment of follow-on biologics. *J Biopharm Stat* 2010a Jan;20(1):31–45.

Chow SC, Liu JP. Statistical assessment of biosimilar products. *J Biopharm Stat* 2010 Jan;20(1):10–30.

Chow SC, Lu Q, Tse SK, Chi E. Statistical methods for assessment of biosimilarity using biomarker data. *J Biopharm Stat* 2010b Jan;20(1):90–105.

Chow SC, Wang J, Endrenyi L, Lachenbruch PA. Scientific considerations for assessing biosimilar products. *Stat Med* 2013b Feb 10;32(3):370–381.

Chow SC, Yang LY, Starr A, Chiu ST. Statistical methods for assessing interchangeability of biosimilars. *Stat Med* 2013c Feb 10;32(3):442–448.

Choy E, Jacobs IA. Biosimilar safety considerations in clinical practice. *Semin Oncol* 2014 Feb;41(Suppl. 1):S3–S14.

Colletti KS. Conference Report: Bioanalysis-related topics presented at the International Conference and Exhibition on Biowaivers and Biosimilars. *Bioanalysis* 2013 Mar;5(5): 529–531.

Corbel MJ, Cortes Castillo Mde L. Vaccines and biosimilarity: A solution or a problem? *Expert Rev Vaccines* 2009 Oct;8(10):1439–1449.

Crisino RM, Dulanto B. Bioanalysis-related highlights from the 2011 AAPS National Biotechnology Conference. *Bioanalysis* 2011 Aug;3(16):1809–1814.

Dazzi C et al. Relationships between total CD34+ cells reinfused, CD34+ subsets and engraftment kinetics in breast cancer patients. *Hematologica* 2000;85:396–402. http://www.haematologica.org.

Declerck PJ. Biosimilar monoclonal antibodies: A science-based regulatory challenge. *Expert Opin Biol Ther* 2013 Feb;13(2):153–156.

DeVries JH, Gough SC, Kiljanski J, Heinemann L. Biosimilar insulins: A European perspective. *Diabetes Obes Metab* 2014 Nov 7;17(5):445–451.

Ebbers HC. Biosimilars: In support of extrapolation of indications. *J Crohns Colitis* 2014 May 1;8(5):431–435.

Ebbers HC, van Meer PJ, Moors EH, Mantel-Teeuwisse AK, Leufkens HG, Schellekens H. Measures of biosimilarity in monoclonal antibodies in oncology: The case of bevacizumab. *Drug Discov Today* 2013 Sept;18(17–18):872–879.

EMA. Concept paper on the need for a reflection paper on the statistical methodology for the comparative assessment of quality attributes in drug development, 2013. http://www.ema.europa.eu/ema/doc_index.jsp?curl=pages/includes/document/document_detail.jsp?webContentId=WC500144945&murl=menus/document_library/document_library.jsp&mid=0b01ac058009a3dc (accessed August 31, 2015).

Endrenyi L, Chang C, Chow SC, Tothfalusi L. On the interchangeability of biologic drug products. *Stat Med* 2013 Feb 10;32(3):434–441.

Epstein MS, Ehrenpreis ED, Kulkarni PM; the FDA-Related Matters Committee of the American College of Gastroenterology. Biosimilars: The need, the challenge, the future: The FDA perspective. *Am J Gastroenterol* 2014 Jun 24;109(12):1856–1859.

Evans DR, Romero JK, Westoby M. Concentration of proteins and removal of solutes. *Methods Enzymol* 2009;463:97–120.

Fávero-Retto MP, Palmieri LC, Souza TA, Almeida FC, Lima LM. Structural meta-analysis of regular human insulin in pharmaceutical formulations. *Eur J Pharm Biopharm* 2013 Nov;85(3 Pt B):1112–1121.

FDA. Guidance for industry quality considerations in demonstrating biosimilarity to reference protein product, 2012. http://www.fda.gov/Drugs/GuidanceComplianceRegulatoryInformation/Guidances/ucm290967.htm (accessed May 28, 2015).

FDA. FDA briefing document Oncologic Drugs Advisory Committee meeting, January 7, 2015. http://www.fda.gov/downloads/AdvisoryCommittees/CommitteesMeetingMaterials/Drugs/OncologicDrugsAdvisoryCommittee/UCM428780.pdf.

FDA. http://www.fda.gov/downloads/drugs/guidancecomplianceregulatoryinformation/guidances/ucm070262.pdf.

FDA. Quality considerations in demonstrating biosimilarity of a therapeutic protein product to a reference product: Guidance for industry. http://www.fda.gov/downloads/drugs/guidancecomplianceregulatoryinformation/guidances/ucm291134.pdf (accessed August 2015).

FDA. Guidances (drugs). http://www.fda.gov/Drugs/GuidanceComplianceRegulatoryInformation/Guidances/default.htm.

FDA. U.S.-licensed Neupogen labeling approved on September 13, 2013. http://www.accessdata.fda.gov/drugsatfda docs/label/2013/103353s5157lbl.pdf.

FDA. ZARXIO—Sandoz presentation to the ODAC. http://www.fda.gov/downloads/AdvisoryCommittees/CommitteesMeetingMaterials/Drugs/OncologicDrugsAdvisoryCommittee/UCM428780.pdf.

Fryklund L, Ritzén M, Bertilsson G, Arnlind MH. Is the decision on the use of biosimilar growth hormone based on high quality scientific evidence? *Eur J Clin Pharmacol* 2014 May;70(5):509–517.

Gabrilove JL et al. Phase I study of granulocyte colony-stimulating factor in patients with transitional cell carcinoma of the urothelium. *J Clin Invest* 1988;82:1454–1461.

Gascon P, Fuhr U, Sörgel F, Kinzig-Schippers M, Makhson A, Balser S, Einmahl S, Muenzberg M. Development of a new G-CSF product based on biosimilarity assessment. *Ann Oncol* 2010 Jul;21(7):1419–1429.

Greenbaum AM, Link DC. Mechanisms of G-CSF-mediated hematopoietic stem and progenitor mobilization. *Leukemia* 2011;25:211–217.

Gsteiger S, Bretz F, Liu W. Simultaneous confidence bands for nonlinear regression models with application to population pharmacokinetic analyses. *J Biopharm Stat* 2011 Jul;21(4):708–725.

U.S. Department of Health and Human Services, Food and Drug Administration, Center for Drug Evaluation and Research (CDER), Center for Biologics Evaluation and Research (CBER). Guidance for industry—Clinical pharmacology data to support a demonstration of biosimilarity to a reference product, May 2014. http://www.fda.gov/downloads/Drugs/GuidanceCompliance Regulatory Information /Guidances/UCM397017.pdf (accessed August 2015).

Guidance for industry—Statistical approaches to establishing bioequivalence, January 2001. http://www.fda.gov/downloads/Drugs/Guidance Compliance Regulatory Information/Guidances/UCM070124.pdf.

Hashii N, Harazono A, Kuribayashi R, Takakura D, Kawasaki N. Characterization of N-glycan heterogeneities of erythropoietin products by liquid chromatography/mass spectrometry and multivariate analysis. *Rapid Commun Mass Spectrom* 2014 Apr 30;28(8):921–932.

Haverick M, Mengisen S, Shameem M, Ambrogelly A. Separation of mAbs molecular variants by analytical hydrophobic interaction chromatography HPLC: Overview and applications. *MAbs* 2014 Jul–Aug;6(4):852–858.

Herman AC et al. Characterization, formulation, and stability of Neupogen (Filgrastim), a recombinant human granulocyte-colony stimulating factor. *Pharm Biotechnol* 1996;9:303–328.

Holloway C, Mueller-Berghaus J, Lima BS, Lee SL, Wyatt JS, Nicholas JM, Crommelin DJ. Scientific considerations for complex drugs in light of established and emerging regulatory guidance. *Ann N Y Acad Sci* 2012 Dec;1276:26–36.

Hsieh TC, Chow SC, Liu JP, Hsiao CF, Chi E. Statistical test for evaluation of biosimilarity in variability of follow-on biologics. *J Biopharm Stat* 2010 Jan;20(1):75–89.

Hsieh TC, Chow SC, Yang LY, Chi E. The evaluation of biosimilarity index based on reproducibility probability for assessing follow-on biologics. *Stat Med* 2013 Feb 10;32(3):406–414.

Hulse WL, Gray J, Forbes RT. Evaluating the inter and intra lot variability of protein aggregation behaviour using Taylor dispersion analysis and dynamic light scattering. *Int J Pharm* 2013 Sept 10;453(2):351–357.

Islam R. Bioanalytical challenges of biosimilars. *Bioanalysis* 2014 Feb;6(3):349–56.

Jackisch C, Scappaticci FA, Heinzmann D, Bisordi F, Schreitmüller T, von Minckwitz G, Cortés J. Neoadjuvant breast cancer treatment as a sensitive setting for trastuzumab biosimilar development and extrapolation. *Future Oncol* 2014 Aug;28:1–11.

Kálmán-Szekeres Z, Olajos M, Ganzler K. Analytical aspects of biosimilarity issues of protein drugs. *J Pharm Biomed Anal* 2012 Oct;69:185–195.

Kang SH, Chow SC. Statistical assessment of biosimilarity based on relative distance between follow-on biologics. *Stat Med* 2013 Feb 10;32(3):382–392.

Kang SH, Kim Y. Sample size calculations for the development of biosimilar products. *J Biopharm Stat* 2014;24(6):1215–1224.

Kang YS, Moon HH, Lee SE, Lim YJ, Kang HW. Clinical experience of the use of CT-P13, a biosimilar to infliximab in patients with inflammatory bowel disease: A case series. *Dig Dis Sci* 2014 Oct 18;60(4):951–956.

Keizer RJ, Budde IK, Sprengers PF, Levi M, Beijnen JH, Huitema AD. Model-based evaluation of similarity in pharmacokinetics of two formulations of the blood-derived plasma product c1 esterase inhibitor. *J Clin Pharmacol* 2012 Feb;52(2):204–213.

Kerpel-Fronius S. Clinical pharmacology aspects of development and application of biosimilar antibodies. *Magy Onkol* 2012 May;56(2):104–112.

King G. Quotient bioresearch complete £1.5 million expansion of bioanalytical research facilities. *Bioanalysis* 2012 Nov;4(22):2666.

Koyfman H. Biosimilarity and interchangeability in the biologics price competition and innovation act of 2009 and FDA's 2012 draft guidance for industry. *Biotechnol Law Rep* 2013 Aug;32(4):238–251.

Kuczka K, Harder S, Picard-Willems B, Warnke A, Donath F, Bianchini P, Parma B, Blume H. Biomarkers and coagulation tests for assessing the biosimilarity of a generic low-molecular-weight heparin: Results of a study in healthy subjects with enoxaparin. *J Clin Pharmacol* 2008 Oct;48(10):1189–1196.

Kuhlmann M, Covic A. The protein science of biosimilars. *Nephrol Dial Transplant* 2006 Oct;21(Suppl. 5):v4–v8.

Kumar M et al. Mass spectrometric distinction of in-source and in-solution pyroglutarnate and succinirnide in proteins: A case study on rhG-CSF. *J Am Soc Mass Spectrom* 2012;24(2):202–212. doi:10.1007/s13361-012-0531-7.

Lapadula G, Ferraccioli GF. Biosimilars in rheumatology: Pharmacological and pharmaco-economic issues. *Clin Exp Rheumatol* 2012 Jul–Aug;30(4 Suppl. 73):S102–S106. Epub 2012 Oct 18.

Lee JF, Litten JB, Grampp G. Comparability and biosimilarity: Considerations for the healthcare provider. *Curr Med Res Opin* 2012 Jun;28(6):1053–1058.

Li Y, Liu Q, Wood P, Johri A. Statistical considerations in biosimilar clinical efficacy trials with asymmetrical margins. *Stat Med* 2013 Feb 10;32(3):393–405.

Liao JJ, Darken PF. Comparability of critical quality attributes for establishing biosimilarity. *Stat Med* 2013 Feb 10;32(3):462–469.

Lin JR, Chow SC, Chang CH, Lin YC, Liu JP. Application of the parallel line assay to assessment of biosimilar products based on binary endpoints. *Stat Med* 2013 Feb 10;32(3):449–461.

Lingg N, Tan E, Hintersteiner B, Bardor M, Jungbauer A. Highly linear pH gradients for analyzing monoclonal antibody charge heterogeneity in the alkaline range. *J Chromatogr A* 2013 Dec 6;1319:65–71.

Locatelli F, Roger S. Comparative testing and pharmacovigilance of biosimilars. *Nephrol Dial Transplant* 2006 Oct;21 (Suppl. 5):v13–v6.

Lu Y, Zhang ZZ, Chow SC. Frequency estimator for assessing of follow-on biologics. *J Biopharm Stat* 2014;24(6):1280–1297.

Lubenau H, Bias P, Maly AK, Siegler KE, Mehltretter K. Pharmacokinetic and pharmacodynamic profile of new biosimilar filgrastim XM02 equivalent to marketed filgrastim Neupogen: Single-blind, randomized, crossover trial. *BioDrugs* 2009;23(1):43–51.

Marini JC, Anderson M, Cai XY, Chappell J, Coffey T, Gouty D, Kasinath A et al. Systematic verification of bioanalytical similarity between a biosimilar and a reference biotherapeutic: Committee recommendations for the development and validation of a single ligand-binding assay to support pharmacokinetic assessments. *AAPS J* 2014 Nov;16(6):1149–1158.

McCamish M, Woollett G. The continuum of comparability extends to biosimilarity: How much is enough and what clinical data are necessary? *Clin Pharmacol Ther* 2013 Apr;93(4):315–317.

Minocha M, Gobburu J. Drug development and potential regulatory paths for insulin biosimilars. *J Diabetes Sci Technol* 2014 Jan 1;8(1):14–19.

Mounho B, Phillips A, Holcombe K, Grampp G, Lubiniecki T, Mollerup I, Jones C. Global regulatory standards for the approval of biosimilars. *Food Drug Law J* 2010;65(4):819–837, ii–iii.

Nagasaki M, Ando Y. Clinical development and trial design of biosimilar products: A Japanese perspective. *J Biopharm Stat* 2014;24(6):1165–1172.

Nellore R. Regulatory considerations for biosimilars. *Perspect Clin Res* 2010 Jan;1(1):11–14.

No Authors. 3rd International Regulatory Workshop on A to Z on bioequivalence, bioanalysis, dissolution and biosimilarity. *Acta Pharm Hung* 2012;82(3):121–124.

O'Connor A, Rogge M. Nonclinical development of a biosimilar: The current landscape. *Bioanalysis* 2013 Mar;5(5):537–544.

Oh MJ, Hua S, Kim BJ, Jeong HN, Jeong SH, Grimm R, Yoo JS, An HJ. Analytical platform for glycomic characterization of recombinant erythropoietin biotherapeutics and biosimilars by MS. *Bioanalysis* 2013 Mar;5(5):545–559.

Oldfield P. A wide angle view of biosimilars from a bioanalytical perspective. *Bioanalysis* 2013 Mar;5(5):533–535.

Oldfield P. Differences in bioanalytical method validation for biologically derived macromolecules (biosimilars) compared with small molecules (generics). *Bioanalysis* 2011 Jul;3(14):1551–1553.

Pani L, Montilla S, Pimpinella G, Bertini Malgarini R. Biosimilars: The paradox of sharing the same pharmacological action without full chemical identity. *Expert Opin Biol Ther* 2013 Oct;13(10):1343–1346.

Panopoulus AD, Watowich SS. Granulocyte colony-stimulating factor: Molecular mechanisms of action during steady state and "emergency" hematopoiesis. *Cytokine* 2008;42:277–288.

Parnham MJ, Schindler-Horvat J, Kozlović M. Non-clinical safety studies on biosimilar recombinant human erythropoietin. *Basic Clin Pharmacol Toxicol* 2007 Feb;100(2):73–83.

PHS Act 61 BLA 125553 ODAC Brief EP2006, a proposed biosimilar to Neupogen.

PHS Act Section 351(k)(2)(A)(i)(I) of the PHS Act. As discussed in the Background section, the statute provides that FDA may determine, in FDA's discretion, that certain studies are unnecessary in a 351(k) application (see section 351(k)(2) of the PHS Act).

PHS Act Section 351(k)(2)(A)(i)(IV) of the PHS Act.

Prync AE, Yankilevich P, Barrero PR, Bello R, Marangunich L, Vidal A, Criscuolo M et al. Two recombinant human interferon-beta 1a pharmaceutical preparations produce a similar transcriptional response determined using whole genome microarray analysis. *Int J Clin Pharmacol Ther* 2008 Feb;46(2):64–71.

Pucaj K, Riddle K, Taylor SR, Ledon N, Bolger GT. Safety and biosimilarity of ior(®) EPOCIM compared with Eprex(®) based on toxicologic, pharmacodynamic, and pharmacokinetic studies in the Sprague-Dawley rat. *J Pharm Sci* 2014 Nov;103(11):3432–3441.

Pulsipher MA, Chitphakdithai P, Logan BR et al. Lower risk for serious adverse events and no increased risk for cancer after PBSCs BM donation. *Blood* 2014;123:3655.

Rathore AS, Bhambure R. Establishing analytical comparability for "biosimilars": Filgrastim as a case study. *Anal Bioanal Chem* 2014 Oct;406(26):6569–6576.

Rouse H. A new look at biosimilarity. *Biorheology* 1977;14(5–6):295–298.

Schiestl M. A biosimilar industry view on the implementation of the WHO guidelines on evaluating similar biotherapeutic products. *Biologicals* 2011 Sept;39(5):297–299.

Schiestl M, Li J, Abas A, Vallin A, Millband J, Gao K, Joung J, Pluschkell S, Go T, Kang HN. The role of the quality assessment in the determination of overall biosimilarity: A simulated case study exercise. *Biologicals* 2014 Mar;42(2):128–132.

Schiestl M, Stangler T, Torella C, Cepeljnik T, Toll H, Grau R. Acceptable changes in quality attributes of glycosylated biopharmaceuticals. *Nat Biotechnol* 2011;29(4):310–312.

Schreiber S, Luger T, Mittendorf T, Mrowietz U, Müller-Ladner U, Schröder J, Stallmach A, Bokemeyer B. Evolution of biologicals in inflammation medicine—Biosimilars in gastroenterology, rheumatology and dermatology. *Dtsch Med Wochenschr* 2014 Nov; 139(47):2399–2404.

Section 351(k)(2)(A)(i)(I) of the PHS Act. As discussed in the Background section, the statute provides that FDA may determine, in FDA's discretion, that certain studies are unnecessary in a 351(k) application (see section 351(k)(2) of the PHS Act).

Shaltout EL, Al-Ghobashy MA, Fathalla FA, Salem MY. Chromatographic and electrophoretic assessment of filgrastim biosimilars in pharmaceutical formulations. *J Pharm Biomed Anal* 2014 Aug;97:72–80.

Shin W, Kang SH. Statistical assessment of biosimilarity based on the relative distance between follow-on biologics for binary endpoints. *J Biopharm Stat* 2013;32(3):382–392.

Shirafuji N et al. A new bioassay for human granulocyte colony-stimulating factor (hG-CSF) using murine myeloblastic NFS-60 cells as targets and estimation of its levels in sera from normal healthy persons and patients with infectious and hematological disorders. *Exp Hematol* 1989;17:116–119.

Singleton CA. MS in the analysis of biosimilars. *Bioanalysis* 2014;6(12):1627–1637.

Skrlin A, Radic I, Vuletic M, Schwinke D, Runac D, Kusalic T, Paskvan I, Krsic M, Bratos M, Marinc S. Comparison of the physicochemical properties of a biosimilar filgrastim with those of reference filgrastim. *Biologicals* 2010 Sept;38(5):557–566.

Sörgel F, Lerch H, Lauber T. Physicochemical and biologic comparability of a biosimilar granulocyte colony-stimulating factor with its reference product. *BioDrugs* 2010 Dec 1; 24(6):347–357.

Sörgel F, Thyroff-Friesinger U, Vetter A, Vens-Cappell B, Kinzig M. Biosimilarity of HX575 (human recombinant epoetin alfa) and epoetin beta after multiple subcutaneous administration. *Int J Clin Pharmacol Ther* 2009a Jun;47(6):391–401.

Sörgel F, Thyroff-Friesinger U, Vetter A, Vens-Cappell B, Kinzig M. Bioequivalence of HX575 (recombinant human epoetin alfa) and a comparator epoetin alfa after multiple subcutaneous administrations. *Pharmacology* 2009b;83(2):122–130.

Stroncek DF et al. Treatment of normal individuals with granulocyte-colony stimulating factor: Donor experiences and the effects on peripheral blood CD34+ cell counts and on the collection of peripheral blood stem cells. *Transfusion* 1996;36:601–610.

Su J, Mazzeo J, Subbarao N, Jin T. Pharmaceutical development of biologics: Fundamentals, challenges and recent advances. *Ther Deliv* 2011 Jul;2(7):865–871.

Suh SK, Park Y. Regulatory guideline for biosimilar products in Korea. *Biologicals* 2011 Sept;39(5):336–338.

Sveikata A, Gumbrevičius G, Seštakauskas K, Kregždytė R, Janulionis V, Fokas V. Comparison of the pharmacokinetic and pharmacodynamic properties of two recombinant granulocyte colony-stimulating factor formulations after single subcutaneous administration to healthy volunteers. *Medicina (Kaunas)* 2014;50(3):144–149.

Szeto KJ, Wolanski M. Initial steps in the regulation of generic biological drugs: A comparison of U.S. and Canadian regimes. *Food Drug Law J* 2012;67(2):131–141, i.

Takahashi O, Kirikoshi R, Manabe N. Acetic acid can catalyze succinirnide formation from aspartic acid residues by a concerted bond reorganization mechanism: A computational study. *Int J Mol Sci* 2015;16:1613–1626,

Tan Q, Guo Q, Fang C, Wang C, Li B, Wang H, Li J, Guo Y. Characterization and comparison of commercially available TNF receptor 2-Fc fusion protein products. *MAbs* 2012 Nov–Dec;4(6):761–774.

Tóthfalusi L, Endrényi L, Chow SC. Statistical and regulatory considerations in assessments of interchangeability of biological drug products. *Eur J Health Econ* 2014 May;15(Suppl. 1): S5–S11.

Trnka H, Wu JX, Van De Weert M, Grohganz H, Rantanen J. Fuzzy logic-based expert system for evaluating cake quality of freeze-dried formulations. *J Pharm Sci* 2013 Dec;102(12):4364–4374.

Tsiftsoglou AS, Trouvin JH, Calvo G, Ruiz S. Demonstration of biosimilarity, extrapolation of indications and other challenges related to biosimilars in Europe. *BioDrugs* 2014 Nov 13;28(6):479–486.

UNIPROT. UniProtKB—P09919 (CSF3_HUMAN). http://www.uniprot.org/uniprot/P09919 (accessed August 2015).

van Aerts LA, De Smet K, Reichmann G, Willem van der Laan J, Schneider CK. Biosimilars entering the clinic without animal studies: A paradigm shift in the European Union. *MAbs* 2014 Aug 5;6(5):1155–1162.

173

Visser J, Feuerstein I, Stangler T, Schmiederer T, Fritsch C, Schiestl M. Physicochemical and functional comparability between the proposed biosimilar rituximab GP2013 and originator rituximab. *BioDrugs* 2013 Oct;27(5):495–507.

Wadhwa M. Immunogenicity: What do we know and what can we measure?, PhD thesis, NIBSC–HPA, Hertfordshire, U.K., April 12, 2011.

Wang X, Chen L. Challenges in bioanalytical assays for biosimilars. *Bioanalysis* 2014 Aug;6(16):2111–21113.

Wish JB. The approval process for biosimilar erythropoiesis-stimulating agents. *Clin J Am Soc Nephrol* 2014 Sept 5;9(9):1645–1651.

Yang J, Zhang N, Chow SC, Chi E. An adapted F-test for homogeneity of variability in follow-on biological products. *Stat Med* 2013 Feb 10;32(3):415–423.

Yang LY, Lai CH. Estimation and approximation approaches for biosimilar index based on reproducibility probability. *J Biopharm Stat* 2014;24(6):1298–1311.

Zaia J. Mass spectrometry and glycomics. *OMICS* 2010 Aug;14(4):401–418.

Zhang A, Tzeng JY, Chow SC. Establishment of reference standards in biosimilar studies. *GaBi J* 2013a July 31;2(4):173–177.

Zhang A, Tzeng JY, Chow SC. Statistical considerations in biosimilar assessment using biosimilarity index. *J Bioequiv Availab* 2013b Sept 2;5(5):209–214.

Zhang N, Yang J, Chow SC, Chi E. Nonparametric tests for evaluation of biosimilarity in variability of follow-on biologics. *J Biopharm Stat* 2014;24(6):1239–1253.

Zhang N, Yang J, Chow SC, Endrenyi L, Chi E. Impact of variability on the choice of biosimilarity limits in assessing follow-on biologics. *Stat Med* 2013 Feb 10;32(3):424–433.

Zink T et al. Structure ad dynamics of the human granulocyte colony stimulating factor determined by NMR spectroscopy. Loop mobility in a four helixbundle protein. *Biochemistry* 1994;33: 8453–8463. http://www.brnrb.wisc.edu/datalibrary/surnrnary/index.php?brnrbld=18291.

Chapter 6 Recombinant expression systems

Background

Whereas most biosimilar product developers would prefer to replicate the expression systems used by the originator, this is not a requirement of the regulatory agencies. As a result, the choice of the recombinant cell line needs to be made judiciously keeping in mind three factors: first, the degree of analytical and functional similarity obtained; second, the yield to control the cost of production; and third, the time and cost involved in developing and validating the expression system. Currently, the most promising cell lines involve mammalian cells. Recent developments in expediting their development and increase in the yield are the most significant discussions in the literature.

The starting material for manufacturing recombinant drugs is the bacterial yeast or insect or mammalian cell culture that expresses the protein product or monoclonal antibody of interest. The cell seed lot system is used by manufacturers to assure the identity and purity of the starting raw material. A cell seed lot consists of aliquots of a single culture. The master cell bank (MCB) is derived from a single colony (bacteria, yeast) or a single eukaryotic cell, stored cryogenically to assure genetic stability and is composed of sufficient ampoules of culture to provide source material for the working cell bank (WCB). The WCB is defined as a quantity of cells derived from one or more ampoules of the MCB, stored cryogenically and used to initiate the production batch.

The most common cellular expression systems used to manufacture therapeutic proteins include bacteria (*Escherichia coli*, *Bacillus subtilis*, *Lactococcus lactis*), yeast (*Saccharomyces cerevisiae*, *Pichia pastoris*), mammalian cells (Chinese hamster ovary [CHO], Baby hamster kidney [BHK]), and insect cells where the baculovirus expression system is used to some extent and may prove itself as a future biopharmaceutical expression system. On February 6, 2009, FDA approved an orphan drug ATryn to treat a rare clotting disorder; its first approval for a biological product produced by genetically engineered (GE) animals. The ATryn, is an anticoagulant used for the prevention of blood clots in patients with a rare disease known as hereditary antithrombin (AT) deficiency. ATryn is a therapeutic protein derived from the milk of goats that have been genetically engineered by introducing a segment of DNA into their genes (called a recombinant DNA or rDNA construct) with instructions for the goat to produce human AT in its milk. AT is a protein that naturally occurs in healthy individual and helps to keep blood from clotting in the veins and arteries. The Center for Biologics Evaluation and Research (CBER) approved the human biologic based on its safety and efficacy, and the Center for Veterinary Medicine (CVM) approved the rDNA construct in the goats that produce ATryn.

On January 16, 2013, the U.S. Food and Drug Administration approved Flublok, the first trivalent influenza vaccine made using an insect virus (baculovirus) expression system and recombinant DNA technology. Flublok is approved for the prevention of seasonal influenza in people 18 through 49 years of age.

A list of major cell lines is provided in Table 6.1.

Table 6.1 Cell Lines Used for Commercially Produced Recombinant Products as Approved by the FDA

Cell Line	Expressed Product
African Monkey Kidney Cells (COS-1)	Antihemophilic factor (Advate®)
Baby Hamster Kidney Cells	Antihemophilic factor (Helixate® and Kogenate®FS)
Chinese Hamster Ovary Cell	Adalimumab (Humira®)
	Alefacept (Amevive®)
	Alemtuzumab (Campath®)
	Algasidase beta (Fabrazume®)
	Alteplase (Activase®, Cathflo®)
	Antihemophilic factor (Bioclate®, Recombinate® Rahf®, ReFacto®)
	Choriogonadotropin alfa (Ovidrel®)
	Coagulation factor IX (BeneFix®)
	Darbepoetin alfa (Aranesp®)
	Dornase alfa (Pulmozyme®)
	Drotrecogin alfa (Xigris®)
	Efalizumab (Raptiva®)
	Epoietin alfa (Epogen®, Procrit®)
	Etanercept (Enbrel®)
	Follitropin alfa (Gonal-F®)
	Follitropin beta (Follistim®)
	Ibritumomab tiuxetan (Zevalin®)
	Imiglucerase (Cerezyme®)
	Interferon beta 1-alpha (Avonex®)
	Laronidase (Aldurazyme®)
	Omalizumab (Xolair®)
	Rituximab (Rituxan®)
	Tenecteplase (TNKase®)
	Thyrotropin alfa (Thyrogen®)
	Trastuzumab (Herceptin®)
E. coli	Aldesleukin Proleukin, IL-2®
	Alpha interferon + ribavarin Reberon®
	Alpha-interferon (Intron A®)
	Anakinra (Kineret®)
	Bone morphogenetic protein [rhBMP-2-] device (Infuse®Bone Graft/LT-CAGE®)
	Coagulation factor VIIa (NovoSeven®)
	Denileukin difitox (Ontak®)
	Filgrastim (Neupogen®)
	Growth Hormone (BioTropin®)
	Insulin (Humalog®, Humulin®, Velosulin®, BR Novolin®, Novolin L®, Novloin R®, Novolin® 70/30, Novloin N®)
	Insulin aspart (Novolog®)
	Insulin glargine (Lantus®)
	Insulin glulisine (Apidra®)
	Interferon alfa-2a (Roferon-A®)
	Interferon alfacon-1 (Infergen®)
	Interferon beta-1a (Rebif®)
	Interferon beta-1b (Betaseron®)
	Interferon gamma-1b (Actimmune®)

(*Continued*)

Table 6.1 (*Continued*) Cell Lines Used for Commercially Produced Recombinant Products as Approved by the FDA

Cell Line	Expressed Product
	Nesiritide (Natrecor®)
	Oprelvekin (Neumega®)
	Pegfilgrastim (Neulasta®)
	Peginterferon alfa-2z (Pegasys®)
	Pegvisomant (Somavert®)
	Pegylated inerferon alfa-2b (PEG-Intron®)
	Reteplase (Retavase®)
	Somatotropin (Humatrope®)
	Somatrem (Protropin®)
	Somatropin (Norditropin®, Nutropin®/Nutropin AQ®) (Nutropin Depot® GenoTropin® Geref®)
	Teriparatide (Forteo®)
Lymphocyte Activated	Daclizumab (Zenapax®)
Mouse C 127	Growth Hormone (Saizen®, Serostim®)
Mouse Myeloma	Basiliximab (Simulect®)
Myeloma NSO	Gemtuzumab ozogamicin (Mylotarg®)
Mammalian	Tositumomab with I-131 (Bexxar®)
	Abciximab (ReoPro®)
Prostate Epithelium Cell	Capromab pendetide with In-111 (ProstaScint®)
Saccharomyces cerevisiae	Becaplermin gel (Regranex® Gel)
	Glucagon (GlucaGen®)
	Granulocyte macrophage colony-stimulating factor (Leukine®)
	Haemophilus B conjugate (Comvax®)
	Hepatitis A inactivated and hepatitis B vaccine (Twinrix®)
	Hepatitis B and inactivated polio-virus vaccine (Pediarix®)
	Hepatitis B vaccine (Engerix-B®, Recombivax-HB®)
	Lepirudin (Refludan®)
	OspA lipoprotein (LYMErix®)
	Rasburicase (Elitek®)

Of the currently approved products, the largest numbers of these products are expressed in three hosts: *E. coli*, followed by CHO cells and *S. cerevisiae* (Table 6.1). There are clear advantages and disadvantages of each of the host system described. These are summarized in Figure 6.1.

Figure 6.1 shows the relative prevalence of the use of these cell lines.

Table 6.2 lists the comparative advantages and disadvantages of key expression systems.

The choice of expression system depends on factors such as type of target protein, posttranslational modifications, expression level, intellectual property rights, and economy of manufacture. The *E. coli* expression system offers rapid and cheap expression, but cannot express complex proteins, and include in vitro folding and tag-removal into the downstream process. Yeast generally expresses the target protein in its native form to the medium, but expression levels are very low. Insect cells provide many advantages of the mammalian cell characteristics. Table 6.3 lists more specific comparisons of these expression systems.

While the developer of biosimilar products may find it lucrative to select an expression system based on the various attributes described earlier, often the choice of the expression system will be limited to the expression system used by the originator,

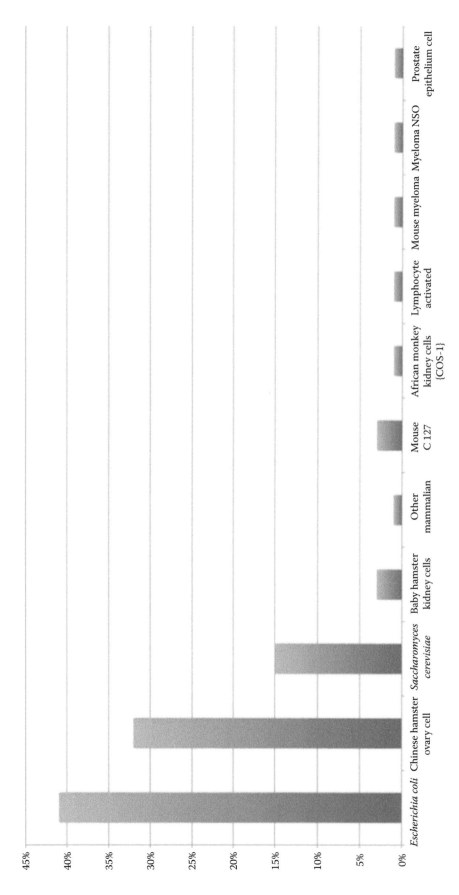

Figure 6.1 Most popular hosts for approved drugs. (From *Physician's Desk Reference* 2014, http://www.pdr.net/.)

Table 6.2 Comparison of Host Systems Advantages and Disadvantages

Host	Advantages	Disadvantages
Bacteria (e.g., *Escherichia coli*)	Many reference and experience available, wide choice of cloning vectors, gene expression easily controlled; easy to grow with high yields, product forming up to 50% of total cell protein, can be designed for secretion into growth media allowing for the removal of unwanted N-terminal methionine groups.	No posttranslational modification, Biological activity and immunogenicity may differ from natural proteins; high endotoxin content in gram-negative. The expressed protein product may cause cellular toxicity; inclusion bodies difficult to process.
Bacteria (e.g., *Staphylococcus aureus*)	Secretes fusion proteins into the growth media.	Does not express such high levels as *E. coli*; pathogenic.
Mammalian cells (e.g., Chinese hamster ovary cells)	Same biologic activity as native proteins, mammalian expression vectors available, can be grown in large-scale cultures.	Cells can be difficult and expensive to grow, cells grow slowly; manipulated cells can be genetically unstable; low productivity as compared to microorganisms.
Yeasts (e.g., *Sacchromyces cervisiae*)	Lack detectable endotoxin, generally regarded as safe (GRAS), fermentation relatively inexpensive, facilitates glycosylation and formation of disulfide bonds, only 0.5% native proteins are secreted so isolation of secreted product is simplified, well established large-scale production and downstream processing.	Gene expression less easily controlled; glycosylation not identical to mammalian systems.
Cultured insect cells (Basculovirus vector)	Facilitates glycosylation and formation of disulfide bonds, safe since few anthropods are adequate host for baculovirus; baculovirus vector received FDA approval for a clinical trial, virus stops host protein amplification, high level of expression of product.	Lack of information on glycosylation mechanism, product not always fully functional; few differences in functional and antigenic properties between product and native protein.
Fungi (e.g., *Aspergills* sp.)	Well-established system for fermentation of filamentous fungi; growth inexpensive.	High level of expression not yet achieved; genetics not always characterized.
Fungi (e.g., *A. niger* sp.)	This is also generally regarded as safe (GRAS), can secrete large quantities of product into growth media, source of many industrial enzymes.	No cloning vectors available.
Transgenic plants	*Large-scale production, very low cost, environmental control.*	Low transformation efficiency, long generation time; often commercially not viable.

Table 6.3 Comparison of Various Expression Systems, Advantages and Disadvantages

Qualifier	Bacteria	Yeast	Insect Cells	Mammalian Cells	Transgenic Animals
Example	*E. coli*	*S. cerevisiae, P. pastoris*	*Lepidopteran*	*Chinese Hamster Ovary*	*Cattle, goat*
Level of expression	High	Medium	Medium	Medium	Very high
Time to produce expression system	Fast	Fast	Medium	Slow	Very slow
	5 days	14 days	4 weeks	4–8 weeks	6–33 months
Cost	Low	Low	High	High	Medium
Extracellular expression	No	Yes	Yes	Yes	Yes
Met-protein expression	Yes	No	No	No	No
Posttranslation modifications	No	No	Yes	Yes	Yes
Major impurities	Endotoxins	Glycosylated products	Viruses	Viruses	Viruses and prions
In vitro protein refolding	Yes	No	No	No	No
Unintended glycosylation	No	Possible	Possible	Possible	Possible
Host cell protein expression	No	No	No	No	Yes
Regulatory track record	Good	Good	N/A	Good	N/A

unless the developer is willing to go through extensive additional safety and efficacy studies. However, in some instances this may be useful. One such selection may be the expression system for insulin.

Expression system development

A typical cell line development scheme using these technologies is illustrated in Figure 6.2.

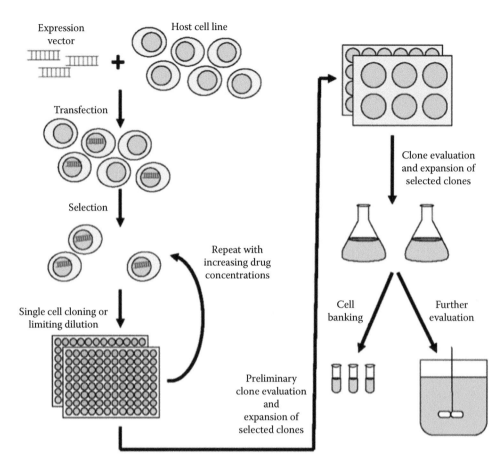

Figure 6.2 Cell line development scheme. (From Lai, T. et al., *Pharmaceuticals*, 6, 579, 2013.)

Bacterial expression systems

The bacterial expression system makes use of gram-negative (e.g., *E. coli*) and gram-positive host cells (*Bacillus* and *Lactococcus*) and allows both intracellular and extracellular expression of target protein, however, without posttranslational modifications. It takes about 5 days from introduced gene to protein production at acceptable levels of a few hundred mg/L to g/L. Bacterial systems are easy to scale up from culture flask to fermenters with capacity into thousands of liters because of simpler nutrition and aeration requirements and the ability to take carrying shear force. Several types of culture process are used such as batch and fed-batch, making this system highly flexible. The major adventitious agents are host cell proteins and endotoxins from *E. coli*; viruses are of little concern, except what may be required to control general exposure during processing. Critical steps including control of impurities arising out of released proteolytic enzymes and endotoxins, besides the handling of inclusion bodies, in vitro folding and cleavage of the N-terminal extension introduced to overcome the problem with expression of Met-protein. As a result, the downstream processing is often complicate, which can limit the choice of this expression system. However, the regulatory record is impressive (Table 6.1); at this stage only gram-negative organisms have been approved.

The typical protein to be expressed in gram-negative bacteria comprises 100–300 amino acids, has no posttranslational modifications, and it possesses a restricted number of cysteine residues to make in vitro refolding possible. The problems of methionine blocking the N-terminus and the intracellular formation of inclusion bodies, which add host cell impurities in the use of gram-negative bacteria, can be resolved by using gram-positive bacteria such as *B. subtilis* and *L. lactis*. Also, the presence of endotoxin is of little concern in gram-positive bacteria. On the negative side, while using *Bacillus*, endogenous proteolytic degradation can be significant, a drawback not found in *Lactococcus*. The extracellular expression of the folded target protein in the use of gram-positive bacteria additionally eliminates significant problems in the use of *E. coli* expression system: the requirements of in vitro folding and performing cleavage of the N-terminal tag. This results in simpler process design comprising of harvest, capture, intermediate purification, polishing, concentration and finishing as described in the following discussion for gram-negative bacteria.

Although bacteria cannot be used to express some large, complex proteins (with eukaryotic posttranslational modifications including glycosylation, acetylation, and amidation) other proteins such as interferons, interleukins, colony-stimulating factors, growth hormones, growth factors, and human serum albumin have been successfully produced. Marketed products produced include human growth hormone, human insulin, α, β, and γ-interferon, and interleukin 2.

The bacterial system is most popular because of its low fermentation cost and high expression yields. The most common form of bacterial expression is cytoplasmatic or intracellular expression where inclusion bodies contain the target protein, which is in an inactive form and thus harmful to bacteria; the yield is generally high and the system requires simpler plasmid construct. The disadvantages include need for in vitro folding, expression of Met-protein, and complex purification steps needed. The cell disruption to recover inclusion bodies releases a large volume of host cell proteins, nucleic acids, and proteolyitc enzymes that can damage target protein. Measures to reduce this proteolysis include use of thioredoxin deficient strains, processing at low temperatures, and coexpression of chaperones, coexpression of PDI, and fusion partners; unfortunately, none of the methods have been developed far enough to be useful in commercial processing operations. Cytosolic expression in *E. coli* also produces methionine blocking of the N-terminus requiring a specific enzymatic cleavage step to produce the native molecule. To eliminate expression of Met-protein, N-terminally extended protein constructs are expressed followed by an enzymatic cleavage step in which the extension is cleaved with an endopeptidase or removed step by step with an exopeptidase. Typical extensions are fusion proteins, histidine tags or, if proline is the second amino acid in the protein, a few amino acids. These disadvantages, particularly the inability of gram-negative organisms to express folded proteins to the periplasm or to the medium has resulted in attempts to use other forms of expression, namely, periplasmic and secretion. Attention has been focused on gram-positive bacteria, for example, *Bacillus* species, which naturally secretes large amounts of protein to the medium, but endogenous proteolytic degradation is of general concern. *Lactococcus*, traditionally used in the dairy industry, has no detectable extracellular proteases or toxins and also offers direct expression to the medium. The periplasmic expression has the advantage of producing native protein with correct N-terminal structure, requiring simpler purification and less extensive incidence of proteolysis. The disadvantages include inefficient translocation through the inner membrane and the fact that inclusion bodies may still form. The extracellular secretion system would be ideal as it results in least extensive proteolysis, provides for simpler purification, and corrects N-terminus situation; however, there is little secretion that results in very low yields; one exception being the expression of β-glucanase.

Genetic modification of bacteria

The genetically modified bacteria are prepared by introducing a recombinant gene into the bacteria on a plasmid, modified to optimize heterologous protein expression comprising complete genetic elements to enhance transcription and translation and to stabilize mRNA. A plasmid is a double-strand circular DNA, small 2–25 kbp (some bigger), bacterial artificial chromosome that has the characteristics of autonomous replication. Plasmids vectors are very useful as a quick way to introduce genes to the cell, and are much easier to manipulate than the chromosome. Components of a gene includes promoter, operator, RBS, CDS, start, and stop codon. All vectors comprise an origin of replication (Ori) that determines the vector copy number. The stability of the expression cassette has a high impact on productivity. Loss of the cassette during the course of the fermentation (the cell tends to minimize the stress by getting rid of the plasmid) results in formation of nonproductive, plasmid free cells, which usually outgrow plasmid carrying cells in fermentation unless selective pressure can be effectively employed. Plasmid stability is increased by using vectors that confer an antibiotic resistance (e.g., ampicillin) to the host or complement some auxotrophic features of the host strain. If periplasmic or extracellular expression is desired, the vector must also include a signal sequence for the transport of the protein.

Great advances have been made in the availability of commercial systems that help resolve some of the problems related to the expressed protein harming the host, formation of inclusion bodies, target proteins appearing translational, and complex purification procedures. The popular commercial systems provide supporting materials and reagents in kit to cone, express, and purify recombinant proteins with a fair amount of ease. These support systems are broadly classified as promoters, stabilizing and optimizing elements, ribosome-binding sites, transcriptional and translational termination sequences.

Promoters

Promoters are DNA sequences to which the endogenous bacterial RNA polymerase will bind to initiate transcription. The promoter has an important function in that it determines the polarity of the transcript by specifying which strand will be transcribed. More important, how tightly regulated a promoter/operator is will greatly affect the ability to express proteins. Most bacterial promoters used in expression vectors consist of two elements located (−35) and (−10) from the actual transcriptional start. These elements comprise the consensus sequences that are bound by a specific transcription factor and the RNA polymerase. One of the most commonly encountered promoters is the Trc promoter.

DNA sequences that act in conjunction with promoters and bind repressor molecules regulate the induction of transcription. One of the most commonly used in *E. coli* is the lacO/lacIq repressor system. In this system, transcription is virtually shut off until the promoter is derepressed by the addition of isopropyl β-D-thiogalactoside (IPTG). At this point, the promoter is freed (i.e., the repressor no longer physically blocks transcription) and transcription is turned on. Inducible elements provide the ability to keep expression of the target gene off, should it produce a product that might be toxic to the host strain. There are a number of different methods of regulation that are available in commercial expression systems. While most are capable of inducing tremendous levels of expression, even slightly leaky expression during culture expansion can limit final product yield.

Terminators

In prokaryotic systems, anti-termination elements are incorporated into vectors to stabilize the RNA polymerase on the DNA template. These elements help ensure optimal transcript elongation during message synthesis. Transcription terminators are used to signal the active RNA polymerase to release the DNA template and halt transcription of the newly transcribed RNA. These terminators are ordinarily positioned downstream of the multicloning site and act to prevent pausing, prevent premature termination, and limit read-through of transcription, which adversely affects plasmid replication. Use of transcription terminators such as rrnB T1 and T2 from *E. coli* 5S rRNA are especially important for use with strong promoters.

Ribosome-binding sites

Ribosome-binding sites (RBS) are small, open reading frames upstream of the coding sequence of interest engineered to encourage ribosomes to bind and translate the sequence of interest. An RBS (Shine-Dalgarno sequence) is required just upstream of the translational start to provide the context for efficient translation initiation. RBS sequences are engineered into vectors to enhance and stabilize ribosome binding. Often RBS elements may be "borrowed" from different sources or are the native RBS for the fusion species. Frequently used RBS elements are obtained from either the *lacZ* gene or the gene 10 of the bacteriophage T7.

Several commercial systems are now available with more appearing on the scene regularly; what used to be a great deal of science and art has been practically reduced to a kit approach with remarkable consistency and proven results. Following are some of these systems; the Novogen pET system remains one of the best systems that we have used:

Arabinose-regulated promoter (Invitrogen pBAD Vector): This system generates very low levels of uninduced expression through glucose-catabolite-dependent repression. The pBAD promoter is induced by arabinose. Adding arabinose into the medium in increasing concentrations induces transcription in a dose-dependent manner.

T7 expression systems (Novagen, Promega, Stratagene): The pET-based vectors utilize the T7 RNA polymerase-based expression vector of Studier and Moffat (1986) to achieve very high levels of protein expression. The power of the pET system is that the T7 RNA polymerase is specific for its own promoter, which is found only on the expression plasmid. During growth and culture expansion, the T7 promoter is under the tight control of the *lacIq* gene, repressing any expression that might adversely affect bacterial growth. When induced with IPTG, rather than directly inducing the T7 promoter 5' to the target gene, a T7 RNA polymerase is expressed in host *E. coli*, allowing transcription from the T7 promoter. Transcription and translation can be accomplished in only a few hours with the expressed protein often comprising the most abundant cellular component. It is important to know that some of these commercial systems are sold under intellectual property rights that may require royalty agreements. For example, the Invitrogen's pET system is licensed from Brookhaven Institute that has no fee when used for development in the laboratory, $5000 if used to test products in humans, and a royalty of 1% if the product derived from using the T7 pET system results in a commercial product used in humans in the United States. The licensing agreement is signed directly with Brookhaven and requires the manufacturer carry a liability insurance. This patent is about to expire and then free use of this system will be possible.

Trc/Tac promoter systems (Clontech, Invitrogen, Kodak, Life Technologies, MBI Fermentas, New England BioLabs, Amersham Biotech, Promega): Trc promoters are IPTG-inducible hybrid promoters. The Trc promoter is a trp/lac fusion, where the (−35) position is derived from trp and the (−10) position is obtained from lacUV5 promoter elements. While extremely strong, some low-level expression of the recombinant protein may occur during growth.

P_L *promoters (Invitrogen pLEX and pTrxFus Vectors):* P_L-based systems place the protein of interest under the tight transcriptional control of the Lambda cI repressor protein. The repressor protein must be engineered into the *E. coli* host or be incorporated into the vector itself. The repressor protein is placed under the control of an inducible promoter. Expression of the cI repressor binds to the operator of the P_L promoter to prevent transcription of the recombinant gene. Induction occurs with the addition of tryptophan, which prevents expression of the repressor, allowing transcription for the P_L promoter.

Lambda PR promoter (Amersham pRIT2T Vector): The PR promoter of the bacteriophage Lambda provides for thermoinducible expression of proteins when temperatures are shifted during growth from 37°C to 42°C. Induction is accomplished in 90 min.

Phage T5 promoter (Qiagen): The phage T5 promoter provides a strong recognition site for *E. coli* RNA polymerase and can direct the expression of targets to levels up to 50% of total cellular protein. The promoter is regulated by two lactose operator elements.

tetA promoter (Biometra pASK75 Vector): The tetA promoter is induced by the addition of anhydrotetracycline in concentrations that are not antibiotically effective. This promoter is not regulated by any endogenous cellular mechanisms and therefore is not influenced by catabolic repression.

Fusion proteins and tags

One-step purification of the recombinant protein by high-affinity binding can be accomplished in some situations using vectors engineered with DNA sequences encoding a specific peptide fused to the expressed protein. One of the most popular systems is the 6xHis system in which six histidine residues enable the tagged-recombinant protein to be purified by a nickel-chelating resin. Often, an endopeptidase recognition sequence is also engineered between the affinity tag and the protein of interest to allow subsequent removal of the leader sequence, peptide-tag, or fusion sequences by enzymatic digestion.

Tags serve several functions providing purification and stabilization of the expressed protein; fusion may act as a tag with the generation of antibodies.

Calmodulin-binding peptide (CBP) tag (Stratagene pCAL Vectors): pCAL expression vectors contain a sequence encoding a CBP. The CBP tag allows the hybrid recombinant protein to bind to a calmodulin resin in the presence of low concentrations of calcium. Elution is accomplished in the presence of 2 mM EGTA and neutral pH. The conditions are milder than in other tag systems. The CAP tag is one of the smaller tags, encoding a 4 kDa tag. Smaller tags have potentially less impact on the protein of interest than larger tags.

Glutathione S-Transferase (GST) tag (Amersham pGEX Vectors): Vectors containing the GST-fusion tag allow encoded proteins to be efficiently purified from bacterial lysates utilizing an affinity matrix containing glutathione. Elution of the purified protein is accomplished under mild, nondenaturing conditions. The GST fusion adds a 26 kDa tag to the recombinant protein, which can be removed when an endopeptidase cleavage site sequence is incorporated between the tag and the protein.

6xHIS tag (Invitrogen, Kodak, Life Technologies, New England BioLab, Pierce, Amersham (GE), Promega, Qiagen, Novagen): 6xHis-tagging vectors fuse a six histidine-peptide to the recombinant protein. This small addition rarely affects protein structure to a significant degree and therefore usually does not require removal following purification of the protein. The 6-His residues impart a remarkable affinity for matrices containing nickel. The fact that binding can occur under native as well as under denaturing conditions distinguishes this affinity purification method from the others. Recombinant proteins, frequently encountered in inclusion bodies in bacterial expression systems, can be solubilized under denaturing conditions using urea or guanidine hydrochloride. Binding to nickel ions on the matrix can then purify the solubilized 6xHis-tagged proteins. The strong affinity of the 6xHis tag tolerates denaturing conditions that facilitate the removal of nonspecific contaminants often associated with recombinant proteins expressed in bacteria. Elution is accomplished under mild conditions by either reducing the pH or adding imidazole as a competitor. With great advances taking place in the field of commercial kits and systems, the reader is referred to market leaders to obtain information on a system suitable for their use.

Dihydrofolate reductase (QIAGEN): Short peptide sequences from the murine dihydrofolate reductase (DHFR) are fused to target proteins to increase stability of the target protein, as well as enhance its antigenicity.

Thioredoxin fusion sequences (Invitrogen, pTrxFUS and pThioHis Vectors, Novagen pET-32 Vectors): Designed to maximize the accumulation of fusion proteins in the soluble fraction, a number of additional advantages are imparted when recombinant proteins are fused with sequences encoding thioredoxin. Fusion with thioredoxin helps overcome insolubility encountered with some bacterial systems. Additionally, in *E. coli*, thioredoxin accumulates in adhesion zones, which can be selectively released by rapid osmotic shock. The temperature stability of thioredoxin also allows temperatures as high as 80°C to be tolerated during purification. These features can be exploited to simplify purification of expressed fusion proteins. Finally, thioredoxin possesses a unique active dithiol, which directs high-affinity binding to phenylarsine oxide, allowing rapid purification of expressed proteins.

Protein A (Amersham pEZZ 18 and pRIT2T): The ability of Protein A to bind to IgG immunoglobulins continues to be a workhorse in biotechnology. pEZZ 18 contains two synthetic "Z" domains of the "B" IgG-binding domains of Protein A and pRIT2T contains the natural IgG-binding domains of protein A. Fusion proteins are easily purified using IgG Sepharose resins. The two Z domains add a 14 kDa peptide to the recombinant protein.

Biotinylation (Promega PinPoint™ Vector): Biotinylation-based epitopes encode peptide sequences that become biotinylated in vivo during expression. Relying on the strong affinity of streptavidin for biotin, the biotinylated peptide acts as a purification tag for the fusion protein. Promega provides a unique monomeric avidin which allows elution of the fused proteins under mild nondenaturing conditions (5 mM biotin), not possible with native avidin.

Cellulose-binding domain (CBD) (Novagen, pET CBD Vectors): CBDs allow for the purification of expressed proteins with cellulose or chitin matrices. These materials are generally very low cost and quite stable and may be found in a variety of forms including beads, powders, fibers, membranes, filters, and sheets. Additionally, the CBD fusion often increases thermostability of the recombinant protein, which may be useful in purification strategies or in simply increasing the stability of expressed proteins.

Maltose-binding protein (MBP) (New England BioLabs, pMAL Vectors): The *malE* gene encodes the MBP, which has the capability to bind tightly to amylose. The MBP is separated from the desired expressed sequence by a polylinker encoding 10 asparagine residues. The linker is designed to assure the MBP can adequately bind amylose that is immobilized on resin. In addition to aiding in expression and purification, the MBP fusion partner also helps keep proteins soluble and provides a choice of folding pathways.

S-peptide tag (Novagen, selected pET Vectors): The 15 amino acid peptide (S-Tag) encoded by these vectors is a small, hydrophilic tag that has a strong affinity ($Kd = 109$ M) for a 104 amino acid protein (S-protein) derived from pancreatic ribonuclease A. When associated, the S-protein: S-peptide complex possesses ribonuclease activity (ribonuclease S). This strong association and activity have been exploited to measure S-Tag fusion protein (ribonuclease assay, which can detect as little as 5 fmol target) and can be used with reagents and resins for detection and purification purposes. For example, the S-protein itself can be directly conjugated to alkaline phosphatase or horseradish peroxidase and used as a probe to detect S-tag fusion protein by Western blot.

Strep-tag (Biometra pASK75 Vector): A C-terminal fusion tag, the Strep-tag encodes a 10 amino acid sequence that binds streptavidin. Affinity of the tag for streptavidin allows purification of recombinant proteins under mild conditions. Along with affinity-purification applications, the strep-tag can be utilized to directly detect recombinant proteins by Western blot or ELISA assays using streptavidin-enzyme (alkaline phosphatase) conjugates.

Intei-mediated purification with affinity chitin-binding tag (New England BioLabs, pCYB Vectors/IMPACT System): This chimeric system produces a C-terminally tagged protein in which a small 5kDa chitin-binding domain from *Bacillus circulans* is physically separated from the target protein by the protein splicing element derived from the *S. cerevisiae VMA* gene 1. The splicing element, or Intein, has been modified such that it undergoes a self-cleavage reaction at its N-terminus at low temperatures in the presence of thiols. Passing extracts through a chitin column purifies the fusion protein; the purified protein is then induced to undergo the Intein-mediated self-cleavage on the column by overnight incubation at 4°C in the presence of DTT or b-mercaptoethanol. The target protein without additional residues is released while the chitin-binding fusion domain remains adhered to the resin.

Immunoreactive epitopes (Invitrogen, Novagen, Kodak): While many vectors utilize fusion or epitope tags for multifunctional purposes, some vectors encode fragments that are essentially immunoreactive epitopes. Immunoreactive tags provide a rapid means of detecting tag-hybrid proteins with high specificity and affinity without needing to generate specific antisera to each protein of interest. The monoclonal anti-myc antibody detects recombinant proteins containing the myc epitope (GluGln LysLeuIleSerGluGluAspLeuAsn). An alternative is the 11 amino acid epitope tag (GlnProGluLeuAlaProGluAspProGluAsp) derived from Herpes Simplex (HSV) glycoprotein D that can be recognized by an anti-HSV monoclonal antibody.

Kinase sequences for in vitro labeling (Stratagene, Amersham): To enable rapid in vitro labeling of proteins with 32P, some vectors incorporate the recognition sequence for the catalytic subunit of cAMP-dependent protein kinase as a fusion epitope. The site encoding the sequence (ArgArgAlaSerVal) is usually located between a primary fusion sequence and the sequence for the gene of interest (GOI). Expressed proteins can be directly labeled using protein kinase and [gamma 32P]-ATP (Kelin et al. 1992) protein A signal sequence (Amersham pEZZ18). The protein A signal sequence of this vector results in the secretion of the expressed protein into the aqueous culture.

ompT and ompA leader signal peptides (Biometra, New England BioLabs, Kodak): The ompT and ompA leader sequence directs secretion of proteins into the periplasmic space. This periplasmic localization can simplify recovery of expressed proteins since osmotic shock treatments can enrich for proteins localized within the periplasm.

malE signal sequence (New England BioLabs pMAL-p2): The normal malE signal sequence contains residues that direct the expressed fusion protein through the cytoplasmic membrane to the periplasm. This provides an alternate folding pathway that particularly helps in the formation of disulfide bonds.

T7 gene 10 leader peptide (Novagen, Stratagene, Promega, Invitrogen): An 11 amino acid leader peptide (MetAlaSerMetThrGlyGlyGlnGlnMetGly) derived from the T7 major capsid protein (gene 10) is incorporated into vectors to provide an ATG source for fused proteins. This leader peptide is often used in T7 expression systems. Antibodies generated against this epitope can be used to detect the production of expressed protein in Western or immunoprecipitation assays.

Fusion and tagged protein cleavage systems

Whereas the use of tagged or fusion proteins allows for rapid quantitation and purification of recombinant proteins, this requires an addition to the vector-derived epitope to provide a cleavage system. While some epitopes such as the 6xHis tag are small and may be inconsequential, others are significantly larger and may affect the downstream use of the recombinant protein. By incorporating site-specific protease cleavage sites between the epitope/fusion and the target protein, these sequences can be removed. One complication arises when the recombinant protein has proteolytic cleavage site within its own sequence making it difficult for the protease to recognize its target site intended. Using a protease to mediate cleavage eventually requires that the protease itself must be removed following the reaction in order to obtain a truly purified recombinant protein. To address this issue, some manufacturers have incorporated novel methods of releasing the target protein from the fusion product while simultaneously including a strategy of immobilizing the protease on a resin through use of an engineered tag on a recombinant endopeptidase. The PreScission™ Protease from Amersham is a genetically engineered fusion protein consisting of the 3C protease of the human rhinovirus type 14 and glutathione *S*-tranferase. This protease can be used either following purification of the recombinant protein, or while the target protein is bound to glutathione sepharose. Since the protease itself contains a GST tag, it remains bound to the sepharose

187

matrix allowing rapid purification of the target away from the protease. New England BioLabs has recently introduced a vector that relies on Intein-mediated self-cleavage rather than proteolytic cleavage of the fusion. The Intein sequence mediates self-cleavage between the tag-epitope and the target sequence and liberates the target protein from the fusion species without the introduction of additional proteins. The affinity tag can remain immobilized and purification of the target is relatively easy.

Some commonly encountered protease/cleavage sites are

- Thrombin (KeyValProArg/GlySer)
- Factor Xa protease (IleGluGlyArg)
- Enterokinase (AspAspAspAspLys)
- rTEV (GluAsnLeuTyrPheGln/Gly): rTEV is a recombinant endopeptidase from the tobacco etch virus
- Intein-mediated self-cleavage (New England BioLabs)
- 3C human rhinovirus protease (Amersham Biotech) (LeuGluValLeuPhe Gln/GlyPro)

Polylinkers or multiple cloning sites (MCSs) are synthetic DNA sequences encoding a series of restriction endonuclease recognition sites that are engineered for convenient cloning of DNA into a vector at a specific position. MCS sites can range in size from two restriction sites and up. To accommodate reading frame differences with translational start sequences, many vectors are offered in an "A, B, C" format, meaning there is a single base shift between each vector, allowing easy cloning into each of the three reading frames. Additionally, some vectors are provided where the MCSs are in opposite orientations. For sequences where the amino-terminus has not been precisely defined, Kodak has developed a FLAG-Shift vector. These vectors contain a "shift" sequence that allows expression of an open reading frame without regard to the reading frame in which it was originally cloned. This shift is facilitated by ribosomal slippage caused by a run of A/Ts.

LIC Vectors (ligation-independent cloning, Novagen, Stratagene) allow directional cloning of PCR products without restriction digestion or ligation reactions of the amplified products. LIC vectors rely on sequence complimentarity between the primer used for amplification and the ends of the vector. These overhangs, following treatment with T4 polymerase, allow specific and efficient annealing and subsequent transformation directly into bacteria. Primers for the overhang can also be engineered to accomplish additional tasks. Novagen, for example, uses primers that encode the recognition site for Enterokinase. All vectors must contain a DNA sequence that directs binding of DNA polymerase and associated factors in order to maintain copies of the vector. Most vectors utilize elements from pBR322. Many vectors also provide an M13 origin of DNA replication (f1) allowing the production of single-stranded DNA for sequencing and mutagenesis.

Selection pressure

All vectors contain sequences that provide a means to select only those cells containing a vector. As a selection pressure, an antibiotic is usually added to the medium. The following markers function by either conveying drug resistance on the host or enabling the host to compensate for the absence of an essential component in the media (auxotrophic markers):

Ampicillin: Interferes with a terminal reaction in bacterial cell wall synthesis. The resistance gene (*bla*) encoding beta-lactamase cleaves the beta-lactam ring of ampicillin. Ampicillin is rapidly degraded by the extracellular enzyme β-lactamase secreted by the plasmid-carrying cells. This can be a major problem decreasing yields significantly due to the outgrowing of plasmid carrying cells by non-carrying ones. This problem has been addressed adding ampicillin during the fermentation or using carbenicillin that is more stable but expensive.

Tetracycline: Prevents bacterial protein synthesis by binding to the 30S ribosomal subunit. The resistance gene (*tet*) specifies a protein that modifies the bacterial membrane and prevents transport of the antibiotic into the cell.

Kanamycin: Binds to the 70S ribosomes and causes misreading of messenger RNA. The resistant gene (*Km*) modifies the antibiotic and prevents interaction with the ribosome.

With the ever-increasing rate of discovering new genes comes the equal challenge of understanding how the products of these genes are used and regulated, when they appear, and how they interact within cells. While much attention has been given to high-throughput PCR and sequencing, numerous advancements have been made in lower profile areas such as developing tools to rapidly express and purify proteins.

Manipulations to improve yield

In recent years, a number of genetic strategies have designed to improve production yields, to simplify the recovery processes, to facilitate in vitro refolding, provide site-specific cleavage of gene fusion product, and to create tailor-made product-specific affinity ligands.

Gene fusion

Currently, there are two types of expression vector in *E. coli* for expressing mammalian genes. One is the nonfusion expression vector (as described earlier in detail), and the other is the fusion expression vector. The former is easier to use. However, certain genes are not expressed well or even not expressed at all in nonfusion expression vectors. In such cases, the only choice for expressing genes is to use a fusion expression vector. The features of fusion protein amplification that may influence the final choice of vector are listed in Table 6.4.

Several kinds of fusion expression vectors have been constructed, characterized, and commercialized. The fusion partners include glutathione *S*-transferase (GST), MBP, staphylococcal protein A, and thioredoxin. The greatest advantage of fusion

Table 6.4 Features of Fusion Tags

Fusion Proteins	Targeting information can be incorporated into a tag, provides a marker for expression, simpler purification using affinity chromatography under denaturating and non-denaturating conditions, easy detection, refolding achievable on a chromatography column ideal for secreted proteins as the product is easily isolated from growth media	The Tag may interfere with protein structure and affect folding and biological activity, the cleavage site is not 100% specific if the tag needs to be removed.
Nonfusion Proteins	No cleavage steps necessary	Purification and detection not as simple reducing potential yield, problem with solubility may be difficult to overcome

Table 6.5 Key Features of GST and His Tags

GST Tag	(His)6 Tag
Can be used in any expression system	Can be used in any expression system
Purification procedure gives high yields of pure product	Purification procedure gives high yields of pure product
Selection of purification products available for any scale	Selection of purification products available for any scale
pGEX6P PreScission protease vectors enable cleavage and purification in a single step	Small tag may not need to be removed, e.g., tag is poorly immunogenic so fusion partner can be used directly as an antigen in antibody production
Site-specific proteases enable cleavage of tag if required	Site-specific proteases enable cleavage of tag if required. N.B. Enterokinase sites that enable tag cleavage without leaving behind extra amino acids are preferable
GST tag easily detected using an enzyme assay or an immunoassay	(His)6 tag easily detected using an immunoassay
Simple purification. Very mild elution conditions minimize the risk of damage to functionality and antigenicity of the target protein.	Simple purification, but elution conditions are not as mild as for GST fusion proteins. Purification can be performed under denaturing conditions if required. N.B. Neutral pH but imidazole may cause precipitation. Desalting to remove imidazole may be necessary.
GST tag can help stabilize folding of recombinant proteins.	(His)6-dihydrofolate reductase tag stabilizes small peptides during expression.
Fusion proteins from dimers.	A small tag is less likely to interfere with structure and function of fusion partner; mass determination by mass spectrometry not always accurate for some (His)6 fusion proteins.

expression vectors is that the inserted genes can usually be expressed well. The major shortcoming of current fusion expression vectors is that chemicals, such as IPTG are expensive and in short supply. The two most commonly used tags are glutathione *S*-transferase (GST tag) and 6x histidine residues (His)6 tag. As for the selection of host and vectors, the decision to use either a GST or a (His)6 tag must be made according to the needs of the specific application. Following are the key features of these tags that should be considered (Table 6.5).

Polyhistidine tags such as (His)4 or (His)10 are also used. They may provide useful alternatives to (His)6 if there are specific requirements for purification.

When choosing an affinity-fusion system, it is important to remember that all systems have their own characteristics, and no single system is ideal for all applications. For example, if secretion of the gene product is desired, it is necessary to choose a system with a secretable affinity tag. If the gene product needs to be purified under denatured conditions, a system that has a tag that can bind under those conditions must be chosen, for example, the polyhistidine affinity tag. The polyhistidine tag is suitable for purification of gene products accumulated as inclusion bodies because the fusion protein can be directly applied to an immobilized-metal-ion-affinity-chromatography (IMAC) column after being solubilized with a suitable denaturing agent. An additional advantage with the small polyhistidine affinity tag is that it can easily be genetically fused to a target protein by PCR techniques. It is also important to choose an affinity-fusion system with elution conditions under which the target protein does not get denatured. For large-scale pharmaceutical production, however, affinity fusions have not been as extensively utilized, despite the ability to replace multiple steps with one step. The main reason is most probably that, for most applications, the affinity tag needs to be removed

Table 6.6 Examples of Fusion Protein Vectors

Vector Family	Tag
pGEX	Glutathione *S*-transferase
PQE	6x Histidine
pET	6x Histidine
pEZZ 18 (non-inducible expression)	IgG-binding domain of protein A
pRIT2T (expression inducible by temperature change)	IgB-binding domain of protein A

afterward. Furthermore, proteinaceous ligands may leak from the column during elution, making it necessary to remove the ligand from the eluate. If the ligand originates from a mammalian source, there is also the risk of viral contamination. Questions concerning the possibility of column sanitation, column lifetime, capacity, and cost must also be considered.

Examples of vectors for fusion proteins together with suggested purification products are listed in Table 6.6.

Cleavage of fusion proteins

It is necessary to remove the affinity tag after the affinity purification step. There are several methods, based on chemical or enzymic treatment, available for site-specific cleavage of fusion proteins. Advantages of the chemical cleavage methods are that the reagents used are inexpensive and widely available, and the reactions are generally easy to scale up. However, the harsh reaction conditions often required can lead to amino-acid-side-chain modifications or denaturation of the target protein. Furthermore, the selectivity is often rather poor, and cleavage can occur on additional sites within the target protein. Therefore, chemical cleavage methods are usually only suitable for the release of peptides and smaller proteins. For many applications, enzymic cleavage methods are preferred to chemical ones because of their higher selectivity and also the cleavage often can be performed under physiological conditions. Disadvantages of enzymic cleavage methods are that some enzymes are very expensive and that not all enzymes are widely available. Furthermore, if the enzyme is of mammalian origin, virus removal, and virus clearance validations need to be performed if the target protein is to be used as a pharmaceutical. Recombinant proteases, produced in bacteria or yeast, are for that reason preferred.

Improved recovery

Examples of gene fusions that have been used to improve initial recovery steps include fusions of hydrophobic tails to the target proteins to favor the partitioning into the top phase in aqueous two-phase systems and fusions of aspartic acid residues to the protein to enhance polyelectrolyte precipitation efficiency. Increased efficiency in anion-exchange chromatography in the EBA format was achieved by the fusion of the target protein to the ZZ domains from Protein A, whereby the pI was lowered. By fusion of a stretch of arginine residues, glutamic acid residues, and phenylalanine residues, the efficiency of ion-exchange chromatography and hydrophobic interaction chromatography was increased. One example of a tailor-made fusion partner is the engineered basic variant of the Z domain (Z_{basic}), enabling cation-exchange-chromatography separations to be performed at high pH values. Since almost no other host-cell proteins were found to bind under such conditions, very efficient purification could be achieved. Utilizing the features of the charged Z_{basic}, an integrated production strategy for Klenow DNA polymerase was developed. The Klenow DNA polymerase was produced as a Z_{basic}–Klenow

191

fusion protein that could be efficiently recovered by cation-exchange chromatography in the EBA model. The Z_{basic}–Klenow fusion was subsequently cleaved to release free Klenow polymerase, with the help of a Z_{basic}-tagged viral protease 3C, whereafter fused Klenow could be recovered from the reaction mix by separating Z_{basic}-protease 3C and Z_{basic} fusion partner using cation-exchange chromatography.

Facilitated in vitro refolding

Fusions of target proteins to highly soluble fusion tags have been shown to enhance in vitro refolding. For example, a high refolding yield at high protein concentration was obtained by fusion of a moderately soluble target protein to ZZ from protein A. By fusion of a target protein to a histidine tag, immobilization of the fusion protein on an IMAC column can be made under denaturing conditions. A subsequent on-column refolding step typically gives a high yield of the renatured target protein. A related example, in which a hexa-arginine polypeptide extension was fused to the target protein, the fusion protein was immobilized on a cation-exchange column and renatured target protein was obtained after on-column refolding.

Gene multimerization

When expressing peptides in *E. coli*, low yields are often obtained. One reason could be the susceptibility of the peptides to proteolysis. A common strategy to improve the stability is to produce the peptide as a fusion. A major disadvantage of this strategy is that the desired product only constitutes a small portion of the fusion protein, often resulting in low yields of the target peptide. One way of increasing the molar ratio, and hence increasing the amount of peptide produced is to produce a fusion protein with multiple copies of the target peptide. An additional beneficial effect is often obtained by this strategy since the gene multimerization has also been shown to increase the proteolytic stability of the produced peptides. When the gene multimerization strategy is employed to increase the production yield, subsequent processing of the gene product to obtain the native peptide is needed. By flanking a peptide gene with codons encoding methionine, CNBr cleavage of the fusion protein, containing multiple repeats of the peptide, has successfully been used for obtaining native peptide at high yield. Takasuga and co-workers produced a pentapeptide multimerized to 3, 14, and 28 copies, fused to dihydrofolate reductase, engineered to be separated by trypsin cleavage. A similar strategy was used to produce a peptide hormone of 28 residues. Eight copies of the peptide gene were linked in tandem, separated by codons specifying lysine residues flanking the peptide, and the construct was fused to a gene fragment encoding a portion of β-galactosidase. Endoproteinase Lys-C, an enzyme which specifically cleaves on the C-terminal side of lysine residues, was used instead of trypsin, together with carboxypeptidase B, to release the native peptide. Similarly, a multimerization strategy was used to improve the yields of the 31-amino-acid human proinsulin C-peptide. The C-peptide was expressed intracellularly in *E. coli* as one, three, or seven copies of parts of fusion proteins. Since it was found that the three different fusion proteins were expressed at equal levels and that they all were efficiently processed by trypsin/carboxypeptidase B treatment to release native C-peptide, the seven-copy construct was used to generate a recombinant production process.

Simplified site-specific removal of fusion partners

Genetically designed recombinant proteases have been used to simplify the removal of proteases after site-specific cleavage of fusion proteins. By fusing the protease to the same affinity tag as the target protein, an efficient removal of the affinity-tagged protease, the released affinity tag, and uncleaved fusion protein can be achieved using affinity chromatography. This principle is commercially available; examples being the systems based on His-tagged tobacco-etch-virus protease and human rhinovirus 3C protease fused to a glutathione *S*-transferase tag (PreScission protease). An affinity-tagged protease can, as an alternative to covalent coupling, also be immobilized to an affinity matrix and be utilized for on-column cleavage. On-column cleavage, in which the produced fusion proteins are site-specifically cleaved while still immobilized on the affinity column, has also been described. An affinity-fusion system, consisting of a protein splicing intein domain from *S. cerevisiae* and a chitin-binding domain, allows simultaneous affinity purification and on-column cleavage. Different immobilizing approaches are especially important for large-scale applications since they can reduce the protease consumption and help to avoid additional contamination by the added protease.

Tailor-made product-specific affinity ligands

Powerful in vitro selection technologies, such as phage display, have proven efficient for the isolation of novel binding proteins from large collections (libraries) of peptides or proteins constructed, for example, by combinatorial protein engineering. One example of such binding proteins is the so-called "affibodies," selected from libraries constructed by random mutagenesis of the Z domain derived from streptococcal protein A (SPA). The Z domain, used as scaffold during library constructions, is proteolytically stable, highly soluble, small (6 kDa), and has a compact and robust structure devoid of intramolecular disulfide bridges, making it an ideal domain for ligand development. Using phage-display technology, affibody ligands to a wide range of targets have been successfully selected. Recently, such affibody ligands showed selective binding in authentic affinity-chromatographic applications involving the purification of target proteins from *E. coli* total cell lysates. Such tailor-made product-specific affinity ligands have also been generated and used for highly efficient recovery of recombinant human factor VIII produced in CHO cells and a recombinant vaccine candidate, derived from the RSV G protein, produced in BHK cells.

The obvious advantage of using a ligand selected to bind to the target protein instead of fusing the target protein to an affinity tag is that no cleavage step to obtaining the native protein is needed. The disadvantage is that a new high-affinity ligand must be selected and produced for every new recombinant protein needed to be purified. It is nevertheless likely that this strategy will be attractive in recombinant bioprocesses, since highly selective affinity matrices can be created that potentially even could discriminate between different folding forms of the target protein and could thus replace several other chromatographic steps in the recovery process. Interestingly, no loss of column capacity or selectivity for the target protein was obtained even after repeated cycles of low pH elution and column sanitation protocols, including 0.5 M NaOH. This might suggest that affinity chromatography using protein ligands could become increasingly used also in industrial-scale recombinant-proteins recovery processes in the future.

Molecular chaperons

It is now well established that the efficient posttranslational folding of proteins, the assembly of polypeptides into oligomeric structures, and the localization of proteins are mediated by specialized proteins termed molecular chaperones. The demonstration that efficient production and assembly of prokaryotic ribulose bisphosphate carboxylase in *E. coli* require both GroES and GroEL proteins led to an increasing interest in the use of molecular chaperones for high-level gene expression in *E. coli*. In addition to their utility in purification and detection, specific fusion peptides may confer advantages to the target protein during expression, such as increased solubility, protection from proteolysis, improved folding, increased yield, and secretion. The engineering of specific protease sites in many fusion proteins facilitates the cleavage and removal of the fusion partner(s).

Normally, protein folding proceeds toward a thermodynamically stable end product. Proteins that are drastically destabilized will probably fold incorrectly, even in the presence of chaperones. Thus, the truncation of polypeptides, the production of single domains from multisubunit protein complexes, the lack of formation of disulfide bonds, which ordinarily contribute to protein structure, or the absence of posttranslational modifications such as glycosylation may make it impossible to attain thermodynamic stability. Moreover, it is now clear that different types of chaperones normally act in concert. Therefore, the overproduction of a single chaperone may be ineffective. For example, the overproduction of DnaK alone resulted in plasmid instability, which was alleviated by the coproduction of DnaJ. Similarly, the coexpression of three chaperone genes in *E. coli* increased the solubility of several kinases. In some cases, it may be necessary to coexpress chaperones cloned from the same source as the target protein. Still another variable to consider is growth temperature. For example, GroES–GroEL coexpression increased the production of β-galactosidase at 30°C but not 37°C or 42°C, whereas DnaK and DnaJ were effective at all temperatures tested. Finally, the overexpression of chaperones can lead to phenotypic changes, such as cell filamentation, that can be detrimental to cell viability and protein production.

Codon usage

Genes in both prokaryotes and eukaryotes show a nonrandom usage of synonymous codons. The systematic analysis of codon usage patterns in *E. coli* led to the following observations:

- There is a bias for one or two codons for almost all degenerate codon families.
- Certain codons are most frequently used by all different genes irrespective of the abundance of the protein; for example, CCG is the preferred triplet encoding proline.
- Highly expressed genes exhibit a greater degree of codon bias than do poorly expressed ones.
- The frequency of use of synonymous codons usually reflects the abundance of their cognate tRNAs. These observations imply that heterologous genes enriched with codons that are rarely used by *E. coli* may not be expressed efficiently in *E. coli*.

The minor arginine tRNA has been shown to be a limiting factor in the bacterial expression of several mammalian genes because the codons AGA and AGG are infrequently used in *E. coli*. The coexpression of the *argU* (*dnaY*) gene that codes for tRNA results in the high-level production of the target protein. The production of β-galactosidase decreases when AGG codons are inserted before the 10th codon from the initiation codon of the *lacZ* gene. To date, however, it has not been possible to formulate general and unambiguous "rules" to predict whether the content of low-usage codons in a specific gene might adversely affect the efficiency of its expression in *E. coli*. Nevertheless, from a practical point of view, it is clear that the codon context of specific genes can have adverse effects on both the quantity and quality of protein levels. Usually, this problem can be rectified by the alteration of the codons in question, or by the coexpression of the cognate tRNA genes.

Mammalian cells expression systems

The main disadvantage in the *de novo* synthesis of recombinant eukaryotic proteins in a prokaryotic system is the improper protein folding and assembly and the lack of posttranslational modification, principally glycosylation and phosphorylation. This leads to utilization of eukaryotic cells wherein viruses are used as vectors. Eukaryotic expression systems fall into four distinct classes based on host type: yeast, drosophila, insect (nondrosophila), and mammalian.

Mammalian cell lines

Mammalian cell culture systems offer the distinct advantages of extracellular expression in native form including complex posttranslational modified proteins; the expression vectors are commercially available with large-scale production batches (2–5000 L) and the track record for FDA and other regulatory body approvals have been very good. On the negative side, mammalian cells take longer time (4–6 months) to develop from introduction of gene to protein production; they grow slowly with density not exceeding beyond 100 million cells/mL, the culture media used is expensive, provides low yields (generally less than 100 mg/L; recently, systems have begun to yield gram quantities per liter) and may contain bovine products and allergens; the ICH requirements for characterization are elaborate and extensive as there is a possibility of carrying virus contamination. The process is subject to sheer stress and thus difficult to use as suspension culture or even in Wave bioreactors. Cell lines are also sensitive to osmolarity changes and finally there are safety issues in the management of cell lines.

Therapeutic antibodies are mainly produced in mammalian host cell lines, including murine myeloma cells, PER.C6 human cells, and CHO cells. The choice of which expression system to choose is determined by an individual company's experience and by the host cell's ability to deliver high productivity with acceptable product quality attributes.

NS0 expressions system Murine NS0 cells are nonimmunoglobulin secreting myeloma cells and are cholesterol auxotrophs requiring the presence of cholesterol in culture medium for growth. However, cholesterol-independent NS0 cells also have been established. NS0 cells lack endogenous glutamine synthetase (GS) enzyme activity, making them suitable for use with GS as a selectable marker for recombinant antibody expression. However, antibody production up to 3 g/L has been reported from non-GS NS0 cell lines as well. Most mouse-derived lines, including NS0, produce *N*-glycosylneuraminic acid (NGNA), a sialic acid that does not exist in human antibodies and that might be immunogenic in humans. Although NS0 cells have been used in industry to produce therapeutic antibodies, immunogenicity concern might at least partially make NS0 cells not as widely used CHO cells for therapeutic antibody production.

PER.C6 expression system Compared with NS0 and CHO cells, PER.C6 cells are a relatively new technology and are derived from human retina cells that have been immortalized by transfecting the E1 genes from adenovirus five DNA. PER.C6 cells can proliferate indefinitely under serum-free conditions similar to NS0 and CHO cells. Their human might favor the argument that it is advantageous to produce therapeutic antibodies PER.C6 cells because the posttranslational modifications are human-like and are less immunogenic. However, the human origin may make cells more vulnerable to adventitious agents that can infect humans and cause diseases. This may raise some safety concerns and require more stringent requirements for cell line characterization and downstream 1 viral removal capability.

Chinese hamster ovary expression system CHO cells are the predominant host used to produce therapeutic proteins. About 70% of all recombinant proteins produced today are made in CHO cells, including DUXB 11, DG44, and CHOK1 cells. Whereas DUXB 11 and DG44 cells do not have DHFR activity, CHOK1 cells have endogenous DHFR activity.

Other popular cell lines used include the human cervix (HeLa), African green monkey kidney (COS), BHK cells, and hybridomas.

Well-characterized cell lines can be obtained from the American Type Culture Collection (ATCC) and the European Collection of Cell Cultures (ECACC). These continuous cell lines have the potential of an infinite life span and can usually be cultivated as perfusion cultures. A cell bank system comprising the MCB bank and the WCB provides the means for the production of well-characterized and standardized cells. These details are provided later in this chapter.

Chinese hamster ovary cells

The approval of CHO-derived tissue plasminogen activator (tPA, Activase BLA 103172) in 1986 validated the recombinant technology and created a large value of mammalian cell culture for the manufacturing of protein therapeutic products. Today, CHO cells remain at the forefront for several reasons including their adaptability, lesser risk of virus contamination, and the ability to grow in the serum-free environment. The fact that CHO cells can allow posttranslational modifications to recombinant proteins makes it possible to use them for complex proteins like monoclonal antibodies. In this regard, the CHO cells act like human cells without the immunogenic α-galactose epitope. The CHO cells are also good for gene amplification that can result in higher yields of recombinant proteins. The titers of CHO cell cultures have improved continuously over the past two decades to 100 times more than what was the norm when these cells were first used for recombinant expression. It is for this reason that the biosimilar product developer must look into the recent advances in the science of cell line development using the techniques that were not available at the time when the originators developed their cell lines. Today, multi-gram yields of recombinant proteins are easily possible using the CHO cells.

Since the first approval, over a hundred new recombinant protein therapeutics produced from mammalian cells have been approved and the numbers continue to grow as on a 15 new approvals per year by U.S. Food and Drug Administration (FDA).

The cell line development technologies used mostly involve the use of either the methotrexate (MTX) amplification or the Lonza's glutamine synthetase (GS) system. Both systems make use of a specific drug to inhibit a selectable enzyme marker essential for cellular metabolism: MTX inhibits DHFR in the MTX amplification system, and methionine sulfoximine (MSX) inhibits GS in the GS system. In addition, cell lines that are deficient in these enzymes such as DHFR are also available.

After transfection with expression vectors containing the expression cassettes for the recombinant protein and selection marker genes, the cells are selected and gene amplified with the selection drug, for example, MTX or MSX. Here, gene amplification describes the increase in recombinant gene copy number in the cells commonly associated with, but not limited to, the applications of MTX and MSX. MTX or MSX concentration can also be increased step-wise to further increase cell protein productivity by further gene amplification. Single cell cloning or limiting dilution is then performed to ensure that the selected cells for further processing are producing the recombinant protein. Analyses of protein titers are subsequently used to choose the clones for progressive expansions. Finally, selected clones are evaluated in controlled bioreactors and banked for future use.

After transfection of the host cell line with the expression vector containing the GOI and selection marker, the cells undergo drug selection and cloning to derive cells that are producing the GOI. When gene amplification systems are used, concentrations of selection drug (e.g., MTX or MSX) can be increased step-wise to derive cell clones that are more productive. Cell clones with high recombinant protein titer are chosen for progressive expansions before cell banking and further clone evaluations, such as production stability of the cell clones and quality of recombinant protein.

Vectors

Eukaryotic expression vectors are of two basic types: virion or virion-plasmid hybrids. Virion-type vectors are most commonly used for the delivery of foreign genes, or a replacement for a defective host gene, into mammalian cell hosts. The virion-plasmid hybrid vectors are used to facilitate the overexpression of protein in native form. Additionally, the availability of authentic pure protein has hastened the development of structure-dependent epitope-specific antibodies to native proteins, the crystallization and subsequent x-ray/NMR analysis of proteins in their native, correctly folded, posttranslationally modified state, the characterization of gene products for genes whose phenotype may or may not be known, and the isolation of proteins whose native form is in such low abundance that they are very difficult to purify from the original natural organism or tissue. The main features of eukaryotic expression vectors include various sequence elements that explicitly define the level of expression, the transcriptional start and stop points, postprocessing (transcript splicing, polyadenylation, etc.) transport, selectable markers, and in some cases a peptide tag to facilitate isolation and purification of the gene product. The earliest expression vectors, pSV2, is a composite of sequence elements from the papova virus, Simian Virus 40 and the prokaryotic cloning vector, pBR322. SV40 sequence provided transcriptional enhancers, promoter (early region), splicing signal (small t-antigen gene), and the polyadenylation signal element, and pBR322 provided the origin of replication (ori) element. The presence of a selectable marker, pBR322 *AmpR* gene, completes this model eukaryotic expression vector. Practically all modern eukaryotic expression vectors possess one or more of these elements. Other viruses used include cytomegalovirus (CMV), Murine sarcoma virus (MSV), Rous sarcoma virus (RSV), mouse mammary tumor virus (MMTV), and Semliki Forest virus (SFV).

Generally, recombinant proteins are expressed in a constitutive manner in most eukaryotic expression systems and frequently with inducible promoters heat shock protein, metallothionien, and human and mouse growth hormone, MMTV-LTR, and inducible enhancer elements ecdysone, muristerone A, and tetracycline/doxycycline.

Two principal strategies have been employed to increase cell productivity: gene regulation and gene amplification. Gene regulation aims to increase the number of times a single gene is transcribed and then translated into product. Gene amplification aims to increase the number of genes that are available for transcription to produce the product (e.g., use of BPV virus, use of DNA amplifying drugs such as methotrexate).

Commercial systems that incorporate the entire gene sequences to support the expression of desired proteins are available from Clontech, Invitrogen, Novagen, Life Technologies, Promega, Pharmacia Biotech (GE), Strategene, Quantum

Biotech and many more. Selectable markers generally are either recessive or dominant; the recessive markers are usually genes that encode products that are not produced in the host cells (cells that lack the "marker" product or function). Marker genes for thymidine kinase (TK), DHFR, adenine phosphoribosyl transferase (APRT), and hypoxanthine–guanine phosphoribosyl transferase (HGPRT) are in this category. Dominant markers include genes that encode products that confer resistance to growth-suppressing compounds (antibiotics, drugs) and/or permit growth of the host cells in metabolically restrictive environments. Commonly used markers within this category include a mutant *DHFR* gene that confers resistance to methotrexate; the *gpt* gene for xanthine–guanine phosphoribosyl transferase, which permits host cell growth in mycophenolic acid/xanthine containing media; and the *neo* gene for aminoglycoside 3′-phosphotransferase, which can confer resistance to G418, gentamycin, kanamycin, and neomycin. Practically all eukaryotic expression vectors used today possess at least one marker from either or both of these categories, although dominant markers are the most common.

Host-range specificity for virion vectors is determined primarily by the presence of recognizable host cell surface receptors, whereas for virion-plasmid hybrid vectors, the determinant is the degree to which the host's cellular machinery recognizes the transcriptional control signals of the hybrid vector. Transcriptional enhancers appear to be the primary determinants of cell-type specificity.

After the target gene has been engineered into the appropriate cloning vector, the vector will need to be introduced into the host cell. Given the infectious nature of the viruses from which most eukaryotic expression vectors are derived, it would seem that the introduction of the vector into the host would be rather straightforward. Unfortunately, this process can be complicated.

The introduction of the vector into mammalian cells can be effected by microinjection, electroporation, and calcium phosphate-, DEAE-dextran-, polybrene-, DMSO-, or cationic lipid-mediated transfection. Although the method of choice usually depends on the cell type and cloning application, cationic lipids or liposome-mediated transfection protocols generally yield the highest and most consistent transfection efficiencies in mammalian cell systems. Thus, many corporate suppliers of expression systems recommend a lipid-based transfection protocol, with many providing proprietary liposome reagents (e.g., Clonetech's Clonfectin Transfection Reagent, Novogen's GeneJuice, Invitrogen's PerFect Transfection Kit, Promega's Transfection and Tfx-50 reagents, or Life Technologies Lipofectin Reagent).

Mammalian cell expression improvements

Productivity of cell culture titer can be increased through the modulation of transcriptional activity via expression vector engineering (Figure 6.3a and b) by modulating the coexpression of product and selection marker genes, the stringency of the selection marker, the DNA regulatory elements carried on the vector, and targeting its integration site on the host cell genome. Productivity can also be increased by improving cell culture characteristics via cell line engineering.

A successful recombination event will also place the promoterless selection marker gene downstream of a promoter and activate its expression to facilitate cell clone selection while disrupting the expression of the reporter gene. A site-specific recombination to integrate the GOI into a previously determined genomic hot spot is thus promoted with the use of the recombinase and its corresponding target sequences. Identification of such genomic hot spot is dependent on the expression level of a single copy of reporter gene randomly integrated into the genome.

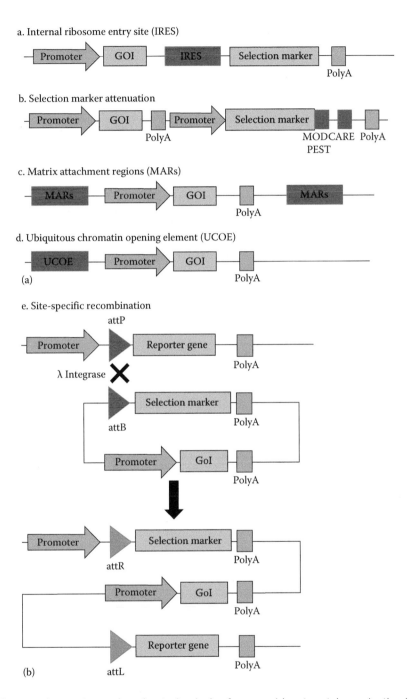

Figure 6.3 Illustration of expression vector engineering technologies for recombinant protein production in mammalian cells. (a) Selection marker attenuation increases the selection stringency, which leads to increased probability of isolating cell clones with high productivity of the recombinant gene of interest (GOI). One of the strategies for selection marker attenuation involves the use of mRNA and protein destabilizing elements such as AU-rich elements (ARE) and murine ornithine decarboxylase (MODC) PEST region, respectively, to reduce the expression of the selection marker. (b) The IRES is used to link expression of multiple genes. Placement of the selection marker gene downstream of the gene of interest ensures that expression of selection marker is dependent on the successful transcription of the GOI. (c) Matrix attachment regions (MARs) flanking the gene of interest promote gene expression by the creation of chromatin loops, which maintain a transcriptionally active chromatin structure. (d) Ubiquitous chromatin opening elements (UCOEs) flanking the gene of interest augment gene expression by sustaining the chromatin in an "open" configuration. (e) An example of site-specific recombination illustrated here uses mutant λ integrase to integrate an expression cassette into the genome. The recombination event is irreversible as attP and attB target sequences are changed to attR and attL sites upon recombination. (Cell line development scheme: From Lai, T. et al., *Pharmaceuticals*, 6, 579, 2013.)

Internal ribosome entry site

Product and selection genes can be coexpressed by cotransfection of mammalian cells with separate vectors. The strategy is limited by the inefficiency of cotransfection and the reliability of product expression based on the selection of the cotransfected marker gene can be very low. Expression of the product and marker genes on the same vector partially improves the reliability of selection of product expression. However, the use of multiple promoters in one vector may result in transcriptional interference and suppression of one active transcriptional unit on another unit in stable transfections.

These problems can be solved with the applications of internal ribosome entry site (IRES) elements. There are many reported IRES elements that can be broadly categorized into cellular or viral IRESes. Expression of multiple genes such as selection marker and gene of interest can be linked to insertion of an IRES element between the two genes. This allows both genes to be dependent on the same promoter for transcription into a single mRNA. The IRES on the mRNA then allows for the 5′ cap-independent translation initiation of the downstream gene, while the transcription initiation of the upstream gene is 5′ cap-dependent. Hence, two different proteins can be translated from the single mRNA.

There are several advantages to linking the expression of multiple genes through the use of IRES. First, a single promoter can be used to drive the transcription of the polycistronic mRNA and ensure a more consistent expression ratio of the linked genes. This is demonstrated to be advantageous for the successful expression of heterodimeric protein like antibody, which is dependent on the balanced expression of the heavy and light chains. Second, by designing the selection marker gene as the downstream gene in an IRES expression vector, the expression of the selection marker is made dependent on the successful transcription of the upstream GOI. As such, this reduces or eliminates the occurrence of selection marker expression without that of the GOI, which can occur due to gene fragmentation when a dual promoter dicistronic vector is used. This concept has been applied to improve the odds of picking a high producer cell clone even when gene amplification is used. The application of IRES also allows for high recombinant protein production from MTX amplified cell pools without the need for cloning.

Selection marker attenuation

When selection stringency is high, the surviving clones will have high transcript levels as a result of gene amplification or integration of the expression vector in a transcriptionally active spot in the genome, since the rate-limiting step of recombinant protein expression occurs in the transcription process. While selection stringency can be increased by increasing the drug concentrations in the cell culture, this approach is typically limited by the slower growth of cells when drug concentration is very high. An alternative approach that has been explored is the attenuation of the selection marker. Theoretically, this allows higher selection stringency at lower drug concentrations since cells with low productivity of the selection marker gene will be selected against. Surviving cells are thereby forced to be more productive in the locus of the selection marker, thereby resulting in the high productivity of the adjacent GOI.

Two strategies have been employed to attenuate the selection marker. The first strategy involves the mutation of the selection maker to reduce its activity. This is demonstrated by the mutation of a selection marker, neomycin phosphotransferase II, to reduce its affinity to neomycin, which leads to a subsequent improvement in the specific monoclonal antibody productivity of 1.4–16.8-fold. The second strategy for selection marker attenuation is through the modulation of gene expression level of the selection marker. A variety of methods have been used to achieve the objective. Codon deoptimization of the selection marker gene through the use of least preferred codons of the expression host lowers the translation efficiency of the selection marker gene and hence leads to a reduction in protein expression. Alternatively, the level of transcription of the selection marker gene can be moderated through the use of a weak Herpes simplex virus thymidine kinase (HSV-tf) promoter. In addition, the use of AU-rich elements (ARE) and murine ornithine decarboxylase (MODC) PEST region as respective mRNA and protein destabilizing elements have been shown to successfully weaken the selection marker, which results in improvements in recombinant protein productivity using the MTX amplification system. In a follow-up study, an attenuated IRES element to reduce the expression of the downstream selection marker gene has also been employed to substitute ARE, resulting in high recombinant protein titers.

Matrix attachment regions

Matrix attachment regions (MARs) are genomic DNA sequences, which serve as attachment points within the DNA that facilitate the anchoring of chromatin structure to the nuclear matrix during interphase. Thus, MARs maintain a transcriptonally active chromatin structure through chromatin loops creation. Furthermore, MARs are associated with increased histone hyperacetylation, which indirectly lead to demethylation of DNA to make it accessible to transcription machinery. In addition to its chromatin modeling function, MARs also serve as binding sites for transcription factors like CCCTC-binding factor (CTCF) and nuclear matrix proteins (NMP) to augment gene expression. When used as *cis* acting elements or by flanking the transgene with MARs, the human β-globin MAR, chicken lysozyme MAR, and β-interferon scaffold attachment region (SAR) promote gene expression and increase the occurrence of high producing clones.

Ubiquitous chromatin opening element

Ubiquitous chromatin opening element (UCOE) is an insulator element against heterochromatin expansion, which is marketed by Merck Millipore (http://www.emdmillipore.com/US/en/support/licensing/ucoe-expression/V_2b.qB. JroAAAFA9l0QWTWn,nav). It is a methylation-free CpG island that abolishes integration position-dependent effects and maintains the chromatin in an "open" configuration to increase accessibility of the DNA region to transcription machinery. Antibody production in CHO cells increased significantly upon the incorporation of UCOEs into expression vectors. It has also been reported that UCOE increases the proportion of high producers and hence improving the expression of antibody by sixfold in CHO stable transfection pools. An alternative noncoding GC-rich DNA fragment is proposed to be a novel UCOE as flanking the GOI with the GC-rich fragment augments recombinant protein expression. It was subsequently proposed that the rigidity of the GC bonds in a DNA double-helix allows the formation of DNA secondary structure that may affect methylation of DNA and histones, which will influence the configuration of the chromatin.

Site-specific recombination

While traditional stable transfection strategies typically involve the random integration of the foreign gene into chromosomes, site-specific recombination offers an alternative strategy to develop high producing and stable clones in a reproducible and predictable manner. This is made possible through the use of recombinases that greatly improve the recombination efficiency in mammalian cell lines in contrast

201

to the low recombination efficiency of traditional homologous recombination. This method, commonly called site-specific recombination, requires the initial generation of a marked host cell line, prior to the introduction of the GOI and recombinase for targeted integration into the marked genomic site of the host cell line. To generate the marked host cell line, a reporter cassette flanked by short *cis*-acting DNA target sequences recognized by specific recombinases are randomly integrated into different loci in the genome via stable transfection. Subsequently, the transfected cell clones are screened for high expression of the reporter gene and single copy integration. Effectively, this will select for clones that have the reporter gene integrated into genomic loci, which promote high transcription rate of the reporter gene. The chance of the reporter gene integrating into these genomic loci (also known as genomic hot spots) is low as only 0.1% of the genomic DNA contains transcriptionally active sequences. Nevertheless, once marked host cell lines that are high producers of the reporter gene are identified, a vector containing the GOI and the same or corresponding DNA target sequences, and a separate expression vector for the recombinase, are cotransfected into the marked host cell line. This leads to a strand recombination between the integrated reporter sequences and that of the GOI, thereby improving the odds of the GOI integrating into a genomic hot spot in the marked host cell line.

Two site-specific tyrosine recombinases from the P1 phage and the yeast *S. cerevisiae*, Cre and Flp, respectively, are commonly used to recognize and recombine their respective short cis-acting DNA target sequences: 34 bp loxP sites and 48 bp Flp Recombination Target (FRT). The Cre/loxP system was first used for human monoclonal antibody production in CHO cells. Recently, artificially caused gene amplification has been used through the use of multiple Cre-mediated integration processes with mutated loxPs sequences to repeatedly insert multiple genes into one target site. Similarly, the Flp-In™ cell line (Life Technologies, Carlsbad, CA; http://www.lifetechnologies.com/us/en/home/about-us.html) is used for Flp-mediated integration of 25 individual antibody expression cassette into specific FRT tagged sites in the genome to express human polyclonal anti RhD antibody. Nevertheless, the reversibility of site-specific recombination is a common drawback of the Cre and Flp recombinases as the recognition sites are recreated upon cassette exchange.

Another site-specific recombination technology makes use of integrase enzymes, such as λ integrase and φC31 integrase, which target two different sequences typically called attP and attB (attachment sites on the phage and bacteria, respectively). Upon recognition of an attachment site previously integrated into the genome of the marked host cell, the integrase catalyze a recombination event, which alters the attB and attP sites upon cassette exchange. Since the integrase is unable to recognize the altered sites, the recombination is irreversible.

Artificial chromosome expression system

The artificial chromosome expression (ACE) system consists of a mammalian-based artificial chromosome known as Platform ACE, an ACE targeting vector (ATV) and a mutant λ integrase (ACE integrase) for targeted recombination. Platform ACE consists of mainly tandem repeated ribosomal genes and repetitive satellite sequences that form the pericentromeric heterochromatin. It also has natural centromeres and telomeres to enable DNA replication without the need of integration into the host cell genome, reducing the probability of chromosomal aberration and clonal heterogeneity. Due to a higher proportion of AT base pairs to GC base pairs in the Platform ACE nucleotide composition, Platform ACE can be purified by high-speed flow cell sorting and subsequently be transfected to different cell types. Platform ACE is pre-engineered to contain 50–70 attP recombination acceptor sites, thus allowing the incorporation of multiple copies of the GOI. In a typical transfection, the platform ACE cell line is cotransfected with the ATV and the ACE integrase plasmid. The recombination event activates the promoterless selection marker on the ATV by integrating the gene downstream of the SV40 promoter in the platform ACE. Hence, cells that survive under application of respective selection pressure are identified as clones that have undergone correct recombination event. In addition, consecutive transfections using different selection markers can be carried out to saturate the recombination acceptor sites on the platform ACE and hence, the high copy number of the GOI can be achieved without gene amplification.

Using this system, high expressing clones are reportedly selected from 100 to 200 cell clones, and monoclonal antibodies expressing cell line achieved yields greater than 500 mg/L in batch terminal shake flask cultures.

Cell line engineering

The quantity of recombinant protein expressed in a cell culture is dependent on the time integral of viable cell density (IVCD) and specific protein productivity (q) of the cells. To improve IVCD, cell line engineering strategies focus on extending the longevity of cell culture, accelerating the specific growth rate and increasing the maximum viable cell density. Similarly, cell line engineering has been employed to improve the folding, transport, and secretion of the recombinant protein to enhance q. A variety of cell line engineering strategies that target diverse cellular functions of CHO cells such as apoptosis, autophagy, proliferation, regulation of cell cycle, protein folding, protein secretion and metabolites production IVCD have been comprehensively reviewed recently, and thus will not be covered in this review. Interestingly, cell engineering can also be used to simultaneously improve IVCD and q. For example, a combinatorial strategy of anti-apoptosis engineering and secretion engineering has produced a CHO cell line, which expressed X-box-binding protein 1 (XBP-1) and caspase-inhibitor, an x-linked inhibitor of apoptosis (XIAP). XBP-1 is a potent transcription factor, which binds to ER stress response element to stimulate promoters of the secretory pathway genes. This leads to an increase in overall protein synthesis which enhances q. Nevertheless, it has been observed that expression of XBP-1 is correlated with reduced viability and stability of the engineered cell line. Thus, an expression vector containing the *XIAP* gene can be transfected into the cell line to inhibit apoptosis. Subsequently, overexpression of XIAP helps rescue the negative effects of XBP-1 on the cell line, and it also leads to a 60% increase in titers.

Cell engineering effort is further boosted by the discovery of zinc finger proteins and transcription activator-like effectors (TALEs), which are protein domains that can be designed to recognize specific DNA sequences; through varying the combination of the types and number of zinc finger proteins, DNA recognition modules that target unique sites (18–36 bp) on the genome can be created. As for TALEs, the DNA-binding domain is a tandem array of repeating units, each of which targets one DNA base as determined by the amino acid residues at two specific positions in the highly conserved unit. By fusing the DNA-binding domain of these proteins to an endonuclease domain, zinc finger nucleases (ZFNs) and transcription activator-like effector nucleases (TALENs) can be created to target and cut specific DNA sequences. The endonuclease domain of restriction enzyme Fok I can be used for this purpose, since it does not have a specific cleavage site and it requires dimerization to cleave DNA. Hence, a pair of ZFNs or TALENs targeting adjacent DNA sequences can position two of these Fok I endonuclease domains in the proximity of each other to allow dimerization and thus DNA cleavage at the targeted DNA sequence. The resulting double-stranded DNA break at the targeted gene loci is subsequently repaired by nonhomologous end joining which often perturbs gene function. Alternatively, the cotransfection of a transgene with homologous region to the cut site can result in transgene integration at the nuclease targeted cleavage site because the presence of the double-stranded break greatly stimulates homologous gene targeting via homologous end joining pathway. As the recognition module can be customized to target any DNA sequence, multiple genes can be targeted using this method to develop an optimized cell line. For example, a triple gene knockout (dihydrofolate reductase [DHFR], glutamine synthase [GS], and $\alpha1,6$-fucosyltransferase8 [FUT8]) CHO cells can be obtained through the use of ZFN. The absence of DHFR and GS allows for the selection of clones with high gene copy while the absence of FUT8 allows the production of mAbs with increased antibody-dependent cell-mediated cytotoxicity (ADCC) for higher treatment efficacy. Using the ZFN technology, a GS-knockout CHO cell line can be generated as offered by Lonza (http://www.lonza.com/products-services/bio-research/transfection.aspx). This is an improvement over the older GS system whereby a GS-containing CHOK1SV cell line is used. The absence of the GS gene in the new host cell line allows for faster cell line development.

Another class of gene editing tools for targeted mutagenesis or transgene integration are meganucleases. These are sequence-specific endonucleases that recognize DNA sites comprising of 12 or more base pairs. Due to their specificity, these enzymes are also used to create double-strand break at targeted DNA site. Meganuclease has also been applied to cell line development for targeted transgene integration, which improves the efficiency in obtaining stably expressing cell lines.

Improvement in genome-wide *in silico* modeling of mammalian systems has also identified novel pathway targets for modification in a mammalian cell line. Coupled with the availability of genome data and advancement of -omics tools, the field of mammalian cell line engineering has the potential to advance to an equivalent level of microbial cell line engineering. Thus, the creation of optimized mammalian cell line through multiple genetic modifications to enhance stability and high expression of recombinant proteins is no longer a far-fetched concept.

Clone screening technologies

As a result of the random integration of foreign genes of interest and subsequent disruption of the genome by gene amplification systems, the cell clones obtained during cell line development are highly heterogeneous. Furthermore, high producing clones are typically rare in a population of transfected cells because the active region supporting high gene expression in the chromosome is rare and these high producer cell clones typically have lower growth rates since a significant portion of resources are being used for expression of the recombinant protein. Therefore, the screening of a large number of cell clones is commonly required to isolate the high producing clones.

Traditionally, serial limiting dilution method is most commonly performed to screen for high producer cell clones due to its simple operation, despite being time, labor and capital intensive. In this method, cells are sequentially diluted on well plates to obtain dilutions at which a portion of the wells is devoid of cells. At the dilution, the wells containing cells will have expanded from a small subset of clones from the original cell pool. To ensure monoclonality, multiple rounds of serial subcloning steps is thus necessary. More importantly, additional steps of cultivating the cells and protein tittering typically by enzyme-linked immunosorbent assay (ELISA) are necessary to determine the protein productivity of the clones. Advancement in clone screening technologies can reduce the time and effort in this endeavor to find rare high-producing cell clones.

Fluorescence-activated cell sorting–based screening

Fluorescence-activated cell sorting (FACS) sorters are equipment that can simultaneously monitor the levels of multiple fluorescence wavelengths associated with a cell at a rate of 108/h. Cells to be analyzed enter the FACS sorter singularly as a moving, focused stream and one or more laser beams interrogate them. The relevant optical detector measures the resulting fluorescence from the cell, and the collected data is quantified and analyzed. The machine then applies a charge to the droplet containing the cell to sort it into specific collection tube or well plates. Depending on the fluorescence signal, cell parameters such as granularity and cell size can also be obtained. However, the accuracy of the FACS-based screening of high producer cell clones is dependent on the fluorescence signal that remains associated with the cell. Hence, it is more suited for selection of high producing cell clones that do not secrete its recombinant protein (Figure 6.4).

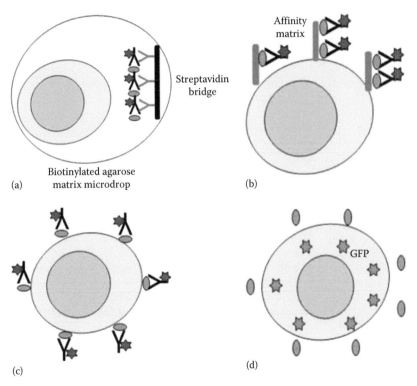

Figure 6.4 Fluorescence labeling strategies for different clone screening technologies to identify high producer cell clones. The coloring scheme for cell clone, primary antibodies, fluorescence agent, and recombinant protein is yellow, black, red, and purple, respectively. All figures follow the same coloring scheme unless it is stated otherwise. (a) Gel microdrop secretion assay used in FACS that encapsulate individual cells in biotinylated agarose matrix. Primary antibodies labeled with fluorescence agent binds to the recombinant protein and the complexes are subsequently bound to secondary antibodies (blue) immobilized on streptavidin bridge (black). (b) The affinity matrix attachment method used in FACS that cross-linked the matrix to the cell surface within a gelatin-based low permeability medium. Secreted recombinant protein remain bound to the affinity matrix (blue), which are subsequently probed by fluorescently labeled antibodies. (c) The cold capture method used in FACS where fluorescently labeled antibodies bind to secreted recombinant protein that remains associated with the cell surface at low temperature. (d) The expression of secreted recombinant protein is linked to an intracellular selection marker like green fluorescene protein (green), and the cell clones are processed by FACS. (*Continued*)

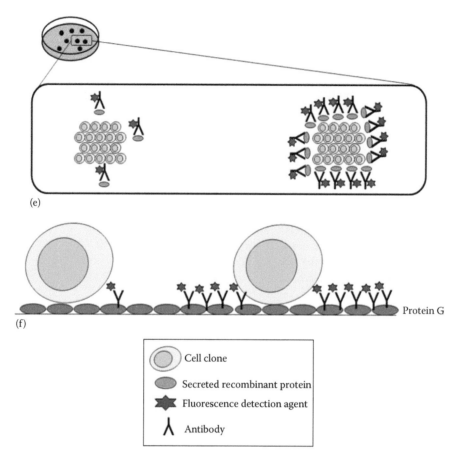

(e)

(f)

Protein G

⊙	Cell clone
⬭	Secreted recombinant protein
✦	Fluorescence detection agent
λ	Antibody

Figure 6.4 (*Continued*) Fluorescence labeling strategies for different clone screening technologies to identify high producer cell clones. The coloring scheme for cell clone, primary antibodies, fluorescence agent, and recombinant protein is yellow, black, red, and purple, respectively. All figures follow the same coloring scheme unless it is stated otherwise. (e) In Clonepix system, cell clones are grown in semisolid media to limit the diffusion of the secreted recombinant protein. Fluorescently labeled antibodies capture the secreted protein where they form a halo structure surrounding the cell colony. In this figure, the cell colony on the right is depicted as a high producer clone as compared to the cell colony on the left. After ranking by the Clonepix system, cell clones are transferred by micro-pins to a new well plate for further characterization. (f) For the Cell Xpress™ system, cell clone expressing recombinant protein such as therapeutic antibodies (black) are captured by protein G (blue) in the well. The captured antibodies will be subsequently probed by fluorescence detection agent and screened by the system. In the figure, a high producer cell clone (right) and low producer cell clone (left) are shown. After ranking the cell clones in the same well, all cell clones except the highest-ranking clone will be subjected to photomechanical lysis. (Cell line development scheme: From Lai, T. et al., *Pharmaceuticals*, 6, 579, 2013.)

Nevertheless, several strategies can be used to select high producer cell clones that secrete its recombinant protein. The first strategy is known as the gel micro-drop secretion assay, which encapsulates the cell in a biotinylated agarose matrix with a diameter of 35 μm. Fluorescently labeled primary antibody is added to the microdrop to bind to the secreted recombinant protein. Subsequently, the secondary biotinylated antibody is used to capture the primary antibody and the complex will bind to a streptavidin bridge immobilized on the biotinylated agarose matrix. Hence, the cell and its secreted protein remain in the microdrop for subsequent processing by FACS. The second strategy involves the cross-linking of an affinity matrix on the cell surface within a gelatin-based low permeability medium to capture the secreted recombinant protein. Subsequently, the immobilized recombinant proteins are detected with fluorescent-labeled antibodies and processed with FACS. While the two strategies have improved the ability of the FACS method in identifying and isolating high producers, the protocols for immobilizing the secreted recombinant proteins are often technically challenging and time consuming. Consequentially, it was hypothesized that the amount of secreted recombinant protein, which associates transiently with the cell surface, should correlate with the total amount of protein being secreted from the cell clone. As such, a cold capture method was proposed. This involved the use of fluorescently tagged antibodies that binds to surface associated recombinant protein at low temperatures of 0°C–4°C. Subsequently, the cell clones were subjected to three rounds of reiterative sorting by FACS to isolate the high producer clones. Using this method, clones with 20-fold increase in specific productivity as compared to the unsorted cell population were isolated.

Alternatively, strategies that measure the level of intracellular selection marker has been proposed as an indirect screen for high producing cell clones that secrete its recombinant protein. In one study, fluorescein isothiocyanate-labeled methotrexate (F-MTX) was used to bind to intracellular DHFR selection marker. The study demonstrated that the distribution of high producer cell clones is highest at a median level of F-MTX fluorescence intensity. In another study, green fluorescent protein (GFP) was used as a second selection marker, and the correlation of GFP fluorescence and recombinant protein productivity is demonstrated with a correlation coefficient range of 0.52–0.70. Consequentially, depending on the intensity of the fluorescence signal in the heterogeneous cell pool, FACS will be carried out to isolate high producer cell clones.

ClonePix

ClonePix FL system (Molecular Devices, http://www.moleculardevices.com/) is an automated colony picker, which is capable of screening large number of clones and identifying high producer clones within a short period of time. The process of ClonePix FL is initiated by cultivating single cells in a semisolid media to allow the formation of individual colonies. The high viscosity of the media allows the progeny of the single cell to remain as a single colony and also trap the secreted recombinant proteins in close proximity to the secreting colonies. These secreted recombinant proteins are captured by fluorescein isothiocyanante (FITC) conjugated antibodies that were previously added to the semisolid media. Upon capturing of the expressed recombinant proteins by the antibodies, they will be deposited as immunoprecipitates around the secreting cell clone and hence, forming a halo fluorescence structure. Consequentially, clones are ranked according to the fluorescent intensity of the halo structure. High-ranking high producer cell clones are then aspirated by micro-pins and transferred to a new well plate for further characterization. The entire process of imaging 10,000 cell clones and selection of high producer cell clones is completed within an hour, and it is sensitive enough to isolate rare high producing clones that formed 0.003% of the population. Due to the high throughput nature of this method, the ClonePix FL system has been used in different studies to consistently select high producer cell clones expressing a range of recombinant proteins like green fluorescent protein and humanized antibody.

Cell Xpress

Cell Xpress technology (http://cellxpress.org/) consists of the Cell Xpress software module and the laser-enabled analysis and processing (LEAP™) platform, which are based on the principle of laser-mediated semiconductor manufacturing technologies. An advantage of the LEAP platform is that it is built for high throughput operation, and hence the entire screening process is fully automated and accomplished by robotics. The Cell Xpress technology combines live cell imaging and laser-mediated cell manipulation to identify and purify the highest producer cell clone in a sample well. Multicolor live cell imaging of Cell Xpress technology is achieved through the simultaneous use of fluorescence detection reagents that associate specifically with either the cell clones or expressed recombinant protein. Furthermore, the sample wells are coated with capture matrix to mediate in situ capture of expressed recombinant protein as most of the recombinant proteins are secreted out of the cell. For example, if therapeutic antibodies are expressed, protein G is commonly used to capture the secreted antibodies. Typically, within the same well, custom software algorithms will locate the cell clones and create a kernel surrounding the individual cell. The area of the kernel will then be expanded to include the secreted recombinant protein until adjacent kernels are encountered. Through measurement of the fluorescent intensity of the respective detection reagents, each cell in the same well will be ranked based on the amount of secreted recombinant protein. Additional criteria for cell growth rate, cell area, and proximity to other cells will also affect the ranking. Hence, the highest ranking cell in a well will be identified as the most suitable high producer cell clone. Subsequently, laser-mediated cell purification will commence whereby the laser beam is directed to the lower ranking cell clones in the same well via large field-of-view optics and galvanometer steering to induce photomechanical lysis. A high producing cell, which remains in the well, will be allowed to grow and finally be transferred to a larger well for expansion.

Since the purification process is a closed system, the probability of contamination is reduced. Cells obtained from a sample size range of 10–108 cells have commonly reached 99.5% purity. Furthermore, as the entire process requires less than 30 s to screen a single well, large numbers of cell clones can be screened within a short time to identify high producers. It has also been reported that the analysis result from Cell Xpress technology shares a good correlation ($R^2 = 0.84$) with the peak IgG volumetric productivities in shake flask growth and expression experiments. In general, Cell Xpress technology has routinely picked cell lines with specific antibody secretion rates of >50 pg/cell/day. Nevertheless, it has been reported that the laser used in LEAP may damage high producing cell clone.

Future prospects

The probability of breakthroughs in cell line development technologies is boosted by the availability of the CHO genome. With this information, the use of comparative transcriptome analysis using completed CHO cell DNA microarray is facilitated, whereas research had been performed using incomplete CHO cell microarray or non-CHO-derived DNA arrays prior to the availability of CHO genomic information. In a recent study, data from large-scale proteomic analysis was complemented by the genome data of CHO cells to discover a total of 6164 grouped proteins, which is an eightfold increase in the identified proteins in the CHO cells proteome. Furthermore, the codon bias of CHO cells, which is distinct from human, was solved. These data will facilitate the expression of human proteins in CHO cells in future. With the availability of techniques in the analysis of metabolites in CHO cells, combined data from genomics, transcriptomics, proteomics, and metabolomics can identify novel genes that affect the growth and protein production rate of CHO cells.

Bioinformatics analysis is also boosted by the availability of the CHO genome. For example, the genomic data of CHO cells has facilitated in silico identification of CHO microRNA (miRNA) loci. Thus, the profiling and subsequent use of microRNA (miRNA) to regulate gene expression has also increased in CHO cell line development in recent years. This is because miRNA can be easily introduced into the cells, and it can efficiently regulate multiple gene targets through mRNA cleavage or translation repression by interaction with 3′ untranslated region (UTR) of the mRNA. Furthermore, as miRNAs are non-coding RNAs, they present no additional translational burden to the cells. Alternatively, bioinformatics analysis of the CHO genome may also reveal new genomic hot spots for site-specific integration of the GOI to generate high producing clones. This has been previously accomplished in the human genome.

Besides the focus on the increased production of protein therapeutics, there will also be a need to improve the quality of the recombinant protein product, which entails metabolic engineering of CHO cells to perform posttranslational protein modification. For example, the *N*-acetylglucosaminyltransferase-III gene has been overexpressed in CHO cells to ensure accurate protein glycosylation pattern in protein therapeutics. Taking it further, there has been an attempt to produce non-protein therapeutic like heparin in CHO cells. While the composition of disaccharide species from expressed heparin sulfate differs from pharmaceutical heparin in the study, it was proposed that fine tuning the expression of transgenes involved in heparin synthesis pathway may solve the problem. In conclusion, with the new technologies discussed earlier, new tools in cell line development can be generated, and the process can be further streamlined to facilitate biopharmaceutical drug discovery and development.

Yeast expression systems

The use of nonmammalian hosts has a distinct generic advantage that these hosts are generally regarded as safe since they are not pathogenic to humans. Yeasts are particularly attractive as they can be rapidly grown on minimal (inexpensive) media. Recombinants can be easily selected by complementation, using any one of a number of selectable (complementation) markers. Expressed proteins can be specifically engineered for cytoplasmic localization or for extracellular export. And finally yeasts are exceedingly well suited for large-scale fermentation to produce large quantities of heterologous protein. Classical studies in yeast genetics have generated a wide array of potential cloning vectors and, in the process, defined which plasmid and host genomic sequences are important in expression technology.

In a summary the yeast culture system has many significant advantages such as it is easy and cheap to grow in large scale, has good regulatory track record, the genetics of yeast is well understood, allows some posttranslational modifications, short doubling time, high cell densities and yield is achievable, no endotoxin release from host cell organism, extracellular expression to low viscosity medium is possible, minor secretion of host cell proteins, no specialized bioreactor required, and it is safer than working with mammalian tissue or cell lines.

The disadvantages of yeast system include problems with correct glycosylation, overglycosylation can ruin protein bioactivity, glycosylation is not identical to mammalian glycosylation, extensive proteolysis of target protein, nonnative proteins are not always correctly folded, fewer cloning vectors are available, gene expression is less easy to control, *S. cerevisiae* is unable to excise introns in gene transcripts of higher eukaryotes

The use of yeast expression systems involves an entirely different set of techniques and principles than those used for other eukaryotic or prokaryotic systems. Commonly used yeast hosts are *S. cerevisiae*, *Schizosaccharomyces pombe*,

P. pastoris, Hansela polymorpha, Kluyveromyces lactis, and *Yarrowia lipolytica*. Newer research tools like lithium acetate and electroporation-mediated transformation of intact yeast cells and the creation of 2 μm yeast episomal plasmids have helped yeast rise to the most favorite list of expression systems.

Wild-type yeasts are prototrophic, that is, they are nutritionally self-sufficient, capable of growing on minimal media. Classical genetic studies have created auxotrophic strains—those that require specific nutritional supplements to grow in minimal media. The nutritional requirements of the auxotrophic strains are the basis for selection of successfully transformed strains. By including a gene in the plasmid expression cassette that complements one or more defective genes in the host auxotroph, one can easily select recombinants on minimal media. Hence, strains requiring leucine will grow on minimal media if they harbor a plasmid expressing the *LEU2* gene. The most commonly used selectable markers found in yeast are for leucine, uracil, histidine, and tryptophan deficiencies.

While the number and a variety of *S. cerevisiae* and *S. pombe* strains possessing nutrition selectable markers make these yeasts attractive hosts, their limitations include hyperglycosylation; weak, poorly regulated promoters; and biomass fermentation. Many of these and other problems are circumvented using, *P. pastoris, K. lactis,* and *Y. lipolytica* that have been extensively utilized for the industrial-scale production of metabolites and native proteins (e.g., β-galactosidase). Vector-host genetic incompatibilities and a relatively undefined biology have limited their use as heterologous expression hosts.

The methylotrophic yeast, *H. polymorpha*, and to a greater extent *P. pastoris*—unique in that they will grow using methanol as the sole carbon source—are becoming a favorite expression host alternative for many researchers. *P. pastoris* has produced some of the highest heterologous protein yields to date (12 g/L fermentation culture), 10–100-fold higher than in *S. cerevisiae*. In *P. pastoris*, growth in methanol is mediated by alcohol oxidase, an enzyme whose de novo synthesis is tightly regulated by the alcohol oxidase promoter. The enzyme has a very low specific activity. To compensate for this, it is overproduced, accounting for more than 30% of the total soluble protein in methanol-induced cells. Thus, by engineering a heterologous protein gene downstream of the genomic AOX1 promoter, one can induce its overproduction. This is the basis for the *P. pastoris* expression system. *H. polymorpha* produces the methanol oxidase (MOX) protein under the control of the MOX1 promoter. A complete *P. pastoris* expression system is available from Invitrogen.

Most yeast vectors for protein expression contain one or more of these basic elements: the *S. cerevisiae* 2 μm plasmid origin of replication, a ColE1 element, an antibiotic resistance "marker" gene (to aid development and screening of plasmid constructs in *E. coli*), a heterologous (constitutive or inducible) promoter, a termination signal, signal sequence (encoding secretion leader peptides), and occasionally fusion protein genes (to facilitate purification).

Constitutive gene expression by the yeast plasmid cassette is commonly mediated (in *S. cerevisiae* and *S. pombe*) by the promoters for genes to the glycolytic enzymes: glyceraldehyde-3-phosphate dehydrogenase (TDH3), triose phosphate isomerase (TPI1), or phosphoglycerate isomerase (PGK1). Protein expression can also be regulated (induced) using the alcohol dehydrogenase isozyme II (ADH2) gene promoter (glucose-repressed), glucocorticoid responsive elements (GREs, induced with deoxycorticosterone), GAL1 and GAL10 promoters (to control galactose utilization pathway enzymes, which are glucose-repressed and galactose-induced), the metallothionein promoter from the *CUP1* gene (induced by copper sulfate), and the PHO5 promoter (induced by phosphate limitation). Most native yeast gene termination signals, when included in the plasmid expression cassette, will provide proper termination of RNA transcripts. The most commonly used are terminator signals for the MF-alpha-1, TPI1, CYC1, and PGK1 genes.

Complete packaged system for quickly developing yeast expression systems are available from Clontech, Invitrogen or Stratagene, and others. For example, Invitrogen sells an expression vector for *S. cerevisiae* and a complete system for *P. pastoris* expression cloning. The Easy Select Pichia Expression Kit includes vectors (pPICZ series), *P. pastoris* strains, reagents for transformation, sequencing primers, media, and a comprehensive manual. Researchers can clone their protein gene into any reading frame contained in each of the three different vectors, select recombinants by Zeocin resistance, induce protein expression with methanol (sole carbon source), and identify expression using antibody to a C-terminal c-myc peptide tag. These vectors also harbor a *S. cerevisiae* alpha-factor secretion gene and polyHIS-encoding element, thus expressed protein is easily recovered from culture extract supernatant and purified using metal-chelate chromatography (e.g., ProBond resin packaged in Invitrogen's Xpress Purification System).

The yeast species *S. cerevisiae* and *P. pastoris* are the most widely used species for therapeutic protein manufacturing as they offer high efficiency, lower cost, and large yields (10G+/L). The development time from introduced gene to protein is about 14 days (in comparison *E. coli* is 5 days and mammalian cells from 4 to 5 months). In contrast to *E. coli*, yeast can express correctly folded proteins directly to the medium, which greatly facilitates purification (low level of host cell proteins, nonviscous solution, low DNA level). The rigid cell wall renders the use of all sorts of bioreactors possible regardless of stirring and shaking mechanisms. Yeast cell cultures have been widely used for the production of biopharmaceuticals such as insulin, streptokinase, hirudin, interferons, tissue necrosis factor, tissue plasminogen activator, hepatitis B vaccine and epidermal growth factor.

Large-scale operations are associated with a number of restrictions related to hosting strain physiology and fermentation technology. Many promotor systems that work well in small scale cannot be implemented in production processes demanding a substantial number of generations. Use of low-expression cassettes reduces the loss of plasmid to some extent, but a profound effect on productivity may be observed due to lack of stability. Expression systems characterized by a high specific rate of product formation at low specific growth rates are highly favorable for large-scale operations. Oxygen demand and temperature control are a key factor in controlling yeast fermentations, which are mainly carried out as fed-batch cultivations. Proteolysis of the target protein can be a major problem.

Saccharomyces cerevisiae

The baker's yeast *S. cerevisiae* is the most extensively studied yeast strain. Its genetics are well known and it has obtained a generally regarded as safe (GRAS) status. Protease deficient strains such as BT150, deficient in proteases A and B, carboxypeptidase Y and carboxypeptidase S, are available. *S. cerevisiae* exhibits alcoholic fermentation under aerobic conditions unless the sugar supply rate is low. This response occurs at glucose concentrations of 0.15 g/L, hence the demand for low sugar supply rate (which is growth limiting). To prevent this, the use of nonrepressing substrates such as raffinose or continuous culture are employed. It is common practice to measure the ethanol concentration or the respiratory quotient (RQ).

An indirect feedback control of RQ data has been applied to assure effective production ($RQ < 1.3$), but it must be kept in mind that the low growth rate may affect target protein expression.

A commonly employed strategy when expressing recombinant proteins in yeast is to put the GOI under the control of an inducible promotor. The use of inducible promotors (e.g., GAL1 induced by galactose) provides for the separation of growth and recombinant protein production. There is no strong inducible promotor available in *S. cerevisiae*, and the expression of recombinant protein will only amount to about 1%–5% of total cellular protein, as compared to 35% in *P. pastoris*. An example is the expression of recombinant Hirudin, a thrombin inhibitor and therapeutic for cardiovascular diseases. If separation of growth and production is not desired (e.g., if protein expression is growth associated) the use of a constitutive promotor such as PHO5 can be employed.

The stability of the expression cassette has a high impact on productivity. Loss of the cassette during the course of the fermentation results in the formation of nonproductive, plasmid free cells, which usually outgrow plasmid carrying cells in fermentation unless selective pressure can be effectively employed. More than 30 generations are required from stock culture to the final stage, and if the expression cassette is lost during the process, this will result in loss of productivity.

To target the recombinant protein for secretion a signal sequence is needed. This can be derived directly from the protein of interest if it is recognized and correctly processed in yeast. If not, the signal sequence of the *S. cerevisiae* α-mating pre-pro leader sequence or invertase may be used. This signal directs the recombinant protein to the endoplasmatic reticulum for further processing and secretion.

While *S. cerevisiae* has been fermented in several different ways, the fermentation protocol of *P. pastoris* cultures is, broadly speaking, nearly always the same. Generating biomass by growing on excess glycerol then inducing protein production with methanol is a three-stage combination of batch and fed batch modes. The first stage is a 24 h glycerol batch usually conducted with 4% v/v glycerol. The end of this stage is indicated by a sharp rise in dissolved oxygen (DO) after which a glycerol fed-batch phase is started, typically with a 50% v/v glycerol solution at a feed rate of 15 mL/L. This phase derepresses the AOX1 promotor and must last for at least 1 h. Examples of glycerol fed-batch stages lasting all the way up to 32 h have been employed. The third stage, a methanol fed-batch, is initiated either with a 100% methanol feed when the last batch of glycerol is exhausted or by slowly adapting the cells to growing on methanol typically on 5% v/v methanol while still feeding glycerol. This induction phase triggers the production of heterologous protein and usually lasts about 80–90 h. Tight control of the methanol concentration is obtained by regularly performing "spike tests" to make sure the methanol is present in limiting amounts. The rapid proliferation of *P. pastoris* cells set high demands on oxygen supply and cooling.

The nutritional needs of yeast are simple and depend on the yeast species, strain and growth conditions. The main culture media components are a carbon source (which also functions as the energy source), salts, trace elements, nitrogen, and growth factors.

For *S. cerevisiae* fermentations, glucose is often chosen as the carbon source because of its inexpensiveness and relative high solubility, whereas for *P. pastoris* (see following section) fermentations glycerol and methanol are used. The growth factor requirements vary in a case-to-case manner. *S. cerevisiae* has in some cases a requirement for inositol, pantothenate, pyridoxine, thiamine, nicotinic acid, and biotin whereas *P. pastoris* has been reported to require biotin.

Pichia pastoris

Pichia pastoris is a methylotrophic yeast which means that it can grow on methanol as the only carbon source. The knowledge of this yeast's genetics stems from the extensive research on the production of single-cell protein (SCP). It is characterized as glucose insensitive yeast. *P. pastoris* has a high oxygen demand and so the fermentation of this yeast generates considerable amounts of heat. It has the advantage of a very strong inducible promotor, the AOX1 promotor. The *AOX1* gene encodes one of the two alcohol oxidases expressed in the presence of methanol. When growing on glucose, transcription from the AOX1 promotor is completely repressed even in the presence of methanol. Growth on glycerol allows for induction with methanol, but the total effect of glycerol is not completely understood. It has become evident that glycerol even at low levels also inhibits expression from the AOX1 promotor, though not as pronounced as glucose. The availability of the strong AOX1 has provided for protein expression levels as high as 22 g/L for intracellular expressed proteins and 11 g/L for secreted ones. In contrast to *S. cerevisiae*, this species exhibits relatively short glycosylation chain length, and the major problem is enzymic deglycosylation and varying glycosylation patterns.

Insect cells

Background

Insect cells constitute a promising, yet unproven, alternative to bacterial and yeast expression systems for a wide range of target proteins requiring proper posttranslational modifications. The difficulties in scale up arise because of difficulties in aeration and type of infection needed for high-level expression. The use of baculovirus system is becoming accepted very fast; it has been used to transform *lepidopteran* insect cells into high-level expression systems in the range of 1–600 mg/L. Preparation and purification of the recombinant virus are faster than the process in mammalian cells and can be completed in about 4 weeks. In the fermentation cycle, the insect cells grow 50-fold in about a week but only in single batches or in semicontinuous batches because of the sensitivity of the cells to shear force. The costs for culture media are moderate (serum free media) to expensive compared to bacteria and yeast media. The system is suited for the expression of cell toxic products since the cells can be grown in a healthy state before infection. Insect cells lack the ability to properly process proteins that are initially synthesized as larger inactive precursor proteins (e.g., peptide hormones, neuropeptides, growth factors, matrix metalloproteases). Baculoviruses are not infectious to vertebrates and, therefore, do not pose a health threat, though the risk of adventitious viruses is still not settled requiring virus inactivation and active filtration. Also, the cyclic killing and lysis of the host cell releases intracellular proteins and nucleic acids into the medium severely straining the downstream purification steps (like in the case of bacterial cells with inclusion bodies). The regulatory record of insect cells remains poor as no products have been approved by the FDA yet.

Insect cells

The nonmammalian hosts like insect cells have several advantages:

- The protein is secreted to the medium in its native form.
- Expresses posttranslational modified proteins including lycosylation, phosphorylation, palmitoylation, myristoylation, and glycosyl-phosphatidylinositol anchors.

- Expression vectors are commercially available.
- The system is suited for expression of cell toxic products since the cells can be grown in a healthy state before infection.
- The baculovirus vectors are harmless to humans.
- Safer to handle.
- Provide high yields (1–600 mg/L).
- Easy to develop (taking about a month to prepare and purify a recombinant virus, compared to more than 6 months for mammalian cells).
- Less expensive to operate.

The main drawbacks are as follows:

- Minimal regulatory track record; FDA is yet to approve the first product.
- Semi-expensive culture media.
- Cell line stability.
- Cells are killed during infection releasing intracellular proteins (*Lepidoterin* species).
- Inactivation of the secretory pathway results in low expression yields due to aeration requirements not met.
- Presence of immunogenic host cell proteins due to release of proteolytic enzymes upon cell distruption.
- Inability to produce eukaryoic glycoproteins with complex N-linked glycans.
- Inability to proper processing of proteins that are initially synthesized as larger inactive precursor proteins (e.g., peptide hormones, neuropeptides, growth factors, matrix metalloproteases).
- Risk of infection with mammalian viruses.
- Sensitive to sheer forces.
- Difficulties in scale up.

The classical strategy for production of proteins in insect cells involves the distinct stages of growing insect cells (*Lepidoterin* species) to mid-exponential growth phase, infecting the cells with the vector, *Autographa californica* nuclear polyhedrosis virus (AcNPV) or *Bombyx mori* nuclear polyhedrosis virus (BmNPV), containing the gene cloning for the target protein and finally harvest and purification of the expressed protein. The reason for using the AcNPV infection step is the ability of baculovirus to replicate in established insect cell lines, where the polyhedrin gene is replaced with a gene of choice (under the control of the strong polyhedrin promotor). The result is high level expression of the gene insert and the accumulation of the target protein. In contrast to other microbial or mammalian cell expression systems, the host cells are killed during each infection cycle.

Eukaryotic expression systems employing insect cell hosts are based upon one of the two vector types: plasmid or plasmid-virion hybrids. Although the latter is the most commonly used, plasmid-based systems offer methodological advantages. The typical insect host is the common fruit fly, *Drosophila melanogaster*. Other insect hosts include mosquito (*Aedes albopictus*), fall army worm (*Spodoptera frugiperda*), cabbage looper (*Trichoplusia ni*), salt marsh caterpillar (*Estigmene acrea*), and silkworm (*B. mori*). In almost all cases, heterologous protein overexpression occurs in suspension cell cultures. The exception, and one of the advantages of plasmid-virion systems is that the recombinant virus may also be injected into larval host hemocel or literally fed to the mature host. Three basic options are available for protein expression in insect cells: vectors that enable high level transient expression; vectors that enable continuous expression from stably transfected cells; and the lytic baculovirus system.

Transient expression systems

The transient expression is the rapidest method requiring simple transfection of insect cell expression vectors containing appropriate promoters (e.g., ie1 and gp64 promoters) in the absence of selection. The expression is optimized when peaks between 24 and 48 h after transfection. Novogen's InsectDirect® System (http://www.novogen.com) is a complete system based on several vectors featuring an enhanced ie1 promoter. The plasmid-based vector systems provide a mechanism for both transient and long-term expression of the recombinant protein. This expression system is exemplified by the Drosophila expression system (DES) available from Invitrogen. The transfection of competent *D. melanogaster* cells with engineered plasmid will mediate the transient (2–7 days) expression of the heterologous protein.

Continuous expression systems

Continuous expression using stably transfected insect cells lines is useful for the study of glycoproteins, secreted proteins, and membrane proteins such as receptors. An alternative to the discontinuous insect cell system (cells are killed by the bacoluvirus infection; see following discussion) is a continuous culture of permanently transfected cells. Cell cultures of the fruit fly, *D. melanogaster* grown to cell densities of 5.7×10^7 cells/mL in a low-cost media makes this system a promising candidate for future recombinant protein expression.

The GOI is cloned into a vector that utilizes a promoter recognized by the insect cell transcription machinery (e.g., baculovirus ie1 or gp64 promoters). The majority of resistant cells will express the target protein in various intervals. Such cell lines maintain stable expression for many passages (more than 50), enabling long-term culture for the accumulation and study of expressed protein. Establishing transformed cells that will express protein for longer time periods requires that the host cells be cotransfected with a "selection" vector, which results in the stable integration of the expression cassette into the host genome. This system offers two advantages over plasmid-virion systems: methodological simplicity, saving the researcher time, effort, and materials, and a choice of expression regimes, constitutive or inducible. Constitutive expression is mediated using the Ac5 Drosophila promoter, whereas a metallothionein promoter guides copper-inducible expression. The DES vectors are designed with multiple cloning sites for insertion of the heterologous protein gene in any of three reading frames. A choice of vectors also provides for the expression of a variety of C-terminal fusion tags: V5 epitope for identification of expressed protein with V5 epitope antibody, polyhistidine peptide for simplified purification with metal chelate affinity resin, and the BiP secretion leader peptide. The DES system also includes media for maintenance of the host cell line and expression of protein, as well as reagents to facilitate transfection. The commonly used stable cell lines include *Aedes aegypti, A. albopictus,* and *Anopheles Gambia*; in addition, the following lines can be stable cell lines or used with baculovirus system (see following discussion): *Drosophilia melangolaster* (Schneider S2 and S3), *S. frugiperda* (Sf9), *S. frugiperda* (Sf21), *T. ni* (HighFive™), *T. ni* (BTI-TN-5B1-4).

Baculovirus expression systems

The baculovirus expression cassette contains all the genetic information needed for propagation of progeny virus, so no helper virus is needed in the transfection process. The biology of the virus provides a simple means, using plaque morphology, to identify transformed host cells. The virus does not appear to be transmissible to vertebrate species; therefore, this virus-based system is safe for human handlers. Since with many virus vectors, heterologous protein genes are under the control of the late-stage baculovirus p10 and polyhedrin promoters, recombinant protein is, in most cases, the sole product produced. Hence, cells harboring the baculovirus expression cassette integrated in their genomes can produce relatively high amounts of heterologous protein. Most of this protein is easily extracted from the cytoplasm (no inclusion bodies characteristic of prokaryotic systems) or harvested from extracellular culture filtrate (when the expression cassette includes a secretory leader fusion peptide engineered to the recombinant protein). However, the cell machinery may be starting to shut down late in infection, which can impact, in particular, proteins requiring processing. Hence, some companies have introduced viral vectors with hybrid early/late promoters that permits the still functioning cell to process glycosylated or secreted proteins. The commonly used cell lines are: *Drosophilia melangolaster* (Schneider S2 and S3), *S. frugiperda* (Sf9), *S. frugiperda* (Sf21), *T. ni* (HighFive), *T. ni* (BTI-TN-5B1-4; HighFive).

Baculovirus expression system provides one of the highest levels of target protein expression using baculovirus expression vector system such as BacVector offered by Novogen. In this system, insect cells are infected with a recombinant baculovirus bearing the GOI. The infected cells undergo a burst of protein expression, after which the cells die and may lyse. Expression at earlier times can be advantageous to allow more complete protein modification such as glycoprotein processing and is obtained by using alternative baculovirus promoters, for example, ie1 and gp84 promoters. The process of creating and expressing heterologous protein with the plasmid-virion system is rather straightforward, in theory, but does require a bit of technical finesse and close attention to detail. The process begins with the engineering of the heterologous protein gene into a "transfer plasmid." This plasmid contains all the elements for autonomous replication in *E. coli*, a bacterial selection marker (usually an ampicillin resistance gene), and elements of the baculovirus genome. The heterologous protein gene is inserted in a specific orientation and location into the plasmid, so it is flanked by elements of the baculovirus genome. Successfully engineered plasmids are then cotransfected with viral expression vector (essentially wild-type baculovirus DNA with p10 and/or polyhedrin genes removed) into permissive host cells. Cell-mediated double recombination between viral sequences flanking the heterologous protein gene and the corresponding sequences of the viral expression vector results in the incorporation of the heterologous protein gene into the viral genome. Hence, recombinant progeny viruses will produce heterologous protein late in their life cycle. Novagen's pIE vectors are based on the baculovirus immediate early promotor ie1. These plasmids can be used with the G418 selection to generate stable cell lines from Sf0 or Sf21 cell lines. Other suppliers of vectors and complete Baculovirus sytems are Clontech, Invitrogen, Life Technologies, Novagen, Pharmingen, Quantum Biotechnologies, and Stratagene.

The baculovirus system has several drawbacks:

- The cyclic killing and lysis of the host cell releases intracellular proteins to the medium adversely affecting the purification of the target protein.
- The production can only be achieved in batches or at best, semi-continuously.
- The expression yield is often much lower than expected due to inactivation of the secretory pathway during the late phase of infection.
- Inactivation of secretory path affects posttranslational events such as glycosylation (the sugar chain will end in a mannose and not contain galactose or terminal sialic acid) rendering the expressed protein unsuitable for in vivo applications.

- Baculovirus infected cells do not efficiently excise introns from expressed genomic DNA, thus limiting foreign protein expression from cDNAs.

Though currently not significant, insect cell technology is likely to produce many dramatic advances in the near future.

Transgenic animals

A transgenic animal is one that carries a foreign gene that has been deliberately inserted into its genome. The foreign gene is constructed using recombinant DNA methodology. In addition to a structural gene, the DNA usually includes other sequences to enable it to be incorporated into the DNA of the host and to be expressed correctly by the cells of the host. Transgenic sheep and goats have been produced that express foreign proteins in their milk. Transgenic chickens are now able to synthesize human proteins in the "white" of the eggs. These animals should eventually prove to be valuable sources of proteins for human therapy. In July 2000, researchers from the team that produced Dolly reported success in producing transgenic lambs in which the transgene had been inserted at a specific site in the genome and functioned well. Transgenic mice have provided the tools for exploring many biological questions.

Until recently, the transgenes introduced into sheep inserted randomly in the genome and often worked poorly. However, in July 2000, success at inserting a transgene into a specific gene locus was reported. The gene was the human gene for alpha 1-antitrypsin, and two of the animals expressed large quantities of the human protein in their milk. The method used to create transgenic sheep is as follows. Sheep fibroblasts (connective tissue cells) growing in tissue culture are treated with a vector that contained the segments of DNA: two regions homologous to the sheep COL1A1– gene. This gene encodes Type 1 collagen. This locus is chosen because fibroblasts secrete large amounts of collagen and thus one would expect the gene to be easily accessible in the chromatin. Also inserted is a neomycin-resistance gene to aid in isolating those cells that successfully incorporates the vector and the human gene encoding alpha 1-antitrypsin. The vector further contains promoter sites from beta lactoglobulin gene to promote hormone-driven gene expression milk producing cells and also binding sites for ribosomes for efficient translation of the mRNAs. Successfully transformed cells are then fused with enucleated sheep eggs and implanted in the uterus of an ewe (female sheep). The off spring secretes milk that contains large amounts of alpha1-antitrypsin (650 µg/mL; 50 times higher than previous results using random insertion of the transgene). This project has now been abandoned because of its high cost despite remarkable success.

Chickens grow faster than sheep, and large numbers can be grown in close quarters; synthesize several grams of protein in the "white" of their eggs. Two methods have succeeded in producing chickens carrying and expressing foreign genes: infecting embryos with a viral vector carrying the human gene for a therapeutic protein, and the promoter sequences that will respond to the signals for making proteins such as lysozyme in egg white. This is followed by transforming rooster sperm with a human gene and the appropriate promoters and checking for any transgenic offspring. Initial results from both methods indicate that it may be possible for chickens to produce as much as 0.1 g of human protein in each egg that they lay and these proteins are likely to have correct sugars to glycosylate proteins, something not possible when using E. coli.

Transgenic pigs have also been produced by fertilizing normal eggs with sperm cells that have incorporated foreign DNA. This procedure called sperm-mediated gene transfer (SMGT) may someday be able to produce transgenic pigs that can serve as a source of transplanted organs for humans. Progress is being made on several fronts to introduce new traits into plants using recombinant DNA. The genetic manipulation of plants has been going on since the dawn of agriculture, but until recently this has required the slow and tedious process of cross-breeding varieties. Genetic engineering promises to speed the process and broaden the scope of what can be done. There are several methods of introducing genes into plants, including infecting plant cells with plasmids as vectors carrying the desired gene and shooting microscopic pellets containing the gene directly into the cell. In contrast to animals, there is no real distinction between somatic cells and germ line cells. Somatic tissues of plants, for example, root cells grown in culture can be transformed into the laboratory with the desired gene and grown into mature plants with flowers. Therapeutic protein genes can be inserted into plants and expressed by them with glycosylation, reduce dangers inherent in tissue culture techniques, and offer simple purification potential. Corn is the most popular plant for these purposes, but tobacco, tomatoes, potatoes, and rice are also being used. Some of the proteins that are being produced by transgenic crop plants: human growth hormone with the gene inserted into the chloroplast DNA of tobacco plants, humanized antibodies against such infectious agents as HIV, respiratory syncytial virus (RSV), sperm (a possible contraceptive), herpes simplex virus (HSV), the cause of "cold sores"; protein antigens to be used in vaccines and other useful proteins like lysozymes and trypsin.

Transgenic animals are one of the most promising recombinant protein expression systems making it possible to produce plasma proteins, human antibodies, and other proteins not easily derived from other sources; however, the development process is long as it takes 18–33 months from introduction of gene to production at usable levels. Two products have been approved by the FDA from the transgenic source.

The second one is a product of Protalix and Pfizer that is branded as Elelyso, a drug to treat Gaucher's disease that was approved in 2012. The Protalix product will compete with the expensive genzyme product, Cerezyme. What makes this approval especially noteworthy is that Elelyso is produced in carrots, making it the first FDA-approved transgenic drug. This is a major step that takes away the legendary reluctance by the regulatory agencies in accepting transgenic route as a viable route for the production of recombinant drugs.

In 2009, the U.S. FDA approved the first transgenic drug of GTC Biotherapeutics and Ovation Pharmaceuticals. The approved drug ATryn (Antithrombin [Recombinant]) is used for the prevention of perioperative and peripartum thromboembolic events in hereditary AT deficient patients. ATryn is the first ever transgenically produced therapeutic protein and the first recombinant AT approved in the United States. It is produced in transgenic goat milk.

Cell banks

Characterization of cell and virus banks

The establishment of cell banks for newly developed cell lines is critical to the successful development of many biological products. The cell bank system assures that the cell line is preserved, its integrity is maintained, and a sufficient supply is readily available. A well characterized cell bank is the consistent source of production cells throughout the life of the product. Worldwide regulatory authorities require screening cell cultures for the presence of contaminating agents by testing for adventitious viral or microbial agents using both in vivo and in vitro methodologies.

The cGMP compliant preparation of MCBs and MWCBs is performed in a class 100 environment with in-phase testing for mycoplasma and sterility. Cell banks are stored in validated and continuously monitored liquid nitrogen dewars.

The ICH guideline describes characterization of cells used in recombinant DNA work (http://www.ema.europa.eu/docs/en_GB/document_library/Scientific_guideline/2009/09/WC500003280.pdf). A cell bank is a collection of vials containing cells stored under defined conditions, with uniform composition and obtained from pooled cells derived from a single cell clone, such as bacteria or mammalian tissue. The cell bank system usually consists of an MCB and a WCB, although more tiers are possible. The MCB is produced in accordance with CGMP and preferably obtained from a qualified repository source (source free from adventitious agents) whose history is known and documented. The WCB is produced or derived by expanding one or more vials of the MCB. The WCB, or MCB in early trials, becomes the source of cells for every batch produced for human use. Cell bank systems contribute greatly to the consistency of production of clinical or licensed product batches, because the starting cell material is always the same. Mammalian and bacterial cell sources are used for establishing cell bank systems. The master virus bank (MVB) is similar to the MCB in that it is derived from a single production run and is uniform in composition. The working virus bank (WVB) is derived directly from the MVB. As with the cell banks, the focus of virus bank usage is to have a consistent source of virus, shown to be free of adventitious agents, to use in production of clinical or product batches. In keeping with CGMP guidelines, testing of the cell bank to be used for production of the virus banks, including quality assurance testing, should be completed prior to the use of this cell bank for production of virus banks. Virus banks are used to provide testing of various therapeutic proteins such as interferon.

Cell and viral bank characterization is an important step toward obtaining a uniform final product with lot-to-lot consistency and freedom from adventitious agents. Testing to qualify the MCB or MVB is performed once and can be done on an aliquot of the banked material or on cell cultures derived from the cell bank. Specifications for qualification of the MCB or MVB should be established. It is important to document the MCB and MVB history, methods and reagents used to produce the bank, and storage conditions. All the raw materials required for production of the banks, namely, media, sera, trypsin, and the like, must also be tested for adventitious agents.

Testing to qualify the MCB includes the following: (1) testing to demonstrate freedom from adventitious agents and endogenous viruses and (2) identity testing. The testing for adventitious agents may include tests for nonhost microbes, mycoplasma, bacteriophage, and viruses. Freedom from adventitious viruses should be demonstrated using both in vitro and in vivo virus tests, and appropriate species-specific tests such as the mouse antibody production (MAP) test. Identity testing of the cell bank should establish the properties of the cells and the stability of these properties during manufacture. Cell banks should be characterized with respect to cellular isoenzyme expression and cellular phenotype and genotype, which could include expression of a gene insert or presence of a gene-transfer vector. Suitable techniques, including restriction endonuclease mapping or nucleic acid sequencing, should be used to analyze the cell bank for vector copy number and the physical state of the vector (vector integrity and integration). The cell bank should also be characterized for the quality and quantity of the gene product produced.

Testing of the MVB is similar to that of the MCB and should include testing for freedom from adventitious agents in general (such as, bacteria, fungi, mycoplasma, or viruses) and for organisms specific to the production cell line, including RCV. Identity testing of the MVB should establish the properties of the virus and the stability of these properties during manufacture. Characterization of the WCB or WVB is generally less extensive, requiring the following: (1) testing for freedom from adventitious agents that may have been introduced from the culture medium, (2) testing for RCV, if relevant, (3) routine identity tests to check for cell line cross-contamination, and (4) demonstration that aliquots can consistently be used for final product production.

The following are suggested for testing MCBs and end of production cells:

- Microbial contamination: Sterility (USP 25): Determines the presence of aerobic and anaerobic bacteria and fungi, mycoplasma (1993 PTC CBER/FDA)
- Cell line authenticity: Karyology and isoenzyme analysis determines the species of origin of the cell line
- Transmission electron microscopy (TEM): Determines cellular morphology, presence of microbial contaminants, enumeration of retroviral and retrovirus-like particles
- Endogenous retroviruses: Reverse transcriptase assays and retroviral infectivity assays detects the presence of retroviruses
- In vitro and in vivo adventitious virus testing
- Mouse, rat, and hamster antibody production assays (MAP, RAP, and HAP): Detects species-specific viruses in rodent cell lines
- Bovine and porcine virus assays (9 CFR)
- Human viruses HIV-I and -II, HTLV-I and -II, HAV, HBV, HCV, CMV, EBV, HSV-2, HHV-6, AAV-2, and B-19
- Primate viruses SIV, STLV, foamy agent, and SMRV

The following tests are required for the WCBs:

- Microbial contamination: Sterility (USP 24) determines the presence of aerobic and anaerobic bacteria and fungi, mycoplasma (1993 PTC CBER/FDA)
- In vitro adventitious virus testing

Table 6.7 Suggested Methods for Test of Microbial Cell Banks (Bacteria and Yeast)

Test	Comments
Auxotrophic markers	Conformation of key markers (yeast).
Characterization	Characterization of the insert by Southern Blot and eventually DNA sequence of the integrated expression cassette.
Identity	Phenotyping or genotyping may be used to confirm species and strain. Growth on selective media is considered adequate to confirm host cell identity for most microbial cells. For E. coli, phage typing should be considered as a supplementary test.
Purity	Freedom from adventitious microbial agents and adventitious cellular contaminants must be assessed (bacteria, bacteriophages, fungi). Visual examination of the characteristics of the isolated colonies using different media is suggested.
Resistance to antibiotics	When antibiotic selection markers have been used.
Stability	It must be demonstrated that the cell line produces the desired product in a consistent quality and quantity. Studies shall be performed to determine whether manipulation of the cell line changes it characteristics significantly.

Table 6.8 Suggested Methods for Test of Insect and Metazoan Cell Banks

Test	Comments
Identity	Phenotyping or genotyping may be used. In most cases, isoenzyme analysis is sufficient to confirm the species of origin. Other technologies include banding cytogenetics and use of species-specific anti-sera. An alternative strategy is to demonstrate the presence of unique markers.
Purity	Freedom from adventitious microbial agents and adventitious cellular contaminants must be assessed. Test for the presence of bacteria, fungi, and mycoplasma should be performed. Freedom of contaminating cell lines of the same or different species must be demonstrated.
Purity, viral agents using cell cultures	Mono-layer cultures of the following cell types: • Cultures of the same species and tissue type as the cell line • Cultures of a human diploid cell line • Cultures of another cell line from a different species The PCR technology is receiving increasing attention from regulatory authorities especially after new validated quantitative tests have emerged.
Purity, viral agents using animals and eggs	Test for pathogen viruses not able to grow in cell cultures in both animals and eggs (e.g., suckling mice, adult mice, guinea pigs, fertilized eggs). The cell banks are suitable for production if none of the animals or eggs show evidence of the presence of any viral agent (*European Pharmacopoeia* V.2.2.12).
Purity, test for retroviruses, endogenous viruses or viral nucleic acid	The test shall include infective assays, transmission electron microscopy (TEM)4, and reverse transcriptase (Rtase) of cells cultured up to or beyond in vitro cell age. Induction studies have not been found to be useful.
Purity test for selected viruses	Murine cell lines shall be tested species-specific using mouse, rat, and hamster antibody production tests (MAP, RAP, HAP). In vivo testing for lymphocytic chorimeningitis virus is required. PCR techniques may be used as well. Human cell lines shall be screened for human viral pathogens (Epstein-Barr virus, cytomegalovirus, human retroviruses, hepatitis B/C viruses with appropriate in vitro techniques.
Stability	It must be demonstrated that the cell line produces the desired product in a consistent quality and quantity. Studies shall be performed to determine whether manipulation of the cell line changes it characteristics significantly.
Serum	Freedom from cultivable bacteria, fungi, mycoplasma, and infectious viruses must be demonstrated. Human serum should not be used, and use of bovine serum shall meet specified requirements for biological substances.
Trypsin	Trypsin shall be tested and found free of cultivable bacteria, fungi, mycoplasma, and infectious viruses, especially bovine or porcine parvoviruses, as appropriate.
	Contaminating viruses are of major concern. Viruses can be introduced into the cell bank by several routes such as derivation of cell lines from infected animals, use of viruses to establish the cell line, use of contaminated reagents (e.g., animal serum), and contaminants during handling of cells. Although continuous cell lines are extensively characterized, viral contaminants will not be cytolytic. However, chronic or latent viruses may be present. Examples of viruses harbored in cell substrates are retroviruses (oncogenic), hantaviruses (CHO cells), hepatitis viruses (human cells), human papilloma virus (human cells), and cytomegalovirus (human cells). Additionally, cell line establishment or cell transformation is achieved using Epstein-Barr or Sendai viruses.

Master cell banks

An MCB is a homogeneous pool of the production cell line dispensed in multiple containers, by which each aliquot is representative for each other vial, stored under defined conditions (liquid nitrogen). A tested and released MCB provides material for the WCB using one more tubes from MCB. The MCB and WCB may differ from each other in certain respects (e.g., culture components and culture conditions) but generally there is no need to extensively characterize WCB unless records indicate wide variability between MCB and WCB. The methods used to characterize cell banks are derived from the WHO requirements for the use of in vitro substrates for the production of biologicals (Requirements for biological substances No. 50 [WHO 1998] and the ICH guidelines Q5A, B, and D); (Tables 6.7 and 6.8).

Bibliography

Abrahmsen L, Moks T, Nilsson B, Uhlén M. Secretion of heterologous gene products to the culture medium of *Escherichia coli. Nucleic Acids Res* 1986;14:7487–7500.

Adams JM. On the release of the formyl group from nascent protein. *J Mol Biol* 1968;33:571–589.

Adams TE, MacIntosh B, Brandon MR, Wordsworth P, Puri NK. Production of methionyl-minus ovine growth hormone in *Escherichia coli* and one-step purification. *Gene* 1992;122:371–375.

Adari H, Andrews B, Ford PJ, Hannig G, Brosius J, Makrides SC. Expression of the human T-cell receptor V5.3 in *Escherichia coli* by thermal induction of the trc promoter: Nucleotide sequence of the lacIts gene. *DNA Cell Biol* 1995;14:945–950.

Adhya S, Gottesman M. Promoter occlusion: Transcription through a promoter may inhibit its activity. *Cell* 1982;29:939–944.

Airenne KJ, Kulomaa MS. Rapid purification of recombinant proteins fused to chicken avidin. *Gene* 1995;167:63–68.

Amann E, Brosius J, Ptashne M. Vectors are bearing a hybrid trp-lac promoter useful for regulated expression of cloned genes in *Escherichia coli. Gene* 1983;25:167–178.

Amrein KE, Takacs B, Stieger M, Molnos J, Flint NA, Burn P. Purification and characterization of recombinant human p50csk pro-tein-tyrosine kinase from an *Escherichia coli* expression system overproducing the bacterial chaperones GroES and GroEL. *Proc Natl Acad Sci USA* 1995;92:1048–1052.

Anderson KP, Low MA, Lie YS, Keller GA, Dinowitz M. Endogenous origin of defective retroviruslike particles from a recombinant Chinese hamster ovary cell line. *Virology* 1991;181(1):305–311.

Andrews B, Adari H, Hannig G, Lahue E, Gosselin M, Martin S, Ahmed A, Ford PJ, Hayman EG, Makrides SC. A tightly regulated high level expression vector that utilizes a thermosensitive lac repressor: Production of the human T cell receptor V5.3 in *Escherichia coli. Gene* 1996;182(1–2):101–109.

Aranha H. Viral clearance strategies for biopharmaceutical safety. Part 1: General considerations. *BioPharm* 2001;14(1):28–35.

Ariga O, Andoh Y, Fujishita Y, Watari T, Sano Y. Production of thermophilic-amylase using immobilized transformed *Escherichia coli* by the addition of glycine. *J Ferment Bioeng* 1991;71:397–402.

Aristidou AA, San K-Y, Bennett GN. Modification of the central metabolic pathway in *Escherichia coli* to reduce acetate accumulation by heterologous expression of the *Bacillus subtilis* acetolactate synthase gene. *Biotechnol Bioeng* 1994;44:944–951.

Aristidou AA, San K-Y, Bennett GN. Metabolic engineering of *Escherichia coli* to enhance recombinant protein production through acetate reduction. *Biotechnol Prog* 1995;11:475–478.

Aristidou AA, Yu P, San K-Y. Effects of glycine supplement on protein production and release in recombinant *Escherichia coli. Biotechnol Lett* 1993;15:331–336.

Astley K, Al-Rubeai M. The role of Bcl-2 and its combined effect with p21CIP1 in adaptation of CHO cells to suspension and protein-free culture. *Appl Microbiol Biotechnol* 2008;78:391–399.

Atochina O, Mylvaganam R, Akselband Y, McGrath P. Comparison of results using the gel microdrop cytokine secretion assay with ELISPOT and intracellular cytokine staining assay. *Cytokine* 2004;27:120–128.

Bachmair A, Finley D, Varshavsky A. In vivo half-life of a protein is a function of its amino-terminal residue. *Science* 1986;234:179–186.

Bachmair A, Varshavsky A. The degradation signal in a short-lived protein. *Cell* 1989;56:1019–1032.

Backman K, O'Connor MJ, Maruya A, Erfle M. Use of synchronous site-specific recombination in vivo to regulate gene expression. *Bio/Technology* 1984;2:1045–1049.

Backman K, Ptashne M. Maximizing gene expression on a plasmid using recombination in vitro. *Cell* 1978;13:65–71.

Backman K, Ptashne M, Gilbert W. Construction of plasmids carrying the cI gene of bacteriophage. *Proc Natl Acad Sci USA* 1976;73:4174–4178.

Baik JY, Gasimli L, Yang B, Datta P, Zhang F, Glass CA, Esko JD, Linhardt RJ, Sharfstein ST. Metabolic engineering of Chinese hamster ovary cells: Towards a bioengineered heparin. *Metab Eng* 2012;14:81–90.

Baik JY, Lee MS, An SR, Yoon SK, Joo EJ, Kim YH, Park HW, Lee GM. Initial transcriptome and proteome analyses of low culture temperature-induced expression in CHO cells producing erythropoietin. *Biotechnol Bioeng* 2006;93:361–371.

Baird SD, Turcotte M, Korneluk RG, Holcik M. Searching for IRES. *RNA* 2006;12:1755–1785.

Baker RT, Smith SA, Marano R, McKee J, Board PG. Protein expression using cotranslational fusion and cleavage of ubiquitin. Mutagenesis of the glutathione-binding site of human Pi class glutathione S-transferase. *J Biol Chem* 1994;269:25381–25386.

Baker TA, Grossman AD, Gross CA. A gene regulating the heat shock response in *Escherichia coli* also affects proteolysis. *Proc Natl Acad Sci USA* 1984;81:6779–6783.

Balakrishnan R, Bolten B, Backman KC. A gene cassette for adapting *Escherichia coli* strains as hosts for att-Int-mediated rearrangement and pL expression vectors. *Gene* 1994;138:101–104.

Balbas P, Bolivar F. Design and construction of expression plasmid vectors in *Escherichia coli*. *Methods Enzymol* 1990;185:14–37.

Baneyx F, Georgiou G. Construction and characterization of *Escherichia coli* strains deficient in multiple secreted proteases: Protease III degrades high-molecular-weight substrates in vivo. *J Bacteriol* 1991;173:2696–2703.

Baneyx F, Georgiou G. Degradation of secreted proteins in *Escherichia coli*. *Ann N Y Acad Sci* 1992a;665:301–308.

Baneyx F, Georgiou G. Expression of proteolytically sensitive polypeptides in *Escherichia coli*. In: *Stability of Protein Pharmaceuticals. A. Chemical and Physical Pathways of Protein Degradation*, Ahern TJ, Manning MC, eds. Plenum Press, New York, 1992b, pp. 69–108.

Bardwell JCA. Building bridges: Disulfide bond formation in the cell. *Mol Microbiol* 1994; 14:199–205.

Bardwell JCA, McGovern K, Beckwith J. Identification of a protein required for disulfide bond formation in vivo. *Cell* 1991;67:581–589.

Barrick D, Villanueba K, Childs J, Kalil R, Schneider TD, Lawrence CE, Gold L, Stormo GD. Quantitative analysis of ribosome binding sites in *E. coli*. *Nucleic Acids Res* 1994;22:1287–1295.

Barron N, Kumar N, Sanchez N, Doolan P, Clarke C, Meleady P, O'Sullivan F, Clynes M. Engineering CHO cell growth and recombinant protein productivity by overexpression of miR-7. *J Biotechnol* 2011;151:204–211.

Battistoni A, Carri MT, Steinkuhler C, Rotilio G. Chaperonins dependent increase of Cu, Zn superoxide dismutase production in *Escherichia coli*. *FEBS Lett* 1993;322:6–9.

Bauer KA, Ben-Bassat A, Dawson M, de la Puente VT, Neway JO. Improved expression of human interleukin-2 in high-cell-density fermentor cultures of *Escherichia coli* K-12 by a phosphotransacetylase mutant. *Appl Environ Microbiol* 1990;56:1296–1302.

Baycin-Hizal D, Tabb DL, Chaerkady R et al. Proteomic analysis of Chinese hamster ovary cells. *J Proteome Res* 2012;11:5265–5276.

Bebbington CR, Renner G, Thomson S, King D, Abrams D, Yarranton GT. High-level expression of a recombinant antibody from myeloma cells using a glutamine synthetase gene as an amplifiable selectable marker. *Biotechnology (NY)* 1992;10:169–175.

Bechhofer D. 5 mRNA stabilizers. In: *Control of Messenger RNA Stability*, Belasco JG, Brawerman G, eds. Academic Press, Inc., San Diego, CA, 1993, pp. 31–52.

Bechhofer DH, Dubnau D. Induced mRNA stability in *Bacillus subtilis*. *Proc Natl Acad Sci USA* 1987;84:498–502.

Becker E, Florin L, Pfizenmaier K, Kaufmann H. Evaluation of a combinatorial cell engineering approach to overcome apoptotic effects in XBP-1(s) expressing cells. *J Biotechnol* 2010;146:198–206.

Becker J, Craig EA. Heat-shock proteins as molecular chaperones. *Eur J Biochem* 1994;219:11–23.

Bedouelle H, Duplay P. Production in *Escherichia coli* and one-step purification of bifunctional hybrid proteins which bind maltose. Export of the Klenow polymerase into the periplasmic space. *Eur J Bio-Chem* 1988;171:541–549.

Belasco JG. mRNA degradation in prokaryotic cells: An overview. In: *Control of Messenger RNA Stability*, Belasco JG, Brawerman G, eds. Academic Press, Inc., San Diego, CA, 1993, pp. 3–12.

Belasco JG, Brawerman G, eds. *Control of Messenger RNA Stability*. Academic Press, Inc., San Diego, CA, 1993.

Belasco JG, Higgins CF. Mechanisms of mRNA decay in bacteria: A perspective. *Gene* 1988;72:15–23.

Belasco JG, Nilsson G, von Gabain A, Cohen SN. The stability of *E. coli* gene transcripts is dependent on determinants localized to specific mRNA segments. *Cell* 1986;46:245–251.

Bell AC, West AG, Felsenfeld G. The protein CTCF is required for the enhancer blocking activity of vertebrate insulators. *Cell* 1999;98:387–396.

Belt A. Characterization of cultures used for biotechnology and industry. In: *Maintaining Cultures for Biotechnology and Industry*, Hunter-Cevera JC, Belt A, eds. Academic Press, Inc., San Diego, CA, 1996, pp. 251–258.

Ben-Bassat A, Bauer K, Chang S-Y, Myambo K, Boosman A, Chang S. Processing of the initiation methionine from proteins: Properties of the *Escherichia coli* methionine aminopeptidase and its gene structure. *J Bacteriol* 1987;169:751–757.

Bentley WE, Mirjalili N, Andersen DC, Davis RH, Kompala DS. Plasmid-encoded protein: The principal factor in the "metabolic burden" associated with recombinant bacteria. *Biotechnol Bioeng* 1990;35:668–681.

Benton T, Chen T, McEntee M, Fox B, King D, Crombie R, Thomas TC, Bebbington C. The use of UCOE vectors in combination with a preadapted serum free, suspension cell line allows for rapid production of large quantities of protein. *Cytotechnology* 2002;38:43–46.

Berg KL, Squires C, Squires CL. Ribosomal RNA operon antitermination. Function of leader and spacer region boxB-boxA sequences and their conservation in diverse microorganisms. *J Mol Biol* 1989;209:345–358.

Berkow R, ed. *The Merck Manual of Diagnosis and Therapy*, 16th edn. Merck Research Laboratories, Rahway, NJ, 1992, pp. 24–30.

Bernard H-U, Remaut E, Hershfield MV, Das HK, Helinski DR, Yanofsky C, Franklin N. Construction of plasmid cloning vehicles that promote gene expression from the bacteriophage lambda pL promoter. *Gene* 1979;5:59–76.

Better M, Chang CP, Robinson RR, Horwitz AH. *Escherichia coli* secretion of an active chimeric antibody fragment. *Science* 1988;240:1041–1043.

Betton J-M, Hofnung M. Folding of a mutant maltose-binding protein of *Escherichia coli* which forms inclusion bodies. *J Biol Chem* 1996;271:8046–8052.

Bidwell JP, Torrungruang K, Alvarez M, Rhodes SJ, Shah R, Jones DR, Charoonpatrapong K, Hock JM, Watt AJ. Involvement of the nuclear matrix in the control of skeletal genes: The NMP1 (YY1), NMP2 (Cbfa1), and NMP4 (Nmp4/CIZ) transcription factors. *Crit Rev Eukaryot Gene Exp* 2001;11:279–297.

Birikh KR, Lebedenko EN, Boni IV, Berlin YA. A high-level prokaryotic expression system: Synthesis of human interleukin 1 and its receptor antagonist. *Gene* 1995;164:341–345.

Bishai WR, Rappuoli R, Murphy JR. High-level expression of a proteolytically sensitive diphtheria toxin fragment in *Escherichia coli*. *J Bacteriol* 1987;169:5140–5151.

Björnsson A, Mottagui-Tabar S, Isaksson LA. Structure of the C-terminal end of the nascent peptide influences translation termination. *EMBO J* 1996;15:1696–1704.

Black CB, Duensing TD, Trinkle LS, Dunlay RT. Cell-based screening using high-throughput flow cytometry. *Assay Drug Dev Technol* 2011;9:13–20.

Blackwell JR, Horgan R. A novel strategy for production of a highly expressed recombinant protein in an active form. *FEBS Lett* 1991;295:10–12.

Blight MA, Chervaux C, Holland IB. Protein secretion pathways in *Escherichia coli*. *Curr Opin Biotechnol* 1994;5:468–474.

Blondel A, Nageotte R, Bedouelle H. Destabilizing interactions between the partners of a bifunctional fusion protein. *Protein Eng* 1996;9:231–238.

Blum P, Ory J, Bauernfeind J, Krska J. Physiological consequences of DnaK and DnaJ overproduction in *Escherichia coli*. *J Bacteriol* 1992a;174:7436–7444.

Blum P, Velligan M, Lin N, Matin A. DnaK-mediated alterations in human growth hormone protein inclusion bodies. *Bio/Technology* 1992b;10:301–304.

Boch J, Scholze H, Schornack S, Landgraf A, Hahn S, Kay S, Lahaye T, Nickstadt A, Bonas U. Breaking the code of DNA binding specificity of TAL-type III effectors. *Science* 2009;326:1509–1512.

Boeger H, Bushnell DA, Davis R, Griesenbeck J, Lorch Y, Strattan JS, Westover KD, Kornberg RD. Structural basis of eukaryotic gene transcription. *FEBS Lett* 2005;579:899–903.

Boni IV, Isaeva DM, Musychenko ML, Tzareva NV. Ribosome-messenger recognition: mRNA target sites for ribosomal protein S1. *Nucleic Acids Res* 1991;19:155–162.

Bowden GA, Baneyx F, Georgiou G. Abnormal fractionation of β-lactamase in *Escherichia coli*: Evidence for an interaction of β-lactamase with the inner membrane in the absence of a leader peptide. *J Bacteriol* 1992;174:3407–3410.

Bowden GA, Georgiou G. The effect of sugars on β-lactamase aggregation in *Escherichia coli*. *Biotechnol Prog* 1988;4:97–101.

Bowden GA, Georgiou G. Folding and aggregation of β-lactamase in the periplasmic space of *Escherichia coli*. *J Biol Chem* 1990;265:16760–16766.

Bowden GA, Paredes AM, Georgiou G. Structure and morphology of protein inclusion bodies in *Escherichia coli*. *Bio/Technology* 1991;9:725–730.

Bowie JU, Sauer RT. Identification of C-terminal extensions that protect proteins from intracellular proteolysis. *J Biol Chem* 1989;264:7596–7602.

Branda CS, Dymecki SM. Talking about a revolution: The impact of site-specific recombinases on genetic analyses in mice. *Dev Cell* 2004;6:7–28.

Brenner S, Jacob F, Meselson M. An unstable intermediate carrying information from genes to ribosomes for protein synthesis. *Nature (London)* 1961;190:576–581.

Brewer SJ, Sassenfeld HM. The purification of recombinant proteins using C-terminal polyarginine fusions. *Trends Biotechnol* 1985;3:119–122.

Brezinsky SC, Chiang GG, Szilvasi A, Mohan S, Shapiro RI, MacLean A, Sisk W, Thill G. A simple method for enriching populations of transfected CHO cells for cells of higher specific productivity. *J Immunol Methods* 2003;277:141–155.

Brinkmann U, Mattes RE, Buckel P. High-level expression of recombinant genes in *Escherichia coli* is dependent on the availability of the *dnaY* gene product. *Gene* 1989;85:109–114.

Brizzard BL, Chubet RG, Vizard DL. Immunoaffinity purification of FLAG epitope-tagged bacterial alkaline phosphatase using a novel monoclonal antibody and peptide elution. *BioTechniques* 1994;16:730–734.

Brosius J. Compilation of superlinker vectors. *Methods Enzymol* 1992;216:469–483.

Brosius J, Erfle M, Storella J. Spacing of the 10 and 35 regions in the tac promoter. Effect on its in vivo activity. *J Biol Chem* 1985;260:3539–3541.

Brosius J, Holy A. Regulation of ribosomal RNA promoters with a synthetic lac operator. *Proc Natl Acad Sci USA* 1984;81:6929–6933.

Brosius J, Ullrich A, Raker MA, Gray A, Dull TJ, Gutell RG, Noller F. Construction and fine mapping of recombinant plasmids containing the rrnB ribosomal RNA operon of *E. coli*. *Plasmid* 1981;6:112–118.

Brown WC, Campbell JL. A new cloning vector and expression strategy for genes encoding proteins toxic to *Escherichia coli*. *Gene* 1993;127:99–103.

Browne SM, Al-Rubeai M. Selection methods for high-producing mammalian cell lines. *Trends Biotechnol* 2007;25:425–432.

Buchner J. Supervising the fold: Functional principles of molecular chaperones. *FASEB J* 1996;10:10–19.

Buell G, Schulz M-F, Selzer G, Chollet A, Movva NR, Semon D, Escanez S, Kawashima E. Optimizing the expression in *E. coli* of a synthetic gene encoding somatomedin-C (IGF-I). *Nucleic Acids Res* 1985;13:1923–1938.

Bujard H, Gentz R, Lanzer M, Stueber D, Mueller M, Ibrahimi I, Haeuptle M-T, Dobberstein B. A T5 promoter-based transcription-translation system for the analysis of proteins in vitro and in vivo. *Methods Enzymol* 1987;155:416–433.

Bukrinsky MI, Barsov EV, Shilov AA. Multicopy expression vector based on temperature-regulated lac repressor: Expression of human immunodeficiency virus env gene in *Escherichia coli*. *Gene* 1988;70:415–417.

Bula C, Wilcox KW. Negative effect of sequential serine codons on expression of foreign genes in *Escherichia coli*. *Protein Expression Purif* 1996;7:92–103.

Bulmer M. Codon usage and intragenic position. *J Theor Biol* 1988;133:67–71.

Butler JS, Springer M, Grunberg-Manago M. AUU-to-AUG mutation in the initiator codon of the translation initiator factor IF3 abolishes translational autocontrol of its own gene (infC) in vivo. *Proc Natl Acad Sci USA* 1987;84:4022–4025.

Cabaniols JP, Ouvry C, Lamamy V et al. Meganuclease-driven targeted integration in CHO-K1 cells for the fast generation of HTS-compatible cell-based assays. *J Biomol Screen* 2010;15:956–967.

CabillyS. Growth at sub-optimal temperatures allows the production of functional, antigen-binding Fab fragments in *Escherichia coli*. *Gene* 1989;85:553–557.

Campbell M, Corisdeo S, McGee C, Kraichely D. Utilization of site-specific recombination for generating therapeutic protein producing cell lines. *Mol Biotechnol* 2010;45:199–202.

Cao H, Widlund HR, Simonsson T, Kubista M. TGGA repeats impair nucleosome formation. *J Mol Biol* 1998;281:253–260.

Carter P. Site-specific proteolysis of fusion proteins. In: *Protein Purification: From Molecular Mechanisms to Large-Scale Processes*, Ladisch MR, Willson RC, Painton C-C, Builder SE, eds. American Chemical Society Symposium Series No. 427. American Chemical Society, Washington, DC, 1990, pp. 181–193.

Caspers P, Stieger M, Burn P. Overproduction of bacterial chaperones improves the solubility of recombinant protein tyrosine kinases in *Escherichia coli*. *Cell Mol Biol* 1994;40:635–644.

Caulcott CA, Rhodes M. Temperature-induced synthesis of recombinant proteins. *Trends Biotechnol* 1986;4:142–146.

Chalfie M, Tu Y, Euskirchen G, Ward WW, Prasher DC. Green fluorescent protein as a marker for gene expression. *Science* 1994;263:802–805.

Chalmers JJ, Kim E, Telford JN, Wong EY, Tacon WC, Shuler ML, Wilson DB. Effects of temperature on *Escherichia coli* overproducing β-lactamase or human epidermal growth factor. *Appl Environ Microbiol* 1990;56:104–111.

225

Chamberlin MJ. New models for the mechanism of transcription elongation and its regulation. *Harvey Lect* 1994;88:1–21.

Chang CN, Kuang W-J, Chen EY. Nucleotide sequence of the alkaline phosphatase gene of *Escherichia coli. Gene* 1986;44:121–125.

Charbit A, Molla A, Saurin W, Hofnung M. Versatility of a vector for expressing foreign polypeptides at the surface of Gram-negative bacteria. *Gene* 1988;70:181–189.

Chau V, Tobias JW, Bachmair A, Marriott D, Ecker D, Gonda DK, Varshavsky A. A multiubiquitin chain is confined to a specific lysine in a targeted short-lived protein. *Science* 1989;243:1576–1583.

Cheah KC, Harrison S, King R, Crocker L, Wells JRE, Robins A. Secretion of eukaryotic growth hormones in *Escherichia coli* is influenced by the sequence of the mature proteins. *Gene* 1994;138:9–15.

Chen BPC, Hai TW. Expression vectors for affinity purification and radiolabeling of proteins using *Escherichia coli* as host. *Gene* 1994;139:73–75.

Chen C-YA, Beatty JT, Cohen SN, Belasco JG. An intercistronic stem-loop structure functions as an mRNA decay terminator necessary but insufficient for puf mRNA stability. *Cell* 1988;52:609–619.

Chen C-YA, Belasco JG. Degradation of pufLMX mRNA in *Rhodobacter capsulatus* is initiated by nonrandom endonucleolytic cleavage. *J Bacteriol* 1990;172:4578–4586.

Chen G-FT, Inouye M. Suppression of the negative effect of minor arginine codons on gene expression: Preferential usage of minor codons within the first 25 codons of the *Escherichia coli* genes. *Nucleic Acids Res* 1990;18:1465–1473.

Chen G-FT, Inouye M. Role of the AGA/AGG codons, the rarest codons in global gene expression in *Escherichia coli. Genes Dev* 1994;8:2641–2652.

Chen HY, Bjerknes M, Kumar R, Jay E. Determination of the optimal aligned spacing between the Shine-Dalgarno sequence and the translation initiation codon of *Escherichia coli* mRNAs. *Nucleic Acids Res* 1994a;22:4953–4957.

Chen HY, Pomeroy-Cloney L, Bjerknes M, Tam J, Jay E. The influence of adenine-rich motifs in the 3 portion of the ribosome binding site on human IFN-gene expression in *Escherichia coli. J Mol Biol* 1994b;240:20–27.

Chen K-S, Peters TC, Walker JR. A minor arginine tRNA mutant limits translation preferentially of a protein dependent on the cognate codon. *J Bacteriol* 1990;172:2504–2510.

Chen L-H, Emory SA, Bricker AL, Bouvet P, Belasco JG. Structure and function of a bacterial mRNA stabilizer: Analysis of the 5 untranslated region of ompA mRNA. *J Bacteriol* 1991;173:4578–4586.

Chen W, Kallio PT, Bailey JE. Construction and characterization of a novel cross-regulation system for regulating cloned gene expression in *Escherichia coli. Gene* 1993;130:15–22.

Chen W, Kallio PT, Bailey JE. Process characterization of a novel cross-regulation system for cloned protein production in *Escherichia coli. Biotechnol Prog* 1995;11:397–402.

Cheng Y-SE, Kwoh DY, Kwoh TJ, Soltvedt BC, Zipser D. Stabilization of a degradable protein by its overexpression in *Escherichia coli. Gene* 1981;14:121–130.

Chesshyre JA, Hipkiss AR. Low temperatures stabilize interferon-2 against proteolysis in *Methylophilus methylotrophus* and *Escherichia coli. Appl Microbiol Biotechnol* 1989;31:158–162.

Chevalier BS, Stoddard BL. Homing endonucleases: Structural and functional insight into the catalysts of intron/intein mobility. *Nucleic Acids Res* 2001;29:3757–3774.

Chong WP, Goh LT, Reddy SG, Yusufi FN, Lee DY, Wong NS, Heng CK, Yap MG, Ho YS. Metabolomics profiling of extracellular metabolites in recombinant Chinese Hamster Ovary fed-batch culture. *Rapid Commun Mass Spectrom* 2009;23:3763–3771.

Chopra AK, Brasier AR, Das M, Xu X-J, Peterson JW. Improved synthesis of *Salmonella typhimurium* enterotoxin using gene fusion expression systems. *Gene* 1994;144:81–85.

Chou C-H, Aristidou AA, Meng S-Y, Bennett GN, San K-Y. Characterization of a pH-inducible promoter system for high-level expression of recombinant proteins in *Escherichia coli. Biotechnol Bioeng* 1995;47:186–192.

Chou C-H, Bennett GN, San K-Y. Effect of modified glucose uptake using genetic engineering techniques on high-level recombinant protein production in *Escherichia coli* dense cultures. *Biotechnol Bioeng* 1994;44:952–960.

Chu L, Robinson DK. Industrial choices for protein production by large-scale cell culture. *Curr Opin Biotechnol* 2001;12:180–187.

Chusainow J, Yang YS, Yeo JH, Toh PC, Asvadi P, Wong NS, Yap MG. A study of monoclonal antibody-producing CHO cell lines: What makes a stable high producer? *Biotechnol Bioeng* 2009;102:1182–1196.

Clarke AR. Molecular chaperones in protein folding and translocation. *Curr Opin Struct Biol* 1996;6:43–50.

Cloney LP, Bekkaoui DR, Hemmingsen SM. Coexpression of plastid chaperonin genes and a synthetic plant Rubisco operon in *Escherichia coli*. *Plant Mol Biol* 1993;23:1285–1290.

Cole PA. Chaperone-assisted protein expression. *Structure* 1996;4:239–242.

Coleman J, Inouye M, Nakamura K. Mutations upstream of the ribosome-binding site affect translational efficiency. *J Mol Biol* 1985;181:139–143.

Collier DN, Strobel SM, Bassford Jr. PJ. SecB-independent export of *Escherichia coli* ribose-binding protein (RBP): Some comparisons with export of maltose-binding protein (MBP) and studies with RBP-MBP hybrid proteins. *J Bacteriol* 1990;172:6875–6884.

Collins-Racie LA, McColgan JM, Grant KL, DiBlasio-Smith EA, Coy J, LaVallie ER. Production of recombinant bovine enterokinase catalytic subunit in *Escherichia coli* using the novel secretory fusion partner DsbA. *Bio/Technology* 1995;13:982–987.

Condon C, Squires C, Squires CL. Control of rRNA transcription in *Escherichia coli*. *Microbiol Rev* 1995;59:623–645.

Cornelis P, Sierra JC, Lim Jr. A et al. Development of new cloning vectors for the production of immunogenic outer membrane fusion proteins in *Escherichia coli*. *Bio/Technology* 1996;14:203–208.

Cost GJ, Freyvert Y, Vafiadis A, Santiago Y, Miller JC, Rebar E, Collingwood TN, Snowden A, Gregory PD. BAK and BAX deletion using zinc-finger nucleases yields apoptosis-resistant CHO cells. *Biotechnol Bioeng* 2010;105:330–340.

Craigen WJ, Lee CC, Caskey CT. Recent advances in peptide chain termination. *Mol Microbiol* 1990;4:861–865.

Crameri A, Whitehorn EA, Tate E, Stemmer WPC. Improved green fluorescent protein by molecular evolution using DNA shuffling. *Nat Biotechnol* 1996;14:315–319.

Cronan Jr. JE. Biotination of proteins in vivo. A post-translational modification to label, purify, and study proteins. *J Biol Chem* 1990;265:10327–10333.

Cull MG, Miller JF, Schatz PJ. Screening for receptor ligands using large libraries of peptides linked to the C terminus of the lac repressor. *Proc Natl Acad Sci USA* 1992;89: 1865–1869.

Cumming DA. Improper glycosylation and the cellular editing of nascent proteins. In: *Stability of Protein Pharmaceuticals. B. In Vivo Pathways of Degradation and Strategies for Protein Stabilization*, Ahern TJ, Manning MC, eds. Plenum Press, New York, 1992, pp. 1–42.

d'Aubenton Carafa Y, Brody E, Thermes C. Prediction of rho-independent *Escherichia coli* transcription terminators. A statistical analysis of their RNA stem-loop structures. *J Mol Biol* 1990;216:835–858.

Dalbøge H, Dahl H-HM, Pedersen J, Hansen JW, Christensen T. A novel enzymatic method for production of authentic hGH from an *Escherichia coli* produced hGH-precursor. *Bio/Technology* 1987;5:161–164.

Dale GE, Broger C, Langen H, D'Arcy A, Stuber D. Improving protein solubility through rationally designed amino acid replacements: Solubilization of the trimethoprim-resistant type S1 dihydrofolate reductase. *Protein Eng* 1994a;7:933–939.

Dale GE, Schönfeld H-J, Langen H, Stieger M. Increased solubility of trimethoprim-resistant type S1 DHFR from *Staphylococcus aureus* in *Escherichia coli* cells overproducing the chaperonins GroEL and GroES. *Protein Eng* 1994b;7:925–931.

Das A. Overproduction of proteins in *Escherichia coli*: Vectors, hosts, and strategies. *Methods Enzymol* 1990;182:93–112.

Datar RV, Cartwright T, Rosen C-G. Process economics of animal cell and bacterial fermentations: A case study analysis of tissue plasminogen activator. *Bio/Technology* 1993; 11:349–357.

Datta P, Linhardt RJ, Sharfstein ST. An 'omics' approach towards CHO cell engineering. *Biotechnol Bioeng* 2013;110:1255–1271.

Davies SL, Lovelady CS, Grainger RK, Racher AJ, Young RJ, James DC. Functional heterogeneity and heritability in CHO cell populations. *Biotechnol Bioeng* 2013;110:260–274.

de Boer HA, Comstock LJ, Vasser M. The tac promoter: A functional hybrid derived from the trp and lac promoters. *Proc Natl Acad Sci USA* 1983;80:21–25.

de Boer HA, Kastelein RA. Biased codon usage: An exploration of its role in optimization of translation. In: *Maximizing Gene Expression*, Reznikoff WS, Gold L, eds. Butterworths, Boston, MA, 1986, pp. 225–285 .

de la Torre JC, Ortin J, Domingo E, Delamarter J, Allet B, Davies J, Bertrand KP, Wray Jr. LV, Reznikoff WS. Plasmid vectors based on Tn10 DNA: Gene expression regulated by tetracycline. *Plasmid* 1984;12:103–110.

De Leon Gatti M, Wlaschin KF, Nissom PM, Yap M, Hu WS. Comparative transcriptional analysis of mouse hybridoma and recombinant Chinese hamster ovary cells undergoing butyrate treatment. *J Biosci Bioeng* 2007;103:82–91.

De Oliveira Dal'Molin CG, Quek LE, Palfreyman RW, Brumbley SM, Nielsen LK. AraGEM, a genome-scale reconstruction of the primary metabolic network in *Arabidopsis*. *Plant Physiol* 2010;152:579–589.

De Poorter JJ, Lipinski KS, Nelissen RG, Huizinga TW, Hoeben RC. Optimization of short-term transgene expression by sodium butyrate and ubiquitous chromatin opening elements (UCOEs). *J Gene Med* 2007;9:639–648.

de Smit MH, van Duin J. Secondary structure of the ribosome binding site determines translational efficiency: A quantitative analysis. *Proc Natl Acad Sci USA* 1990;87:7668–7672.

de Smit MH, van Duin J. Control of translation by mRNA secondary structure in *Escherichia coli*. A quantitative analysis of literature data. *J Mol Biol* 1994a;244:144–150.

de Smit MH, van Duin J. Translational initiation on structured messengers. Another role for the Shine-Dalgarno interaction. *J Mol Biol* 1994b;235:173–184.

De Sutter K, Hostens K, Vandekerckhove J, Fiers W. Production of enzymatically active rat protein disulfide isomerase in *Escherichia coli*. *Gene* 1994;141:163–170.

Dejong G, Telenius AH, Telenius H, Perez CF, Drayer JI, Hadlaczky G. Mammalian artificial chromosome pilot production facility: Large-scale isolation of functional satellite DNA-based artificial chromosomes. *Cytometry* 1999;35:129–133.

Del Tito Jr. BJ, Ward JM, Hodgson J, Gershater CJL, Edwards H, Wysocki LA, Watson FA, Sathe G, Kane JF. Effects of a minor isoleucyl tRNA on heterologous protein translation in *Escherichia coli*. *J Bacteriol* 1995;177:7086–7091.

Denèfle P, Kovarik S, Ciora T, Gosselet N, Bénichou J-C, Latta M, Guinet F, Ryter A, Mayaux J-F. Heterologous protein export in *Escherichia coli*: Influence of bacterial signal peptides on the export of human interleukin 1. *Gene* 1989;85:499–510.

Derman AI, Prinz WA, Belin D, Beckwith J. Mutations that allow disulfide bond formation in the cytoplasm of *Escherichia coli*. *Science* 1993;262:1744–1747.

Derom C, Gheysen D, Fiers W. High-level synthesis in *Escherichia coli* of the SV40 small-t antigen under the control of the bacteriophage lambda pL promoter. *Gene* 1982;17:45–54.

Derouazi M, Martinet D, Besuchet Schmutz N, Flaction R, Wicht M, Bertschinger M, Hacker DL, Beckmann JS, Wurm FM. Genetic characterization of CHO production host DG44 and derivative recombinant cell lines. *Biochem Biophys Res Commun* 2006;340:1069–1077.

Derynck R, Remaut E, Saman E, Stanssens P, De Clercq E, Content J, Fiers W. Expression of human fibroblast interferon gene in *Escherichia coli*. *Nature (London)* 1980;287:193–197.

Deuschle U, Kammerer W, Gentz R, Bujard H. Promoters of *Escherichia coli*: A hierarchy of in vivo strength indicates alternate structures. *EMBO J* 1986;5:2987–2994.

Devlin PE, Drummond RJ, Toy P, Mark DF, Watt KWK, Devlin JJ. Alteration of amino-terminal codons of human granulo-cyte-colony-stimulating factor increases expression levels and allows efficient processing by methionine aminopeptidase in *Escherichia coli*. *Gene* 1988;65:13–22.

Dharshanan S, Chong H, Hung CS, Zamrod Z, Kamal N. Rapid automated selection of mammalian cell line secreting high level of humanized monoclonal antibody using Clone Pix FL system and the correlation between exterior median intensity and antibody productivity. *Electron J Biotechnol* 2011;14(2).

di Guan C, Li P, Riggs PD, Inouye H. Vectors that facilitate the expression and purification of foreign peptides in *Escherichia coli* by fusion to maltose-binding protein. *Gene* 1988;67:21–30.

Dietmair S, Hodson MP, Quek LE, Timmins NE, Chrysanthopoulos P, Jacob SS, Gray P, Nielsen LK. Metabolite profiling of CHO cells with different growth characteristics. *Biotechnol Bioeng* 2012;109:1404–1414.

Doherty AJ, Connolly BA, Worrall AF. Overproduction of the toxic protein, bovine pancreatic DNaseI, in *Escherichia coli* using a tightly controlled T7-promoter-based vector. *Gene* 1993;136:337–340.

Donovan WP, Kushner SR. Amplification of ribonuclease II (rnb) activity in *Escherichia coli* K-12. *Nucleic Acids Res* 1983;11:265–275.

Donovan WP, Kushner SR. Polynucleotide phosphorylase and ribonuclease II are required for cell viability and mRNA turnover in *Escherichia coli* K-12. *Proc Natl Acad Sci USA* 1986;83:120–124.

Dreesen IA, Fussenegger M. Ectopic expression of human mTOR increases viability, robustness, cell size, proliferation, and antibody production of chinese hamster ovary cells. *Biotechnol Bioeng* 2011;108:853–866.

Dreyfus M. What constitutes the signal for the initiation of protein synthesis on *Escherichia coli* mRNAs? *J Mol Biol* 1988;204:79–94.

Dubendorff JW, Studier FW. Controlling basal expression in an inducible T7 expression system by blocking the target T7 promoter with lac repressor. *J Mol Biol* 1991;219:45–59.

Duffaud GD, March PE, Inouye M. Expression and secretion of foreign proteins in *Escherichia coli*. *Methods Enzymol* 1987;153:492–507.

Duvoisin RM, Belin D, Krisch HM. A plasmid expression vector that permits stabilization of both mRNAs and proteins encoded by the cloned genes. *Gene* 1986;45:193–201.

Dykes CW, Bookless AB, Coomber BA, Noble SA, Humber DC, Hobden AN. Expression of atrial natriuretic factor as a cleavable fusion protein with chloramphenicol acetyltransferase in *Escherichia coli*. *Eur J Biochem* 1988;174:411–416.

Easton AM, Gierse JK, Seetharam R, Klein BK, Kotts CE. Production of bovine insulin-like growth factor 2 (bIGF2) in *Escherichia coli*. *Gene* 1991;101:291–295.

Edalji R, Pilot-Matias TJ, Pratt SD, Egan DA, Severin JM, Gubbins EG, Petros AM, Fesik SW, Burres NS, Holzman TF. High-level expression of recombinant human FK-binding protein from a fusion precursor. *J Protein Chem* 1992;11:213–223.

Ehretsmann CP, Carpousis AJ, Krisch HM. mRNA degradation in procaryotes. *FASEB J* 1992;6:3186–3192.

Eliasson M, Olsson A, Palmcrantz E, Wiberg K, Inganas M, Guss B, Lindberg M, Uhlén M. Chimeric IgG-binding receptors engineered from staphylococcal protein A and streptococcal protein G. *J Biol Chem* 1988;263:4323–4327.

Ellis RJ, Hartl FU. Protein folding in the cell: Competing models of chaperonin function. *FASEB J* 1996;10:20–26.

Elvin CM, Thompson PR, Argall ME, Hendry P, Stamford NPJ, Lilley E, Dixon NE. Modified bacteriophage lambda promoter vectors for overproduction of proteins in *Escherichia coli*. *Gene* 1990;87:123–126.

Emory SA, Belasco JG. The ompA 5 untranslated RNA segment functions in *Escherichia coli* as a growth-rate-regulated mRNA stabilizer whose activity is unrelated to translational efficiency. *J Bacteriol* 1990;172:4472–4481.

Emory SA, Bouvet P, Belasco JG. A 5-terminal stem-loop structure can stabilize mRNA in *Escherichia coli*. *Genes Dev* 1992;6:135–148.

Enfors S-O. Control of in vivo proteolysis in the production of recombinant proteins. *Trends Biotechnol* 1992;10:310–315.

Ernst JF, Kawashima E. Variations in codon usage are not correlated with heterologous gene expression in *Saccharomyces cerevisiae* and *Escherichia coli*. *J Biotechnol* 1988;7:1–9.

Eyre-Walker A, Bulmer M. Reduced synonymous substitution rate at the start of enterobacterial genes. *Nucleic Acids Res* 1993;21:4599–4603.

Fahey RC, Hunt JS, Windham GC. On the cysteine and cystine content of proteins. Differences between intracellular and extracellular proteins. *J Mol Evol* 1977;10:155–160.

Falkenberg C, Björck L, Åkerström B. Localization of the binding site for streptococcal protein G on human serum albumin. Identification of a 5.5-kilodalton protein G binding albumin fragment. *Biochemistry* 1992;31:1451–1457.

Fan L, Kadura I, Krebs LE, Hatfield CC, Shaw MM, Frye CC. Improving the efficiency of CHO cell line generation using glutamine synthetase gene knockout cells. *Biotechnol Bioeng* 2012;109:1007–1015.

Faxen M, Plumbridge J, Isaksson LA. Codon choice and potential complementarity between mRNA downstream of the initiation codon and bases 1471–1480 in 16S ribosomal RNA affects expression of glnS. *Nucleic Acids Res* 1991;19:5247–5251.

Figge J, Wright C, Collins CJ, Roberts TM, Livingston DM. Stringent regulation of stably integrated chloramphenicol acetyl transferase genes by *E. coli* lac repressor in monkey cells. *Cell* 1988;52:713–722.

Firpo MA, Connelly MB, Goss DJ, Dahlberg AE. Mutations at two invariant nucleotides in the 3-minor domain of *Escherichia coli* 16 S rRNA affecting translational initiation and initiation factor 3 function. *J Biol Chem* 1996;271:4693–4698.

Ford CF, Suominen I, Glatz CE. Fusion tails for the recovery and purification of recombinant proteins. *Protein Expression Purif* 1991;2:95–107.

Forsberg G. Site specific cleavage of recombinant fusion proteins expressed in *Escherichia coli* and characterization of the products. PhD dissertation. Royal Institute of Technology, Stockholm, Sweden, 1992.

Forsberg G, Baastrup B, Rondahl H, Holmgren E, Pohl G, Hartmanis M, Lake M. An evaluation of different enzymatic cleavage methods for recombinant fusion proteins, applied on des(1–3)insulin-like growth factor I. *J Protein Chem* 1992;11:201–211.

Freundlich M, Ramani N, Mathew E, Sirko A, Tsui P. The role of integration host factor in gene expression in *Escherichia coli*. *Mol Microbiol* 1992;6:2557–2563.

Friedman DI. Integration host factor: A protein for all reasons. *Cell* 1988;55:545–554.

Friefeld BR, Korn R, de Jong PJ, Sninsky JJ, Horwitz MS. The 140-kDa adenovirus DNA polymerase is recognized by antibodies to *Escherichia coli*-synthesized determinants predicted from an open reading frame on the adenovirus genome. *Proc Natl Acad Sci USA* 1985;82:2652–2656.

Frorath B, Abney CC, Berthold H, Scanarini M, Northemann W. Production of recombinant rat interleukin-6 in *Escherichia coli* using a novel highly efficient expression vector pGEX-3T. *BioTechniques* 1992;12:558–563.

Fuchs J. Isolation of an *Escherichia coli* mutant deficient in thioredoxin reductase. *J Bacteriol* 1977;129:967–972.

Fuchs P, Breitling F, Dubel S, Seehaus T, Little M. Targeting recombinant antibodies to the surface of *Escherichia coli*: Fusion to a peptidoglycan associated lipoprotein. *Bio/Technology* 1991;9:1369–1372.

Fuh G, Mulkerrin MG, Bass S, McFarland N, Brochier M, Bourell JH, Light DR, Wells JA. The human growth hormone receptor. Secretion from *Escherichia coli* and disulfide bonding pattern of the extracellular binding domain. *J Biol Chem* 1990;265:3111–3115.

Fujimoto K, Fukuda T, Marumoto R. Expression and secretion of human epidermal growth factor by *Escherichia coli* using enterotoxin signal sequences. *J Biotechnol* 1988;8:77–86.

Fussenegger M, Bailey JE, Hauser H, Mueller PP. Genetic optimization of recombinant glycoprotein production by mammalian cells. *Trends Biotechnol* 1999;17:35–42.

Gaal T, Barkei J, Dickson RR, de Boer HA, de Haseth PL, Alavi H, Gourse RL. Saturation mutagenesis of an *Escherichia coli* rRNA promoter and initial characterization of promoter variants. *J Bacteriol* 1989;171:4852–4861.

Gafny R, Cohen S, Nachaliel N, Glaser G. Isolated P2 rRNA promoters of *Escherichia coli* are strong promoters that are subject to stringent control. *J Mol Biol* 1994;243:152–156.

Galas DJ, Eggert M, Waterman MS. Rigorous pattern-recognition methods for DNA sequences. Analysis of promoter sequences from *Escherichia coli*. *J Mol Biol* 1985;186:117–128.

Garcia GM, Mar PK, Mullin DA, Walker JR, Prather NE. The *E. colidnaY* gene encodes an arginine transfer RNA. *Cell* 1986;45:453–459.

Gardella TJ, Rubin D, Abou-Samra A-B, Keutmann HT, Potts Jr. JT, Kronenberg HM, Nussbaum SR. Expression of human parathyroid hormone-(1–84) in *Escherichia coli* as a factor X-cleavable fusion protein. *J Biol Chem* 1990;265:15854–15859.

Gates CM, Stemmer WPC, Kaptein R, Schatz PJ. Affinity selective isolation of ligands from peptide libraries through display on a lac repressor "headpiece dimer." *J Mol Biol* 1996;255:373–386.

Georgiou G. Expression of proteins in bacteria. In: *Protein Engineering: Principles and Practice*, Cleland JL, Craik CS, eds. Wiley Liss, New York, 1996, pp. 101–127.

Georgiou G, Poetschke HL, Stathopoulos C, Francisco JA. Practical applications of engineering Gram-negative bacterial cell surfaces. *Trends Biotechnol* 1993;11:6–10.

Georgiou G, Stephens DL, Stathopoulos C, Poetschke HL, Mendenhall J, Earhart CF. Display of β-lactamase on the *Escherichia coli* surface: Outer membrane phenotypes conferred by Lpp-OmpA-β-lactamase fusions. *Protein Eng* 1996;9:239–247.

Georgiou G, Valax P. Expression of correctly folded proteins in *Escherichia coli*. *Curr Opin Biotechnol* 1996;7:190–197.

Germino J, Bastia D. Rapid purification of a cloned gene product by genetic fusion and site specific proteolysis. *Proc Natl Acad Sci USA* 1984;81:4692–4696.

Germino J, Gray JG, Charbonneau H, Vanaman T, Bastia D. Use of gene fusions and protein-protein interaction in the isolation of a biologically active regulatory protein: The replication initiator protein of plasmid R6K. *Proc Natl Acad Sci USA* 1983;80:6848–6852.

Gething M-J, Sambrook J. Protein folding in the cell. *Nature (London)* 1992;355:33–45.

Ghaderi D, Zhang M, Hurtado-Ziola N, Varki A. Production platforms for biotherapeutic glycoproteins. Occurrence, impact, and challenges of non-human sialylation. *Biotechnol Genet Eng Rev* 2012;28:147–175.

Gheysen D, Iserentant D, Derom C, Fiers W. Systematic alteration of the nucleotide sequence preceding the translation initiation codon and the effects on bacterial expression of the cloned SV40 small-t antigen gene. *Gene* 1982;17:55–63.

Ghrayeb J, Kimura H, Takahara M, Hsiung H, Masui Y, Inouye M. Secretion cloning vectors in *Escherichia coli*. *EMBO J* 1984;3:2437–2442.

Giacomini A, Ollero FJ, Squartini A, Nuti MP. Construction of multipurpose gene cartridges based on a novel synthetic promoter for high-level gene expression in Gram-negative bacteria. *Gene* 1994;144:17–24.

Giladi H, Goldenberg D, Koby S, Oppenheim AB. Enhanced activity of the bacteriophage PL promoter at low temperature. *Proc Natl Acad Sci USA* 1995;92:2184–2188.

Giladi H, Koby S, Gottesman ME, Oppenheim AB. Supercoiling, integration host factor, and a dual promoter system, participate in the control of the bacteriophage pL promoter. *J Mol Biol* 1992;224:937–948.

Gilbert HF. Protein chaperones and protein folding. *Curr Opin Biotechnol* 1994;5:534–539.

Giordano TJ, Deuschle U, Bujard H, McAllister WT. Regulation of coliphage T3 and T7 RNA polymerases by the lac repressor-operator system. *Gene* 1989;84:209–219.

Girod PA, Nguyen DQ, Calabrese D et al. Genome-wide prediction of matrix attachment regions that increase gene expression in mammalian cells. *Nat Methods* 2007;4:747–753.

Girod PA, Zahn-Zabal M, Mermod N. Use of the chicken lysozyme 5′ matrix attachment region to generate high producer CHO cell lines. *Biotechnol Bioeng* 2005;91:1–11.

Goeddel DV. Systems for heterologous gene expression. *Methods Enzymol* 1990;185:3–7.

Goeddel DV, Kleid DG, Bolivar F, Heyneker HL, Yansura DG, Crea R, Hirose T, Kraszewski A, Itakura K, Riggs AD. Expression in *Escherichia coli* of chemically synthesized genes for human insulin. *Proc Natl Acad Sci USA* 1979;76:106–110.

Goff SA, Casson LP, Goldberg AL. Heat shock regulatory gene htpR influences rates of protein degradation and expression of the lon gene in *Escherichia coli*. *Proc Natl Acad Sci USA* 1984;81:6647–6651.

Goff SA, Goldberg AL. Production of abnormal proteins in *E. coli* stimulates transcription of lon and other heat shock genes. *Cell* 1985;41:587–595.

Goff SA, Goldberg AL. An increased content of protease La, the lon gene product, increases protein degradation and blocks growth in *E. coli*. *J Biol Chem* 1987;262:4508–4515.

Gold L. Posttranscriptional regulatory mechanisms in *Escherichia coli*. *Annu Rev Biochem* 1988;57:199–233.

Gold L. Expression of heterologous proteins in *Escherichia coli*. *Methods Enzymol* 1990;185: 11–14.

Gold L, Stormo GD. High-level translation initiation. *Methods Enzymol* 1990;185:89–93.

Goldberg AL. The mechanism and functions of ATP-dependent proteases in bacterial and animal cells. *Eur J Biochem* 1992;203:9–23.

Goldberg AL, Dice JF. Intracellular protein degradation in mammalian and bacterial cells. *Annu Rev Biochem* 1974;43:835–869.

Goldberg AL, Goff SA. The selective degradation of abnormal proteins in bacteria. In: *Maximizing Gene Expression*, Reznikoff W, Gold L, eds. Butterworths, Boston, MA, 1986, pp. 287–314.

Goldberg AL, Goff SA, Casson LP. Hosts and methods for producing recombinant products in high yields. U.S. patent 4,758,512, July 1988.

Goldberg AL, St. John AC. Intracellular protein degradation in mammalian and bacterial cells. Part 2. *Annu Rev Biochem* 1976;45:747–803.

Goldman E, Rosenberg AH, Zubay G, Studier FW. Consecutive low-usage leucine codons block translation only when near the 5 end of a message in *Escherichia coli*. *J Mol Biol* 1995;245:467–473.

Goldstein J, Lehnhardt S, Inouye M. Enhancement of protein translocation across the membrane by specific mutations in the hydrophobic region of the signal peptide. *J Bacteriol* 1990a;172:1225–1231.

Goldstein J, Pollitt NS, Inouye M. Major cold shock protein of *Escherichia coli*. *Proc Natl Acad Sci USA* 1990b;87:283–287.

Goldstein MA, Doi RH. Prokaryotic promoters in biotechnology. *Biotechnol Annu Rev* 1995;1:105–128.

Golic MM, Rong YS, Petersen RB, Lindquist SL, Golic KG. FLP-mediated DNA mobilization to specific target sites in *Drosophila* chromosomes. *Nucleic Acids Res* 1997;25:3665–3671.

Goloubinoff P, Gatenby AA, Lorimer GH. GroE heat-shock proteins promote assembly of foreign prokaryotic ribulose bisphosphate carboxylase oligomers in *Escherichia coli*. *Nature (London)* 1989;337:44–47.

Gonda DK, Bachmair A, Wünning I, Tobias JW, Lane WS, Varshavsky A. Universality and structure of the N-end rule. *J Biol Chem* 1989;264:16700–16712.

Gorski K, Roch J-M, Prentki P, Krisch HM. The stability of bacteriophage T4 gene 32 mRNA: A 5 leader sequence that can stabilize mRNA transcripts. *Cell* 1985;43:461–469.

Gottesman S. Minimizing proteolysis in *Escherichia coli*: Genetic solutions. *Methods Enzymol* 1990;185:119–129.

Gottesman S, Maurizi MR. Regulation by proteolysis: Energy-dependent proteases and their targets. *Microbiol Rev* 1992;56:592–621.

Gourse RL, de Boer HA, Nomura M. DNA determinations of rRNA synthesis in *E. coli*: Growth rate dependent regulation, feedback inhibition, upstream activation, antitermination. *Cell* 1986;44:197–205.

Gouy M, Gautier C. Codon usage in bacteria: Correlation with gene expressivity. *Nucleic Acids Res* 1982;10:7055–7074.

Gram H, Ramage P, Memmert K, Gamse R, Kocher HP. A novel approach for high level production of a recombinant human parathyroid hormone fragment in *Escherichia coli*. *Bio/Technology* 1994;12:1017–1023.

Grauschopf U, Winther JR, Korber P, Zander T, Dallinger P, Bardwell AJC. Why is DsbA such an oxidizing disulfide catalyst? *Cell* 1995;83:947–955.

Gray F, Kenney JS, Dunne JF. Secretion capture and report web: Use of affinity derivatized agarose microdroplets for the selection of hybridoma cells. *J Immunol Methods* 1995;182:155–163.

Gray GL, Baldridge JS, McKeown KS, Heyneker HL, Chang CN. Periplasmic production of correctly processed human growth hormone in *Escherichia coli*: Natural and bacterial signal sequences are interchangeable. *Gene* 1985;39:247–254.

Gren EJ. Recognition of messenger RNA during translational initiation in *Escherichia coli*. *Biochimie* 1984;66:1–29.

Grentzmann G, Brechemier-Baey D, Heurgué V, Mora L, Buckingham RH. Localization and characterization of the gene encoding release factor RF3 in *Escherichia coli*. *Proc Natl Acad Sci USA* 1994;91:5848–5852.

Grisshammer R, Duckworth R, Henderson R. Expression of a rat neurotensin receptor in *Escherichia coli*. *Biochem J* 1993;295:571–576.

Gronenborn B. Overproduction of phage lambda repressor under control of the lac promoter of *Escherichia coli*. *Mol Gen Genet* 1976;148:243–250.

Gros F, Hiatt H, Gilbert W, Kurland CG, Risebrough RW, Watson JD. Unstable ribonucleic acid revealed by pulse labelling of *Escherichia coli*. *Nature (London)* 1961;190:581–585.

Gross G, Hauser H. Heterologous expression as a tool for gene identification and analysis. *J Biotechnol* 1995;41:91–110.

Gross G, Mielke C, Hollatz I, Blöcker H, Frank R. RNA primary sequence or secondary structure in the translational initiation region controls expression of two variant interferon-genes in *Escherichia coli*. *J Biol Chem* 1990;265:17627–17636.

Groth AC, Fish M, Nusse R, Calos MP. Construction of transgenic *Drosophila* by using the site-specific integrase from phage phiC31. *Genetics* 2004;166:1775–1782.

Gualerzi CO, Pon CL. Initiation of mRNA translation in prokaryotes. *Biochemistry* 1990;29:5881–5889.

Guan K, Dixon JE. Eukaryotic proteins expressed in *Escherichia coli*: An improved thrombin cleavage and purification procedure of fusion proteins with glutathione S-transferase. *Anal Biochem* 1991;192:262–267.

Guan X, Wurtele ES. Reduction of growth and acetyl-CoA carboxylase activity by expression of a chimeric streptavidin gene in *Escherichia coli*. *Appl Microbiol Biotechnol* 1996;44:753–758.

Guarneros G, Montanez C, Hernandez T, Court D. Posttranscriptional control of bacteriophage int gene expression from a site distal to the gene. *Proc Natl Acad Sci USA* 1982;79:238–242.

Guilhot C, Jander G, Martin NL, Beckwith J. Evidence that the pathway of disulfide bond formation in *Escherichia coli* involves interactions between the cysteines of DsbB and DsbA. *Proc Natl Acad Sci USA* 1995;92:9895–9899.

Gurtu V, Yan G, Zhang G. IRES bicistronic expression vectors for efficient creation of stable mammalian cell lines. *Biochem Biophys Res Commun* 1996;229:295–298.

Gutman GA, Hatfield GW. Nonrandom utilization of codon pairs in *Escherichia coli*. *Proc Natl Acad Sci USA* 1989;86:3699–3703.

Hackl M, Jadhav V, Jakobi T, Rupp O, Brinkrolf K, Goesmann A, Puhler A, Noll T, Borth N, Grillari J. Computational identification of microRNA gene loci and precursor microRNA sequences in CHO cell lines. *J Biotechnol* 2012;158:151–155.

Hackl M, Jakobi T, Blom J et al. Next-generation sequencing of the Chinese hamster ovary microRNA transcriptome: Identification, annotation and profiling of microRNAs as targets for cellular engineering. *J Biotechnol* 2011;153:62–75.

Hall MN, Gabay J, Débarbouillé M, Schwartz M. A role for mRNA secondary structure in the control of translation initiation. *Nature (London)* 1982;295:616–618.

Hammarberg B, Nygren P-Å, Holmgren E, Elmblad A, Tally M, Hellman U, Moks T, Uhlén M. Dual affinity fusion approach and its use to express recombinant human insulin-like growth factor II. *Proc Natl Acad Sci USA* 1989;86:4367–4371.

Hammill L, Welles J, Carson GR. The gel microdrop secretion assay: Identification of a low productivity subpopulation arising during the production of human antibody in CHO cells. *Cytotechnology* 2000;34:27–37.

Hammond S, Swanberg JC, Polson SW, Lee KH. Profiling conserved microRNA expression in recombinant CHO cell lines using illumina sequencing. *Biotechnol Bioeng* 2012;109:1371–1375.

Hanania EG, Fieck A, Stevens J, Bodzin LJ, Palsson BO, Koller MR. Automated in situ measurement of cell-specific antibody secretion and laser-mediated purification for rapid cloning of highly-secreting producers. *Biotechnol Bioeng* 2005;91:872–876.

Hansson M, Ståhl S, Hjorth R, Uhlén M, Moks T. Single-step recovery of a secreted recombinant protein by expanded bed adsorption. *Bio/Technology* 1994;12:285–288.

Harley CB, Reynolds RP. Analysis of *E. coli* promoter sequences. *Nucleic Acids Res* 1987; 15:2343–2361.

Hartz D, McPheeters DS, Gold L. Influence of mRNA determinants on translation initiation in *Escherichia coli*. *J Mol Biol* 1991;218:83–97.

Hasan N, Szybalski W. Control of cloned gene expression by promoter inversion in vivo: Construction of improved vectors with a multiple cloning site and the ptac promoter. *Gene* 1987;56:145–151.

Hasan N, Szybalski W. Construction of lacIts and lacIqts expression plasmids and evaluation of the thermosensitive lac repressor. *Gene* 1995;163:35–40.

Hawley DK, McClure WR. Compilation and analysis of *Escherichia coli* promoter DNA sequences. *Nucleic Acids Res* 1983;11:2237–2255.

Hay RJ. Animal cells in culture. In: *Maintaining Cultures for Biotechnology and Industry*. Hunter-Cevera JC, Belt A, eds. Academic Press, 1996, pp. 161–178.

Hayashi MN, Hayashi M. Cloned DNA sequences that determine mRNA stability of bacteriophage X174 in vivo are functional. *Nucleic Acids Res* 1985;13:5937–5948.

Hayes SA, Dice JF. Roles of molecular chaperones in protein degradation. *J Cell Biol* 1996;132:255–258.

He BA, McAllister WT, Durbin RK. Phage RNA polymerase vectors that allow efficient gene expression in both prokaryotic and eukaryotic cells. *Gene* 1995;164:75–79.

Hedgpeth J, Ballivet M, Eisen H. Lambda phage promoter used to enhance expression of a plasmid-cloned gene. *Mol Gen Genet* 1978;163:197–203.

Hein R, Tsien RY. Engineering green fluorescent protein for improved brightness, longer wavelengths and fluorescence resonance energy transfer. *Curr Biol* 1996;6:178–182.

Helke A, Geisen RM, Vollmer M, Sprengart ML, Fuchs E. An unstructured messenger RNA region and a 5 hairpin represent important elements of the *E. coli* translation initiation signal determined by using the bacteriophage T7 gene 1 translation start site. *Nucleic Acids Res* 1993;21:5705–5711.

Hellebust H, Murby M, Abrahmsén L, Uhlén M, Enfors S-O. Different approaches to stabilize a recombinant fusion protein. *Bio/Technology* 1989;7:165–168.

Hellman J, Mäntsä P. Construction of an *Escherichia coli* export-affinity vector for expression and purification of foreign proteins by fusion to cyclomaltodextrin glucanotransferase. *J Biotechnol* 1992;23:19–34.

Hénaut A, Danchin A. Analysis and predictions from *Escherichia coli* sequences, or *E. coli* in silico. In: *Escherichia coli and Salmonella: Cellular and Molecular Biology*, Vol. 2, Neidhardtrtiss III FC, Ingraham JL, Lin ECC, Low KB, Magasanik B, Wznikoff, Riley M, Schaechter M, Umbarger HE, eds. ASM Press, Washington, DC, 1996, pp. 2047–2066.

Hendrick JP, Hartl F-U. The role of molecular chaperones in protein folding. *FASEB J* 1995;9:1559–1569.

Herbst B, Kneip S, Bremer E. pOSEX: Vectors for osmotically controlled and finely tuned gene expression in *Escherichia coli*. *Gene* 1994;151:137–142.

Hernan RA, Hui HL, Andracki ME, Noble RW, Sligar SG, Walder JA, Walder RY. Human hemoglobin expression in *Escherichia coli*: Importance of optimal codon usage. *Biochemistry* 1992;31:8619–8628.

Higgins CF, Causton HC, Dance GSC, Mudd EA. The role of the 3 end in mRNA stability and decay. In: *Control of Messenger RNA Stability*, Belasco JG, Brawerman G, eds. Academic Press, Inc., San Diego, CA, 1993, pp. 13–30.

Hirel P-H, Schmitter J-M, Dessen P, Fayat G, Blanquet S. Extent of N-terminal methionine excision from *Escherichia coli* proteins is governed by the side-chain length of the penultimate amino acid. *Proc Natl Acad Sci USA* 1989;86:8247–8251.

Ho SC, Bardor M, Feng H, Mariati, Tong YW, Song Z, Yap MG, Yang Y. IRES-mediated Tricistronic vectors for enhancing generation of high monoclonal antibody expressing CHO cell lines. *J Biotechnol* 2012;157:130–139.

Hochuli E, Döbeli H, Schacher A. New metal chelate adsorbent selective for proteins and peptides containing neighbouring histidine residues. *J Chromatogr* 1987;411:177–184.

Hochuli E, Bannwarth W, Dobeli H, Gentz R, Stuber D. Genetic approach to facilitate purification of recombinant proteins with a novel metal chelate adsorbent. *Nat Biotech* 1988;6:1321–1325.

Hockney RC. Recent developments in heterologous protein production in *Escherichia coli*. *Trends Biotechnol* 1994;12:456–463.

Hodgson J. Expression systems: A user's guide. *Bio/Technology* 1993;11:887–893.

Hoffman CS, Wright A. Fusions of secreted proteins to alkaline phosphatase: An approach for studying protein secretion. *Proc Natl Acad Sci USA* 1985;82:5107–5111.

Høgset A, Blingsmo OR, Saether O, Gautvik VT, Holmgren E, Josephson S, Gabrielsen OS, Gordeladze JO, Alestrøm P, Gautvik KM. Expression and characterization of a recombinant human parathyroid hormone secreted by *Escherichia coli* employing the staphylococcal protein A promoter and signal sequence. *J Biol Chem* 1990;265:7338–7344.

Holland IB, Kenny B, Steipe B, A. Plückthun. Secretion of heterologous proteins in *Escherichia coli*. *Methods Enzymol* 1990;182:132–143.

Holmes P. Al-Rubeai M. Improved cell line development by a high throughput affinity capture surface display technique to select for high secretors. *J Immunol Methods* 1999;230: 141–147.

Holmgren A. Thioredoxin and glutaredoxin systems. *J Biol Chem* 1989;264:13963–13966.

Hopp TP, Prickett KS, Price VL, Libby RT, March CJ, Cerretti DP, Urdal DL, Conlon PJ. A short polypeptide marker sequence useful for recombinant protein identification and purification. *Bio/Technology* 1988;6:1204–1210.

Horii T, Ogawa T, Ogawa H. Organization of the recA gene of *Escherichia coli*. *Proc Natl Acad Sci USA* 1980;77:313–317.

Hsiung HM, Cantrell A, Luirink J, Oudega B, Veros AJ, Becker GW. Use of bacteriocin release protein in *E. coli* for excretion of human growth hormone into the culture medium. *Bio/Technology* 1989;7:267–271.

Hsiung HM, MacKellar WC. Expression of bovine growth hormone derivatives in *Escherichia coli* and the use of the derivatives to produce natural sequence growth hormone by cathepsin C cleavage. *Methods Enzymol* 1987;153:390.

Hsiung HM, Mayne NG, Becker GW. High-level expression, efficient secretion and folding of human growth hormone in *Escherichia coli*. *Bio/Technology* 1986;4:991–995.

Hsu LM, Giannini JK, Leung T-WC, Crosthwaite JC. Upstream sequence activation of *Escherichia coli* argT promoter in vivo and in vitro. *Biochemistry* 1991;30:813–822.

Huang Y, Li Y, Wang YG, Gu X, Wang Y, Shen BF. An efficient and targeted gene integration system for high-level antibody expression. *J Immunol Methods* 2007;322:28–39.

Huh KR, Cho EH, Lee SO, Na DS. High level expression of human lipocortin (annexin) 1 in *Escherichia coli*. *Biotechnol Lett* 1996;18: 163–168.

Hui A, Hayflick J, Dinkelspiel K, de Boer HA. Mutagenesis of the three bases preceding the start codon of the β-galactosidase mRNA and its effect on translation in *Escherichia coli*. *EMBO J* 1984;3:623–629.

Hummel M, Herbst H, Stein H. Gene synthesis, expression in *Escherichia coli* and purification of immunoreactive human insulin-like growth factors I and II. Application of a modified HPLC separation technique for hydrophobic proteins. *Eur J Biochem* 1989;180: 555–561.

Humphreys DP, Weir N, Mountain A, Lund PA. Human protein disulfide isomerase functionally complements a dsbA mutation and enhances the yield of pectate lyase C in *Escherichia coli*. *J Biol Chem* 1995;270:28210–28215.

Hüttenhofer A, Noller HF. Footprinting mRNA-ribosome complexes with chemical probes. *EMBO J* 1994;13:3892–3901.

Hwang C, Sinskey AJ, Lodish HF. Oxidized redox state of glutathione in the endoplasmic reticulum. *Science* 1992;257:1496–1502.

Hwang SO, Lee GM. Effect of Akt overexpression on programmed cell death in antibody-producing Chinese hamster ovary cells. *J Biotechnol* 2009;139:89–94.

ICH. Quality of biotechnological products: Viral safety evaluation of biotechnology products derived from cell lines of human or animal origin. ICH Harmonised Tripartite Guideline. *Dev Biol Stand* 1998;93:177–201.

Ikehara M, Ohtsuka E, Tokunaga T et al. Inquiries into the structure-function relationship of ribonuclease T1 using chemically synthesized coding sequences. *Proc Natl Acad Sci USA* 1986;83:4695–4699.

Ikemura T. Codon usage and tRNA content in unicellular and multicellular organisms. *Mol Biol Evol* 1985;2:13–34.

Ingram LO, Conway T, Clark DP, Sewell GW, Preston JF. Genetic engineering of ethanol production in *Escherichia coli*. *Appl Environ Microbiol* 1987;53:2420–2425.

Inouye H, Michaelis S, Wright A, Beckwith J. Cloning and restriction mapping of the alkaline phosphatase structural gene (phoA)of *Escherichia coli* and generation of deletion mutants in vitro. *J Bacteriol* 1981;146:668–675.

234

Inouye S, Inouye M. Up-promoter mutations in the lpp gene of *Escherichia coli*. *Nucleic Acids Res* 1985;13:3101–3110.

Irwin B, Heck JD, Hatfield GW. Codon pair utilization biases influence translational elongation step times. *J Biol Chem* 1995;270:22801–22806.

Iserentant D, Fiers W. Secondary structure of mRNA and efficiency of translation initiation. *Gene* 1980;9:1–12.

Itakura K, Hirose T, Crea R, Riggs AD, Heyneker HL, Bolivar F, Boyer HW. Expression in *Escherichia coli* of a chemically synthesized gene for the hormone somatostatin. *Science* 1977;198:1056–1063.

Ito K, Kawakami K, Nakamura Y. Multiple control of *Escherichia coli* lysyl-tRNA synthetase expression involves a transcriptional repressor and a translational enhancer element. *Proc Natl Acad Sci USA* 1993;90:302–306.

Ivanov I, Alexandrova R, Dragulev B, Saraffova A, AbouHaidar MG. Effect of tandemly repeated AGG triplets on the translation of CAT-mRNA in *E. coli*. *FEBS Lett* 1992;307:173–176.

Iwakura M, Obara K, Kokubu T, Ohashi S, Izutsu H. Expression and purification of growth hormone-releasing factor with the aid of dihydrofolate reductase handle. *J Biochem* 1992;112:57–62.

Izard J, Parker MW, Chartier M, Duché D, Baty D. A single amino acid substitution can restore the solubility of aggregated colicin A mutants in *Escherichia coli*. *Protein Eng* 1994;7:1495–1500.

Jacob F, Monod J. Genetic regulatory mechanisms in the synthesis of proteins. *J Mol Biol* 1961;3:318–356.

Jacques N, Guillerez J, Dreyfus M. Culture conditions differentially affect the translation of individual *Escherichia coli* mRNAs. *J Mol Biol* 1992;226:597–608.

Jadhav V, Hackl M, Bort JA, Wieser M, Harreither E, Kunert R, Borth N, Grillari J. A screening method to assess biological effects of microRNA overexpression in Chinese hamster ovary cells. *Biotechnol Bioeng* 2012;109:1376–1385.

Jensen EB, Carlsen S. Production of recombinant human growth hormone in *Escherichia coli*: Expression of different precursors and physiological effects of glucose, acetate, salts. *Biotechnol Bioeng* 1990;36: 1–11.

Jia Q, Wu H, Zhou X et al. A "GC-rich" method for mammalian gene expression: A dominant role of non-coding DNA GC content in regulation of mammalian gene expression. *Sci China Life Sci* 2010;53:94–100.

Johnson ES, Gonda DK, Varshavsky A. *cis-trans* recognition and subunit-specific degradation of short-lived proteins. *Nature (London)* 1990;346:287–291.

Johnson DL, Middleton SA, McMahon F, Barbone FP, Kroon D, Tsao E, Lee WH, Mulcahy LS, Jolliffe LK. Refolding, purification, and characterization of human erythropoietin binding protein produced in *Escherichia coli*. *Protein Expression Purif* 1996;7:104–113.

Jones PG, Krah R, Tafuri SR, Wolffe AP. DNA gyrase, CS7.4, and the cold shock response in *Escherichia coli*. *J Bacteriol* 1992;174:5798–5802.

Josaitis CA, Gaal T, Gourse RL. Stringent control and growth-rate-dependent control have non-identical promoter sequence requirements. *Proc Natl Acad Sci USA* 1995;92:1117–1121.

Josaitis CA, Gaal T, Ross W, Gourse RL. Sequences upstream of the 35 hexamer of rrnB P1 affect promoter strength and upstream activation. *Biochim Biophys Acta* 1990;1050:307–311.

Jost JP, Oakeley EJ, Zhu B, Benjamin D, Thiry S, Siegmann M, Jost YC. 5-Methylcytosine DNA glycosylase participates in the genome-wide loss of DNA methylation occurring during mouse myoblast differentiation. *Nucleic Acids Res* 2001;29:4452–4461.

Kadokura H, Yoda K, Watanabe S, Kikuchi Y, Tamura G, Yamasaki M. Enhancement of protein secretion by optimizing protein synthesis: Isolation and characterization of *Escherichia coli* mutants with increased secretion ability of alkaline phosphatase. *Appl Microbiol Biotechnol* 1994;41:163–169.

Kadonaga JT, Gautier AE, Straus DR, Charles AD, Edge MD, Knowles JR. The role of the β-lactamase signal sequence in the secretion of proteins by *Escherichia coli*. *J Biol Chem* 1984;259:2149–2154.

Kaelin WG Jr, Krek W, Sellers WR et al. Expression cloning of a cDNA encoding a retinoblastoma-binding protein with E2F-like properties. *Cell* 1992;70(2):351–364.

Kameyama Y, Kawabe Y, Ito A, Kamihira M. An accumulative site-specific gene integration system using Cre recombinase-mediated cassette exchange. *Biotechnol Bioeng* 2010;105:1106–1114.

Kane JF. Effects of rare codon clusters on high-level expression of heterologous proteins in *Escherichia coli*. *Curr Opin Biotechnol* 1995;6:494–500.

Kane JF, Hartley DL. Formation of recombinant protein inclusion bodies in *Escherichia coli*. *Trends Biotechnol* 1988;6:95–101.

Kastelein RA, Berkhout B, van Duin J. Opening the closed ribosome-binding site of the lysis cistron of bacteriophage MS2. *Nature (London)* 1983;305:741–743.

Kato C, Kobayashi T, Kudo T et al. Construction of an excretion vector and extracellular production of human growth hormone from *Escherichia coli*. *Gene* 1987;54:197–202.

Kaufman RJ, Davies MV, Wasley LC, Michnick D. Improved vectors for stable expression of foreign genes in mammalian cells by use of the untranslated leader sequence from EMC virus. *Nucleic Acids Res* 1991;19:4485–4490.

Kaufman RJ, Sharp PA. Amplification and expression of sequences cotransfected with a modular dihydrofolate reductase complementary dna gene. *J Mol Biol* 1982;159:601–621.

Kaufman RJ, Wasley LC, Spiliotes AJ, Gossels SD, Latt SA, Larsen GR, Kay RM. Coamplification and coexpression of human tissue-type plasminogen activator and murine dihydrofolate reductase sequences in Chinese hamster ovary cells. *Mol Cell Biol* 1985;5:1750–1759.

Kaufmann A, Stierhof Y-D, Henning U. New outer membrane-associated protease of *Escherichia coli* K-12. *J Bacteriol* 1994;176:359–367.

Kavanaugh JS, Rogers PH, Arnone A. High-resolution X-ray study of deoxy recombinant human hemoglobins synthesized from beta-globins having mutated amino termini. *Biochemistry* 1992;31:8640–8647.

Keiler KC, Waller PRH, Sauer RT. Role of a peptide tagging system in degradation of proteins synthesized from damaged messenger RNA. *Science* 1996;271:990–993.

Kelman Z, Yao N, O'Donnell M. *Escherichia coli* expression vectors containing a protein kinase recognition motif, His6-tag and hemagglutinin epitope. *Gene* 1995;166:177–178.

Kendall RL, Yamada R, Bradshaw RA. Cotranslational amino-terminal processing. *Methods Enzymol* 1990;185:398–407.

Kenealy WR, Gray JE, Ivanoff LA, Tribe DE, Reed DL, Korant BD, Petteway Jr. SR. Solubility of proteins overexpressed in *Escherichia coli*. *Dev Ind Microbiol* 1987;28:45–52.

Kennard ML. Engineered mammalian chromosomes in cellular protein production: Future prospects. *Methods Mol Biol* 2011;738:217–238.

Kennard ML, Goosney DL, Monteith D, Zhang L, Moffat M, Fischer D, Mott J. The generation of stable, high MAb expressing CHO cell lines based on the artificial chromosome expression (ACE) technology. *Biotechnol Bioeng* 2009;104:540–553.

Kern I, Ceglowski P. Secretion of streptokinase fusion proteins from *Escherichia coli* cells through the hemolysin transporter. *Gene* 1995;163: 53–57.

Khosla C, Bailey JE. Characterization of the oxygen-depen-dent promoter of the Vitreoscilla hemoglobin gene in *Escherichia coli*. *J Bacteriol* 1989;171:5995–6004.

Khosla C, Curtis JE, Bydalek P, Swartz JR, Bailey JE. Expression of recombinant proteins in *Escherichia coli* using an oxygen-responsive promoter. *Bio/Technology* 1990;8:554–558.

Kikuchi Y, Yoda K, Yamasaki M, Tamura G. The nucleotide sequence of the promoter and the amino-terminal region of alkaline phosphatase structural gene (phoA)of *Escherichia coli*. *Nucleic Acids Res* 1981;9:5671–5678.

Kim JD, Yoon Y, Hwang HY, Park JS, Yu S, Lee J, Baek K, Yoon J. Efficient selection of stable chinese hamster ovary (CHO) cell lines for expression of recombinant proteins by using human interferon beta SAR element. *Biotechnol Prog* 2005;21:933–937.

Kim JM, Kim JS, Park DH, Kang HS, Yoon J, Baek K, Yoon Y. Improved recombinant gene expression in CHO cells using matrix attachment regions. *J Biotechnol* 2004;107:95–105.

Kim J-S, Raines RT. Ribonuclease S-peptide as a carrier in fusion proteins. *Protein Sci* 1993;2:348–356.

Kim J-S, Raines RT. Peptide tags for a dual affinity fusion system. *Anal Biochem* 1994;219:165–166.

Kim JY, Kim YG, Lee GM. CHO cells in biotechnology for production of recombinant proteins: Current state and further potential. *Appl Microbiol Biotechnol* 2012;93:917–930.

Kim NS, Byun TH, Lee GM. Key determinants in the occurrence of clonal variation in humanized antibody expression of cho cells during dihydrofolate reductase mediated gene amplification. *Biotechnol Prog* 2001;17:69–75.

Kim SJ, Kim NS, Ryu CJ, Hong HJ, Lee GM. Characterization of chimeric antibody producing CHO cells in the course of dihydrofolate reductase-mediated gene amplification and their stability in the absence of selective pressure. *Biotechnol Bioeng* 1998;58:73–84.

Kitai K, Kudo T, Nakamura S, Masegi T, Ichikawa Y, Horikoshi K. Extracellular production of human immunoglobulin G Fc region (hIgG-Fc) by *Escherichia coli*. *Appl Microbiol Biotechnol* 1988;28:52–56.

Kitano K, Fujimoto S, Nakao M, Watanabe T, Nakao Y. Intracellular degradation of recombinant proteins in relation to their location in *Escherichia coli* cells. *J Biotechnol* 1987;5:77–86.

Kito M, Itami S, Fukano Y, Yamana K, Shibui T. Construction of engineered CHO strains for high-level production of recombinant proteins. *Appl Microbiol Biotechnol* 2002;60: 442–448.

Kleerebezem M, Tommassen J. Expression of the pspA gene stimulates efficient protein export in *Escherichia coli. Mol Microbiol* 1993;7:947–956.

Knappik A, Krebber C, Plückthun A. The effect of folding catalysts on the in vivo folding process of different antibody fragments expressed in *Escherichia coli. Bio/Technology* 1993;11:77–83.

Knappik A, Plückthun A. An improved affinity tag based on the FLAG peptide for the detection and purification of recombinant antibody fragments. *BioTechniques* 1994;17:754–761.

Knappik A, Plückthun A. Engineered turns of a recombinant antibody improve its in vivo folding. *Protein Eng* 1995;8:81–89.

Knott JA, Sullivan CA, Weston A. The isolation and characterization of human atrial natriuretic factor produced as a fusion protein in *Escherichia coli. Eur J Biochem* 1988;174:405–410.

Ko JH, Park DK, Kim IC, Lee SH, Byun SM. High-level expression and secretion of streptokinase in *Escherichia coli. Biotechnol Lett* 1995;17:1019–1024.

Kobayashi M, Nagata K, Ishihama A. Promoter selectivity of *Escherichia coli* RNA polymerase: Effect of base substitutions in the promoter 35 region on promoter strength. *Nucleic Acids Res* 1990;18:7367–7372.

Köhler K, Ljungquist C, Kondo A, Veide A, Nilsson B. Engineering proteins to enhance their partition coefficients in aqueous two-phase systems. *Bio/Technology* 1991;9:642–646.

Koken MHM, Odijk HHM, van Duin M, Fornerod M, Hoeijmakers JHJ. Augmentation of protein production by a combination of the T7 RNA polymerase system and ubiquitin fusion: Overproduction of the human DNA repair protein, ERCC1, as a ubiquitin fusion protein in *Escherichia coli. Biochem Biophys Res Commun* 1993;195:643–653.

Kolb AF, Siddell SG. Expression of a recombinant monoclonal antibody from a bicistronic mRNA. *Hybridoma* 1997;16:421–426.

Koller MR, Hanania EG, Stevens J, Eisfeld TM, Sasaki GC, Fieck A, Palsson BO. High-throughput laser-mediated in situ cell purification with high purity and yield. *Cytometry A* 2004;61:153–161.

Kosen PA. Disulfide bonds in proteins. In: Ahern TJ, 1992, pp. 31–67.

Kushner SR. mRNA decay. In: *Escherichia coli* and *Salmonella*: *Cellular and Molecular Biology*, Vol. 1, Neidhardt FC, Curtiss III R, Ingraham JL, Lin ECC, Low KB, Magasanik B, Reznikoff WS, Riley M, Schaechter M, Umbarger HE, eds. ASM Press Washington, DC, 1996, pp. 849–860.

Kwon S, Kim S, Kim E. Effects of glycerol on β-lactamase production during high cell density cultivation of recombinant *Escherichia coli. Biotechnol Prog* 1996;12:205–208.

Lai T, Yang Y, Ng S. Advances in mammalian cell line development technologies for recombinant protein production. *Pharmaceuticals* 2013;6:579–603.

Landick R, Turnbough Jr. CL, Yanofsky C. Transcription attenuation. In: *Escherichia coli* and *Salmonella*: *Cellular and Molecular Biology*, Vol. 1, Neidhardt FC, Curtiss III R, Ingraham JL, Lin ECC, Low KB, Magasanik B, Reznikoff WS, Riley M, Schaechter M, Umbarger HE, eds. ASM Press, Washington, DC, 1996, pp. 1263–1286.

Lange R, Hengge-Aronis R. The cellular concentration of the s subunit of RNA polymerase in *Escherichia coli* is controlled at the levels of transcription, translation, and protein stability. *Genes Dev* 1994;8:1600–1612.

Langer T, Lu C, Echols H, Flanagan J, Hayer MK, Hartl FU. Successive action of DnaK, DnaJ and GroEL along the pathway of chaperone-mediated protein folding. *Nature (London)* 1992;356:683–689.

Lanthier M, Behrman R, Nardinelli C. Economic issues with follow-on protein products. *Nat Rev Drug Discov* 2008;7:733–737.

Lanzer M, Bujard H. Promoters largely determine the efficiency of repressor action. *Proc Natl Acad Sci USA* 1988;85:8973–8977.

Lattenmayer C, Loeschel M, Schriebl K, Steinfellner W, Sterovsky T, Trummer E, Vorauer-Uhl K, Muller D, Katinger H, Kunert R. Protein-free transfection of CHO host cells with an IgG-fusion protein: Selection and characterization of stable high producers and comparison to conventionally transfected clones. *Biotechnol Bioeng* 2007;96:1118–1126.

LaVallie ER, DiBlasio EA, Kovacic S, Grant KL, Schendel PF, McCoy JM. A thioredoxin gene fusion expression system that circumvents inclusion body formation in the *E. coli* cytoplasm. *Bio/Technology* 1993;11:187–193.

LaVallie ER, McCoy JM. Gene fusion expression systems in *Escherichia coli. Curr Opin Biotechnol* 1995;6:501–506.

Le Calvez H, Green JM, Baty D. Increased efficiency of alkaline phosphatase production levels in *Escherichia coli* using a degenerate PelB signal sequence. *Gene* 1996;170:51–55.

Lee C, Li P, Inouye H, Brickman ER, Beckwith J. Genetic studies on the instability of -galactosidase to be translocated across the *Escherichia coli* cytoplasmic membrane. *J Bacteriol* 1989;171:4609–4616.

Lee C, Ly C, Sauerwald T, Kelly T, Moore G. High-throughput screening of cell lines expressing monoclonal antibodies. *BioProcess Int* 2006;4:32–35.

Lee H-W, Joo J-H, Kang S, Song I-S, Kwon J-B, Han MH, Na DS. Expression of human interleukin-2 from native and synthetic genes in *E. coli*: No correlation between major codon bias and high level expression. *Biotechnol Lett* 1992;14:653–658.

Lee J, Cho MW, Hong E-K, Kim K-S, Lee J. Characterization of the nar promoter to use as an inducible promoter. *Biotechnol Lett* 1996;18:129–134.

Lee N, Zhang S-Q, Cozzitorto J, Yang J-S, Testa D. Modification of mRNA secondary structure and alteration of the expression of human interferon 1 in *Escherichia coli*. *Gene* 1987;58:77–86.

Lee SC, Olins PO. Effect of overproduction of heat shock chaperones GroESL and DnaK on human procollagenase production in *Escherichia coli*. *J Biol Chem* 1992;267:2849–2852.

Lee SY. High cell-density culture of *Escherichia coli*. *Trends Biotechnol* 1996;14:98–105.

Lehnhardt S, Pollitt S, Inouye M. The differential effect on two hybrid proteins of deletion mutations within the hydrophobic region of the *Escherichia coli* ompA signal peptide. *J Biol Chem* 1987;262:1716–1719.

Lei S-P, Lin H-C, Wang S-S, Callaway J, Wilcox G. Characterization of the Erwinia carotovora pelB gene and its product pectate lyase. *J Bacteriol* 1987;169:4379–4383.

Levitsky VG. RECON: A program for prediction of nucleosome formation potential. *Nucleic Acids Res* 2004;32:W346–W349.

Li SC, Squires CL, Squires C. Antitermination of *E. coli* rRNA transcription is caused by a control region segment containing lambda nut-like sequences. *Cell* 1984;38:851–860.

Liang S-M, Allet B, Rose K, Hirschi M, Liang C-M, Thatcher DR. Characterization of human interleukin 2 derived from *Escherichia coli*. *Biochem J* 1985;229:429–439.

Lindsey DF, Mullin DA, Walker JR. Characterization of the cryptic lambdoid prophage DLP12 of *Escherichia coli* and overlap of the DLP12 integrase gene with the tRNA gene argU. *J Bacteriol* 1989;171:6197–6205.

Lisser S, Margalit H. Compilation of *E. coli* mRNA promoter sequences. *Nucleic Acids Res* 1993;21:1507–1516.

Little M, Fuchs P, Breitling F, Dübel S. Bacterial surface presentation of proteins and peptides: An alternative to phage technology? *Trends Biotechnol* 1993;11:3–5.

Little P. Genetics. Small and perfectly formed. *Nature* 1993;366:204–205.

Little S, Campbell CJ, Evans IJ, Hayward EC, Lilley RJ. A short N-proximal region of prochymosin inhibits the secretion of hybrid proteins from *Escherichia coli*. *Gene* 1989;83:321–329.

Liu PQ, Chan EM, Cost GJ et al. Generation of a triple-gene knockout mammalian cell line using engineered zinc-finger nucleases. *Biotechnol Bioeng* 2010;106:97–105.

Ljungquist C, Lundeberg J, Rasmussen A-M, Hornes E, Uhlén M. Immobilization and recovery of fusion proteins and B-lymphocyte cells using magnetic separation. *DNA Cell Biol* 1993;12:191–197.

Lo AC, MacKay RM, Seligy VL, Willick GE. *Bacillus subtilis* β-1,4-endoglucanase products from intact and truncated genes are secreted into the extracellular medium by *Escherichia coli*. *Appl Environ Microbiol* 1988;54:2287–2292.

Lobito AA, Ramani SR, Tom I, Bazan JF, Luis E, Fairbrother WJ, Ouyang W, Gonzalez LC. Murine insulin growth factor-like (IGFL) and human IGFL1 proteins are induced in inflammatory skin conditions and bind to a novel tumor necrosis factor receptor family member, IGFLR1. *J Biol Chem* 2011;286:18969–18981.

Lofdahl S, Guss B, Uhlén M, Philipson L, Lindberg M. Gene for staphylococcal protein A. *Proc Natl Acad Sci USA* 1983;80:697–701.

Lonza launches next generation GS gene expression system. Available online: http://www.lonza.com/about-lonza/media-center/news/2012/120710-GS-System-e.aspx.

Lorimer GH. A quantitative assessment of the role of chaperonin proteins in protein folding in vivo. *FASEB J* 1996;10:5–9.

Lovatt A, McMutrie D, Black J, Doherty I. Validation of quantitative PCR assays. Addressing virus contamination concerns. *BioPharm* 2002;15(3):22–32.

Lowary PT, Widom J. New DNA sequence rules for high affinity binding to histone octamer and sequence-directed nucleosome positioning. *J Mol Biol* 1998;276:19–42.

Lu ZJ, Murray KS, Van Cleave V, LaVallie ER, Stahl ML, McCoy JM. Expression of thioredoxin random peptide libraries on the *Escherichia coli* cell surface as functional fusions to flagellin: A system designed for exploring protein-protein interactions. *Bio/Technology* 1995;13:366–372.

Luli GW, Strohl WR. Comparison of growth, acetate production and acetate inhibition of *Escherichia coli* strains in batch and fed-batch fermentations. *Appl Environ Microbiol* 1990;56:1004–1011.

Lundeberg J, Wahlberg J, Uhlén M. Affinity purification of specific DNA fragments using a lac repressor fusion protein. *Genet Anal Tech Appl* 1990;7:47–52.

MacFerrin KD, Chen L, Terranova MP, Schreiber SL, Verdine GL. Overproduction of proteins using expression-cassette polymerase chain reaction. *Methods Enzymol* 1993;217:79–102.

MacIntyre S, Henning U. The role of the mature part of secretory proteins in translocation across the plasma membrane and in regulation of their synthesis in *Escherichia coli*. *Biochimie* 1990;72:157–167.

Mackman N, Baker K, Gray L, Haigh R, Nicaud J-M, Holland IB. Release of a chimeric protein into the medium from *Escherichia coli* using therminal secretion signal of haemolysin. *EMBO J* 1987;6:2835–2841.

Makoff AJ, Smallwood AE. The use of two-cistron constructions in improving the expression of a heterologous gene in *E. coli*. *Nucleic Acids Res* 1990;18:1711–1718.

Makrides S. Strategies for achieving high level expression of genes in *E. coli*. *Microbiol Rev* 1996;60(3):512–538.

Makrides SC, Nygren P-Å, Andrews B, Ford PJ, Evans KS, Hayman EG, Adari H, Levin J, Uhlén M, Toth CA. Extended in vivo half-life of human soluble complement receptor type 1 fused to albumin-binding receptor. *J Pharmacol Exp Ther* 1996;277:534–542.

Malke H, Ferretti JJ. Streptokinase: Cloning, expression and secretion by *Escherichia coli*. *Proc Natl Acad Sci USA* 1984;81:3557–3561.

Manning MC, ed., *Stability of Protein Pharmaceuticals. A. Chemical and Physical Pathways of Protein Degradation*. Plenum Press, New York.

Kozak M. Comparison of initiation of protein synthesis in prokaryotes, eukaryotes, and organelles. *Microbiol Rev* 1983;47:1–45.

Krueger JK, Kulke MN, Schutt C, Stock J. Protein inclusion body formation and purification. *BioPharm* 1989;2:40–45.

Manz R, Assenmacher M, Pfluger E, Miltenyi S, Radbruch A. Analysis and sorting of live cells according to secreted molecules, relocated to a cell-surface affinity matrix. *Proc Natl Acad Sci USA* 1995;92:1921–1925.

Marino MH. Expression systems for heterologous protein production. *BioPharm* 1989;2:18–33.

Marston FAO. The purification of eukaryotic polypeptides synthesized in *Escherichia coli*. *Biochem J* 1986;240:1–12.

Martin J, Hartl FU. Molecular chaperones in cellular protein folding. *Bioessays* 1994;16:689–692.

Masuda K, Kamimura T, Kaneshaki M, Ishii K, Imaizumi A, Sugiyama T, Suzuki Y, Ohtsuka E. Efficient production of the C-terminal domain of secretory leukoprotease inhibitor as a thrombin-cleavable fusion protein in *Escherichia coli*. *Protein Eng* 1996;9:101–106.

Matin A. Starvation promoters of *Escherichia coli*. Their function, regulation, and use in bioprocessing and bioremediation. *Ann N Y Acad Sci* 1994;721:277–291.

Maurizi MR. Proteases and protein degradation in *Escherichia coli*. *Experientia* 1992;48:178–201.

McCarthy JEG, Brimacombe R. Prokaryotic translation: The interactive pathway leading to initiation. *Trends Genet* 1994;10:402–407.

McCarthy JEG, Gualerzi C. Translational control of prokaryotic gene expression. *Trends Genet* 1990;6:78–85.

McCarthy JEG, Schairer HU, Sebald W. Translational initiation frequency of atp genes from *Escherichia coli*: Identification of an intercistronic sequence that enhances translation. *EMBO J* 1985;4:519–526.

McCarthy JEG, Sebald W, Gross G, Lammers R. Enhancement of translational efficiency by the *Escherichia coli* atpE translational initiation region: Its fusion with two human genes. *Gene* 1986;41:201–206.

Meerman HJ, Georgiou G. Construction and characterization of a set of *E. coli* strains deficient in all known loci affecting the proteolytic stability of secreted recombinant proteins. *Bio/Technology* 1994a;12:1107–1110.

Meerman HJ, Georgiou G. High-level production of proteolytically sensitive secreted proteins in *Escherichia coli* strains impaired in the heat-shock response. *Ann N Y Acad Sci* 1994b;721:292–302.

Melville M, Doolan P, Mounts W, Barron N, Hann L, Leonard M, Clynes M, Charlebois T. Development and characterization of a Chinese hamster ovary cell-specific oligonucleotide microarray. *Biotechnol Lett* 2011;33:1773–1779.

Meng YG, Liang J, Wong WL, Chisholm V. Green fluorescent protein as a second selectable marker for selection of high producing clones from transfected CHO cells. *Gene* 2000;242:201–207.

Mertens N, Remaut E, Fiers W. Tight transcriptional control mechanism ensures stable high-level expression from T7 promoter-based expression plasmids. *Bio/Technology* 1995a;13:175–179.

Mertens N, Remaut E, Fiers W. Versatile, multi-featured plasmids for high-level expression of heterologous genes in *Escherichia coli*: Overproduction of human and murine cytokines. *Gene* 1995b;164:9–15.

Michaelis S, Beckwith J. Mechanism of incorporation of cell envelope proteins in *Escherichia coli*. *Annu Rev Microbiol* 1982;36:435–465.

Mikuni O, Ito K, Moffat J, Matsumura K, McCaughan K, Nobukuni T, Tate W, Nakamura Y. Identification of the prfC gene, which encodes peptide-chain-release factor 3 of *Escherichia coli*. *Proc Natl Acad Sci USA* 1994;91:5798–5802.

Miller CG. Protein degradation and proteolytic modification. In: *Escherichia coli and Salmonella: Cellular and Molecular Biology*, Vol. 1, Neidhardt FC, Curtiss III R, Ingraham JL, Lin ECC, Low KB, Magasanik B, Reznikoff WS, Riley M, Schaechter M, Umbarger HE, eds. ASM Press, Washington, DC, 1996, pp. 938–954.

Miller JC, Holmes MC, Wang J et al. An improved zinc-finger nuclease architecture for highly specific genome editing. *Nat Biotechnol* 2007;25:778–785.

Miller JC, Tan S, Qiao G et al. A TALE nuclease architecture for efficient genome editing. *Nat Biotechnol* 2011;29:143–148.

Minas W, Bailey JE. Co-overexpression of prlF increases cell viability and enzyme yields in recombinant *Escherichia coli* expressing *Bacillus stearothermophilus* α-amylase. *Biotechnol Prog* 1995;11:403–411.

Minor PD. Ensuring safety and consistency in cell culture production processes: Viral screening and inactivation. *Trends Biotechnol* 1994a;12(7):257–261.

Minor PD. Significance of contamination with viruses of cell lines used in the production of biological medicinal products. In: *Animal Cell Technology. Products of Today, Prospects for Tomorrow*, Spier RE, Griffiths JB, Berthold W, eds. Butterworth Heinemann, Oxford, U.K., 1994b, pp. 741–748.

Mirkovitch J, Mirault ME, Laemmli UK. Organization of the higher-order chromatin loop: Specific DNA attachment sites on nuclear scaffold. *Cell* 1984;39:223–232.

Misoka F, Fuwa T, Yoda K, Yamasaki M, Tamura G. Secretion of human interferon-α induced by using secretion vectors containing a promoter and signal sequence of alkaline phosphatase gene of *Escherichia coli*. *J Biochem* 1985;97:1429–1436.

Misoka F, Miyake T, Miyoshi K-I, Sugiyama M, Sakamoto S, Fuwa T. Overproduction of human insulin-like growth factor-II in *Escherichia coli*. *Biotechnol Lett* 1989;11:839–844.

Mitraki A, King J. Protein folding intermediates and inclusion body formation. *Bio/Technology* 1989;7:690–697.

Mohan C, Kim YG, Koo J, Lee GM. Assessment of cell engineering strategies for improved therapeutic protein production in CHO cells. *Biotechnol J* 2008;3:624–630.

Mohsen A-WA, Vockley J. High-level expression of an altered cDNA encoding human isovaleryl-CoA dehydrogenase in *Escherichia coli*. *Gene* 1995;160:263–267.

Moks T, Abrahmsén L, Holmgren E et al. Expression of human insulin-like growth factor I in bacteria: Use of optimized gene fusion vectors to facilitate protein purification. *Biochemistry* 1987a;26:5239–5244.

Moks T, Abrahmsén L, Osterlöf B, Josephson S, Ostling M, Enfors S-O, Persson I, Nilsson B, Uhlén M. Large-scale affinity purification of human insulin-like growth factor I from culture medium of *Escherichia coli*. *Bio/Technology* 1987b;5:379–382.

Moore JT, Uppal A, Maley F, Maley GF. Overcoming inclusion body formation in a high-level expression system. *Protein Expression Purif* 1993;4:160–163.

Morino T, Morita M, Seya K, Sukenaga Y, Kato K, Nakamura T. Construction of a runaway vector and its use for a high-level expression of a cloned human superoxide dismutase gene. *Appl Microbiol Biotechnol* 1988;28:170–175.

Morioka-Fujimoto K, Marumoto R, Fukuda T. Modified enterotoxin signal sequences increase secretion level of the recombinant human epidermal growth factor in *Escherichia coli*. *J Biol Chem* 1991;266:1728–1732.

Moscou MJ, Bogdanove AJ. A simple cipher governs DNA recognition by TAL effectors. *Science* 2009;326:1501.

Mottagui-Tabar S, Björnsson A, Isaksson LA. The second to last amino acid in the nascent peptide as a codon context determinant. *EMBO J* 1994;13:249–257.

Mukhija R, Rupa P, Pillai D, Garg LC. High-level production and one-step purification of biologically active human growth hormone in *Escherichia coli*. *Gene* 1995;165:303–306.

Muller D, Katinger H, Grillari J. MicroRNAs as targets for engineering of CHO cell factories. *Trends Biotechnol* 2008;26:359–365.

Müller-Hill B, Crapo L, Gilbert W. Mutants that make more lac repressor. *Proc Natl Acad Sci USA* 1968;59:1259–1264.

Munro S, Pelham HRB. An Hsp70-like protein in the ER: Identity with the 78 kd glucose-regulated protein and immunoglobulin heavy chain binding protein. *Cell* 1986;46:291–300.

Murby M, Cedergren L, Nilsson J, Nygren P-Å, Hammarberg B, Nilsson B, Enfors S-O, Uhlén M. Stabilization of recombinant proteins from proteolytic degradation in *Escherichia coli* using a dual affinity fusion strategy. *Biotechnol Appl Biochem* 1991;14:336–346.

Murby M, Samuelsson E, Nguyen TN, Mignard L, Power U, Binz H, Uhlén M, Ståhl S. Hydrophobicity engineering to increase solubility and stability of a recombinant protein from respiratory syncytial virus. *Eur J Biochem* 1995;230:38–44.

Murby M, Uhlén M, Ståhl S. Upstream strategies to minimize proteolytic degradation upon recombinant production in *Escherichia coli*. *Protein Expression Purif* 1996;7:129–136.

Mussolino C, Cathomen T. TALE nucleases: Tailored genome engineering made easy. *Curr Opin Biotechnol* 2012;23:644–650.

Nagahari K, Kanaya S, Munakata K, Aoyagi Y, Mizushima S. Secretion into the culture medium of a foreign gene product from *Escherichia coli*: Use of the ompF gene for secretion of human β-endorphin. *EMBO J* 1985;4:3589–3592.

Nagai K, Perutz MF, Poyart C. Oxygen binding properties of human mutant hemoglobins synthesized in *Escherichia coli*. *Proc Natl Acad Sci USA* 1985;82:7252–7255.

Nagai K, Thøgersen HC. Generation of β-globin by sequence-specific proteolysis of a hybrid protein produced in *Escherichia coli*. *Nature (London)* 1984;309:810–812.

Nagai K, Thøgersen HC. Synthesis and sequence-specific proteolysis of hybrid proteins produced in *Escherichia coli*. *Methods Enzymol* 1987;153:461–481.

Nagai H, Yuzawa H, Yura T. Interplay of two cis-acting mRNA regions in translational control of 32 synthesis during the heat shock response of *Escherichia coli*. *Proc Natl Acad Sci USA* 1991;88:10515–10519.

Nakamura K, Inouye M. Construction of versatile expression cloning vehicles using the lipoprotein gene of *Escherichia coli*. *EMBO J* 1982;1:771–775.

Nakashima K, Kanamaru K, Mizuno T, Horikoshi K. A novel member of the cspA family of genes that is induced by cold shock in *Escherichia coli*. *J Bacteriol* 1996;178:2994–2997.

Neri D, De Lalla C, Petrul H, Neri P, Winter G. Calmodulin as a versatile tag for antibody fragments. *Bio/Technology* 1995;13:373–377.

Newbury SF, Smith NH, Robinson EC, Hiles ID, Higgins CF. Stabilization of translationally active mRNA by prokaryotic REP sequences. *Cell* 1987;48:297–310.

Ng SK, Lin W, Sachdeva R, Wang DI, Yap MG. Vector fragmentation: Characterizing vector integrity in transfected clones by Southern blotting. *Biotechnol Prog* 2010;26:11–20.

Ng SK, Tan TR, Wang Y, Ng D, Goh LT, Bardor M, Wong VV, Lam KP. Production of functional soluble Dectin-1 glycoprotein using an IRES-linked destabilized-dihydrofolate reductase expression vector. *PLoS One* 2012;7:e52785.

Ng SK, Wang DI, Yap MG. Application of destabilizing sequences on selection marker for improved recombinant protein productivity in CHO-DG44. *Metab Eng* 2007;9:304–316.

Nguyen TN, Hansson M, Ståhl S, Bächi T, Robert A, Domzig W, Binz H, Uhlén M. Cell-surface display of heterologous epitopes on *Staphylococcus xylosus* as a potential delivery system for oral vaccination. *Gene* 1993;128:89–94.

Nicaud J-M, Mackman N, Holland IB. Current status of secretion of foreign proteins by microorganisms. *J Biotechnol* 1986;3:255–270.

Nierlich DP, Murakawa GJ. The decay of bacterial messenger RNA. *Prog Nucleic Acid Res Mol Biol* 1996;52:153–216.

Nilsson B, Abrahmsén L. Fusions to staphylococcal protein A. *Methods Enzymol* 1990;185:144–161.

Nilsson B, Forsberg G, Moks T, Hartmanis M, Uhlén M. Fusion proteins in biotechnology and structural biology. *Curr Opin Struct Biol* 1992;2:569–575.

Nilsson J, Nilsson P, Williams Y, Pettersson L, Uhlén M, Nygren P-Å. Competitive elution of protein A fusion proteins allows specific recovery under mild conditions. *Eur J Biochem* 1994;224:103–108.

Nishi T, Itoh S. Enhancement of transcriptional activity of the *Escherichia coli* trp promoter by upstream AT-rich regions. *Gene* 1986;44:29–36.

Nishihara T, Iwabuchi T, Nohno T. A T7 promoter vector with a transcriptional terminator for stringent expression of foreign genes. *Gene* 1994;145:145–146.

Niwa H, Yamamura K, Miyazaki J. Efficient selection for high-expression transfectants with a novel eukaryotic vector. *Gene* 1991;108:193–199.

Nomura M, Gourse R, Baughman G. Regulation of the synthesis of ribosomes and ribosomal components. *Annu Rev Biochem* 1984;53:75–118.

Nordström K, Uhlin BE. Runaway-replication plasmids as tools to produce large quantities of proteins from cloned genes in bacteria. *Bio/Technology* 1992;10:661–666.

Nossal NG, Heppel LA. The release of enzymes by osmotic shock from *Escherichia coli* in exponential phase. *J Biol Chem* 1966;241:3055–3062.

Novo JB, Morganti L, Moro AM, Paes Leme AF, Serrano SM, Raw I, Ho PL. Generation of a Chinese hamster ovary cell line producing recombinant human glucocerebrosidase. *J Biomed Biotechnol* 2012;2012:875383.

Novotny J, Ganju RK, Smiley ST, Hussey RE, Luther MA, Recny MA, Siliciano RF, Reinherz EL. A soluble, single-chain T-cell receptor fragment endowed with antigen-combining properties. *Proc Natl Acad Sci USA* 1991;88:8646–8650.

Nygren P-Å, Ljungquist C, Tromborg H, Nustad K, Uhlén M. Species-dependent binding of serum albumins to the streptococcal receptor protein G. *Eur J Biochem* 1990;193:143–148.

Nygren P-Å, Ståhl S, Uhlén M. Engineering proteins to facilitate bioprocessing. *Trends Biotechnol* 1994;12:184–188.

Nygren P-Å, Uhlén M, Flodby P, Andersson R, Wigzell H. In vivo stabilization of a human recombinant CD4 derivative by fusion to a serum-albumin-binding receptor. *Vaccines* 1991;91:363–368.

O'Connor CD, Timmis KN. Highly repressible expression system for cloning genes that specify potentially toxic proteins. *J Bacteriol* 1987;169:4457–4462.

O'Gorman S, Fox DT, Wahl GM. Recombinase-mediated gene activation and site-specific integration in mammalian cells. *Science* 1991;251:1351–1355.

Obukowicz MG, Staten NR, Krivi GG. Enhanced heterologous gene expression in novel rpoH mutants of *Escherichia coli. Appl Environ Microbiol* 1992;58:1511–1523.

Obukowicz MG, Turner MA, Wong EY, Tacon WC. Secretion and export of IGF-1 in *Escherichia coli* strain JM101. *Mol Gen Genet* 1988;215:19–25.

Oka T, Sakamoto S, Miyoshi K-I, Fuwa T, Yoda K, Yamasaki M, Tamura G, Miyake T. Synthesis and secretion of human epidermal growth factor by *Escherichia coli. Proc Natl Acad Sci USA* 1985;82:7212–7216.

Olejniczak ET, Ruan Q, Ziemann RN, Birkenmeyer LG, Saldana SC, Tetin SY. Rapid determination of antigenic epitopes in human NGAL using NMR. *Biopolymers* 2010;93:657–667.

Olins PO, Devine CS, Rangwala SH, Kavka KS. The T7 phage gene 10 leader RNA, a ribosome-binding site that dramatically enhances the expression of foreign genes in *Escherichia coli. Gene* 1988;73:227–235.

Olins PO, Lee SC. Recent advances in heterologous gene expression in *Escherichia coli. Curr Opin Biotechnol* 1993;4:520–525.

Olins PO, Rangwala SH. A novel sequence element derived from bacteriophage T7 mRNA acts as an enhancer of translation of the lacZ gene in *Escherichia coli. J Biol Chem* 1989;264:16973–16976.

Olins PO, Rangwala SH. Vector for enhanced translation of foreign genes in *Escherichia coli. Methods Enzymol* 1990;185:115–119.

Olsen MK, Rockenbach SK, Curry KA, Tomich C-SC. Enhancement of heterologous polypeptide expression by alterations in the ribosome-binding-site sequence. *J Biotechnol* 1989;9:179–190.

Omer CA, Diehl RE, Kral AM. Bacterial expression and purification of human protein prenyltransferases using epitope-tagged, translationally coupled systems. *Methods Enzymol* 1995;250:3–12.

Ong E, Gilkes NR, Warren RAJ, Miller Jr. RC, Kilburn DG. Enzyme immobilization using the cellulose-binding domain of a *Cellulomonas fimi* exoglucanase. *Bio/Technology* 1989a; 7:604–607.

Ong E, Greenwood JM, Gilkes NR, Kilburn DG, Miller Jr. RC, Warren RAJ. The cellulose-binding domains of cellulases: Tools for biotechnology. *Trends Biotechnol* 1989b;7:239–243.

Oppenheim AB, Giladi H, Goldenberg D, Kobi S, Azar I. Vectors and transformed host cells for recombinant protein production at reduced temperatures. International patent application WO 96/03521, February 1996.

Ostermeier M, De Sutter K, Georgiou G. Eukaryotic protein disulfide isomerase complements *Escherichia coli* dsbA mutants and increases the yield of a heterologous secreted protein with disulfide bonds. *J Biol Chem* 1996;271:10616–10622.

Pace CN, Shirley BA, McNutt M, Gajiwala K. Forces contributing to the conformational stability of proteins. *FASEB J* 1996;10:75–83.

Peng RW, Abellan E, Fussenegger M. Differential effect of exocytic SNAREs on the production of recombinant proteins in mammalian cells. *Biotechnol Bioeng* 2011;108:611–620.

Pérez-Pérez J, Márquez G, Barbero J-L, Gutiérrez J. Increasing the efficiency of protein export in *Escherichia coli. Bio/Technology* 1994;12:178–180.

Persson M, Bergstrand MG, Bülow L, Mosbach K. Enzyme purification by genetically attached polycysteine and polyphenylalanine affinity tails. *Anal Biochem* 1988;172:330–337.

Petersen C. Control of functional mRNA stability in bacteria: Multiple mechanisms of nucleolytic and non-nucleolytic inactivation. *Mol Microbiol* 1992;6:277–282.

Pilbrough W, Munro TP, Gray P. Intraclonal protein expression heterogeneity in recombinant CHO cells. *PLoS One* 2009;4:e8432.

Pilot-Matias TJ, Pratt SD, Lane BC. High-level synthesis of the 12-kDa human FK506-binding protein in *Escherichia coli* using translational coupling. *Gene* 1993;128:219–225.

Pipelinereview.com. TOP 30 Biologics. *La Merie Daily* 2011. http://www.pipelinereview.com/index.php/archive/view/listid-1-la-merie-daily/mailid-35-La-Merie-Daily-TOP-30-Biologics-2011-new-free-report/tmpl-component.

Platt T. Transcriptional termination and the regulation of gene expression. *Annu Rev Biochem* 1986;55:339–372.

Plückthun A. Mono-and bivalent antibody fragments produced in *Escherichia coli*: Engineering, folding and antigen binding. *Immunol Rev* 1992;130:151–188.

Podhajska AJ, Hasan N, Szybalski W. Control of cloned gene expression by promoter inversion in vivo: Construction of the heat-pulse-activated att-nutL-p-att-N module. *Gene* 1985;40:163–168.

Pohlner J, Krämer J, Meyer TF. A plasmid system for high-level expression and in vitro processing of recombinant proteins. *Gene* 1993;130:121–126.

Pollitt S, Zalkin H. Role of primary structure and disulfide bond formation in β-lactamase secretion. *J Bacteriol* 1983;153:27–32.

Pollock MR, Richmond MH. Low cyst(e)ine content of bacterial extracellular proteins: Its possible physiological significance. *Nature (London)* 1962;194:446–449.

Poole ES, Brown CM, Tate WP. The identity of the base following the stop codon determines the efficiency of in vivo translational termination in *Escherichia coli*. *EMBO J* 1995;14:151–158.

Powell KT, Weaver JC. Gel microdroplets and flow cytometry: Rapid determination of antibody secretion by individual cells within a cell population. *Biotechnology (NY)* 1990;8:333–337.

Prickett KS, Amberg DC, Hopp TP. A calcium-dependent antibody for identification and purification of recombinant proteins. *Bio-Techniques* 1989;7:580–589.

Proba K, Ge LM, Plückthun A. Functional antibody single-chain fragments from the cytoplasm of *Escherichia coli*: Influence of thioredoxin reductase (TrxB). *Gene* 1995;159:203–207.

Proudfoot AEI, Power CA, Hoogewerf AJ, Montjovent M-O, Borlat F, Offord RE, Wells TNC. Extension of recombinant human RANTES by the retention of the initiating methionine produces a potent antagonist. *J Biol Chem* 1996;271:2599–2603.

Pugsley AP. The complete general secretory pathway in gram-negative bacteria. *Microbiol Rev* 1993;57:50–108.

Pugsley AP, Schwartz M. Export and secretion of proteins by bacteria. *FEMS Microbiol Rev* 1985;32:3–38.

Quek LE, Nielsen LK. On the reconstruction of the *Mus musculus* genome-scale metabolic network model. *Genome Inform* 2008;21:89–100.

Ramesh V, De A, Nagaraja V. Engineering hyperexpression of bacteriophage Mu C protein by removal of secondary structure at the translation initiation region. *Protein Eng* 1994;7:1053–1057.

Rangwala SH, Finn RF, Smith CE, Berberich SA, Salsgiver WJ, Stallings WC, Glover GI, Olins PO. High-level production of active HIV-1 protease in *Escherichia coli*. *Gene* 1992;122:263–269.

Rao L, Ross W, Appleman JA, Gaal T, Leirmo S, Schlax PJ, Record MT, Gourse RL. Factor independent activation of rrnB P1—An "extended" promoter with an upstream element that dramatically increases promoter strength. *J Mol Biol* 1994;235:1421–1435.

Rees S, Coote J, Stables J, Goodson S, Harris S, Lee MG. Bicistronic vector for the creation of stable mammalian cell lines that predisposes all antibiotic-resistant cells to express recombinant protein. *BioTechniques* 1996;20:102–104, 106, 108–110.

Remaut E, Stanssens P, Fiers W. Plasmid vectors for high-efficiency expression controlled by the pL promoter of coliphage lambda. *Gene* 1981;15:81–93.

Richardson G, Lin N, Lacy K, Davis L, Gray M, Cresswell J, Gerber M, Caple M, Kayser K. Cell Xpress™ technology facilitates high-producing Chinese hamster ovary cell line generation using glutamine synthetase gene expression system. In: *Cells and Culture*, Vol. 4, Noll T, ed. Noll T, Berlin, Germany, 2010, pp. 45–4.

Richardson JP. Transcription termination. *Crit Rev Biochem Mol Biol* 1993;28:1–30.

Richardson JP, Greenblatt J. Control of RNA chain elongation and termination. In: *Escherichia coli and Salmonella: Cellular and Molecular Biology*, Vol. 1, Neidhardt FC, Curtiss III R, Ingraham JL, C EC. Lin, Low KB, Magasanik B, Reznikoff WS, Riley M, Schaechter M, Umbarger HE, eds. ASM Press, Washington, DC, 1996, pp. 822–848.

Rinas U, Tsai LB, Lyons D, Fox GM, Stearns G, Fieschko J, Fenton D, Bailey JE. Cysteine to serine substitutions in basic fibroblast growth factor: Effect on inclusion body formation and proteolytic susceptibility during in vitro refolding. *Bio/Technology* 1992;10:435–440.

Ringquist S, Shinedling S, Barrick D, Green L, Binkley J, Stormo GD, Gold L. Translation initiation in *Escherichia coli*: Sequences within the ribosome-binding site. *Mol Microbiol* 1992;6:1219–1229.

Robben J, Massie G, Bosmans E, Wellens B, Volckaert G. An *Escherichia coli* plasmid vector system for high-level production and purification of heterologous peptides fused to active chloramphenicol acetyl-transferase. *Gene* 1993;126:109–113.

Roberts TM, Kacich R, Ptashne M. A general method for maximizing the expression of a cloned gene. *Proc Natl Acad Sci USA* 1979;76:760–764.

Robertson JS. Strategy for adventitious agent assays. *Dev Biol Stand* 1996;88:37–40.

Rogers S, Wells R, Rechsteiner M. Amino acid sequences common to rapidly degraded proteins: The PEST hypothesis. *Science* 1986;234:364–368.

Ron D, Dressler H. pGSTag—A versatile bacterial expression plasmid for enzymatic labeling of recombinant proteins. *BioTechniques* 1992;13:866–869.

Rosenberg AH, Goldman E, Dunn JJ, Studier FW, Zubay G. Effects of consecutive AGG codons on translation in *Escherichia coli*, demonstrated with a versatile codon test system. *J Bacteriol* 1993;175:716–722.

Rosenberg AH, Studier FW. T7 RNA polymerase can direct expression of influenza virus cap-binding protein (PB2) in *Escherichia coli*. *Gene* 1987;59:191–200.

Rosenberg M, Court D. Regulatory sequences involved in the promotion and termination of RNA transcription. *Annu Rev Genet* 1979;13:319–353.

Rosenwasser TA, Hogquist KA, Nothwehr SF, Bradford-Goldberg S, Olins PO, Chaplin DD, Gordon JI. Compartmentalization of mammalian proteins produced in *Escherichia coli*. *J Biol Chem* 1990;265:13066–13073.

Ross J. mRNA stability in mammalian cells. *Microbiol Rev* 1995;59:423–450.

Ross W, Gosink KK, Salomon J, Igarashi K, Zou C, Ishihama A, Severinov K, Gourse RL. A third recognition element in bacterial promoters: DNA binding by the subunit of RNA polymerase. *Science* 1993;262:1407–1413.

Rudolph R, Lilie H. In vitro folding of inclusion body proteins. *FASEB J* 1996;10:49–56.

Russell DR, Bennett GN. Cloning of small DNA fragments containing the *Escherichia coli* tryptophan operon promoter and operator. *Gene* 1982a;17:9–18.

Russell DR, Bennett GN. Construction and analysis of in vivo activity of *E. coli* promoter hybrids and promoter mutants that alter the 35 to 10 spacing. *Gene* 1982b;20:231–243.

Russell JP, Chang DW, Tretiakova A, Padidam M. Phage Bxb1 integrase mediates highly efficient site-specific recombination in mammalian cells. *BioTechniques* 2006;40:460, 462, 464.

Sagawa H, Ohshima A, Kato I. A tightly regulated expression system in *Escherichia coli* with SP6 RNA polymerase. *Gene* 1996;168:37–41.

Saier Jr. MH. Differential codon usage: A safeguard against inappropriate expression of specialized genes? *FEBS Lett* 1995;362:1–4.

Saier Jr. MH, Werner PK, Muller M. Insertion of proteins into bacterial membranes: Mechanism, characteristics, and comparisons with the eucaryotic process. *Microbiol Rev* 1989;53:333–366.

Sali A, Shakhnovich E, Karplus M. How does a protein fold? *Nature (London)* 1994;369:248–251.

Samuelsson E, Moks T, Nilsson B, Uhlén M. Enhanced in vitro refolding of insulin-like growth factor I using a solubilizing fusion partner. *Biochemistry* 1994;33:4207–4211.

Samuelsson E, Wadensten H, Hartmanis M, Moks T, Uhlén M. Facilitated in vitro refolding of human recombinant insulin-like growth factor I using a solubilizing fusion partner. *Bio/Technology* 1991;9:363–366.

San K-Y, Bennett GN, Aristidou AA, Chou CH. Strategies in high-level expression of recombinant protein in *Escherichia coli*. *Ann N Y Acad Sci* 1994a;721:257–267.

San K-Y, Bennett GN, Chou C-H, Aristidou AA. An optimization study of a pH-inducible promoter system for high-level recombinant protein production in *Escherichia coli*. *Ann N Y Acad Sci* 1994b;721:268–276.

Sandler P, Weisblum B. Erythromycin-induced stabilization of ermA messenger RNA in *Staphylococcus aureus* and *Bacillus subtilis*. *J Mol Biol* 1988;203:905–915.

Sandler P, Weisblum B. Erythromycin-induced ribosome stall in the ermA leader: A barricade to 5-to-3 nucleolytic cleavage of the ermA transcript. *J Bacteriol* 1989;171:6680–6688.

Sandman K, Grayling RA, Reeve JN. Improved N-terminal processing of recombinant proteins synthesized in *Escherichia coli*. *Bio/Technology* 1995;13:504–506.

Sano T, Cantor CR. Expression vectors for streptavidincontaining chimeric proteins. *Biochem Biophys Res Commun* 1991;176:571–577.

Sano T, Glazer AN, Cantor CR. A streptavidin-metallothio-nein chimera that allows specific labeling of biological materials with many different heavy metal ions. *Proc Natl Acad Sci USA* 1992;89:1534–1538.

Sarmientos P, Duchesne M, Denèfle P, Boiziau J, Fromage N, Del-porte N, Parker F, Lelievre Y, Mayaux J-F, Cartwright T. Synthesis and purification of active human tissue plasminogen activator from *Escherichia coli. Bio/Technology* 1989;7:495–501.

Sassenfeld HM. Engineering proteins for purification. *Trends Biotechnol* 1990;8:88–93.

Sassenfeld HM, Brewer SJ. A polypeptide fusion designed for the purification of recombinant proteins. *Bio/Technology* 1984;2:76–81.

Sato K, Sato MH, Yamaguchi A, Yoshida M. Tetracy-cline/H antiporter was degraded rapidly in *Escherichia coli* cells when truncated at last transmembrane helix and this degradation was protected by overproduced GroEL/ES. *Biochem Biophys Res Commun* 1994;202:258–264.

Sautter K, Enenkel B. Selection of high-producing CHO cells using NPT selection marker with reduced enzyme activity. *Biotechnol Bioeng* 2005;89:530–538.

Schatz G, Dobberstein B. Common principles of protein translocation across membranes. *Science* 1996;271:1519–1526.

Schatz PJ. Use of peptide libraries to map the substrate specificity of a peptide-modifying enzyme: A 13 residue consensus peptide specifies biotinylation in *Escherichia coli. Bio/Technology* 1993;11:1138–1143.

Schatz PJ, Beckwith J. Genetic analysis of protein export in *Escherichia coli. Annu Rev Genet* 1990;24:215–248.

Schauder B, Blöcker H, Frank R, McCarthy JEG. Inducible expression vectors incorporating the *Escherichia coli* atpE translational initiation region. *Gene* 1987;52:279–283.

Schauder B, McCarthy JEG. The role of bases upstream of the Shine-Dalgarno region and in the coding sequence in the control of gene expression in *Escherichia coli*: Translation and stability of mRNAs in vivo. *Gene* 1989;78:59–72.

Schein CH. Production of soluble recombinant proteins in bacteria. *Bio/Technology* 1989;7:1141–1149.

Schein CH. Optimizing protein folding to the native state in bacteria. *Curr Opin Biotechnol* 1991;2:746–750.

Schein CH. Solubility and secretability. *Curr Opin Biotechnol* 1993;4:456–461.

Schein CH, Boix E, Haugg M, Holliger KP, Hemmi S, Frank G. Secretion of mammalian ribonucleases from *Escherichia coli* using the signal sequence of murine spleen ribonuclease. *Biochem J* 1992;283:137–144.

Scherer GFE, Walkinshaw MD, Arnott S, Morré DJ. The ribosome binding sites recognized by *E. coli* ribosomes have regions with signal character in both the leader and protein coding segments. *Nucleic Acids Res* 1980;8:3895–3907.

Schertler GFX. Overproduction of membrane proteins. *Curr Opin Struct Biol* 1992;2:534–544.

Schimke RT. Gene amplification in cultured animal cells. *Cell* 1984;37:705–713.

Schmidt TGM, Skerra A. The random peptide library-assisted engineering of a C-terminal affinity peptide, useful for the detection and purification of a functional Ig Fv fragment. *Protein Eng* 1993;6:109–122.

Schneider TD, Stormo GD, Gold L, Ehrenfeucht A. Information content of binding sites on nucleotide sequences. *J Mol Biol* 1986;188:415–431.

Schoner BE, Belagaje RM, Schoner RG. Translation of a synthetic two-cistron mRNA in *Escherichia coli. Proc Natl Acad Sci USA* 1986;83:8506–8510.

Schoner BE, Belagaje RM, Schoner RG. Enhanced translational efficiency with two-cistron expression system. *Methods Enzymol* 1990;185:94–103.

Schoner BE, Hsiung HM, Belagaje RM, Mayne NG, Schoner RG. Role of mRNA translational efficiency in bovine growth hormone expression in *Escherichia coli. Proc Natl Acad Sci USA* 1984;81:5403–5407.

Schümperli D, McKenney K, Sobieski DA, Rosenberg M. Translational coupling at an intercistronic boundary of the *Escherichia coli* galactose operon. *Cell* 1982;30:865–871.

Scolnik E, Tompkins R, Caskey T, Nirenberg M. Release factors differing in specificity for terminator codons. *Proc Natl Acad Sci USA* 1968;61:768–774.

Selvarasu S, Ho YS, Chong WP, Wong NS, Yusufi FN, Lee YY, Yap MG, Lee DY. Combined in silico modeling and metabolomics analysis to characterize fed-batch CHO cell culture. *Biotechnol Bioeng* 2012;109:1415–1429.

Serpieri F, Inocencio A, de Oliveira JM, Pimenta Jr. AA, Garbuio A, Kalil J, Brigido MM, Moro AM. Comparison of humanized IgG and FvFc anti-CD3 monoclonal antibodies expressed in CHO cells. *Mol Biotechnol* 2010;45:218–225.

Shaffer AL, Shapiro-Shelef M, Iwakoshi NN et al. XBP1, downstream of Blimp-1, expands the secretory apparatus and other organelles, and increases protein synthesis in plasma cell differentiation. *Immunity* 2004;21:81–93.

Sharp PM, Bulmer M. Selective differences among translation termination codons. *Gene* 1988;63:141–145.

Sharp PM, Cowe E, Higgins DG, Shields DC, Wolfe KH, Wright F. Codon usage patterns in *Escherichia coli, Bacillus subtilis, Saccharomyces cerevisiae, Schizosaccharomyces pombe, Drosophila melanogaster* and *Homo sapiens*: A review of the considerable within-species diversity. *Nucleic Acids Res* 1988;16:8207–8211.

Shatzman AR. Expression systems. *Curr Opin Biotechnol* 1995;6:491–493.

Shean CS, Gottesman ME. Translation of the prophage cI transcript. *Cell* 1992;70:513–522.

Shen S-H. Multiple joined genes prevent product degradation in *Escherichia coli*. *Proc Natl Acad Sci USA* 1984;81:4627–4631.

Shen T-J, Ho NT, Simplaceanu V, Zou M, Green BN, Tam MF, Ho C. Production of unmodified human adult hemoglobin in *Escherichia coli*. *Proc Natl Acad Sci USA* 1993;90:8108–8112.

Shine J, Dalgarno L. The 3-terminal sequence of *Escherichia coli* 16S ribosomal RNA: Complementarity to nonsense triplets and ribosome binding sites. *Proc Natl Acad Sci USA* 1974;71:1342–1346.

Shine J, Dalgarno L. Determinant of cistron specificity in bacterial ribosomes. *Nature (London)* 1975;254:34–38.

Shirakawa M, Tsurimoto T, Matsubara K. Plasmid vectors designed for high-efficiency expression controlled by the portable recA promoter-operator of *Escherichia coli*. *Gene* 1984;28:127–132.

Shirano Y, Shibata D. Low temperature cultivation of *Escherichia coli* carrying a rice lipoxygenase L-2 cDNA produces a soluble and active enzyme at a high level. *FEBS Lett* 1990;271:128–130.

Shuman HA, Silhavy TJ, Beckwith JR. Labeling of proteins with β-galactosidase by gene fusion. Identification of a cytoplasmic membrane component of the *Escherichia coli* maltose transport system. *J Biol Chem* 1980;255:168–174.

Simon LD, Randolph B, Irwin N, Binkowski G. Stabilization of proteins by a bacteriophage T4 gene cloned in *Escherichia coli*. *Proc Natl Acad Sci USA* 1983;80:2059–2062.

Simon LD, Tomczak K, St. John AC. Bacteriophages inhibit degradation of abnormal proteins in *E. coli*. *Nature (London)* 1978;275:424–428.

Singer BS, Gold L. Phage T4 expression vector: Protection from proteolysis. *Gene* 1991;106:1–6.

Skerra A. Bacterial expression of immunoglobulin fragments. *Curr Opin Immunol* 1993;5:256–262.

Skerra A. Use of the tetracycline promoter for the tightly regulated production of a murine antibody fragment in *Escherichia coli*. *Gene* 1994;151:131–135.

Smith DB, Johnson KS. Single-step purification of polypeptides expressed in *Escherichia coli* as fusions with glutathione S-transferase. *Gene* 1988;67:31–40.

Smith MC, Thorpe HM. Diversity in the serine recombinases. *Mol Microbiol* 2002;44:299–307.

Snyder WB, Silhavy TJ. Enhanced export of β-galactosidase fusion proteins in prlF mutants is Lon dependent. *J Bacteriol* 1992;174:5661–5668.

Sprengart ML, Fatscher HP, Fuchs E. The initiation of translation in *E. coli*: Apparent base pairing between the 16S rRNA and downstream sequences of the mRNA. *Nucleic Acids Res* 1990;18:1719–1723.

Sprengart ML, Fuchs E, Porter AG. The downstream box: An efficient and independent translation initiation signal in *Escherichia coli*. *EMBO J* 1996;15:665–674.

Stadel JM, Ecker DJ, Crooke ST. Ubiquitin fusion augments the yield of cloned gene products in *Escherichia coli*. *Proc Natl Acad Sci USA* 1989;86:2540–2544.

Stader JA, Silhavy TJ. Engineering *Escherichia coli* to secrete heterologous gene products. *Methods Enzymol* 1990;185:166–187.

Ståhl S, Nygren P-Å, Sjolander A, Uhlén M. Engineered bacterial receptors in immunology. *Curr Opin Immunol* 1993;5:272–277.

Stanssens P, Remaut E, Fiers W. Alterations upstream from the Shine-Dalgarno region and their effect on bacterial gene expression. *Gene* 1985;36:211–223.

Stark MJR. Multicopy expression vectors carrying the lac repressor gene for regulated high-level expression of genes in *Escherichia coli*. *Gene* 1987;51:255–267.

Steitz JA, Jakes K. How ribosomes select initiator regions in mRNA: Base pair formation between the 3 terminus of 16S rRNA and the mRNA during initiation of protein synthesis in *Escherichia coli*. *Proc Natl Acad Sci USA* 1975;72:4734–4738.

Stempfer G, Höll-Neugebauer B, Kopetzki E, Rudolph R. A fusion protein designed for noncovalent immobilization: Stability, enzymatic activity, and use in an enzyme reactor. *Nat Biotechnol* 1996;14:481–484.

Stempfer G, Höll-Neugebauer B, Rudolph R. Improved refolding of an immobilized fusion protein. *Nat Biotechnol* 1996;14:329–334.

Stormo GD, Schneider TD, Gold LM. Characterization of translational initiation sites in *E. coli*. *Nucleic Acids Res* 1982;10:2971–2996.

Strandberg L, Enfors S-O. Factors influencing inclusion body formation in the production of a fused protein in *Escherichia coli*. *Appl Environ Microbiol* 1991;57:1669–1674.

Studier FW, Moffatt BA. Use of bacteriophage T7 RNA polymerase to direct selective high-level expression of cloned genes. *J Mol Biol* 1986;189:113–130.

Studier FW, Rosenberg AH, Dunn JJ, Dubendorf JW. Use of T7 RNA polymerase to direct expression of cloned genes. *Methods Enzymol* 1990;185:60–89.

Stueber D, Bujard H. Transcription from efficient promoters can interfere with plasmid replication and diminish expression of plasmid specified genes. *EMBO J* 1982;1:1399–1404.

Su X, Prestwood AK, McGraw RA. Production of recombinant porcine tumor necrosis factor alpha in a novel *E. coli* expression system. *BioTechniques* 1992;13:756–762.

Sugimoto S, Yokoo Y, Hatakeyama N, Yotsuji A, Teshiba S, Hagino H. Higher culture pH is preferable for inclusion body formation of recombinant salmon growth hormone in *Escherichia coli*. *Biotechnol Lett* 1991;13:385–388.

Summers RG, Knowles JR. Illicit secretion of a cytoplasmic protein into the periplasm of *Escherichia coli* requires a signal peptide plus a portion of the cognate secreted protein. Demarcation of the critical region of the mature protein. *J Biol Chem* 1989;264:20074–20081.

Suominen I, Karp M, Lähde M, Kopio A, Glumoff T, Meyer P, Mäntsälä P. Extracellular production of cloned α-amylase by *Escherichia coli*. *Gene* 1987;61:165–176.

Suter-Crazzolara C, Unsicker K. Improved expression of toxic proteins in *E. coli*. *BioTechniques* 1995;19:202–204.

Swamy KHS, Goldberg AL. *E. coli* contains eight soluble proteolytic activities, one being ATP dependent. *Nature (London)* 1981;292:652–654.

Swamy KHS, Goldberg AL. Subcellular distribution of various proteases in *Escherichia coli*. *J Bacteriol* 1982;149:1027–1033.

Szekely M. *From DNA to Protein. The Transfer of Genetic Information*. John Wiley & Sons, Inc., New York, 1980, pp. 13–17.

Tabor S, Richardson CC. A bacteriophage T7 RNA polymer-ase/promoter system for controlled exclusive expression of specific genes. *Proc Natl Acad Sci USA* 1985;82:1074–1078.

Tacon W, Carey N, Emtage S. The construction and characterization of plasmid vectors suitable for the expression of all DNA phases under the control of the *E. coli* tryptophan promoter. *Mol Gen Genet* 1980;177:427–438.

Talmadge K, Gilbert W. Cellular location affects protein stability in *Escherichia coli*. *Proc Natl Acad Sci USA* 1982;79:1830–1833.

Tanabe H, Goldstein J, Yang M, Inouye M. Identification of the promoter region of the *Escherichia coli* major cold shock gene, cspA. *J Bacteriol* 1992;174:3867–3873.

Tarragona-Fiol A, Taylorson CJ, Ward JM, Rabin BR. Production of mature bovine pancreatic ribonuclease in *Escherichia coli*. *Gene* 1992;118:239–245.

Tate WP, Brown CM. Translational termination: "stop" for protein synthesis or "pause" for regulation of gene expression. *Biochemistry* 1992;31:2443–2450.

Taylor A, Brown DP, Kadam S, Maus M, Kohlbrenner WE, Weigl D, Turon MC, Katz L. High-level expression and purification of mature HIV-1 protease in *Escherichia coli* under control of the araBAD promoter. *Appl Microbiol Biotechnol* 1992;37:205–210.

Taylor ME, Drickamer K. Carbohydrate-recognition domains as tools for rapid purification of recombinant eukaryotic proteins. *Biochem J* 1991;4:575–580.

Tessier L-H, Sondermeyer P, Faure T, Dreyer D, Benavente A, Villeval D, Courtney M, Lecocq J-P. The influence of mRNA primary and secondary structure on human IFN-γ gene expression in *E. coli*. *Nucleic Acids Res* 1984;12:7663–7675.

Thomann H-U, Ibba M, Hong K-W, Söll D. Homologous expression and purification of mutants of an essential protein by reverse epitope-tagging. *Bio/Technology* 1996;14:50–55.

Thomas CD, Modha J, Razzaq TM, Cullis PM, Rivett AJ. Controlled high-level expression of the lon gene of *Escherichia coli* allows overproduction of Lon protease. *Gene* 1993;136:237–242.

Thornton JM. Disulfide bridges in globular proteins. *J Mol Biol* 1981;151:261–287.

Tobias JW, Shrader TE, Rocap G, Varshavsky A. The N-end rule in bacteria. *Science* 1991;254:1374–1377.

Tolentino GJ, Meng S-Y, Bennett GN, San K-Y. A pH-regulated promoter for the expression of recombinant proteins in *Escherichia coli*. *Biotechnol Lett* 1992;14:157–162.

Torriani A. Influence of inorganic phosphate in the formation of phosphatases by *Escherichia coli*. *Biochim Biophys Acta* 1960;38:460–479.

Trill JJ, Shatzman AR, Ganguly S. Production of monoclonal antibodies in COS and CHO cells. *Curr Opin Biotechnol* 1995;6:553–560.

Trudel P, Provost S, Massie B, Chartrand P, Wall L. pGATA: A positive selection vector based on the toxicity of the transcription factor GATA-1 to bacteria. *BioTechniques* 1996;20:684–693.

Tunner JR, Robertson CR, Schippa S, Matin A. Use of glucose starvation to limit growth and induce protein production in *Escherichia coli*. *Biotechnol Bioeng* 1992;40:271–279.

Tzareva NV, Makhno VI, Boni IV. Ribosome-messenger recognition in the absence of the Shine-Dalgarno interactions. *FEBS Lett* 1994;337:189–194.

Uhlén M, Forsberg G, Moks T, Hartmanis M, Nilsson B. Fusion proteins in biotechnology. *Curr Opin Biotechnol* 1992;3:363–369.

Uhlén M, Moks T. Gene fusions for purpose of expression: An introduction. *Methods Enzymol* 1990;185:129–143.

Uhlén M, Nilsson B, Guss B, Lindberg M, Gatenbeck S, Philipson L. Gene fusion vectors based on the gene for staphylococcal protein A. *Gene* 1983;23:369–378.

Ullmann A. One-step purification of hybrid proteins which have β-galactosidase activity. *Gene* 1984;29:27–31.

Umana P, Jean-Mairet J, Moudry R, Amstutz H, Bailey JE. Engineered glycoforms of an anti-neuroblastoma IgG1 with optimized antibody-dependent cellular cytotoxic activity. *Nat Biotechnol* 1999;17:176–180.

Underwood PA, Bean PA. Hazards of the limiting-dilution method of cloning hybridomas. *J Immunol Methods* 1988;107:119–128.

Urlaub G, Chasin LA. Isolation of Chinese hamster cell mutants deficient in dihydrofolate reductase activity. *Proc Natl Acad Sci USA* 1980;77:4216–4220.

van Dijl JM, de Jong A, Smith H, Bron S, Venema G. Signal peptidase I overproduction results in increased efficiencies of export and maturation of hybrid secretory proteins in *Escherichia coli*. *Mol Gen Genet* 1991;227:40–48.

Varshavsky A. The N-end rule. *Cell* 1992;69:725–735.

Vasquez JR, Evnin LB, Higaki JN, Craik CS. An expression system for trypsin. *J Cell Biochem* 1989;39:265–276.

Vellanoweth RL, Rabinowitz JC. The influence of ribosome-binding-site elements on translational efficiency in *Bacillus subtilis* and *Escherichia coli* in vivo. *Mol Microbiol* 1992;6:1105–1114.

Villa-Komaroff L, Efstratiadis A, Broome S, Lomedico P, Tizard R, Chick SWL, Gilbert W. A bacterial clone synthesizing proinsulin. *Proc Natl Acad Sci USA* 1978;75:3727–3731.

von Heijne G. Transcending the impenetrable: How proteins come to with membranes. *Biochim Biophys Acta* 1988;947:307–333.

von Heijne G. The signal peptide. *J Membr Biol* 1990;115:195–201.

von Heijne G, Abrahmsén L. Species-specific variation in signal peptide design. Implications for protein secretion in foreign hosts. *FEBS Lett* 1989;244:439–446.

von Strandmann EP, Zoidl C, Nakhei H, Holewa B, von Strand-mann RP, Lorenz P, Klein-Hitpass L, Ryffel GU. A highly specific and sensitive monoclonal antibody detecting histidine-tagged recombinant proteins. *Protein Eng* 1995;8:733–735.

Voorma HO. Control of translation initiation in prokaryotes. In: *Translational Control*, Hershey JWB, Mathews MB, Sonenberg N, eds. Cold Spring Harbor Laboratory Press, Cold Spring Harbor, NY, 1996, pp. 759–777.

Voziyanov Y, Konieczka JH, Stewart AF, Jayaram M. Stepwise manipulation of DNA specificity in Flp recombinase: Progressively adapting Flp to individual and combinatorial mutations in its target site. *J Mol Biol* 2003;326:65–76.

Voziyanov Y, Pathania S, Jayaram M. A general model for site-specific recombination by the integrase family recombinases. *Nucleic Acids Res* 1999;27:930–941.

Wall JG, Plu A.

Wada K-N, Wada Y, Ishibashi F, Gojobori T, Ikemura T. Codon usage tabulated from the GenBank genetic sequence data. *Nucleic Acids Res* 1992;20(Suppl.):2111–2118.

Wang L-F, Yu M, White JR, Eaton BT. BTag: A novel six-residue epitope tag for surveillance and purification of recombinant proteins. *Gene* 1996;169:53–58.

Warburton N, Boseley PG, Porter AG. Increased expression of a cloned gene by local mutagenesis of its promoter and ribosome binding site. *Nucleic Acids Res* 1983;11:5837–5854.

Ward ES. Expression and secretion of T-cell receptor V and V domains using *Escherichia coli* as a host. *Scand J Immunol* 1991;34:215–220.

Ward ES, Gupta D, Griffiths AD, Jones PT, Winter G. Binding activities of a repertoire of single immunoglobulin variable domains secreted from *Escherichia coli*. *Nature (London)* 1989;341:544–546.

Ward GA, Stover CK, Moss B, Fuerst TR. Stringent chemical and thermal regulation of recombinant gene expression by vaccinia virus vectors in mammalian cells. *Proc Natl Acad Sci USA* 1995;92:6773–6777.

Warne SR, Thomas CM, Nugent ME, Tacon WCA. Use of a modified *Escherichia coli* trpR gene to obtain tight regulation of high-copy-number expression vectors. *Gene* 1986;46:103–112.

West AG, Fraser P. Remote control of gene transcription. *Hum Mol Genet* 2005;14:R101–R111.

Westwood AD, Rowe DA, Clarke HR. Improved recombinant protein yield using a codon deoptimized DHFR selectable marker in a CHEF1 expression plasmid. *Biotechnol Prog* 2010;26:1558–1566.

Wetzel R. Protein aggregation in vivo. Bacterial inclusion bodies and mammalian amyloid. In: *Stability of Protein Pharmaceuticals. B. In Vivo Pathways of Degradation and Strategies for Protein Stabilization*, Ahern TJ, Manning MC, eds. Plenum Press, New York, 1992, pp. 43–48.

WHO Expert Committee on Biological Standardization. Requirements for the collection, processing and quality control of blood, blood components and plasma derivatives (Requirements for biological substances No. 27, revised 1992). 43rd Report, Annex 2. 1994. World Health Organization, Geneva, Switzerland. WHO Technical Report Series, No. 840.

WHO. Report of a WHO consultation on medicinal and other products in relation to human and animal transmissible spongiform encephalopathies. WHO/BLG/97.2. 1997. World Health Organization, Geneva, Switzerland.

WHO. WHO requirements for the use of animal cells as in vitro substrates for the production of biologicals (Requirements for biological substances No. 50). *Dev Biol Stand* 1998;93:141–171.

Wiberg FC, Rasmussen SK, Frandsen TP et al. Production of target-specific recombinant human polyclonal antibodies in mammalian cells. *Biotechnol Bioeng* 2006;94:396–405.

Wickner W. Assembly of proteins into membranes. *Science* 1980;210:861–868.

Wigler M, Perucho M, Kurtz D, Dana S, Pellicer A, Axel R, Silverstein S. Transformation of mammalian cells with an amplifiable dominant-acting gene. *Proc Natl Acad Sci USA* 1980;77:3567–3570.

Wikstrom PM, Lind LK, Berg DE, Björk GR. Importance of mRNA folding and start codon accessibility in the expression of genes in a ribosomal protein operon of *Escherichia coli*. *J Mol Biol* 1992;224:949–966.

Wilkinson DL, Harrison RG. Predicting the solubility of recombinant proteins in *Escherichia coli*. *Bio/Technology* 1991;9:443–448.

Wilkinson DL, Ma NT, Haught C, Harrison RG. Purification by immobilized metal affinity chromatography of human atrial natriuretic peptide expressed in a novel thioredoxin fusion protein. *Biotechnol Prog* 1995;11:265–269.

Williams KL, Emslie KR, Slade MB. Recombinant glycoprotein production in the slime mould *Dictyostelium discoideum*. *Curr Opin Biotechnol* 1995;6:538–542.

Wilson BS, Kautzer CR, Antelman DE. Increased protein expression through improved ribosome-binding sites obtained by library mutagenesis. *BioTechniques* 1994;17:944.

Wilson KS, von Hippel PH. Transcription termination at intrinsic terminators: The role of the RNA hairpin. *Proc Natl Acad Sci USA* 1995;92:8793–8797.

Wirth D, Gama-Norton L, Riemer P, Sandhu U, Schucht R, Hauser H. Road to precision: Recombinase-based targeting technologies for genome engineering. *Curr Opin Biotechnol* 2007;18:411–419.

Wittliff JL, Wenz LL, Dong J, Nawaz Z, Butt TR. Expression and characterization of an active human estrogen receptor as a ubiquitin fusion protein from *Escherichia coli*. *J Biol Chem* 1990;265:22016–22022.

Wlaschin KF, Nissom PM, Gatti Mde L et al. EST sequencing for gene discovery in Chinese hamster ovary cells. *Biotechnol Bioeng* 2005;91:592–606.

Wolber V, Maeda K, Schumann R, Brandmeier B, Wiesmüller L, Wittinghofer A. A universal expression-purification system based on the coiled-coil interaction of myosin heavy chain. *Bio/Technology* 1992;10:900–904.

Wong HC, Chang S. Identification of a positive retroregulator that stabilizes mRNAs in bacteria. *Proc Natl Acad Sci USA* 1986;83:3233–3237.

Wong HC, Chang S. 3-Expression enhancing fragments and method. U.S. patent 4,910,141, March 1990.

Wückthun C, Plückthun A. A versatile and highly repressible *Escherichia coli* expression system based on invertible promoters: Expression of a gene encoding a toxic product. *Gene* 1993;136:199–203.

Wückthun C, Plückthun A. Correctly folded T-cell receptor fragments in the periplasm of *Escherichia coli*—Influence of folding catalysts. *Mol J Biol* 1994a;242:655–669.

Wückthun C, Plückthun A. Protein folding in the periplasm of *Escherichia coli*. *Mol Microbiol* 1994b;12:685–692.

Wurm FM. Production of recombinant protein therapeutics in cultivated mammalian cells. *Nat Biotechnol* 2004;22:1393–1398.

Wurtele H, Little KC, Chartrand P. Illegitimate DNA integration in mammalian cells. *Gene Ther* 2003;10:1791–1799.

Xu X, Nagarajan H, Lewis NE et al. The genomic sequence of the Chinese hamster ovary (CHO)-K1 cell line. *Nat Biotechnol* 2011;29:735–741.

Xue G-P, Johnson JS, Smyth DJ, Dierens LM, Wang X, Simpson GD, Gobius KS, Aylward JH. Temperature regulated expression of the tac/lacI system for overproduction of a fungal xylanase in *Escherichia coli. Appl Microbiol Biotechnol* 1996;45:120–126.

Xue G-P, Plückthun A. Cultivation process and constructs for use therein. International patent application WO 95/11981, May 1995.

Yabuta M, Onai-Miura S, Ohsuye K. Thermo-inducible expression of a recombinant fusion protein by *Escherichia coli* lac repressor mutants. *Biotechnol J* 1995;39:67–73.

Yamada M, Kubo M, Miyake T, Sakaguchi R, Higo Y, Imanaka T. Promoter sequence analysis in *Bacillus* and *Escherichia*: Construction of strong promoters in *E. coli. Gene* 1991;99:109–114.

Yamamoto T, Imamoto F. Differential stability of trp messenger RNA synthesized originating at the trp promoter and pL promoter of lambda trp phage. *J Mol Biol* 1975;92:289–309.

Yamane T, Shimizu S. Fed-batch techniques in microbial processes. *Adv Biochem Eng* 1984;30: 147–194.

Yamano N, Kawata Y, Kojima H, Yoda K, Yamasaki M. In vivo biotinylation of fusion proteins expressed in *Escherichia coli* with a sequence of *Propionibacterium freudenreichii* trans-carboxylase 1.3S biotin subunit. *Biosci Biotechnol Biochem* 1992;56:1017–1026.

Yang G, Withers SG. Ultrahigh-throughput FACS-based screening for directed enzyme evolution. *Chembiochem* 2009;10:2704–2715.

Yang M-T, Scott II HB, Gardner JF. Transcription termination at the thr attenuator. Evidence that the adenine residues upstream of the stem and loop structure are not required for termination. *J Biol Chem* 1995;270:23330–23336.

Yang W, Xia W, Mao J et al. High level expression, purification and activation of human dipeptidyl peptidase I from mammalian cells. *Protein Expression Purif* 2011;76:59–64.

Yanisch-Perron C, Vieira J, Messing J. Improved M13 phage cloning vectors and host strains: Nucleotide sequences of the M13mp18 and pUC19 vectors. *Gene* 1985;33:103–119.

Yansura DG. Expression as trpE fusion. *Methods Enzymol* 1990;185:161–166.

Yansura DG, Henner DJ. Use of *Escherichia coli* trp promoter for direct expression of proteins. *Methods Enzymol* 1990;185:54–60.

Yasukawa T, Kaneiishii C, Maekawa T, Fujimoto J, Yamamoto T. Increase of solubility of foreign proteins in *Escherichia coli* by coproduction of the bacterial thioredoxin. *J Biol Chem* 1995;270:25328–25331.

Ye J, Alvin K, Latif H, Hsu A, Parikh V, Whitmer T, Tellers M, de la Cruz Edmonds MC, Ly J, Salmon P, Markusen JF. Rapid protein production using CHO stable transfection pools. *Biotechnol Prog* 2010;26:1431–1437.

Yee L, Blanch HW. Recombinant protein expression in high cell density fed-batch cultures of *Escherichia coli. Bio/Technology* 1992;10:1550–1556.

Yike I, Zhang Y, Ye J, Dearborn DG. Expression in *Escherichia coli* of cytoplasmic portions of the cystic fibrosis transmembrane conductance regulator: Apparent bacterial toxicity of peptides containing R-domain sequences. *Protein Expression Purif* 1996;7:45–50.

Yoshikawa T, Nakanishi F, Ogura Y, Oi D, Omasa T, Katakura Y, Kishimoto M, Suga KI. Flow cytometry: An improved method for the selection of highly productive gene-amplified CHO cells using flow cytometry. *Biotechnol Bioeng* 2001;74:435–442.

Young JF, Dusselberger U, Palese P, Ferguson B, Shatzman AR. Efficient expression of influenza virus NS1 nonstructural proteins in *Escherichia coli. Proc Natl Acad Sci USA* 1983;80:6105–6109.

Yu P, Aristidou AA, San K-Y. Synergistic effects of glycine and bacteriocin release protein in the release of periplasmic protein in recombinant *E. coli. Biotechnol Lett* 1991;13:311–316.

Zacharias M, Goringer HU, Wagner R. Analysis of the Fis-dependent and Fis-independent transcription activation mechanisms of the *Escherichia coli* ribosomal RNA P1 promoter. *Biochemistry* 1992;31:2621–2628.

Zahn-Zabal M, Kobr M, Girod PA, Imhof M, Chatellard P, de Jesus M, Wurm F, Mermod N. Development of stable cell lines for production or regulated expression using matrix attachment regions. *J Biotechnol* 2001;87:29–42.

Zentgraf H, Frey M, Schwinn S, Tessmer C, Willemann B, Samstag Y, Velhagen I. Detection of histidine-tagged fusion proteins by using a high-specific mouse monoclonal anti-histidine tag antibody. *Nucleic Acids Res* 1995;23:3347–3348.

Zhang J, Deutscher MP. *Escherichia coli* RNase D: Sequencing of the rnd structural gene and purification of the overexpressed protein. *Nucleic Acids Res* 1988;16:6265–6278.

Zhang J, Deutscher MP. Analysis of the upstream region of the *Escherichia coli* rnd gene encoding RNase D. Evidence for translational regulation of a putative tRNA processing enzyme. *J Biol Chem* 1989;264:18228–18233.

Zhang J, Deutscher MP. A uridine-rich sequence required for isolation of prokaryotic mRNA. *Proc Natl Acad Sci USA* 1992;89:2605–2609.

Zhang S, Zubay G, Goldman E. Low usage codons in *Escherichia coli*, yeast, fruit fly and primates. *Gene* 1991;105:61–72.

Zhang Z, Yang X, Yang H, Yu X, Li Y, Xing J, Chen Z. New strategy for large-scale preparation of the extracellular domain of tumor-associated antigen HAb18G/CD147 (HAb18GED). *J Biosci Bioeng* 2011;111:1–6.

Zhou M, Crawford Y, Ng D et al. Decreasing lactate level and increasing antibody production in Chinese Hamster Ovary cells (CHO) by reducing the expression of lactate dehydrogenase and pyruvate dehydrogenase kinases. *J Biotechnol* 2011;153:27–34.

Zhu B, Benjamin D, Zheng Y, Angliker H, Thiry S, Siegmann M, Jost JP. Overexpression of 5-methylcytosine DNA glycosylase in human embryonic kidney cells EcR293 demethylates the promoter of a hormone-regulated reporter gene. *Proc Natl Acad Sci USA* 2001;98:5031–5036.

Chapter 7 Upstream systems optimization

Background

Biosimilar products are almost universally manufactured using recombinant technology that comprises distinct upstream and downstream stages. The upstream phase consists of expressing a recombinant protein or antibody in a culture medium and harvesting it; the downstream phase consists of separating and purifying the target product. It is noteworthy that the scale of production bears a significant consideration on the upstream system optimization process since the dosing of biosimilar products may vary from a few micrograms to a few grams, making the choice of the upstream process critical to the success of biosimilar products. Hence, these products must be manufactured at a reasonable cost considering their market price will inevitably be lower, reducing profit margins significantly. This is particularly true for monoclonal antibodies (mAbs); mAbs doses are high, and the process methods are generally more intricate and complex. Developers of biosimilar products are faced with the challenge of the large investment requirement when embarking on the development of monoclonal antibodies, which is why biosimilar products developers' focus is rather on using innovative systems with lower capital requirements, high titer cell lines, and simplified purification processes, etc. The manufacturing of recombinant DNA products, for example, is a process that involves the two stages discussed earlier:

1. Upstream processing consists of a pure *culture* of the chosen organism, in sufficient quantity and in the correct physiological state; *sterilized*, properly formulated media, a seed bioreactor to develop inoculum to initiate the process in the main bioreactor; the production scale bioreactor.
2. Downstream processing consists of equipment for drawing the culture medium in steady state, cell separation, a collection of cell-free supernatant, product purification, and effluent treatment.

The key considerations in upstream recombinant systems include the cell line (mainly bacterial, but mammalian cells are also used), the type of media used, the process of expression and its conditions, and the type of bioreactor. Growing cells in supporting media under optimal conditions results in a wide range of products; the process requires control of precise conditions to provide a commercially feasible yield of the target protein, which is achieved in several forms: secretion of media, inclusion bodies in bacteria, etc. The fermentation process (or bioprocess) is also used to generate biomass, various amino acids and vitamins, and alcohol or even to modify compounds in addition to producing recombinant products, which is the focus of this book. Microorganisms (prokaryotic) and eukaryotic cell cultures are grown in volumes ranging from a few hundred milliliters to several thousand liters using a variety of cell growth methods and bioreactor designs. The large range of bioreactor sizes are due to recombinant protein quantity needs that vary widely, from hundreds of kilograms for albumin, hemoglobin, and insulin to perhaps a few grams for drugs like erythropoietins, interleukins, etc. As a result, it is unlikely that the initial biosimilar therapeutic protein will require a large volume due mainly to the high initial investment required.

Bioreactors

A bioreactor is an apparatus used to grow biological cells and organisms. Historically, the word fermenter was used when microorganisms were grown and bioreactor when cells other than microorganisms are grown; however, these terms are now frequently used interchangeably, with preference for bioreactor in the recombinant production industry. A large variety of bioreactors are available such as shake flasks, roller bottles, spinner flasks, flexible cell culture flasks, wave bags, stirred tanks, and airlift reactors to support suspension, microcarrier and cell encapsulation cultures. Many of these systems work at laboratory scale, yet a large array of them is used in the manufacturing operations if the required growth conditions are not reproduced in larger more efficient reactors. Biosimilar products manufacturers must be wary of replicating the system of production used by the originator and as approved by the regulatory authorities. At times, the biosimilar manufacturer may have to opt for a less efficient system merely to keep from introducing additional variable factors like using cube system for cells that require stationary media vis-à-vis the use of roller bottles. The extent to which the biosimilar manufacturer will be able to modify the production technique depends on the nature of validation required by the FDA in its comparability protocol guidelines for biosimilar product manufacturing; these guidelines are not yet released but are likely to be released in the coming months. The manufacturer may decide to forego the original method of production used and instead go for a more efficient and innovative system to obtain better yields even though this may require a larger investment upfront. The chapter on single-use technology in this book offers an attractive option to startups in the field of biosimilar products.

The Wave technology (General Electric) offers systems up to 500 L cultures with the advantage of using disposable materials. The technology is based on disposable plastic containers with ports to provide media and utility and monitoring probes; the bag rests on a plate that flips back and forth to create a wave motion inside the bag. The wave bioreactor is not free of shear forces and cell damage has been observed though improvement in the system are rapidly appearing and it is likely that this may turn out to be the premier system for mammalian cells and whenever contamination can be a problem, for example, using the same suite of multiple products.

The stirred tanks consist of stainless steel containers with rotating impellers of various configurations and devices that maintain constant agitation and control gas transfer, temperature, pH, and fluid level. Most stirred tanks use some form of sintered material or perforated tubing in order to sparge air or specific gas mixtures into the cell suspension. Figure 7.1 shows a typical and perhaps one of the most popular bioreactors.

Airlift bioreactors have no moving parts or mechanical seals and offers low hydrodynamic shear forces with a low power input per unit volume (10–15 W/m^3). This kind of bioreactors offers gentle gas circulation and good oxygen transfer. However, as the volume of the reactor increases, mixing becomes a limiting factor for the productivity (i.e., the amount of product formed per unit volume per unit time).

Batch culture

A batch culture is a closed system with a fixed culture volume in which the cells grow until the maximum cell density is reached, depending on medium nutrients,

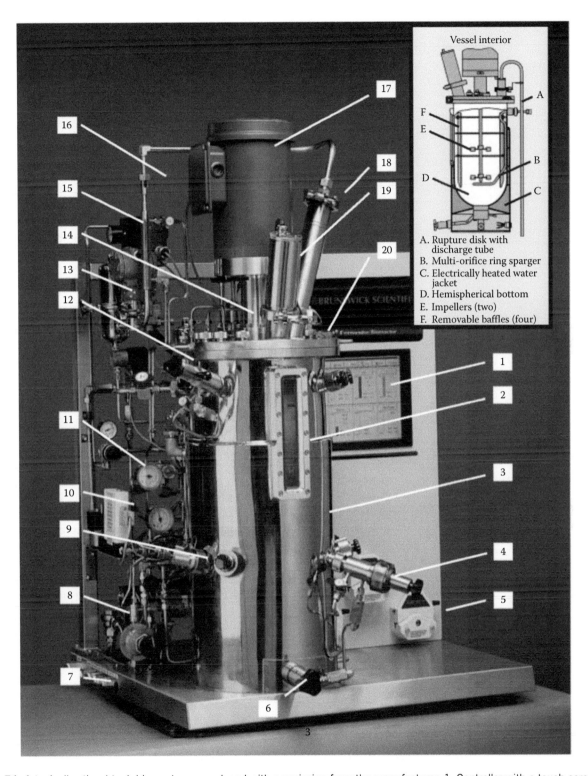

Figure 7.1 A typically stirred tank bioreactors reproduced with permission from the manufacturer. 1. Controller with a touch-screen interface, 2. Viewing window, 3. Sterilizer-In-Place system, 4. Resterilizable sample valve, 5. Peristaltic pump, 6. Resterilizable harvest/drain valve, 7. Services, water, air, clean and house steam, water return, and drain, 8. Steam taps, 9. Ports to allow the addition of RTD, pH, DO, and other sensors, 10. Thermal mass flow controller, 11. Open frame is piping to facilitate access for cleaning maintenance and servicing, 12. Resterilizable inoculation/addition valves, 13. Filters, 14. Rupture disk to prevent over pressurization, 15. Automatic back pressure regulator, 16. Exhaust line with heat exchanger and view glass, 17. Motor, 18. Exhaust condenser, 19. Combination light and fill port, 20. Headplate ports for sampling, insertion of sensors and other devices.

product toxicity, waste product toxicity, and other essential factors. When a particular organism is introduced into a selected growth medium, the growth of the inoculum does not occur immediately, rather it takes a pause, the period of adaptation, called the lag phase. Following the lag phase, the rate of growth of the organism steadily increases, for a certain period called the logarithmic or exponential phase. After a certain time based on a variety of nutritional and cell characteristics, the rate of growth slows down, due to the continuously falling concentrations of nutrients and/or a continuously increasing (accumulating) concentrations of toxic substances. This is called the deceleration phase. After the deceleration phase, growth ceases and the culture enters a stationary phase or a steady state. The biomass (or the total quantity of cell mass) remains constant, except when certain accumulated chemicals in the culture begin to lyse the cells (chemolysis). Unless other microorganisms contaminate the culture, the chemical constitution remains unchanged. Mutation of the organism in the culture can also be a source of contamination called internal contamination. If the desired product is produced in the log phase, it can be prolonged by manipulating the growth conditions but only for a very limited time, and if the desired amount is not produced quickly, it will be reasonable to choose another culture method.

Batch processing or batch culture manufacturing thus relies on disbanding a culture at the onset of the stationary phase for the recovery of its biomass (cells and organism) or the compounds (expressed proteins) that accumulated in the medium. A significant advantage of batch processing is the optimum levels of product recovery, control of growth conditions, and better regulatory compliance. The disadvantages are the wastage of unutilized nutrients, high labor costs, and the time lost in batch preparation. Batch cultivation, however, remains the simplest way to produce a recombinant protein. In batch cultivation, all the nutrients required for cell growth are supplied from the start, and the growth is initially unrestricted. However, the unrestricted growth commonly leads to unfavorable changes in the growth medium such as oxygen limitation and pH changes. Also, certain metabolic pathways in the cell will be saturated, which potentially leads to the accumulation of inhibitory by-products in the medium. Therefore, only moderate cell densities and production levels can normally be obtained with batch cultivations. Acetate is produced when the culture is growing in the presence of excess glucose or under oxygen-limiting conditions. A high concentration of acetate reduces the growth rate, maximum obtainable cell density, and the level of production of the recombinant protein. It is, therefore, important to maintain the acetate concentration below the inhibitory level. This can be achieved by controlling the cultivation in several ways; the growth rate could be controlled by limiting nutrients, such as sources of carbon or nitrogen, by using glycerol or fructose instead of glucose as the carbon/energy source, by addition of glycine and methionine or by lowering the cultivation temperature, or by metabolic engineering (http://www.babonline.org/bab/035/0091/bab0350091.htm-REF48#REF48). Other problems concerning growth to high cell densities are oxygen limitation, reduced mixing efficiency, heat generation, and high partial pressure of carbon dioxide.

Continuous culture

In continuous processing, the growth is limited by the availability of one or two components of the medium. As the initial quantity of a critical component is exhausted, growth ceases and a steady state is reached; growth is renewed by the addition of the limiting component. A certain amount of the whole culture medium can also be added periodically after the steady state sets in. These additions increase the volume of the medium in the fermentation vessel, which is arranged such that the excess volume drains off as an overflow, which is collected and used for recovery of products. At each step of the addition of the medium, the medium dilutes the concentration of the biomass and the products. New growth, stimulated by the added medium increases the biomass and the products, to another steady state set in, and another aliquot of medium reverses the process. It is called continuous culture or processing since the growth of the organism is controlled by the availability of growth limiting chemical component of the medium; this system is called a chemostat. The rate at which aliquots are added or the dilution rate determines the growth rate.

Commercial adaptation of continuous processing is confined to biomass production, and to a limited extent to the production of potable and industrial alcohol. The production of growth-associated products like ethanol is more efficient in continuous processing, particularly for industrial use.

Fed-batch culture

In the fed-batch medium, a fresh aliquot of the medium is continuously or periodically added, without the removal of the culture fluid. The bioreactor is designed to accommodate the increasing volumes. The system is always in a quasi-steady state. Fed-batch processing requires a greater degree of process and product control. A low but constantly replenished medium has the advantages of maintaining conditions in the culture within the aeration capacity of the bioreactor, removing the repressive effects of medium components such as rapidly used carbon and nitrogen sources and phosphate, avoiding the toxic effects of a medium component and providing limiting level of a required nutrient for an auxotrophic strain. Classically, the fed-batch culture is used in the production of baker's yeast, where biomass is the desired product. Diluting the culture with a batch of fresh medium prevents the production of ethanol, which affects the yield; in the production of yeast, the traces of ethanol were detected in the exhaust gas and the processing steps adjusted accordingly. Another classic fed-batch process product is penicillin, a secondary metabolite. Penicillin process has two stages: an initial growth phase followed by the production phase called the "idiophase." The culture is maintained at low levels of biomass and phenyl acetic acid, the precursor of penicillin is fed into the fermenter continuously, but at a low rate, as the precursor is toxic to the organism at higher concentrations.

In fed-batch cultivation, the carbon/energy source is added in proportion to the consumption rate. Thereby overflow metabolism and the accumulation of inhibitory by-products is minimized. Moreover, the growth rate can be balanced to achieve a maximal production level. The bioreactor should preferentially be equipped to maintain an optimal oxygen concentration, pH, and temperature. Defined media are generally used in fed-batch cultivation. As the concentrations of the nutrients are known and can be controlled by the cultivation, the cultivation is also more reproducible compared with the use of a complex growth medium. However, the addition of complex media, such as yeast extract, is sometimes necessary to obtain a high level of the desired recombinant protein.

In fed-batch processes, neither cells nor medium leave the bioreactor keeping the sugar levels low for a long time, and it is possible to switch from one substrate to another, thus rendering the use of inducible promoters possible. The process is usually performed at low growth rates adjusting the feed rate whose upper limit is dictated by the oxygen transfer limit and cooling strategies available. The feed rate can be subjected to direct feedback control using substrate concentration and indirect feedback control parameters include cell concentration, culture fluorescence, carbon dioxide evolution rate, pDO, and pH-stat; constant, exponential, or increasing rate feeding are fixed and not subject to any feedback control as they are determined in the scale-up stage.

A variation of the classical fed-batch process is the semifed-batch process, where nutrients are added in dry form without changing the culture volume. In contrast to the batch mode, the operation of large-scale bioreactors in the fed-batch mode is subject to much variability. The substrate gradients formed may result in overflow metabolism locally in the bioreactor. The resulting product might be inhomogeneous as the cells produce variants of the target protein during different life cycle phases. Product stability might also be challenged by the presence of proteolytic enzymes, as the product cannot be removed from the reactor before the end of production.

A mathematical model of a fed-batch reactor can be derived from a material balance across the reactor. Despite the apparent similarity between the fed-batch reactor model and the continuous culture model, they are very different. Whereas the chemostat equation for biomass accumulation is composed of a growth and a removal component:

$$\frac{dX}{dt} = \underset{\substack{\text{Biomass} \\ \text{growth}}}{\mu X} \cdot \underset{\substack{\text{Biomass} \\ \text{removal}}}{DX} \tag{7.1}$$

The fed-batch equation is composed of a growth and a dilution component (F/V):

$$\frac{dX}{dt} = \underset{\text{Growth}}{\mu X} - \underset{\substack{\frac{F}{V}X \\ \text{Dilution}}}{\frac{F}{V}X} \tag{7.2}$$

The fed-batch reactor model contains an additional time-dependent equation:

$$\frac{dV}{dt} = F \tag{7.3}$$

The concept of the steady state cannot be as easily applied to a fed-batch reactor. The equations must, therefore, be solved numerically.

Perfusion culture

In perfusion batch cultivation, fresh media is added to the bioreactor, and an equivalent amount is removed, with or without cells. A controlled perfusion bioreactor offers tight control of the growth conditions, and cells can be kept in their productive phase for several months if required. A significant risk in this process is contamination as the bioreactor is frequently handled in addition to the accumulation of nonproducing variants that can affect productivity. The advantages include short harvesting volumes, use of smaller bioreactor vessels, reducing initial capital costs despite lower yields obtained. Since the cells are maintained at the steady state, the resulting product is often more homogeneous. The isolation of product can be controlled as the harvest can be selected from the steady-state production phase of the cell culture resulting in a more homogeneous product. The higher quality of the product obtained from the steady-state culture in perfusion system makes downstream processing more efficient; there is also an option for batch processing of multiple harvests to reduce costs. These differences are often considered in making a choice for the perfusion system when considering the large media volume used in perfusion culture systems.

Table 7.1 summarizes the advantages and disadvantages of the various types of culture processes used.

Table 7.1 Comparative Advantages and Disadvantages of Various Culture Systems

Culture	Advantages	Disadvantages
Batch	Well-tested technology, less contamination risk, and less expensive.	Limited cell density, downtime between batches, nonhomogeneous product, and variable quality between batches.
Fed batch	High cell densities and longer culture periods using low medium viscosity.	Nonhomogeneous product due to changing medium, large reactor size required, more susceptible to contamination, difficult to control, and needs sophisticated monitoring systems.
Perfusion	Long production phase, short down time, reducing cleaning and sterilization, waster removal, continuous expression, dilution of the toxic medium, and smaller reactor vessel.	Susceptible to contamination due to frequent handling, difficult to control, requires sophisticated monitoring, and nutrient gradients.

Suspension culture

A large variety of cells requires adhesion to the stationary surface to express proteins, for example, Chinese hamster ovary (CHO) cells in the production of erythropoietin. Classically, roller bottles are used to immobilize these cells whereby they secrete the protein in the medium that is frequently harvested. These cells can also be immobilized onto the surface of microcarriers or to macroporous particles giving two advantages: easier control and high cell densities, for example, upward of 10^8 cells/mL. Microcarrier technology is used both in the batch and in the perfusion mode. See below for details on the use of microcarrier technology. Several bioreactor systems are based on immobilized cell technology (hollow fiber, ceramic matrix, packed bed, and fluidized bed reactors). Since suspension culture technology is relatively newer and the products approved by the FDA decades ago did not have the option of adopting them, many approved processes still rely on validated methods like the use of roller bottle. There are newer issues with suspension cultures such as cell aggregation and lack of system homogeneity when using hollow fiber systems. These remain to be resolved.

The absence of adventitious organisms in cell cultures is critical. In addition to demonstrating that bacteria, yeast, and molds are not present in cell cultures, the manufacturer must provide evidence that mycoplasmas and adventitious viruses are not present for each individual culture. It is important to recognize that certain hybridomas used for monoclonal antibody production may contain endogenous retroviruses. However, it must be demonstrated that any viruses present in the culture are removed from the final product. This requires the development of suitable analytical techniques to ensure the absence of contamination by mycoplasmas or human and animal adventitious viruses.

The degree and type of glycosylation may be important in the design of cell culture conditions for the production of glycosylated proteins. The degree of glycosylation present may affect the half-life of the product in vivo as well as its potency and antigenicity. Although the glycosylation status of a cell culture product is difficult to determine, it can be verified to be consistent if the culture conditions are highly reproducible.

Cell culture conditions are dependent upon the host system. Before performing a large-scale purification, it is important to check protein amplification in a small pilot experiment to establish optimum conditions for expression. The expression is monitored during growth and induction phase by retaining small samples at key steps in all procedures for the analysis of the purification method. The yield of fusion proteins is highly variable and is affected by the nature of the fusion protein, the host cell, and the culture conditions.

Microcarrier support

Microcarrier culture is a technique that makes possible the high yield culture of anchorage-dependent cells. By using macroporous microcarriers, it is possible to immobilize semiadherent and suspension cell lines, even in protein-free media. Growing cells on microcarriers can dramatically improve yield, reduce serum and media costs, decrease the risks of contamination, and reduce the number of handling steps. The microcarrier surface supplies focal adhesion sites that support cellular traction, the formation of the cytoskeleton, and orientation of organelles. Cells modify and lay down their own extracellular matrix on the microcarrier surface. Cells that form tight junctions can create a uniform cellular sheet around a microporous microcarrrier and generate a specific microenvironment inside the carrier. Cells in such sheets are polarized with typical apical and basolateral side. Microporous carriers allow cell-to-cell communication of low-to-medium molecular weight media components through the microcarrier. The volumetric cell densities in microcarriers allow the use of serum or protein-free media, simplifying downstream processing. There is an additional protection of cells from the shear force stress due to aeration by sparging, spin filters, and impellers. The use of microcarriers allows inoculation and growth of one type of cells inside and another type outside where necessary such as in creating "artificial organs." Compared to spin filters, immobilization of cells allows running perfusion cultures, which in turn allows the use of serum-free media and allows the switch from growth to production media; high densities cause faster rate of exchange of nutrient minimizing retention times of secreted products and speeding harvesting of product at lower temperature—this can significantly improve product degradation profile and build up of toxic substances from the cells.

Amersham Company (now GE) was the first to produce these microporous microcarriers and offers several kinds such as Cytodex for use in animal cells; it is a transparent, hydrophilic, and hydrated, cross-linked dextran for use in stirred cultures. Cytodex 1 is formed by cross-linking dextran matrix with positively charged DEAE groups making it more suitable for established cell lines for production from cultures of primary, and normal diploid cells. Cytodex 3 is formed by coupling a thin layer of denatured pigskin collagen type 1 to the cross-linked dextran matrix; this is more suitable for difficult to cultivate cells with an epithelial-type morphology. The collagen surface is susceptible to digestion by a variety of proteolytic enzymes that provide the opportunity to harvest cells from the microcarriers while maintaining maximum cell viability and membrane integrity.

Cytopore is a transparent, hydrophilic, and hydrated microporous cross-linked cellulose microcarrier with positively charged N,N-ethyl-aminoethyl groups; it is more rigid than Cytodex. Cytopores can be used to immobilize insect cells, yeast, and bacteria besides the CHO cells. Cytopore 1 is designed for use in suspension culture systems for growth of recombinant CHO cells with charge capacity of 1.00 meq/g; Cytopore 2 is optimized for anchorage-dependent cells where the optima charge density required is about 1.8 meq/g.

Cytoline is a range of weighted macroporous carriers for use in stirred tank, packed bed, and fluidized bed cultures; it is nontransparent and composed of medical quality polyethylene, which is weighted by silica. Cytoline offers increased protection from shear forces, improves nutrition and aeration while reducing the need to use the serum. Cytoline 1 is optimized for use in fluidized bed cultures of CHO cells; its high sedimentation rate (120–220 cm/min) enables a high recirculation rate to allow a better supply of oxygen. Cytoline 2 is a lower sedimentation rate (25–75 cm/min) microcarrier for use in stirred cultures but mainly in the culture of hybridomas and other stress-sensitive cells in fluidized beds. Cytopilot Mini offered by Amersham is a laboratory-scale fluidized bed reactor, designed specifically to exploit the potential of Cytoline.

Other suppliers of microporous microcarrier include Cultispher by Percell Biolytica AB (http://www.percell.se/) and Hillex microcarriers by SoloHill (http://www.solohill.com/hillex.html).

Generally, microcarrier cultures can be contained in virtually any type of culture vessel. However, the best results are obtained with equipment that gives even suspension of the microcarriers with gentle stirring. The most suitable vessels for general-purpose microcarrier culture are those with efficient gassing and mixing systems that do not generate high shear forces and provide a homogeneous culture environment. For really high cell densities, a perfusion culture system is needed. However, when selecting a vessel for a perfusion culture, some design criteria need consideration. The stirrer should never come into contact with the inside surface of the vessel during culture because it may damage the microcarriers. Similarly, spinner vessels with a bearing immersed in the culture medium are not suitable because the microcarriers can circulate through the bearing and get crushed. Alternatives to fermenters for perfusion culture exist for laboratory, pilot, and production scale applications.

Note: *Glass culture vessels should be siliconized before use.*

The exact culture procedure depends on the type of cell and on the culture vessel. Macroporous microcarrier cultures normally contain 1–2 g Cytopore 1 and are usually inoculated with about 2×10^6 cells/mL. Perfused cultures may contain much higher microcarrier concentrations. In such instances, the inoculum should be increased proportionally. Successful microcarrier culture depends on the state of the inoculum and correct operation during the initial stages. Conditions vary with the cell type and the culture conditions. Anchorage-dependent cells cannot survive unattached in suspension for very long. The easy access to the interior of the carriers facilitates the initiation of the culture at full culture volume and enables continuous stirring at 30 rpm from the commencement of the culture.

For a stationary culture, cover the bottom of a bacteriological Petri dish with microcarriers. The suggested starting concentration of microcarriers for a 60 mm Ø Petri dish is approximately 2 mg/mL (about 0.1 cc/mL). For stirred cultures, the optimum concentration varies from cell to cell. The concentration of CHO cells is approximately 1–2 mg/mL. However, this very much depends on the feeding strategy for the culture. The high cell density experienced with macroporous carriers means that the culture rapidly consumes any available metabolites. A steady state should be maintained, toxic metabolites should not be allowed to accumulate, and pH values should be maintained at the set level. Rapid changes in pH cause cell peeling and a reduction in the final cell yield. CO_2 should also be kept at the desired level. Cell growth can be monitored by glucose consumption, CO_2 consumption, lactate build-up, and cell counting.

Roller bottle culture system

This used to be the most commonly used method for initial scale-up of attached cells also known as anchorage-dependent cell lines. Roller bottles are cylindrical vessels that revolve slowly (between 5 and 60 revolutions per hour) and which bathes the cells that are attached to the inner surface with the medium. Roller bottles are available typically with surface areas of 1050 cm². The size of some of the roller bottles presents problems since they are difficult to handle in the confined space of a microbiological safety cabinet. Recently, roller bottles with expanded inner surfaces have become available, which has made handling large surface area bottles more manageable, but repeated manipulations and subculture with roller bottles should be avoided if possible. A further problem with roller bottles is with the attachment of cells since some cell lines do not attach evenly. This is a particular problem with epithelial cells. This may be partially overcome a little by optimizing the speed of rotation, generally by decreasing the speed, during the period of attachment for cells with low attachment efficiency.

Spinner flask culture

This is the method of choice for suspension lines including hybridomas and attached lines that have been adapted to growth in suspension, for example, HeLa S3. Spinner flasks are either plastic or glass bottles with a central magnetic stirrer shaft and side arms for the addition and removal of cells and medium, and gassing with CO_2-enriched air. Inoculated spinner flasks are placed on a stirrer and incubated under the culture conditions appropriate for the cell line. Cultures should be stirred at 100–250 revolutions per minute. Spinner flask systems designed to handle culture volumes of 1–12 L are available from Techne (http://www.techne.com), Sigma (http://www.sigmaaldrich.com), and Bellco (http://www.bellcoglass.com/).

Other scale-up options

The next stage of scale-up for both suspension and attached cell lines is the bioreactor that is used for large culture volumes (in the range 100–10,000 L). For suspension cell lines, the cells are kept in suspension by either a propeller in the base of the chamber vessel or by air bubbling through the culture vessel. However, both of these methods of agitation give rise to mechanical stresses. A further problem with suspension lines is that the density obtained is relatively low in the order of 2×10^6 cells/mL.

For attached cell lines, the cell densities obtained are increased by the addition of microcarrier beads. These small beads are 30–100 μm in diameter and can be made of dextran, cellulose, gelatin, glass, or silica and increase the surface area available for cell attachment considerably. The range of microcarriers available means that it is possible to grow most cell types in this system.

A recent advance has been the development of porous microcarriers that has increased the surface area available for cell attachment by a further 10–100 fold. The surface area of 2 g of beads is equivalent to 15 small roller bottles.

The Roller Cell® system utilizes a rotary system of multiple bottles attached to a central collection duct avoiding labor costs associated with periodic harvesting of media. It can be used for a simple culture protocol with a "cell harvest" or multiharvest system (i.e., 10 × refeed/harvest). A comparison of the time taken to process the equivalent of 200 standard bottles manually, with a robotic automated system, and the Roller Cell 40™ shows that for simple harvesting the total man-hours are 15.5 in regular roller bottles, 5.7 h in robotic systems, and only 1.5 h in Roller Cell™ system. In multiple harvest system, if the manual hours are 37 in regular roller bottle system and 33.3 in robotic systems, it takes only 6.3 h in Roller Cell system. The Roller Cell™ system (http://www.synthecon.com/Cellon/rc2.shtml) is certainly worth considering when scaling up production of suspension cell culture systems.

Wave bioreactor

This is a disposable bioreactor that has many advantages including no cleaning, cross-contamination, or validation issues. Cells stay in contact with a disposable sterile biocompatible plastic that conforms to USP Class VI and ISO 10993. Bioreactors, including all fittings and filters, are delivered sterile and ready for use. These are ideal for cGMP applications, and no biosafety cabinet is required for their use. These can be used in an incubator or on the bench with integral heater and optional CO_2/air mixing unit. For suspension, microcarrier or perfusion culture, spinners, roller bottles, and similar systems are not scalable due to the inherently limited mass transfer surface area. The Wave Bioreactor® has no such limits and operation up to 580 L has been demonstrated, with cell densities over 6×10^7 cells/mL. Studies have shown excellent validation for CHO cells from 1 to 500 L capacity Wave bioreactors (http://www.wavebiotech.com).

The biosimilar recombinant protein manufacturers are advised to consider this system as their choice system to avoid many cGMP issues that are inevitable in any bioprocess scaling. The Wave Bioreactor system is an excellent option for suspension clones or attachment-dependent lines using microcarriers. It is not usually a good option for attachment-dependent cell lines that have hitherto been grown on rigid surfaces in roller bottles (e.g., currently available erythropoietin cell line). The surface of the bag is usually not designed for attachment. The Waver Bioreactor provides still too much disturbance during periodic harvesting that can cause a high degree of stress on the cells and probably cause them to slough off the surface and clump rather than maintain the confluent monolayer. Given below are some of the present applications of Wave Bioreactors:

Monoclonal antibodies: The Wave Bioreactor has been used extensively for monoclonal antibody production. Culture can be started at low volume and then fresh medium is added whenever the cell count is sufficiently high. This enables inoculum scale-up without transfers. Batches ranging from 100 mL to 580 L have been run with cell densities over 6×10^6 cells/mL, and the productivity was comparable to stirred tank bioreactors. Dissolved oxygen concentrations were not limiting and remained above 50% saturation.

Insect cells/baculovirus: The high oxygen supply capability of the Wave Bioreactor makes it ideal for insect cell culture. Ten-liter batch volumes are routine with cell densities over 9×10^6 cells/mL. Baculovirus yields are higher than with conventional bioreactors. The Wave Bioreactor system is extremely easy to operate, and inoculum scale-up, and infection can be done inside the bioreactor, reducing the need for transfers.

Anchorage-dependent cells: Agitation in the Wave Bioreactor is powerful enough to mix and aerate the culture; yet, it is gentle enough to cultivate anchorage-dependent cells on various microcarriers. Some reports indicate displacement and rupture of cells not specifically designed for the purpose.

Perfusion culture: Unique internal perfusion filter equipped Cellbags make perfusion culture easy. Bioreactors can be operated for weeks, and cell densities up to 6×10^7 cells/mL have been reported. Applications include high-density culture and patient-specific cell therapy. Wave Bioreactors are in use in GMP applications producing inoculum for large conventional bioreactors, and also for clinical and commercial production of human therapeutics. Reduced cleaning and validation requirements make this an ideal system for GMP applications.

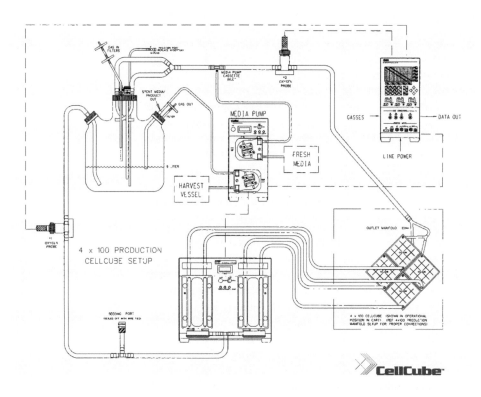

Figure 7.2 The CellCube® culture system. (Courtesy of Corning.)

Other uses: The Wave Bioreactor has many other uses, for example, keeping in-process inoculum pools agitated and aerated prior to use; bead-to-bead transfer; thawing, and mediamixing. Custom Cellbags can be provided for special applications.

Cell cube technology

The CellCube® System offered by Corning (http://www.corning.com/lifesciences/news_center/press_releases/electronic_e-cube.asp) provides a fast, simple, and compact method for the mass culture of attachment-dependent cells. Disposable CellCube modules have a polystyrene tissue culture treated growth surface for cell attachment where the production can range from the module with 8,500 cm² cell growth surface to one with 340,000 cm² using the same control package. The system continually perfuses the cells with fresh media for increased cell productivity. The CellCube System is comprised of four pieces of capital equipment—the system controller, oxygenator, circulation, and media pumps. The cell cubes consist of a series of parallel rigid plates designed for attached monolayers. One set of plates has the equivalent area of 200 roller bottles. The media is contained in a large reservoir that could be replaced on a daily basis (Figure 7.2).

Rotary culture system

A newer entry into culture systems is the rotary cell culture system (e.g., Synthecon's RCCS™; http://www.synthecon.com), which is different from all other cell culture systems. The cylindrical culture vessel is filled with culture fluid, and the cells or tissue particles are added. All air bubbles are removed from the culture vessel. The vessel is attached to the rotator base and rotated about the horizontal axis.

Cells establish a fluid orbit within the culture medium in the horizontally rotating cylindrical vessel. They do not collide with the walls or any other parts of the vessel and often appear as if embedded in gelatin. As cells grow in size, the rotation speed is adjusted to compensate for the increased settling rates of the larger particles. The tissue particles do move enough within the fluid culture medium to exchange nutrients, wastes, and dissolved gases and make contact with other tissue particles. The cells and/or tissue particles often join to form larger tissue particles that continue the differentiation process.

Oxygen supply and carbon dioxide removal are achieved through a gas permeable silicone rubber membrane, which acts very much as lung membranes. Since The Rotary Cell Culture System™ has no impellers, airlifts, bubbles, or agitators, tissue damage from impact and turbulence is significantly decreased as compared to conventional bioreactors. Shear stress and damage are so low that it is essentially insignificant. Under these conditions, cells communicate and synthesize tissue as they would in the body rather than concentrate their energy on repair.

Unlike cell and tissue cultures grown in two-dimensional flat plate systems, cells grown in the Rotary Cell Culture System™ are functionally similar to tissues in the human body. You will be able to grow three-dimensional tissues in vitro that mimic the structure and function of the same tissue in vivo.

Media

The integral component of upstream processing is the media used; its selection depends on the type of bioreactor used and type of culture system adopted. Whereas a large volume of data are available in the scientific literature on the selection of media, the best advice is available from the patent applications that described the original product as well as the information available from the media suppliers. Several environmental and regulatory considerations have changed the selection of media and the manufacturers are strongly advised to develop a good working relationship with media suppliers, who are more than cooperative in offering advice and frequently offer to run test batches to optimize the selection process. The search for media should begin with the following companies:

- http://www.bd.com
- http://www.cambrex.com
- http://www.specialtymedia.com
- http://www.hynetics.com
- http://www.invitrogen.com
- http://www.jrhbio.com
- http://www.irvinesci.com
- http://www.cellgro.com
- http://www.pharma-ingredients.questint.com
- http://www.serologicals.com
- http://www.sial.com

The composition of the cell growth medium is very important as it significantly affects both the cells and the protein expression. For example, the translation of different mRNAs is differentially affected by temperature as well as changes in the culture medium. Nutrient composition and fermentation variables, such as temperature, pH, and other parameters, can affect proteolytic activity, secretion, and production levels. Specific manipulations of the culture medium have been shown to enhance protein release into the medium. Thus, supplementation of the growth medium with glycine enhances the release of periplasmic proteins into the medium without causing significant cell lysis. Similarly, the growth of cells under osmotic stress in the presence of sorbitol and glycyl betaine causes more than a 400-fold increase in the production of soluble, active protein.

Scaling and production costs

The upstream process is linked to downstream process, and the selection of each step in the two phases is determined based on cost optimization considerations, even though the expertise required in each of these steps is highly specialized. The downstream processing adds 80%–90% of the total cost of production; compare this with the cost of recovery in other biological fermentation productions: 5% for whole-cell yeast biomass, 10%–50% for bulk chemicals, 10% for extracellular enzymes, and 20%–50% for antibiotics. The high cost of recombinant DNA products arises from the low yields in aqueous fermentation broths and high purification regulatory requirements. As a result, it is not unusual to find the market price for rDNA products in the range of $100,000 plus while the biomass and chemicals can be bought for pennies per kg.

Cost reduction in the manufacturing of recombinant products is an integrated approach wherein the upstream and downstream processing are developed to minimize waste, use of raw materials, capital, and energy. The large-scale process development for the upstream process takes into consideration several key factors, such as

- Organism selection, with regard to substrate versatility, by-product formation characteristics, robustness of the organism, for example, to process upsets, viability with regard to cell recycling, physiological characteristics (maximum growth rate, aeration requirements, etc.), and genetic accessibility.
- Metabolic and cellular engineering to improve existing properties of the organism, to introduce novel functions, for example, by simplifying product recovery, expanding substrate and product ranges, and enabling fermentation to occur under nonstandard conditions.
- Fermentation process development to achieve culture and media optimization (from complex to defined minimal media), optimization of cultivation parameters that take into account product recovery and purification (minimize byproduct formation, minimize chemical inputs, and develop high cell density cultivation) and incorporation of cell retention/recycling. Several specific steps can be taken to minimize cost.
- Simplification of broth to remove whatever is nonessential, albeit at reduced efficiency, is a good general rule; inevitably, any added component burdens the downstream processing.
- Selecting alternate product form that is easier to separate in downstream processing.
- Reusing broth components, for example, recycling cells, although a technical challenge, holds promise for improving fermentation efficiency. The CHO cells show declining yield over the 7-day period; however, adding fresh cells to already present cells may present a cost-reducing possibility. This cannot be done for processes where the drug is contained in inclusion body. Another strategy is reusing some or all of the broth after product separation. Often, optimum product synthesis and biomass growth take place when medium nutrients are present in excess. However, this results in nutrients being left over at the end of fermentation.
- Removing the product during fermentation improves the yield as the possible inhibitory effects of the product on production are reduced. Using continuous extraction, a side stream can be routed out of the unit, and

the extracted broth returned to conserve broth as well. Further, two-phase fermentations have been developed to extract the product from a biomass-containing aqueous phase into an organic phase, which can then be removed online.

- Reducing the water content, which is typically as high as 90%, reduces downstream processing cost as well as the cost of purified water; this is accomplished by increasing the biomass concentration (i.e., high cell density [HCD] fermentation), engineering the organism to tolerate higher product concentrations, and removing inhibitory elements from the fermentation media composition.
- Use of microcarriers in bioreactors for cells that requires stationary surface is an excellent approach to improve overall yield.
- The introduction of downstream unit operations within a fermentation process reduces the cost substantially, for example, extractive fermentation, electrodialysis, and inline membrane separation technologies.

Problem resolution in fusion protein expression

A large expense should be budgeted for problem resolution during scale-up of upstream processes. The inherent nature of cells, how they interact with media, the role of contamination, etc., make it almost impossible to predict the fate of any scale-up batch. It is this knowledge and experience that the originator companies purportedly adduce as the reason why a biological product cannot be produced as a biosimilar equivalent. It is, therefore, imperative that the scale-up should be fully validated. Following are some of the noted problems that may arise in the expression of fusion proteins and their solutions:

- Too high a level of expression.
  - Add 2% glucose to the growth medium. This will decrease the basal expression level associated with the upstream *lac* promoter but will not affect basal level expression from the *tac* promoter. The presence of glucose should not significantly affect overall expression following induction with IPTG.
  - Basal level expression (i.e., expression in the absence of an inducer, such as IPTG), present with most inducible promoters, can affect the outcome of cloning experiments for toxic inserts; it can select against inserts cloned in the proper orientation. The basal level expression can be minimized by catabolite repression (e.g., growth in the presence of glucose). The *tac* promoter is not subject to catabolite repression. However, with the pGEX vector system there is a *lac* promoter located upstream between the 3′-end of the *lacI*q gene and the *tac* promoter. This *lac* promoter may contribute to the basal level of expression of inserts cloned into the pGEX multiple cloning sites, and it is subject to catabolite repression.
- No protein detected in bacterial sonicate.
  - Check DNA sequences. It is essential that protein-coding DNA sequences are cloned in the proper translation frame in the vectors. Cloning junctions should be sequenced to verify that inserts are in frame.
  - Optimize culture conditions to improve yield. Investigate the effect of cell strain, medium composition, incubation temperature, and induction conditions. Exact conditions will vary for each fusion protein expressed.
  - Analyze a small aliquot of an overnight culture by SDS-PAGE. Generally, a highly expressed protein will be visible by Coomassie™ blue staining when 5–10 μL of an induced culture whose A600 is ~1.0 is loaded on the gel. Nontransformed host *Escherichia coli* cells and

cells transformed with the parental vector should be run in parallel as negative and positive controls, respectively. The presence of the fusion protein in this total cell preparation and its absence from a clarified sonicate may indicate the presence of inclusion bodies.

- Check for expression by immunoblotting. Some fusion proteins may be masked on an SDS-PAGE by a bacterial protein of approximately the same molecular weight. Immunoblotting can be used to identify fusion proteins in these cases. Run an SDS-PAGE of induced cells and transfer the proteins to a nitrocellulose or PVDF membrane (such as Hybond™-C or Hybond-P). Detect fusion protein using anti-GST or anti-His antibody.

- Most of fusion protein is in the postsonicate pellet.
  - Check cell disruption procedure. Cell disruption is seen by partial clearing of the suspension or by microscopic examination. The addition of lysozyme (0.1 volume of a 10 mg/mL lysozyme solution in 25 mM Tris–HCl, pH 8.0) prior to sonication may improve results. Avoid frothing as this may denature the fusion protein.
  - Reduce sonication since oversonication can lead to copurification of host proteins with the fusion protein.
  - Fusion protein may be produced as insoluble inclusion bodies. Try altering the growth conditions to slow the rate of translation, as suggested below. It may be necessary to combine these approaches. Exact conditions must be determined empirically for each fusion protein.
  - Lower the growth temperature (within the range of +20°C to +30°C) to improve solubility.
  - Decrease IPTG concentration to <0.1 mM to alter induction level.
  - Alter time of induction.
  - Induce for a shorter period of time.
  - Induce at a higher cell density for a short period of time.
  - Increase aeration. High oxygen transport can help prevent the formation of inclusion bodies. It may be necessary to combine the above approaches. Exact conditions must be determined empirically for each fusion protein.
  - Alter extraction conditions to improve solubilization of inclusion bodies.

Bacterial manufacturing systems

Bacterial process optimization

Commercial success of a bacterial cell production system depends on a large number of factors, from the cell density to type of media to the conditions of process; optimization remains trial and error based though some basic principles help us avoid obvious pitfalls.

Cell density and viability As most proteins are intracellular accumulated in *E. coli*, productivity is proportional to the final cell density and the specific productivity, that is, the amount of product formed per unit cell mass per unit time. High cell densities require sufficient oxygen supply to avoid formation of acetate, lactate, or pyruvate when *E. coli* is grown under anaerobic or oxygen-limiting conditions (mixed acid fermentation). A further complication of *E. coli* growth on glucose under aerobic conditions is incomplete glucose oxidation resulting in accumulation of acetate, sometimes referred to as the bacterial "Crab Tree" effect; therefore, excess glucose in the medium should be avoided.

Proteolysis is often a major problem especially when the microorganism is stressed, where a high proteolytic protein turnover can be expected. Variations between shake flasks and large-scale bioreactors have been observed challenging the scale-up and limiting the value of down-scale studies.

Higher cell densities are also obtained when using *E. coli* by separating growth and production phases taking advantage of regulated promoters to achieve high cell densities in the first phase (promoter off) and high rate of target protein expression in the second phase (promoter on). Inducers, such as 3,β-indolacrylic acid (IAA), isopropyl-β-D-thiogalactoside (IPTG), and lactose, are used to turn the promoter on (trp and Lac promoters, respectively).

High cell density growth is inevitably faced with problems of oxygen supply, formation of toxic products, high carbon dioxide, and heat generation; one way to resolve this would be to switch to modes other than batch mode since merely increasing the concentration of nutrients is not advised. For example, the maximum concentration of glucose in media should not exceed 50 g/L; ammonia by 3 g/L; iron(II) by 1.15 g/L; magnesium by 8.7 g/L; manganese by 68 mg/L; phosphate by 10 g/L; zinc by 0.038 g/L; molybdenum by 0.8 g/L; boron by 44 mg/L; and cobalt by 0.5 mg/L. Even in fed-batch mode, and in some situations perfusion mode, the longevity of continuous culturing produces an accumulation of nonproducing variants of the microbial cells resulting in decreasing productivity.

Cell density is calculated from OD_{470} on the basis of a calibration curve or by laser turbidimetry and expressed as unit g/L. The viscosity of culture broth increases sharply when the cell concentration exceeds 200 g/L, which is regarded as the maximum attainable cell density of *E. coli*. The cell dry weight is calculated from OD_{470} or OD_{500} on the basis of a calibration curve or by centrifugation and weight of the pellet. Viable cells can be identified by spread out of the cell suspension on nutrient agar plates with or without kanamycin or like selection markers. Cell lysis can be measured from the DNA content in the supernatant following centrifugation. Plasmid instability influences the productivity. Instability may result from defective partitioning during cell division (loss) or undesired modifications (insertion, deletion, and rearrangement of DNA). Most plasmids of industrial interest are lost at frequencies of 10^{-2} to 10^{-5} per cell generation. The plasmid-free cells will have a higher specific growth rate, and they will, in time, reach a higher concentration. Plasmid instability is influenced by plasmid construction, plasmid copy number, cultivation conditions, and bioreactor's configuration. The stability can be estimated by replica plating to ampicillin, kanamycin, or like selection marker agar plates. The method sensitivity is 1%. The number of plasmid molecules per genome in a single cell. PCN is a criterion for the strength of the expression system on the DNA level. The production of target protein depends on the PCN. The PCN is calculated from the measurement of the total DNA content and the plasmid DNA content of the biomass by quantitative DNA assays (e.g., agarose slab gel electrophoresis and capillary electrophoresis) after cell disintegration.

Media

The common media include

- Complex systems containing chemically undefined nutrients such as yeast extract, peptone, tryptone, and casamino acids; it typically comprises tryptone, yeast extract, mineral salts (e.g., sodium choride, potassium hydrogen phosphate, and magnesium sulfate), glucose, and an antibiotic (e.g., kanamycin, ampicillin, and chloramphenicol).
- Defined media comprising solely defined nutrients, it typically comprises mineral salts including ammonium sulfate, trace elements, foam

controlling agents, glucose, and eventually additional components such as thiamine and ampicillin.
- Semicomplex media that is a combination of the above two types.

Control parameters The in-process control parameters have well-defined lower and upper parameter limits and measurement of a variety of output parameters (responses):

- pH is typically set at 6.9; repeated sterilization of probe may delay response time.
- Dissolved oxygen must exceed 20%; because of low solubility in media (approx. 7.6 mg/L), the oxygen transfer rate is a limiting factor when scaling up bacterial cultures. The dissolved oxygen is measured by a polarographic oxygen electrode.
- Temperature lowering to 26°C–30°C not only reduces nutrient uptake and growth rate as well as inclusion bodies but also reduces toxic by-product formation. Heat generation during culture in large-scale fermenters requires cooling.
- Agitation rate is adjusted to keep dissolved oxygen above critical value; generally, 300–1000 rpm are maintained.
- Aeration rate of 1 vvm at 30–70 kPa proves optimal.
- Ammonia feed is maintained to keep ammonium concentrations below 170 mM; ammonia is determined by Kjeldahl analysis.
- Glucose feed is adjusted to not exceed optimal glucose levels using feedback mechanisms. The levels of glucose are measured by HPLC, RI, or commercial kits.
- Acetate forms in complex and defined media when the specific growth rate exceeds 0.2 and 0.35 h^{-1}, respectively; $pCO_2 > 0.3$ atm. decreases growth rate and stimulates acetate formation, which is reduced by using glycerol as a carbon source instead of glucose and by using the exponential feeding method. A concentration of less than 20 mM of acetate maintains an exponential growth. Acetate concentration is monitored by gas chromatography or with an acetate kit.
- Foam level should be minimized by adding antifoam agents.
- By-products are routinely monitored; ethanol detected by gas chromatography (flame ionization detector) or ethanol sensor; propionate and lactate detected by gas chromatography (flame ionization detector); amino acids detected by HPLC, and the product by specific assay (see BP, EP, or USP for assay methods).

Mammalian manufacturing systems

Mammalian cell systems

Mammalian cells like CHO, human cervix (HeLa), African green monkey kidney (COS), baby hamster kidney (BHK) cells, and hybridomas are widely used for the production of monoclonal antibodies and complex posttranslational eukaryotic proteins. The target protein is generally expressed directly to the medium in its native form. The development timeline from gene construct to production is 4–5 months, and the yields obtained range from a few mg/L to g/L. Since the protein is expressed in the low-viscosity medium and the content of host cell proteins is usually low, the purification processes are relative simple; however, these must remove components like peptone, antifoam reagents, growth factors, etc., when used and the released

Table 7.2 Important Parameters in Yeast Cultures

Parameter	Comments
Aeration rate	The aeration rate is commonly set to a fixed value or to increase as the biomass increases. Typical values are 1–2 vvm (liter oxygen per liter of culture per min).
Agitation rate	The stirrer speed is either fixed on a value known to be sufficient to keep the DO above the critical 20% (common values are 500–600 rpm) or established empirically in response to the DO9.
Dissolved oxygen	The dissolved oxygen must exceed 20% and is most commonly set to 30%. Since oxygen solubility in aqueous solution is very low (7.6 µg/mL) the oxygen transfer rate (OTR) is a main limiting factor when scaling up yeast cultures. The oxygen level can be controlled via the aeration rate, vessel top pressure, and agitation rate. The DO is a valuable tool for analyzing the actual culture composition via the "spike test." The "spike tests" confirms that the compound in question (e.g., methanol, glycerol, and glucose) is the rate-limiting factor. The feeding of the said compound is stopped, and the DO is measured. If the response time is long (1–2 min) or the rise in DO is small (~10%), the fermentation is not limited by the compound. A typical rate limited culture will have spike times of 15–30 s.
Foam level	The foam level is controlled by the addition of antifoam agents.
Glucose and glycerol feed	These carbon substrates can be determined online with a near-infrared analyzer or offline with HPLC. Glucose is often determined offline using an enzymatic analytical kit. Spike tests are used to see if these substrates are present in limiting amounts.
Methanol and ethanol	Since high levels (2%–3%) of methanol are toxic to the cells, it is important to keep strict control with the concentration in the cell culture and adjust the feed rate often. Traditionally, the culture is kept methanol limited by performing spike tests regularly. Since a limiting concentration of methanol may not be optimal, it can be advantageous to perform online control with an alcohol sensor in the exhaust gas or a near infrared analyzer. Alternatively, the alcohol content can be analyzed offline with gas chromatography. It has been reported that a methanol concentration of 1% is optimal for *Pichia pastoris* fermentations. The same analytical methods apply to ethanol measurement in *Saccharomyces cerevisiae* fermentations.
pCO_2	It is not certain, that pCO_2 have any effect on yeast cultures, but a negative effect on cell growth has been observed at pressures above 350 mbar.[1] pCO_2 is used to calculate the respiration quotient (RQ), an important parameter, as described below.
pH	Yeast grows well at pH 5–6 although it has been reported that lowering pH to 4 has no effect on the growth of *S. cerevisiae*. A pH decrease to 3 has been employed to reduce the action of proteases. pH 4 has been reported to increase plasmid stability and thereby protein production and has been employed to lower the risk of bacterial contamination. The metabolism of actively growing yeast will result in a decrease in pH; this is opposed to the automatic addition of ammonia hydroxide that can also be a nitrogen source. pH is measured with permanently sealed gel-filled glass combination electrodes or with pressurized electrodes. They are standardized against commercially available standard solutions. Repeated sterilization over a prolonged period of time depletes the outer gel layer of the glass membrane increasing the response time. Pilot-scale and larger vessels are sterilized with steam in situ. NaOH or HCl is often used to adjust pH. Care should be taken to avoid locally high pH during the addition. It is recommended to use less concentrated NaOH solutions (0.1 M).
RQ	The RQ is defined as the ratio of CO_2 production to oxygen consumption. It is a valuable guide to the metabolic state of the yeast cell culture. Aerobic growth yields RQ values above one, whereas values below one indicate a switch to anaerobic fermentation with the concomitant decrease in biomass. For *S. cerevisiae* BT150, an RQ of 1 is optimal for biomass production. The optimal value for protein production must be determined from case to case.
Temperature	The optimal temperature is 30°C. Temperature is sometimes shifted to 28°C or lower to prevent overheating. Respiration generates heat and especially *P. pastoris* fermentations, demanding high amounts of oxygen, generate substantial heat.

host cell proteins and nucleic acids due to apoptosis and cell sensitivity to sheer forces, and the process of virus inactivation/removal. Mammalian cells are difficult and slow to grow and are more fragile than microbial cells making them very sensitive to sheer forces; batch or fed-batch cultures are often used for antibody production, while other recombinant protein also may be produced in continuous cultures over 4–8 weeks. Culture media are expensive relative to those used for microbial and yeast protein expression. Since viral contamination is a real risk, inactivation and removal must be designed into the downstream process, and extensive control procedures established (e.g., end of production testing and virus validation). The relative low expression levels combined with high prices on culture media and expensive quality control programs make it generally more expensive to produce recombinant proteins in animal cells than in microbial systems. However, complex proteins cannot be expressed in microbial systems leaving transgenic animals or plants as the only alternative. The regulatory record of the use of mammalian cells is very good (Table 7.2).

Monoclonal antibodies production

Monoclonal antibodies are expressed in mammalian cells and also in mouse ascites. In the earlier period, the typical expression level of the target protein was from 50 to 1000 mg/L but recently results as high as 20 g/L have been reported. The protein purification strategy takes into consideration the poor stability at pH below 4.5 and irreversible structural changes that may result in loss of immunoactivity. This consideration is different from the manufacturing of other classes of therapeutic proteins where the goal is to keep the protein as nonimmunogenic as possible through structural modifications. IgM tends to precipitate at low conductivity where the protein is in its pentameric form. Generally, IgMs are less stable than IgGs. Most IgGs are stable and soluble in medium to high pH and low conductivity, a condition often used in application samples for chromatographic procedures. However, some mAbs (up to 20% of IgMs) are cryoglobulins with reduced solubility below 37°C. mAbs being highly basic, form stable ionic complexes with polyvalent anions (phosphate, citrate, sulfate, and borate). The complexes easily aggregate. mAbs also form complexes with nucleic acids; the reaction can be reversed in the presence of 0.3–1.0 M NaCl. mAbs strongly bind to divalent metal ions; the resulting change in the net charge of the molecule leads to destabilization. Table 7.1 lists the principles used to remove adventitious agents in the manufacturing of human monoclonal antibodies. Many of these principles apply to the removal of adventitious agents in general as well.

Upstream process optimization

Commercial scale manufacturing is carried out in closed reactors with no or limited supply of medium (batch or rarely fed-batch mode) or in bioreactors allowing media throughput with cell retention (perfusion systems). Immobilized cells can be utilized in both modes resulting in cell densities from 10^6 cells/mL (suspension cultures) to 10^8 cells/mL (immobilized cells). The productivity of batch cultures is often relatively high (up to 2 g/L for antibodies), but product stability, toxicity, and bioreactor volume reduction make perfusion bioreactors fairly competitive. The total process yield following downstream processing may be better using the perfusion mode (fewer purification problems due to a higher quality of the starting material), despite the slightly lower productivity.

There are two types of media used, one as growth media to support the multiplication of cells and production media to maintain cells in the most productive phase. The culture media used for establishing master cell banks still use serum, but the fact that MCB is extensively characterized, the use of serum is justified. However, the rest of the media used should be free of adventitious agents (e.g., viruses and prions), which restricts the use of fetal calf serum, peptones of animal origin, and porcine trypsin. Several commercials supply these types of media and the claims made regarding their suitability should be evaluated vis-à-vis the process requirements. The serum-free media consist of salts, vitamins, amino acids, and carbohydrates supplemented with specific growth and attachment factors such as insulin, IGF-1, transferrin, epidermal growth factor, somatostatin, fibronectin, and collagen. In addition, cells often need steroid-type hormones such as dexamethasone, testosterone, progesterone, and hydrocortisone and lipid-based factors including phospholipids, cholesterol, and sphingomyelin. Protein additives (e.g., albumin, transferrin, growth factors, and peptones) are often of recombinant or plant origin. Albumin, which is the most important protein of all animal sera, is included in many sera-free media. It exhibits a number of functions (transport, detoxification, buffering, and mechanical protection against sheer forces). However, the growth stimulatory effect seems to be associated with a factor not present in the recombinant form indicating that recombinant albumin can only partially replace the natural source. Transferrin, which transports Fe ions to the cells, is used in almost any serum-free medium or is replaced with iron complexes (e.g., ferri-sulfate, iminodiacetic acid complexes, ferri-sulfate–glycine–glycine complexes, ferri-citrate, ferri-tropolone, and phosphate compounds). Peptones are often produced from Soya by enzymatic degradation. Typical concentrations of additives are 5 mg/L of insulin, 5–35 mg/L of transferring, 20 μM of ethanolamine, and 5 μg of selenium. Sometimes, sodium carboxymethyl cellulose (0.1%) is added to prevent mechanical damage to cells. Pluronic F68 (0.1%) is used to reduce foaming and to protect cells from bubble sheer forces in sparged cultures.

Cell detachment has been achieved by the use of trypsin or trypsin/EDTA derived from porcine or bovine pancreas. Serum-free media are devoid of the antitrypsin activity of serum and trypsin has been sought to be replaced by dispase I/II, papain, or pronase or by use of thermoresponsive polymer surfaces.

N-linked glycosylation is a posttranslational modification commonly performed on proteins by eukaryotic cells, and it can significantly alter the efficacy of human therapeutic protein. The use of mammalian cells, such as Chinese hamster ovary (CHO) cells, for the commercial production of recombinant human proteins, is often attributed to their ability to impart desired glycosylation features on proteins. An essential step in N-linked glycosylation is the transfer of oligosaccharide from a lipid in the endoplasmic reticulum membrane to an asparagine residue within a specific amino acid consensus sequence on a nascent polypeptide. In cultured cells, this reaction does not occur at every identical potential glycosylation site on different molecules of the same protein. The resulting variation in the extent of glycosylation for a given protein is known as site occupancy heterogeneity. Although current regulatory practice permits product heterogeneity, demonstration of specific and reproducible glycosylation is required. Hence, heterogeneity in protein glycosylation presents special challenges to the development and production of a candidate therapeutic with consistent properties. In view of the inevitable occurrence and significance of glycosylation heterogeneity in protein therapeutics derived from mammalian cells, much research has been directed toward understanding factors that influence glycosylation heterogeneity during a bioprocess. When using CHO cells, there occurs a gradual decline in glycosylation site occupancy over the course of batch and fed-batch cultures of recombinant CHO cells; the proportion of fully glycosylated molecule can decrease by 9%–25% during the exponential growth phase. This deterioration in glycosylation does neither arise from the extracellular degradation of the product, nor could it be overcome by supplementation of the cultures with extra nutrients, such as nucleotide sugars, glucose, and glutamine. Certain lipid supplements can minimize the glycosylation changes; since lipid-linked oligosaccharides (LLOs) are the oligosaccharide donors in N-linked glycosylation, their availability may be a key regulatory mechanism for controlling the extent of protein glycosylation. Inadequate formation or excessive degradation of LLOs can result in LLO shortages and consequently limit cellular glycosylation capacity. Under subsaturating LLO levels, a gradual decrease in the intracellular pool of LLOs would lead to a corresponding decrease in protein glycosylation. The CHO cells glycosylate their proteins to a gradually increasing extent as culture progresses, until the onset of massive cell death (15%–25%). The glycosylation site occupancy of different proteins may undergo distinct changes over the length of culture even though the net glycosylation efficiency in CHO cells improves with cultivation time. The glycosylation pattern of each individual glycoprotein product needs to be tracked over the course of culture because different proteins may exhibit different glycosylation variations with time, even when the same culture method is used.

Process parameters of importance in optimizing mammalian cell yields include the following:

- pH, which is optimal at 6.8–7.2; lower pH inhibits growth; since CO_2 is generated, control of pH throughout fermentation may be needed due to the formation of lactic acid. pH must be controlled to compensate lactic acid formation by adjusting the supply of CO_2. The problems associated with deterioration of pH measuring devices as described in bacterial cell culture apply here as well.
- Redox potential optimal value is +75 mV, which equals a pO_2 of 8%–10%, which is approximately 50% of air saturation. The redox potential depends on the concentration of reducing and oxidizing agents, the temperature and the pH of the solution. The redox potential falls under logarithmic growth and is at its lowest 24 h before the onset of stationary phase.
- Dissolved oxygen should be in the range 0.06–0.3 µmol oxygen/106 cells/h. As oxygen solubility in aqueous solutions is very low (7.6 µg/mL), the oxygen transfer rate (OTR) is a main limiting factor when scaling up cell cultures.
- Temperature, the optimal temperature range, is from 33°C to 38°C. The relative low level of metabolic activity makes it easier to control the temperature in animal cell bioreactors.
- Osmolality in the culture medium used with lepidopteran cell lines is 345–380 mOsm/kg.
- Agitation rate ranging from 50 to 200 rpm is common but needs optimization depending on the size of the equipment.
- Ammonia production rate (mmol/10^6 cells/h) is an indicator of metabolic rate and should be continuously measured and kept under control. The specific ammonia production rate (mmol/10^6 cells/h) is an indicator of metabolic rate. Analysis of ammonia can be performed using a BioProfile 100 analyzer or similar instrument. Ammonia may also be determined using an enzyme-based assay kit.
- Glucose levels should be monitored to record glucose utilization (mmol/10^6 cells/h) is an indicator of metabolic rate.
- Glutamine, an essential amino acid for cell growth, should not be used as an energy source. The metabolic product, ammonia, which is toxic to the cells, should be controlled. The glutamine concentration should be kept in the range of 2–8 mM.
- Lactate is formed by the conversion of glucose—a balance between glycolysis and oxidation of glucose is maintained.
- pCO_2 is used to adjust the CO_2 flow to regulate the pH of the cell culture, and the flow is maintained only as long as the pH is above the set point. In late stage cultures, the CO_2–HCO_3 buffer system is no longer sufficient to maintain pH and sodium hydrogen carbonate or NaOH is used instead.
- Foam level is controlled by the addition of antifoam agents.

Growth rates are measured in terms of cell density (using a hemocytometer); cell viability is tested using trypan blue, erythrosine staining, or electronic counting (e.g., Coulter counter); and specific production rates are monitored using a cell-hour approach, which expresses the relationship between protein productivity and cell population dynamics. LDH activity reflects the extent of cell lyses since mammalian cells do not secrete LDH. The concentration of the reduced form of nicotinamide adenine dinucleotide or its phosphorylated form correlates with cell mass during the lag phase and exponential growth phase and provides information about the physiological state of the culture. The reduced form fluoresces at 460 nm when irradiated with light at 340 nm while the oxidized form does not fluoresce. Various amino acids can be analyzed by HP-RPC after derivatization with *o*-phthalaldehyde, 3-mercaptopropionic acid, and 9-fluorenylmethylchloroformate using a fluorescence detector. Glutamine can be analyzed using a BioProfile 100 analyzer or similar instrument.

Adventitious agents are tested at the end of the cell culture in the unprocessed bulk and where multiple harvest pools are prepared at different times, the culture is tested at the time of the collection of each pool. The test program should include a test for bacteria, fungi, mycoplasma, and viruses. Test for adventitious viruses in continuous cell line cultures used for expression of recombinant proteins should include inoculation onto monolayer cultures of the same species and tissue type as that used for production, of human diploid cell line, of another cell line from a different species. If appropriate, a PCR test or another suitable method may be used. When cells are readily accessible (e.g., hollow fiber), the unprocessed bulk would constitute harvest from the fermenter.

Yeast cell manufacturing systems

The yeast species *Saccharomyces cerevisiae* and *Pichia pastoris* are also used and their use is growing fast (Tables 7.2 and 7.3) as these systems offer high efficiency (short doubling times, high cell densities, and high yields from better mass transfer of nutrients in unicellular growth morphology) and low fermentation costs. The media cost and scale-up considerations are similar to those encountered in bacterial systems; however, the development time frame is a bit longer, about 14 days from gene construct to production. Both batch and fed-batch culture methods are used. The target protein is usually expressed directly to the medium in its native form although some proteins tend to undergo degradation upon expression (e.g., proinsulin). A low redox potential of the medium/harvest is sometimes observed resulting in cleavage of disulfide bonds. Yeast has GRAS (Generally Regarded As Safe) FDA status, and host-related impurities are very low since the organism has a rigid cell wall, and the expressed protein is secreted into the medium. Like the bacterial cultures, viruses are of little significance as contaminants. The purification process is usually simpler compared to the bacterial system (since the product is secreted in yeast) and no extra host cell–related operations are required. However, the disadvantages include lack of posttranslational modifications and unexpected formation of mono- and diglycosylated forms of the target proteins that may be difficult to remove.

Yeast cell cultures

Yeast can grow both anaerobically and aerobically. In anaerobic growth, most of the carbon substrate is metabolized into inorganic substances like ethanol and carbon dioxide, making this growth less desirable for protein production. The preferred large-scale aerobic fermentation yields more biomass. However, the presence of high levels of readily metabolically available sugars (e.g., glucose) represses aerobic respiration (the Crab-tree effect) in certain yeast species (e.g., *S. cerevisiae*) and it can be difficult in practice to completely suppress the anaerobic growth. The use of a nonrepressing substrate (e.g., raffinose) or continuous culture is often employed to overcome this problem.

Table 7.3 Important Responses in Yeast Cultures

Response	Comments
Amino acids	Detected by HPLC.
Ammonia	Can be determined by Kjeldahl analysis.
Biomass and cell density	The biomass states the amount of wet cell weight (WCW) that can be converted to the dry cell weight (DCW). DCW can be measured gravimetrically. DCW has a linear relationship with the absorbance at a given wavelength (A_{590}–A_{660}) over a certain range, a relationship that must be established on case by case since it varies with size and shape of the cells. For a typical diploid strain $A_{660} = 2$ equals 10^7 cells and for a typical haploid strain $A_{660} = 2$ equals 2×10^7 cells. The linear range is often very short (usually within 0.1–0.3 absorbance units at 660 nm). The actual cell number can be calculated electronically (Coulter counter) or visually by coupling a hemocytometer to a light microscope. The normal convention with budding yeast cells is to consider a daughter cell only an individual cell when completely separated from the mother cell. A brief sonification of the cells before counting will separate cells that have completed cytokinesis. Cell aggregations should not be a frequently encountered problem since the yeast strains used in the industry usually aggregate only when contaminated with bacteria or fungi. In addition, indirect methods measuring the metabolic activities such as glucose or oxygen consumption, RQ, and increase in product formation can be applied.
Cell viability	A wide variety of marker genes is currently available. Use of marker genes that encode resistance to antibiotics is generally not recommended.
Methanol and ethanol	Since high levels (2%–3%) of methanol are toxic to the cells, it is important to keep strict control with the concentration in the cell culture and adjust the feed rate often. Traditionally, the culture is kept methanol limited by performing spike tests regularly. Since a limiting concentration of methanol may not be optimal, it can be advantageous to perform online control with an alcohol sensor in the exhaust gas or a near-infrared analyzer. Alternatively, the alcohol content can be analyzed offline with gas chromatography. It has been reported that a methanol concentration of 1% is optimal for *P. pastoris* fermentations. The same analytical methods apply to ethanol measurement in *S. cerevisiae* fermentations.
Product	Specific assay.

Basically, the fermentation of yeast can be divided into two modes: batch systems with its different variants and continuous systems. The classical batch systems offer minor risks, relative inexpensiveness, and well-tested procedures. The Crab-tree effect of *S. cerevisiae* and other limiting factors of batch fermentations have led to the development of fed-batch systems, and this is now the most commonly employed fermentation strategy. Continuous modes offer the possibility of prolonged fermentation, high productivity, and greater flexibility than do batch modes. However, the higher risk of contamination and greater expense make it less amenable. When choosing fermentation mode, the volumetric productivity, final product concentration, stability, and reproducibility must be considered in a case-to-case manner.

Batch culture

This is a closed system with a definite amount of nutrients and one single harvest. The cells follow classic kinetics with a log phase of rapid proliferation where some products are produced and a stationary phase where the amount of cells does not change and where other products are produced. If the desired product is produced in the log phase, it can be prolonged by manipulating the growth conditions but only for a very limited time. If the desired amount is not produced that quickly, it will be reasonable to choose another culture method. When fermenting *S. cerevisiae*, the cells will start by producing ethanol due to the Crab-tree effect and will not enter the log phase of aerobic respiration until the sugar level is low enough. The osmotic sensitivity of yeast cells puts another constraint on the initial sugar concentration. In addition to these disadvantages, batch culture does not allow for control of the growth rate and the culture becomes rapidly limited by oxygen. All together these factors restrict the duration of batch fermentations. Higher yield of biomass can be achieved in batch fermentations by keeping the sugar level low, using a glucose insensitive yeast like *Candida* or replacing glucose with a nonrepressive substrate such as acetate, galactose, or glycerol.

Fed-batch culture

Another solution to the problems with the Crab-tree effect and osmotic sensitivity is to conduct the fermentation in a fed-batch mode. The medium is added in fixed volumes throughout the process thus increasing the volume of the cell culture with time. Neither cells nor medium leaves the bioreactor. In this way, the sugar levels can be kept low for a long time, and it is possible to switch from one substrate to another thus rendering the use of inducible promoters possible. The process is usually performed at low growth rates, the key control parameter being the feed rate whose upper limit is dictated by the oxygen transfer limit and cooling strategies available. The feed rate is often subjected to feedback control strategies using, for instance, measurement of the respiratory quotient (RQ), biomass production, or heat generation. In *S. cerevisiae*, this feedback includes online analysis of glucose and ethanol, whereas *P. pastoris* fermentations often are controlled by either direct online measurement of methanol but more frequently by "spike tests" (the "spike tests" confirms that the compound in question [e.g., methanol, glycerol, and glucose] is the rate-limiting factor). Variants of the classical fed-batch strategy such as semifed batch where nutrients and vitamins are added in a dry form not changing the culture volume have also been employed with success.

Perfusion culture

An alternative approach to batch cultivation is to continuously add fresh medium to the bioreactor and to remove equivalent amounts of medium with cells. Continuous cultures are performed at low dilution rates usually up to 0.3 h^{-1}. It has been reported that higher dilution rates cause washout even though successful fermentation has been performed with dilution rates up to 0.6 h^{-1} after a time of adaptation. Low dilution rates usually promote a high yield of biomass and an RQ of about 1. In *S. cerevisiae* fermentations, high dilution rates promote a switch to anaerobic fermentation because of too much accessible sugar. Continuous cultures are conducted in chemostats having all the sophisticated monitoring and control apparatus necessary to maintain a successful continuous culture. In the chemostat, the cell density will not be as high as in batch fermentors resulting in a lower productivity, but the long-term continuous expression of the product will in many cases be advantageous to the batch mode. In contrast to batch cultures, a steady state is reached, where the cell density, the substrate concentration, and the product concentration are constants. The disadvantages of continuous cultures are high risk of contamination with faster growing organisms and demand for very complex and expensive apparatus. The longevity of continuous culturing can cause accumulation of nonproducing variants of the yeast cells resulting in decreasing productivity.

Cell immobilization strategies

It has been shown that immobilizing cells increase the achievable cell density and plasmid stability enabling longer cultivation time and higher productivity. The reason for this is not yet clear. Conventional immobilizing methods, such as chemical cross-linking or entrapment of cells in a gel matrix, can alter cell physiology, increase contamination risks, and decrease efficiency, reasons for why these methods have not become very widespread. However, it has been reported, that immobilization of *S. cerevisiae* strain XV2181 cells in a fibrous bed bioreactor promoted a stable long-term production of GM-CSF with a relatively high volumetric productivity (0.98 mg/L/h) performed for 4 weeks without any contamination or cell physiology alteration. The specific protein production and total product yield was, however, much lower than the control batch fermentation with cells in suspension.

Media

Media can be divided into two categories: complex and defined. The complex media are rich broths comprising yeast extract or yeast nitrogen base along with the carbon source and can be supplemented with biotin and peptone. They can be buffered with potassium phosphate. Typical concentrations are 1% yeast extract, 2% peptone, and 0.00004% biotin and the carbon source can be 2% glucose, 1% glycerol, or 0.5% methanol. The defined media comprise salts, trace elements, and optionally amino acids. Often the trace elements are added after sterilization along with the carbon substrate and the required vitamins. Commonly used additives are casein hydrolysates that can prevent proteolytic degradation of the final product by inhibiting extracellular proteases and antifoam agents.

In-process control

In-process control is becoming an increasingly important part of the safety measurements taken in biopharmaceutical development and manufacture. The control program comprises strict parameter control using defined lower and upper parameter limits and measurement of a variety of output parameters (responses). Critical parameters and interactions should be identified from the data collected. The aim is to assure a robust and reproducible process. Below is listed a number of important parameters and responses related to bacterial cell cultures (Table 7.2).

The responses monitored while yeast culture are listed in Table 7.3.

The regulatory issues relating to yeast systems are relatively simpler. The risk related directly to the cells fall into three categories: viruses and other transmissible agents, cellular DNA, and host cell proteins (e.g., growth factors). The potential introduction of adventitious agents such as fungi, bacteria, viruses, and prions in cell culture media is of major concern and has let the biotech community to constantly search for safe media and raw materials. Thus, raw materials possessing high contamination risk (e.g., hydrolysates and peptones produced from animals or by means of animal-derived enzymes) should be avoided. Use of antibiotics is discouraged. Yeast is generally regarded as a safe expression system (GRAS status). Its use is not associated with the release of endotoxins and adventitious agents such as viruses or prions are not present. The major concern is the presence of DNA and host cell proteins in the final product.

Insect cell systems

Insect cells Insect cells constitute a promising, yet unproven, alternative to bacterial and yeast expression systems for a wide range of target proteins requiring proper posttranslational modifications. The difficulties in scale-up arise because of difficulties in aeration and the type of infection needed for high-level expression. The use of baculovirus system is becoming accepted very fast; it has been used to transform *lepidopteran* insect cells into high-level expression systems in the range of 1–600 mg/L. Preparation and purification of the recombinant virus are faster than the process in mammalian cells and can be completed in about 4 weeks. In the fermentation cycle, the insect cells grow 50-fold in about a week but only in single batches or in semicontinuous batches because of the sensitivity of the cells to shear force. The costs for culture media are moderate (serum-free media) to expensive compared to bacteria and yeast media. The system is suited for the expression of cell toxic products since the cells can be grown in a healthy state before infection. Insect cells lack the ability to properly process proteins that are initially synthesized as larger inactive precursor proteins (e.g., peptide hormones, neuropeptides, growth factors, and matrix metalloproteases). Baculoviruses are not infectious to vertebrates and, therefore, do not pose a health threat though the risk of adventitious viruses is still not settled requiring virus inactivation and active filtration. Also, the cyclic killing and lysis of the host cell releases intracellular proteins and nucleic acids into the medium severely straining the downstream purification steps (like in the case of bacterial cells with inclusion bodies). The regulatory record of insect cells remains poor as no products have been approved by the FDA yet.

Process optimization Large-scale insect cell cultures are processed in a batch or fed-batch suspension culture grown in a serum-free medium to a cell density of $1–3 \times 10^6$ cells/mL before the viral infection is conducted at early or middle exponential phase. Late exponential phase infection can be done if cells are resuspended in fresh medium leading since production is limited either by depletion of nutrients or by the accumulation of toxic compounds.

Suspension cultures are the preferred choice for large-scale insect cell processes although in some cases relative higher yields have been reported in static cultures.

Growing the culture to higher cell densities in spinner flasks ($2–3 \times 10^6$ cells/mL in uninfected cells) often results in decreased target protein expression due to an unusually high oxygen demand; as a result, culture volumes above 500 mL requires additional oxygen sparging that can damage the shear-sensitive insect cells. As a result, airlift fermentors are recommended that provide good agitation at low shear. Polymers (e.g., 0.1% w/v Pluronic polyol F-68) are added to reduce foam formation and thus provide protection of the cells. Typical doubling times are 20–40 h.

If the fed-batch mode is used, the addition of nutrients such as yeastolate ultrafiltrate, lipids, amino acids, vitamins, trace elements, and glucose is done under controlled conditions.

The recombinant baculovirus is generated by homologous recombination to the stock solution of a titer of 1×10^7 to 1×10^8 pfu/mL, which is used to infect the insect cells around day 40–47. Protein expression usually takes place shortly after infection. The product yield decreases sharply when cultures are infected later than an optimal time of infection (TOI), which is in the early- or midexponential phase of a culture.

Replacement of growth media prior to infection and feeding with glucose, glutamine, and yeastolate in later stages of the infection improves yield.

The addition of human nerve growth factor also improves yield if the cells are grown in the fed-batch mode feeding with a mixture of glutamine, yeastolate, and lipids.

Use of efficient medium like YPR for *sf9* and HighFive™ cells.

Assure that all components, oxygen, glucose, and glutamine feeding are available in ample supply during protein synthesis and also viral replication stages.

Use of microemulsions to introduce lipids to culture to avoid the presence of insoluble lipid droplets in the culture medium.

Where serum-free media cannot be used, assure that it is free of TSE; serum-free media may contain discrete proteins or bulk protein fractions but not of animal origin; also there should be no components of unknown origin.

A pH range of 6.0–6.4 is optimal for most *Lepidopteran* cell lines.

Dissolved oxygen (DO) corresponding to 10%–50% of air saturation is needed in large-scale bioreactors. Supply of pure oxygen may be needed for high-density cell cultures. A standard oxygen probe (e.g., Ingold) is used for oxygen measurements.

Temperature range is from 25°C to 30°C for optimal operations; lower temperatures (20°C) are useful for keeping the cells in a slower growing stock.

Osmolality is optimally maintained at 345–380 mOsm/kg for culture medium used with lepidopteran cell lines.

Agitation rates are determined empirically; a range of 50–200 rpm is most common.

Ammonia production rate (mmol/10^6 cells/h) is an indicator of metabolic rate.

Glucose utilization (mmol/10^6 cells/h) is an indicator of metabolic rate.

pCO_2 is monitored through pH changes; the CO_2 flow is used to regulate the pH of the cell culture, and the flow is maintained only as long as the pH is above the set point. In late stage cultures, the CO_2–HCO_3 buffer system is no longer sufficient to maintain pH and sodium hydrogen carbonate or NaOH is used instead.

Lactate dehydrogenase activity reflects the extent of cell lysis since insect cells do not secrete LDH.

MOI has a limited effect on the maximum achievable yield; it is generally in the interval between 0.1 and 1.0 pfu/cell for most efficient operation; lower MOI is used for fed-batch production.

TOI is typically initiated at cell densities of $1–3 \times 10^6$ cells/mL.

Adventitious agent contamination is best detected at the end of the cell culture in the unprocessed bulk (if multiple harvest pools are prepared at different times, the culture shall be tested at the time of the collection of each pool). The test program should include a test for bacteria, fungi, mycoplasma, and viruses.

Transgenic animal systems

Transgenic animals are one of the most promising recombinant protein expression systems in the biopharmaceutical industry. The first product, an antibody expressed in goats, is about to be approved by the FDA. In this system, even complex posttranslational modified proteins are successfully expressed in their native biologically active form, thus making it possible to produce plasma proteins, human antibodies, and other proteins not easily derived from other sources at the industrial scale. It takes 6–33 months (depending on host organism) from gene construct to product expression, usually in the mammary gland, often at high protein concentrations (50 mg/mL), resulting in up to a yearly production of 10–100 kg per animal (cows). Transgenic animals coexpress species-specific target protein, which can be difficult to separate from the recombinant target protein. Virus inactivation and removal procedures must be included. Even though rarely recorded, prions as well as the most likely viruses undergo virus inactivation and removal and with extensive control procedures must be established (e.g., end of production test and virus validation). The skim milk fraction is an excellent starting material after the capture of fat and casein. For large-scale production (>100 kg/year), the costs of raw products are one-tenth for transgenic animals compared to mammalian cell cultures, mainly because of reduction in capital investment. The regulatory record in the use of transgenic animals is poor (Table 7.2).

The target protein is usually expressed in the mammary gland, often at high protein concentrations (50 mg/mL), resulting in a yearly production of 10–100 kg per animal (cows). The animal husbandry and milking procedures are known technologies upgraded to Good Agricultural Practices (GAP). The whey fraction can be processed by known chromatographic procedures, resulting in high-quality pathogen-free drug substance bulk materials—a prerequisite for preclinical and clinical trials. For large-scale production (>100 kg/year), the costs of raw products are approximately 10 times lower for transgenic animals compared to mammalian cell cultures, mainly because of reduction in capital investment.

Examples of major products undergoing clinical trials are Alpha-1 antitrypsin (Cystic fibrosis), Alpha-glucosidase (Pompe's disease), and Antithrombin III (Coronary artery bypass grafting).

The considerations in the use of transgenic animals are different from those of cell culture techniques:

- Need to redefine the master/working cell bank concept.
- Consider the variation of milk composition with lactation period.
- Control of sick animals and use of medications.
- The presence of pathogenic agents in the expressed proteins.
- Virus inactivation and documented clearance are required. There is a link between bovine spongiform encephalopathy (BSE) and Creutzfeldt-Jakob disease (CJD). Both types of CJD and other forms of transmissible spongiform encephalopathy (TSE) are probably caused by aberrant protein agents, named prions. Prions are notoriously difficult to inactivate without denaturing the protein product, but specific filters are entering the market. Prions have not been found in milk from bovine spongiform encephalopathy (BSE) infected cows, and use of good breeding practice and pathogen-free purification facilities should reduce the risk of infections to a minimum.
- Strategic methodologies, such as milking pigs and rabbits, can be very arduous besides other problems of herd control.
- Coexpression of the animal protein in milk, often possessing close physical and chemical properties with the target protein (e.g., bovine serum albumin has 76% homology with serum albumin); process design should include powerful purification procedures for removal of the coexpressed protein.
- Regulatory controls are poorly defined.
- High bioburden (levels up to 10,000 cfu/mL are common) can affect stability; raw milk cannot be stored for more than 12 h, and some expressed proteins may not tolerate pasteurization; freezing and low temperature storage is therefore required.

Cell lines and characterization

Because the genetic stability of the cell bank during storage and propagation is a major concern, it is important to know the origin and history (number of passages) of both the MCB and WCB. An MCB ampoule is kept frozen or lyophilized and only used once. Occasionally, a new MCB may be generated from a WCB. The new MCB should be tested and properly characterized. For biological products, a product license application or amendment must be submitted and approved before a new MCB can be generated from a WCB. Information about the construction of the expression vector, the fragment containing the genetic material that encodes the desired product, and the relevant genotype and phenotype of the host cell(s) are submitted as part of a product application. The major concerns of biological systems are the genetic stability of cell banks during production and storage, contaminating microorganisms, and the presence of endogenous viruses in some mammalian cell lines. As part of the application document, manufacturers are required to submit a description of all tests performed to characterize and qualify a cell bank.

It must be emphasized that the tests required to characterize a cell bank will depend on the intended use of the final product, the host/expression system, and the method of production including the techniques employed for purification of the product. In addition, the types of tests may change as technology advances. The MCB is rigorously tested using the following tests though the testing may not be limited to these tests:

- Genotypic characterization by DNA fingerprinting
- Phenotypic characterization by nutrient requirements, isoenzyme analysis, growth and morphological characteristics
- Reproducible production of desired product
- Molecular characterization of vector/cloned fragment by restriction enzyme mapping, sequence analysis
- Assays to detect viral contamination
- Reverse transcriptase assay to detect retroviruses
- Sterility test and mycoplasma test to detect other microbial contaminants

It is not necessary to test the WCB as extensively as the MCB; however, limited characterization of a WCB is necessary. The following tests are generally performed on the WCB, but this list is not inclusive:

- Phenotypic characterization
- Restriction enzyme mapping
- Sterility and mycoplasma testing
- Testing the reproducible production of desired product

The MCB and WCB must be stored in conditions that assure genetic stability. Generally, cells stored in liquid nitrogen (or its vapor phase) are stable longer than cells stored at −70°C. In addition, it is recommended that the MCB and WCB be stored in more than one location in the event that a freezer malfunctions.

Future prospects

The largest focus of recombinant manufacturing is now on antibodies; currently, all approved mAbs are produced in mammalian cells such as CHO, mouse myeloma (NSO), BHK, human embryonic kidney (HEK-293), and PER.C6. Secreted antibodies from these cell lines are glycosylated in a similar, but not identical, way to their human counterparts. The glycan structures in mAbs can influence their interaction with immune natural killer cells that bind to the constant region (Fc) and destroy antibody-targeted cells. This process known as antibody-dependent cell toxicity (ADCC) depends on a specific N-glycosylation site at Asn 297 in the Fe domain of the heavy chain of IgG1s. In addition to glycoform structure, the lack of fucose or presence of N-acetylglucosamine can impact ADCC potency.

Although many mAbs do not exhibit ADCC activity, glycosylation is difficult to control precisely in mammalian cells and may be dependent on a variety of factors such as clonal variations, media, and culture conditions.

Recently, several alternative expression systems, such as *P. pastoris* and *E. coli*, have emerged as promising hosts for mAb secretion when it may be possible engineer-specific antibodies. Researchers at Glycofi (a subsidiary of Merck & Co. Inc., Whitehouse Station, NJ) have successfully demonstrated the feasibility of glycol-engineered *P. pastoris* cell lines to produce mAbs with highly specific glycoforms. Several different glycoforms of commercially available rituximab (manufactured as Rituxan by Genentech Inc., CA) were generated, and binding to Fcγ receptors and ADCC activity were measured. This study demonstrated a 10-fold increase in binding affinity as well as enhanced ADCC activity with the glycanengineered proteins compared with Rituxan. Thus, controlling the glycan composition and structure of IgGs appears to be a promising method for improving the efficacy of therapeutic mAbs when ADCC activity is deemed important. Coupled with the use of well-established *P. pastoris* as a platform, including high cell density cultures, scalability, cost effectiveness, and existing large-scale fermentation capacity, can allow for the high-fidelity production of human glycosylated therapeutic proteins.

Escherichia coli has been most commonly used for the production of antibody fragments such as Fabs when Fc-mediated effector functions are not required or deleterious. Efficient secretion of heavy and light chains in a favorable ratio has been demonstrated to result in the high-level expression and assembly of full-length IgGs in the *E. coli* periplasm. The technology described offers a rapid and potentially inexpensive method for the production of full-length aglycosylated therapeutic antibodies when ADCC function is not required.

However, for the developers of biosimilar antibodies, it is imperative that the originator systems are replicated wherever possible. These novel methods can become useful when the intellectual property constraints make it difficult to overcome infringement.

The past two decades have seen significant advances in cell culture technology that have helped increase the expression of recombinant proteins from 100 mg/L to several g/L concentrations. These advances have resulted from intensive research in cell line engineering (vector improvements, high-throughput screening methods for high-producer clones, and introduction of antiapoptotic genes in cell lines to prolong culture viability), media development, feeding strategies, cell metabolism, and better process understanding and their impact on product quality and scale-up studies. Further increases in titers will result from a combination of the above. In particular, media improvement, including the development of high-yielding chemically defined media with replacement of hydrosylates with specific peptides, is an important area for further work.

Current fed-batch technology typically does not yield bioreactor cultures in excess of 10% packed cell volume. In contrast, in microbial and fungal fed-batch cultures, the packed cell volume can be as high as 30%–40% of the total bioreactor volume and, therefore, presents an opportunity to target further increases in cell concentration. Application of robust online sensors (e.g., capacitance sensors and NIR) for cell stoichiometric-based online feeding strategies with lower byproduct formation can be used achieving higher cell density cultures. However, prolongation of culture viability is also required. Better understanding of cell biology including the application of gene expression analysis using CHO chips and genomic-scale models will also be key. The factors contributing to optimal producer cell lines are often complex and not simply due to one gene, which necessitates an understanding of system-wide properties for better engineering of cell lines. Advances in genomics, proteomics, and metabolomics will fuel these advances; current application of these technologies is still in its infancy. A complete genomic scale-model requires the full-annotated gene sequences of CHO cells, and this is currently unavailable. As a result, a gene expression study by cross-species hybridization using human or mouse DNA microarrays is an option, although it is not very reliable.

Another area of research has focused on the use of inducible expression systems in mammalian cells. This strategy has the advantage of decoupling cell growth from product formation.

Integration of high-throughput technologies, online monitoring, and control capability and automation will allow researchers to broaden the experimental design space as well as lower the cost of process development. From an operational perspective, disposable technologies should continue to see widespread adoption because of the advantages discussed earlier (operational simplicity, reduced turnaround time, facility change-over time, lower capital costs, and lower resource requirement).

Bibliography

Airenne KJ, Laitinen OH, Alenius H, Mikkola J, Kalkkinen N, Arif SA, Yeang HY, Palosuo T, Kulomaa MS. Avidin is a promising tag for fusion proteins produced in baculovirus-infected insect cells. *Protein Expr Purif* 1999;17(1):139–145.

Airenne KJ, Oker-Blom C, Marjomaki VS, Bayer EA, Wilchek M, Kulomaa MS. Production of biologically active recombinant avidin in baculovirus-infected insect cells. *Protein Expr Purif* 1997;9(1):100–108.

Altmann F, Staudacher E, Wilson IB, Marz L. Insect cells as hosts for the expression of recombinant glycoproteins. *Glycoconj J* 1999;16(2):109–123.

Ambesi-Impiombato FS, Parks LA, Coon HG. The culture of hormone-dependent functional epithelial cells from rat thyroids. *Proc Natl Acad Sci USA* 1980;77(6):3455–3459.

Andrews B, Adari H, Hannig G, Lahue E, Gosselin M, Martin S, Ahmed A, Ford PJ, Hayman EG, Makrides SC. A tightly regulated high level expression vector that utilizes a thermo-sensitive lac repressor: Production of the human T cell receptor V beta 5.3 in *Escherichia coli*. *Gene* 1996;182(1–2):101–109.

Arthur PM, Duckworth B, Seidman M. High level expression of interleukin-1 beta in a recombinant *Escherichia coli* strain for use in a controlled bioreactor. *J Biotechnol* 1990;13(1):29–46.

Bae CS, Yang DS, Lee J, Park YH. An improved process for the production of recombinant yeast-derived monomeric human G-CSF. *Appl Microbiol Biotechnol* 1999;52(3): 338–344.

Baneyx F. Recombinant protein expression in *Escherichia coli*. *Curr Opin Biotechnol* 1999;10(5):411–421.

Barnes D, Sato G. Methods for growth of cultured cells in serum-free medium. *Anal Biochem* 1980;102(2):255–270.

Batas B, Chaudhuri JB. Protein refolding at high concentration using size-exclusion chromatography. *Biotechnol Bioeng* 1996;50(1):16–23.

Batt BC, Yabannavar VM, Singh V. Expanded bed adsorption process for protein recovery from whole mammalian cell culture broth. *Bioseparation* 1995;5(1):41–52.

Bech Jensen E, Carlsen S. Production of recombinant human growth hormone in *Escherichia coli*: Expression of different precursors and physiological effects of glucose, acetate and salts. *Biotechnol Bioeng* 1990;36(1):1–11.

Bermudez-Humaran LG, Langella P, Miyoshi A, Gruss A, Guerra RT, Montes dO-L, Le Loir Y. Production of human papillomavirus type 16 E7 protein in *Lactococcus lactis*. *Appl Environ Microbiol* 2002;68(2):917–922.

Berry DR. Growth of yeast. In: *Fermentation Process Development of Industrial Organisms*, Neway JO, ed. Marcel Dekker Inc., New York, 1989, pp. 277–311.

Bettger WJ, Ham RG. The nutrient requirements of cultured mammalian cells. *Adv Nutr Res* 1982;4:249–286.

Billman-Jacobe H. Expression in bacteria other than *Escherichia coli*. *Curr Opin Biotechnol* 1996;7(5):500–504.

Birnboim HC, Doly J. A rapid alkaline extraction procedure for screening recombinant plasmid DNA. *Nucleic Acids Res* 1979;7(6):1513–1523.

Bova GS, Eltoum IA, Kiernan JA, Siegal GP, Frost AR, Best CJ, Gillespie JW, Su GH, Emmert-Buck MR. Optimal molecular profiling of tissue and tissue components: Defining the best processing and microdissection methods for biomedical applications. *Mol Biotechnol.* 2005 Feb;29(2):119–152.

Breuer S, Marzban G, Cserjan-Puschmann M, Durrschmid E, Bayer K. Off-line quantitative monitoring of plasmid copy number in bacterial fermentation by capillary electrophoresis. *Electrophoresis* 1998;19(14):2474–2478.

Broad D, Boraston R, Rhodes M. Production of recombinant proteins in serum-free media. *Cytotechnology* 1991;5(1):47–55.

Buntemeyer H, Lutkemeyer D, Lehmann J. Optimization of serum-free fermentation processes for antibody production. *Cytotechnology* 1991;5(1):57–67.

Bylund F, Castan A, Mikkola R, Veide A, Larsson G. Influence of scale-up on the quality of recombinant human growth hormone. *Biotechnol Bioeng* 2000;69(2):119–128.

Bylund F, Collet E, Enfors SO, Larsson G. Substrate gradient formation in the large-scale bioreactor lowers cell yield and increases by-product formation. *Bioprocess Eng* 1998; 18(3):171–180.

Caron AW, Archambault J, Massie B. High level recombinant protein production in bioreactors using the baculovirus-insect cell expression system. *Biotechnol Bioeng* 1990;36(11): 1133–1140.

Carver A. Recombinant protein production in lactating animals—Transgenic livestock represent an ideal route for the production of recombinant proteins. *Pharmaceut Biotechnol Int* 1996;3pp.

Carver A, Wright G, Cottom D, Cooper J, Dalrymple M, Temperley S, Udell M, Reeves D, Percy J, Scott A. Expression of human alpha 1 antitrypsin in transgenic sheep. *Cytotechnology* 1992;9(1–3):77–84.

Carver AS, Dalrymple MA, Wright G et al. Transgenic livestock as bioreactors: Stable expression of human alpha-1-antitrypsin by a flock of sheep. *Biotechnology* 1993;11(Nov):1263–1270.

Cassiman JJ, Brugmans M, Van den BH. Growth and surface properties of dispase dissociated human fibroblasts. *Cell Biol Int Rep* 1981;5(2):125–132.

CBER/FDA. Points to Consider in the Characterization of Cell Lines Used to Produce Biologicals, 1993; Department of Health and Human Services memorandum dated July 12, 1993; from CBER to biological product manufacturers.

CBER/FDA. Points to Consider in the Manufacture and Testing of Monoclonal Antibody Products for Human Use, 1997; U.S. Department of Health and Human Services memorandum dated February 27, 1997; to biological product manufacturers.

CBER/FDA. Points to Consider in the Production and Testing of New Drugs and Biologicals Produced by Recombinant DNA Technology, 1985; Draft report from the Office of Biologics Research and Review Center for Drugs and Biologics dated April 10, 1985.

Cereghino GP, Cereghino JL, Ilgen C, Cregg JM. Production of recombinant proteins in fermenter cultures of the yeast *Pichia pastoris*. *Curr Opin Biotechnol* 2002;13(4):329–332.

Chan LC, Greenfield PF, Reid S. Optimising fed-batch production of recombinant proteins using the baculovirus expression vector system. *Biotechnol Bioeng* 1998;59(2):178–188.

Ciaccia AV, Cunningham EL, Church FC. Characterization of recombinant heparin cofactor II expressed in insect cells. *Protein Expr Purif* 1995;6(6):806–812.

Clark AJ. The mammary gland as a bioreactor: Expression, processing, and production of recombinant proteins. *J Mammary Gland Biol Neoplasia* 1998;3(3):337–350.

Clark AJ, Bessos H, Bishop JO, Brown P, Harris S, Lathe R, McClenaghan M, Prowse C, Simons JP, Whitelaw CBA, Wilmut I. Expression of human anti-hemophilic factor IX in the milk of transgenic sheep. *Biotechnology* 1989;7(5):487–492.

Cleland JL, Builder SE, Swartz JR, Winkler M, Chang JY, Wang DI. Polyethylene glycol enhanced protein refolding. *Biotechnology (NY)* 1992;10(9):1013–1019.

Collodi P, Rawson C, Barnes D. Serum-free culture of carcinoma cell lines. *Cytotechnology* 1991;5(1):31–46.

Dahlgren ME, Powell AL, Greasham RL, George HA. Development of scale-down techniques for investigation of recombinant *Escherichia coli* fermentations: Acid metabolites in shake flasks and stirred bioreactors. *Biotechnol Prog* 1993;9(6):580–586.

Dale C, Allen A, Fogerty S. *Pichia pastoris*: A eukaryotic system for the large-scale production of biopharmaceuticals. *BioPharm* 1999;12(11):36–42.

285

Dalton JC, Bruley DF, Kang KA, Drohan WN. Separation of recombinant human protein C from transgenic animal milk using immobilized metal affinity chromatography. *Adv Exp Med Biol* 1997;411:419–428.

Das T, Johns PW, Goffin V, Kelly P, Kelder B, Kopchick J, Buxton K, Mukerji P. High-level expression of biologically active human prolactin from recombinant baculovirus in insect cells. *Protein Expr Purif* 2000;20(2):265–273.

Dave AS, Bruley DF. Separation of human protein C from components of transgenic milk using immobilized metal affinity chromatography. *Adv Exp Med Biol* 1999;471:639–647.

de Oliveira JE, Soares CR, Peroni CN et al. High-yield purification of biosynthetic human growth hormone secreted in *Escherichia coli* periplasmic space. *J Chromatogr A* 1999;852(2):441–450.

de Vos WM. Gene expression systems for lactic acid bacteria. *Curr Opin Microbiol* 1999; 2(3):289–295.

Dieye Y, Usai S, Clier F, Gruss A, Piard JC. Design of a protein-targeting system for lactic acid bacteria. *J Bacteriol* 2001;183(14):4157–4166.

Doerfler W, Bohm P, eds. *The Molecular Biology of Baculoviruses. Current Topics in Microbiology and Immunology*, Vol. 131. Springer Verlag, New York, 1987.

Drohan WN, Zhang DW, Paleyanda RK, Chang R, Wroble M, Velander W, Lubon H. Inefficient processing of human protein C in the mouse mammary gland. *Transgenic Res* 1994;3(6):355–364.

Dutton PL. Redox potentiometry: Determination of midpoint potentials of oxidation-reduction components of biological electron-transfer systems. *Methods Enzymol* 1978;54:411–435.

Dutton RL, Scharer JM, Moo-Young M. Descriptive parameter evaluation in mammalian cell culture. *Cytotechnology* 1998;26(2):139–152.

Ebert KM, Selgrath JP, DiTullio P, Denman J, Smith TE, Memon MA, Schindler JE, Monastersky GM, Vitale JA, Gordon K. Transgenic production of a variant of human tissue-type plasminogen activator in goat milk: Generation of transgenic goats and analysis of expression. *Biotechnology (NY)* 1991;9(9):835–838.

Echelard Y. Recombinant protein production in transgenic animals. *Curr Opin Biotechnol* 1996;7(5):536–540.

Elia JA, Floudas CA. Energy supply chain optimization of hybrid feedstock processes: A review. *Annu Rev Chem Biomol Eng* 2014;5:147–179.

Faber KN, Harder W, Ab G, Veenhuis M. Review: Methylotrophic yeasts as factories for the production of foreign proteins. *Yeast* 1995;11(14):1331–1344.

Fahey EM, Chaudhuri JB, Binding P. Refolding of low molecular weight urokinase plasminogen activator by dilution and size exclusion chromatography—A comparative study. *Sep Sci Technol* 2000;35:1743–1760.

Fahrner RL, Blank GS, Zapata GA. Expanded bed protein A affinity chromatography of a recombinant humanized monoclonal antibody: Process development, operation, and comparison with a packed bed method. *J Biotechnol* 1999;75(2–3):273–280.

Farrell PJ, Lu M, Prevost J, Brown C, Behie L, Iatrou K. High-level expression of secreted glycoproteins in transformed lepidopteran insect cells using a novel expression vector. *Biotechnol Bioeng* 1998;60(6):656–663.

Feliu JX, Cubarsi R, Villaverde A. Optimized release of recombinant proteins by ultrasonication of E. coli cells. *Biotechnol Bioeng* 1998;58(5):536–540.

Feng M, Glassey J. Physiological state-specific models in estimation of recombinant *Escherichia coli* fermentation performance. *Biotechnol Bioeng* 2000;69(5):495–503.

Feng W, Graumann K, Hahn R, Jungbauer A. Affinity chromatography of human estrogen receptor-alpha expressed in *Saccharomyces cerevisiae*. Combination of heparin- and 17beta-estradiol-affinity chromatography. *J Chromatogr A* 1999;852(1):161–173.

Franchi E, Maisano F, Testori SA, Galli G, Toma S, Parente L, de Ferra F, Grandi G. A new human growth hormone production process using a recombinant *Bacillus subtilis* strain. *J Biotechnol* 1991;18(1–2):41–54.

Friehs K, Reardon KF. Parameters influencing the productivity of recombinant E. coli cultivations. *Adv Biochem Eng Biotechnol* 1993;48:53–77.

Fritzsch FS, Dusny C, Frick O, Schmid A. Single-cell analysis in biotechnology, systems biology, and biocatalysis. *Annu Rev Chem Biomol Eng.* 2012;3:129–155.

Gaeng S, Scherer S, Neve H, Loessner MJ. Gene cloning and expression and secretion of Listeria monocytogenes bacteriophage-lytic enzymes in *Lactococcus lactis*. *Appl Environ Microbiol* 2000;66(7):2951–2958.

Georgiou G, Valax P. Isolating inclusion bodies from bacteria. *Methods Enzymol* 1999;309:48–58.

Goldberg ME, Expert-Bezancon N, Vuillard L, Rabilloud T. Non-detergent sulphobetaines: A new class of molecules that facilitate in vitro protein renaturation. *Folding Design* 1996;1(1):21–27.

Goldman M. Processing challenges for transgenic milk products. *BioProcess Int* 2003;1(10):60–63.

Goosen MF. Large-scale insect cell culture: Methods, applications and products. *Curr Opin Biotechnol* 1991;2(3):365–369.

Gorfien SF, Paul W, Judd D, Tescione L, Jayme DW. Optimized nutrient additives for fed-batch cultures. *BioPharm International* 2003;16(4):34–40.

Griffiths B. The use of oxidation-reduction potential (ORP) to monitor growth during a cell culture. *Dev Biol Stand* 1983;55:113–116.

Grossmann M, Wong R, Teh NG, Tropea JE, East-Palmer J, Weintraub BD, Szkudlinski MW. Expression of biologically active human thyrotropin (hTSH) in a baculovirus system: Effect of insect cell glycosylation on hTSH activity in vitro and in vivo. *Endocrinology* 1997;138(1):92–100.

Guilbert LJ, Iscove NN. Partial replacement of serum by selenite, transferrin, albumin and lecithin in haemopoietic cell cultures. *Nature* 1976;263(5578):594–595.

Handa-Corrigan A. Bioreactors for mammalian cells. In: *Mammalian cell biotechnology. A practical approach*, Butler M, ed. IRL Press, Oxford, U.K., 1991, pp. 139–158.

Hannig G, Makrides SC. Strategies for optimizing heterologous protein expression in *Escherichia coli. Trends Biotechnol* 1998;16(2):54–60.

Harder MP, Sanders EA, Wingender E, Deckwer WD. Studies on the production of human parathyroid hormone by recombinant *Escherichia coli. Appl Microbiol Biotechnol* 1993;39(3):329–334.

Hardy E, Martinez E, Diago D, Diaz R, Gonzalez D, Herrera L. Large-scale production of recombinant hepatitis B surface antigen from *Pichia pastoris. J Biotechnol* 2000;77(2–3):157–167.

Hasegawa M, Kawano Y, Matsumoto Y, Hidaka Y, Fujii J, Taniguchi N, Wada A, Hirayama T, Shimonishi Y. Expression and characterization of the extracellular domain of guanylyl cyclase C from a baculovirus and Sf21 insect cells. *Protein Expr Purif* 1999;15(3):271–281.

Hawthorne TR, Burgi R, Grossenbacher H, Heim J. Isolation and characterization of recombinant annexin V expressed in *Saccharomyces cerevisiae. J Biotechnol* 1994;36(2):129–143.

Hellwig S, Emde F, Raven NP, Henke M, van Der LP, Fischer R. Analysis of single-chain antibody production in *Pichia pastoris* using on-line methanol control in fed-batch and mixed-feed fermentations. *Biotechnol Bioeng* 2001;74(4):344–352.

Hennighausen L. The mammary gland as a bioreactor: Production of foreign proteins in milk. *Protein Expr Purif* 1990;1(1):3–8.

Hensing MC, Rouwenhorst RJ, Heijnen JJ, van Dijken JP, Pronk JT. Physiological and technological aspects of large-scale heterologous-protein production with yeasts. *Antonie Van Leeuwenhoek* 1995;67(3):261–279.

Hewlett G. Strategies for optimising serum-free media. *Cytotechnology* 1991;5(1):3–14.

Hodgson J. Expression systems: A user's guide. Emphasis has shifted from the vector construct to the host organism. *Biotechnology (NY)* 1993;11(8):887–893.

Hoenicka M, Becker EM, Apeler H, Sirichoke T, Schroder H, Gerzer R, Stasch JP. Purified soluble guanylyl cyclase expressed in a baculovirus/Sf9 system: Stimulation by YC-1, nitric oxide, and carbon monoxide. *J Mol Med* 1999;77(1):14–23.

Hoffman BJ, Broadwater JA, Johnson P, Harper J, Fox BG, Kenealy WR. Lactose fed-batch overexpression of recombinant metalloproteins in *Escherichia coli* BL21 (DE3): Process control yielding high levels of metal-incorporated, soluble protein. *Protein Expr Purif* 1995;6(5):646–654.

Hofman V, Ilie M, Gavric-Tanga V et al. Role of the surgical pathology laboratory in the preanalytical approach of molecular biology techniques. *Ann Pathol* 2010 Apr;30(2):85–93.

Holms WH. The central metabolic pathways of *Escherichia coli*: Relationship between flux and control at a branch point, efficiency of conversion to biomass, and excretion of acetate. *Curr Top Cell Regul* 1986;28:69–105.

Hoshi H, Kan M, Yamane I, Minamoto Y. Hydrocortisone potentiates cell proliferation and promotes cell spreading on tissue culture substrata of human diploid fibroblasts in a serum-free hormone supplemented medium. *Biomed Res* 1982;3(5):546–552.

Hossler P, Khattak SF, Li ZJ. Optimal and consistent protein glycosylation in mammalian cell culture. *Glycobiology* 2009 Sep;19(9):936–49.

Houdebine LM. Expression of recombinant proteins in the milk of transgenic animals. *Rev Fr Transf Hemobiol* 1993;36(1):49–72.

Houdebine LM. Production of pharmaceutical proteins from transgenic animals. *J Biotechnol* 1994;34(3):269–287.

Houdebine LM. The production of pharmaceutical proteins from the milk of transgenic animals. *Reprod Nutr Dev* 1995;35(6):609–617.

Houdebine LM. Transgenic animal bioreactors. *Transgenic Res* 2000;9(4–5):305–320.

Hoyer LW, Drohan WN, Lubon H. Production of human therapeutic proteins in transgenic animals. *Vox Sang* 1994;67(Suppl 3):217–220.

Hu WS, Aunins JG. Large-scale mammalian cell culture. *Curr Opin Biotechnol* 1997;8(2):148–153.

Hughes PR, Wood HA. In vivo and in vitro bioassay methods for baculoviruses. In: *The Biology of Baculoviruses*, Vol. II, Granados RR, Federici BA, eds. CRC Press Inc., Boca Raton, FL, 1986, pp. 1–30.

Hunninghake D, Grisolia S. A sensitive and convenient micromethod for estimation of urea, citrulline, and carbamyl derivatives. *Anal Biochem* 1966;16(2):200–205.

ICH. Quality of biotechnological products: Viral safety evaluation of biotechnology products derived from cell lines of human or animal origin. ICH Harmonised Tripartite Guideline. *Dev Biol Stand* 1998;93:177–201.

ICH Guidance. Derivation and Characterization of Cell Substrates Used for Production of Biotechnological/Biological Products, 1997.

Ignoffo CM. Evaluation of in vivo specificity of insect viruses. In: *Baculoviruses for Insect Pest Control: Safety Considerations,* Summers M, Engler R, Falcon LA, Vail PV, eds. American Society for Microbiology, Washington, DC, 1975, pp. 52–57.

Ikonomou L, Bastin G, Schneider YJ, Agathos SN. Design of an efficient medium for insect cell growth and recombinant protein production. *In Vitro Cell Dev Biol Anim* 2001;37(9):549–559.

Invitrogen. *Guide to Baculovirus Expression Vector Systems (BEVS) and Insect Cell Culture Techniques.* Invitrogen life technologies, Carlsbad, CA, 2002.

Ishihara K, Satoh I, Nittoh T, Kanaya T, Okazaki H, Suzuki T, Koyama T, Sakamoto T, Ide T, Ohuchi K. Preparation of recombinant rat interleukin-5 by baculovirus expression system and analysis of its biological activities. *Biochim Biophys Acta* 1999;1451(1):48–58.

Jarvis DL, Kawar ZS, Hollister JR. Engineering N-glycosylation pathways in the baculovirus-insect cell system. *Curr Opin Biotechnol* 1998;9(5):528–533.

Jayme DW. Nutrient optimization for high density biological production applications. *Cytotechnology* 1991;5(1):15–30.

Jazwinski SM. Preparation of extracts from yeast. *Methods Enzymol* 1990;182(13):154–174.

Jones EW. Tackling the protease problem in *Saccharomyces cerevisiae. Methods Enzymol* 1991;194:428–453.

Jung G, Denefle P, Becquart J, Mayaux JF. High-cell density fermentation studies of recombinant *Escherichia coli* strains expressing human interleukin-1 beta. *Ann Inst Pasteur Microbiol* 1988;139(1):129–146.

Kannan R, Tomasetto C, Staub A, Bossenmeyer-Pourie C, Thim L, Nielsen PF, Rio M. Human pS2/trefoil factor 1: Production and characterization in *Pichia pastoris. Protein Expr Purif* 2001;21(1):92–98.

Karuppiah N, Sharma A. Cyclodextrins as protein folding aids. *Biochem Biophys Res Commun* 1995;211(1):60–66.

Keith MB, Farrell PJ, Iatrou K, Behie LA. Screening of transformed insect cell lines for recombinant protein production. *Biotechnol Prog* 1999;15(6):1046–1052.

Kilburn DG. Monitoring and control of bioreactors. In: *Mammalian cell biotechnology. A practical approach*, Butler M, ed. IRL Press, Oxford, U.K., 1991, pp. 159–185.

King DJ, Walton F, Smith BW, Dunn M, Yarranton GT. Recovery of recombinant proteins from yeast. *Biochem Soc Trans* 1988;16(6):1083–1086.

King LA, Possee RD. Insect cell culture media and maintenance of insect cell lines. In: *The Baculovirus Expression System. A Laboratory Guide.* Chapman & Hall, London, U.K., 1992a, pp. 75–105.

King LA, Possee RD. Production and selection of recombinant virus. In: *The Baculovirus Expression System. A Laboratory Guide.* Chapman & Hall, London, U.K., 1992b, pp. 127–140.

King LA, Possee RD. Propagation of baculoviruses in insect larvae. In: *The Baculovirus Expression System. A Laboratory Guide.* Chapman & Hall, London, U.K., 1992c, pp. 180–194.

King LA, Possee RD. Scaling up the production of recombinant protein in insect cells; laboratory bench level. In: *The Baculovirus Expression System. A Laboratory Guide.* Chapman & Hall, London, U.K., 1992d, pp. 171–179.

King LA, Possee RD. The baculoviruses. In: *The Baculovirus Expression System. A Laboratory Guide.* Chapman & Hall, London, U.K., 1992e, pp. 1–15.

Kost TA, Condreay JP. Recombinant baculoviruses as expression vectors for insect and mammalian cells. *Curr Opin Biotechnol* 1999;10(5):428–433.

Kramer W, Elmecker G, Weik R, Mattanovich D, Bayer K. Kinetics studies for the optimization of recombinant protein formation. *Ann N Y Acad Sci* 1996;782:323–333.

Ladisch MR, Rudge SR, Ruettimann KW, Lin JK. Bioseparations of milk proteins. In: *Bioproducts and Bioprocesses*, Fiechter A, Okada H, Tanner RD, eds. Springer-Verlag, New York, 1989, pp. 209–221.

Langella P, Le Loir Y. Heterologous protein secretion in *Lactococcus lactis*: A novel antigen delivery system. *Braz J Med Biol Res* 1999;32(2):191–198.

Larrick JW, Thomas DW. Producing proteins in transgenic plants and animals. *Curr Opin Biotechnol* 2001;12(4):411–418.

Larsson G, Törnquist M, Ståhl Wernesson E, Trägårdh C, Noorman H, Enfors SO. Substrate gradients in bioreactors: Origin and consequences. *Bioprocess Eng* 1996;14(6):281–289.

Lauritzen C, Pedersen J, Madsen MT, Justesen J, Martensen PM, Dahl SW. Active recombinant rat dipeptidyl aminopeptidase I (cathepsin C) produced using the baculovirus expression system. *Protein Expr Purif* 1998;14(3):434–442.

Le Loir Y, Gruss A, Ehrlich SD, Langella P. A nine-residue synthetic propeptide enhances secretion efficiency of heterologous proteins in *Lactococcus lactis. J Bacteriol* 1998; 180(7):1895–1903.

Le Loir Y, Nouaille S, Commissaire J, Bretigny L, Gruss A, Langella P. Signal peptide and propeptide optimization for heterologous protein secretion in *Lactococcus lactis. Appl Environ Microbiol* 2001;67(9):4119–4127.

Lee SY. High cell-density culture of *Escherichia coli. Trends Biotechnol* 1996;14(3):98–105.

Lee JY, Yoon CS, Chung IY, Lee YS, Lee EK. Scale-up process for expression and renaturation of recombinant human epidermal growth factor from *Escherichia coli* inclusion bodies. *Biotechnol Appl Biochem* 2000;31(Pt 3):245–248.

Lenhard T, Reilander H. Engineering the folding pathway of insect cells: Generation of a stably transformed insect cell line showing improved folding of a recombinant membrane protein. *Biochem Biophys Res Commun* 1997;238(3):823–830.

Levin W, Daniel RF, Stoner CR, Stoller TJ, Wardwell-Swanson JA, Angelillo YM, Familletti PC, Crowl RM. Purification of recombinant human secretory phospholipase A2 (group II) produced in long-term immobilized cell culture. *Protein Expr Purif* 1992;3(1):27–35.

Lin HY, Neubauer P. Influence of controlled glucose oscillations on a fed-batch process of recombinant *Escherichia coli. J Biotechnol* 2000;79(1):27–37.

Lindsay D, Betenbaugh M. Quantification of cell culture factors affecting recombinant protein yields in baculovirus-infected insect cells. *Biotechnol Bioeng* 1992;39(6):614–618.

Liu S, Tobias R, McClure S, Styba G, Shi Q, Jackowski G. Removal of endotoxin from recombinant protein preparations. *Clin Biochem* 1997;30(6):455–463.

Logan JS, Martin MJ. Transgenic swine as a recombinant production system for human hemoglobin. *Methods Enzymol* 1994;231:435–445.

Looker D, Mathews AJ, Neway JO, Stetler GL. Expression of recombinant human hemoglobin in *Escherichia coli. Methods Enzymol* 1994;231:364–374.

Lorincz Z, Kalabay L, Cseh S, Zavodszky P, Arnaud P, Jakab L. Isolation of human alpha 2HS-glycoprotein synthesized by Sf9 cells. *Acta Microbiol Immunol Hung* 1998;45(3–4): 419–424.

Lubiniecki AS, Lupker JH. Purified protein products of rDNA technology expressed in animal cell culture. *Biologicals* 1994;22(2):161–169.

Lubon H. Transgenic animal bioreactors in biotechnology and production of blood proteins. *Biotechnol Annu Rev* 1998;4:1–54.

Lubon H, Paleyanda RK, Velander WH, Drohan WN. Blood proteins from transgenic animal bioreactors. *Transf Med Rev* 1996;10(2):131–143.

Luckow VA, Summers MD. Trends in the development of baculovirus expression vectors. *Nat Biotechnol* 1988;6(1):47–55.

Luli GW, Strohl WR. Comparison of growth, acetate production, and acetate inhibition of *Escherichia coli* strains in batch and fed-batch fermentations. *Appl Environ Microbiol* 1990;56(4):1004–1011.

Lüllau E, Marison IW, von Stockar U. Ceramic hydroxyapatite: A new tool for separation and analysis of IgA monoclonal antibodies. In: *Animal Cell Technology: From Vaccines to Genetic Medicine*, Carrondo MJT, Griffiths JB, Moreira JLP, eds. Kluwer Academic Publishers, Dordrecht, the Netherlands, 1997, pp. 265–269.

Lundgren B, Blüml G. Microcarriers in cell culture production. In: *Bioseparation and Bioprocessing. Volume II: Processing, Quality and Characterization, Economics, Safety and Hygiene*, Subramanian G, ed. Wiley-VCH, New York, 1998, pp. 165–222.

289

MacDonald HL, Neway JO. Effects of medium quality on the expression of human interleukin-2 at high cell density in fermentor cultures of *Escherichia coli* K-12. *Appl Environ Microbiol* 1990;56(3):640–645.

Maiorella B, Inlow D, Shauger A, Harano D. Large-scale insect cell-culture for recombinant protein production. *Nat Biotechnol* 1988;6(12):1406–1410.

Manousos M, Ahmed M, Torchio C, Wolff J, Shibley G, Stephens R, Mayyasi S. Feasibility studies of oncornavirus production in microcarrier cultures. *In Vitro* 1980;16(6):507–515.

Marston FA, Hartley DL. Solubilization of protein aggregates. *Methods Enzymol* 1990;182(20): 264–276.

Mather JP, Sato GH. The growth of mouse melanoma cells in hormone-supplemented, serum-free medium. *Exp Cell Res* 1979;120(1):191–200.

Maurer HR. Serum-free media and cell cultures. Introductory remarks. *Cytotechnology* 1991;5(1):1.

McCarroll L, King LA. Stable insect cell cultures for recombinant protein production. *Curr Opin Biotechnol* 1997;8(5):590–594.

Merten OW. Safety issues of animal products used in serum-free media. *Dev Biol Stand* 1999;99:167–180.

Merten OW, Litwin J. Serum-free medium for fermentor cultures of hybridomas. *Cytotechnology* 1991;5(1):69–82.

Metcalfe H, Field RP, Froud SJ. The use of 2-hydroxy-2,4,6-cycloheptarin-1-one (tropolone) as a replacement for transferrin. In: *Animal Cell Technology, Products of Today, Prospects for Tomorrow*, Spier RE, Griffiths JB, Berthold W, eds. Butterworth-Heinemann, Oxford, U.K., 1994, pp. 88–90.

Miksch G, Neitzel R, Fiedler E, Friehs K, Flaschel E. Extracellular production of a hybrid beta-glucanase from Bacillus by *Escherichia coli* under different cultivation conditions in shaking cultures and bioreactors. *Appl Microbiol Biotechnol* 1997;47(2):120–126.

Miller DW, Safer P, Miller LK. An insect baculovirus host-vector system for high-level expression of foreign genes. In: *Genetic Engineering. Principles and Methods*, Vol. 8, Setlow JK, Hollander A, eds. Plenum Publishing Corp., New York, 1986, pp. 277–298.

Miller LK. Insect viruses. In: *Fields—Virology*, Fields BN, Knipe DM, Howley PM, eds. Lippincott-Raven Publishers, Philadelphia, PA, 1996, pp. 533–556.

Miltenburger HG, David P. Mass production of insect cells in suspension. *Dev Biol Stand* 1980;46:183–186.

Miyoshi A, Poquet I, Azevedo V, Commissaire J, Bermudez-Humaran L, Domakova E, Le Loir Y, Oliveira SC, Gruss A, Langella P. Controlled production of stable heterologous proteins in *Lactococcus lactis Appl Environ Microbiol* 2002;68(6):3141–3146.

Mizutani S, Mori H, Shimizu S, Sakaguchi S, Kobayashi T. Effect of amino acid supplement on cell yield and gene product in *Escherichia coli* horboring plasmid. *Biotechnol Bioeng* 1986;28:204–209.

Moody AJ, Hejnaes KR, Marshall MO, Larsen FS, Boel E, Svendsen I, Mortensen E, Dyrberg T. Isolation by anion-exchange of immunologically and enzymatically active human islet glutamic acid decarboxylase 65 overexpressed in Sf9 insect cells. *Diabetologia* 1995;38(1):14–23.

Mori H, Yano T, Kobayashi T, Shimizu S. High density cultivation of biomass in fed-batch system with DO-Stat. *J Chem Eng* Jpn 1979;12:313–319.

Muller C, Rinas U. Renaturation of heterodimeric platelet-derived growth factor from inclusion bodies of recombinant *Escherichia coli* using size-exclusion chromatography. *J Chromatogr A* 1999;855(1):203–213.

Munshi CB, Fryxell KB, Lee HC, Branton WD. Large-scale production of human CD38 in yeast by fermentation. *Methods Enzymol* 1997;280:318–330.

Murakami H, Yamada K, Shirahata S, Enomoto A, Kaminogawa S. Physiological enhancement of immunoglobulin production of hybridomas in serum-free media. *Cytotechnology* 1991;5(1):83–94.

Murby M, Uhlén M, Ståhl S. Upstream strategies to minimize proteolytic degradation upon recombinant production in *Escherichia coli*. *Protein Expr Purif* 1996 Mar;7(2):129–136.

Murhammer DW, Goochee CF. Scale-up of insect cultures: Protective effects of pluronic F-68. *Nat Biotechnol* 1988;6(12):1411–1418.

Nagai K, Thogersen HC, Luisi BF. Refolding and crystallographic studies of eukaryotic proteins produced in *Escherichia coli*. *Biochem Soc Trans* 1988;16(2):108–110.

Nagata M, Matsumura T. Action of the bacterial neutral protease, dispase, on cultured cells and its application to fluid suspension culture with a review on biomedical application of this protease. *Jpn J Exp Med* 1986;56(6):297–307.

Neubauer P, Hofmann K, Holst O, Mattiasson B, Kruschke P. Maximizing the expression of a recombinant gene in *Escherichia coli* by manipulation of induction time using lactose as inducer. *Appl Microbiol Biotechnol* 1992;36(6):739–744.

Nguyen B, Jarnagin K, Williams S, Chan H, Barnett J. Fed-batch culture of insect cells: A method to increase the yield of recombinant human nerve growth factor (rhNGF) in the baculovirus expression system. *J Biotechnol* 1993;31(2):205–217.

Nielsen LK, Smyth GK, Greenfield PF. Accuracy of the endpoint assay for virus titration. *Cytotechnology* 1992;8(3):231–236.

Niemann H, Halter R, Carnwath JW, Herrmann D, Lemme E, Paul D. Expression of human blood clotting factor VIII in the mammary gland of transgenic sheep. *Transg Res* 1999;8(3):237–247.

Niu PD, Lefevre F, Mege D, La Bonnardiere C. Atypical porcine type I interferon. Biochemical and biological characterization of the recombinant protein expressed in insect cells. *Eur J Biochem* 1995;230(1):200–206.

Novotny J, Ganju RK, Smiley ST, Hussey RE, Luther MA, Recny MA, Siliciano RF, Reinherz EL. A soluble, single-chain T-cell receptor fragment endowed with antigen-combining properties. *Proc Natl Acad Sci USA* 1991;88(19):8646–8650.

O'Donnell JK, Martin MJ, Logan JS, Kumar R. Production of human hemoglobin in transgenic swine: An approach to a blood substitute. *Cancer Detect Prev* 1993;17(2):307–312.

O'Reilly DR, Miller LK, Luckow VA. *Baculovirus Expression Vectors: A Laboratory Manual.* W.H. Freeman and Co., New York, 1992.

Overton LK, Patel I, Becherer JD, Chandra G, Kost TA. Expression of tissue inhibitor of metalloproteinases by recombinant baculovirus-infected insect cells cultured in an airlift fermentor. In: *Baculovirus Expression Protocols*, Richardson CD, ed. Humana Press, Totowa, NJ, 1995, pp. 225–242.

Patra AK, Mukhopadhyay R, Mukhija R, Krishnan A, Garg LC, Panda AK. Optimization of inclusion body solubilization and renaturation of recombinant human growth hormone from *Escherichia coli*. *Protein Expr Purif* 2000;18(2):182–192.

Pedersen J, Lauritzen C, Madsen MT, Weis DS. Removal of N-terminal polyhistidine tags from recombinant proteins using engineered aminopeptidases. *Protein Expr Purif* 1999;15(3):389–400.

Peng S, Sommerfelt M, Logan J, Huang Z, Jilling T, Kirk K, Hunter E, Sorscher E. One-step affinity isolation of recombinant protein using the baculovirus/insect cell expression system. *Protein Expr Purif* 1993;4(2):95–100.

Pennock GD, Shoemaker C, Miller LK. Strong and regulated expression of *Escherichia coli* beta-galactosidase in insect cells with a baculovirus vector. *Mol Cell Biol* 1984;4(3):399–406.

Petsch D, Anspach FB. Endotoxin removal from protein solutions. *J Biotechnol* 2000;76(2–3): 97–119.

Pfeifer TA. Expression of heterologous proteins in stable insect cell culture. *Curr Opin Biotechnol* 1998;9(5):518–521.

Piard JC, Jimenez-Diaz R, Fischetti VA, Ehrlich SD, Gruss A. The M6 protein of Streptococcus pyogenes and its potential as a tool to anchor biologically active molecules at the surface of lactic acid bacteria. *Adv Exp Med Biol* 1997;418:545–550.

Pluckthun A, Krebber A, Krebber C et al. Producing antibodies in *Escherichia coli*: From PCR to fermentation. In: *Antibody Engineering*, McCafferty J, Hoogenboom H, Chiswell D, eds. JRL Press, Oxford, U.K., 1996: 203–252.

Pollock DP, Kutzko JP, Birck-Wilson E, Williams JL, Echelard Y, Meade HM. Transgenic milk as a method for the production of recombinant antibodies. *J Immunol Methods* 1999;231(1–2):147–157.

Poncet N, Taylor PM. The role of amino acid transporters in nutrition. *Curr Opin Clin Nutr Metab Care* 2013 Jan;16(1):57–65.

Poquet I, Ehrlich SD, Gruss A. An export-specific reporter designed for gram-positive bacteria: Application to *Lactococcus lactis*. *J Bacteriol* 1998;180(7):1904–1912.

Power JF, Reid S, Radford KM, Greenfield PF, Nielsen LK. Modelling and optimization of the baculovirus expression system in batch suspension culture. *Biotechnol Bioeng* 1994;44(6):710–719.

Propst CL, Von Wedel RJ, Lubiniecki AS. Using mammalian cells to produce products. In: *Fermentation Process Development of Industrial Organisms*, Neway JO, ed. Marcel Dekker Inc., New York, 1989, pp. 221–276.

Prunkard D, Cottingham I, Garner I, Bruce S, Dalrymple M, Lasser G, Bishop P, Foster D. High-level expression of recombinant human fibrinogen in the milk of transgenic mice. *Nat Biotechnol* 1996;14(July):867–871.

Quirk AV, Geisow MJ, Woodrow JR, Burton SJ, Wood PC, Sutton AD, Johnson RA, Dodsworth N. Production of recombinant human serum albumin from *Saccharomyces cerevisiae*. *Biotechnol Appl Biochem* 1989;11(3):273–287.

Qureshi GA, Fohlin L, Bergstrom J. Application of high-performance liquid chromatography to the determination of free amino acids in physiological fluids. *J Chromatogr* 1984;297:91–100.

Rasmussen L, Toftlund H. Phosphate compounds as iron chelators in animal cell cultures. *In Vitro Cell Dev Biol* 1986;22(4):177–179.

Reichert JM. Monoclonal antibodies in the clinic. *Nat Biotechnol* 2001;19(9):819–822.

Reiser J, Glumoff V, Kalin M, Ochsner U. Transfer and expression of heterologous genes in yeasts other than *Saccharomyces cerevisiae*. *Adv Biochem Eng Biotechnol* 1990;43:75–102.

Ren Y, Busch R, Durban E, Taylor C, Gustafson WC, Valdez B, Li YP, Smetana K, Busch H. Overexpression of human nucleolar proteins in insect cells: Characterization of nucleolar protein p120. *Protein Expr Purif* 1996;7(2):212–219.

Reuveny S, Kim YJ, Kemp CW, Shiloach J. Production of recombinant proteins in high density insect cultures. *Biotechnol Bioeng* 1993;42(2):235–239.

Rice JW, Rankl NB, Gurganus TM, Marr CM, Barna JB, Walters MM, Burns DJ. A comparison of large-scale Sf9 insect cell growth and protein production: Stirred vessel vs. airlift. *Biotechniques* 1993;15(6):1052–1059.

Riesenberg D, Menzel K, Schulz V, Schumann K, Veith G, Zuber G, Knorre WA. High cell density fermentation of recombinant *Escherichia coli* expressing human interferon alpha 1. *Appl Microbiol Biotechnol* 1990;34(1):77–82.

Riesenberg D. High-cell-density cultivation of *Escherichia coli*. *Curr Opin Biotechnol* 1991;2(3):380–384.

Rinas U, Kracke-Helm HA, Schügerl K. Glucose as substrate in recombinant strain fermentation technology. *Appl Microbiol Biotechnol* 1989;31(2):163–167.

Roddie PH, Ludlam CA. Recombinant coagulation factors. *Blood Rev* 1997;11(4):169–177.

Rodewald HR, Langhorne J, Eichmann K, Kupsch J. Production of murine interleukin-4 and interleukin-5 by recombinant baculovirus. *J Immunol Methods* 1990;132(2):221–226.

Rohricht P. Transgenic protein production. Part 2: Process economics. *BioPharm* 1999a;12(9):52–54.

Rohricht P. Transgenic protein production. The technology and major players. *BioPharm* 1999b;12(3):46–49.

Romanos MA, Scorer CA, Clare JJ. Foreign gene expression in yeast: A review. *Yeast* 1992; 8(6):423–488.

Rosenfeld SA, Brandis JW, Ditullio DF, Lee JF, Armiger WB. High-cell-density fermentations based on culture fluorescence. In: *Expression Systems and Processes for rDNA Products*, Hatch RT, Goochee C, Moreira A, Alroy Y, eds. ACS, Washington, DC, 1991, pp. 23–33.

Roush DJ, Lu Y. Advances in primary recovery: Centrifugation and membrane technology. *Biotechnol Prog* 2008 May–Jun;24(3):488–495.

Rudolph N. Technologies and economics for protein production in transgenic animal milk. *Gen Eng News* 1997;17(18):16.

Ruiz Jr. LP, Binion SB, Clark DR, Cady ME. Production of pharmaceutical products from milk—Computer model application in assessment of economics of large-scale recombinant protein production in cattle transgenic animal milk. *J Cell Biochem* 1990;(Suppl. 14D):17.

Rymaszewski Z, Abplanalp WA, Cohen RM, Chomczynski P. Estimation of cellular DNA content in cell lysates suitable for RNA isolation. *Anal Biochem* 1990;188(1):91–96.

Sandkvist M, Bagdasarian M. Secretion of recombinant proteins by Gram-negative bacteria. *Curr Opin Biotechnol* 1996;7(5):505–511.

Savijoki K, Kahala M, Palva A. High level heterologous protein production in Lactococcus and Lactobacillus using a new secretion system based on the *Lactobacillus brevis* S-layer signals. *Gene* 1997;186(2):255–262.

Schein CH. Production of soluble recombinant proteins in bacteria. *Biotechnology (NY)* 1989; 7(11):1141–1149.

Sellick I. Improve product recovery during cell harvesting. Enhanced TFF may reduce the capacity crunch. *BioProcess Int* 2003;1(4):62–65.

Skerra A. Bacterial expression of immunoglobulin fragments. *Curr Opin Immunol* 1993;5(2): 256–262.

Smith GE, Summers MD, Fraser MJ. Production of human beta interferon in insect cells infected with a baculovirus expression vector. *Mol Cell Biol* 1983;3(12):2156–2165.

So AD, Gupta N, Brahmachari SK, Chopra I, Munos B, Nathan C, Outterson K, Paccaud JP, Payne DJ, Peeling RW, Spigelman M, Weigelt J. Towards new business models for R&D for novel antibiotics. *Drug Resist Updat* 2011 Apr;14(2):88–94.

Sondergaard L. Drosophila cells can be grown to high cell densities in a bioreactor. *Biotechnol Tech* 1996;10(3):161–166.

Spivak JL, Avedissian LS, Pierce JH, Williams D, Hankins WD, Jensen RA. Isolation of the full-length murine erythropoietin receptor using a baculovirus expression system. *Blood* 1996;87(3):926–937.

Steidler L, Robinson K, Chamberlain L, Schofield KM, Remaut E, Le Page RW, Wells JM. Mucosal delivery of murine interleukin-2 (IL-2) and IL-6 by recombinant strains of *Lactococcus lactis* coexpressing antigen and cytokine. *Infect Immun* 1998;66(7): 3183–3189.

Stratton J, Chiruvolu V, Meagher M. High cell-density fermentation. *Methods Mol Biol* 1998;103:107–120.

Strijker R. Production of proteins in milk of transgenic cows—Human recombinant lactoferrin production via mamma tissue-specific gene expression in cattle transgenic animal milk. *Meded Fac Landbouwwet Rijksuniv Gent* 1994;59(4a):1733–1736.

Stromqvist M, Houdebine M, Andersson JO, Edlund A, Johansson T, Viglietta C, Puissant C, Hansson L. Recombinant human extracellular superoxide dismutase produced in milk of transgenic rabbits. *Transgenic Res* 1997;6(4):271–278.

Sudbery PE. The expression of recombinant proteins in yeasts. *Curr Opin Biotechnol* 1996;7(5):517–524.

Sumathy S, Palhan VB, Gopinathan KP. Expression of human growth hormone in silkworm larvae through recombinant Bombyx mori nuclear polyhedrosis virus. *Protein Expr Purif* 1996;7(3):262–268.

Summers DK, Sherratt DJ. Multimerization of high copy number plasmids causes instability: ColE1 encodes a determinant essential for plasmid monomerization and stability. *Cell* 1984;36(4):1097–1103.

Summers MD, Smith GE. A manual of methods for baculovirus virus vectors and insect cell culture procedures. Texas Agricultural Experimental Station Bulletin No. 1555, Texas A&M University, College Station, TX, 1987.

Swanson ME, Martin MJ, O'Donnell JK, Hoover K, Lago W, Huntress V, Parsons CT, Pinkert CA, Pilder S, Logan JS. Production of functional human hemoglobin in transgenic swine. *Biotechnology (N Y)* 1992;10(5):557–559.

Swartz JR. Advances in *Escherichia coli* production of therapeutic proteins. *Curr Opin Biotechnol* 2001;12(2):195–201.

Szablowska-Gadomska I, Zayat V, Buzanska L. Influence of low oxygen tensions on expression of pluripotency genes in stem cells. *Acta Neurobiol Exp (Wars)* 2011;71(1):86–93.

Tandon S, Horowitz P. The effects of lauryl maltoside on the reactivation of several enzymes after treatment with guanidinium chloride. *Biochim Biophys Acta* 1988;955(1):19–25.

Tejeda-Mansir A, Montesinos RM. Upstream processing of plasmid DNA for vaccine and gene therapy applications. *Recent Pat Biotechnol* 2008;2(3):156–72.

Thatcher DR. Recovery of therapeutic proteins from inclusion bodies: Problems and process strategies. *Biochem Soc Trans* 1990;18(2):234–235.

Thøgersen C, Holtet TL, Etzerodt M, Denzyme Aps, Assignee. Iterative method of at least five cycles for the refolding of proteins. U.S. Patent 5917018. Issued 29-6-1999. Filed 18-9-1995.

Thompson BG, Kole M, Gerson DF. Control of ammonium concentration in *Escherichia coli* fermentations. *Biotechnol Bioeng* 1985;27:818–824.

Tom RL, Caron AW, Massie B, Kamen AA. Scale-up of recombinant virus and protein production in stirred-tank reactors. In: *Baculovirus Expression Protocols*, Richardson CD, ed. Humana Press, Totowa, NJ, 1995, pp. 203–224.

Tramper J, Williams JB, Joustra D. Shear sensitivity of insect cells in suspension. *Enzyme Microb Technol* 1986;8:33–36.

van Berkel PH, Welling MM, Geerts M, van Veen HA, Ravensbergen B, Salaheddine M, Pauwels EK, Pieper F, Nuijens JH, Nibbering PH. Large scale production of recombinant human lactoferrin in the milk of transgenic cows. *Nat Biotechnol* 2002;20(5):484–487.

Van Cott KE. Transgenic animals as drug factories: A new source of recombinant protein therapeutics. *Expert Opin Invest Drugs* 1998;7(10):1683–1690.

van Urk H, Voll WSL, Scheffers WA, van Dijken JP. Transient-state analysis of metabolic fluxes in crabtree-positive and crabtree-negative yeasts. *Appl Environ Microbiol* 1990;56:281–287.

Velander WH. Recovering recombinant proteins from milk of transgenic animals is feasible. *Genet Technol News* 1989;9(12):4.

Velander WH, Johnson JL, Page RL, Russell CG, Subramanian A, Wilkins TD, Gwazdauskas FC, Pittius C, Drohan WN. High-level expression of a heterologous protein in the milk of transgenic swine using the cDNA encoding human protein C. *Proc Natl Acad Sci USA* 1992;89(24):12003–12007.

293

Velander WH, Page RL, Morcol T, Russell CG, Canseco R, Young JM, Drohan WN, Gwazdauskas FC, Wilkins TD, Johnson JL. Production of biologically active human protein C in the milk of transgenic mice. *Ann N Y Acad Sci* 1992;665:391–403.

Wall RJ, Pursel VG, Shamay A, McKnight RA, Pittius CW, Hennighausen L. High-level synthesis of a heterologous milk protein in the mammary glands of transgenic swine. *Proc Natl Acad Sci USA* 1991;88(5):1696–1700.

Walsh G. Biopharmaceutical benchmarks. *Nat Biotechnol* 2000;18(8):831–833.

Wang MY, Kwong S, Bentley WE. Effects of oxygen/glucose/glutamine feeding on insect cell baculovirus protein expression: A study on epoxide hydrolase production. *Biotechnol Prog* 1993;9(4):355–361.

Weickert MJ, Doherty DH, Best EA, Olins PO. Optimization of heterologous protein production in *Escherichia coli. Curr Opin Biotechnol* 1996;7(5):494–499.

Weiss SA, Godwin GP, Gorfien SF, Whitford WG. Insect cell culture in serum-free media. In: Baculovirus expression protocols, Richardson CD, ed. Humana Press, Totowa, NJ, 1995, pp. 79–95.

Werner MH, Clore GM, Gronenborn AM, Kondoh A, Fisher RJ. Refolding proteins by gel filtration chromatography. *FEBS Lett* 1994;345(2–3):125–130.

Werten MW, van den Bosch TJ, Wind RD, Mooibroek H, de Wolf FA. High-yield secretion of recombinant gelatins by *Pichia pastoris. Yeast* 1999;15(11):1087–1096.

Wetlaufer DB, Xie Y. Control of aggregation in protein refolding: A variety of surfactants promote renaturation of carbonic anhydrase II. *Protein Sci* 1995;4(8):1535–1543.

WHO. WHO requirements for the use of animal cells as in vitro substrates for the production of biologicals (Requirements for biological substances No. 50). *Dev Biol Stand* 1998;93:141–171.

Wilkins TD, Velander W. Isolation of recombinant proteins from milk. *J Cell Biochem* 1992;49(4):333–338.

Wright G. Manufacturing from transgenics: Scale up and production Issues. *J Biotechnol Health Care* 1997;4(3):247–254.

Wright G, Carver A, Cottom D, Reeves D, Scott A, Simons P, Wilmut I, Garner I, Colman A. High level expression of active human alpha-1-antitrypsin in the milk of transgenic sheep—Expression as beta-lactoglobulin fusion protein in transgenic animal milk. *Biotechnology* 1991;9(9):830–834.

Wright G, Colman A, Cottom D, Williams M. Licensing of protein products from the milk of transgenic animals. Validation for pathogen removal—A strategy. *Dev Biol Stand* 1996;88:269–276.

Wright G, Noble J. Production of transgenic protein. In: *Bioseparation and Bioprocessing*, Vol. II. *Processing, Quality and Characterization, Economics, Safety and Hygiene*, Subramanian G, ed. Wiley-VCH, 1998, pp. 67–79.

Xu B, Jahic M, Blomsten G, Enfors SO. Glucose overflow metabolism and mixed-acid fermentation in aerobic large-scale fed-batch processes with *Escherichia coli. Appl Microbiol Biotechnol* 1999;51(5):564–571.

Yabe N, Kato M, Matsuya Y, Yamane I, Iizuka M, Takayoshi H, Suzuki K. Role of iron chelators in growth-promoting effect on mouse hybridoma cells in a chemically defined medium. *In Vitro Cell Dev Biol* 1987;23(12):815–820.

Yamada KM, Olden K. Fibronectins - adhesive glycoproteins of cell surface and blood. *Nature* 1978;275(5677):179–184.

Yamane T, Shimizu S. Fed-batch techniques in microbial processes. *Adv Biochem Eng* 1984;30:147–194.

Yang S. *Influence of Proteolysis on Production of Recombinant Proteins in Escherichia coli.* Royal Institute of Technology (KHT), Stockholm, Sweden, 1995.

Yang ST, Shu CH. Kinetics and stability of GM-CSF production by recombinant yeast cells immobilized in a fibrous-bed bioreactor. *Biotechnol Prog* 1996;12(4):449–456.

Yang XM. Optimization of a cultivation process for recombinant protein production by *Escherichia coli. J Biotechnol* 1992;23(3):271–289.

Yee L, Blanch HW. Recombinant protein expression in high cell density fed-batch cultures of *Escherichia coli. Biotechnology (N Y)* 1992;10(12):1550–1556.

Yoon SK, Ahn YH, Han K. Enhancement of recombinant erythropoietin production in CHO cells in an incubator without CO_2 addition. *Cytotechnology* 2001;37(2):119–132.

Young MW et al. Production of biopharmaceutical proteins in the milk of transgenic dairy animals. *BioPharm* 1997;10(6):34–38.

Zardeneta G, Horowitz PM. Micelle-assisted protein folding. Denatured rhodanese binding to cardiolipin-containing lauryl maltoside micelles results in slower refolding kinetics but greater enzyme reactivation. *J Biol Chem* 1992;267(9):5811–5816.

Zhang J, Alfonso P, Thotakura NR et al. Expression, purification, and bioassay of human stanniocalcin from baculovirus-infected insect cells and recombinant CHO cells. *Protein Expr Purif* 1998;12(3):390–398.

Zhang J, Kalogerakis N, Behie LA, Iatrou K. A two-stage bioreactor system for the production of recombinant proteins using genetically engineered baculovirus/insect cell system. *Biotechnol Bioeng* 1993;42:357–366.

Zhang W, Bevins MA, Plantz BA, Smith LA, Meagher MM. Modeling *Pichia pastoris* growth on methanol and optimizing the production of a recombinant protein, the heavy-chain fragment C of botulinum neurotoxin, serotype A. *Biotechnol Bioeng* 2000;70(1):1–8.

Zhou WB, Zhou XS, Zhang YX. [Decolorization and isolation of recombinant hirudin expressed in the methylotrophic yeast *Pichia pastoris*]. *Sheng Wu Gong Cheng Xue Bao* 2001;17(6):683–687.

Zigova J. Effect of RQ and pre-seed conditions on biomass and galactosyl transferase production during fed-batch culture of *S. cerevisiae* BT150. *J Biotechnol* 2000;80(1):55–62.

Zigova J, Mahle M, Paschold H, Malissard M, Berger EG, Weuster-Botz D. Fed-batch production of a soluble β-1,4-galactosyltransferase with *Saccharomyces cerevisiae*. *Enzyme Microb Technol* 1999;25(3–5):201–207.

Ziomek CA. Minimization of viral contamination in human pharmaceuticals produced in the milk of transgenic goats. *Dev Biol Stand* 1996;88:265–268.

Chapter 8 Downstream systems optimization

Introduction

Downstream processing comprises the steps in the manufacturing of biological drugs once the biological entity such as a recombinant bacteria or a recombinant mammalian cell has expressed or produced the target protein or antibody. Generally, this requires harvesting, to collect the product, the concentration of the product, and subsequent purification, leading to the concentrated drug substance.

Bacterial downstream processing

The downstream process comprises harvest, capture, in vitro refolding, intermediary purification, enzymatic cleavage, and polishing after the step of generating starting material from fermentation and extraction. The most commonly used techniques used in downstream processing include chromatography, filtration, microfiltration, ultrafiltration, active filtration, centrifugation, precipitation, and crystallization. The downstream process is designed for effective and efficient removal of cell debris, host cell proteins, nucleic acids, and endotoxins that arise from the disruption of the cells. It is also important to remove the closely related target protein derivatives such as cleaved, deamidated, oxidized, and scrambled forms as much as possible. As a result of these considerations, a minimum of three different chromatographic steps is generally required. In most instances, the last step would be size exclusion chromatography (and thus four steps). Virus removal is required in instances where animal-derived raw materials are used; this would add an additional step to the total processing scheme.

The biosimilar manufacturing client should make a thorough search of the intellectual property infringement in using any technique since a large number of patents often protect the originator's molecules. (See Chapter 2 on Intellectual Property Issues, Strategic Elements.)

Harvest

Harvesting step generally uses techniques like centrifugation, filtration, microfiltration, or ultrafiltration. The purpose of the harvesting step is to isolate the inclusion bodies from cell culture impurities such as cell debris and to condition the sample for the following chromatographic capture step. Generally, centrifugation separates the cells from media; the cells are washed and their disruption is a mechanical (e.g., French press), chemical (e.g., urea and cysteine), or enzymatic nature (e.g., lysozyme) process to release host cell proteins (and therefore proteolytic enzymes also) to the medium. If the cells are exposed directly to the extraction buffer, cell debris is removed by centrifugation. It is recommended to use urea (7–8 M) as the denaturing agent and cysteine (50–100 mM, of nonanimal origin) as the reducing agent. Note that urea solution must be free of cyanate to prevent carbamylation of primary amino groups (for further information, see the carbamylation document). EDTA is often added to the extraction buffer in 1–2 mM concentrations in order to inhibit metalloproteases. The typical result of the harvest and extraction procedure is a viscous sample at 4C with high levels of target protein, host cell proteins, DNA and endotoxins in a 7–8 M urea, 50–100 mM cysteine, and 1–2 mM EDTA buffer at pH 7.5–9.0. The use of expanded bed adsorption (EBA) technique recovers proteins directly from preparations of broken cells, saving one or several centrifugation and/or filtration steps. Since during cell lysis and extraction the recombinant product is exposed to a variety of proteolytic enzymes, enzyme inhibitors should not be used at this stage. The enzymatic activity can be reduced by using protease negative mutant hosts, low temperatures, and fast procedures. Inevitably, the presence of denaturing agents leads to the formation of minor amounts of cleaved products, such as split and truncated forms.

Capture

Capture step follows the harvesting and generally uses chromatographic techniques like AC, IEC, and IMAC; expanded bed, packed bed. The purpose of the capture step is to remove cell culture and process-related impurities and to prepare for in vitro folding. The nature of the capture step depends on the protein construct expressed. For example, histidine tags are exceptional, binding to metal chelating matrices under denaturing conditions. Affinity chromatography can be used for fusion proteins if in vitro refolding has been carried out prior to the capture step. For nontagged proteins with short N-terminal extensions, ion exchange chromatography (IEC) can be used under denaturing conditions provided urea is used as a denaturant. The particle size of media should be in the range of 100–300 nm. It may be an advantage to using expanded bed technology. The typical result of the capture procedure is a purified sample comprising the target protein (1–5 mg/mL) and low to medium levels of impurities in 7–8 M urea, 50–100 mM cysteine, and 1–2 mM EDTA buffer at pH 7.5–9.0.

In vitro refolding

As mentioned earlier, inclusion bodies do not produce target protein in its natural form requiring refolding that is often chaotrope mediated, co-solvent assisted, requires dilution, desalting, use of hollow fiber, SEC, or immobilized protein. In some instance, prior to refolding of the proteins, some purification can be made by means of dilution, size exclusion chromatography, dialysis, or immobilization techniques ahead of the addition of refolding buffer. The purpose of the in vitro folding step is to bring the denatured target protein into its native biological active form. From a regulatory point, it is important to keep this procedure as a separate unit operation as random in vitro refolding is very difficult to control and impossible to document. Note that whereas the inclusion bodies from different sub-batches can be pooled, refolding step requires the use of single container and uniform conditions. The in vitro refolding comprises transfer from the initial denaturation buffer (e.g., in 7–8 M urea, 50–100 mM cysteine, 1–2 mM EDTA buffer at pH 7.5–9.0) by means of dilution, desalting, dialysis, size exclusion chromatography, or immobilization technologies to the renaturation buffer (e.g., 1–5 mM cysteine, 1–2 mM EDTA, co-solvent, buffer pH 7–9.5), which is also the composition of the sample for intermediary purification. The sample should be directly taken to intermediate purification column, avoiding excessive handling to prevent aggregation of the newly folded target protein.

The refolding is always carried out at a low protein concentration (0.1–0.5 mg/mL) to prevent protein aggregation; as a result, the volume of buffer used at this stage is large. Additionally, this may require the use of co-solvents added to the folding buffer, typically salts, sugars, detergents, sulfobetaines, or short chain alcohols. The most difficult refolding, the formation of disulfide bonds between cysteine residues, is managed by adjusting the redox potential, pH, temperature, conductivity, and protein concentration. Whereas refolding is often spontaneous when the environment is more oxidizing, a validated system is needed to reduce the losses of proteins at this stage. It is not uncommon to lose 50% of protein at this stage and, therefore, to make a commercial yield possible, this step must be most carefully monitored.

Enzymatic cleavage

The purpose of the cleavage procedure is to remove the N-terminal extension. The enzyme used to remove the tag must be of nonanimal origin. The enzyme may be immobilized in order to reduce its amount in the final product. If immobilized folding is being used, the cleavage enzyme may be added after refolding, thus combining the two operations. A specific immobilized refolding method using the FXa cleavage site has been described and used in the small scale. The folded protein is eluted from the column by means of FXa, leaving the tag immobilized to the column. The composition of the sample for polishing depends on the cleavage procedure used. Typical tags are fusion proteins or repeated histidine sequences. If the second amino acid (N-terminally) is proline, this amino acid will act as a stop codon for specific di-amino acid exo-peptidases and the tag can be removed without the enzymatic cleavage site insert. There are several types of cleavage sites, for example, where there is an enzymatic cleavage site insert between a fusion protein or tag and the mature protein and thus the cleavage is performed with an endopeptidase, such as Factor Xa (FXa), which recognizes the sequence -Ile-Glu-Gly-Arg-, resulting in a correct N-terminal sequence of the mature protein as FXa cleaves after the arginine residue. More than 10 eukaryotic proteins have been expressed by this method. Another construct type makes use of the exopeptidase, dipeptidyl amino peptidase I (DPPI), which cleaves every second peptide bond from the N-terminal end of the molecule except for the amino acid residues Glu or Pro acting as terminators for proteolytic degradation. The Glu stop terminator is used in constructs where the second amino acid residue is not proline. The N-terminal extension, be it a fusion protein, a His-tag, or a few selected amino acids, must be removed using specific bioprocess grade enzymes of recombinant origin. Well-characterized bioprocess enzymes for this purpose are rare (an exception is DPPI), limiting the number of protein constructs to be used in biopharmaceutical processes.

Intermediary purification

This step utilized techniques like IEC, IMAC, HAC, HIC, or packed bed to remove cell culture and process-derived impurities and to prepare the sample for polishing. Anion exchange, cation exchange, or hydroxyapatite chromatography will apply for the purification step following in vitro folding. Chromatography based on hydrophobic interaction should be avoided as the renatured protein tends to aggregate during such procedures. The particle size of media should be in the range of 50–120 nm. An important part of the process design is to assure a smooth transfer between unit operations making use of the ability of ion exchangers to bind proteins at low salt concentrations and hydrophobic interaction media to bind proteins at high salt concentrations. However, in certain cases, buffer exchange or sample concentration is required. It is recommended to use chromatographic desalting as the buffer exchange principle (no shear forces) and ultra-filtration (or precipitation) for sample concentration. Note that variations in the composition of eluted samples (e.g., protein concentration, conductivity, pH) may influence the proceeding sample application and thus interfere with process robustness.

Polishing

This step typically used techniques like RPC, IEC, or packed bed. This is the most refined step in purification to remove host cell, process, and product-related impurities. Powerful chromatographic methods (HP-RPC or HP-IEC) should be used to separate the protein from its derivatives (enzymatic cleaved forms, des-amido forms, oxidized forms, scrambled forms, etc.). To obtain an optimal resolution, the particle size of media should be in the range of 15–60 nm.

Concentration

In most instances, the eluent from the polishing step would provide the final product in a concentrated form. However, in some instances, reduction of volume may be required. This can be accomplished by ultrafiltration. See comments under intermediate purification relating to buffer exchange.

Finishing

This is generally the final step using SEC, desalting, or packed bed. SEC (packed bed) removes di- and polymer forms, small molecules (e.g., endotoxins, co-solvents, leachables), and prepares the drug substance for formulation into drug product. Whereas the host cell proteins are removed all along wherein DNA is removed by means of anion exchange or hydroxyapatite chromatography (DNA binds strongly to both types), endotoxins are removed by binding to anion exchange matrices, polymyxin, sepharose or histamine-sepharose; additionally, endotoxins are removed by ultrafiltration, gel filtration, or phase separation with the detergent Triton X114, the final step of finishing assure that the drug substance is ready to be used in the drug product formulation.

Mammalian downstream

Since the target protein is typically expressed directly in the medium, the typical procedures of harvest capture, intermediary purification, and polishing are used, in addition to viral clearance steps where required (when used, it is advisable to clear virus as early as possible despite the difficulties in handling early samples of media due to their viscosity, etc.). Since mammalian cell lines are subject to shear, milder bioreactor conditions are used; roller bottles are still widely used despite the high cost; similarly, the downstream processes chosen should be of such nature as to maintain the structural integrity of protein. For example, the use of chromatographic desalting as the buffer exchange principle (no shear forces) and ultrafiltration (or precipitation) for sample concentration is recommended. The process design should assure a smooth transfer between unit operations making use of the ability of ion exchangers to bind proteins at low salt concentrations and hydrophobic interaction media to bind proteins at high salt concentrations fully coordinated.

Harvest

This stage conditions the sample for the capture step. Cells and cell debris are typically removed by means of centrifugation, microfiltration, or expanded bed technology. The expanded bed technology may prove impractical due to cell aggregation. Several new membrane technologies obviate the problems of cell aggregation and improve the efficiency of tangential flow filtration. Mammalian cell cultures are sometimes carried out as continuous cultures running over several weeks (typically from 4 to 8 weeks). The harvest is collected into sub-batches at regular intervals and sometimes even processed further before pooling. It should be emphasized that each subbatch must undergo testing before being added to the pool of batches in order to demonstrate process control.

Capture

This step removes cell culture and process-derived impurities and prepares the sample for intermediary purification. Both packed bed and expanded bed technology are used. The typical chromatographic capture procedure used are IEC, although affinity or hydrophobic interaction chromatography (HIC) work for some proteins. The presence of antifoam agents in the harvest may eliminate the use of HIC due to their relative high hydrophobicity. The particle size of chromatographic media is ideally in the range of 100–300 nm.

Several purification principles are suited for capture purification of IgG. Protein A affinity chromatography has been used in a number of cases, but the less common protein G or IMAC technology should offer an advantage as well as many monoclonal antibodies bind strongly to metal affinity matrices. Note that monoclonal antibodies bind to nucleic acids and divalent metal ions. Protein A binds to all subclasses of human IgG (except for IgG class 3) in the Fc region at neutral to alkaline pH with low to high conductivity. Generally, only slight adjustment of the culture medium is needed for application and binding, although high ionic strength increases the binding capacity of the column. The binding is unaffected by the nature and variations of the glycosylation pattern. The mAb is eluted by decreasing pH, but the instability at low pH must seriously be taken into consideration. The addition of co-solvents, such as 30%–60% ethylene glycol or 1–2 M urea to the buffer, is sometimes used to elute the protein at a higher pH. Further, a stabilizing effect is obtained if the mAb is eluted into a neutral pH buffer. If fetal calf serum has been added to the culture medium, bovine IgG may amount up to 50% of the IgG eluted. Several IgMs bind to protein A, but elution often requires lower pH than 4.5, which seriously challenges the protein stability. Leakage of the protein A ligand is a serious drawback, and specific analytical assays must be introduced to prove efficient removal. IMAC is a very attractive capture purification principle for IgGs, which binds to metal chelating columns at neutral to alkaline pH (buffers containing chelating agents, free ammonium ions, free amino acids, amines or aminated zwitterion should not be used). The binding is relatively independent of salt concentration, the presence of nonionic detergents, and of urea in low concentrations. The cell culture medium can, therefore, be applied to the column with little or no conditioning. Elution is performed by decreasing pH in the range of 8–4, mainly reflecting titration of histidyl residues or with chelating agents such as EDTA displacing the metal from the column. Leakage of Ni^{2+} or similar metal ions must be expected, and thus further purification is needed in subsequent steps. Although IMAC can be used virtually anywhere in the downstream process with only minor adjustment of the application sample, the capture features dominate because there is little or no adjustment of the large volume of cell culture application sample required, the majority of contaminants pass through the column at loading conditions, and that metal ions can be removed in later steps. IMAC binds IgG from more species and subclasses than do protein A, and it operates under far milder conditions. It is less expensive and does not leach cytotoxic biological material into the product. IMAC will selectively recover intact IgG from supernatants in which the light chain is in surplus. It should be emphasized that the expanded bed technology offers the advantages of very gentle isolation of the expressed mAb, saving at least one centrifugation or filtration step.

Virus inactivation

Inactivation is typically (but not necessarily) performed after the capture step, where the sample volume has been severely reduced. The virus inactivation program should be linked to the infectious viruses used for transfection of the cells.

Intermediary purification

This step removes cell culture and process-derived impurities and prepares the sample for polishing. The intermediary purification step offers a wide range of chromatographic principles (IMAC, HIC, IEC, HAC). The particle size of chromatographic media should be in the range of 50–120 nm. Both cation and anion exchange chromatography (IEC) are suited purification principles for all mAbs, although the former is restricted in that very few mAbs are soluble at low pH and low conductivity. Binding conditions and choice of ion exchanger depend on the pI of the antibody, and thus some IgG class 3 mAb may be too basic to support high binding capacities of anion exchangers. HIC is applicable to all mAbs. Application in highly concentrated salt solutions and elution in low conductivity buffers may result in precipitation of the antibody and care should, therefore, be taken to investigate the stability of the mAb under these conditions. Mixed mode ion exchangers, such as hydroxyapatite, is a suited purification principle, also for IgM. However, chelating agents cannot be tolerated in the application buffer, and the matrix does not withstand acidic pH buffers.

Virus filtration

The techniques used include pH modification, the addition of detergents, and use of microwave heating. Membrane filtration can contribute to the overall virus reduction in a reliable and controlled manner without damaging the target protein. In many cases, large viruses could be removed to the detection limit of the assay used. Small viruses (e.g., parvoviruses) may require filtration through special membranes that have limited protein transmission. cGMP facilities are typically divided into an area where viruses are expected to be present (harvest and capture) and nonvirus areas potentially free of viruses. An early virus filtration step is recommended if the viscosity and particulates levels would allow an effective process (Table 8.1).

Polishing

This step removes host cell, process, and product-related impurities. Powerful chromatographic methods (HP-IEC) should be used to separate the protein from its derivatives (enzymatic cleaved forms, des-amido forms, oxidized forms, scrambled forms, etc.). The polishing step makes use of the most highly developed purification methods to obtain optimal resolution. The particle size of chromatographic media should be in the range of 15–60 nm.

Table 8.1 Removal of Adventitious Agents

Principle	Endotoxins	Nucleic Acids	Viruses	Note
Anion exchange	++	+++	+++	Strongly basic IgGs and IgMs form stable complexes with DNA, especially at low conductivity reducing the clearance.
Cation exchange	+	+++	++	Nucleic acids do not bind to cation exchangers.
HIC	+	+++	++	Nucleic acids do not bind to HIC media. Antibody-nucleic acid complexes are dissociated at high salt concentrations. Endotoxins may form micelles or higher secondary structures in aqueous solutions, especially at high salt concentrations and thus get excluded from the matrix.
Hydroxyapatite	+	+	+	Nucleic acid clearance is variable in phosphate gradients.
IMAC	++	++	+++	Assuming similar purification properties as for protein A.
Protein A	++	++	+++	Complexes between monoclonal antibodies and nucleic acids are dissociated at high salt concentrations tolerated in the protein A application.
Size exclusion	+	+	+	Nucleic acids form complexes with monoclonal antibodies at low conductivity resulting in a reduced clearance factor. The endotoxin clearance factor is less for IgM.

Finishing

This step removes di- and polymer forms and small molecules (e.g., endotoxins, co-solvents, leachables) and allows for easy reformulation of the drug substance. Techniques used include SEC, desalting; packed bed. The high molecular weight of IgM at 900 kDa makes SEC a powerful purification tool for IgM antibodies.

Yeast downstream processing

Downstream processing

In contrast to the bacterial expression systems, yeast does express proteins with the correct N-terminal amino acid residue. Consequently, use of tagged proteins is rarely observed with this expression system. The protein is normally expressed in the medium in its native form, although disulfide rearrangement or reduction may appear due to the low redox potential of the fermentation broth. Proteins are often expressed in yields ranging from 50 to 300 mg/L. Besides being regarded as generally safe to use, the yeast expression system does not raise any concerns related to endotoxin release from cell walls and possible virus infections. The cell wall is rigid, and separation of cells from the medium is relatively easy using centrifugation or expanded bed technology.

The purification techniques generally used are similar to those used for bacterial expression systems except there are no refolding and cleavage steps involved. There are no specific purification methods for removal of host cell proteins, but the use of three to four different chromatographic methods during downstream processing will, in most cases, result in an acceptable level. Unintended glycosylation products may be coexpressed together with the target protein (e.g., mono- and di-glycosylated forms). They can be difficult to separate from the target molecule even during polishing. Given in Table 8.2 are the major steps in the downstream processing of yeast-derived proteins.

Table 8.2 Steps in Yeast-Derived Systems

Harvest (centrifugation filtration microfiltration ultrafiltration)	The purpose of the harvesting step is to condition the sample for the capture step. Cells and cell debris are typically removed by means of centrifugation or expanded bed technology. The latter may in some cases prove difficult to operate due to cell aggregation. Several new membrane technologies have been developed to improve the efficiency of tangential flow filtration. The redox potential of the solution may be close to reducing conditions (<100 mV), resulting in cleavage of the disulfide bonds. This parameter should, therefore, be closely monitored during operation.
Capture (AC, IEC, IMAC; expanded bed packed bed)	The purpose of the capture step is to remove cell culture and process derived impurities and to prepare the sample for intermediary purification. Both packed bed and expanded bed technology will apply. The typical chromatographic capture procedure is ion exchange chromatography, although affinity or hydrophobic interaction chromatography may apply for some proteins. The presence of anti-foam agents in the harvest may eliminate the use of HIC due to their relative high hydrophobicity. The particle size of chromatographic media should be in the range of 100–300 nm. The sample composition depends on the chromatographic principle used.
Intermediary purification (IEC, IMAC, HAC, HIC; packed bed)	The purpose of the intermediary purification step is to remove cell culture and process derived impurities and to prepare the sample for polishing. The intermediary purification step offers a wide range of chromatographic principles (IMAC, HIC, IEC, and HAC). The choice of media does not only include the protein purification ability, but also the ability to "receive" pooled fraction from the capture step and to "deliver" a suitable application sample to the polishing step. The particle size of chromatographic media should be in the range of 50–120 nm. The nature of the sample for polishing depends on the chromatographic principle used.
Polishing (RPC, IEC, HIC; packed bed)	The purpose of the polishing step is to remove host cell, process, and product-related impurities. Powerful chromatographic methods (HP-RPC or HP-IEC) should be used to separate the protein from its derivatives (enzymatic cleaved forms, des-amido forms, oxidized forms, scrambled forms, etc.). The polishing part of the process makes use of the most highly developed purification methods to obtain optimal resolution. Minor amounts of glycosylated products may be present from the fermentation broth, and specific care should be taken to remove these products during polishing. The particle size of chromatographic media should be in the range of 15–60 nm.
Concentration (ultrafiltration)	This step may be required to concentrate the output to meet the product formulation requirements.
Finish (SEC, desalting; packed bed)	The purpose of the SEC step is to remove di- and polymer forms, small molecules (e.g., endotoxins, co-solvents, leachable) and to allow for easy reformulation of the drug substance.

Insect cell processing

Downstream processing

The use of the standard *baculo* virus system requires cyclic killing and lysis of the host cell that releases intracellular proteins to the medium complicates the purification of the target protein; as a result, alternate systems like transfection with appropriate plasmids are often preferred. The typical expression level of the target protein is from 50 to 500 mg/L and tags are not necessary. Insect cells are sensitive to shear forces, a factor that is not relevant when using baculovirus systems. Since there is a release of proteolytic enzymes, harvesting and capturing should be done under conditions where the enzymatic activity is decreased (low temperature and fast procedures) without using any enzyme inhibitors. The standard procedures of harvesting, capturing, intermediate purification, and polishing apply to insect cells as well as to other systems described earlier, and additionally, a step of virus decontamination is introduced. It is recommended to use this virus inactivation step early in the purification process despite the difficulties in the filtration in the early stages of processing.

Harvest   Most insect cell cultures are run as batch or fed batch cultures; hollow fiber modules prove useful.

Capture   The typical chromatographic capture procedure is IEC, although affinity or HIC may apply for some proteins. The presence of antifoam agents in the harvest may eliminate the use of HIC due to their relative high hydrophobicity. The particle size of chromatographic media should be in the range of 100–300 nm.

Virus inactivation   Since pathogenic or infectious viruses are not used to transform, it is difficult to define a suitable inactivation program. Generic inactivation methods (e.g., microwave) may be considered, but it may be enough to highlight inactivation measures taken as part of the downstream process (e.g., low pH, the presence of co-solvents).

Intermediary purification   The intermediary purification step offers a wide range of chromatographic principles (IMAC, HIC, IEC, and HAC).

Virus filtration   Small viruses (e.g., parvoviruses) may require filtration through special membranes that have limited protein transmission. Harvest and capture are the areas where viruses can be expected; use early viral filtration where possible taking into account viscosity and particulate load.

Polishing   Chromatographic methods (HP-RPC or HP-IEC) are used to separate the protein from its derivatives (enzymatic cleaved forms, des-amido forms, oxidized forms, scrambled forms, etc.).

Finishing   SEC removes di- and polymer forms, small molecules (e.g., endotoxins, co-solvents, leachable). DNA is removed by means of anion exchange or hydroxyapatite chromatography (DNA binds strongly to both types).

Transgenic animals processing

Downstream processing

Assuming the starting material is raw milk, it is normally converted to skim milk by centrifugation using techniques common in the dairy industry. After removal of casein, the capture, intermediary purification, and the polishing concept can be applied using the same purification principles as for cell culture harvests. Special attention should be paid to proteases and to the separation of the target protein from its animal counterpart coexpressed in milk.

Harvest   This step consists of removing fat micelles and casein; centrifugation, filtration, microfiltration, and possibly the use of expanded bed. Variations in the starting material composition are associated with the lactation cycle, the presence of subclinical infections, etc., requiring testing after each milking to establish conditions of separation of therapeutic proteins. The most dominant proteins present in cow or sheep milk is α- and β-lactoglobulin, immunoglobulin, and serum albumin, but a number of plasma proteins may also be found. Skimming removes 95%–98% of lipid, still enough fat to affect the useful life of chromatographic capture column and to block filters used.

Capture   This step removes cell culture and process-derived impurities and prepares the sample for intermediary purification. The shear bulk of casein (40 g/L) can be removed by precipitation at pH 4.5 or by adding precipitating agents such as polyethylene glycol. Use of low pH should be avoided as many proteins lose their biological activity such as glycoproteins may lose sialic acid residues. The casein micelles are solubilized by chelation of the calcium with EDTA or citrate prior to chromatography. Ceramic and organic membranes are used to remove casein micelles by micro- or ultrafiltration. Finally, clarification of skim milk using EDTA followed by addition calcium phosphate–based particles has been used to reform casein micelles away from the target protein. Many proteins such as protein C bind to the casein micelles, and if the micelles are not dissolved, the protein C can be lost. The process is suitable for expanded bed or big bead technology in the capture step, assuming that the protein does not bind to casein micelles. In a simple two-step procedure (de-creaming and capture), the protein solution is made ready for virus inactivation and purification. The typical chromatographic capture procedure is IEC, although affinity or HIC may apply for some proteins. The particle size of chromatographic media should be in the range of 100–300 nm.

Virus inactivation   Inactivation is typically (but not necessarily) performed after the capture step, where the sample volume has been severely reduced. Details provided for virus inactivation in other chapters on cell cultures apply here as well.

Intermediary purification   This separation removes cell culture and process-derived impurities and prepares the sample for polishing. The intermediary purification step offers a wide range of chromatographic principles (IMAC, HIC, IEC, and HAC).

Virus filtration   Membrane filtration can contribute to the overall virus reduction in a reliable and controlled manner without damaging the target protein.

Polishing   This step removes host cells, process, and product-related impurities. Powerful chromatographic methods (HP-IEC) are used to separate the protein from its derivatives (enzymatic cleaved forms, des-amido forms, oxidized forms, scrambled forms, etc.). Reversed phase chromatography is probably not an option taking the type of molecules expressed in transgenic animals into consideration (monoclonal antibodies, complex proteins). The particle size of chromatographic media should be in the range of 15–60 nm.

Finishing   The use of SEC removes di- and polymer forms, small molecules (e.g., endotoxins, co-solvents, leachable). The coexpressed animal target protein must be efficiently removed.

Optimization of downstream processing

Extraction, isolation, and purification   Several techniques have been used to condition the sample for the first chromatographic capture step: centrifugation, filtration, and microfiltration. In some cases, the harvest and capture steps have been united by means of expanded bed technology. Due to the large volumes handled, major changes in ionic strength or pH are not recommended. Instead, a purification principle matching the characteristics of the application sample should be selected. The recovery process begins with the isolation of the desired protein from the fermentation or cell culture medium, often in a very impure form. The advantage of cell culture and yeast-derived products is that many of these proteins are secreted directly into the medium, thus requiring only cell separation to obtain a significant purification. For *Escherichia coli*–derived products, lysis of the bacteria is often necessary to recover the desired protein. It is important in each case to achieve rapid purification of the desired protein because proteases released by the lysed organisms may cleave the desired product. Such trace proteases are a major concern for the purification of biotechnology-derived products because they can be very difficult to remove, may complicate the recovery process, and can significantly affect final product stability. The recovery process is usually designed to purify the final product to a high level. The purity requirement for a product depends on many factors, although chronic use products may be required to have much higher purity than those intended for single-use purposes. Biotechnology products contain certain impurities that the recovery processes are specifically designed to eliminate or minimize. These impurities include trace amounts of DNA, growth factors, residual host proteins, endotoxins, and residual cellular proteins from the media.

Capture Once the fermentation process is completed, the desired product is separated, and if necessary, refolded to restore configuration integrity, and purified. The first part of the downstream process, capture,

- Separates the expressed protein from major impurities (water, cell debris, lipoproteins, lipids, carbohydrates, proteases, glycosidases, colored compounds, adventitious agents, fermentation additives, fermentation by-products)
- Conditions the sample for further intermediary purification steps

The target protein should be concentrated and transferred to an environment, which will conserve the biological activity. The main purpose of the capture step is to get rid of water, host, and process-related impurities. Due to the large volumes handled and the nature of the application sample, the first chromatographic purification step is built on the on/off principle aiming for selective binding of the protein to the matrix. Large particle sizes are used to avoid clotting of the column, and low resolution should be expected. One of the fundamental principles for capture operations is to focus on the target protein properties to achieve effective binding and pay less attention toward contaminants since purity is not the issue at this stage. The high selectivity of affinity chromatography can be an attractive approach for capture provided that the affinity ligand stability does not put severe restrictions on the use of efficient cleaning and sterilization regimes. The most frequent purification principles used are packed bed or expanded bed affinity, and IEC. In expanded bed technology, direct application of the cell culture is possible, thereby reducing the number of unit operations. Capture operations should make use of simple technologies, broad parameter intervals, high flow rates, cheap chromatographic media, and large particle sizes to assure process robustness, consistency, and economy. The outcome of the capture operation is a severe reduction in sample volume, reduction of the amount of impurities, and a sample, which is conditioned for the next step.

For recovery of intracellular proteins, cells must be disrupted after fermentation. This is done by chemical, enzymatic, or physical methods. Following a disruption, cellular debris can be removed by centrifugation or filtration. For recovery of extracellular protein, the primary separation of product from producing organisms is accomplished by centrifugation or membrane filtration. Initial separation methods, such as ammonium sulfate precipitation and aqueous two-phase separation, can be employed following centrifugation to concentrate the products. Further purification steps primarily involve chromatographic methods to remove impurities and bring the product closer to final specifications. Extraction and isolation require either filtration or centrifugation:

- *Filtration*: Ultrafiltration is commonly used to remove the desired product from the cell debris. The porosity of the membrane filter is calibrated to a specific molecular weight, allowing molecules below that weight to pass through while retaining molecules above that weight. Filtration is an integrated part of every downstream process in order to condition the sample for chromatographic purification and to perform sterile filtrations.

307

The filtration techniques comprise conventional filtration, microfiltration, ultrafiltration, and the use of specific filters for removal of defined impurities such as endotoxins, viruses, or prions. In-depth filtration consists of the separation of particles from the solute using large pore size filters. Most samples are filtered before entering the chromatographic column to prevent an increase in backpressure. Microfiltration consists of separation of particles from the solute using specifically designed tangential flowover membranes with pore sizes ranging from 0.1 to 0.3 pm. Sheer forces may destabilize the protein. Ultrafiltration separates the protein from low molecular solvent molecules using specifically designed tangential flowover membranes with a cut-off range from 1 to 100,000 kDa. Sheer forces may destabilize the protein. Filters with specific ligands are used specific filters for virus and prion removal are entering the market. They may be used in future processes to increase product safety. Filtration is used throughout for sample conditioning. Microfiltration is used for clarification of the harvest. Ultrafiltration is used for buffer exchange and/or sample concentration. It is not recommended to use the technique as the final step in the downstream process, as the sheer forces may affect protein stability.

- Precipitation is rarely used as a purification technique, but rather as an intermediary step between two chromatographic unit operations. The trend is to avoid precipitation for reasons of economy, time, compliance issues, and convenience. However, it may be useful to precipitate the protein for the purpose of storage. Precipitation is rarely used in capture due to the large volumes handled.

- *Centrifugation*: Centrifugation can be open or closed. The environment where centrifugation is performed must be controlled. Centrifugation is commonly applied to remove cells, cell debris, and precipitates from the harvest. The technique is labor demanding and is today sought replaced with microfiltration techniques or expanded bed technology.

- Crystallization is protein specific. It offers both purification and the ability to store the protein in a convenient way. Crystallization may be used for some intermediary products.

- Virus inactivation and/or active filtration is added as an extra unit operation when using mammalian cells or transgenic animals as the expression system.

Purification Chromatography is one of the most powerful protein purification techniques. Over the years, the technique has been developed into an industrial tool, where most biopharmaceutical processes comprise at least three different chromatographic steps carefully adjusted to the capture, intermediary, and polishing principle. The following table gives an overview of the applications areas for different chromatographic purification methods. The selection of chromatographic medium depends on the composition of the sample. The table lists some of the general features of the different chromatographic principles. Co-solvents mentioned will, generally, not affect the binding of the protein to the column. For every chromatographic technique used, there is a balance between resolution, speed, capacity, and recovery. Choose logical combinations of chromatographic techniques, based on the main benefits of the technique and the condition of the sample at the beginning or end of each step. Techniques may be combined but mainly when they are complimentary to each other. Given next is an explanation of various types of chromatography practices.

The purification process is primarily achieved by one or more column chromatography techniques.

- *AC (affinity chromatography)*: AC works on the basis of structural epitome recognition. Low protein concentration; large application volumes; any pH that applies; low to medium ionic strength. Affinity chromatography is a selective technique wherein the on/off principle fits very well into the capture mode, where large amounts of impurities are removed. Proteins normally elute from the column at low pH; common practice to apply the sample at neutral to slightly alkaline pH.

- *AEC (anion exchange chromatography)*: AEC works on the principle of electrostatic interactions. Low protein concentration; large application volumes; pH > pI; low ionic strength; ethanol, urea, and nonionic detergents tolerated. IEC is the most common chromatographic technique in use. It suits capture, intermediate purification, and polishing, making use of the highly specific media developed with defined particle sizes, spacers, and ligands. HIC offers, generally, less resolution than IEC, but selectivity can be high when the proper ligand has been defined.

- *CEC (cation exchange chromatography)*: Low protein concentration; large application volumes; pH < pI; low ionic strength; ethanol, urea, and nonionic detergents tolerated.

- *HAC (hydroxyapatite chromatography)*: Low protein concentration; large application volume; pH > 7.0; the matrix does not tolerate acidic pH; low to medium ionic strength; medium to high ionic strength; ethanol tolerated. HAC often provides excellent solutions to purification challenges. The high-resolution RPC technique is primarily used for proteins of molecular weight lower than 25 kDa, as the binding constant for higher molecular weight proteins is too high. Elution with organic solvents is common, and protein stability in these buffers is a major issue.

- *HIC (hydrophobic interaction chromatography)*: Low protein concentration; large application volume; any pH that applies; medium to high ionic strength. HIC is often used in combination with ion exchange, making use of the binding of proteins at high salt concentrations and elution at low salt concentrations. The presence of hydrophobic antifoam agents in fermentation broth and cell cultures may lower the binding capacity if HIC is used in the capture step. It may also be a less attractive capture technique if the feed volume is large and the addition of salt is needed in order to increase the ionic strength of the solution. The large salt quantities thus needed add to the manufacture costs and causes a waste disposal problem.

- *IMAC (immobilized metal affinity chromatography)*: IMAC works on complex binding between metal-ligand and proteins. Low protein concentration; large application volumes; neutral pH; high ionic strength; denaturants, detergents, and ethanol tolerated. IMAC is rarely exploited in industrial downstream processing. It is a powerful capture technique for a number of proteins and a unique tool for histidine tagged proteins. It is also a salt-tolerant technique being a useful feature considering the typical ionic strength in biological starting materials. Selectivity can be very high and depends on the combined efforts of the primary, secondary, and tertiary structure of proteins.

- *RPC (reversed phase chromatography)*: Hydrophobic interaction; silica-based matrices do not tolerate alkaline pH. Low protein concentration; large application volumes; any pH that applies; organic solvents can be used.

Table 8.3 Relative Importance of Various Chromatographic Methods in Various Downstream Processes

Principle	Capture	Intermediary Purification	Polishing
Affinity	+++	+	+
Crystallization	+	++	+++
Filtration	+++	+++	+++
HAC	+	+++	+
HIC	++	+++	+
IEC	+++	+++	+++
IMAC	+++	++	+
Microfiltration	+++	+	+
Precipitation	+	+++	+++
RPC	+	+	+++
SEC	+	+	+++
Ultrafiltration	+++	+++	+++

- *SEC (size exclusion chromatography)*: Steric exclusion from the intra-particle volume; the sample volume is restricted to maximum 5% of the column volume. High protein concentration; small application volumes; any pH; any ionic strength; any solvent. SEC is recommended as the final purification step. Although the sample volume rarely exceeds 5% v/v of the column volume, the technique offers removal of di- and polymeric compounds, transfer to a well-defined buffer and, in many cases, enhanced protein stability. The relative importance of each of the unit operations is given in Table 8.3.

Impurity removal   A variety of impurities that these chromatographic techniques are expected to remove and the testing methods used to ascertain this removal include

- Aggregated proteins: SDS-PAGE, HPSEC
- Amino acid substitutions: amino acid analysis, peptide mapping, MS, Edman degradation analysis
- Deamidation: IEF, HPLC, MS, Edman degradation analysis
- DNA: DNA hybridization, UV spectrophotometry, protein binding
- Endotoxin: Bacterial Endotoxins Test (pyrogen test)
- Formyl methionine: peptide mapping, HPLC, MS
- Host cell proteins: SDS-PAGE, immunoassays
- Microbial (bacteria, yeast, fungi): microbial limit tests, sterility tests, microbiological testing
- Monoclonal antibodies: SDS-PAGE, immunoassays
- Mycoplasma: modified 21 CFR method, DNAF
- Other protein impurities (media): SDS-PAGE, HPLCb, immunoassays
- Oxidized methionines: peptide mapping, amino acid analysis, HPLC, Edman degradation analysis, MS
- Protein mutants: peptide mapping, HPLC, IEF, MS
- Proteolytic cleavage: IEF, SDS-PAGE (reduced), HPLC, Edman degradation analysis, MSS
- Viruses (endogenous and adventitious): CPE and HAd (exogenous virus only), reverse transcriptase activity, MAP

Validation Chemicals used in chromatography methods, either in the stationary (bonded) phase or in the mobile phase, may become impurities in the final product and the burden of validation (i.e., demonstrating removal of potentially harmful chemicals) lies with the manufacturer. A column material supplier certification regarding leaching of chemicals is not sufficient since the contamination is process and product dependent. Validation is necessary when isolating end product monoclonal antibodies or using a technique that contains a monoclonal antibody purification step. The process must demonstrate the removal of leaching antibody or antibody fragments. It is also required to ensure the absence of adventitious agents such as viruses and mycoplasmas in the cell line that is the source of the monoclonal antibodies. The main concern is the possibility of contamination of the product with an antigenic substance whose administration could be detrimental to patients. Continuous monitoring of the process is necessary to avoid or limit such contamination. The problem of antigenicity related to the active, as well as host proteins, is one that is unique to biotechnology-derived products in contrast to traditional pharmaceuticals. Manufacturing methods that use certain solvents should be monitored if these solvents are able to cause chemical rearrangements that could alter the antigenic profile of the drug substance. The manufacturer is also obligated to produce evidence regarding performance consistency of novel chromatographic columns. Considerations for single-use products such as vaccines may differ because they are not administered continuously and, in this case, antigenicity is desirable. On the other hand, validating the removal of ligand or extraneous protein contamination is necessary. Unlike drugs derived from natural sources, manufacturers of biotechnology-derived products have been required to provide validation of the removal of nucleic acids during purification. Vaccines may again be different in this regard because of the accumulated clinical history on these products.

Intermediate purification

The intermediate purification step:

- Removes the majority of key impurities cellular proteins, culture media components, DNA, viruses, endotoxins, etc.
- Is the first stage purification.
- Using *medium size* particles in a variety of chromatographic techniques.
- For chemical and/or enzymatic modifications of the protein.
- For higher resolution between related compounds by
  - Applying more selective desorption principles such as multistep or continuous gradient elution procedures.
  - More specific chromatographic matrices, using smaller particles (offering better resolution) and technically more advanced solutions (gradients); made possible by higher purity and generally lower viscosity of the samples applied.
- Ideal for packed bed technology.

Polishing

The polishing step(s):

- Are high-resolution chromatographic methods to separate even closely related compounds (des-amido forms, oxidized forms, etc.).
- Require expensive, small uniform media particles with reversed phase and IEC as the dominant chromatographic principles.

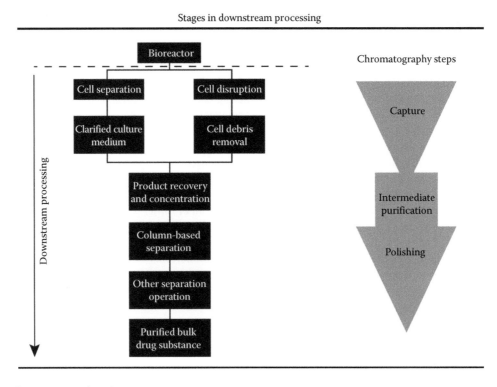

Figure 8.1 Downstream processing stages.

- Involve recommended size exclusion chromatographic or desalting operation as the final polishing step. Use of SEC as the final step may serve a number of purposes:
  - Removal of di- and polymers
  - Possible buffer exchange to any buffer requested for the formulation of drug substance bulk
  - The increase in protein stability
  - The uniform composition of the bulk material

The eluent from the polishing operation should be a stable, well-defined bulk material (drug substance) meeting specifications and with a composition acceptable for further formulation.

Overview

Whereas upstream process generates the crude protein, and much depends on the nature of gene construct and the conditions of upstream processing that determines the nature of the protein, it is the downstream processing that defines the final product.

Downstream processing schemes include several stages (Figure 8.1), each serving a specific function; primarily this consists of capture, intermediate purification, and polishing or final purification. Highest efficiency and cost reduction is achieved when the number of steps is reduced to a minimum, without appreciably affecting the final yield. The methods used in downstream processing are crucial to the safety of the product and, as a result, the processing area is required to be of the higher air quality standard than the upstream area (10,000 vs. 100,000). For recombinant proteins intended for use as human drugs, the purity must often exceed 99%, and some impurities, such as endotoxins and DNA, are limited to an

Table 8.4 Protein Properties and Downstream Processing Systems

Property	Affects
Charge	Choice of purification methods, precipitation, crystallization, and chromatography.
Co-factors	The choice to add cofactors to stabilize proteins.
Co-solvents	Choice of cosolvents depends on the function of pH, ionic strength, temperature, and redox potential on stability and solubility.
Detergent requirement	Choice and concentration of detergents depends on the hydrophobicity of proteins during separation and purification.
Disulfide bonds	Control of redox potential; a shift of one pH unit results in a 60 mV change in redox potential (toward the reducing side when pH is raised); use of reducing agents such as dithiothreitol or cysteine to reduce the cysteine residues to cysteine conversion.
Free cysteines	Control of reducing agents and condition (mM amounts of reducing agents are added to the buffer to prevent air oxidation) to maintain free cysteine residues.
Hydrophobicity	Use of detergents in the buffers to prevent aggregation and binding to surfaces. Hydrophobic proteins are sticky and tend to bind to surfaces and to other proteins and hydrophobic co-solvents (detergents); choice of reverse phase chromatography depends on hydrophobicity. Appendix lists hydrophobiticity of various amino acids.
Ionic strength	The solubility of the protein depends on the ionic strength of the solution.
Isoelectric point	pH adjustment to prevent unintended precipitation, particularly at low ionic concentration, during elution from chromatographic columns.
Metal ion sensitivity	The addition of EDTA to remove divalent metal ions.
Molecular weight	Selection of ultra- and microfiltration membranes and chromatographic columns.
Posttranslational modifications	Choice of methods including translation medium to control glycosylation and phosphorylation as essential properties for regulatory approvals.
Protease sensitivity	Storage of the harvest, or during downstream processing to prevent protease activity.
Solubility	Conditions that cause unintended precipitation by electrostatic interaction at low ionic strength, especially at pH values close to the isoelectric point (isoelectric precipitation), hydrophobic interactions under conditions of high salt concentration (salting out) in the preparation of feed stock for hydropohobic interaction chromatography. [Precipitation of even small particles blocks filters, columns, and valves in preparative loadings to chromatographic columns.]
Stability	Control of pH, conductivity, temperature, redox potential, protein concentration, di-valent metal ions, proteolytic enzymes, co-factors, and presence of co-solvents to prevent degradation and aggregation. The biological activity correlates with stability.

upper level in the range of parts per million. Chromatographic methods available provide adequate purification and purity levels required and are easily scalable. The most frequently used chromatographic methods are ion-exchange chromatography, size-exclusion chromatography, hydrophobic-interaction chromatography, reversed-phase chromatography, and affinity chromatography.

A careful selection of the processes in any downstream processing begins with an understanding of the characteristics of the protein in question, its stability profile, and the factors that may alter its structure. The most important criteria of the design of the downstream process are the physical and chemical properties of the protein manufactured (Table 8.4).

Capture The capture step may include a number of different unit operations such as cell harvesting, product release, feedstock clarification, concentration, and initial purification. EBA technology is specifically designed to address the problems related to the beginning of the downstream sequence and may serve as the ultimate capture step since it combines clarification, concentration, and initial purification into one single operation. Details of expanded bed operations are provided later in the chapter.

The overall purpose of the capture stage is to rapidly isolate the target molecule from critical contaminants such as proteases, glycosidases, remove particulate matter, concentrate, and transfer to an environment that conserves potency/activity.

At this stage, high throughput (i.e., capacity and speed) and short process times are very important. Processing time is critical because the fermentation broths and crude cell homogenates contain proteases and glycosidases that reduce product recovery and produce degradation products that may be difficult to remove later. Adsorption of the target molecule on a solid adsorbent as done in expanded bed systems and adsorption chromatography decreases the likelihood of interaction between degradative enzymes and susceptible intramolecular bonds in the target molecule. The first step is a capturing step, where the product binds to the adsorbent while the impurities do not. The product is often eluted with a step gradient, giving a high concentration of the product but a moderate degree of purification. The main requirements of this first step are high capacity, a high degree of product recovery, and high chemical and physical stability. IEC, and to some extent hydrophobic-interaction chromatography, is frequently used as the first chromatographic step.

Intermediate purification   After the capturing step, the bulk of impurities, such as host-cell proteins, nucleic acids, and endotoxins, are typically removed with high-resolution techniques such as hydrophobic-interaction chromatography, IEC, reversed-phase chromatography, or affinity chromatography. Lower flow rates, gradient elution, and matrices with particles of smaller size are used for enhanced resolution. After these steps, the product purity is typically at a level of 99%.

Polishing   The last step is a polishing step to remove any aggregates, degradation products, or target protein molecules that may have been modified during the purification procedure. It also serves to condition the purified product for its use or storage. Commonly used techniques for the final step are size-exclusion chromatography and reversed-phase chromatography.

System suitability   The purification schemes should enable the eluted sample from one step to be applied directly on to the next step, avoiding buffer changes and concentration steps. It is also important to keep the number of steps as low as possible since the total recovery decreases rapidly with the increasing number of steps. A convenient way to reduce the number of steps in the purification scheme without reducing the purity is to include a step with high selectivity, such as affinity chromatography, at each stage. Finally, the sanitization and cleanability of equipment is important. Cleaning chromatography columns is particularly challenging because of the inability to achieve the recommended linear velocities necessary to efficiently clean bioprocess equipment. Consequently, the mechanical design of the column plays a large role in its cleanability. The column's internal flow geometries and seal configurations must prevent dead flow areas that allow liquid to remain static during normal operation. All areas of the column must be swept with sufficient and consistent velocity to ensure effective cleaning and sanitization. The prospective manufacturer is advised to seek from manufacturer's certification that the column is sanitizable, and the regulatory authorities do accept this claim. Examples of sanitizable column include the QuikScale® biochromatography column is designed to distribute the solution evenly across the packed media bed to prevent any areas of stagnant liquid. This column can be easily and consistently sanitized and cleaned to FDA guidelines using typical sanitization protocols employing NaOH. Given next is a list of columns that can be used for separation of products intended for human consumption. As an example, the BioProcess™ HPLC columns from Amersham (now GE) comprise a family of stainless steel, high-pressure chromatography columns with a wide range of bed heights and inner diameters and feature dynamic axial compression that uses solvent as the compression medium. In addition, dynamic axial piston pressure eliminates the formation of voids or channels in packed beds. Together with a specially designed flow distribution system fitted with

Table 8.5 Commonly Used Downstream Buffers and the Conditions of Their Use

Buffer	pKa	dpKa/dT	Comments
Acetic acid	4.76	−0.0002	Supplied as glacial acetic acid ($d = 0$ 1.058 g/mL at 20°C). Acetic acid is corrosive and fumes irritant.
Ammonia	9.25	−31	Volatile. Fumes are harmful. Buffers are often prepared from ammonium hydrogen chloride or bicarbonate.
Boric acid	9.23	−0.008	Often used in combination with Tris and not so often as a standalone buffer.
Carbonic acid	6.35	−0.0055	CO_2 exchange with the environment. Carbonate buffers are commonly used in cell cultures to sustain
	10.33	−0.009	CO_2 levels. Work above pH 7.5 to avoid the release of CO_2.
Citric acid	3.14		Commonly used buffer in the pH range 3–7.
	4.76		
	6.39		
Ethanolamine	9.50	−0.029	Smelly harmful liquid. Reacts with many amine-modifying agents.
Glycine	2.35	−0.002,	Make sure the amino acid is not of animal origin. It is a zwitterion, a reactive primary amine. Also used
	9.78	−0.025	as a carbon and nitrogen source in cell cultures.
HEPES	7.66	−0.014	4-(2-hydroxyethyl)piperazine-ι-ethanesulfonic acid. Low UV absorbance, commonly used in cell culture media.
Phosphoric acid	2.15	0.0044	Reacts with divalent metal ions. pH changes may occur if freezing samples are stored in phosphate
	7.20	−0.0028	buffers.
	12.33		
Tris	8.06	−0.028	Tris(hydroxymethyl-aminomethane. High temperature sensitivity, reactive primary amine, influences pH measurements, undesirable effects on some biological systems.

Source: Beynon, R.J. and Easterby, J.S., Properties of common buffers, in: *Buffer Solutions: The Basics*, IRL Press at Oxford University Press, Oxford, U.K., 1996, pp. 67–82.

both the piston and the bottom of the column, this ensures a uniform distribution of the mobile phase across the whole bed. These columns comply with the technical performance demands placed on equipment operated at pressures up to 100 bar in industrial bioprocessing. In addition, they withstand temperatures up to 50°C, which allows effective cleaning with warm water. Construction is in stainless steel AISI 316L, and the sanitary design includes Quick connections. Sealing materials are resistant to organic solvents, and a leakage detection system is included.

Downstream processing systems

The choice of process equipment, chemicals used, columns selected, and the conditions under which processing is made are all critical. This section provides a general discussion of these elements. Details of processing can be found in the discussion of cell-specific processes described elsewhere in the book.

Buffers and solvents   A large number of buffers and solvents are used in the various unit operations associated with downstream processing. The largest range of buffers should normally be pKa ± 1. The solvents are selected based on their incompatibility with the proteins and the resins used. Given next is a list of the most commonly used buffers (Table 8.5) and co-solvents (Table 8.6) along with the general conditions of their use.

Sample preparation   Each unit operation processes a prepared sample, which is generally a solution of the target protein with variable levels of impurities and has specific chemical and physical properties suitable for the processing desired. Conditioning of samples includes such steps as centrifugation, filtration, microfiltration, ultrafiltration, desalting, and precipitation (Table 8.7).

An optimized sample preparation will include the least number of steps; for samples prepared for initial handling, use of expanded bed technology is advised

Table 8.6 Commonly Used Downstream Solvents and the Conditions of Their Use

Cosolvent	Conditions for Use	Comments
Ammonium sulfate	1–3 M	Precipitating and stabilizing agent at concentrations above 0.5 mg/mL and to increase ionic strength in HIC.
Benzonase		The enzyme used to degrade DNA and RNA.
Cysteine	1–100 mM	Reducing agent. Used in low mM concentrations in protein refolding, and in 50–100 mM concentrations in the reduction of protein disulfide bonds. Relatively cheap.
Dextran sulfate	10%	Precipitates lipoproteins.
1,4-Dithiothreitol (DTT)	1–100 mM	Reducing agent. Used in low mM concentrations in protein refolding, and in 50–100 mM concentrations in the reduction of protein disulfide bonds. Relatively expensive.
Ethylenediaminotetraacetic acid (EDTA)	1–10 mM	Binds strongly to divalent metal ions
EGTA	1–10 mM	Binds strongly to divalent metal ions
Dodecyl-β-D-maltoside	0.5%–1.0% v/v	Nonionic detergent for membrane protein solubilization. May absorb at 280 nm. Expensive.
Glucose	20–50 mM	Protein stabilizer.
Glycerol	5%–50% v/v	Protein stabilizer.
Guanidinium hydrochloride	1–6 M	Protein denaturant used to dissolve inclusion bodies. Expensive.
Mannose	20–50 mM	Protein stabilizer.
NaCl	0.1–1 M	Maintains ionic strength. Corrosive to stainless steel.
Nonidet P40	0.05%–2.0%	Nonionic detergent. May absorb at 280 nm. Expensive.
Octyl-β-D-glucoside	0.05%–1.5%	Nonionic detergent for membrane protein solubilization. May absorb at 280 nm. Expensive.
Polyethylene glycol	Up to 20% v/v	A precipitating agent with no denaturing effect. Please check if the phrase "Mr usually less than 6000" is correct. Complete removal may be difficult; does not affect AC or IEC.
Polyethylene imine	0.1% v/v	Precipitates aggregated nucleoproteins.
1-Propanol	0.05%–1.0%	Organic modifier.
Protamine sulfate	1%	Precipitates aggregated nucleoproteins.
Sodium dodecyl sulfate (SDS)	0.1%–0.5%	Ionic detergent and denaturant. Expensive.
Sucrose	20–50 mM	Protein stabilizer.
Tris	0.01–0.2 M	Maintains pH.
Triton X-100	1%–2% v/v	Nonionic denaturing detergent absorbs at 280 nm. Expensive.
Urea	1–8 M	Protein denaturant used to dissolve inclusion bodies. Cheap.

Table 8.7 Sample Conditioning Steps

Technique	Comments
Filtration	To clarify the sample; the sample must pass a 0.45 μm filter before being applied to the column. Filtration should be done just prior to chromatography as its composition may change with time due to precipitation or viscosity changes. These steps must be properly validated.
Ultrafiltration	For buffer exchange or to concentrate the sample. Sheer forces may result in protein destabilization.
Microfiltration	To remove cells or cell debris. Sheer forces may result in protein destabilization.
Centrifugation	To remove cells, cell debris, or particulate matter.
Dilution	To reduce conductivity and/or viscosity when sample volume is not extremely large.
Removal of divalent metal ions	With EDTA or similar chelating agents.
Desalting	Chromatographic desalting is a very efficient buffer exchange method; used for removal of GuHCl or urea.
Hollow fiber dialysis	For buffer exchange.
Bioprocessing aids	Normally added batch-wise to crude protein solutions in order to remove particulate matter and hydrophobic compounds.
Benzonase	The enzyme used to degrade DNA and RNA.
Pre-column	Safety device to protect the main column.
Cyclodextrins	Bind strongly to detergents; used to remove detergents from protein surfaces.
Precipitation	Separation and concentration step; better to use other newer techniques allowing continuous operations.

Table 8.8 Sample Characteristics and Their Effect on Protein Stability

Parameter	Comments
Concentration	Very low protein concentration may limit the equilibrium capacity in adsorption chromatography.
Conductivity	The ionic strength of the solution affects protein solubility and binding to ion exchangers. Precipitation may occur slowly over time gradually transforming the clear solution to a filter-blocking sample.
Holding times	The sample stability is a function of time. Note that holding times in large-scale operations are much longer than in laboratory scale.
pH	The pH of the solution will affect the solubility of the protein, its ability to bind to ion exchangers, and protein stability. pH will also influence the redox potential of the solution.
Redox potential	The redox potential of the solution influences disulfide bond stability. Under reducing conditions, the disulfide bonds will open up resulting in free cysteinyl residues. The redox potential is a function of pH.
Temperature	The temperature influences pH, ionic strength, and the redox potential. The protein stability is a function of temperature. Proteolytic enzymes are less active at low temperatures.
Viscosity	Viscous samples are difficult to filter and apply to chromatographic columns.
Volume	When working with SEC or other isocratic techniques a large volume will call for a larger column to satisfy throughput.

Table 8.9 Cosolvents and Impurities That Can Affect Sample Stability

Cosolvent and Impurities	Comments
Cell debris	Increases column back pressure and blocks of filters.
Nucleic acids	Increase viscosity.
Enzymes	Proteolytic enzymes degrade the target protein.
Divalent metal ions	Bind to proteins and ion exchangers and affect in vitro renaturation. EDTA binds strongly to divalent metal ions.
Salts	Increase the conductivity (ionic strength) of the solution which may produce salting-in and salting-out effects; significantly affects affinity, ion exchange, and hydrophobic interaction chromatography, etc.
Nonionic detergents	Affects affinity, hydrophobic interaction, reversed phase chromatography. Bind strongly to the protein.
Ionic detergents	Affects ion exchange, hydrophobic interaction, reversed phase, and affinity chromatography. Bind strongly to the protein.
Zwitterionic detergents	Affects ion exchange, hydrophobic interaction, reversed phase, and affinity chromatography. Bind strongly to the protein.
Organic solvents	May cause precipitation. Affects hydrophobic and reversed phase chromatography. Large volumes require costly explosion proof facilities.
Poly-alcohols	May cause precipitation.
Antifoam agents	Reduces formation of foam. These agents are hydrophobic in nature and bind to hydrophobic matrices.
Carbohydrates	Are present from the fermentation process. Rarely affect the subsequent purification step.
Lipids/lipoproteins	Hydrophobic in nature. May affect all forms of chromatography due to irreversible binding to the matrix.
Colored compounds	Very often bind irreversibly to chromatographic matrices.
Urea	Is present from E. coli inclusion body extraction buffers. Does not affect ion exchange chromatography.
GuHCl	Is present from E. coli inclusion body extraction buffers. Does affect ion exchange chromatography.

(described later in the chapter). Any of these processes used can significantly affect sample stability (Table 8.8). Of greater importance is the physical stability of a sample comprising of aggregation and precipitation that change in the secondary and tertiary structure of the protein. Ideally, the protein should retain its biologically active form throughout the processing though it is often not practical. The sample stability is monitored by the biological activity of sample (e.g., IU/mg) throughout the process development stages. A comprehensive stability assurance plan would include testing to study effects of overnight storage of sample, changes in pH (2–10), ionic strength (0.05–2.0 M), solvents, temperature (1°C–40°C), and sheer force. A factorial design is most optimal to evaluate these data.

The stability of samples is further affected by the presence of co-solvents and other impurities (Table 8.9).

Protein folding

When using *E. coli*, the protein is retained in the inclusion bodies, which are brought into solution prior to downstream processing; at this stage, the protein is likely to have incorrect disulfide pattern as *E. coli* does not carry out the intra-cellular folding of the cloned protein. The amino acid cysteine plays a dominant role in protein folding. Under oxidizing conditions, this amino acid will form a disulfide bridge to another cysteine residue making the redox potential of the solution an essential (but rarely measured) parameter in protein folding. It is a difficult task to reestablish the secondary and tertiary structure of proteins with disulfide bonds. Carefully designed folding buffers and fine-tuned parameter intervals (protein concentration, pH, conductivity, temperature, and redox potential) are needed to assure even reasonable yields and to reduce the amount of incorrectly folded (scrambled) forms, which are often very difficult to separate from the native form.

During the renaturation process, proteins tend to aggregate either as a result of electrostatic and/or hydrophobic interactions or as a result of intermolecular disulfide formation. As the intra-molecular folding reaction is of first order and the inter-molecular aggregation reaction is of second order, it is often necessary to carry out the folding at low protein concentrations (0.05–0.2 mg/mL), resulting in large tank volumes, when folding takes place in an industrial environment. The composition of the buffer used to extract proteins produces first a denaturation (and reducing), and the buffer used to refold the protein (producing oxidation) are different. The final buffer used is the greatest importance. Final yields can be improved often by adding specific co-solvents; if detergent-assisted renaturation is used, the detergents can be removed by the addition of cyclodextrins. The protein refolding generally takes place at alkaline pH (usually the interval 7.5–9.5 is used), as the deprotonated form of cysteine is required.

There are several methods for transfer from the initial denaturing or reducing buffer to the folding or oxidizing buffer. The dialysis in the bag is a common laboratory method, not suited for manufacturing scale production. Dilution is often used even in the large sale, but it requires larger tanks and careful addition of denatured protein into the folding buffer to assure proper mixing to obtain low protein concentrations uniformly throughout the tank. Dilution folding can also be accomplished using a method of pulse renaturation where aliquots of the denatured protein are added at successive time intervals. Desalting assures a very efficient transfer to the folding buffer, thus assuring well-defined folding conditions. If aggregation can be prevented by means of additional co-solvents, the method can be very efficient (moderate protein concentrations). The size exclusion chromatography allows high protein concentrations in the application sample as polymeric protein is removed from the folding zone due to size exclusion. The fast removal of aggregates will decrease the rate of the second-order aggregation reaction. Ultrafiltration is used for buffer exchange by using methods like hollow fiber dialysis that operates at low protein concentrations; the high capacity of hollow fiber devices makes this technique suited for an industrial scale. The application sample is dialyzed against the folding buffer. In the technique of immobilized folding, the protein is immobilized on a chromatographic matrix in the presence of the initial buffer. Freedom for structure formation is facilitated by binding through N-terminal extensions (His-tags or cellulose tags), which are later removed by enzymatic cleavage. Folding is achieved by applying the renaturation buffer to the column. Aggregation is severely reduced and in situ purification can be achieved by eluting the protein.

The first buffer system used is for the purpose of denaturing a protein prior to renaturing. Commonly used denaturants are urea, guanidinium chloride, or detergents. Urea is cheap, but the presence of cyanate will result in carbamylation of free amino groups. Therefore, cyanate must be removed (using a mixed anion exchange matrix) before use of the urea solution. The temperature is often kept at 4°C–8°C during the operation to prevent the formation of cyanate. A concentration of 7–8 M urea is recommended. Guanidinium chloride is a strong denaturant, but is very expensive. A concentration of 4–6 M is recommended. Most detergents are also very efficient denaturants at 1%–2% solutions. They may bind strongly to the protein, and certain chromatographic techniques cannot be carried out in the presence of detergents. EDTA is often added to the buffer (to bind divalent metal ions) in order to prevent oxidative side reactions during folding. A concentration of 1–5 mM is recommended. Dithiothreitol (DTT) or cysteine is commonly used as the reducing agent. Cysteine is the cheaper of the two and does not smell. A concentration of 50–100 mM is recommended. A recommended buffer for denaturation includes 7 M urea or 6 M GuHCl or 0.1%–2.0% detergent, 1–5 mM EDTA, 50–100 mM reducing agent (dithiothreitol or cysteine), and 50 mM buffer in the pH interval 7.5–9.5 (Tris or glycine). The buffer should have pH range 7.5–9.5; protein concentration 1–50 mg/mL; conductivity 10–50 mS/cm; temperature (urea) 4°C–8°C; temperature 4°C–25°C; and redox potential −300 to −100 mV.

In the oxidation step or renaturation step, the typical conditions are low protein concentration, alkaline pH, low to medium denaturant concentration, presence of reducing or oxidizing disulfide agent, presence of EDTA, and often a specific co-solvent to facilitate folding. The folding is a slow process and up to 24 h reaction time should be expected for folding of proteins with disulfide bonds. Specific enzymes (protein disulfide isomerase) are rarely used in the industrial scale. Refolding strategies depend on whether there is a concomitant disulfide bond formation or not. Where such bond is formed, the most common methods include air oxidation, the use of mixed disulfides, and the use of low molecular weight thiols. The latter method is well suited for large-scale operations partly because of its simplicity and partly because of the few side reactions observed. The cysteinyl amino acid residue will form a disulfide bond with another cysteinyl residue at oxidizing conditions (some disulfide bonds form with a redox potential of −50 mV). The reagents used are 2-mercaptoethanol, glutathion, or cysteine, which are added to the reaction mixture. It may be useful to take advantage of the presence of oxygen (approximately 0.4 mM) in normal aqueous buffers and add instead the reduced mercapto reagents. Cysteine is a cheap, nonsmelling, reagent well suited for large-scale operations (make sure the amino acid is of nonanimal origin). It has been common practice to make use of a specified ratio between the reduced and oxidized form of the mercapto reagent. However, in industrial application, it is strongly recommended to adjust the ratio of the reduced and oxidized form by controlling the redox potential of the solution. A typical folding buffer comprises 50–100 mM Tris or glycine, 0–3 M of denaturant, 1–5 mM EDTA, 2–5 mM mercapto reagent, a co-solvent. The conditions of reaction should be at pH 9.0–9.5, protein concentration of 0.1–0.5 mg/mL (G25 or hollow fiber), 3–15 mg/mL (GPC), conductivity 10–30 mS/cm, temperature 4°C–25°C, and redox potential of −50 to +50 mV. Several co-solvents have been used to facilitate protein folding mainly by suppressing aggregation in the inhibition of intermolecular hydrophobic interactions. These include acetamide, albumin, L-arginine hydrochloride, Brij 30, 35, and 58, carboxymethylcellulose, cetyltrimethylammonium bromide, CHAPS 3-(3-chloramidopropyl)dimethylammonia-1-propane sulfonate), CHAPSO, CTAB cetyltrimethylammonium bromide, cyclodextrins, cyclohexanol, deoxycholate, dodecyl maltoside, dodecyltrimethyl ammonium bromide, ethanol, ethylurea, formamide, glycerol, *n*-hexanol, hexadecyldimethylethyl ammonium bromide, hexadecylpyridinium chloride monohydrate, laurylmaltoside, methylformamide, methylurea, myristyltrimethyl ammonium bromide, NP40, *n*-pentanol, POE(10)L (CH$_3$(CH$_2$)$_{11}$(OCH$_2$CH$_2$)$_{10}$OH), potassium sulfate, SB3–14 (*N*-tetradecyl-*N*,*N*-dimethyl-3-ammonio-1-propane sulfonate), SB12, sodium dodecyl sulfate, sodium sulfate, sorbitol, sulfobetaines, tauro cholate, tetradecyl trimethyl ammonium bromide, Tris, Triton X-100, Tween 20, 40, 60, 80, and 81, and ZW3–14 (CH$_3$(CH$_2$)$_{13}$(N(CH$_3$)$_2$CH$_2$CH$_2$SO$_3$).

Another co-solvent approach, called the dilution additive strategy, has been to employ small molecules to promote protein folding in which the interaction between the small molecule and the protein is transient. In another co-solvent approach, named artificial chaperone-assisted folding, aggregation is prevented by the formation of protein-detergent complexes. In the second step, the detergent is removed with cyclodextrins having a higher binding constant for the detergent than that of the protein. The stripping of the protein facilitates intra-molecular folding. The naturally occurring chaperones GroEL and GroES are rarely used in industrial applications. Compounds used for chaperone-assisted refolding include cycloamylose, cyclodextrins, and linear dextrins.

In those instances where folding of proteins does not involve the formation of disulfide bonds, an optimization for disulfide formation is omitted. Further, those conditions that promote disulfide formation such as a correct redox potential are important. As most proteins comprise cysteinyl residues, it is common practice to add 1–10 mM reduced mercapto reagent to the folding mixture in order to prevent unintended disulfide bond formation.

Tables 8.10 and 8.11 show typical initial and final buffer conditions for two proteins, lysozyme, and IGF-1.

Table 8.10 Buffer Conditons for Lysozyme

Parameter	Initial Conditions	Final Conditions
Protein concentration	50 mg/mL	0.2 mg/mL
Denaturant	6 M GuHCl Guanidiniumhydrochloride	30 mM GuHCl Guanidiniumhydrochloride
Buffer	0.1 M Tris-sulfate pH 8.5	0.1 M Tris-sulfate pH 8.5
EDTA		2 mM EDTA
Reducing agent	30 mM Dithiothreitol	
Oxidizing agent		4 mM GSH glutathione/GSSG oxidized glutathione
Co-solvent		4 mM detergent
Cyclodextrin		16.5 mM methyl-beta-cyclodextrin

Table 8.11 Buffer Conditions for IGF-1

Parameter	Initial Conditions	Final Conditions
Protein concentration	1.5 mg/mL	
Denaturant	7 M urea	
Buffer	50–100 mM Tris	50 mM Tris
EDTA (ethylenediaminotetraacetic acid)	1–2 mM	2 mM EDTA
Reducing agent	50–100 mM cysteine	
Oxidizing agent		2 mM Cys/Cys-Cys
Co-solvent		25% v/v ethanol

Filtration

Filtration unit process is used for

- Harvesting, washing, or clarification of cell cultures, lysates, colloidal suspensions; microfiltration of cell cultures (prior to the capture operations) is an alternative to centrifugation or expanded bed technology
- Removal of aggregates and precipitated proteins
- Removal of particulate matter
- Prechromatographic clarification to remove colloidal particles (chromatographic columns act as filters; a medium of particle size 50–100 μm is equivalent to a 0.45 μm filter); use a prefilter to columns to prolong their life
- Buffer filtration
- Sterile filtration and depyrogenation of small molecules
- Specific virus removal
- Buffer exchange; diafiltration can be used and tangential flow or hollow fibers are commonly used in parallel with chromatographic desalting procedures; force applied here can affect protein stability
- Concentration, clarification, and desalting of proteins; ultrafiltration is used for sample concentration (e.g., prior to size exclusion chromatography or precipitation)

Alternates to filtration unit process are the techniques like expanded bed, precipitation, or desalting. The choice of filtration vis-à-vis, other techniques is made by comparing the losses of the target protein with the losses incurred in using other methods. The common losses during filtration consists of retention (0.4%–10%), adsorption (0.02%–2%), aggregation (0.1%–20%), and hold-up volumes (0.2%–10%). The wide range of losses is carefully optimized by choosing membranes with appropriate retention characteristics and adjusting various chemical and physical parameters such as pH and ionic strength. The adsorption depends on the membrane area vis-à-vis the amount of protein present in the sample feed, and the volume passed; generally, hydrophilic membranes will exhibit lower protein binding than hydrophobic membranes. Hold-up volumes can be reduced by careful piping design, optimization of membrane area, and flushing the system (and thereby diluting the filtrate or retentate). Aggregation in micro- and ultrafiltration affects protein stability by inducing aggregation.

Filtration is a pressure-driven separation process that uses membranes to separate components in a liquid solution or suspension based on their size and charge differences:

- Prefiltration (5 μm)
- Clarification (1 μm)
- Filtration (0.45 μm)
- Sterilization (0.22 μm)

In conventional flow, the fluid is pushed across the membrane; in tangential flow filtration, the fluid is pumped tangentially to the surface of the membrane such as microfiltration and ultrafiltration.

Microfiltration is usually used upstream in the recovery process to separate intact cells or cell debris from the harvest. Membrane pore size cut-off values are typically in the range from 0.05 to 1.0 μm. Ultrafiltration is used to separate protein from buffer components (buffer exchange), sample concentration, or virus filtration. Depending on the protein to be retained, the nominal molecular weight limits are from 1 to 1000 kDa or even up to 0.05 μm (virus filtration). In tangential flow filtration, the liquid is divided into two streams, namely, permeate and concentrate, the latter is too large to pass.

The ultrafiltration technique has traditionally been used to separate solutes that differ by more than 10 fold in size, making it ideal for buffer exchange and protein concentration. The low selectivity offered has mainly been attributed to the wide pore size distributions in commercial membranes, the presence of significant bulk mass transfer limitations, and membrane fouling phenomena. However, by proper selection of solute wall concentrations, in combination with the intrinsic sieving coefficients, the separation characteristics of the system can be altered toward better separation of the protein solute molecules.

The microfiltration technique is used early in the process to separate cells or cell debris from the harvest in order to clarify the sample before the first chromatographic capture unit operation. Ultrafiltration may be used throughout the downstream process with the purpose of buffer exchange or protein concentration. However, the sheer forces applied to the target protein may affect its stability, and tangential flow operations should always be carefully controlled. It is common practice to use ultrafiltration as the final downstream unit operation in order to assure correct bulk buffer composition, but protein stability may be affected. Alternatively, desalting using size exclusion chromatography should be considered.

The most commonly used types of membranes are as follows:

- Flat membranes in a cartridge where the feed is applied tangentially to the membrane. The incoming solution flows in the channels between the elements (filter surfaces), and the permeate filters through the elements. The channels can be open or equipped with a turbulent promoter (net). The concentrate, which is recycled, flows through the cartridge or modular elements. The system is easy to clean.
- Tubular membranes are used with a porous support assembly that is perforated inside and supports the tube. The input feed flows through the tube, and the permeate flows through the membrane outside the tube. It has a greater surface area than a flat plate and high capital costs.
- Spiral membranes are rolled up flat sheets. Alternate layers of membrane, porous support, membrane, and a spacer are wound around a perforated tube. The surface is greater than tubular and flat plate. It may be difficult to clean deposited material on the surface of the spacer.
- Hollow fiber filters are small porous fibers bundled together and sealed in a chamber. The feed is pumped through the fibers, and the permeate flows through the tubing in the chamber. Hollow fiber filters have very large surface area; however, they cannot handle solids in the feed stream.

Filtration optimization considerations The tangential flow filtration is controlled by adjusting the constant cross-flow rate to adjust changes in viscosity of sample, the pressure drop is maintained when cross-flow rates are adjusted, the adjustment of retentate pressure and the transmembrane pressure, which is often kept constant. The process conditions that affect these adjustments include the following:

- Configuration of the membrane is important to obtain correct flux and pressure. The flux depends on the membrane configuration, where the energy input per membrane unit area may vary considerably. Typically, filtration is run at a flux of 25–250 L/m^2 and pressure of 0.2–4 bar. The selection of the optimal membrane material/type is not straightforward and depends on fouling, membrane rejection, flux, pressure, time, energy consumption, etc.

- Molecular weight cut-off of the membrane must match the need. A membrane with a higher molecular weight cutoff has a higher permeability and flux. Sometimes, the difference in rejection is not significant, and there can be considerable economical advantages in optimizing the cut-off range.
- Intact cells/cell debris are retained on 0.05–1 μm membrane passing colloidal material, viruses, proteins, and salts.
- Viruses are retained on 100 kDa–0.05 μm membrane passing proteins and salts.
- Proteins are retained on 10–300 kDa or 1–1000 kDa membrane passing other proteins including small peptides and salts.

Physicochemical characteristics such as the following:

- pH and ionic strength, which affect protein solubility; precipitation of proteins reduces the permeability of the membrane.
- Temperature increases the diffusivity and decreases the viscosity. The effect is significant resulting in an increased flux. If the temperature is not adjusted, an increase in temperature should be expected as the filtration proceeds due to the energy input. Note that the protein stability might be affected by the temperature change.
- Protein concentration is a limiting factor for flux decreases with increasing protein concentration in the retentate; there is always a maximum concentration to which the feed can be concentrated.
- Diffusivity directly but not linearly determines process flux; higher diffusivity results in a higher flux because solids are removed from the gel layer at a faster rate.
- Viscosity increased decrease flux.
- Transmembrane pressure increases compress the gel, lowering its permeability. At low protein concentrations, the flux is a function of the transmembrane pressure while, at high protein concentration, the flux becomes less dependent on transmembrane pressure. Membranes with relatively high water fluxes, such as polysulfones, are less pressure dependent than membranes that have low water fluxes (e.g., cellulose-based membranes).
- Fouling of membranes results from adsorption of solutes to the membrane surface or polymerization of solutes, which is an irreversible process leading to decreased flux.
- Cross-flow velocity (tangential flow) increases flux at equal transmembrane pressure, but it can also affect protein stability.
- Filtrate velocity is normally uncontrolled (the operation depends on cross-flow velocity and transmembrane pressure); transmembrane pressure can be controlled by filtrate velocity.

Precipitation

It is often advantageous to precipitate the protein

- As an intermediary product for storage
- As final drug substance bulk material
- For buffer exchange and/or protein concentration

The advantages of protein precipitation include mainly the volume reduction, removal of specific impurities, and enhanced stability when using certain co-solvents. The main drawbacks to using this unit process are that the precipitated protein may be difficult to redissolve, disposal of waste may create environmental issues, the high cost (large space, labor intensive, the use of explosion-proof environment when solvents are used, inefficiency at low protein concentration), and decreased protein stability when some solvents are used. A volume reduction remains the primary reason for using precipitation techniques.

Protein precipitation depends on the distribution of hydrophobic and charged patches on the surface and the properties of the surrounding aqueous phase. The most common technique of precipitation is salting out using high concentrations of salts, largely depending on the hydrophobicity of the protein (Table 8.12). The nature of the salt is of importance and salts, such as ammonium sulfate, which encourage hydration of polar regions and dehydration of the hydrophobic regions without interacting with the protein surface, are favored. The salting-out effect of anions follows the Hofmeister series (phosphate > sulfate > acetate > chloride) and the most effective cations are ammonium > potassium > sodium. The heat of solution and change in solvent viscosity should be taken into account when choosing a precipitating salt as should be the presence of other contaminants in the salts that can also affect protein structure and stability (e.g., heavy metals in ammonium sulfate).

A globular protein will exhibit minimum solubility near its isoelectric point, as an overall charge close to zero minimizes the electrostatic repulsion between the solute molecules. This is called isoelectric precipitation, which is often carried out in the presence of polyalcohols in order to enhance the precipitation yield.

The addition of organic solvents (e.g., acetone, ethanol) to the aqueous phase reduces the water activity and decreases the dielectric constant of the solvent. It has been suggested that the precipitating forces are electrostatic in nature, much in the same way as under isoelectric precipitation, rather than through hydrophobic interaction. Larger proteins tend to precipitate at lower concentrations of organic solvent.

Table 8.12 Precipitation Agents in Downstream Processing

Precipitating Agent	Typical Conditions for Use	Comments
Acetone	0%–80% v/v at 2°C–8°C	Denatures irreversibly; explosion-proof area (10 L+) required; volume contraction calculations required.
Ammonium sulfate	1–3 M	May damage proteins; solid use produces uncontrolled precipitation locally; use saturated solution only; the ideally protein concentration of 1 mg/mL is required to obtain acceptable yields.
Caprylic acid	Sample volume/15 g	Precipitates bulk of proteins from sera and ascites, leaving IgG in solution.
Dextran sulfate	0%–0.5% v/v	Precipitates lipoproteins.
Ethanol	0%–60% v/v	Denatures irreversibly; explosion-proof area (10 L+) required; volume contraction calculations required.
Polyethylene glycol (3,000–20,000)	0–20 w/v	Rarely denatures; difficult to remove. The residual polymer will rarely interfere with the purification procedures used in industrial downstream processing.
Polyethylene imine	0%–0.1% w/v	Precipitates aggregated nucleoproteins.
Polyvinylpyrrolidine	0%–3% w/v	Precipitates lipoproteins. Alternative to dextran sulfate.
Protamine sulfate	0%–1% w/v	Precipitates aggregated nucleoproteins.

324

Polyethylene glycol of molecular weight 4,000–20,000 is used to precipitate proteins (in concentrations up to 20% w/v). The mechanism is similar to that of organic solvents, and PEG can be regarded as a polymerized organic solvent. It should be noted that PEG is not easy to remove from the protein, although its presence rarely affects chromatographic techniques.

Expanded bed adsorption system

The initial purification of the target molecule is traditionally treated with adsorption chromatography using a packed bed of adsorbent (see the chapter on Downstream Chromatography). However, this requires clarification of the crude feed before application to the chromatography column. The cells and/or cell debris are removed by centrifugation and/or microfiltration. Both of these traditional methods of separation have several disadvantages. For example, the efficiency of a centrifugation step depends on particle size, the density difference between the particles and the surrounding liquid, and viscosity of the feedstock. When handling small cells, such as *E. coli*, or cell homogenates, small particle size, and high viscosity reduce the feed capacity during centrifugation and sometimes make it difficult to obtain a completely particle-free liquid. To obtain a particle-free solution, centrifugation is usually combined with microfiltration. Although microfiltration yields cell free solutions, the disadvantages of this combination system include

- Reduction in the flux of liquid per unit membrane area due to fouling of the membrane
- Long process times
- Use of large units and thus large capital costs
- Recurrent cost of equipment maintenance
- Product loss due to degradation

The EBA technology provides a fluidized adsorption resin; adsorption of the target molecule to an adsorbent in a fluidized bed also eliminates the need for particulate removal done by centrifugation and/or microfiltration. The properties of EBA make it the ultimate capture step for the initial recovery of target proteins from crude feedstock. The process steps of clarification, concentration, and initial purification can be combined into one unit operation, providing increased process economy due to a decreased number of process steps, increased yield, shorter overall process time, reduced labor cost, and reduced running cost and capital expenditure. Successful processing by EBA has been reported in many commercial processes, for example, *E. coli* homogenate, lysate, inclusion bodies, and secreted products; yeast cell homogenate and secreted products; whole hybridoma fermentation broth; myeloma cell culture; whole mammalian cell culture broth; milk; animal tissue extracts, etc. The EBA technology has been pioneered by Amersham (GE Healthcare) through their Streamline® product line, which is fully scalable and widely used in the manufacturing of therapeutic proteins.

As the name describes, the adsorbent in the EBA system is expanded and equilibrated by applying an upward liquid flow to the column. A stable fluidized bed is formed when the adsorbent particles are suspended in equilibrium due to the balance between particle sedimentation velocity and upward liquid flow velocity. The column adaptor is positioned in the upper part of the column during this phase. Crude, the unclarified feed is applied to the expanded bed with the same upward flow as used during expansion and equilibration. Target proteins are bound to the adsorbent while cell debris, cells, particulates, and contaminants pass through unhindered. Weakly bound material, such as residual cells, cell debris, and another type of particulate material, is washed out from the expanded bed using upward liquid flow.

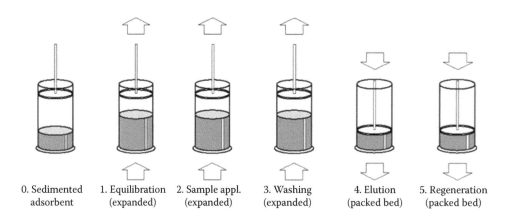

0. Sedimented adsorbent   1. Equilibration (expanded)   2. Sample appl. (expanded)   3. Washing (expanded)   4. Elution (packed bed)   5. Regeneration (packed bed)

Figure 8.2 Schematic presentation of the steps of expanded bed adsorption.

When all weakly retained material has been washed out from the bed, the liquid flow is stopped, and the adsorbent particles quickly settle in the column. The column adaptor is then lowered to the surface of the sedimented bed. Flow is reversed, and the captured proteins are eluted from the sedimented bed using suitable buffer conditions. The eluate contains the target protein, which is clarified and partly purified, ready for further purification by packed bed chromatography. After elution, the bed is regenerated by washing it with the downward flow in sedimented bed mode using buffers specific for the type of chromatographic principle applied. This regeneration removes the more strongly bound proteins that are not removed during the elution phase (Figure 8.2).

Finally, a cleaning-in-place procedure is applied to remove nonspecifically bound, precipitated, or denatured substances from the bed, and restore it to its original performance. During this phase, a moderate upward flow is used with the column adaptor positioned at approximately twice the sedimented bed height. Tailoring the chromatographic characteristics of an adsorbent for use in EBA includes careful control of the sedimentation velocity of the adsorbent beads. The sedimentation velocity is proportional to the density difference between the adsorbent and the surrounding fluid multiplied by the square of the adsorbent particle diameter. To achieve the high throughput required in industrial applications of adsorption chromatography, flow velocities must be high throughout the complete purification cycle. Most EBA systems are based on adsorbent agarose, a material proven to work well for industrial scale chromatography. The macroporous structure of the highly cross-linked agarose matrices combines good binding capacities for large molecules, such as proteins, with high chemical and mechanical stability. High mechanical stability is an important property of a matrix to be used in expanded bed mode to reduce the effects of attrition when particles are moving freely in the expanded bed. Agarose is also modified to make it less brittle to improve performance. The column also has a significant impact on the formation of stable expanded beds. The columns are equipped with a specially designed liquid distribution system to allow the formation of a stable expanded bed. The need for a specially designed liquid distribution system for expanded beds derives from the low pressure drop over the expanded bed. Usually, the flow through a packed bed generates such a high-pressure drop over the bed that it can assist the distributor in producing plug flow through the column. Since the pressure drop over an expanded bed is much smaller, the distributor in an expanded bed column must produce a plug flow itself. Consequently, it is necessary to build in an additional pressure drop into the distribution system. Besides generating a pressure drop, the distributor also has to direct the flow in a vertical direction only. Any flow in a radial direction of the bed will cause turbulence that propagates through the column. Shear stress associated with flow constrictions also requires consideration when designing the liquid distributor. Shear stress should be kept to a minimum to reduce the risk of molecular degradation. Another function of the distribution system is to prevent the adsorbent from leaving the column. This is usually accomplished by a net mounted on that side of the distributor that is facing the adsorbent. The net must have a mesh size that allows particulate material to pass through and yet at the same time confine the adsorbent to the column. The distributor must also have a sanitary design, which means that it should be free from stagnant zones where cells/cell debris can accumulate.

Understanding the hydrodynamics of the expanded bed is critical for the performance of an EBA operation. The hydrodynamics of a stable expanded bed, run under well-defined process conditions, are characterized by a high degree of reproducibility, which allows the use of simple and efficient test principles to verify the stability (i.e., functionality) of the expanded bed before the feed is applied to the column. The same type of test principles used to verify the functionality of a packed chromatography column is used in EBA. The bed is stable when only small circulatory movements of the adsorbent beads are observed. Other movements may indicate turbulent flow or channeling, which leads to inefficient adsorption. Large circular movements of beads in the upper part of the bed usually indicate that the column is not in a vertical position. Channelings in the lower part of the bed usually indicates air under the distributor plate or a partially clogged distribution system. These visual patterns are illustrated in Figure 8.3.

Besides visual inspection, bed stability is evaluated by the degree of expansion and number of theoretical plates, before each run. The degree of expansion is determined by the ratio of expanded bed height to sedimented bed height, H/H0. If the degree of expansion differs from the expected value, it may indicate an unstable bed. Absolute values for the degree of expansion can only be compared if the buffer system (liquid density and viscosity) and temperature are constant between runs. A significant decrease in the degree of expansion may indicate poor stability or channeling due to trapped air under the distributor plate, infection or fouling of the adsorbent, the column not being in a vertical position, or a blocked distributor plate.

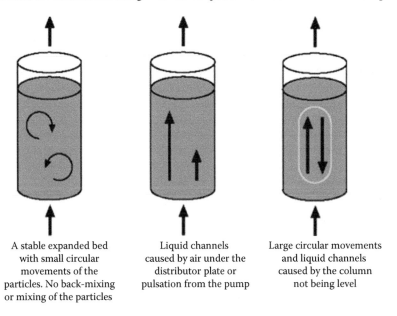

| A stable expanded bed with small circular movements of the particles. No back-mixing or mixing of the particles | Liquid channels caused by air under the distributor plate or pulsation from the pump | Large circular movements and liquid channels caused by the column not being level |

Figure 8.3 Visual patterns of movement of adsorbent beads in an expanded bed.

The Residence Time Distribution (RTD) test is a tracer stimulus method that can be used to assess the degree of longitudinal axial mixing (dispersion) in the expanded bed by defining the number of theoretical plates. A dilute acetone solution is used as a tracer input into the fluid entering the column. The UV absorbance of the acetone is measured in the exit stream from the column. The number of theoretical plates is calculated from the mean residence time of the tracer in the column and the variance of the tracer output signal, representing the standard band broadening of a sample zone. The RTD test is a simple but efficient tool for function testing complete systems. If used to test systems before feed application, the risk of wasting valuable feed is reduced considerably. The test should be performed with the buffer and flow rate that are to be used during process operation. Note that when using a small tracer molecule (such as acetone) with a porous adsorbent the measurement of RTD is a function of tracer permeation in the matrix pores in addition to the actual dispersion in the liquid phase. The Plate number of 170–200 N/m should be easily reached.

The critical parameters in EBA can be divided into chemical parameters, physical parameters, and the nature of the feedstock.

Chemical parameters relate to the selectivity and capacity of the separation process and include pH, ionic strength, types of ions, and buffers used. The influence on separation performance of these parameters is virtually the same in EBA as in traditional packed bed chromatography. For example, high-conductivity feedstock applied directly to an ion exchange adsorbent would reduce capacity requiring dilution prior to the application. If the conductivity is minimized at the end of the fermentation step, dilution is unnecessary. This results in less feed volume and shorter feed application time. In an intracellular system, the conductivity of feedstock can be reduced by running the homogenization step in water or a dilute buffer. The pH range defined during method scouting should also be verified in expanded bed mode since reduced pH in some systems may cause aggregation of biomass. This aggregation can block the column distribution system causing poor flow distribution and an unstable bed.

Physical parameters relate to the hydrodynamics and stability of a homogeneous fluidization in the expanded bed. Some physical parameters are related to the broth composition, for example, cell density, biomass content, and viscosity. Others are related to operating conditions such as temperature, flow velocity, and bed height. Cell density and biomass content both affect viscosity, which may reduce the maximum operational flow rate by over-expanding the bed. Temperature also affects the viscosity and, hence, the operational flow rate in the system. Increased temperature can improve binding kinetics. Optimization experiments are usually carried out at room temperature, but a broth taken directly from the fermentor may have a higher temperature. This difference in temperature must be considered when basing decisions on results of small-scale experiments.

Feedstock characteristics widely determine the application and efficiency of EBA systems (Table 8.13). Secretion systems generate dilute, low viscosity feedstock that contains rather low amounts of protein and intracellular contaminants, thus providing favorable conditions for downstream processing. Intracellular systems, on the other hand, generate feedstocks rich in intracellular contaminants and cell wall/cell membrane constituents. Along with the nutrient broth, these contaminants pose a greater challenge during the optimization phase of EBA. Much of the nutrient broth and associated contamination can be removed prior to cell lyses by

Table 8.13 Characteristics of Feed-Stocks according to the Location of the Product in the Recombinant Organism

E. coli	Yeast	Mammalian Cells
Secreted—dilute, low-viscosity feed containing low amounts of protein. Proteases, bacterial cells, and endotoxins are present. Cell lysis often occurs with handling and at low pH. DNA can be released and cause high viscosity.	Secreted—dilute, low-viscosity feed containing low amounts of protein. Proteases and yeast cells are present.	Secreted—dilute, low-viscosity feed containing low amounts of protein. Proteases and mammalian cells are present. Cell lyses often occurs with handling and at low pH. DNA can be released and cause high viscosity. Cell lysis can also release significant amounts of lipids. Agglomeration of cells can occur.
Cytoplasmic—Cell debris, high content of protein. Lipid, DNA, and proteases are present. Very thick feedstock which needs dilution. Intact bacterial cells and endotoxins are present.	Cytoplasmic—Cell debris, high content of protein. Lipid, DNA, and proteases are present. Very thick feedstock which needs dilution. Intact yeast cells are present.	Cytoplasmic—Unusual location for product accumulation.
Periplasmic—cell debris, high content of protein. Lipid and proteases are present. Thick feedstock which needs dilution. DNA is present if the cytoplasmic membrane is pierced. Intact bacterial cells and endotoxin are present.	Periplasmic—not applicable to yeast cells.	Periplasmic—not applicable to mammalian cells.
Inclusion body—Cell debris, high content of protein. Lipid and proteases are present. Very diluted solutions after renaturation. Intact bacterial cells, DNA, and endotoxin are present. Precipitation of misfolded variants occurs in a time-dependent manner.	Inclusion body—Not applicable to yeast cells.	Inclusion body—Not applicable to mammalian cells.

thorough washing of the cells, but such steps introduce additional costs to the process. The main source of contaminants in feed where the target molecule is located within the host cell is the complex cell membrane that has to be disrupted to release the target molecule. Bacterial and yeast cell walls have a high polysaccharide content that can nucleate into larger structures that foul solid surfaces. Proteins and phospholipids are other integral parts of such cell walls that will be released upon cell disintegration. Bacterial cell walls are particularly rich in phospholipids, lipopolysaccharides, peptidoglycans, lipoproteins, and other types of large molecules that are associated with the outer membrane of a bacterial cell. These contaminants may complicate downstream processing by fouling the chromatographic adsorbent.

The contaminant may also be present as charged particulates that can act as ion exchangers and adsorb proteins, especially basic ones if the ionic strength of the homogenate is low. This problem is, however, not specifically related to EBA and should be addressed when selecting conditions for cell disruption.

The main concern when processing a feed based on a secretion system would be to maintain intact cells, thereby avoiding the release of cell membrane components and intracellular contaminants such as DNA, lipids, and intracellular proteins that may foul the adsorbent or block the inlet distribution system of the column. Release of intracellular proteases is a further concern since it will have a negative impact on the recovery of biologically active material.

Animal cells lack a cell wall, which makes them more sensitive to shearing forces than microbial cells. The mammalian cell membrane is composed mainly of proteins and lipids. It is particularly rich in lipids, composing a central layer covered by protein layers and a thin mucopolysaccharide layer on the outside surface. Due to the high membrane content of mammalian cells, lysis can complicate the downstream process by causing extensive lipid fouling of the adsorbent. Another consequence of cell lysis is the release of large fragments of nucleic acids, which can cause a significant increase in the viscosity of the feedstock or disturb the flow due to clogging the column inlet distribution system. Nucleic acids may also bind to cells and adsorbent causing aggregation in the expanded bed. These types of contamination also lead to problems in traditional processing where they cause severe fouling during microfiltration.

Hybridoma cells are generally considered to be particularly sensitive to shear forces resulting from vigorous agitation or sparging. In contrast, CHO cells have relatively high resistance to shear rates and good tolerance to changes in osmotic pressure.

The use of EBA reduces the amount of cell lysis that occurs, as compared with traditional centrifugation and cross-flow filtration unit operations, since the cells are maintained in a freely flowing, low shear environment during the entire capture step. Nevertheless, it is important to actively prevent cell lysis during processing, for instance, by avoiding exposure to osmotic pressure shocks during dilution of the feedstock and by minimizing the sample application time.

Nonsecreted products sometimes accumulate intracellularly as inclusion bodies, which are precipitated protein aggregates that result from overexpression of heterologous genes. Inclusion bodies are generally insoluble, and recovery of the biologically active protein requires denaturation by exposure to high concentration of chaotropic salts such as guanidine hydrochloride or dissociants such as urea. The subsequent renaturation by dilution provides very large feedstock volumes. EBA can be advantageous since precipitation of misfolded variants increases with time, which usually causes problems for traditional packed bed chromatography. Even after extensive centrifugation of the feedstock, precipitation continues and may finally block a packed chromatography bed.

When a nonsecreted product accumulates in the periplasmic compartment, it can be released by disrupting the outer membrane without disturbing the cytoplasmic membrane. Accumulation in the periplasmic space can thus reduce both the total volume of liquid to be processed and the amount of contamination from intracellular components. However, it is usually very difficult to release the product from the periplasmic space without piercing the cytoplasmic membrane and thereby releasing intracellular contaminants such as large fragments of nucleic acids, which may significantly increase the viscosity of the feedstock.

Method scouting, that is, defining the most suitable adsorbent and the optimal conditions for binding and elution, is performed at small scale using clarified feed in packed bed mode. Selection of the adsorbent is based on the same principles as in packed bed chromatography. The medium showing strongest binding to the target protein while binding as few as possible of the contaminating proteins, that is, the medium with the highest selectivity and/or capacity for the protein of interest, will be the medium of choice. Regardless of the binding selectivity for the target protein, adsorbents are compatible with any type of feed material. The flow velocity during method scouting should be similar to the flow velocity to be used during the subsequent experiments in expanded mode. The nominal flow velocity for EBA is 300 cm/h. This may need adjustment during optimization, depending on the properties of the feedstock.

Elution can be performed stepwise or by applying a gradient. Linear gradients are applied in the initial experiments to reveal the relative binding of the target molecule versus the contaminants. This information can be used to optimize selectivity for the target molecule, that is, to avoid binding less strongly bound contaminants. It can also be used to define the stepwise elution to be used in the final expanded bed.

When selectivity has been optimized, the maximum dynamic binding capacity is determined by performing breakthrough capacity tests using the previously determined binding conditions. The breakthrough capacity determined at this stage will give a good indication of the breakthrough capacity in the final process in the expanded bed.

The purpose of the method optimization in expanded mode is to examine the effects of the crude feed on the stability of the expanded bed and on the chromatographic performance. If necessary, adjustments are made to achieve stable bed expansion with the highest possible recovery, purity, and throughput.

The principle for scale-up is similar to that used in packed bed chromatography. Scale-up is performed by increasing the column diameter and maintaining the sedimented bed height, flow velocity, and expanded bed height. This preserves both the hydrodynamic and chromatographic properties of the system.

In any type of adsorption chromatography, the washing stage removes nonbound and weakly bound soluble contaminants from the chromatographic bed. In EBA, washing also removes remaining particulate material from the bed. Since EBA combines clarification, concentration, and initial purification, the particulate removal efficiency is a critical functional parameter for the optimal utilization of the technique. Washing may also be performed with a buffer containing a viscosity enhancer such as glycerol, which may reduce the number of bed volumes needed to clear the particulates from the bed. A viscous wash solution follows the feedstock through the bed in a plug-like manner, increasing the efficiency of particulate removal. Even if the clarification efficiency of an EBA step is very high, some interaction between cell/cell debris material and adsorbent beads can be expected, which retain small amounts of cells and/or cell debris on the adsorbent. Such particulates may be removed from the bed during regeneration, for instance, when running a high salt buffer through an ion exchanger, or during cleaning between cycles using a well-defined CIP protocol.

Cells retained on the adsorbent may be subjected to lysis during the washing stage. Such cell lysis can be promoted by reduced ionic strength when wash buffer is introduced into the expanded bed. Nucleic acids released due to cell lysis can cause significant aggregation and clog owing to the "glueing" effect of nucleic acids forming networks of cells and adsorbent beads. If not corrected during the washing stage, wash volume/time may increase due to channeling in the bed. Other problems may also arise during later phases of the purification cycle, such as high back pressure during elution in packed bed mode and increased particulate content in the final product pool. If such effects are noted during washing, a modified wash procedure containing Benzonase (Merck, Nycomed Pharma A/S) can be applied to degrade and remove nucleic acids from the expanded bed.

Stepwise elution is often preferred to continuous gradients since it allows the target protein to be eluted in a more concentrated form, reduces buffer consumption, and gives shorter cycle times. Being a typical capture step, separation from impurities in EBA is usually achieved by selective binding of the product, which can simply be eluted from the column at high concentration with a single elution step.

The efficiency of the CIP protocol should be verified by running repetitive purification cycles and testing several functional parameters such as degree of expansion, number of theoretical plates in the expanded bed, and breakthrough capacity. If the nature of the coupled ligand allows it, an efficient CIP protocol would be based on

0.5–1.0 M NaOH as the main cleaning agent. If the medium to be cleaned is an ion exchange medium, the column should always be washed with a concentrated aqueous solution of a neutral salt, for example, 1–2 M NaCl, before cleaning with NaOH.

Occasionally, the presence of nucleic acids in the feed is the cause of fouling the adsorbent and in such a case, treating the adsorbent with a nuclease (e.g., Benzonase, Merck, Nycomed Pharma A/S) could restore performance. Benzonase can be pumped into the bed and be left standing for some hours before washing it out. Where the delicate nature of the attached ligand prevents the use of harsh chemicals such as NaOH, 6 M guanidine hydrochloride, 6 M urea and 1 M acetic acid can be used.

Virus inactivation and removal

Mammalian cell cultures and monoclonal antibodies derived from hybridoma cell cultures pose an essential risk of contamination with retroviral particles or adventitious viruses. Despite the preventive actions (master cell bank characterization, use of raw materials of nonanimal origin, end of production procedures), viral contamination remains a problem for all recombinant products as adventitious viruses can also be introduced during production (e.g., raw materials, cross-contamination) with the risk of contaminating the final product. Virus inactivation and virus removal during protein purification is an important step, particularly in insect cells, mammalian cells, or transgenic animals. Though it is possible to inactivate the viruses without appreciably affecting the target protein, only prevention and partial inactivation techniques suitable for maintaining the purity of proteins is a viable choice. Besides inactivation, chromatographic and filtration procedures are used to clear viruses during downstream processing. The overall level of clearance is the ratio of the viral contamination per unit volume in the pretreatment suspension to the concentration per unit volume in the post-treatment suspension. It is usually expressed in terms of the sum of the logarithm of the clearance found for individual steps possessing a significant reduction factor. Reduction of the virus titer of one $\log_{10}$ or less is considered negligible. Clearance should be demonstrated for viruses (or viruses of the same species) known to be present in the master cell bank (*relevant viruses*) or when a relevant virus is not available a *specific model virus* may be used as a substitute to challenge the samples and calculate clearance ratios. The choice of the viruses selected should be fully justified.

The accepted methods of virus clearance are: virus inactivation and virus removal; inactivation is the irreversible loss of any viral infectivity, while the virus removal is the physical reduction of viral particles in number achieved by such methods as depth and ultrafiltration and chromatography techniques exploiting electrostatic, hydrophobic, and hydrophilic surface characteristics. Inactivation of virus is achieved by the following:

- *Radiation*: γ- and UV-irradiation destroy the virus genome; Short-wavelength UV treatment can cause the formation of free radicals leading to protein damage. This can be minimized by the use of antioxidants or filters excluding the 185 nm wavelength.
- Heat inactivation contributes significantly to the inactivation of viruses. The introduction of high temperature short time (HTST) heat treatment offers substantial inactivation of small nonenveloped viruses while fully maintaining the integrity of the protein product. This method features the unique opportunity to spike and recollect a virus sample of a volume as low as 20–30 mL into the fluid pathway using a designed sample applicator under operational conditions for the manufacturing process

(flow rate 35–80 L/h, peak temperature 60°C–165°C) at full scale. The complete pathway is disposable, hence offering an extraordinary validation opportunity as well as a multiproduct use and avoiding any potential cross-contamination.

- *pH*: Acid treatment results in destruction in the nucleocapsid and genome. Acid treatment works well with larger virus particles. Particles < 40 nm may require a combination of pH < 3 and 3 M urea. Acid treatment at pH > 3 does not inactivate nonenveloped viruses. Enveloped viruses are not inactivated in a reliable manner.
- *Pressure*: High-pressure procedures at near-zero temperatures may be useful in inactivation of viruses.
- *Co-solvents*: Organic solvents dissolve the virus envelope and/or disintegrate the nucleocapsid.
- *Detergents*: Detergents dissolve the virus envelope and/or disintegrate the nucleocapsid. The method is very efficient for lipid-enveloped viruses but is ineffective against nonenveloped viruses.
- *Denaturation*: The method selected for inactivation depends to a great degree on the nature of virus; Results in the disintegration of the nucleocapsid. Urea treatment does not inactivate small nonenveloped viruses. Large enveloped viruses are inactivated to some extent. Urea treatment works well with larger virus particles. Particles < 40 nm may require a combination of pH < 3 and 3 M urea. Chaotropic agents dissolve the virus envelope and/or disintegrate the nucleocapsid.
- *Fatty acids*: Treatment with caprylate is a useful approach to remove the risk of lipid-enveloped viruses from protein pharmaceuticals.
- β-Propiolactone is an effective virucidal agent; 3.5–5 log reduction of viral infectivity is observed but it is toxic and removal must be assured by analytical testing validation.
- *Inactine*: The technology is based on the disruption of nucleic acid replication while preserving the integrity of lipids and protein.
- *Biosurfactants*: Surfactin, a cyclic lipopeptide antibioic with a molecular weight of 1036, possessing antiviral effect (enveloped RNA and DNA viruses) and antimycoplasma properties. The activity decreases with increasing protein concentration, and the reagent is not effective in solutions of high protein concentration. The in vivo toxicity is low.
- *Imines*: 0.05% v/v N-acetylethylenimine (AEI) inactivates infectious units of poliovirus and foot-and-mouth disease virus at 4°C or 37°C without damaging a variety of proteins tested.
- *Quaternary ammonium chloride*: 3-(trimethyloxysilyl)-propyldimethyloctadecylammonium chloride (Si-QAC) covalently binds to the alginate and removes viruses from protein solutions.

Because of the differences in the nature of viruses and their in vitro reactivity, it is difficult to adopt general methods for their clearance, for example, solvent-detergent treatment is highly effective for inactivation of enveloped viruses, but has no or little effect on nonenveloped viruses. Low pH is effective in mammalian cell systems but has little effect on parvoviruses. It is important to include at least one virus inactivation step in the downstream process scheme. The process should be properly validated keeping in mind that the inactivation is a biphasic process. The sample is ordinarily spiked with the virus in small amounts so as not to change the sample characteristics (e.g., buffer composition, protein concentration, conductivity). Parallel control assays should be included to assess loss of infectivity due to dilution, concentration, or storage. Buffers and products should be evaluated independently for toxicity or interference in assays used to determine the virus titer as these components may adversely affect the indicator cells. Virus escaping a first inactivation step may be more resistant to subsequent steps (e.g., virus aggregates).

Most downstream processes include at least three different chromatographic procedures that also help remove adventitious agents. The virus capsid is a shell consisting of different proteins, and the virus particle might behave like any other protein present in the sample to be purified. Also, the envelope of enveloped viruses contains a variety of proteins. Given the diversity in the nature of viruses, these methods cannot be relied upon to definitely remove viruses. Generally, anion exchange chromatography at basic pH and low conductivity proves most useful.

Filtration works through two mechanisms, size exclusion and absorptive retention of particulates. Size exclusion, which occurs due to geometric or spatial constraint, provides a predictable method of particle removal as it is not directly influenced by process and filtration conditions. In contrast, a number of process-dependent factors (e.g., charge, hydrophobicity, pH, ionic strength) influence the absorption of particulates in a much less controlled manner. Size exclusion and adsorptive filtration are not mutually exclusive. Depth filters with anion exchanging characteristics are very efficient virus removers.

The viruses of interest, such as retroviruses (100–140 nm) and small viruses (about 30 nm), can be removed by use of nanofiltration or by ultrafiltration. Nanofilters featuring pore sizes in the range of 20–70 nm utilize the classical depth structures, which are typical for commercial microfiltration membranes. Tighter pore sizes results in high back pressure (>0.3 MPa). Nanofilters are inexpensive in practical use and can be used anywhere in the downstream process as long as the load solution is 0.1 μm pre-filtered. A significant feature of nanofilters is the possibility of measuring their integrity. Complete clearance (to the limits of detection) was demonstrated in the removal of viruses above 35 nm from human globulin by 35 nm nanofiltration.

Ultrafiltration offers the possibility to operate at much smaller pore sizes (>1 kDa cut-off) more or less depending on the stokes radius of the protein. A cut-off of 200 kDa allows for the passage of IgG-type antibodies (M_w around 150 kDa), resulting in excellent clearance factors for virus particles >40 nm. Depending on process parameters, even a 100 kDa filter may be used successfully with significant retention of particles <40 nm[3].

Membrane filtration is an effective, reliable, and controllable technique to remove viruses. In many cases, large viruses could be removed to the detection limit of the assay used. Small viruses (e.g., parvoviruses) may require filtration through special membranes that have limited protein transmission. The cGMP facilities are typically divided into an area where viruses are expected to be present (harvest and capture) and nonvirus areas potentially free of viruses. An early virus filtration step is therefore of advantage but is challenged by the composition (viscosity, particulates) of the sample material in early downstream processing. Virus filtration should also be used whenever cell culture media or buffers are believed to include a virus contamination risk.

Process flow

The entire downstream process is presented in a process flow diagram that specifies the details of processes in a logical order, properly sized to meet the requirements of each operation. Given next is a process flow for the manufacturing of erythropoietin (Table 8.14).

Table 8.14 Process Flow for the Manufacturing of Erythropoietin

Equipment Used		Process and Material		Process Description
• XX L S.S. vessel • Sartorius membrane filter 2.0 and 0.45 µm	←	Growing media preparation: using fetal bovine serum, Gentamicin, DMEM, high glucose ↓	→	• Add X L of DMEM, high glucose in XX L vessel. • Add X L fetal bovine serum and X g Gentamicin, mix till dissolve. • QS to XX L with high glucose. • Filter the solution and keep at X°C for XX h.
• X L Sterile S.S. Vessel • Sartorius membrane filter 2.0 and 0.45 µm	←	Harvesting media preparation using human insulin Gentamicin DMEM, high glucose ↓	→	• Add X L of DMEM, high glucose in XX L sterile vessel. • Add XX mg human insulin and X g Gentamicin, mix for XX min. • QS to XX L with high glucose. • Filter the solution and keep at X°C for XX h.
• Water bath • Centrifuge • CO₂ incubator • Rolling bottle apparatus	←	Cell subculture for XX frozen vials of CHO EPO ↓	→	• Thaw the vials at X°C in water bath, wash vials with X% ethanol, transfer the contents of each vial into XX mL centrifuge tube. • Wash the cell by centrifugation at XX rpm for X min. with X–Y mL growing media for X times. • Disperse the cellular pellets in X mL of growing media in X mL flask and divide the dispersion into X flasks and complete the volume to X mL with growing media. • Repeat the above steps for all vials. • Incubate all flasks at X°C for X days. • Split the contents of each flask into X flasks and complete the volume to X mL with growing media. • Incubate all flasks at X°C for X days. • Split the contents of each flask into X roller bottles and complete the volume to X mL with growing media.
• Centrifuge • Sartorius membrane filter 2.0 and 0.45 µm	←	Harvesting for X days ↓	→	• Transfer the contents of each bottle to a centrifuge flask and wash the cell by centrifugation two times with PBS 15–20 mL. • Inoculate the cell dispersion of each flask with harvesting medium into roller bottles. • Media containing EPO collected every 24 h, filtered through 2.0 and 0.45 µm. • Collected medium replaced with a new medium for X days.
• Column Blue Sepharose XK	←	Purification: affinity chromatography ↓	→	• Filtered cellular supernatant rh-EPO is purified through an affinity column. • Filtered collected media is loaded in the column equilibrated in PBS. • The column is then washed with X column volumes with PBS and elute the bounded protein with PBS-1.4 M NaCl (X g of NaCl/L of PBS). The sample obtained are stored at −4°C. • After the EPO elution, the column is washed first with X mL of NaOH 0.1 M and later with 2 column volume of PBS-X M NaCl. Finally, the column is equilibrated in PBS for the next day.
• Column Sephadex Fine	←	Desalting ↓	→	• The rh-EPO recovered from the previous step is desalted using the column. • The column is equilibrated in X mM Tris-HCl pH X with an increasing flow going from X mL/min to X mL/min with a limit pressure of X MPa. The maximum volume loaded is around X mL. • The elution of the protein is controlled at X nm, and the salt concentration is followed by the conductimeter in line. • The column is cleaned with X column volumes of X M NaOH and re-equilibrated with X mM Tris-HCl pH X for the next day.
• Column Hi Trap Q Sepharose	←	Anion exchange ↓	→	• The desalted fraction containing the rh-EPO is purified using a Hi Trap Q Sepharose. • The column is equilibrated with X volumes of X mM Tris-HCl pH X and later with X volumes of 50 mM Tris-HCl pH X–X mM NaCl (X% buffer B, X mM Tris-HCl pH X–NaCl XM)

(Continued)

Table 8.14 (*Continued*) Process Flow for the Manufacturing of Erythropoietin

Equipment Used		Process and Material		Process Description
• Column Hi Prep	←	Desalting ↓	→	• The rh-EPO recovered from the previous step is desalted using column Hi Prep. • The column is equilibrated in X mM NaAC pH X at X mL/min. The maximum volume loaded into the column is X mL. • The solution is monitored at X nm controlling the elution of the protein and of the salt with the conductimeter line. The flow through is collected, and the salt is left to complete its elution. • The column is cleaned with X volumes of XM NaOH and re-equilibrated with X mM NaAC pH 6.0 for the next day.
• Column Hi Trap SP Sepharose	←	Cation exchange ↓	→	• The rh-EPO containing fractions eluted from the Q-Sepharose column and desalted is loaded in a Hi Trap SP-sepharose column. • This SP-column retains isoforms of incomplete glycosylation enriching the specific activity and another high molecular weight contaminant products. • The fraction containing the EPO molecules with high glycolisation elute with the flow through (FT) of the column using X mM NaAC, pH X. • The column is first equilibrated with 10 volumes of X mM NaAC pH X. • The fraction containing the rh-EPO is loaded in the column at X mL/min. • Measure the optical density with a spectrophotometer to determine the quantity of protein present in the sample. • Measure the volume obtained. • Filter the flow through in a Millipore Stericup 0.22 μm • Store the product obtained at 2°C–8°C • The column is then washed with X volumes of X mM NaAC, pH X-XM NaCl and re-equilibrated in X mM NaAC, pH X.
• Gel filtration column, Superose	←	Gel formation ↓	→	• The final polishing of the purified rh-EPO is performed using a gel filtration column. • The fraction is loaded at X mL/min. following the spectrum at X nm, collecting the peak containing the purified rh-EPO.
• Polyethylene storage container		Final product	→	• Erythropoietin concentrated solution has a concentration of X mg/mL to X mg/mL and potency of not less than 100,000 international units (IU) per milligram of an active substance determined.

Bibliography

Ahmed AK, Schaffer SW, Wetlaufer DB. Nonenzymic reactivation of reduced bovine pancreatic ribonuclease by air oxidation and by glutathione oxidoreduction buffers. *J Biol Chem* 1975;250(21):8477–8482.

Ambrosius D, Rudolph R. - Boehringer Mannheim GmbH, Assignee. The process for the reactivation of denatured protein. U.S. Patent 5618927. Issued 8-4-1997. Filed 2-7-1992.

Arakawa T, Timasheff SN. Preferential interactions of proteins with salts in concentrated solutions. *Biochemistry* 1982;21(25):6545–6552.

Aranha H. Viral clearance strategies for biopharmaceutical safety. Part 2: Filtration for viral clearance. *BioPharm* 2001;14(2):32–43.

Aranha-Creado H, Brandwein H. Application of bacteriophages as surrogates for mammalian viruses: A case for use in filter validation based on precedents and current practices in medical and environmental virology. *PDA J Pharm Sci Technol* 1999;53(2):75–82.

Aranha-Creado H, Fennington GJ, Jr. Cumulative viral titer reduction demonstrated by the sequential challenge of a tangential flow membrane filtration system and a direct flow pleated filter cartridge. *PDA J Pharm Sci Technol* 1997;51(5):208–212.

Aranha-Creado H, Oshima K, Jafari S, Howard G, Jr., Brandwein H. Virus retention by a hydrophilic triple-layer PVDF microporous membrane filter. *PDA J Pharm Sci Technol* 1997;51(3):119–124.

Aranha-Creado H, Peterson J, Huang PY. Clearance of murine leukaemia virus from monoclonal antibody solutions by a hydrophilic PVDF microporous membrane filter. *Biologicals* 1998;26(2):167–172.

Arsenault PR, Wobbe KK, Weathers PJ. Recent advances in artemisinin production through heterologous expression. *Curr Med Chem* 2008;15(27):2886–2896.

Asenjo JA, Andrews BA. Enzymatic cell lysis for product release. *Bioprocess Technol* 1990;143–175.

Ayroldi E, Macchiarulo A, Riccardi C. Targeting glucocorticoid side effects: Selective glucocorticoid receptor modulator or glucocorticoid-induced leucine zipper? A perspective. *FASEB J* 2014 Dec;28(12):5055–5070.

Ayyappan S, Prabhakar D, Sharma N. Epidermal growth factor receptor (EGFR)-targeted therapies in esophagogastric cancer. *Anticancer Res* 2013 Oct;33(10):4139–4155.

Bachrach HL. Reactivity of viruses in vitro. *Prog Med Virol* 1966;8:214–313.

Baldock D, Graham B, Akhlaq M, Graff P, Jones CE, Menear K. Purification and characterization of human Syk produced using a baculovirus expression system. *Protein Expr Purif* 2000;18(1):86–94.

Barnfield Frej A-K, Hjorth R, Hammarström Å. Pilot scale recovery of recombinant annexin V from unclarified *Escherichia coli* homogenate using expanded bed adsorption. *Biotech Bioeng* 1994;44:922–929.

Barnfield Frej A-K, Johansson HJ, Johansson S, Leijon P. Expanded bed adsorption at production scale: Scale-up verification, process example and sanitization of column and adsorbent. *Bioprocess Eng* 1997;16:57–63.

Barton S, Swanton C. Recent developments in treatment stratification for metastatic breast cancer. *Drugs* 2011 Nov 12;71(16):2099–2113.

Batas B, Chaudhuri JB. Protein refolding at high concentration using size-exclusion chromatography. *Biotechnol Bioeng* 1996;50(1):16–23.

Batt BC, Yabannavar VM, Singh V. Expanded bed adsorption process for protein recovery from whole mammalian cell culture broth. *Bioseparation* 1995;5;41–52.

Beck JT, Williamson B, Tipton B. Direct coupling of expanded bed adsorption with a downstream purification step. *Bioseparation* 1999;8(1–5):201–207.

Bellara SR, Cui Z, MacDonald SL, Pepper DS. Virus removal from bioproducts using ultrafiltration membranes modified with latex particle pretreatment. *Bioseparation* 1998;7(2):79–88.

Berdichevsky Y, Lamed R, Frenkel D, Gophna U, Bayer EA, Yaron S, Shoham Y, Benhar I. Matrix-assisted refolding of single-chain Fv-cellulose binding domain fusion proteins. *Protein Expr Purif* 1999;17(2):249–259.

Berthold W, Kempken R. Interaction of cell culture with downstream purification: A case study. *Cytotechnology* 1994;15(1–3):229–242.

Berthold W, Werz W, Walter JK. Relationship between nature and source of risk and process validation. *Dev Biol Stand* 1996;88:59–71.

Beynon RJ, Easterby JS. Properties of common buffers. In: *Buffer Solutions: The Basics*. IRL Press at Oxford University Press, Oxford, U.K., 1996, pp. 67–82.

Bezzine S, Ferrato F, Lopez V, de Caro A, Verger R, Carriere F. One-step purification and biochemical characterization of recombinant pancreatic lipases expressed in insect cells. *Methods Mol Biol* 1999;109:187–202.

Bleicher KH, Böhm HJ, Müller K, Alanine AI. Hit and lead generation: Beyond high-throughput screening. *Nat Rev Drug Discov* 2003 May;2(5):369–378.

Bödeker BGD, Potere E, Dove G. Production of recombinant factor VIII from perfusion cultures: II. Large-scale purification. In: *Animal Cell Technology. Products of Today, Prospects for Tomorrow*, Spier RE, Griffiths JB, Berthold W, eds. Butterworth Heinemann, Oxford, U.K., 1994, pp. 584–587.

Bova GS, Eltoum IA, Kiernan JA, Siegal GP, Frost AR, Best CJ, Gillespie JW, Su GH, Emmert-Buck MR. Optimal molecular profiling of tissue and tissue components: Defining the best processing and microdissection methods for biomedical applications. *Mol Biotechnol* 2005 Feb;29(2):119–152.

Bradley DW, Hess RA, Tao F, Sciaba-Lentz L, Remaley AT, Laugharn JA, Jr., Manak M. Pressure cycling technology: A novel approach to virus inactivation in plasma. *Transfusion* 2000;40(2):193–200.

Brandwein H, Aranha-Creado H. Membrane filtration for virus removal. *Dev Biol (Basel)* 2000;102:157–163.

Brantley JD, Martin J. Integrity Testing of Sterilizing Grade Filters. Sci. Tech. R., PBB-STR-28. 1997. Pall Ultrafine Filtration Company, East Hills, NY.

Brinkmann U, Buchner J, Pastan I. Independent domain folding of Pseudomonas exotoxin and single-chain immunotoxins: Influence of interdomain connections. *Proc Natl Acad Sci USA* 1992;89(7):3075–3079.

337

Brown F, Meyer RF, Law M, Kramer E, Newman JF. A universal virus inactivant for decontaminating blood and biopharmaceutical products. *Biologicals* 1998;26(1):39–47.

Brushia RJ, Forte TM, Oda MN, La Du BN, Bielicki JK. Baculovirus-mediated expression and purification of human serum paraoxonase 1A. *J Lipid Res* 2001;42(6):951–958.

Buchner J, Rudolph R. Renaturation, purification and characterization of recombinant Fab-fragments produced in *Escherichia coli*. *Biotechnology (NY)* 1991;9(2):157–162.

Budde RJ, Ramdas L, Ke S. Recombinant pp60c-src from baculovirus-infected insect cells: Purification and characterization. *Prep Biochem* 1993;23(4):493–515.

Buijs A, Wesselingh JA. Batch fluidized ion-exchange column for streams containing suspended particles. *J Chromatogr* 1980;201:319–327.

Burns MA, Graves DJ. Continuous affinity chromatography using a magnetically stabilized fluidized bed. *Biotechnol Prog* 1985;1:95–103.

Caldwell SR, Varghese J, Puri NK. Large scale purification process for recombinant NS1-OspA as a candidate vaccine for Lyme disease. *Bioseparation* 1996;6(2):115–123.

Cameron R, Davies J, Adcock W, MacGregor A, Barford JP, Cossart Y, Harbour C. The removal of model viruses, poliovirus type 1 and canine parvovirus, during the purification of human albumin using ion-exchange chromatographic procedures. *Biologicals* 1997;25(4):391–401.

Cerione RA, Leonard D, Zheng Y. Purification of baculovirus-expressed Cdc42Hs. *Methods Enzymol* 1995;256:11–15.

Cerletti N, McMaster GK, Cox D, Schmitz A, Meyhack B., Ciba-Geigy Corporation, Assignee. Process for refolding recombinantly produced TGF-.beta.-like proteins. U.S. Patent 5650494. Issued 22-7-1997. Filed 7-6-1995.

Cerletti N, McMaster GK, Cox D, Schmitz A, Meyhack B, Novartis AG, Assignee. Process for the production of biologically active protein (e.g. TGF). EP Patent 0433225. Issued 7-4-1999. Filed 27-11-1990.

Cha HJ, Dalal NG, Pham MQ, Bentley WE. Purification of human interleukin-2 fusion protein produced in insect larvae is facilitated by fusion with green fluorescent protein and metal affinity ligand. *Biotechnol Prog* 1999a;15(2):283–286.

Cha HJ, Dalal NG, Vakharia VN, Bentley WE. Expression and purification of human interleukin-2 simplified as a fusion with green fluorescent protein in suspended Sf-9 insect cells. *J Biotechnol* 1999b;69(1):9–17.

Chang Y-K, Chase HA. Development of operating conditions for protein purification using expanded bed techniques: The effect of the degree of bed expansion on adsorption performance. *Biotech Bioeng* 1996a;49:512–526.

Chang Y-K, Chase HA. Ion exchange purification of G6PDH from unclarified yeast cell homogenates using expanded bed adsorption. *Biotech Bioeng* 1996b;49:204–216.

Chang Y-K, McCreath GE, Chase HA. Development of an expanded bed technique for an affinity purification of G6PDH from unclarified yeast cell homogenates. *Biotech Bioeng* 1995;48:355–366.

Chapman J. Progress in improving the pathogen safety of red cell concentrates. *Vox Sang* 2000;78 Suppl 2:203–204.

Charm SE, Landau SH. Thermalizer. High-temperature short-time sterilization of heat-sensitive biological materials. *Ann N Y Acad Sci* 1987;506:608–612.

Charm SE, Landau S, Williams B, Horowitz B, Prince AM, Pascual D. High-temperature short-time heat inactivation of HIV and other viruses in human blood plasma. *Vox Sang* 1992;62(1):12–20.

Chase HA. Purification of proteins by adsorption chromatography in expanded beds. *Trends Biotech* 1994;12;296–303.

Chase HA, Draeger MN. Affinity purification of proteins using expanded beds. *J Chromatogr* 1992a;597:129–145.

Chase HA, Draeger MN. Expanded bed adsorption of proteins using ion-exchangers. *Sep Sci Technol* 1992b;27:2021–2039.

Churgay LM, Kovacevic S, Tinsley FC, Kussow CM, Millican RL, Miller JR, Hale JE. Purification and characterization of secreted human leptin produced in baculovirus-infected insect cells. *Gene* 1997;190(1):131–137.

Cleland JL, Builder SE, Swartz JR, Winkler M, Chang JY, Wang DI. Polyethylene glycol enhanced protein refolding. *Biotechnology (NY)* 1992;10(9):1013–1019.

Clemmitt RH, Chase HA. Facilitated downstream processing of a histidine-tagged protein from unclarified *E. coli* homogenates using immobilized metal affinity expanded-bed adsorption. *Biotechnol Bioeng* 2000;67(2):206–216.

Cole KD, Lee TK, Lubon H. Aqueous two-phase partitioning of milk proteins: Application to human protein-C secreted in pig milk—Recombinant protein-C purification from transgenic pig milk two-phase system extraction. *Appl Biochem Biotechnol* 1997;67(1–2):97–112.

Creighton TE. Process for the production of a protein. U.S. Patent 4977248. Issued 11-12-1990. Filed 10-8-1989.

Dasari G, Prince I, Hearn MTW. High-performance liquid chromatography of amino acids, peptides and proteins. CXXIV. Physical characterization of fluidized-bed behaviour of chromatographic packing materials. *J Chromatogr* 1993;631:115–124.

Daugherty DL, Rozema D, Hanson PE, Gellman SH. Artificial chaperone-assisted refolding of citrate synthase. *J Biol Chem* 1998;273(51):33961–33971.

De Bernardez Clark E. Refolding of recombinant proteins. *Curr Opin Biotechnol* 1998;9(2): 157–163.

De Bernardez Clark E. Protein refolding for industrial processes. *Curr Opin Biotechnol* 2001;12(2):202–207.

De Bernardez Clark E, Schwarz E, Rudolph R. Inhibition of aggregation side reactions during in vitro protein folding. *Methods Enzymol* 1999;309:217–236.

Debanne MT, Pacheco-Oliver MC, O'Connor-McCourt MD. Purification of the extracellular domain of the epidermal growth factor receptor produced by recombinant baculovirus-infected insect cells in a 10-L reactor. In: *Baculovirus Expression Protocols*, Richardson CD, ed. Humana Press, Totowa, NJ, 1995a, pp. 349–361.

Debanne MT, Pacheco-Oliver MC, O'Connor-McCourt MD. Purification of the extracellular domain of the epidermal growth factor receptor produced by recombinant baculovirus-infected insect cells in a 10-L reactor. In: *Baculovirus Expression Protocols*, Richardson CD, ed. Humana Press, Totowa, NJ, 1995b, pp. 349–361.

Degener A, Belew M, Velander WH. Expanded bed purification of a recombinant protein from the milk of transgenic livestock. *Abstr Papers Am Chem Soc* 1996 Mar 24;211(1):P87.

Degener A, Belew M, Velander WH. Zn(2+)-selective purification of recombinant proteins from the milk of transgenic animals. *J Chromatogr A* 1998;799(1–2):125–137.

Degener A, Belew M, Velander WH. Expanded bed purification of a recombinant protein from the milk of transgenic livestock. Presented at *211th American Chemical Society National Meeting*, New Orleans, LA, March 24–28, 1996.

Denman J, Hayes M, O'Day C, Edmunds T, Bartlett C, Hirani S, Ebert KM, Gordon K, McPherson JM. Transgenic expression of a variant of human tissue-type plasminogen activator in goat milk: Purification and characterization of the recombinant enzyme. *Biotechnology* 1991;9(9):839–843.

Dichtelmuller H, Rudnick D, Breuer B, Ganshirt KH. Validation of virus inactivation and removal for the manufacturing procedure of two immunoglobulins and a 5% serum protein solution treated with beta-propiolactone. *Biologicals* 1993;21(3):259–268.

Dobeli H, Andres H, Breyer N, Draeger N, Sizmann D, Zuber MT, Weinert B, Wipf B. Recombinant fusion proteins for the industrial production of disulfide bridge containing peptides: Purification, oxidation without concatamer formation, and selective cleavage. *Protein Expr Purif* 1998;12(3):404–414.

Dobers J, Zimmermann-Kordmann M, Leddermann M, Schewe T, Reutter W, Fan H. Expression, purification, and characterization of human dipeptidyl peptidase IV/CD26 in Sf9 insect cells. *Protein Expr Purif* 2002;25(3):527–532.

Draeger MN, Chase HA. Liquid fluidized bed adsorption of proteins in the presence of cells. *Bioseparations* 1991;2:67–80.

Drohan WN, Wilkins TD, Latimer E, Zhou D, Velander W, Lee TK, Lubon H. A scalable method for the purification of recombinant human protein C from the milk of transgenic swine. *Adv Bioproc Eng* 1994:501–507.

Ebeling F, Baer M, Hormila P, Jarventie G, Koistinen P, Katka K, Oksanen K, Perkkio M, Ruutu T, Soppi E. Tolerability and kinetics of a solvent-detergent-treated intravenous immunoglobulin preparation in hypogammaglobulinaemia patients. *Vox Sang* 1995;69(2):91–94.

Einhäuser W, König P. Getting real-sensory processing of natural stimuli. *Curr Opin Neurobiol* 2010 Jun;20(3):389–395.

Ejima D, Watanabe M, Sato Y, Date M, Yamada N, Takahara Y. High yield refolding and purification process for recombinant human interleukin-6 expressed in *Escherichia coli*. *Biotechnol Bioeng* 1999;62(3):301–310.

Elia JA, Floudas CA. Energy supply chain optimization of hybrid feedstock processes: A review. *Annu Rev Chem Biomol Eng* 2014;147–179.

Emerson MV, Lauer AK. Emerging therapies for the treatment of neovascular age-related macular degeneration and diabetic macular edema. *BioDrugs* 2007;21(4):245–257.

Erickson JC, Finch JD, Greene DC. Direct capture of recombinant proteins from animal cell culture media using a fluidized bed adsorber. In: *Animal Cell Technology: Products for Today, Prospects for Tomorrow*, Griffiths B, Spier RE, Berthold W, eds. Butterworth & Heinemann, Oxford, U.K., 1994, pp. 557–560.

Fabbro D, Batt D, Rose P, Schacher B, Roberts TM, Ferrari S. Homogeneous purification of human recombinant GST-Akt/PKB from Sf9 cells. *Protein Expr Purif* 1999;17(1):83–88.

Fahey EM, Chaudhuri JB, Binding P. Refolding and purification of a urokinase plasminogen activator fragment by chromatography. *J Chromatogr B Biomed Sci Appl* 2000a;737(1–2):225–235.

Fahey EM, Chaudhuri JB, Binding P. Refolding of low molecular weight urokinase plasminogen activator by dilution and size exclusion chromatography—A comparative study. *Sep Sci Technol* 2000b;35:1743–1760.

Fahrner RL, Knudsen HL, Basey CD, Galan W, Feuerhelm D, Vanderlaan M, Blank GS. Industrial purification of pharmaceutical antibodies: Development, operation, and validation of chromatography processes. *Biotechnol Genet Eng Rev* 2001;18:301–327.

Flamand M, Chevalier M, Henchal E, Girard M, Deubel V. Purification and renaturation of Japanese encephalitis virus nonstructural glycoprotein NS1 overproduced by insect cells. *Protein Expr Purif* 1995;6(4):519–527.

Frech M, Cussac D, Chardin P, Bar-Sagi D. Purification of baculovirus-expressed human Sos1 protein. *Methods Enzymol* 1995;255:125–129.

Freitag R, Horváth C. Chromatography in the downstream processing of biotechnological products. *Adv Biochem Eng Biotechnol* 1996;5:17–59.

Fritzsch FS, Dusny C, Frick O, Schmid A. Single-cell analysis in biotechnology, systems biology, and biocatalysis. *Annu Rev Chem Biomol Eng* 2012;129–155.

Funaba M, Mathews LS. Recombinant expression and purification of smad proteins. *Protein Expr Purif* 2000;20(3):507–513.

Gagnon P. *Purification Tools for Monoclonal Antibodies*. Validated Biosystems Inc., Tucson, AZ, 1996.

Gailliot FP, Gleason C, Wilson JJ, Zwarick J. Fluidized bed adsorption for whole broth extraction. *Biotechnol Prog* 1990;6:370–375.

Galés C. G-protein-coupled receptors plasticity and signalling. *Med Sci (Paris)* 2012 Oct; 28(10):883–885.

Georgiev MI, Weber J, Maciuk A. Bioprocessing of plant cell cultures for mass production of targeted compounds. *Appl Microbiol Biotechnol* 2009 Jul;83(5):809–823.

Gerba CP, Staggs CH, Alondie MG. Characterization of sewage solid-associated virus and behavior in natural waters. *Water Res* 1978;12:805–812.

Gibert S, Bakalara N, Santarelli X. Three-step chromatographic purification procedure for the production of a his-tag recombinant kinesin overexpressed in *E. coli*. *J Chromatogr B Biomed Sci Appl* 2000;737(1–2):143–150.

Goldberg ME, Expert-Bezancon N, Vuillard L, Rabilloud T. Non-detergent sulphobetaines: A new class of molecules that facilitate in vitro protein renaturation. *Fold Des* 1996;1(1):21–27.

Graber SG, Figler RA, Garrison JC. Expression and purification of G-protein alpha subunits using baculovirus expression system. *Methods Enzymol* 1994;237:212–226.

Graf EG, Jander E, West A, Pora H, Aranha-Creado H. Virus removal by filtration. *Dev Biol Stand* 1999;99:89–94.

Graf H, Rabaud JN, Egly JM. Ion exchange resins for the purification of monoclonal antibodies from animal cell culture. *Bioseparation* 1994;4(1):7–20.

Gray SP, Jandeleit-Dahm K. The pathobiology of diabetic vascular complications—cardiovascular and kidney disease. *J Mol Med (Berl)* 2014 May;92(5):441–452.

Guiochon G, Beaver LA. Separation science is the key to successful biopharmaceuticals. *J Chromatogr A* 2011 Dec 9;1218(49):8836–8858.

Hagel P, Gerding JJ, Fieggen W, Bloemendal H. Cyanate formation in solutions of urea. I. Calculation of cyanate concentrations at different temperature and pH. *Biochim Biophys Acta* 1971;243(3):366–373.

Hagel L, Sofer G. Cleaning, sanitization and storage. In: *Handbook of Process Chromatography. A Guide to Optimization, Scale up, and Validation*, Academic Press, London, U.K., ISBN: 0-12-654266.

Hale JE, Beidler DE. Purification of humanized murine and murine monoclonal antibodies using immobilized metal-affinity chromatography. *Anal Biochem* 1994;222(1):29–33.

Hama S, Kondo A. Enzymatic biodiesel production: An overview of potential feedstocks and process development. *Bioresour Technol* 2013 May;13:386–395.

Hamalainen E, Suomela H, Ukkonen P. Virus inactivation during intravenous immunoglobulin production. *Vox Sang* 1992;63(1):6–11.

Hansson M, Ståhl S, Hjorth R, Uhlén M, Moks T. Single-step recovery of a secreted recombinant protein by expanded bed adsorption. *Bio/Technol* 1994;12:285–288.

Harakas NK, Schaumann JP, Conolly DT, Wittwer AJ, Olander JV, Feder J. Large-scale purification of tissue type plasminogen activator from cultured human cells. *Biotechnol Progress* 2002;4(3):149–158.

Harris DP, Andrews AT, Wright G, Pyle DL, Asenjo JA. The application of aqueous two-phase systems to the purification of pharmaceutical proteins from transgenic sheep milk. *Bioseparation* 1997;7(1):31–37.

Harrison RG, Jr. Purification of recombinant proteins from yeast. *Bioprocess Technol* 1991; 12:183–191.

Hartleib J, Ruterjans H. High-yield expression, purification, and characterization of the recombinant diisopropylfluorophosphatase from Loligo vulgaris. *Protein Expr Purif* 2001;21(1):210–219.

Hejnaes KR, Bayne S, Norskov L, Sorensen HH, Thomsen J, Schaffer L, Wollmer A, Skriver L. Development of an optimized refolding process for recombinant Ala-Glu-IGF-1. *Protein Eng* 1992;5(8):797–806.

Henzler HJ, Kaiser K. Avoiding viral contamination in biotechnological and pharmaceutical processes. *Nature Biotechnol* 1998;16(11):1077–1079.

Hepler JR, Kozasa T, Gilman AG. Purification of recombinant Gq alpha, G11 alpha, and G16 alpha from Sf9 cells. *Methods Enzymol* 1994;237:191–212.

Hermanson IL, Turchi JJ. Overexpression and purification of human XPA using a baculovirus expression system. *Protein Expr Purif* 2000;19(1):1–11.

Hidalgo M, Bloedow D. Pharmacokinetics and pharmacodynamics: Maximizing the clinical potential of Erlotinib (Tarceva). *Semin Oncol* 2003 Jun;30(3 Suppl 7):25–33.

Hill RM, Brennan SO, Birch NP. Expression, purification, and functional characterization of the serine protease inhibitor neuroserpin expressed in Drosophila S2 cells. *Protein Expr Purif* 2001;22(3):406–413.

Hjorth R, Kämpe S, Carlsson M. Analysis of some operating parameters of novel adsorbents for recovery of proteins in expanded beds. *Bioseparation* 1995;5:217–223.

Holtet TL, Etzerodt M, Thøgersen HC, Denzyme Aps, Holtet TL, Etzerodt M, Thøgersen HC, Assignees. Improved method for the refolding of proteins. WO Patent 9418227. Issued 18-8-1994. Filed 4-2-1993.

Horowitz B. Investigations into the application of tri(n-butyl)phosphate/detergent mixtures to blood derivatives. *Curr Stud Hematol Blood Transfus* 1989;(56):83–96.

Hu YJ, Zhu BQ. Research progress on strain improvement of Acremonium chrysogenum by genetic engineering. *Yi Chuan* 2011 Oct;33(10):1079–1086. Review. Chinese.

Huang YW, Lu ML, Qi H, Lin SX. Membrane-bound human 3beta-hydroxysteroid dehydrogenase: Overexpression with His-tag using a baculovirus system and single-step purification. *Protein Expr Purif* 2000;18(2):169–174.

Huggins DJ, Venkitaraman AR, Spring DR. Rational methods for the selection of diverse screening compounds. *ACS Chem Biol* 2011 Mar 18;6(3):208–217.

Hughes B, Bradburne A, Sheppard A, Young D. Evaluation of anti-viral filters. *Dev Biol Stand* 1996;88:91–98.

Huxtable S, Zhou H, Wong S, Li N. Renaturation of 1-aminocyclopropane-1-carboxylate synthase expressed in *Escherichia coli* in the form of inclusion bodies into a dimeric and catalytically active enzyme. *Protein Expr Purif* 1998;12(3):305–314.

ICH. Quality of biotechnological products: Viral safety evaluation of biotechnology products derived from cell lines of human or animal origin. ICH Harmonised Tripartite Guideline. *Dev Biol Stand* 1998;93:177–201.

Iizuka M, Fukuda K. Purification of the bovine nicotinic acetylcholine receptor alpha-subunit expressed in baculovirus-infected insect cells. *J Biochem (Tokyo)* 1993;114(1):140–147.

Josic D, Kovac S. Application of proteomics in biotechnology—Microbial proteomics. *Biotechnol J* 2008 Apr;3(4):496–509.

Josic D, Loster K, Kuhl R, Noll F, Reusch J. Purification of monoclonal antibodies by hydroxylapatite HPLC and size exclusion HPLC. *Biol Chem Hoppe Seyler* 1991;372(3):149–156.

Juge N, Andersen JS, Tull D, Roepstorff P, Svensson B. Overexpression, purification, and characterization of recombinant barley alpha-amylases 1 and 2 secreted by the methylotrophic yeast Pichia pastoris. *Protein Expr Purif* 1996;8(2):204–214.

Juntunen K, Rochel N, Moras D, Vihko P. Large-scale expression and purification of the human vitamin D receptor and its ligand-binding domain for structural studies. *Biochem J* 1999;344 Pt 2:297–303.

Kariv I, Rourick RA, Kassel DB, Chung TD. Improvement of "hit-to-lead" optimization by integration of in vitro HTS experimental models for early determination of pharmacokinetic properties. *Comb Chem High Throughput Screen* 2002 Sep;5(6):459–472.

341

Karuppiah N, Sharma A. Cyclodextrins as protein folding aids. *Biochem Biophys Res Commun* 1995;211(1):60–66.

Kaszubska W, Zhang H, Patterson RL et al. Expression, purification, and characterization of human recombinant thrombopoietin in Chinese hamster ovary cells. *Protein Expr Purif* 2000;18(2):213–220.

Kelley BD, Ramelmeier RA, eds. *Validation of Biopharmaceutical Manufacturing Processes.* American Chemical Society, Washington, DC, 1998, pp. 114–124.

Kiefhaber T, Rudolph R, Kohler HH, Buchner J. Protein aggregation in vitro and in vivo: A quantitative model of the kinetic competition between folding and aggregation. *Biotechnology (NY)* 1991;9(9):825–829.

Kim CS, Lee EK. Effect of operating parameters in in vitro renaturation of a fusion protein of human growth hormone and glutathione S-transferase from inclusion body. *Process Biochem* 2000;36:111–117.

King DJ. *Applications and Engineering of Monoclonal Antibodies.* Taylor & Francis, London, U.K., 1998.

King LA, Possee RD. Propagation, titration and purification of AcMNPV in cell culture. In: *The Baculovirus Expression System. A Laboratory Guide.* Chapman & Hall, London, U.K., 1992, pp. 106–126.

Klaassen CH, Bovee-Geurts PH, Decaluwe GL, DeGrip WJ. Large-scale production and purification of functional recombinant bovine rhodopsin with the use of the baculovirus expression system. *Biochem J* 1999;342 (Pt 2):293–300.

Klein MT, Vinson PN, Niswender CM. Approaches for probing allosteric interactions at 7 transmembrane spanning receptors. *Prog Mol Biol Transl Sci* 2013;11:1–59.

Kozasa T, Gilman AG. Purification of recombinant G proteins from Sf9 cells by hexahistidine tagging of associated subunits. Characterization of alpha 12 and inhibition of adenylyl cyclase by alpha z. *J Biol Chem* 1995;270(4):1734–1741.

Kunimoto DY, Allison KC, Watson C, Fuerst T, Armstrong GD, Paul W, Strober W. High-level production of murine interleukin-5 (IL-5) utilizing recombinant baculovirus expression. Purification of the rIL-5 and its use in assessing the biologic role of IL-5 glycosylation. *Cytokine* 1991;3(3):224–230.

Kutzko JP, Sherman LT, Hayes ML, Genzyme Transgenics Corporation, Assignee. Purification of biological active peptides from milk. U.S. Patent 6268487. Issued 31-7-2001. Filed 13-5-1996.

Labrou NE. Protein purification: An overview. *Methods Mol Biol* 2014;112:3–10.

Lansmann S, Bartelsen O, Sandhoff K. Purification and characterization of recombinant human acid sphingomyelinase expressed in insect Sf21 cells. *Methods Enzymol* 2000;311:149–156.

Laurent G. Olfactory processing: Maps, time and codes. *Curr Opin Neurobiol* 1997 Aug; 7(4):547–553.

Lee VW, Antonsen KP, PPL Therapeutics Scotland Ltd., Assignee. Purification of alpha-1 proteinase inhibitor. U.S. Patent 6194553. Issued 27-2-2001. Filed 9-3-1998.

Lehr RV, Elefante LC, Kikly KK, O'Brien SP, Kirkpatrick RB. A modified metal-ion affinity chromatography procedure for the purification of histidine-tagged recombinant proteins expressed in Drosophila S2 cells. *Protein Expr Purif* 2000;19(3):362–368.

Lilie H, Schwarz E, Rudolph R. Advances in refolding of proteins produced in *E. coli. Curr Opin Biotechnol* 1998;9(5):497–501.

Lin LS, Yamamoto R, Drummond RJ. Purification of recombinant human interferon E expressed in *Escherichia coli. Methods Enzymol* 1986;119:183–192.

Listrom CD, Morizono H, Rajagopal BS, McCann MT, Tuchman M, Allewell NM. Expression, purification, and characterization of recombinant human glutamine synthetase. *Biochem J* 1997;328(Pt 1):159–163.

Lowe ME. Human pancreatic procolipase expressed in insect cells: Purification and characterization. *Protein Expr Purif* 1994;5(6):583–586.

Luellau E, von Stockar U, Vogt S, Freitag R. Development of a downstream process for the isolation and separation of monoclonal immunoglobulin A monomers, dimers and polymers from cell culture supernatant. *J Chromatogr A* 1998;796(1):165–175.

Lundblad JL, Seng RL. Inactivation of lipid-enveloped viruses in proteins by caprylate. *Vox Sang* 1991;60(2):75–81.

Lutkemeyer D, Ameskamp N, Priesner C, Bartsch EM, Lehmann J. Capture of proteins from mammalian cells in pilot scale using different STREAMLINE adsorbents. *Bioseparation* 2001;10(1–3):57–63.

Lutkemeyer D, Bretschneider M, Buntemeyer H, Lehmann J. Membrane chromatography for rapid purification of recombinant antithrombin III and monoclonal antibodies from cell culture supernatant. *J Chromatogr* 1993;639(1):57–66.

Machida S, Ogawa S, Xiaohua S, Takaha T, Fujii K, Hayashi K. Cycloamylose as an efficient artificial chaperone for protein refolding. *FEBS Lett* 2000;486(2):131–135.

Madhunapantula SV, Mosca PJ, Robertson GP. The Akt signaling pathway: An emerging therapeutic target in malignant melanoma. *Cancer Biol Ther* 2011 Dec 15;12(12):1032–1049.

Madri JA. Modeling the neurovascular niche: Implications for recovery from CNS injury. *J Physiol Pharmacol* 2009 Oct;60(Suppl):95–104.

Mannucci PM, Colombo M. Virucidal treatment of clotting factor concentrates. *Lancet* 1988;2(8614):782–785.

Manzke O, Tesch H, Diehl V, Bohlen H. Single-step purification of bispecific monoclonal antibodies for immunotherapeutic use by hydrophobic interaction chromatography. *J Immunol Methods* 1997; 208(1):65–73.

Marczinovits I, Somogyi C, Patthy A, Nemeth P, Molnar J. An alternative purification protocol for producing hepatitis B virus X antigen on a preparative scale in *Escherichia coli*. *J Biotechnol* 1997; 56(2):81–88.

Marston FA, Hartley DL. Solubilization of protein aggregates. *Method Enzymol* 1990; 182(20):264–276.

Marston FA. The purification of eukaryotic polypeptides synthesized in *Escherichia coli*. *Biochem J* 1986;240(1):1–12.

Masschelein CA, Callegari JP, Laurent M, Simon JP, Taeymans D. Current biotechnological developments in Belgium. *Crit Rev Biotechnol* 1989;8(4):275–303.

Matejtschuk P, Baker RM, Chapman GE. Purification and characterization of monoclonal antibodies. In: *Bioseparation and Bioprocessing*. Vol. II: *Processing, Quality and Characterization, Economics, Safety and Hygiene*, Subramanian G, ed. Wiley-VCH, New York, 1998, pp. 223–252.

Mayank R, Ranjan A, Moholkar VS. Mathematical models of ABE fermentation: Review and analysis. *Crit Rev Biotechnol* 2013 Dec;33(4):419–447.

McCormick DK. Expanded bed adsorption. The first new unit process operation in decades. *Bio/Technol* 1993;11:1059.

McDonald JR, Ong M, Shen C, Parandoosh Z, Sosnowski B, Bussel S, Houston LL. Large-scale purification and characterization of recombinant fibroblast growth factor-saporin mitotoxin. *Protein Expr Purif* 1996;8:97–108.

Michaelis U, Rudolph R, Jarsch M, Kopetzki E, Burtscher H, Schumacher G, Boehringer Mannheim GmbH, Assignee. Process for the production and renaturation of recombinant, biologically active, eukaryotic alkaline phosphatase. U.S. Patent 5434067. Issued 18-7-1995. Filed 30-7-1993.

Miekka SI, Busby TF, Reid B, Pollock R, Ralston A, Drohan WN. New methods for inactivation of lipid-enveloped and non-enveloped viruses. *Haemophilia* 1998;4(4):402–408.

Mirnezami R, Kinross JM, Vorkas PA, Goldin R, Holmes E, Nicholson J, Darzi A. Implementation of molecular phenotyping approaches in the personalized surgical patient journey. *Ann Surg* 2012 May;255(5):881–889.

Misawa S, Kumagai I. Refolding of therapeutic proteins produced in *Escherichia coli* as inclusion bodies. *Biopolymers* 1999;51(4):297–307.

Mitchell DL, Young MA, Entwisle C, Davies AN, Cook RM, Dodd I. Purification and characterisation of recombinant murine interleukin-5 glycoprotein, from a Baculovirus expression system. *Biochem Soc Trans* 1993;21(4):332S.

Mochizuki S, Zydney AL. Theoretical analysis of pore size distribution effects on membrane transport. *J Membr Sci* 1993;82(3):211–227.

Morton HC, Atkin JD, Owens RJ, Woof JM. Purification and characterization of chimeric human IgA1 and IgA2 expressed in COS and Chinese hamster ovary cells. *J Immunol* 1993;151(9):4743–4752.

Muller C, Rinas U. Renaturation of heterodimeric platelet-derived growth factor from inclusion bodies of recombinant *Escherichia coli* using size-exclusion chromatography. *J Chromatogr A* 1999;855(1):203–213.

Murby M, Uhlén M, Ståhl S. Upstream strategies to minimize proteolytic degradation upon recombinant production in *Escherichia coli*. *Protein Expr Purif* 1996 Mar;7(2):129–136.

Nair MP, Kudchodkar BJ, Pritchard PH, Lacko AG. Purification of recombinant lecithin: Cholesterol acyltransferase. *Protein Expr Purif* 1997;10(1):38–41.

Nakatsuka S, Michaels AS. Transport and separation of proteins by ultrafiltration through sorptive and non-sorptive membranes. *J Membr Sci* 1992;69(3):189–211.

Niles AL, Maffitt M, Haak-Frendscho M, Wheeless CJ, Johnson DA. Recombinant human mast cell tryptase beta: Stable expression in Pichia pastoris and purification of fully active enzyme. *Biotechnol Appl Biochem* 1998;28 (Pt 2):125–131.

343

Nilsson J, Stahl S, Lundeberg J, Uhlen M, Nygren PA. Affinity fusion strategies for detection, purification, and immobilization of recombinant proteins. *Protein Expr Purif* 1997;11(1):1–16.

Niu G, Chen X. Has molecular and cellular imaging enhanced drug discovery and drug development? *Drugs R D* 2008;9(6):351–368.

Nixon L, Koval CA, Xu L, Noble RD, Slaff GS. The effects of magnetic stabilization on the structure and performance of fluidized beds. *Bioseparation* 1991;2:217–230.

Noppe W, Hanssens I, De Cuyper M. Simple two-step procedure for the preparation of highly active pure equine milk lysozyme. *J Chromatogr* 1996;719:327–331.

Nozaki Y. The preparation of guanidine hydrochloride. *Methods Enzymol* 1972;26 PtC:43–50.

Odorzynski TW, Light A. Refolding of the mixed disulfide of bovine trypsinogen and glutathione. *J Biol Chem* 1979;254(10):4291–4295.

Ogez JR, Builder SE. Downstream processing of proteins from mammalian cells. *Bioprocess Technol* 1990;10:393–416.

Ohtaki T, Ogi K, Kitada C, Hinuma S, Onda H. Purification of recombinant human pituitary adenylate cyclase-activating polypeptide receptor expressed in Sf9 insect cells. *Ann N Y Acad Sci* 1996;805:590–594.

Ohtaki T, Ogi K, Masuda Y, Mitsuoka K, Fujiyoshi Y, Kitada C, Sawada H, Onda H, Fujino M. Expression, purification, and reconstitution of receptor for pituitary adenylate cyclase-activating polypeptide. Large-scale purification of a functionally active G protein-coupled receptor produced in Sf9 insect cells. *J Biol Chem* 1998;273(25):15464–15473.

O'Riordan CR, Erickson A, Bear C, Li C, Manavalan P, Wang KX, Marshall J, Scheule RK, McPherson JM, Cheng SH. Purification and characterization of recombinant cystic fibrosis transmembrane conductance regulator from Chinese hamster ovary and insect cells. *J Biol Chem* 1995;270(28):17033–17043.

Oshima KH, Evans-Strickfaden T, Highsmith A. Comparison of filtration properties of hepatitis B virus (HBV), hepatitis C virus(HCV) and simian virus 40 SV40) using a polyvinylidene fluoride (PVDF) membrane filter. *Vox Sang* 1998;75(3):181–188.

Oshima KH, Evans-Strickfaden TT, Highsmith AK, Ades EW. The use of a microporous polyvinylidene fluoride (PVDF) membrane filter to separate contaminating viral particles from biologically important proteins. *Biologicals* 1996;24(2):137–145.

Oshima KH, Highsmith AK, Ades EW. Removal of influenza A virus, phage T1, and PP7 from fluids with a nylon 0.04-μm membrane filter. *Environ Toxicol Water Qual* 1994;9:165–170.

Page M, Thorpe R. Purification of monoclonal antibodies. *Methods Mol Biol* 1998;80:113–119.

Pall. Validation guide for Pall UltiporR VFTM Grade DV50 UltipleatTM AB Style virus removal filter cartridges. Pall Ultrafine Filtration Company, Pall Corporation, New York, 1995.

Perreault P, Brantley JD. Characterization of ultrafiltration membranes using mixed dextrans. Pall Ultrafine Filtration Company, Pall Corporation, New York, 1997.

Phillips MW, DiLeo AJ. A validatible porosimetric technique for verifying the integrity of virus-retentive membranes. *Biologicals* 1996;24(3):243–253.

Porfiri E, Evans T, Bollag G, Clark R, Hancock JF. Purification of baculovirus-expressed recombinant Ras and Rap proteins. *Methods Enzymol* 1995;255:13–21.

Price AE, Logvinenko KB, Higgins EA, Cole ES, Richards SM. Studies on the microheterogeneity and in vitro activity of glycosylated and nonglycosylated recombinant human prolactin separated using a novel purification process. *Endocrinology* 1995;136(11):4827–4833.

Prince AM, Stephan W, Brotman B. Beta-propiolactone/ultraviolet irradiation: A review of its effectiveness for inactivation of viruses in blood derivatives. *Rev Infect Dis* 1983;5(1):92–107.

Prior CP. Large-scale process purification of clinical product from animal cell cultures. *Biotechnology* 1991;17:445–478.

Qi ZH, Sikorski CT. Controlled delivery using cyclodextrin technology. *Pharmaceut Technol Eur* 2001;13(11):17–27.

Quality of biotechnological products: Viral safety evaluation of biotechnology products derived from cell lines of human or animal origin. ICH Harmonised Tripartite Guideline. *Dev Biol Stand* 1998;93:177–201.

Ransohoff TC, Levine HL. Purification of monoclonal antibodies. *Bioprocess Technol* 1991;12:213–235.

Roberts P. Efficient removal of viruses by a novel polyvinylidene fluoride membrane filter. *J Virol Methods* 1997;65(1):27–31.

Rogelj B, Strukelj B, Bosch D, Jongsma MA. Expression, purification, and characterization of equistatin in Pichia pastoris. *Protein Expr Purif* 2000;19(3):329–334.

Rogl H, Kosemund K, Kuhlbrandt W, Collinson I. Refolding of Escherichia coli produced membrane protein inclusion bodies immobilised by nickel chelating chromatography. *FEBS Lett* 1998;432(1–2):21–26.

Roper DK, Lightfoot EN. Estimating plate-heights in stacked membrane chromatography by flow reversal. *J Chromatogr A* 1995;702:69–80.

Rosenfeld SA, Nadeau D, Tirado J, Hollis GF, Knabb RM, Jia S. Production and purification of recombinant hirudin expressed in the methylotrophic yeast Pichia pastoris. *Protein Expr Purif* 1996;8(4):476–482.

Roush DJ, Lu Y. Advances in primary recovery: Centrifugation and membrane technology. *Biotechnol Prog* 2008 May–Jun;24(3):488–495.

Rozema D, Gellman SH. Artificial chaperones: Protein refolding via sequential use of detergent and cyclodextrin. *J Am Chem Soc* 1995;117(8):2373–2374.

Rozema D, Gellman SH. Artificial chaperone-assisted refolding of denatured-reduced lysozyme: Modulation of the competition between renaturation and aggregation. *Biochemistry* 1996a;35(49):15760–15771.

Rozema D, Gellman SH. Artificial chaperone-assisted refolding of carbonic anhydrase B. *J Biol Chem* 1996b;271(7):3478–3487.

Rubinfeld B, Polakis P. Purification of baculovirus-produced Rap1 GTPase-activating protein. *Methods Enzymol* 1995;255:31–38.

Rudolph R, Fischer S, Mattes R, Boehringer Mannheim GmbH, Assignee. Process for the activating of gene-technologically produced, heterologous, disulphide bridge-containing eukaryotic proteins after expression in prokaryotes. U.S. Patent 5593865. Issued 14-1-1997. Filed 1-6-1995.

Rudolph R, Fischer S, Boehringer Mannheim GmbH, Assignee. Process for obtaining renatured proteins. U.S. Patent 4933434. Issued 12-6-1990. Filed 13-1-1989.

Rudolph R, Lilie H. In vitro folding of inclusion body proteins. *FASEB J* 1996;10(1):49–56.

Sack M, Hofbauer A, Fischer R, Stoger E. The increasing value of plant-made proteins. *Curr Opin Biotechnol* 2015 Jan 8;32C:163–170.

Saksena S, Zydney AL. Effect of solution pH and ionic strength on the separation of albumin from immunoglobulins (IgG) by selective filtration. *Biotechnol Bioeng* 1994; 43(10):960–968.

Sanghani PC, Moran RG. Purification and characteristics of recombinant human folylpoly-γ-glutamate synthetase expressed at high levels in insect cells. *Protein Expr Purif* 2000;18(1):36–45.

Sarker D, Workman P. Pharmacodynamic biomarkers for molecular cancer therapeutics. *Adv Cancer Res* 2007;9 213–268.

Sato T et al. Integrity test of virus removal membranes through gold particle method and liquid forward flow test method. In: *Animal Cell Technology: Basic and Applied Aspects*, Kobayashi T et al., eds. Kluwer Academic Publishers, Norwell, MA, 1994, pp. 517–522.

Schillberg S, Raven N, Fischer R, Twyman RM, Schiermeyer A. Molecular farming of pharmaceutical proteins using plant suspension cell and tissue cultures. *Curr Pharm Des* 2013;19(31):5531–5542.

Schmidt FR. Recombinant expression systems in the pharmaceutical industry. *Appl Microbiol Biotechnol* 2004 Sep;65(4):363–372. Epub 2004 Jul 24.

Schmidt HH, Genschel J, Haas R, Buttner C, Manns MP. Expression and purification of recombinant human apolipoprotein A-I in Chinese hamster ovary cells. *Protein Expr Purif* 1997;10(2):226–236.

Schorr J, Moritz P, Breul A, Scheef M. Production of plasmid DNA in industrial quantities according to cGMP guidelines. *Methods Mol Med* 2006;12:339–350.

Scopes RK. *Protein Purification: Principles and Practice*, 3rd edn. Springer Verlag, New York, 1994, pp. 1–380.

Scopes RK. Separation by precipitation. In: *Protein Purification: Principles and Practice*. Springer Verlag, New York, 1987, pp. 41–71.

Seeger A, Rinas U. Two-step chromatographic procedure for purification of basic fibroblast growth factor from recombinant *Escherichia coli* and characterization of the equilibrium parameters of adsorption. *J Chromatogr A* 1996;746(1):17–24.

Segall M. Advances in multiparameter optimization methods for de novo drug design. *Expert Opin Drug Discov* 2014 Jul;9(7):803–817.

Seidel-Morgenstern, A. Analysis of boundary conditions in the axial dispersion model by application of numerical laplace inversion. *Chem Eng Sci* 1991;46:2567–2571.

Settleman J, Foster R. Purification and GTPase-activating protein activity of baculovirus expressed p190. *Methods Enzymol* 1995;256:105–113.

345

Sharp JA, Whitley PH, Cunnion KM, Krishna NK. Peptide inhibitor of complement c1, a novel suppressor of classical pathway activation: Mechanistic studies and clinical potential. *Front Immunol* 2014 Aug 22;406.

Shepard SR, Boucher R, Johnston J, Boerner R, Koch G, Madsen JW, Grella D, Sim BK, Schrimsher JL. Large-scale purification of recombinant human angiostatin. *Protein Expr Purif* 2000;20(2):216–227.

Sobsey MD. Methods for detecting enteric viruses in water and wastewater. In: *Viruses in Water*, Berg G, Bodily HL, Lennette EH, eds. American Public Health Association, Washington, DC, 1976, pp. 89–127.

Sobsey MD, Glass JS. Influence of water quality on enteric virus concentration by microporous filter methods. *Appl Environ Microbiol* 1984;47(5):956–960.

Sobsey MD, Moore RS, Glass JS. Evaluating adsorbent filter performance for enteric virus concentrations in tap water. *J Am Water Works Assoc* 1981;73:542–548.

Sofer G. Virus inactivation in the 1990s and into the 21st century. Part 4, Culture media, biotechnology products, and vaccines. *BioPharm* 2003;16(1):50–57.

Stark GR, Stein WH, Moore S. Reactions of the cyanate present in aqueous urea with amino acids and proteins. *J Biol Chem* 1960;235(4):3177–3181.

Steinlein LM, Graf TN, Ikeda RA. Production and purification of N-terminal half-transferrin in Pichia pastoris. *Protein Expr Purif* 1995;6(5):619–624.

Stockel J, Doring K, Malotka J, Jahnig F, Dornmair K. Pathway of detergent-mediated and peptide ligand-mediated refolding of heterodimeric class II major histocompatibility complex (MHC) molecules. *Eur J Biochem* 1997;248(3):684–691.

Sundari CS, Raman B, Balasubramanian D. Artificial chaperoning of insulin, human carbonic anhydrase and hen egg lysozyme using linear dextrin chains—A sweet route to the native state of globular proteins. *FEBS Lett* 1999;443(2):215–219.

Susa M, Luong-Nguyen NH, Crespo J, Maier R, Missbach M, McMaster G. Active recombinant human tyrosine kinase c-Yes: Expression in baculovirus system, purification, comparison to c-Src, and inhibition by a c-Src inhibitor. *Protein Expr Purif* 2000;19(1):99–106.

Swaine DE, Daugulis A J. Review of liquid mixing in packed bed biological reactors. *Biotechnol. Progr* 1988;4:134–148.

Tan AL, Forbes JM, Cooper ME. AGE, RAGE, and ROS in diabetic nephropathy. *Semin Nephrol* 2007 Mar;27(2):130–143.

Tandon S, Horowitz P. The effects of lauryl maltoside on the reactivation of several enzymes after treatment with guanidinium chloride. *Biochim Biophys Acta* 1988;955(1):19–25.

Tandon S, Horowitz PM. Detergent-assisted refolding of guanidinium chloride-denatured rhodanese. The effects of the concentration and type of detergent. *J Biol Chem* 1987; 262(10):4486–4491.

Tao CZ, Cameron R, Harbour C, Barford JP. The development of appropriate viral models for the validation of viral inactivation procedures. In: *Animal Cell Technology. Products of Today, Prospects for Tomorrow*, Spier RE, Griffiths JB, Berthold W, eds, Butterworth Heinemann, Oxford, 1994: 754–756.

Tarnowski SJ, Roy SK, Liptak RA, Lee DK, Ning RY. Large-scale purification of recombinant human leukocyte interferons. *Methods Enzymol* 2002;119:153–165.

Tennagels N, Hube-Magg C, Wirth A, Noelle V, Klein HW. Expression, purification, and characterization of the cytoplasmic domain of the human IGF-1 receptor using a baculovirus expression system. *Biochem Biophys Res Commun* 1999;260(3):724–728.

Thibonnier M, Coles P, Thibonnier A, Shoham M. Molecular pharmacology and modeling of vasopressin receptors. *Prog Brain Res* 2002;13:179–196.

Thiele A, Zerweck J, Schutkowski M. Peptide arrays for enzyme profiling. *Methods Mol Biol* 2009;57:19–65.

Thies MJ, Pirkl F. Chromatographic purification of the C(H)2 domain of the monoclonal antibody MAK33. *J Chromatogr B Biomed Sci Appl* 2000;737(1–2):63–69.

Thøgersen C, Holtet TL, Etzerodt M, Denzyme Aps, Assignee. Iterative method of at least five cycles for the refolding of proteins. U.S. Patent 5917018. Issued 29-6-1999. Filed 18-9-1995.

Thömmes J et al. Purification of monoclonal antibodies from whole hybridoma fermentation broth by fluidized bed adsorption. *Biotech Bioeng* 1995;45:205–211.

Thömmes J, Bader A, Halfar M, Karau A, Kula M-R. Isolation of monoclonal antibodies from cell containing hybridoma broth using a protein A coated adsorbent in expanded beds. *J Chromatogr A* 1996;752:111–122.

Tipton B, Boose JA, Larsen W, Beck J, O´Brien T. Retrovirus and parvovirus clearance from an affinity column product using adsorptive depth filtration. *BioPharm* 2002;15(9):43–50.

Troccoli NM, McIver J, Losikoff A, Poiley J. Removal of viruses from human intravenous immune globulin by 35 nm nanofiltration. *Biologicals* 1999;26(4):321–329.

Tsao IF, Wang HY. Removal and inactivation of viruses by a surface-bonded quaternary ammonium chloride. In: *Downstream Processing and Bioseparation: Recovery and Purification of Biological Products*, Hamel JFP, Hunter JB, Sikdar SK, eds. American Chemical Society, Washington, DC, 1990, pp. 250–267.

Tsujimoto M, Adachi H, Kodama S, Tsuruoka N, Yamada Y, Tanaka S, Mita S, Takatsu K. Purification and characterization of recombinant human interleukin 5 expressed in Chinese hamster ovary cells. *J Biochem (Tokyo)* 1989;106(1):23–28.

Tsumoto K, Shinoki K, Kondo H, Uchikawa M, Juji T, Kumagai I. Highly efficient recovery of functional single-chain Fv fragments from inclusion bodies overexpressed in *Escherichia coli* by controlled introduction of oxidizing reagent—Application to a human single-chain Fv fragment. *J Immunol Methods* 1998;219(1–2):119–129.

Tsurumi T, Osawa N, Hitaka H, Hirasaki T, Yamaguchi K, Manabe S, Yamashiki T. Mechanism of removing monodisperse gold particles from a suspension using cuprammonium regenerated cellulose hollow fibre (BMM hollow fibre). *Polymer J* 1990;22:304–311.

Udell M, McCreath G, PPL Therapeutics Scotland Ltd, Assignee. Purification of fibrinogen from fluids by precipitation and hydrophobic chromatography. WO Patent 0017239; EP1115745. Issued 30-3-2000. Filed 24-9-1998.

Valtanen H, Lehti K, Lohi J, Keski-Oja J. Expression and purification of soluble and inactive mutant forms of membrane type 1 matrix metalloproteinase. *Protein Expr Purif* 2000;19(1):66–73.

Van Cott KE, Williams B, Velander WH, Gwazdauskas F, Lee T, Lubon H, Drohan WN. Affinity purification of biologically active and inactive forms of recombinant human protein C produced in porcine mammary gland. *J Mol Recognit* 1996;9(5–6):407–414.

van Reis R, Gadam S, Frautschy LN, Orlando S, Goodrich EM, Saksena S, Kuriyel R, Simpson CM, Pearl S, Zydney AL. High performance tangential flow filtration. *Biotechnol Bioeng* 1997;56(1):71–82.

van Reis R, Zydney AL. Protein ultrafiltration. In: *Encyclopedia of Bioprocess Technology: Fermentation, Biocatalysis, and Bioseparation*, Flickinger MC, Drew SW, eds. Wiley, New York, 1999, 2197–2214.

Velander WH, Morcol T, Akers RM, Boyle PL, Johnson JL, Drohan WN. The recovery of therapeutic proteins from milk—Protein-C gene cloning and expression in pig or cattle transgenic animal; recombinant protein purification. *J Cell Biochem* 1990;(Suppl. 14 D):36.

Velayudhan A, Menon MK. Modeling of purification operations in biotechnology: Enabling process development, optimization, and scale-up. *Biotechnol Prog* 2007 Jan–Feb;23(1):68–73.

Vicente T, Mota JP, Peixoto C, Alves PM, Carrondo MJ. Rational design and optimization of downstream processes of virus particles for biopharmaceutical applications: Current advances. *Biotechnol Adv* 2011 Nov–Dec;29(6):869–878.

Virus removal and inactivation

Vollenbroich D, Ozel M, Vater J, Kamp RM, Pauli G. Mechanism of inactivation of enveloped viruses by the biosurfactant surfactin from Bacillus subtilis. *Biologicals* 1997;25(3):289–297.

Vynckier AK, Dierickx L, Voorspoels J, Gonnissen Y, Remon JP, Vervaet C. Hot-melt co-extrusion: Requirements, challenges and opportunities for pharmaceutical applications. *J Pharm Pharmacol* 2014 Feb;66(2):167–179.

Walker ID. Detection, purification, and utilization of murine monoclonal IgM antibodies. *Methods Mol Biol* 1995;45:183–188.

Walter JK, Nothelfer F, Werz W. Validation of viral safety for pharmaceutical proteins. In: *Bioseparation and Bioprocessing. Vol. I: Biochromatography, Membrane Separations, Modeling, Validation*, Subramanian G, ed. Wiley-VCH, New York, 1998, pp. 465–496.

Walter JK, Nothelfer F, Werz W. Virus removal and inactivation. A decade of validation studies: Critical evaluation of the data set. In: *Validation of Biopharmaceutical Manufacturing Processes. ACS Symposium Series*, Vol. 698. Boehringer Ingelheim Pharma Deutschland KG, Biberach, Germany, 2009, Chapter 9, pp. 114–124.

Warren TC, Miglietta JJ, Shrutkowski A, Rose JM, Rogers SL, Lubbe K, Shih CK, Caviness GO, Ingraham R, Palladino DE. Comparative purification of recombinant HIV-1 and HIV-2 reverse transcriptase: Preparation of heterodimeric enzyme devoid of unprocessed gene product. *Protein Expr Purif* 1992;3(6):479–487.

Weinglass AB, Baldwin SA. Characterisation and purification of recombinant GLUT1 expressed in insect cells. *Biochem Soc Trans* 1996;24(3):478S.

Wells PA, Garlick RL, Lyle SB, Tuls JL, Poorman RA, Brideau RJ, Wathen MW. Purification of a recombinant human respiratory syncytial virus chimeric glycoprotein using reversed-phase chromatography and protein refolding in guanidine hydrochloride. *Protein Expr Purif* 1994;5(4):391–401.

Werner MH, Clore GM, Gronenborn AM, Kondoh A, Fisher RJ. Refolding proteins by gel filtration chromatography. *FEBS Lett* 1994;345(2–3):125–130.

West SM, Chaudhuri JB, Howell JA. Improved protein refolding using hollow-fibre membrane dialysis. *Biotechnol Bioeng* 1998;57(5):590–599.

Wetlaufer DB, Xie Y. Control of aggregation in protein refolding: A variety of surfactants promote renaturation of carbonic anhydrase II. *Protein Sci* 1995;4(8):1535–1543.

WHO requirements for the use of animal cells as in vitro substrates for the production of biologicals (Requirements for biological substances No 50). *Dev Biol Stand* 1998;93:141–171.

Withers BE, Keller PR, Fry DW. Expression, purification and characterization of focal adhesion kinase using a baculovirus system. *Protein Expr Purif* 1996;7(1):12–18.

Wlad H, Ballagi A, Bouakaz L, Gu Z, Janson JC. Rapid two-step purification of a recombinant mouse Fab fragment expressed in *Escherichia coli. Protein Expr Purif* 2001;22(2):325–329.

Wolf MW, Reichl U. Downstream processing of cell culture-derived virus particles. *Expert Rev Vac* 2011 Oct;10(10):1451–1475.

Wright G, Binieda A, Udell M. Protein separation from transgenic milk. *J Chem Tech Biotechnol* 1994;54:110.

Wright G, Colman A. Purification of recombinant proteins from sheep's milk. In: *Transgenic Animals. Generation and Use.* Harwood Academic Publishers, Amsterdam, the Netherlands 1997, pp. 469–471.

Wu N, Ataai MM. Production of viral vectors for gene therapy applications. *Curr Opin Biotechnol* 2000 Apr;11(2):205–208.

Xiao Z, Lu JR. Strategies for enhancing fermentative production of acetoin: A review. *Biotechnol Adv* 2014 Mar–Apr;32(2):492–503.

Xu W, Simons FE, Peng Z. Expression and rapid purification of an Aedes aegypti salivary allergen by a baculovirus system. *Int Arch Allergy Immunol* 1998;115(3):245–251.

Yi C, Maksimoska J, Marmorstein R, Kissil JL. Development of small-molecule inhibitors of the group I p21-activated kinases, emerging therapeutic targets in cancer. *Biochem Pharmacol* 2010 Sep 1;80(5):683–689.

Zardeneta G, Horowitz PM. Micelle-assisted protein folding. Denatured rhodanese binding to cardiolipin-containing lauryl maltoside micelles results in slower refolding kinetics but greater enzyme reactivation. *J Biol Chem* 1992;267(9):5811–5816.

Zeman LJ, Zydney AL. *Microfiltration and Ultrafiltration: Principles and Applications.*

Zhihui Q. Modulating nitric oxide signaling in the CNS for Alzheimer's disease therapy. *Future Med Chem* 2013 Aug;5(12):1451–1468.

Zhu DX, Hua ZC, Liang XF, Zhang XK, Ding Y, Zhu JQ, Han KK. Purification and characterization of the biologically active human truncated macrophage colony-stimulating factor expressed in Saccharomyces cerevisiae. *Biol Chem Hoppe Seyler* 1993;374(9):903–908.

Zurek C, Kubis E, Keup P, Hörlein D, Beunink J, Thömmes J, Kula M-R, Hollenberg CP, Gellissen G. Production of Two Aprotinin variants in Hansenula polymorpha. *Process Biochem* 1996;31:679–689.

Chapter 9 Single-use manufacturing systems (SUMS)

Background

Manufacturing of biosimilar products offers another formidable challenge for the new entrants to the business—the cost of establishing biological manufacturing. While the established manufacturers like Amgen, Merck, Sanofi, Pfizer, and Baxter have also entered the biosimilar markets, their existing infrastructure of traditional manufacturing would likely be adequate for manufacturing biosimilars, yet the new entrants to the business must carefully examine the technology to adopt for their manufacturing systems.

The cost of traditional stainless steel–based manufacturing systems commands very high capital cost and equally high maintenance cost. Realizing that the regulatory agencies require submission of data from commercial-scale lots (unlike small drugs that can be from development lots), there is a need to make the investment in the infrastructure upfront. This investment can run into hundreds of millions of dollars that will accumulate massive depreciation and ultimately end up the cost of manufacturing, leaving little room for the developers to be flexible in the pricing of these products. This model is therefore contrarian to the entire philosophy of creating this category of products—to make them more affordable.

One way, the new developers of biosimilars, as well as established manufacturers, can go around these large investments is to adopt single-use technologies, which is gaining popularity; even the large manufacturers like Amgen have launched an all-out effort to develop single-use systems for their future technology. Amgen and Novartis are both building large manufacturing facilities in Singapore that are likely to be mostly based on single-use systems.

There is a variation in how the European literature lists these technologies as single-use while in the United States the use of disposable systems is widely used. For the purpose of this chapter, we have used the term single use to get away from the stereotype image of disposable being less environmentally friendly. The bibliography in this chapter may still list them as "disposable."

A quick overview of the cost of manufacturing of biological API using the single-use system in comparison with traditional stainless steel technology is given in Table 9.1.

The cost reduction shown about represents the most conservative calculation; with appropriate adjust of product flow, validation cycles, and batch sizes, it is anticipated that the cost of production in a well-coordinated single-use systems should not be more than 50% of the traditional cost.

A summary of pros and cons of the two systems is given in Table 9.2.

Table 9.1 Unitary Comparative Cost of Manufacturing of a Biological API Using Stainless Steel and Single-Use Systems

Category	Stainless Steel	Single Use
Capital charge	37	10
Materials	14	11
Media	3	3
Buffer	1	1
CIP	1	1
QC tests	9	6
Consumables	11	14
Resins/MA	4	4
Bags, single uses	0	3
Filters	7	7
Labor	29	17
Process	11	10
Quality	14	5
Indirect	4	2
Other	10	5
Insurance and other	2	2
Maintenance	2	1
Utilities	6	2
Total	100	57

Source: Data for stainless steel from Sinclair, A., *BioProcess Int.*, 516–519, 2010.
Note: The capital cost is the amortization assuming the facility is built in Europe or the United States.

Table 9.2 Pros and Cons of Using Single-Use Systems

Single Use	Stainless Steel
Pros	
Presterilized, ready to use	Proven technology
Easy setup	Scalability virtually unlimited
Eliminates CIP/SIP	
Low capital outlay	
Low validation requirements	
Increased flexibility	
Cons	
New technology	Hard to clean and maintain
Volume limitations	Large capital investment
	Extensive CIP/SIP cycle
	Extensive validation
	Expensive installation
	Accessories also stainless steel

History

While the barriers to developing new drugs keep getting higher because of the regulatory demands of assuring the safety, the technological barriers to manufacturing these drugs have certainly come down. The current technology for manufacturing recombinant drugs can be traced back to the dawn of civilization; mammalian cell culture technology—the expression system preferred for the most known therapeutic proteins with desirable glycosylation patterns—is relatively new. It took two decades of trials and tribulations to bring cell culture from a bench technique at milligram scales to industrial production at kilogram scales. The era of biopharmaceuticals is manifested in the capability of producing large quantities of biologics in stainless steel bioreactors. Today, those large-scale stirred-tank bioreactors (usually >10,000 L in scale) represent modern mammalian cell culture technology, a major workhorse of the biopharmaceutical industry. Many blockbuster biologics—such as Enbrel etanercept from Immunex Corporation, Avastin bevacizumab from Genentech (Roche), and Humira (adalimumab) from Abbvie—are produced using large-scale bioreactors. The current state of manufacturing thus represents the peak of what we conveniently call the age of stainless steel.

The method of manufacture of biological drugs progressed through an expected route; fermentation in large vats, whether it was done for wine production or industrial chemicals or drugs like penicillin was a well-established technique, so when time came to manufacture recombinant drugs, the same systems were transported over to this new class of drugs—around 30 years ago. Large stainless steel fermenters suited well as their science and technology were well developed. However, lurking in the ambush was a new inquiry at major regulatory agencies—the quest to control cross-contamination and viral clearance, the two most important causes of the side effects of these drugs. The quality guidelines by the FDA and EMA began emphasizing the safety issues or cleaning validation, and viral clearance, and the industry responded with more robust validation plans to prove compliance. The costs of manufacturing soared, but that did not make any difference because all of these molecules were under patent, and the companies were able to get whatever the price they needed to justify these huge investments.

While the stainless steel manufacturers reaped huge profits selling their multistory fermenters and bioreactors, the industry of flex bag drug formulation and administration, the intravenous bags also thrived. However, few saw the need to connect the two for there was no financial incentive to do so.

The first "disruptive" innovation came to the industry when the first single-use Wave bioreactor was introduced in 1996 coincided with the highest ever number of biotechnology drugs approved in 1 year between 1982 and 2007. Almost immediately, the biological manufacturing industry (and more particularly the stainless steel industry) began a debate on the safety and utility of plastic bags to manufacture biological drugs and the greatest fear inculcated in the heart of prospective users was the issue of extractable and leachables, a topic that gets a detailed review in this chapter. Ironically, this issue had long been resolved when the FDA allowed the use of plastic bags to administer drugs of all types, both aqueous and lipid origin and including hyperalimentation solutions. The risks to patients were minimal vis-à-vis the convenience of administration. In reality, the leachables in the biological manufacturing are of little importance as the exposure to this possible chemical comes at a very early stage in the production and the robust downstream purification that removes even the isomers of the compounds is more than adequate to remove these contaminants. The greater risk lies in the interactions in the final dosage forms. A notable incidence was the reporting of PRCA in using erythropoietin, and while many causes were brought to attention, one was the interaction between the rubber stopper and the newly formulated drug containing a new surfactant that might have extracted some extractables from the rubber stopper.

The fear of leachables and the strong presence of a well-established stainless steel industry and a user industry in no rush to learn how to reduce the cost of production slowed down the implementation of plastic containers, more particularly, the single-use containers in the manufacturing of drugs.

First came the changes in the practice as the industry began using single-use filters, flexible containers, membranes, sampling devices, and now have come a wave of single-use bioreactors to address the most critical barriers in the manufacture of biological drugs. The stainless steel industry remains robust today, thanks to the reluctance or perhaps the inability of the big pharma to switch to this disruptive technology for upstream manufacturing.

Single-use bioreactors have since evolved beyond the wave-based design and been adopted both for research purposes and GMP production. Other single-use technologies such as single-use filters, flexible containers, membranes, sampling devices, and chromatography columns have also made a significant headway in being accepted as the standard of manufacturing.

The final decade of twentieth century was good for the biotech industry that raised billions in the public market and a rush for new regulatory filings was on soon; however, many of these companies did not have in-house expertise to manufacture these molecules and that cause a mushrooming of contract research laboratories and contract manufacturing organizations that were ready to fill the gap. It became relatively easy to secure clinical test supplies without having to construct a recombinant manufacturing facility; this eased the financial pressure as well as made up for the dearth of qualified individuals in these newly found science of manufacturing. However, contract research and manufacturing organizations could not afford to install large stainless steel capacity since they would not know which product they would handle the next day—the single use became very popular (because they required so little capital investment) among the CRO/CMO groups as well as research organizations, even if their need for regulatory compliance was less.

The improved efficiency of a company letting them switch over to different products and manufacturing methods pushed the equipment supplier industry to make some quick innovations. The list of single-use items expanded very quickly, and we can readily classify them into three categories:

> Category I includes well-established single uses that came a long time ago and these include analyzer sample caps, culture containers, flasks, titer plates, Petri dishes, pipette and dispensing tips, protective clothing, gloves, syringes, test tubes, and vent and liquid filters.
> Category II includes line items that were necessitated by the problems in cleaning validation; these became fully accepted about a decade ago and included aseptic transfer systems, bags, manifold systems, connectors, tri-clamps, flexible tubing, liquid containment bags, stoppers, tank liners, and valves.
> Category III includes and most recent trends in the past 5 years and include bioprocess containers (though the first one was introduced in 1996 by Wave but only after GE Healthcare bought Wave that it became a main stream offering), bioreactors, centrifuges, chromatography systems, depth filters and systems, isolators, membrane adsorbes, diafiltration devices, mixing systems, and pumps.

There are many published surveys of the industry reported in the literature, and while these statistics can be tainted because the equipment suppliers support most of these, a few general trends that are established include

1. The use of single-use bioprocessing is growing at a rate of 30% per year.
2. The biopharmaceutical or biological manufacturing industry consumes almost one-third of all single-use products used, followed by the biodiagnostic industry.

3. The CROs are least likely to use single uses because of the capital cost investment and the fact that they are used to the adaptability of the hard-walled systems.

4. Most of the adaptations of the single-use technology are in the United States, comprising two-third of all worldwide use and with Europe a far 50% of the United States.

5. The companies with less than 100 employees constitute about one-third of customers, and so are the companies with more than 5000 employees; the mid-size companies are taking longer to evaluate the merits of single-use systems.

6. Three-fourth of the companies using single-use systems are using these for manufacturing, with less than 10% of companies involved in drug discovery using single-use systems.

7. Companies with more than six products account for almost 60% of all single uses used.

8. More than 80% of new products utilize the single-use system, and almost 70% of existing manufacturing processes have been modified to include single-use systems.

9. The main concerns about the use of single-use systems in the order of importance are
    a. The capital investment
    b. Experience in using these systems
    c. Validation and environmental concerns
    d. The concern about leachables and extractable
    e. The integrity of systems

10. European regulatory agencies, as well as European companies, have greater concerns for the leachables and extractables; while the FDA allows greater flexibility in adopting newer systems, the European Medicines Evaluation Agency (EMEA) has drifted away from common acceptance criteria.

11. The most widely used single-use components are the bags and bioprocess containers, followed by filters (which is also the main cost concern), connectors, bioreactors, mixing vessels, chromatography, and sensors. This trend shows that the simplest of the components, which require little problems in validation, are the easiest to adopt; obviously, chromatography and the use of single-use sensors would present a much high barrier to validation.

12. The unmet needs of the industry in adopting single-use processes include
    a. Leachables
    b. GMP compliance of single-use sensors, calibration scale
    c. Robustness of sensors and chromatography equipment
    d. Reliable bioprocessors that are cheaper
    e. Scalability
    f. High volume and flow rates
    g. Lack of a single pressure flow and tep transmitter
    h. Larger scale, greater than 100 L
    i. Standardization
    j. Lab scale, less than 3 L

13. Most companies have allocated less than 100k for single-use products.

14. Main reasons for adopting single-use systems:
    a. Cleaning/sterilization cycle
    b. Convenience
    c. Flexibility
    d. Operating cost

e. Capital cost
f. Turnaround time
g. Reduced process steps
h. Smaller foot print
i. Rapid scale up
j. Improved environment

15. Selection of specific single-use systems depends on
    a. Availability
    b. Price
    c. Quality
    d. Approved supplier
    e. Documentation
    f. Customer service
    g. Product offering
    h. Past purchase history
    i. Engineering Support

16. The single-use systems are disposed of 57% by incineration, 37% by landfill, 20% waste to energy, and 10% converted for the alternate purpose.

17. Over 80% of users are satisfied with their adoption of single-use systems and have demonstrated cost savings.

18. The main misconcepts in the use of single-use systems include
    a. May be more costly over time, specially filtration
    b. Not sure of savings
    c. New investment needed
    d. Have no need to save cost

19. The main regulatory concerns about single uses include
    a. Sterilization and extractable/leachable
    b. Leaking of containers
    c. Bag integrity
    d. Validation of sterility/manufacturing process
    e. Aseptic process validation
    f. Quality of multiple suppliers
    g. Validation, lot-to-lot variability
    h. Material compatibility
    i. Reproducibility of batch process

Next to the cost (total), which is significantly lower, is the attraction of timeliness in the use of single-use components. A readily available component that require little preparation make it easier to switch over applications easily. The newest concept is to offer a complete line of solutions as offered by all of the major suppliers, GE, Sartorius, Pall, and Millipore.

One way to look at the value of the single-use system is to make comparisons of the time it takes to complete a batch. For example, it takes about 50 h for a batch of a monoclonal antibody to reach the end of upstream stage to purification; using single-use systems, this time can be reduced by at least 50% by using the GE ReadyToProcess® systems, an integrated system of single-use components. Similar systems are offered by Sartorius, Pall, and Millipore, the details of which are provided in Table 9.3.

Another advantage of single-use technologies is their portability. The floor-plan layout of a single-uses-based facility can be changed much more easily than that of a traditional facility. Different process requirements can be easily addressed by moving equipment into or out of a production suite. Because of the single-use nature of single-use systems, contamination is less of a concern (especially cross-contamination for multihost, multiproduct facilities).

Table 9.3 Commercial Unit Operation Systems

	Mixers	FlexReady Solutions	CellReady Bioreactor	Assemblies	Tubes and Connectors	Drums	Novaseal Crimper	Lynx Sterile Connector	Bins
Application/unit operation									
Buffer and media preparation		X		X		X			
Mixing	X			X					
Bioreactor			X						
Clarification		X							
Column protection				X					
Trace contaminant removal									
Virus removal		X							
Ultrafiltration/diafiltration		X							
Sterile filtration				X					
Storage				X		X			X
Transport				X		X			X
Sampling				X					
Connectology					X		X	X	

Single-use systems were first accepted for process development and production groups for toxicology studies and early-stage clinical trials. As commercially available systems become more robust and reliable, single uses have been incorporated into process platforms by many manufacturers, and more commercial production facilities now use these technologies as an integral part of their manufacturing processes and their efficiency and productivity improvement tools.

Regulatory barriers

There are no extraordinary regulatory barriers in the deployment of single-use systems except the need to validate leachables. Not surprisingly, even the hard-walled manufacturing systems also used plastic inevitably for such uses as filtration, packaging, and storage. Whether these devices or components are steam cleaned or not is irrelevant. So, there is already a precedence to show the safety of plastics in the manufacturing.

The use of single use is supposed to reduce regulatory barriers substantially. It has been clearly established that if the manufacturer would not have to do any cleaning validation in between batches, the cost will be reduced. Now let us examine other features of a single-use facility and how it can additionally reduce the regulatory barriers.

One of the easiest and perhaps most robust means of assuring lack of cross-contamination is to dedicate a facility for a single molecule; this way only one cell line enters the facility and only molecule goes out; all equipment is dedicated and nothing in the facility is shared with another operation by the manufacturer. The cleaning between batches would remain a task, but its importance is significantly reduced as no new contaminants are expected. Dedicating a facility to one molecule is not possible today for most companies unless they happen to be the originator producing very large quantities of the product. However, the smaller footprint required in a single-use facility can make it possible to dedicate facilities to individual molecules. First, the elimination of SIP/CIP reduced capital cost substantially and so were the requirement of the total space needed. Combining this with adopting newer techniques in producing purified water (e.g., double RO/EDI), the company can altogether eliminate the large engineering infrastructure needed to provide water in the manufacturing area. It is projected that almost 90% of the water used in traditional facilities is for CIP/SIP and autoclave use. Amgen uses 80,000 gal of water per day in its Rhode Island facility, most of it for the CIP/SIP purpose. The cost of water is very high; use of single uses brings the first savings in terms of energy cost reduction (required in water distillation) and the capital cost of distillation and storage. The new RO systems do not have storage vessels and work on an on-demand basis.

Another design element that reduces the footprint of a single-use manufacturing facility is the size of equipment installed; while many suppliers are still following the traditional design of 3D stainless steel bioreactor with a liner approach, the real breakthrough in the field has come from pillow bag bioreactors that do not need an outer container and can be used horizontally, not vertically as most 3D reactors are. Laying down the bioreactors horizontally makes the facility design compact and allows the use of smaller ceiling heights that adds substantially to cost savings in the facility design.

Path forward

Companies have to weigh a lot of issues when making decisions regarding their use of single-use technologies, some of which are economical while others are technical or regulatory in nature. Faced with these issues, the engineers who implement new technologies at biopharmaceutical manufacturing companies often face pushback from corporate management. There are technical reasons why companies have not opted to implement single-use technologies, such as when manufacturing solutions are available for mammalian cell culture. Single-use bags are not practical if you are using *Escherichia coli* or yeast because the biomass levels are so high that mixing, oxygenating, etc., is not easy with current single-use technology. There are also limitations with single-use centrifuges for harvesting the product after fermentation. Customers generally implement new technologies after they are truly tested and well proven.

Another technical issue that can delay the acceptance of single-use solutions is scale. Many of these single-use solutions cannot be scaled up to the very largest scale that the manufacturers want. So, when growing mammalian cell cultures in the GE WAVE bags, the maximum scale is 500 L, which is fine, but if you want to scale up to 2,000 L or 10,000 L, you will have to invest in much more hardware. For example, some bag solutions for bioreactors combine all the advantages—being both ready-to-use and readily single use. They use a rocking table for mixing and require very little handling and intervention. They are ideal when you want to make a process really lean in terms of logistics, installation, preparation, lack of cleaning, dismantling, turnaround, etc. This is certainly a dynamic area in terms of available products, bag solutions at larger scale (up to 2000 L) require more hardware and handling, such as a supportive bag holders, internal mixers, and other components, making the whole system more complex and less plug and play or unplug and throw away. You lose the level of containment, and you have a lot more handling—introducing risks and being less amenable to lean approaches. However, for many applications, smaller scales are enough, and bags offer speed and flexibility beyond what steel can deliver.

Smaller companies find single-use technologies attractive because they can be set up quickly with reduced capital requirements and operated in a relatively inexpensive lab space to produce a drug under GMP conditions. Single-use technologies are powerful strategic tools for smaller companies who need to advance their early-stage drugs to Phase 2 or even Phase 3 clinical trials before partnering for market approval and commercial manufacturing. Single-use technologies allow promising new drug candidates to move through the approval process quickly and at reduced expense when compared to building new stainless steel facilities. One trend that we expect will drive implementation of single-use technologies is the rise of specific drugs for smaller patient populations and personalized medicines. Combining this with the trend for higher upstream yields of an unpurified drug, a 2000 L bioreactor could easily become the manufacturing tool of choice. In this way, single-use technologies are closing the gap between development, clinical trials, and commercial scale manufacturing.

The establishment of BPSA (Bio-Process System Alliance) (www.bpsaliance.org) is an excellent support portal for the newcomers to single-use systems as to seasoned practitioners. The Bio-Process Systems Alliance (BPSA) was formed in 2005 as an industry-led corporate member trade association dedicated to encouraging and accelerating the adoption of single-use manufacturing technologies used in the production of biopharmaceuticals and vaccines. BPSA facilitates education, sharing of best practices, development of consensus guides, and business-to-business networking opportunities among its member company employees.

Given next are the FAQs derived from BPSA archives (Table 9.4).

Integrated lines of single-use systems

Sartorius-Stedim (www.sartorius-stedim.com) offers its single-use technology factory that includes

1. FlexAct is a new system that enables you to custom-configure single-use solutions for entire biomanufacturing steps. The FlexAct system consists of the central operating module that offers the widest variety of configuration options so you can take complete control of practically any steps in upstream and downstream processing. FlexAct CDS offers configurable single-use solutions for buffer preparation (BP), cell harvest (CH), virus inactivation (VI), media preparation (MP), and virus removal (VR). Next in line are UF|DF cross-flow (UD), polishing (PO), form|fill (FF) and form|transfer (FT).
2. Biostat CultiBag STR Plus used for cell culture at all levels from 50 to 1000 L; includes single-use optical DO and pH measurements.
3. Flexboy is a flexible bag system.
4. Flexel 3D mixing bags.
5. LevMixer and Palletank for mixing bases.

Pall Corporation (www.pall.com) also offers an extensive range of single-use products including

1. Allegro™ 2D and 3D Biocontainers
2. Allegro Jacketed Totes
3. Allegro Single-use Systems—Recommended Capsule Filters and Membrane
4. Kleenpak™ Nova Sterilizing-grade and Virus Removal Capsule Filters
5. Kleenpak Sterile Connectors
6. Kleenpak Sterile Disconnectors
7. Kleenpak TFF CapsulesStax™ Single-use Depth Filter Systems
8. Stax™ Single-use Depth Filter w/Seitz® AKS Activated Carbon Media

Table 9.4 FAQs Derived from BPSA

What makes up a "typical" single-use system? (SUS)	Single-use systems consist of fluid path components to replace reusable stainless steel components. The most typical systems are made up of bag chambers, connectors, tubing and filter capsules. For more complex unit operations such as cross-flow filtration or cell culture, the single-use systems will include other functional components such as agitation systems and single-use sensors.
What are the primary benefits of single-use systems?	Single-use systems boast improved productivity, cost structure, and a reduced environmental footprint compared to traditional stainless steel facilities. This is driven by the demanding cleanliness and sanitization standards in the biopharmaceutical industry. *Productivity:* Less time is used since changing out single-use systems then they do clean and sanitizing a traditional stainless steel system. *Cost:* Without reusable parts to clean, there are no chemicals, water, steam, or other utilities used in the cleaning/sanitizing process. Also, a facility engineered for single-use is simplified, using less space, so the total energy consumption is reduced. *Environmental footprint:* While the plastics in the single-use systems are usually incinerated, the footprint contributed by this is less than that contributed by waste water, chemicals, and energy used for cleaning traditional stainless steel systems. *Cost effectiveness:* Single-use bioprocessing can reduce the capital cost for building and retrofitting biopharmaceutical manufacturing facilities. *Manufacturing efficiencies:* Single-use systems can achieve significant reductions in labor, faster batch-turn around, and product change-over; single-use system modularity facilitates scale-up, speeds of integration and accelerates batch changeovers and retrofits. *Quality and safety:* single-use offers reduced risks of cross-contamination, reduced risk of bioburden contamination, reduction of cleaning (and cleaning validation) issues, and other benefits that can help satisfy the requirements of regulatory agencies. *Sustainability:* Single-use bioprocesses systems can provide a range of environmental benefits beyond those of stainless steel systems. Although single-use system may generate additional solid waste, benefits include reduction in the amount of water, chemicals, and energy required for cleaning and sanitizing as well as avoiding the labor-intensive cleaning processes required with stainless steel systems.
Are single-use systems limited to specific stages of the manufacturing process?	Currently, single-use technology can be utilized in a variety of stages as well as unit operations in bioprocessing manufacturing. However, some unit operations and process scales currently have limitations due to existing capital installed and lack of technologies available in SUS formats. Historically, media and buffer preparation were some of the first unit operations to utilize single-use technologies. As technologies have evolved, there are now larger scales and more unit operations such as cell culture, mixing, purification, formulation, and filling. As the single-use industry continues to evolve, and the bioprocessing industry realizes the benefits, further investments will be required from the supply base to develop even larger scales, as well as newer technologies.
Do manufactures need to replace all stainless technology to take advantage of the benefits of single-use systems?	Not necessarily—the benefits of integrating single-use systems into unit operations can be achieved by making either a full conversion to plastic-based solutions or an appropriate combination of single-use *and* stainless technologies. Manufacturers can increase process flexibility and improve efficiencies even with partial conversion to single-use technologies. Examples include cell culture operations integrating single-use bioreactor technology with traditional stainless reactors and final fill operations combining single-use systems with traditional vial filling equipment.
Are single-use systems limited to specific stages of the manufacturing process?	Currently, single-use technology can be utilized in a *variety* of stages as well as unit operations in bio-processing manufacturing. However, some unit operations and process scales currently have limitations due to existing capital installed and lack of technologies available, in SUS formats.
What are the factors to consider when implementing a single-use technology?	There are four basic questions/areas to ask when thinking about implementing single-use technologies: application, capital for investment, data available, and risk tolerance. What is my *application* (product and process)? Will I be performing upstream and/or downstream processing of cell therapy, drug product, or another bio-pharma process? What part or stage of the process will I be working? Is my process/product scope and overview well understood and documented? Is my application mixing and buffer preparation? Is my application fermentation or other bioreactor cell growth process? Will I require significant filtration and refinement? Will I be making final drug product? What environmental control and handling procedures do I have in place and how to they differ with single use? How much *capital* do I have available to invest in initial start-up and/or technology conversion costs? What *data* is available from the suppliers to help answer the questions I will get from the FDA when validating my product and process? Do the suppliers have a DMF? Have the suppliers performed extensive testing on extractables/leachables, USP Class VI, etc.? How can I reduce the amount of *risk* associated with my product/process? Am I the first to market with this process/product? Do others have a roadmap or platform that can be followed instead of reinventing the wheel? What data do I have or is available that can help reduce the amount of risk? Have I implemented QbD to eliminate future problems and issues? There are many case studies available on how companies have addressed the four topics mentioned above. Refer to other specific FAQ questions links or BPSA's landing page: www.bpsalliance.org. For specific inquiries and a referral to an industry expert or liaison, contact ottk@socma.com

(Continued)

Table 9.4 (*Continued*) FAQs Derived from BPSA

What effect can single-use technology have on capital and start up costs associated with designing and commissioning a new manufacturing facility?	Single-use systems typically have lower capital and start up costs. The capital equipment associated with supporting a single-use unit operation is far less expensive than for multiuse stainless systems. One example is that the materials used to build support systems for single-use operations need not be electro polished and passivated, like the product contact surfaces of a multiuse system. Commissioning a single-use system is often *faster and simpler* than a multiuse system. The product contact surfaces or single-use systems can be qualified well in advance of the actual installation, compressing start-up commissioning time lines.
How does single-use technology affect ongoing manufacturing efficiencies and production costs?	Manufacturing efficiencies and production costs will vary by company, by manufacturing site, and by the unit operation. In general, overall resource deployment is less with single-use operations when compared with stainless operations. BPSA member-company experts can help you evaluate the economics and efficiencies. To be put in touch with a company expert, please ottk@socma.com
Are there established standards or guides for single-use systems?	While there are no industry standards specifically for single-use process systems, there are several standards and technical guides that can be applied. Some specific examples are the ASTM Standard F838-05 on Sterilizing Filtration, the ANSI/AAMI/ISO Standard 11137 on Sterilization of Healthcare Products—Radiation and the ANSI/AAMI/ISO Standard 13408 on Aseptic Processing of Healthcare Products. These and other industry standards and technical, as well as government regulations and regulatory agency guidelines, that can be applied to single-use systems, are further described in the BPSA Component Quality Reference Matrix. BPSA is also a leader in developing new best practice guides for single-use systems that are simulating the development of standards or guides by other organizations (e.g., ASTM-BPE, PDA, ISPE). BPSA guides also cover Irradiation and Sterilization Validation, determination of extractables and leachables and disposal of single-use systems. see these at www.bpsalliance.org
Is there information available on irradiation and sterilization/validation?	The American National Standards Institute (ANSI), American Association of Medical Instrumentation (AAMI) and the International Organization for Standardization (ISO) have jointly issued a standard on Sterilization of Healthcare Products (ANSI/AMI/ISO 11137) that is recognized by regulative authorities around the globe. To help biopharmaceutical manufacturers and Single-Use equipment suppliers better understand the application of this standard and its various options to Single-Use systems, BPSA has published a Guide to Irradiation and Sterilization of Single-Use Systems. This BPSA guide explains the basic principles of sterilization validation under ANSI/AAMI/ISO 11137, discusses considerations in choosing among alternate approach options, and suggests where microbial control by irradiation without validation may be applicable and beneficial in terms of development time and cost without compromising safety or quality. See the references at www.bpsalliance.org
What are extractables and leachables and how should they be considered during validation?	Drug developers and regulators are concerned about potentially adverse impact on drug product quality or safety by chemicals that may migrate into the drug product from fluid contact process equipment (including single-use systems), final drug product containers, or secondary packaging. Extractables are chemicals that migrate from fluid contact materials under exaggerated conditions (e.g., solvent, time, temperature) and represent a "worst case" library of chemicals that could potentially contaminate or interact with the drug product. Leachables are chemicals that can be found in final drug product and typically include some extractables from process equipment as well as from the final container/closures or packaging along with any reaction or degradation products of those extractables and the active drug. The BPSA Guide—*Recommendations for Extractables and Leachables Testing*, provides a risk-based approach for determination of extractables and leachables from single-use process systems that has been recognized by US FDA CBER reviewers and applied successfully by several biopharmaceutical manufacturers. See the guide at www.bpsalliance.org
What are the options for disposing of single-use systems and components?	There is a range of disposal options for single-use systems; the best solution will be dependent on the composite and volume of plastics, local regulations, and available waste treatment facilities. Although recycling is viewed as environmentally appealing, it is not amenable to most single-use systems due to low volumes and mixed plastic content. Landfill options for typical systems include treated, untreated, as well as grind and autoclave. Incineration is a widely accepted treatment option in both the United States and Europe that reduces the volume of waste. Cogeneration is an attractive alternative that converts the plastic waste into energy that produces heat or electricity for consumption by individual facilities or entire communities. Pyrolysis is a relatively new technology that converts plastic waste into oil that can be used as fuel. To learn more about the advantages and disadvantages of each option, please reference BPSA's *Guide to Disposal of Single-Use Bioprocess Systems* at www.bpsalliance.org

(*Continued*)

Table 9.4 (*Continued*) FAQs Derived from BPSA

How does the environmental impact of single-use manufacturing compare to traditional stainless manufacturing?	The total environmental footprint of a manufacturing facility is more complex and adds additional factors to the more familiar term carbon footprint. The environmental footprint additionally considers the consumption of water, usage of land, and the impact of humans that operate and travel to, from and within the facility. The primary contributor to the difference between traditional stainless operations and single-use operations is in sanitization and cleaning processes. While there has been growing public concern about the generation of solid waste products from plastics, by comparison the environmental burden of multiuse stainless operations is extreme in the consumption of chemicals, water, and energy. Single-use systems do not require the same sanitization and cleaning rigor due to their disposability. The single-use plastic materials from single-use processes can typically be effectively incinerated often with energy capture potential in cogenerative operations.
Who can be contacted to learn more about implementing single-use technology into my operation?	BPSA is comprised of 40 member organizations within which reside a variety of experts who are willing to assist you in the implementation of single-use systems. The first stop to gathering the knowledge to educate on single-use manufacturing is to contact BPSA Executive Director Kevin Ott, ottk@socma.com who can direct you to the proper resources contained within the membership of both the BPSA and the additional standard-setting and technical organizations that deal all aspects of single-use systems. 15. What is the value of BPSA and the benefits of membership? BPSA's mission is to advance the adoption of single-use systems, worldwide. BPSA promotes this purpose by conducting networking activities, monitoring legislative and regulatory initiatives involving SUS, publishing industry guides on specific topics relevant to the industry, and being the information clearinghouse for education on all aspects of single use. Suppliers of single-use components, end users of single-use systems, ancillary service providers, and contract manufacturers are all eligible for BPSA membership as dues-paying organizations. Membership information and applications for membership can be seen at www.bpsalliance.org. Finally, BPSA is the only national organization that represents industry interests in single use, and that serves as the focal point for education on matters related to this industry and its customer base.

Millipore offers *its extensive line* (www.millipore.com).

With Mobius FlexReady Solutions, you can install equipment, configure applications, and validate your processes quickly and easily. Mobius FlexReady Solutions target key steps in the mAbs processing/purification train, designed and optimized for

- Clarification
- Media & Buffer Preparation
- Tangential Flow Filtration (TFF)
- Virus Filtration

GE Lifesciences (www.gelifesciences.com) offer a ReadyToProcess system to include all steps of upstream and downstream processing; these components include the following:

1. *Bioreactors*: The WAVE Bioreactor™ is a scalable, effective, cost-efficient rocking platform for cell culture. The culture medium and cells contact only a presterile, single-use Cellbag™ ensuring very short setup times. There is no need for cleaning or sterilization, providing easy operation and protection against contamination. The rocking motion of the platform induces waves in the culture fluid to provide efficient mixing and gas transfer, resulting in an environment well-suited for cell growth. The Wave system is completely scaleable across our platform ranging from 200 mL to 500 L.
2. *Single-use bioreactor bags*: Manufactured from multilayered-laminated clear plastic, Cellbag single-use bioreactors are suitable for your specific cell culture process needs for research, development, or cGMP manufacturing operations. Cellbag components are similar to those used for biological storage bags and meet USP Class VI specifications for plastics. Validation data and Cellbag DMF are available to demonstrate biocompatibility. Cellbags can be highly customized to meet specific processing requirements.

3. *Fluid management*: ReadyCircuit assemblies comprise bags, tubing, and connectors. Together with ReadyToProcess filters and sensors, ReadyCircuit assemblies form self-contained bioprocessing modules that maintain an aseptic path and provide convenience by removing time-consuming process steps associated with conventional systems. Bags, tube sets, filters, and related equipment can be secured in appropriate orientations for efficient operation using the ReadyKart mobile processing station. With an array of features and optional accessories, the ReadyKart is designed to support a variety of process-specific, fluid handling needs.

4. *ReadyToProcess Konfigurator*: ReadyToProcess Konfigurator lets you design your fluid-handling circuits with ease online. Enter the parameters to generate the design you need; includes the fast output of P&ID drawings and convenient Bill of Materials for simplified ordering.

5. *Connectivity*: ReadyMate connectors are genderless aseptic connectors that allow simple connection of components is maintaining secure workflows and sterile integrity. Additional accessories such as tube fuser and sealer of thermoplastic tubing support secure aseptic connectivity throughout the manufacturing process.

6. *Filters*: ReadyToProcess filters are a range of preconditioned and ready-to-use cartridges and capsules for both cross-flow and normal-flow filtration operations. Factory prepared to Water for Injection quality for endotoxins, TOC, and conductivity and sterilized via gamma radiation. They enable simpler and faster bioprocessing with maximum safety.

7. *Chromatography columns*: ReadyToProcess columns are high-performance bioprocessing columns that come prepacked, prequalified, and presanitized. ReadyToProcess columns are designed for seamless scalability, delivering the same performance level as available in conventional processing columns such as AxiChrom™ and BPG™.

8. *Chromatography system—ÄKTA™ ready*: ÄKTA ready (www.gehealthcare.com) is a liquid chromatography system built for process scale-up and production for early clinical phases. The system operates with ready-to-use, single-use flow paths, and as a consequence, cleaning products/batches and validation of cleaning procedures is not required. KTA™ ready is a liquid chromatography system built for process scale-up and production for early clinical phases. System meets GLP and cGMP requirements for Phase I–III in drug development and full-scale production, provides improved economy and productivity due to simpler procedures, single use eliminates risk of cross-contamination between products/batches, easy connection to and operation with prepacked ReadyToProcess columns and other process columns, scalable processes using UNICORN™ software.

Safety of single-use systems

All final containers and closures shall be made of material that will not hasten the deterioration of the product or otherwise render it less suitable for the intended use. All final containers and closures shall be clean and free of surface solids, **leachable** contaminants, and other materials that will hasten the deterioration of the product or otherwise render it less suitable for the intended use. (Biologics 21CFR600.11(h))

Single-use devices make extensive use of plastic materials or elastomer systems from filter housings to the lining of bioreactors; today, perhaps the most significant impediment to the wider acceptance of single-use systems is the controversy surrounding the possibility of contaminating the product from the chemicals in the plastic film. So, before entering a broad description of the choices of single-uses available, we should examine this topic in detail.

Leachables are chemicals that migrate from single-use processing equipment into various components of the drug product during manufacturing. Extractables are chemical entities (organic and inorganic) that can be extracted from single uses using common laboratory solvents in controlled experiments. They represent the worst-case scenario and are used as a tool to predict the types of leachables that may be encountered during pharmaceutical production.

The issue of leaching of chemicals from plastic has been the hottest topic not just for the bioprocess industry but also for many other industries including the food industry where issues like the safety of BPA in water bottles keeps rising. How does the use of plastic affect bioprocessing is of great interest to the stainless steel industry.

While regulatory requirements pertain to the toxic effects of leachables, a risk unique to biological drugs arises in the effect of leachables on the three- and four-dimensional structure of protein drugs; such changes can render the drug more immunogenic if not less effective; these side effects are thus of greater importance to the bioprocessing industry. The most well-known problem is that of the high incidence of PRCA reported in patients using commercial erythropoietin formulations—several deaths ensued; while the source of PRCA induction is not clearly settled, it is generally attributed to a change in the drug formulation that included a new surfactant, which caused unexpected leaching of and elastomer compound from the rubber stopper used.

Polymers and additives

The materials used to fabricate single-use processing equipment for biopharmaceutical manufacturing are usually polymers, such as plastic or elastomers (rubber), rather than the traditional metal or glass. Polymers offer more versatility because they are lightweight, flexible, and much more durable than their traditional counterparts. Plastic and rubber are also single use, so issues associated with cleaning and its validation can be avoided. Additives can also be incorporated into polymers to give them clarity of glass or to add color to label or code the parts.

Unlike metals where the risk lies mainly in the oxidation to environment, polymers are affected by heat, light, oxygen, and autoclaving and thus degrade over time if not stabilized and this can adversely affect the mechanical properties. Polymers are thus stabilized by incorporating chemicals that are prone to leaching during the manufacturing process and storage of biological materials.

When a plastic resin is processed, it is often introduced into an extruder, where it is melted at high temperatures and its stability is influenced by its molecular structure, polymerization process, presence of residual catalysts, and finishing steps used in production. Processing conditions during extrusion (e.g., temperature, shear, and residence time in the extruder) can dramatically affect polymer degradation. End-use conditions that expose a polymer to excessive heat or light (such as outdoor applications or sterilization techniques used in medical practices) can foster premature failure of polymer products as well, leading to a loss of flexibility or strength. If left unchecked, the results often can be a total failure of the plastic component.

Polymer degradation is controlled by the use of additives, which are specialty chemicals that provide a desired effect to a polymer. The effect can be stabilization that allows a polymer to maintain its strength and flexibility or performance improvement that adds color or some special characteristic such as antistatic or antimicrobial properties. There are typically three classes of stabilizers:

- Melt processing aids such as phosphites and hindered phenols, antioxidants that protect a polymer during extrusion and molding.
- Long-term thermal stabilizers that provide defense against heat encountered in end-use applications (e.g., hindered phenols and hindered amines).
- Light stabilizers that provide ultraviolet (UV) protection through mechanisms such as radical trapping, UV absorption, or excited state quenching.

One application in which an additive can improve or alter the performance of a polymer is a filler or modifier that affects its mechanical properties. Additives known as plasticizers can affect the stress–strain relationship for a polymer. Polyvinylchloride (PVC) is used for home water pipes and is a very rigid material. With the addition of plasticizers, however, it becomes very flexible and can be used to make intravenous (IV) bags and inflatable devices. Lubricants and processing aids are also used to reduce polymer manufacturing cycle times (e.g., mold-release agents) or facilitate the movement of plastic and elastomeric components that contact each other (e.g., rubber stoppers used in syringes).

Additives are not always single entities. Some are manufactured from naturally occurring raw materials such as tallow and vegetable oils that are themselves composed of many different components and can vary from batch to batch. Others are considered "products-by-process," as they are formed during the processing by adding several starting materials to affect the chemical reaction. The complexity of chemical reactions that take place in the manufacturing of plastic makes the analysis of extractables and leachables very complex and difficult. In extractables and leachables testing, those lesser-known minor chemical species may be the ones that leach into a drug product as this is not predictable being to a greater degree a function of the characteristics of the product.

Stabilizers incorporated into plastics and rubbers are constantly working to provide the much-needed protection to the polymer substrate. This is a dynamic process that changes according to the external stress on the system. For example, good stabilizers are efficient radical scavengers. Generally, a two-tiered approach is used to protect polymers from the heat and shear they encounter during processing—using primary antioxidants—for example, hindered phenols such as butylated hydroxy toluene (BHT) or Ciba's Irganox to protect during processing and afford long-term heat stability. Secondary antioxidants are also added as process stabilizers, typically hydroperoxide decomposers that protect polymers during extrusion and molding and protecting the primary antioxidants against decomposition. All the by-products of these reactions become available to leach from polymers into a drug product.

Elastomers are also used for special stabilization; acid scavengers are used to neutralize traces of halogen anions formed during aging of halogen-containing rubbers (e.g., brominated or chlorinated isobutylene isoprene). If not neutralized, anions cause premature aging and a decrease in the performance of rubber articles over time. Metal oxides can be very efficient acid scavengers. Ions of copper (Cu), iron (Fe), cobalt (Co), nickel (Ni), and other transition metals that have different oxidation states with comparable stability are called "rubber poisons" because they are easily oxidized or reduced by one-electron transfer. They are very active catalysts for hydroperoxide decomposition and contribute to the degradation of rubber vulcanizates. Rubber poisons thus require a specific stabilizer: a "metal deactivator like 2,3-bis[[3-[3,5-di-*tert*-butyl-4-hydroxyphenyl]propionyl]]propionohydrazide, which binds ions into stable complexes and deactivates them.

As a result of the need to add chemicals to elastomer systems, the extractables in a single-use system can include

- Monomer and oligomers from incomplete polymerization reactions
- Additives and their transformation and degradation products
- Lubricants and surface modifiers
- Fillers
- Rubber curing agents and vulcanizates
- Impurities and undesirable reaction products such as polyaromatic hydrocarbons, nitrosamines, and mercaptobenzothiazoles

Unexpected additives can also be present in a polymer system because of the inconsistencies in the process of manufacturing elastomer systems whereby unpredictable reactions can take place.

Despite the risk in the use of additives added to polymers, the utility of polymers in single-use bioprocess equipment (and in all medical or pharmaceutical applications) far outweighs the risks associated with their use. These risks can be managed well by taking three steps: material selection, implementation of a proper testing program, and partnering with vendors.

Material selection

The type of plastic use should match the needed physical and chemical properties and compatibility of the additives used to the product manufactured. For example, phenolic antioxidants, each with the same active site (the hindered phenol moiety) but with the different nature of the remainder of the molecule, makes them soluble or compatible with a given polymer substrate. An antioxidant that is compatible with nylon might not be the best choice for use in polyolefins, as an example.

Ensuring compatibility often lessens the amount of leaching that can occur. It is also important to select polymers and additives that are approved for use by the regulatory authorities for the specific use. Such compounds have already undergone a fair amount of analytical and toxicological testing, so a good amount of information is often available for them. Because of this, most manufacturers are likely to continue using these additives and thus the user may not have to alter the composition at a laser stage as these compounds and the art of using them is likely to survive obviating the need for a change control step as significant changes in the process need to be reported back to the FDA.

Commercially supplied plastic films are proprietary formulations and arrangements; for example, Advanced Scientific (www.advancedscientifics.com) produces its bags utilizing two films. The fluid contact film is a 5.0 mil polyethylene. The outer is a 5 layer 7 mil coextrusion film, which provides barrier and durability. A typical test report is given in Table 9.5.

ATMI (www.atmi.com) offers its proprietary TK8 film, which is constructed from laminated layers of polyamide (PA), ethylene vinyl alcohol polymer (EVOH), and ultralow density polyethylene (ULDPE). The outer PA layer provides robust puncture resistance, strength, and excellent thermal stability. The EVOH layer minimizes gas diffusion across the film while maintaining a very good flex crack resistance. The ULDPE layers provide flexibility, integrity and an ultra-clean, ultra-pure, low-extractables product contacting layer. The combination of these layers results in a film that has outstanding optical clarity, is easy to handle, and performs well in a broad range of bioprocess applications. The inner ULDPE layer used in TK8 is blow-extruded in-house by ATMI under cleanroom conditions ($0.2 \ \mu m$ filtered air),

Table 9.5 Test Report of Leachability, Extractability Testing

Biocompatibility

	USP Accute Systemic Injection Test	Pass	USP <88>
	USP Intracutaneous Injection Test	Pass	USP <88>
	USP Intramuscular Implantation Test	Pass	USP <88>
	USP MEM Elution Method	Noncytotoxic	USP <87>
	Physiochemical Test for Plastics	Pass	USP <661>

Extractables

		TOC after 90 Days (ppm)	pH Shift after 90 Days
	Purified water (pH = 7)	<2	−0.79
	Acidic water (pH < 2)	<3	+0.01
	Basic water (pH > 10)	<4	+0.87

Physical Data

	Water vapor transmission rate (g/100 in.^{2}/24 h)	0.017		ASTM F-1249
	Carbon dioxide transmission rate (cc/100 in.^{2}/24 h)	0.129		ASTM F-2476
	Oxygen transmission rate (cc/100 in.^{2}/24 h)	0.023		ASTM F-1927
	Average Force	**Average MOE**	**Average Elongation**	
Tensile	32.73 lb	25110 psi	1084%	ASTM D882-02
	Min Force	**Average Force**	**Max Force**	
Tear resistance	6.7 lb	7.21 lb	7.74 lb	ASTM D1004-03
Puncture resistance	16.42 lb	18.61 lb	19.51 lb	FTMS 101C

ensuring the cleanest possible product-contacting surface. Lamination is also performed under controlled, ultra-clean conditions. Lastly, TK8 film is converted into ATMI bag products in our ISO Class 5 cleanroom. All of the layers in TK8 are made from "medical grade" materials, meaning that they comply with industry standards and are subject to strict change controls. The entire structure of TK8 is totally free of any animal-derived components (ADCF). ATMI has also created TK8 with dual sourcing and contingency planning in mind to ensure the security of supply.

- TK8 film complies with USP Class VI (USP<87>, USP<88>, and USP<661>).
- ULDPE resin complies with EP 3.1.3.
- The Shelf life is supported by aging validation studies.
- Certified ADCF.
- Bioburden evaluation available (ISO 11737).
- Particle count data available (EP 2.9.19 or USP<788>).
- By performing blow extrusion in-house, ATMI maintains full control and traceability of the contact film composition, from resin through to finished bag product.

Testing

Polymers used in medical and pharmaceutical applications should comply with the appropriate USP guidelines, and it is recommended that they meet USP Class VI testing as documented in chapter <88> of the US Pharmacopeia. Appropriate extractables and leachables testing programs must be implemented for all bioprocessing materials that come into direct contact with the drug.

Table 9.6 Leachables/Extractable Testing Sources

www.pqri.org	www.us.sgs.com	www.pacelabs.com
www.rapra.net	www.cyanta.com	www.asqlongisland.com
www.impactanalytical.com	www.ialab.com	www.intertek.com
www.avomeen.com		

The best-practice guidelines for conducting such testing are provided by the Bio-Process Systems Alliance (BPSA) as a two-part technical guideline for evaluating the risk associated with extractables and leachables, specifically for single-use processing equipment (www.bpsalliance.org). This organization is dedicated to encouraging the use of single-use systems and provided excellent support and assistance; the reader is highly encouraged to visit their website for newer information as well as participate in their many seminars and conventions to stay abreast of the developments in this fast-changing field.

The testing for leachables should not necessarily end once the materials have been qualified. It is necessary to have in the place quality control program instead of testing the product or equipment alone. The level of quality control testing will depend on risk tolerance; fortunately, the manufacturing of recombinant drugs involves an extensive purification step that are likely to remove most of these leachables; also, since the final medium used for protein solutions is aqueous, and many of the leachables are not soluble in water, this further reduces the risk. Greater risk can be seen in the final packaging components, for example, rubber stoppers used in packaging the final dosage form is more likely to be a risk to the protein formulation than any other component in the chain of single use to which the drug is exposed during the manufacturing process.

While it is always a good idea to establish in-house testing of leachables, often it is neither possible nor recommended; several highly reputed laboratories have fully certified programs; given next is a short list of these laboratories (Table 9.6).

Partnering with vendors

Reputable vendors often have extractables data already on hand to share with their customers. In many cases, they will provide a certificate of analysis and toxicological information associated with materials used to fabricate their products. Vendors also should have well-established change control processes for the products they sell to allow sponsors to modify their applications to regulatory agencies accordingly.

Responsibility of sponsors

Companies filing regulatory approval have the responsibility of complying with the requirements of validating the process to minimize the risk from leachables. Extractables and leachables evaluations are part of a validation program for processes using single-use biopharmaceutical systems and components. There is minimal regulatory guidance that directly addresses extractables and leachables in bioprocessing.

The extractables are evaluated by exposing components or systems to conditions that are more severe than normally found in a biopharmaceutical process, typically using a variety of solvents at high temperatures. The goal of an extractable study is to identify as many compounds as possible that have the potential to become leachables. A positive outcome is one where the list of extractables from a material is sizable. Although it is not expected that many of those extractables will actually leach into the drug product at detectable levels, a material extractables profile provides critical information in pursuit of a comprehensive leachables test.

Not all leachables may be found during the extractables evaluation, as drug formulation components or buffers may interact with a polymer or its additives to form a new "leachable" contaminant that was not previously identified during extractables analysis. In addition, leachables that were not identified as extractables also will be found if the drug product formulation and processing conditions are unique and more severe than the conditions at which extractable tests were performed—or when the analytical methodologies used in the two types of studies are different.

Regulatory requirements

There are as yet no specific standards or guidance that reference extractables and leachables from single-use bioprocessing materials. Many references that do apply were written to address the processing materials and equipment without regard to the materials of construction.

United States and Canada

The foundation for the requirement to assess extractables and leachables in the United States is introduced in Title 21 of the Code of Federal Regulations (CFR) Part 211.65, which states that "Equipment shall be constructed so that surfaces that contact components, in-process materials, or drug products shall not be reactive, additive, or absorptive so as to alter the safety, identity, strength, quality, or purity of the drug product beyond the official or other established requirements." This regulation applies to all materials including metals, glass, and plastics.

Extractables and leachables generally would be considered "additive," although it is also possible for leachables to interact with a product to yield new contaminants.

The FDA regulatory guidance for final container–closure systems, though not written for process contact materials, gives direction to the type of final product testing that may be provided regarding extractables and leachables from single-use process components and systems. The May 1999 guidance document from the FDA's Center for Drug Evaluation and Research (CDER) indicates the types of drug products and component dosage form interactions that the FDA considers to be the highest risks for extractables. Generally, the likelihood of the packaging component interacting with dosage form is the highest in injectable dosage forms, mainly because of the low level of leachables that can be allowed in such drug delivery systems.

Drugs that will be administered as injectables or inhalants will have higher levels of regulatory concern than oral or topical drugs. Similarly, liquid dosage forms will have higher regulatory concern than tablets because extractables migrate into liquids more easily than into solids.

In addition, pharmaceutical-grade materials are expected to meet or exceed industry and regulatory standards and requirements such as those listed in the US Pharmacopeia (USP) chapters <87> and <88>. The USP procedures test the biological reactivity of mammalian cell cultures following contact with polymeric materials. Those chapters are helpful for testing the suitability of plastics for use in fabricating a system to process parenteral drug formulations. However, they are not considered sufficient regulatory documentation for extractables and leachables because many toxicological indicators are not evaluated, including subacute and chronic toxicity along with evaluation of carcinogenic, reproductive, developmental, neurological, and immunological effects.

Europe

In the European Union, a related statement to the US 21 CFR 211.65 is found in the rules governing the manufacture of medicinal products. The EU good manufacturing practice document states "Production equipment should not present any hazard to the products. The parts of the production equipment that come into contact with the product must not be reactive, additive or absorptive to such an extent that it will affect the quality of the product and thus present any hazard."

The EMEA published a guideline on plastic immediate packaging materials in December 2005 that also addresses container–closure systems and has been used to provide direction for single-use process contact materials. Data to be included relating to extractables and leachables come from extraction studies ("worst-case leachables"), interaction studies, and migration studies (similar to leachables information for those components, identify what additional information or testing is required, and then set and execute a plan to fill in the gaps).

Risk assessment

Risk assessment is based on following considerations:

- *Compatibility of materials*: Most biological drugs formulations are aqueous-based and, therefore, compatible with the materials used in most single-use processing components. Still, a check to make sure that the process stream and/or formulation do not violate any of the manufacturer's recommendations for chemical compatibility, pH, and operating pressure/temperature is warranted before proceeding. A full analysis of data generated by the vendor should be completed upfront as a preparatory step.
- *The proximity of a component to the final product*: Product contact immediately before the final fill increases the risk of leachables in a final product. For example, tubing or connectors used to transfer starting buffers probably present a lower risk because of their upstream location. Processing steps such as diafiltration or lyophilization that could remove leachables from a process should also be considered because they may reduce associated risk. However, it cannot be assumed that a step that can potentially remove some leachables will remove all leachables. In such cases, supporting data should be obtained.
- *Product composition*: In general, a product stream or formulation that has higher levels of organics, particularly high or low pH, or solubilizing agents such as surfactants (detergents), will increase the regulatory and safety concern for potential leachables. Neutral buffers lower concern about potential leachables.
- *Surface area*: The surface area exposed to a product stream varies widely. It is relatively high for filters, in which the internal surface area is 1000× the filtration area. Conversely, the surface area is relatively small for O-ring seals.
- *Time and temperature*: Longer contact times allow for more potential leachables to be removed from a material until equilibrium is reached. Higher temperatures lead to more rapid migration of leachables from materials into a process stream or formulation.

- *Pretreatment steps*: Sterilization by steam autoclave and/or gamma irradiation may cause higher levels of extractables and leachables depending on the polymer formulation involved in a single-use component. On the other hand, rinsing may lower the concern for extractables and leachables (e.g., when filters are flushed before use).

Here are some highlights relating to risk assessment of extractables and leachables:

1. Regulatory responsibility for overall assessment and understanding of a finished product and process components involved in its production remains with the product sponsor. This includes evaluations of extractables and leachables. Regulatory agencies do not have a guideline available yet to help sponsors.
2. All elastomeric and plastic-based materials contain extractables specific to the formulated and cured material(s) from which they are constructed.
3. Contaminants are also found in stainless steel systems in the form of residues left after cleaning or traces of metals such as iron, nickel, and chromium salts from the stainless steel itself so the problem of contamination from the container is not restricted to single-use containers.
4. Most polymers without certain additives would not work as materials for use in single-use processing; this includes stabilizing the polymer, extruding it and preventing its oxidation and UV degradation; other additives include antistatic agents, impact modifiers, catalysts, release agents, colorants, brighteners, bactericides, and blowing agents. The choice of polymer or method of polymerization (by heat or chemical means) directly affects the levels and types of compounds found as extractables.
5. Fluoropolymers offer the best choice as they are typically processed without additives, stabilizers, or processing aids.
6. A DMF or BMF for process-contact equipment is not explicitly required by US regulatory authorities. However, it represents a way for vendors to share proprietary information about a component or raw material with the FDA and to ensure that such information remains up to date. It is, therefore, important that sponsors work only with the most reputable suppliers of single-use components.
7. The levels and types of compounds found as extractable analytes are directly affected by the type and degree of sterilization performed (e.g., gamma irradiation, ethylene oxide gas, or autoclaving). The leachable analyte and concentration that may be of the issue to one particular drug formulation may have no impact on another. It is for these differences that the responsibility of product sponsors to qualify and demonstrate the applicability of process components within their manufacturing systems. Leachables are always final-product-specific.
8. All component materials should be evaluated that have the potential to come into direct contact with a manufactured drug product. Of greater importance are the components that would contact the product in the post-purification stage.
9. Controlled extraction studies are designed to generate extractables, the presence of extractables is expected. This does not necessarily reflect the degree and concentration of leachables that will be found in contact with a product stream—leachables are a function of the nature of the product, the length of exposure, and the environmental conditions for the storage of the product.
10. Detection of a toxic or otherwise undesirable extractable under aggressive conditions requires testing to ensure that migration to the product is below acceptable limits under actual processing conditions. It is done

by controlled extraction using studies using multiple solvents of varying polarity to fully elucidate the extractable analytes in question. Techniques such as Soxhlet extraction, solvent refluxing, microwave extraction, sonication, and/or acid washing at elevated temperature may also be used. For extractables testing, the contact surface area can be maximized by mechanical methods such as cutting or grinding.

11. For leachables testing, it is most applicable to mimic actual process conditions by leaving test components intact. Controlled extraction studies should use extraction media of varying polarities and physical properties. Ideally, this would come from using two or three solvents that include analysis by high-pressure liquid chromatography (HPLC), GC-MS, and ICP-MS.

12. Toxicology of leachables should be performed using approved protocols. A PQRI (www.pqri.org) document on extractables and leachables suggests an approach to address toxicology using LD50 with a 1,000× or 10,000× safety factor based on the dosage quantity. In addition, several structural activity relationship (SAR) databases are readily available to professional toxicologists. Examples include the "Carcinogenic Potency Database" (CPDB) (http://potency.berkeley.edu/) and the U.S. Environmental Protection Agency's "Distributed Structure-Searchable Toxicity Network" (DSSTOX) database (http://www.epa.gov/ncct/dsstox/). Chances are that most sponsors will not be able to conduct these studies in-house, and it is advised to outsource these evaluations.

13. The classes of compounds extractable include more particularly, n-nitrosamines, polynuclear aromatics (sometimes termed polyaromatic hydrocarbons, PAH), and 2-mercaptobenzothiozole, along with biologically active compounds such as bisphenol-A (BPA). Individual extractable compounds are too numerous to list, but examples include aromatic antioxidants such as butylated hydroxytoluene (BHT), oleamide, bromide, fluoride, chloride, oleic acid, erucamide, eicosane, and stearic acid. Databases on extractables are widely available (www.pqri.org.) Comprehensive extractable data for components can reduce the time and resources needed to qualify leachables from the systems where they are used. When comparing supplied extractables data for components constructed of like materials, end users should carefully review the methods used to generate the data. Less rigorous methods may under-represent the actual levels and extent of extractables, and a report describing more extractables may simply come from using more rigorous methods.

14. For determination of leachables in products, it is currently the industry standard to validate analytical methods according to ICH and USP criteria. This ensures appropriate levels of analytical precision and accuracy.

15. The overall quantity of extractables or leachables can be estimated using nonspecific methods such as total organic carbon (TOC) and nonvolatile residue (NVR) analysis. Such nonspecific quantitation is especially useful in comparing materials before their final selection for a process. These analyses can be used individually or collectively to estimate amounts of extractable material present and to ensure that targeted methods are not missing a major extractable constituent. For instance, nonpolar compounds without chromophores can be identified using Fourier-transform infrared (FTIR) analysis of nonvolatile residues.

16. Organic extractables will leach into formulations at a higher level if the products have higher organic content or if surfactants are present.

17. The toxicity of leachables is frequently estimated based on the amount entering the human body in each dose. Thus, it is often not the quantity

of leachable in a product but how much of it finds its way into the human body. This is somewhat analogous to the limits many regulatory agencies set on residual DNA in a finished product.

18. The component used for an extractables study should be the same one that will be used in a process, and it should have the same pretreatment steps as is intended for that process. For instance, if a process uses gamma irradiation for sterilization, then the component used for extractable testing should be sterilized by that method. Often, a vendor would provide a simulated data based on similar products by extrapolating the data from other components; this would not be acceptable.

19. The solvents used for leachable studies should include water and a low-molecular-weight alcohol such as ethanol or n-propanol. Where appropriate, an organic solvent with the appropriate solubility parameters will help identify additional extractables. Extractions should be performed at the relatively extreme time and temperature conditions. However, the solvents or extraction conditions should not be so extreme as to degrade materials to a point at which they are not mechanically functional (e.g., melting or dissolving). Extreme conditions used should be relative to those under which a material is normally used. For example, one normally used at room temperature might be extracted at an elevated temperature of 50°C or 70°C.

20. Analytical methods should include HPLC and GC-MS methods to detect and identify specific, individual, extractable compounds. HPLC with ultraviolet (HPLC-UV) or mass spectrometer (LC-MS) detector, and GC-MS are the most scientifically robust methods for this purpose. When metals are a concern, inductively coupled plasma analysis is widely used, both with and without mass-spectrometric detection (ICP and ICP-MS).

21. While it is desirable to identify each extractable, for some extractables such as siloxanes and oligomers of base polymers, precise identification is not feasible because of the large number of closely related isomers and oligomers. In such cases, a general classification can be used. Quantitation of identified extractables is informative, but it does not need to be performed at a high level of precision. This is different from recommendations for evaluating extractables for final containers, closures, for which analytical and toxicological limits should be set based on a measured level of extractables.

22. User-specific components may be built by using subcomponents from different vendors such as filters, connecters, tubing, bags, etc. It is unlikely that the composite system would have a complete data on extractables from the vendor assembling the component. Individual data for each subcomponent can be pooled, but it may be easier for the sponsor to conduct the study on the entire component at one time. It is, therefore, advisable that sponsors use the shelf products where possible.

23. Biocompatibility testing is very complex issue. It is a material's lack of interaction with living tissue or a living system by not being toxic, injurious, or physiologically reactive, and not causing immunological rejection. This testing is required and two common test regimens are commonly used to measure biocompatibility, USP <88>, Biological Reactivity Testing (USP Class VI), and ISO 10993, Biological Evaluation of Medical Devices (www.iso.org), which has replaced the USP Class VI test.

24. The ISO 10993 has 20 parts and provides testing requirements in great detail. These parts include evaluation and testing, animal welfare requirements, tests for genotoxicity, carcinogenicity and reproductive toxicity, selection of tests for interactions with blood, tests for in vitro cytotoxicity,

tests for local effects after implantation, ethylene oxide sterilization residuals, clinical investigation of medical devices, framework for identification and quantification of potential degradation products, tests for irritation and delayed-type hypersensitivity, tests for systemic toxicity, sample preparation and reference materials, identification and quantification of degradation products from polymeric medical devices, identification and quantification of degradation products from ceramics, identification and quantification of degradation products from metals and alloys, toxicokinetic study design for degradation products and leachables, establishment of allowable limits for leachable substances, chemical characterization of materials, physicochemical, morphological, and topographical characterization of materials, and principles and methods of immunotoxicology testing of medical devices.

25. The USP <88> protocols are used to classify plastics in Classes I–VI, based on end use, type, and time of exposure of human tissue to plastics, of which Class VI requires the most stringent testing of all the six classes. These tests measure and determine the biological response of animals to the plastic by either direct or indirect contact or by injection of the specific extracts prepared from the material under test. The tests are described as *Systemic Toxicity Test* to determine the irritant effect of toxic leachables present in extracts of test materials; *Intracutaneous Test* to assess the localized reaction of tissue to leachable substances; and *Implantation Test* to evaluate the reaction of living tissue to the plastic. The extracts for the test are prepared at one of three standard temperatures/time: 50°C (122°F) for 72 h, 70°C (158°F) for 24 h, and 121°C (250°F) for 1 h.

26. A typical testing data for single-use bioreactors (as supplied by GE Healthcare) would include
   • Testing is performed on irradiated film (50 kGy)
   • USP XXII plastic class VI and ISO 10993
   • ISO 10993-4 Hemolysis study in vivo extraction method
   • ISO 10993-5 Cytotoxicity study using ISO elution method
   • ISO 10993-6 Muscle implantation study in rabbit
   • ISO 10993-10 Acute intracutaneous reactivity study in rabbit
   • ISO 10993-11 Acute systemic toxicity in mouse

Single-use containers

Single-use containers form the heart of any comprehensive single-use facility. To replace dozens of hard-walled (steel or glass) containers that are used to store media, starting materials, and intermediate and finished products, whether kept at room temperature or kept frozen, there is a great need for containers. Fortunately, single-use bag systems have been very well adopted as alternates to hard-walled containers. And that is because historically, pharmaceutical products have been stored and dispensed in these types of bags such as sterile intravenous solutions, blood, plasma, plasma expanders, hyperalimentation solutions, etc. For blood storage, a single-use bag would have one-layer films made from polyvinyl chloride (PVC) or ethylene vinyl acetate (EVA).

Given next is a listing of major suppliers of single-use containers; most major equipment suppliers would have proprietary bags to fit their equipment only, and while generic bag manufacturers may have alternates to these proprietary bags, there are intellectual property issues involved as many of these bags may have patent protection.

Proprietary bag suppliers

Thermo Scientific (www.thermoscientific.com)
Sartorius-Stedim (www.sartorius-stedim.com)
Pall (www.pall.com)
GE (www.gelifesciences.com)
Millipore (www.millipore.com)
Xcellerex (www.xcellerex.com)
LevTech (ATMI) (www.atmi.com)

Generic bag suppliers

Advanced Scientifics (www.advancedscientifics.com)
   PL-01077: Polyethylene Single Use Bags
   PL-01077 is a 5 layer 7 mil co-extrusion film that provides barrier and
      durability. Utilized on smaller bag sizes up to 1 L it maintains compa-
      rable, extra values to larger PE bags.
   PL-01026/PL-01077
   Polyethylene Single-Use Bags
   Advanced Scientifics' Polyethylene Single Use systems are produced
      utilizing two films. The fluid contact film is a 5.0 mil polyethylene
      (PL-01026). The outer is a 5 layer 7 mil coextrusion film, which pro-
      vides barrier and durability (PL-01077).
   PL-01028 /Ethyl Vinyl Acetate Single-Use Bags
   Advanced Scientifics' Ethyl Vinyl Acetate Single-Use systems are pro-
      duced utilizing a single film. The film is a 4 layer 12.5 mil co-extrusion
      film, which provides barrier and durability (PL-01028).
   Drums and Protective Containers, Tank Liner
   Containers/Fill Port Automatic Aseptic Filling
   Advanced Scientifics' Filling System, when used in conjunction with good
      technique and a laminar flow hood, yields an aseptic bag fill. The semi-
      automatic filling system utilizes a fixture and cap assembly developed
      and manufactured by Advanced Scientifics, and fully controls the fill-
      ing interface with no user interaction required with the fill port. What
      is left after completion is a tamper evident dispensing port. This results
      in a cleaner, more efficient and cost-effective method of filling.
Charter Medical (www.chartermedical.com)
   Bio-Pak® Cell Culture Bio-Containers
   Bio-Pak Cell Culture containers are designed for single-use bioprocess-
      ing applications The Bio-Pak Cell Culture containers incorporate our
      Clear-Pak® film chosen for its superior clarity and excellent perfor-
      mance in promoting cell growth and viability.
   Bio-Pak® 3D Gusset Bio-Containers
   Bio-Pak 3D Gusset style containers are available in a range of sizes from
      50 to 1000 L. The 3D gusset design is ideal for preparation and stor-
      age of media and buffer solutions. Clear-Pak film is a single-web,
      multilayer, co-extruded film, which provides excellent gas barrier
      properties to minimize pH shift for greater product stability.
   Bio-Pak® Small Volume Bio-Containers
   Bio-Pak Small Volume containers are designed for bioprocessing applica-
      tions, storage, and transport of sterile fluids. Bio-Pak Small Volume
      containers are available in sizes ranging from 50 mL to 20 L. Our
      boat port design provides flexibility in tubing interface options and
      facilitates maximum recovery of stored materials.

Figure 9.1 Layers of plastics in PL-01077 bag film offered by Advanced Scientifics.

Bio-Pak® XL & XLPlus Bio-Containers

The Charter Medical Bio-Pak XL series biocontainers are an efficient and cost-effective alternative to large tanks and totes for sterile fluid containment and processing. The single-use Bio-Pak XL bags eliminate the issues surrounding cleaning validation, storage, and sterilization of traditional biocontainers. In addition, our biocontainers are lightweight, and easier to fill, and handle over large glass or steel tanks.

Contour Tank Liners

Charter Medical's Contour Tank Liners are a cost-effective alternative to dedicated tanks and totes. Contour liners reduce cleaning validation and sterilization of traditional containers. Most importantly, because they are single use, the potential of cross-contamination between different products is reduced.

Bio-Pak® Totes

Charter Medical offers application-designed, mobile totes mounted on durable, nonmarking wheels. These stainless steel totes hold the flexible bag plus outlet tubing in a self-contained, wheeled unit that can be safely transported by forklifts. A unique bottom outlet system allows fast flow rates and minimal container holdup volume.

Freeze-Pak™ Cryogenic Bio-Containers

Freeze-Pak cyrogenic biocontainers are designed for use in cryogenic temperature applications under liquid nitrogen conditions. Freeze-Pak cryogenic biocontainers are used predominately for clinical and research applications. Our Freeze-Pak cryogenic film is preferred based on our film performance during the freeze–thaw process. Freeze-Pak film is a single web polyolefin monolayer and is 12 mil in thickness.

Applied Bioprocessing Containers (http://www.appliedbpc.com)

Small-volume containers, 50 mL–20 L with integrated handle, integrated hanging capability, needle-free sampling port, may be used with a sterile welder, and available as a manifold system.

Containers for cylindrical tanks, 50–750 L, 2D and 3D designs, top or bottom drain, available as a liner, fit most cylindrical tanks, and available as a manifold system.

Single-use bags are made from plastic films whose composition is determined by the need for robustness, performance, and often the size of the container. These bags have multiple layers for strengthening the walls. Given next is the construction of the Advanced Scientifics typical bag design (Figure 9.1).

There is a wide choice available from 1 to 3500 L bags that can be had with a variety of shapes, volume, available ports, tubing, inline filters, and any other custom feature besides the standard offering by these manufacturers. Generally, it would be advisable to use an off-the-shelf item, even though the generic manufacturers would offer customs bags readily; the reason for this choice is that there is likely to be a larger volume of data available on off-the-shelf bags, and also it is likely to be available on a need basis.

The typical applications in bioprocessing would use tank liners and two-dimensional (2D) and three-dimensional (3D) bags.

Tank liners

Tank liners are simple, single-use bags used to line container and transportation systems. In most cases, they are not gamma sterilized since these are used in open systems most of the time, such as in the preparation of buffer solutions and culture media at the first stage of preparation. The container within which the liner is inserted is there only to provide a mechanical support.

Commercially available overhead mixers can readily be integrated because these systems are open (www.lightninmixers.com). A broad choice of low-density polyethylene liners is available from vendors that supply to several industries reducing the cost of liners. Single-use equipment suppliers also offer these choices. For example, Thermo Scientific HyClone tank liners are designed for use with commercially available overhead mixers. The chamber is constructed of CX3-9 film with dimensions optimized for Thermo Scientific HyClone standard drums and commonly used industry standard cylindrical tanks. Standard products for maximum recovery using industry standard connection systems in unit volumes of 50, 100, and 200 L. Tanks are supplied sterile to minimize bioburden. A dolly is available to provide mobility of volumes up to 500 L.

The hard-walled containers are necessary for the preparation of buffers and media as this offers the cheapest alternative; however, these containers do not contact any formulation component and, as a result, the cheapest containers should be used, most likely the choice would be a plastic off the shelf-drum, for example, a 55 gal drum. Several major equipment suppliers provide a complete line of mixing systems, and while these do offer an advantage in handling large volumes consistently, one can readily put together a system from off-the-shelf components at a substantially lower cost. It is noteworthy that the more expensive systems come with programming elements that might make the PAT work easier, but at the stage of buffer and media preparation, the challenges are few and readily overcome by implementing the simplest and cheapest systems. This is what is intended in the max-dispo concept—to use only what adds value.

Pillow bag fluid containers

For smaller volumes, pillow bag bags work well, from less than 1 to 50 L, before they become difficult to handle. The largest pillow bag fluid container for bioreaction is provided by GE for their Cellbag operations (Wave Bioreactor) in 1000 L size; other suppliers like Charter Medical can provide containers up to 3500 L in size. These bags are produced from two-layer films, which are welded together at their ends. The result is a flat chamber, which has ports either face welded or end welded. The choice of ports is determined by the user, and most suppliers have standard combinations that might work well in most instances. It is important to iterate here that any custom-designed bag or configuration would require new studies to establish the role of leachables and this may not be necessary if standard off-the-shelf items are used that have already gone into cGMP manufacturing and approvals of products made using them.

Besides their use as bioreactors, the pillow bag bags are utilized in a reclining or hanging position as manifolds for sampling, dispensing, and to hold the product.

Pillow bag powder bags

In some instances, it may be necessary to use bags to store powders (such as buffer salts, API, excipients); these bags would have a funnel shape and equipped with large sanitary fittings or aseptic transfer systems; these are antistatic and free of additives. An example of such bag is the Thermo Scientific HyClone Powdertrainer (www.thermoscientific.com). Large size powder bags are generally custom designed.

3D bags

The pillow bag bags have an interesting problem in their design that at a larger scale it would be difficult to maintain their integrity. The 1000 L bag offered by GE is recommended to be used with no more than 500 L media; beyond that, the seals may not hold since the weight of fluid inside is transferred to the seams of these bags. This becomes particularly problematic when the pillow bag bags are rocked or shaken—adding stress to seams.

Three-dimensional bags as liners for hard-walled containers obviate the problems of the integrity of pillow bag bags; today, these bags are available in the size of 3–4000 L sizes. The 3D design also provides an additional surface to install ports with complex functions and at both top and bottom. The 3D bags are made by welding films and are mostly offered in cylinders, conical, or cube shapes. Often, the shape is determined by the method how these containers are stored (stacked) in outer containers that have the same shape and allow snug fitting of the 3D bags. While a very large liner can always be brought into manufacturing area, the outer containers are at times built before the facility is completed; companies offering modular construction of outer containers would do well in the future if they offer an option of assembling an outer container from smaller pieces.

To facilitate their use such as in buffer preparation, these outer containers may be equipped with weight sensors, recirculation/mixing fluid management, and temperature control if required. The temperature control can be achieved in several ways, the cheapest one being wrapping them in blankets that are temperature controlled and the most expensive being to use jacketed containers with circulating fluids. The weight measurement is of greatest importance, and while most manufacturers would use a floor scale, large-scale production requires the installation of load cells in the outer containers to avoid moving the containers for weighing.

Transportation container

Products at different stages of manufacturing often need to be transported by the company or to remote locations to complete the process; finished products are also shipped out to customers and this requires selection of safe, stable, and closed container systems that would maintain sterility. Examples of these containers include

- Flexboy, Flexel 3D Palletank, and Celsius FFT products (www.sartorius-stedim.com)
- Nalgene (www.nalgene.com)
- Thermo Scientific (www.thermoscientific)
- BioShell™ container system designed to protect single-use bags during storage, handling, and shipping. High-purity, dual-density foam construction can withstand multiple impacts at −70°C (www.bio-shell.com)

Single-use bags can be readily used to transport or store frozen products, from cell culture as WCB for direct introduction into a bioreactor to shipping biological API;

while flexible bags would survive temperature variations, often it is difficult to detect damage to them during transportation and thus require a protective surface around them to obviate this risk.

Summary

- Plastic single-use containers offer the best solution for single-use components utilization as they remove the cleaning requirements and validation.
- Low-density PE liners in a hard-walled plastic container and a standard mixer make the cheapest combination of pieces to prepare buffers and media.
- More complex mixing systems are not necessary, and neither are the expensive proprietary containers to hold these PE liners.
- The pillow bag bags can be used only for smaller size volumes while 3D bags with an outer nonsingle-use container increase the limit of fluids that can be contained to thousands of liters.
- Several novel shapes and sizes are available to fit just about any need of the sponsor.
- Flexible bags can be used for the transportation of biological drugs, and while they survive freeze–thaw cycles, it is often difficult to record breach in their integrity requiring an outer protective surface.
- Custom-designed bags are readily available, but these are very expensive and do not give the sponsor the benefit of the large database provided by the vendors—stick to off-the-shelf products whenever possible.
- Future novel uses of bags may include storage of WCB for direct addition to bioreactors.

Single-use mixing systems

The unit operation of mixing is extensively involved in bioprocessing systems. Some of the keys mixing operations include mixing to dissolve components of a buffer, culture media, refolding solution, dispersion of cell culture in bioreactors, heating or cooling of liquids, and many other such operations.

All mixing operations must be fully validated as part of PAT to assure that optimal mixing has been achieved all the time. While the stainless steel mixing vessels have long been used, and the principles behind mixing and de-mixing of components with traditional mixing devices studied, much remains to be understood about achieving homogeneous mixtures in single-use bags.

In bioprocessing operations, two types of mixings are important—one that leads to the dissolution of solutes and the other provides a homogeneous environment such as in a bioreactor or a refolding tank. How fast a mixture of powdered components in a buffer mixture dissolves will depend, to a great degree, on the solubility of individual components, the agitation applied, temperature, and the length of mixing. In theory, mixing involves distributive, dispersive (breaking of aggregates), or diffusive steps. All of these steps require energy, which is provided by the mechanical motion induced in liquids. A laminar movement of liquid can achieve the mixing or a turbulent movement and the Reynold's number of mixing obtained can predict this.

In fluid mechanics, the Reynolds number Re is a dimensionless number that gives a measure of the ratio of inertial forces to viscous forces and consequently quantifies the relative importance of these two types of forces for given flow conditions. Laminar flow occurs at low Reynolds numbers, where viscous forces are dominant, and is characterized by smooth, constant fluid motion; while turbulent flow occurs at high Reynolds numbers and is dominated by inertial forces, which tend to produce chaotic eddies, vortices, and other flow instabilities. In a cylindrical vessel stirred by a central rotating paddle, turbine or propeller, the characteristic dimension is the diameter of the agitator D. The velocity is ND where N is the rotational speed (revolutions per second), μ is kinematic viscosity, and ρ are the density of fluid. Then, the Reynolds number is

$$\text{Re} = \frac{\rho N D^2}{\mu}$$

The system is fully turbulent for values of Re above 10,000.

In fluid dynamics, mixing length theory is a method attempting to describe momentum transfer by turbulence Reynolds stresses within a fluid boundary layer by means of an eddy viscosity. The mixing length is a distance that a fluid parcel will keep its original characteristics before dispersing them into the surrounding fluid.

Laminar mixing, often encountered in fluids with high viscosities, originates from a longitudinal mixing where fluid motion is dominated by linear viscous forces. Fluid particles flow along parallel streamlines and to obtain homogeneity, radial mixing is necessary, which can be achieved through mechanical forces such as using a stirring bar, an impeller, or rocking the base. Thus, turbulent mixing provides the greatest effectiveness as evidenced by the utility of high-speed stirrers.

Manufacturing processes are validated for their outcome in a cGMP environment; as a result, the desired mixing quality, which in most cases is a homogeneous mixture, is obtained by mixing for a certain period of time (with a range) and with a certain force applied (such as rpm, rocking motions per minute, or other such parameters) and in those instances where a demixing may occur, a time for which the mixture remains homogeneous. Generally, for most of the mixing processes encountered in bioprocessing, these parameters are easy to study and validate; the most difficult one is the mixing of culture in a bioreactor, a topic that would receive greater discussion in the next chapter.

Types of mixing

There are several distinct types of mixing systems currently available in bioprocessing where single-use mixing containers are used. These include

1. Stirrer systems
   a. Rotating stirrer
   b. Tumbling stirrer
2. Oscillating systems
   a. Rocker
   b. Vibrating disc
   c. Orbital shaker
   d. Pedal push
3. Peristaltic system
   a. Recirculating pump

The systems using stirrers can have the stirring element either driven magnetically or connected through a sealed shaft. Oscillating types mix by moving the liquid inside a bag (mostly pillow bag types) by rocking them or shaking using mechanical vibrations or ultrasonic vibrations. Generally, the mixing systems that do not involve any mechanical parts inside the bag (either pillow bag or 3D) are preferred to reduce the cost, the risk of damage to bag from rotating devices, the grinding of bag or the stirrer inside the bag; those stirring systems that use a magnetic field provide better sterility compared to those that are magnetically coupled.

Stirring magnetic mixer

- XDM (Xcellerex), 100–1000 L, the XDM Quad Mixing System comprises an integrated magnetic stirrer with compact motor, a bottom-mounted single-use stirrer; the coupling between the motor and the single-use stirrer is magnetic. The square configuration offers enhanced mixing efficiency through a natural baffling effect and compact storage capability. The bottom is slanted to ensure a low residual volume after discharge.
- The Flexel 3D LevMix System for Palletank, 50–1000 L, combines the LevTech levitated impeller licensed by ATMI and the Sartorius Stedim Flexel 3D Bag. It comprises a stainless steel, a cube-shaped container with a door for ease of bag mounting. In addition, it has windows to enable observation of the mixing process, a drive unit for levitating or rotating the stirrer and a single-use bag with a center-mounted magnetic stirrer.
- Magnetic Mixer (ATMI Life Sciences), 30–2000 L
- Jet-Drive (ATMI Life Sciences), 50–200 L
- Mobius (Millipore), 100–500 L
- LevMixer (ATMI Life Sciences): 30–2000 L, ultraclean as it does not produce any residue from mechanical motion, suitable for downstream operations as well.

Stirring mechanical coupling mixer

- SUM (Thermo Fisher Scientific), 50–2000 L, there are two types of magnetic stirrers driven by a stirring plate available for different mixing applications. Not intended for sterile applications and suitable applications include dissolving solid media and/or buffer components prior to sterile filtration.
- Thermo Fisher Scientific HyClone Mixtainer Systems, impeller linked to an overhead drive and is coupled with a sealed bearing assembly, which maintains the integrity of the system. The mixing stirrer is installed off-center. This mixer is intended for powder–liquid and liquid–liquid mixing and has sterile single-use contact surfaces.

Tumbling mixer

- Pad-Drive (ATMI Life Sciences), 25–1000 L, uses a tumbling stirrer mounted from the top, the wand rotates inside an inert polymer sleeve.
- WandMixer (ATMI Life Sciences), 5–200 L, uses a tumbling stirrer whose axle is built into the bag from the top of the bag.

Oscillating mixer

- WAVE (GE Healthcare), 20–1000 L, horizontal oscillation on a rocker; the rocking motion is very efficient in generating waves, and the wave-induced motion in the bag causes large volumes of fluid to move and facilitates dispersion of solids. The optimum operating parameters depend on the combination of the container geometry, bag support, filling volume, rocking angle, rocking rate, and the characteristics of the mixture (solids, foam, etc.).

- HyNetics (HyNetics Corporation), 30–5000 L, vertical oscillation of a disk or septum. The key feature is the mixing disk, which is fabricated from rigid, engineered polymers. Multiple, evenly distributed slots penetrate the disk. The underside of the disk incorporates pie-shaped flaps. These flaps open as the disk moves up from the bottom of the mixing bag on the drive's upstroke, allowing fluid to flow through the disk's slots. The flaps close on the down stroke, forcing the liquid toward the bottom of the vessel and subsequently up the walls of the vessel. The mixing disk, flaps, polymer mixing shaft, and the shaft rolling diaphragm seal, which attaches to the bag film, are single use.
- SALTUS (Meissner), 5–2000 L, vertical oscillation of a disk or septum; based on a vibrating disk with conical orifices. Due to the oscillating movement and the conical orifices, liquid jets develop at the tapered end of the holes. Thus, an axial fluid flow pattern is achieved. The frequency and amplitude of the vibration can be adjusted to provide either vigorous or gentle mixing. The bag is preassembled with the rigid high-density polyethylene (HDPE) vibrating disks and with tubes, filters, and a sampling port, in addition to a single-use sensor plate for pH, dissolved oxygen (DO), and temperature measurement. Due to the frictionless, oscillating movement of the disk, it can be used where the ultraclean environment is required.

Peristaltic mixer

- The Flexel 3D Palletank for recirculation mixing incorporates one or two recirculation loops and can be equipped with Sartorius Mechatronics load cells to facilitate fluid management.

Summary

- A large number of unit operations in bioprocessing involve mixing. Fortunately, these are relatively simple operations that are easily validated.
- The largest mixing operations involve buffer and media preparation that can involve thousands of liters. Since these components are sterilized, likely by filtration, it is not necessary to use any special proprietary mixing system. An off-the-shelf plastic drum with a PE liner and industry-standard mixers would do the job well at a fraction of the cost. It is not necessary to use any proprietary liners as long as the sponsor is able to qualify a supplier; at this stage, the qualification is relatively simple. Since all of the unit operations in a cGMP operation are validated, once a system has been qualified, it can be used repeatedly.
- Open mixing of media and buffer may be provided with a laminar hood in those environments where there is a risk of cross-contamination to reduce any additional burden on filter systems.
- The mixing systems available today are the same as used in single-use bioreactors; in some instances, the platform can be used for both operations.
- While many reputable suppliers have developed highly sophisticated 3D systems, these are not necessary for buffer and media preparations; the cost of 3D bags with built-in stirring systems can be prohibitive.

- The pillow bag bags offer many advantages including the easy of storage as they are horizontally expanded; the wave motion created inside these bags is extremely efficient.
- The power requirements for operating the mixing systems are the lowest in nonstirring types, such as the oscillating mixers; however, this is not a major consideration in the overall cost of mixing.
- In the future, several novel systems and utilization of existing systems would appear in the market, and there are likely to be greater integration of the various steps of bioprocessing.

Connectors and transfers

Single-use components came into use first in the field of connectors and lines as it was difficult to clean them. Unlike hard piping, the flexible tubing incorporated into single-use transfer lines does not require costly and time-consuming cleaning and validation. This allows manufacturers to manage to quickly change process steps or converting over to a new product. This is a key advantage for multiple product facilities in which process requirements change depending on the drug being produced. Innovative manufacturers now incorporate single-use tubing assemblies throughout the bioprocess from seed trains to final fill applications. Additional cost savings result from reduced labor, chemical, water, and energy demands associated with cleaning and validation.

Modern bioprocessing facilities scale up inoculum from a few million cells in several milliliters of culture to production volumes of thousands of liters. This process requires an aseptic transfer at each point along the seed train. Traditional bioprocessing facilities accomplish scale-up using a dedicated series of stainless steel bioreactors linked together with valves and rigid tubing. For these systems, to prevent contamination between production runs, a clean-in-place (CIP) system is designed into each bioreactor, vessel, and piping line to remove any residual materials. These CIP and steam-in-place (SIP) systems require extensive validation testing, and the valves and piping contained in these systems can create additional validation challenges.

Advances in single-use technology allow bioprocess engineers to replace most storage vessels and fixed piping networks with single-use storage systems and tubing assemblies. Single-use eliminates the need for CIP validation for many components and reduces maintenance and capital expense by eliminating expensive vessels, valves, and sanitary piping assemblies.

While total single-use systems are not always possible, there is a transition taking place and often there is a need to connect a single-use system with stainless steel vessels. Single-use media storage systems are routinely manufactured in volumes from 20 to 2500 L. Media storage systems arrive at the bioprocess facility sterilized by gamma irradiation and often are fitted with integrated filters, sampling systems, and connectors. Using a SIP connector like Colder Products' Steam-Thru® Connection (www.colder.com) allows operators to make sterile connections between these presterilized single-use systems and stainless steel bioreactors for aseptic transfer of media.

Similarly, single-use tubing assemblies may be used to transfer inoculum between bioreactors using either a peristaltic pump or headspace pressure. Such transfer lines can reduce the number of reusable valves required for transfer and eliminate problem areas for CIP and SIP validation. Terminating each presterilized transfer line with a single-use SIP connector provides sterility assurance equal to that of traditional fixed piping, at lower capital costs.

As single-use bioreactors are beginning to appear, companies are using them for both seed trains and small-scale production. These systems are connected to a cell culture media storage bag (either by aseptic welding or aseptic connectors such as AseptiQuik®) (www.colder.com) using flexible tubing. Flexible tubing with aseptic connectors is used as transfer lines between each reactor in the process.

There are also instances when liquids are transferred from a higher ISO environment to a lower ISO environment, and assurance is needed that it does not result in cross-contamination; to assure this, a conduit can be installed in the walls connecting the two areas, with the cleaner room having a higher pressure. A pre-sterilized tube is then inserted from the lower ISO class side to the higher ISO class side and connected to the vessels between which the liquid is transferred by a peristaltic pump; upon completion of transfer, the tube is pulled into higher ISO class area and discarded. This method allows connection between downstream and upstream areas without the risk of transferring any contamination to a lower ISO class area such as downstream area.

Tubing

Flexible tubes are an essential part of all single-use systems and are subject to the safety concerns described in an earlier chapter with regard to the leachables and extractables. Several attributes of flexible tubing require evaluation such as their heat resistance, operating temperature range, chemical resistance, color, density, shore hardness, flexibility, elasticity, surface smoothness, mechanical stability, abrasion resistance, gas permeability, visible and UV light sensitivity, composition of layers, weldability, and sealability and sterilizability by gamma radiation or in an autoclave.

All tubes used in bioprocessing conform to USP Class VI classification and FDA 21 CFR 177.2600 and EP 3A Sanitary Standard. For cGMP manufacturing, these are classified as bulk pharmaceuticals. Most common materials used for the tubing include the following:

- Thermoplastic elastomer (C-Flex, PharmaPure, PharMed BPT, SaniPur 60, Advanta Flex) is a pump tubing, highly biocompatible, with easy sealability and low permeability. Thermoplastic tubes like C-Flex and PharMed (both from Saint Gobain, www.biopharm.saint-gobain.com/) are particularly suitable for aseptic biopharmaceutical applications because of moldability, being free of animal components, and while thermoplastic (which makes sealing and welding easy), sterilizability. C-Flex is a unique, patented thermoplastic elastomer specifically designed to meet the critical demands of the medical, pharmaceutical, research, biotech and diagnostics industries. C-Flex biopharmaceutical tubing has been used by many of the world's leading biotech and pharmaceutical processing companies for over 20 years. Each coil of C-Flex tubing is extruded to precise ID, OD, and wall dimensions. All tubing is formulated to meet the standards of the biopharmaceutical industry and is QA tested before leaving our production facility.
- *Features/benefits*
  - Complies with USP 24/NF19, Class VI, FDA, and USDA standards
  - Manufactured under strict GMPs
  - Nonpyrogenic, noncytotoxic, nonhemolytic
  - Chemically resistant to concentrated acids and alkalis

- Significantly less permeable than silicone
- Low platelet adhesion and protein binding
- Ultra-smooth inner bore
- Superior to PVC for many applications, with significantly fewer TOC extractables
- Longer peristaltic pump life
- Heat-sealable, bondable, and formable
- Remains flexible from −50°F to 275°F
- Sterilizable by radiation, ETO, autoclave, or chemicals
- Available in animal-derived component free (ADCF), clear, and opaque formulations
- Lot traceable
- Safer disposal through incineration
- *Typical applications*
- Cell culture media and fermentation
- Diagnostic equipment
- Pharmaceutical, vaccine, and botanical product production
- Pinch valves
- High-purity water
- Reagent dispensing
- Medical fluid/drug delivery
- Dialysis and cardiac bypass
- Peristaltic pump segments
- Sterile filling and dispensing systems
- PharMed BPT biocompatible tubing is ideal for use in peristaltic pumps and cell cultures. PharMed BPT tubing is less permeable to gases and vapors than silicone tubing and is ideal for protecting sensitive cell cultures, fermentation, synthesis, separation, purification, and process monitoring and control systems. PharMed BPT tubing has been formulated to withstand the rigors of peristaltic pumping action while providing the biocompatible fluid surface required in sensitive applications. With its superior flex life characteristics, PharMed tubing simplifies manufacturing processes by reducing production downtime due to pump tubing failure. The excellent wear properties of PharMed BPT translate to reduced erosion of interior tubing walls, improving the overall efficiency of filtering systems.
- *Features/benefits*
    - Outlasts silicone tubing in peristaltic pumps by up to 30 times
    - Low particulate spallation
    - Autoclavable and sterilizable
    - Temperature resistant from −60°F to 275°F
    - Withstands repeated CIP and SIP cleaning and sterilization
    - Meets USP Class VI and FDA criteria
- *Typical applications*
    - Diagnostic test product manufacturing
    - Cell harvest and media process systems
    - Vaccine manufacturing
    - Bioreactor process lines
    - Production filtration and fermentation
    - Sterile filling
    - Shear-sensitive fluid transfer
- *Platinum-cured silicon*: PureFit, SMP/SBP/SVP, Tygon 3350-3370, APST; biocompatible, no leachable additives, economical.
- *Peroxide-cured silicon*: Versilic SPX; biocompatible, no leachables

- *Modified polyolefin*: Tygon LFL (www.tygon.com); chemically resistant, flexible, long-lasting. Tygon 2275
- *Modified polyvinyl chloride*: Tygon LFL, chemical resistant, long-lasting

There are new products introduced routinely, and the read is referred to current information on these products. One of the best sources to meet just about all needs for tubing is Saint-Gobain Company; consult with them first.

Fittings and accessories

Connections between bags or other process stages are done by fittings that come in a wide range of configurations, materials, and sterility. This includes a straight coupler, Y-coupler, T-coupler, cross-coupler, elbow coupler, and barbed plugs. This is necessary to allow ready solutions to the often-complex routine of liquids in a bioprocessing facility. The size of these connectors ranges from 1/16 to 1 in. in most instances; often-incompatible sizes are downgraded or upgraded by interim connectors called reduction couplers that are available for most type of connectors.

The barbed plug is most convenient as it can be easily patented with ties to provide a very secure connection.

The tube-to-tube fittings can serve to change the size and are available in a variety of connection options. Also available are caps to close the tube end with the connector attached to transport the components. Table 9.7 lists various types of tube fittings available.

Clams are used for blocking or regulating flows and come in a variety of types, the most common being the inch clamp for quick starting and stopping flow; ratchet clamps adjust the flow rates. Special clamps with mechanical power transmission (Biovalves, www.biouretech.com), which maintain the contact pressure via a thread arbor, are available for larger tubes with thicker walls.

The BioPure BioValves™ is a precision restriction flow controller and shut-off valve for silicone tube for use in bioprocessing and pharmaceutical manufacturing applications. It is profiled to minimize flow path turbulence and can be used one-handed. Its thread pitch is calibrated to 2 mm per turn, permitting accurate estimation of flow restriction. It is molded from glass reinforced Nylon USP Class VI. These can be repeatedly autoclaved at 134°C for 5 min or irradiated at 60 kGy (6 Mrad) with no detectable weakening.

Pumps

Pumps are used for fluid transfer by creating hydrostatic pressure or by differential pressure; the maximum allowed pressure would be determined by the weakest part of the bioprocess component exposed to the pressure. Peristaltic pumps, syringe pumps, and diaphragm pumps are all currently used to provide single-use pumping

Table 9.7 Types of Tube Fittings

LuerLok	Male and female parts are connected securely via a thread; suitable for small-volume flow rates (hose barb: 1116–3116 in.).
Sanitary fittings	Also known as tri-clamps (TC) genderless, a clamp connects both parts and secures a gasket between them. A connection with conventional sanitary fittings made of stainless steel is also possible (hose barb: 1/4–1 in.).
Quick (dis) connect fittings	Male and female parts are connected securely via a click mechanism. An O-ring fitted to the male part provides the seal. Pressing a button on the female part breaks the connection (hose barb: 3/8–1.5 in. also with sanitary termination).

solutions. All of these are volume displacement pumps and are easy to use and avoid contact with the product; they can, however, produce stress on the tubing especially when the operations are conducted for an extended period of time. It is for this reason that special peristaltic pump tubes are made available by Saint-Gobain company. The stress on the tube may produce particles from erosion of tube and contaminate the fluids being passed through.

High-end peristaltic dispensing pumps have benefited from improved pulsation-free pump head design, a precise drive motor, and a state-of-the-art calibration algorithm. They are exceptionally accurate at microliter fill volumes. Peristaltic pumps that incorporate single-use tubing eliminate cross-contamination and do not require cleaning because the tubing is the only product-contact part. Likewise, the cleaning validation of peristaltic pumps with single-use tubing is significantly easier than for piston pumps. The cost of labor and supplies for writing and executing protocols, cleaning, and documenting the cleaning process is higher for a multiple-use piston-pump filling system. Adjusting the flow speed, and therefore preventing foaming or splashing, is easier for a peristaltic pump than for a piston pump. Operators can also use a ramp-up and ramp-down feature to determine how fast a peristaltic pump reaches its fill speed. This option helps optimize overall fill time and increase throughput.

Many biological drugs are shear-sensitive, and peristaltic pumps protect them by applying low pressure and providing gentle handling. In contrast, a piston pump's valve system generates fast flow through small orifices, potentially damaging biological products. Even valveless piston pumps apply high pressures and high shear factors that could harm a biological product.

On the other hand, viscous products can be problematic for peristaltic pumps. The pumps apply only approximately 1.3 bar of pressure, and their accuracy suffers when they handle products more viscous than 100 cP.

A diaphragm pump is a positive displacement pump that uses a combination of the reciprocating action of a rubber, thermoplastic or teflon diaphragm, and suitable nonreturn check valves to pump fluid. Sometimes, this type of pump is also called a membrane pump. There are three main types of diaphragm pumps:

1. Those in which the diaphragm is sealed with one side in the fluid to be pumped, and the other in the air or hydraulic fluid. The diaphragm is flexed, causing the volume of the pump chamber to increase and decrease. A pair of nonreturn check valves prevents reverse flow of the fluid.
2. Those employing volumetric positive displacement where the prime mover of the diaphragm is electromechanical, working through a crank or geared motor drive. This method flexes the diaphragm through simple mechanical action, and one side of the diaphragm is open to air.
3. Those employing one or more unsealed diaphragms with the fluid to be pumped on both sides. The diaphragm(s) again are flexed, causing the volume to change.

When the volume of a chamber of either type of pump is increased (the diaphragm moving up), the pressure decreases, and fluid is drawn into the chamber. When the chamber pressure later increases from decreased volume (the diaphragm moving down), the fluid previously drawn in is forced out. Finally, the diaphragm moving up once again draws fluid into the chamber, completing the cycle. This action is similar to that of the cylinder in an internal combustion engine.

Diaphragm pumps have good suction lift characteristics; some are low-pressure pumps with low flow rates; others are capable of higher flows rates, dependent on the effective working diameter of the diaphragm and its stroke length. They can handle sludges and slurries with a relatively high amount of grit and solid content. They are suitable for discharge pressure up to 1200 bar and have good dry running characteristics. Like peristaltic pumps, they are low-shear pumps and can handle highly viscous liquids.

Mini diaphragm pumps operate using two opposing floating discs with seats that respond to the diaphragm motion. This process results in a quiet and reliable pumping action. The higher efficiency of the pump is evident in the longer lifer of the motor pump unit. These DC motor diaphragm pumps have the excellent self-priming capability and can be run dry without damage, rated to 160°F (70°C). No metal parts come in contact with materials being pumped; diaphragms and check valves are available in Viton, Santoprene, or Buna-N construction. So, these mini diaphragm pumps are greatly chemically resistant. The mini diaphragm pumps prime within seconds of turning the pump on; prime is maintained by two check valves (one on either side). Separated from the motor, the pump body contains no machinery parts, so the pump can be in dry running condition for a short while. A built-in pressure switch inside the pump can automatically stop the pump, when the pressure reach a setting data.

The single-use diaphragm pump head must be integrated into the transfer line prior to sterilization. As the pump head is totally closed, no other part of the pump comes into contact with the fluid. After the process, the pump head is disposed of, together with the rest of the transfer line. Flow rates of 0.1–4000 L/h can be achieved with single-use diaphragm pumps such as from Quattroflow (www.quattrowflow.com).

Aseptic coupling

One of the most commonly used methods is to connect the tubes or components using sterile connectors under a laminar flow hood; however, this is not always possible specially when the components like single-use bags are large and cannot be moved.

Some connectors require installation in a laminar hood followed by sterilization. These are called steam-in-place connectors. Two aseptic systems go through sealing using these connectors following sterilization by autoclave, radiation, or chemical treatment. Examples of these steam-in-place connectors are from Coler (www.coler.com) and Millipore (www.millipore). The Lynx ST system from Millipore comprises an integrated valve, which can be opened and closed after sterilization of the connection.

Aseptic connectors

Critical to effective single-use processing operations are aseptic connection devices. Pharmaceutical manufacturers typically make about 25,000 aseptic connections each year, with some large manufacturers making as many as 100,000 aseptic connections annually.

The most convenient connectors are aseptic connectors that allow aseptic connections in an open, uncontrolled environment without using a laminar flow hood. Examples of these aseptic connectors include the offering from Pall, Sartorius-Stedim Biotech, GE Healthcare, Millipore, and Saint-Gobain. The aseptic parts on the connector side are sealed with sterile membrane filters or caps. After coupling, the sterile membrane filters must be withdrawn, and both parts have to be clamped or fixed. These connectors are secure and recommended to save time but offer an expensive choice and at times limitation of sizes of tubes that can be connected. These connectors are also used as aseptic ports in bioreactors.

One of the earliest entries in the field of aseptic filters was Pall Kleenpak Sterile Connector.

AseptiQuik Connectors (www.colder.com) provide quick and easy sterile connections, even in nonsterile environments. AseptiQuik's "CLICK-PULL-TWIST" design enables users to transfer media easily with less risk of operator error. The connector's robust design provides reliable performance without the need for clamps, fixtures, or tube welders. Biopharmaceutical manufacturers can make sterile connections with the quality and market availability they expect from the leader in single-use connection technology.

The Opta® SFT Sterile Connector (www.sartorius-stedim.com) is a single-use device, composed of pre-sterilized female and male coupling body, that allows a sterile connection in biopharmaceutical manufacturing processes. The Opta SFT-I Connector is supplied with Flexel 3D, Flexboy bags and transfer sets as part of integrated Sartorius Stedim Biotech Fluid Management assemblies. Opta SFT-I is available with a 1/4, 3/8 and 1/2 in. Hosebarb. The Opta SFT-D is available as an individual device for end-user assembly with TPE tubing and autoclave sterilization. They are quick, easy to use, and are backed by extensive validation work as well as 100% in-house integrity testing.

Pall Corporation (www.pall.com) is expanding its line of Kleenpak Aseptic Connectors with two new sizes, 1/4 and 3/8 in. The new sizes enable vaccine manufacturers to apply the safety and efficiency benefits of instant aseptic connections throughout more of their single-use operations to help speed time to market and comply with good manufacturing practices. Pall revolutionized the aseptic connection process by shortening the time needed for connection from 15 min to a few seconds when it introduced its 1/2 in. Kleenpak Connector. The addition of the two new Kleenpak Connector sizes increases flexibility to implement aseptic connections in more applications to improve single-use processing efficiency. This is especially important to complex vaccine production, which often requires a greater number of connection steps. The Kleenpak Connector is easy-to-use and projects an audible snap to signify that a sterile connection has been established.

ReadyMate Single-use Aseptic Connector (DAC) (www.gelifesciences.com) provides connections for high-fluid throughput and offers a secure, simple, and economical connection for upstream and downstream applications. DAC connectors can be autoclaved, or gamma radiated, and can be part of a sterile circuit. The connectors can be used to connect unit operations and assemblies. DAC connectors and their components are manufactured in compliance with the current Good Manufacturing Practices of the FDA and ISO 9000-2000. ReadyMate is a genderless, inter-size connectable single-use aseptic connector. There are four Hose Barb Sizes (3/4, 1/2, 3/8, and 1/4 in.), mini TC, and TC that all interphase. It has a genderless design, user-friendly sanitary coupling, easy to use with Tip'n'latch, USP Class VI, and sterilized by radiation or autoclave. The main advantages of using the Bio Quate connector include: simple set-up, rapid connection, direct connection of different tube sizes, direct connection to different tube materials, aseptic on-site manifold fabrication, large, smooth inside bore, no power, calibration or service required, no capital equipment to purchase. These connectors are supplied by Bioquate (www.bioquate.com).

Welding

When it is possible to use a thermoplastic tube, welding offers an easy, inexpensive, and very secure solution. Examples of thermosplastic tubes include C-Flex, PharMed, and Bioprene. To start, both thermoplastic tubes must be aseptic and should have the same dimensions (inner diameter and OD), and their ends capped. The tubes are placed parallel in the opposite direction during a heated blade cuts through them and seals simultaneously. Preheating of the blade is necessary both to achieve the welding temperature and to sterilize and depyrogenize the blade itself prior to the welding process. The depyrogenize procedure normally lasts 30 s at 250°C or 3 s at 320°C. After being cut, the tubes are moved against each other so that the ends of each tube, which are connected to the aseptic systems, are positioned directly opposite each other on either side of the blade. A welding cycle can be between 1 and 4 min, depending on the material and the diameter of the tubes. The main welding systems available today include Sterile Tube Fuser (GE Healthcare), BioWelder (Sartorius-Stedim), Aseptic Sterile Welder 3960 (www. Sebra.com), TSCD (Terumo), and SCD 11B (Terumo) (www.terumotransfusin. com), which supplies its equipment mainly to blood transfusion industry. Both GE Healthcare and Sartorius lead the installations in the bioprocessing industry.

The Hot Lips Tube Sealer (www.gelifesciences.com) is a portable device used to thermally seal thermoplastic tubing for the transport and setup of inoculums, culture, media, and buffers. The seal forms a tamperproof and leakproof closure. Preprogrammed for a wide range of tubing types, diameters, and wall thicknesses, a single button initiates the sealing operation. The instrument is self-calibrating, and a microprocessor-controlled motor ensures repeatable performance without the need for tubing adaptors.

The Sterile Tube Fuser (www.gelifesciences.com) is an automated device for welding together a wide range of tubing types intended for aseptic operation. Operated via a single push-button operator interface, it connects tubing between sterile containers, Cellbag bioreactors, and process equipment for the aseptic transfer of large volumes of fluids such as inoculum, media, buffers, and process intermediates.

Aseptic transfer systems

Moving product across clean rooms may involve, at times, long distance; while transfer tubings between upstream and downstream areas and pass through autoclave are common, larger volume transfer systems like those offered by Sartorius Stedim, ATMI Life Sciences, Getinge, and LaCalhene, which essentially constitute double door systems using single-use containers. In using these systems, the main, reusable port is always permanently fixed in the separating wall (in a clean room or isolator) and represents the containment barrier. The second connecting part is an integral part of the single-use container, which stores or conducts the components, fluids, and powders to be transferred. Both the connecting parts and the reusable containers and transfer systems can be coupled to the main port. After coupling, the ports are opened from inside the cleaner area, and the transfer is started. The single-use container is normally the package for the fluid and the sterile barrier for the fluid conduction.

> Biosafe® Aseptic Transfer Equipment: The Biosafe range of Aseptic Transfer Ports offers reliable and easy-to-use solutions for the secure transfer of components, fluids, and powders while maintaining the integrity of the critical area—isolators, RABS, and cleanrooms.
>
> Biosafe Aseptic Transfer Single-Use Bag: A complete range of Biosafe Aseptic Transfer Bags either gamma sterile or autoclavable and Biosafe RAFT System are designed to best fit your requirements for aseptic transfer of components into clean rooms, isolators, or RABS and for contained transfer of potent powders.

Biosafe® RAFT System: The Biosafe RAFT system provides easy-to-use and reliable through-the-wall aseptic transfer of liquid between clean rooms of different environmental classification while ensuring a total confinement.

SART™ System: The SART System is designed to allow aseptic liquid transfer between two areas with different containment classifications.

Special bags have therefore been developed, for example, the Biosafe RAFT system by Sartorius Stedim Biotech, allowing aseptic coupling to larger fluid containers and ports in addition to fluid conduction.

Tube sealers

When disconnecting an aseptic connection, the ends must be capped with aseptic caps and this can be done under laminar hood or by using tube sealers, the examples of which include offerings from PDC (www.pdcbiz.com), Saint-Gobain (www.saint-gobain.com), Sartorius-Stedim (www.sartorius-stedim.com), GE Healthcare (www.gelifesciences.com), Terumo (www.terumotransfusion.com), and SEBRA (www.sebra.com). Most of these sealers can seal from 0.25 to about 1.5 in. tubes and take from 1 to 4 min to complete the seal. Most operate on the electrical heating element, but electrical and radio frequencies are also used for sealing tubes. There is no need for using a laminar flow hood for these operations. In most instances, applying a crimper in two places and cutting the tube between the crimps offers the cheapest solution.

Sampling

Sampling is a routine during manufacturing to assure compliance by obtaining these in process parameters like pH, DO, OD, pCO_2, etc. Most single-use systems have one or more integrated sampling lines, which are partly equipped with special sampling valves, sampling manifolds, or special sampling systems. A popular single-use sample valve is the Clave connector from ICU-Medical (www.icumed.com), which is also used in intravascular catheters for medical applications. It allows a sample to be taken with a LuerLok syringe. A dynamic seal inside the valve guarantees that the sample is not taken until the syringe is connected, thereby ensuring the sample only comes in contact with the inner, aseptic parts of the valve. However, the samples drawn does not remain sterile.

Manifolds consisting of sampling bags, sampling flasks, or syringes are appropriate for taking aseptic samples in single-use systems. These manifolds can be connected to the systems via aseptic connectors or tube welding. Sampling manifolds allow multiple sampling for quality purposes over a given period of time. The main feature of the manifold is that the number of manipulations in a process is significantly reduced. The manifold systems are delivered ready for process use, preassembled, and sterile. To use them, only one connection has to be made to allow several bags to be filled, thus

Also used for sampling are a manifold system where sample containers of a manifold are arranged in parallel whereby the last one is used as a waste container. Through using Y-, T-, or X-hose barbs and tube clamps, the initial flow, and the subsequent sample are guided to the appropriate containers. Steam-in-place connections, of course, also allow the connection of manifold systems to conventional stainless steel processing equipment (www.sartorius-stedim.com).

Conclusion

The complexity of bioprocessing makes it difficult to design systems that would have any weak links; contamination is indeed the most significant risk that requires that all connectors, tubing, and implements joining various steps of a process and performing sampling remain patent. Single-use connectors and tubing were one of the first components that went single use. Still in hard-walled systems, steam-in-place systems are in use only because there is steam for CIP/SIP operations. Even then, the risk of contamination remains. Since much of the single-use technology in these applications has come from the biomedical field, the device industry had always been ahead of the regulatory requirements. Biocompatibility issue has long been resolved, and vendors are able to provide detailed information on their devices that might be needed by the regulatory agencies. Since the manufacturing of these devices is complex, it is unlikely for a user to request custom devices; however, the diversity of choices available today are enough to modify any system that would be able to use an off-the-shelf item. As before, I have always emphasized the important of off-the-shelf item over custom designs.

The tube connectors and sealers are a newer entry as single-use bags for mixing and bioreactors have become more popular; still there is a limited choice of suppliers of equipment, mainly GE and Sartorius. The cost of this equipment is still high but then the choice comes down to using expensive aseptic connectors. Generally, if a good choice of aseptic connector is available, that should be preferred over tube connectors since it is always possible to make a poor connection using the heat-activated systems; also, the use of aseptic connectors allows connecting tubes that may not be thermolabile.

Filtration

Except for steel meshes in bulk manufacturing of nonsterile dosage forms, filters are rarely reused in the pharmaceutical industry. They take varied forms from muslin cloth to paper filters to membrane cartridges. Single-use filter devices in biological manufacturing were the earliest changes that went single use mainly because of the problems and cleaning them, and also the cost of these parts have always been reasonable.

There is a multitude of filter designs and mechanisms utilized by the biopharmaceutical industry. Prefilters are commonly pleated or wound filter fleeces manufactured from meltblown random fiber matrices. These filters are used to remove a high contaminant content within the fluid. Prefilters have a large band of retention ratings and can be optimized to all necessary applications. The most common application for prefilters is to protect membrane filters, which are tighter and more selective than prefilters. Membrane filters are used to polish or sterilize fluids. These filters need to be integrity testable to assess whether or not they meet the performance criteria. Cross-flow filtration can be utilized with micro- or ultrafiltration membranes. The fluid sweeps over the membrane layer and, therefore, keeps it unblocked. This mode of filtration also allows diafiltration or concentration of fluid streams. Nanofilters are commonly used as viral removal filters. The most common retention rating of these filters is 20 or 50 nm.

> *Dead-end filtration*: Dead-end filtration operates on the principle of passing a fluid feed stream through a filter device by means of a pressure drop, usually applied by either a pump or compressed gas pressure before the filter device. All contaminants larger in size than the pore size of the filter media are retained by the filter material and will finally cause a filter

blockage by plugging its channels or pores. The dead-end filtration is one of the simplest modes of operation for filters and hence requires minimum accessories such as tubing/piping, tanks, controls, and footprint.

Dead-end filters described using microporous membranes manufactured out of synthetic polymers such as polyethersulfonate, polyamide, cyanoacrylate, and polyvinylidene fluoride (PVDF) are used extensively for sterile processing. They are used for adding media to the bioreactor, bioburden reduction in cell harvest clarification, chromatography column protection, and final filtration of the purified bulk drug substance. These filters often come attached to single-use bags and are gamma sterilized.

The most common dead-end filtration devices are filter cartridges (for the reusable process) or capsules for fully single-use processes. They are used in wide ranging applications as pre- and sterilizing grade filters in upstream as well as downstream applications including media filtration; intermediate product pool filtrations; and in form, fill, and finish for the sterilization of drug substance. Dead-end filter devices are also used for sterilizing grade air and vent filtration, for cell harvest and clarification, and, most recently, for viral clearance and membrane chromatography.

Cross-flow filtration: In chemical engineering, biochemical engineering and protein purification, cross-flow filtration (also known as tangential flow filtration) is a type of filtration (a particular unit operation). Cross-flow filtration is different from dead-end filtration in which the feed is passed through a membrane or bed, the solids being trapped in the filter and the filtrate being released at the other end. Cross-flow filtration gets its name because the majority of the feed flow travels tangentially across the surface of the filter, rather than into the filter. The principal advantage of this is that the filter cake (which can blind the filter) is substantially washed away during the filtration process, increasing the length of time that a filter unit can be operational. It can be a continuous process, unlike batch-wise dead-end filtration. This type of filtration is typically selected for feeds containing a high proportion of small particle size solids (where the permeate is of most value) because solid material can quickly block (blind) the filter surface with dead-end filtration. Industrial examples of this include the extraction of soluble antibiotics from fermentation liquors.

Since in cross-flow filtration the feed stream is led across or tangential to the filter material surface and is recycled continuously around the filter, this requires more complex equipment and controls, but the retentate is allowed to pass through the filter device multiple times by recirculation. Thus, it is possible to perform concentration or buffer exchange processes. Additionally, for liquids with heavy load of suspended particles, the filter is kept from clogging as the turbulent flow of the feed across the filter removes deposited materials, something that is not possible in dead-end filtration.

Filtration media

The filter media generally comprises layers of solid materials in a network or mesh with voids, pores, and channels that allow passage of liquid but retain larger particles, larger than the size of the openings, which may be in nanometer dimension.

Depth filters use their entire depth to retain particulate on the basis of sieving compounded by adsorption effects, unlike retentive filters where the filtered material is concentrated on the surface. The depth filter media dominate pre-filtration and clarification applications because of the high solid mass that is generally required to be removed at this stage.

Table 9.8 Pall Filter Offering (www.pall.com)

Filter	Use	Type
Seitz® K-Series Depth Filter Sheets	Active pharmaceutical ingredients, clarification and prefiltration, plasma fractionation	Sheet filters and sheet filter modules
Seitz® K-Series Depth Filter Sheets	Beer, bottled water, dairy, food, soft drinks, spirits, wine	Sheet filters and sheet filter modules
Seitz® P-Series Depth Filter Sheets	Biotechnology, clarification, and prefiltration, plasma fractionation	Sheet filters and sheet filter modules
Seitz® T-Series Depth Filter Sheets	Pre-filtration, production	Sheet filters and sheet filter modules
Seitz® Z-Series Depth Filter Sheets	Active pharmaceutical ingredients, clarification and prefiltration	Sheet filters and sheet filter modules
Supracap™ 100 Depth Filter Capsules	Active pharmaceutical ingredients, biotechnology, cell separation, clarification and prefiltration, plasma fractionation, scale up/process development, vaccines	Capsules, sheet filters, and sheet filter modules
SUPRAcap™ 200 Encapsulated Depth Filter Modules	Active pharmaceutical ingredients, biotechnology, cell separation, clarification and prefiltration, plasma fractionation, scale up/process development, vaccines	Capsules, sheet filters, and sheet filter modules
Supracap™ 60 Depth Filter Capsules	Active pharmaceutical ingredients, biotechnology, cell separation, clarification and prefiltration, plasma fractionation, scale up/process development, vaccines	Capsules, sheet filters, and sheet filter modules
SUPRAdisc™ Depth Filter Modules	Active pharmaceutical ingredients, biotechnology, cell separation, clarification and prefiltration, plasma fractionation, scale up/process development, vaccines	Sheet filters and sheet filter modules
SUPRAdisc™ Depth Filter Modules	BioFuels and biotechnology, chemicals	Sheet filters and sheet filter modules
SUPRAdisc™ HP Depth Filter Modules	Active pharmaceutical ingredients, biotechnology, cell separation, clarification and prefiltration, plasma fractionation, scale up/process development, vaccines	Sheet filters and sheet filter modules
SUPRAdisc™ II Depth Filter Modules	Beer, food, juice, spirits, wine	Sheet filters and sheet filter modules
SUPRAdisc™ II Modules	Active pharmaceutical ingredients, biotechnology, clarification and prefiltration, plasma fractionation, scale up/process development, vaccines	Sheet filters and sheet filter modules
SUPRApak™ Depth Filter Modules	Beer, spirits	Sheet filters and sheet filter modules
SUPRApak™ SW Series Modules	Beer, beer—corporate brewers, beer—microbreweries, food, soft drinks, spirits	Sheet filters and sheet filter modules
T-Series Depth Filter Sheets	Active pharmaceutical ingredients, biotechnology, cell separation, clarification and prefiltration, plasma fractionation, scale up/process development, vaccines	Sheet filters and sheet filter modules
T-Series Depth Filter Sheets	BioFuels and biotechnology, chemicals	Sheet filters and sheet filter modules
T-Series Membrane Cassettes	Active pharmaceutical ingredients, clarification and prefiltration	Sheet filters and sheet filter modules

Sieving or size exclusion have more uniform pore sizes throughout the bed and are thus used to remove selective size of particles; these filters, mostly membrane types, are ideal as sterilizing filters, for example, the commonly used 0.22 μm filter to sterilize liquid. While the main mechanism of their operation is sieving, the chemical nature of these membranes makes them a good base to adsorb organic substances.

Depth filter are made of fibers that are spread out on a substrate to make a mesh just like making paper; special additives activated carbon, ceramic fibers, and other such specific components are embedded with the help of a binder to form the filter.

Sheet filters are also made like paper using milled cellulose fibers and may contain diatomaceous earth or perlite along with a binder to strengthen the filter.

One of the world's largest suppliers of these filters in biopharmaceutical manufacturing is Pall (Table 9.8).

Because of their thickness, the sheet filters provide a slow filtration option yet are extensively used for pre-filtration.

Microglass fibers are also used to filter media; these are nonwoven spun fibers of borosilicate glass whose web is strengthened by a binder allowing for a 3D structure of asymmetric voids as small as 0.2 µm to act as sterilizing filters.

Polypropylene and polyester fibers are also used for spinning from polymer melt and bonded by the polymer itself, giving better chemical compatibility as no binder is added to them. These are always the preferred filters over polyamide and cellulose filters. The convention method of their manufacture leaves pore sizes 20–50 µm, making them unsuitable for sterile filtration; specially blown process is used to reduce the pore size in the range of 5–50 µm; further spinning is needed to reduce the size further to 1–10 µm range.

Polymer membranes

The history of membrane filters goes back to hundreds of years (Table 9.9).

The main advances in membrane technology (1960–1980) began in 1960 with the invention of the first asymmetric integrally skinned cellulose acetate RO membrane. This development simulated both commercial and academic interest, first in desalination by reverse osmosis, and then in other membrane application and processes. During this period, significant progress was made in virtually every phase of membrane technology: applications, research tools, membrane formation processes, chemical and physical structures, configurations and packaging.

Two basic morphology of hollow fiber membrane are *isotropic* and *anisotropic* (Figure 9.2). Membrane separation is achieved by using these morphologies.

Table 9.9 History of Membrane Filters

Year	Important development
1748	Abbe Nollet—water diffuses from dilute to concentrated solution
1846	The first synthetic (or semisynthetic) polymer studied by Schoenbein and produced commercially in 1869.
1855	Fick employed cellulose nitrate membrane in his classic study *Ueber Diffusion.*
1866	Fick, Traube, artificial membranes (nitrocellulose)
1907	Bechhold, pore size control, "ultrafiltration."
1927	Sartorius Company, membranes available commercially
1945	German scientists, methods for bacterial culturing
1957	USPH, officially accepts membrane procedure
1958	Sourirajan, first success in desalinating water

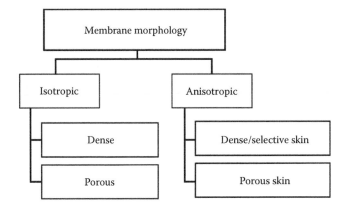

Figure 9.2 Basic membrane morphology.

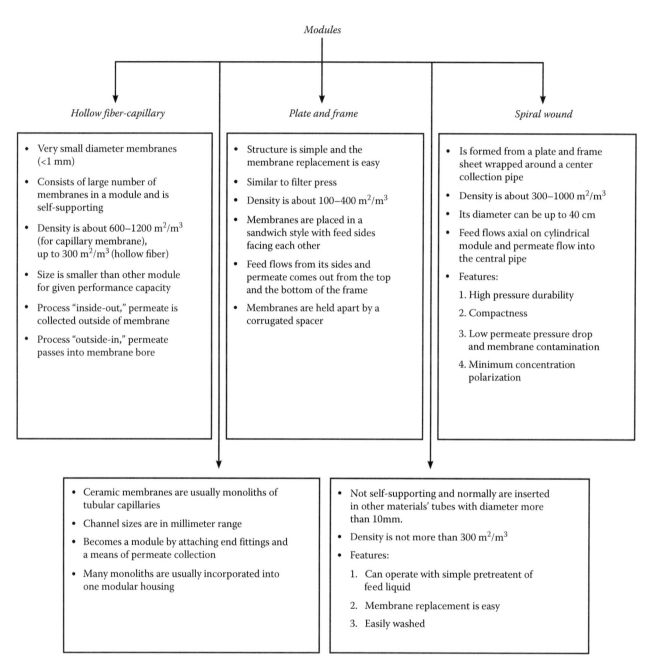

Figure 9.3 Membrane configuration.

The anisotropic configuration is of special value. In the early 1960s, the development of anisotropic membranes exhibiting a dense, ultrathin skin on a porous structure provided a momentum to the progress of membrane separation technology. The semipermeability of the porous morphology is based essentially on the spatial cross-section of the permeating species, that is, small molecules exhibit a higher permeability rate through the fiber wall. The anisotropic morphology of the dense membrane, which exhibit the dense skin, is obtained through the solution-diffusion mechanism. The permeation species chemically interacts with the polymer matrix and selectively dissolves in it, resulting in diffusive mass transport along the chemical potential gradient, as demonstrated in the pervaporation process.

The type of the membrane configuration is given in Figure 9.3.

Hollow fiber is one of the most popular membranes used in industries. Its several beneficial features make it attractive for those industries. Among them are

- *Modest energy requirement*: In the hollow fiber filtration process, no phase change is involved. Consequently, there is need no latent heat. This makes the hollow fiber membrane have the potential to replace some unit operation that consumes heat, such as distillation or evaporation column.
- *No waste products*: Since the basic principal of hollow fiber is filtration, it does not create any waste from its operation except the unwanted component in the feed stream. This can help to decrease the cost of operation to handle the waste.
- *Large surface per unit volume*: Hollow fiber has a large membrane surface per module volume. Hence, the size of hollow fiber is smaller than another type of membrane but can give a higher performance.
- *Flexible*: Hollow fiber is a flexible membrane; it can carry out the filtration by two ways, either "inside-out" or "outside-in."
- *Low operation cost*: Hollow fiber need low operation cost compared to other types of unit operation.

However, it also has some disadvantages that lead to its application constraints. Among the disadvantages are

- *Membrane fouling*: Membrane fouling of hollow fiber is more frequent than another membrane due to its configuration. The contaminated feed will increase the rate of membrane fouling, especially for hollow fiber.
- *Expensive*: Hollow fiber is more expensive than other membranes that are available in the market. It is because of its fabrication method that the expense is higher than the other membranes.
- *Lack of research*: Hollow fiber is a new technology, and, so far, research done on it is less compared to other types of membranes. Hence, more research will be done on it in future because of its potential.

Hollow fiber made of the polymer cannot use corrosive substances and high temperature conditions. Various types of membrane processes can be found in almost all of the literature references. In this text, we will confine ourselves to the few membrane processes that we will encounter in the further discussion of the industrial applications.

There is considerable confusion in the open literature as to the distinction between few membrane separation processes, that is, the microfiltration (MF), ultrafiltration (UF), and reverse osmosis (RO). Occasionally, one will see it referred to by other names such as "hyperfiltration (HF)." In order to distinguish these separation processes clearly, the RO has the separation range of 0.0001–0.001 μm (i.e., 1–10 Å) or <300 mol wt. RO is a liquid-driven membrane process, with the RO membranes capable of passing water while rejecting microsolutes, such as salts or low-molecular-weight organics (<1000 Da). Pressure driving force (1–10 MPa) is needed to overcome the force of osmosis that cause the water to flow from dilute permeate to concentrated feed. The principal use of this membrane process is desalination, which shows its great advantage over the conventional technique of desalination, that is, ion exchange.

The biotechnology industry, which originated in the late 1970s, has become one of the emerging industries that draws the attention of the world, especially with the emergence of the genetic engineering as a means of producing medically important proteins, during the 1980s. Two of the major interesting applications of membrane technology in the biotechnology industry is the separation and purification of the biochemical product, often known as *downstream processing*; and the membrane bioreactor, which was developed for the transformation of certain substrates by enzymes (i.e., biological catalysts).

Since its introduction in the 1970s, membrane bioreactor has generated a lot of attention over other conventional production processes due to the possibility of a high enzyme density and hence high space–time yields. Whereas downstream processing is usually based on discontinuously operated microfiltration, membrane bioreactors are operated continuously and are equipped with UF membranes. Two types of bioreactor designs are possible: dissolved enzymes (as in used with the production of L-alanine from pyruvate) or immobilized enzymes membrane.

Membrane science began emerging as an independent technology only in the mid-1970s, and its engineering concepts still are being defined. Many developments that initially evolved from government-sponsored fundamental studies are now successfully gaining the interest of the industries as membrane separation has emerged as a feasible technology.

Today, membrane polymers used in pharmaceutical processes include PVDF, expanded polytetrafluorethylene (ePTFE), polyethersulfone (PESu), polyamide (PA), cellulose acetate (CA), regenerated cellulose (RC), and mixed cellulose ester (MCE), a mixture of cellulose nitrate (CN) and CA. Membranes provide the highest retention efficiency or the smallest pore sizes; the microfiltration membrane pore sizes range from 10 to 0.1 µm; the ultrafiltration membranes have pore sizes from 0.1 µm to a few nanometers, making them even suitable for virus filtration. The nanofilters are in the range of 50 nm and smaller and rated for the molecular weight that can be retained in kilodaltons. Microfiltration membranes are suitable for prefiltration, while others for sterilization.

Microfiltration cross-flow   The traditional process for adopting cross-flow filtration involves the use of dense ultrafiltration membranes for the purpose of performing a concentration and buffer exchange operation on the target molecule pre- and postcolumn chromatography. When the ultrafiltration media is replaced with a microporous microfiltration membrane, one has the option of performing a wide range of separations involving larger species while usually operating at much lower pressures.

As in the case of the ultrafiltration membrane applications, a suitable cross-flow microfiltration process may involve the concentration and "washing" of a particulate material. For example, as an alternative to the use of centrifugation, one could aseptically recover cells from a cell culture process and proceed to concentrate and wash the cells to remove contaminating macrosolutes and replace the fluid with a solution suitable for freezing the cells. This same type of sequence could be used to aseptically process liposomes and other drug delivery emulsions.

A second widely practiced use of cross-flow microfiltration is also an alternative to centrifugation or normal-flow filtration for the clarification of the target molecule or virus from cells and cell debris. Highly permeable microfiltration membranes allow a large target molecule to pass through the filter while providing complete retention of the contaminating particulates in a single scaleable step.

The filter media needs a device to use; earlier devices were stainless steel holders for flat discs; pleated inserts to provide larger surface area followed this and these were enclosed in plastic containers. While the stainless steel devices could be autoclaved, the plastic components needed gamma radiation to sterilize.

Hollow fiber devices contain bundles of numerous hollow fiber membranes, and they are placed coaxially into a pipe-like perforated cage and sealed using a resin so that either one or both ends of the module are open to give access to the feed or allow exit to the filtrate or permeate. They can be operated from both sides, the tube or the shell side.

Hollow fiber devices are established, for example, in virus clearance by Asahi Kasei.

Medical Planova® Planova filters, the world's first filters designed specifically for virus removal, significantly enhance virus safety in biotherapeutic drug products, such as biopharmaceuticals and plasma derivatives. They exhibit unparalleled performance in removing viruses, ranging from human immunodeficiency virus (HIV) to parvovirus B19 while providing maximum product recovery.

BioOptimal MF-SL™ Designed specifically for use in cell culture clarification applications, BioOptimal MF-SL filters enable biopharmaceutical manufacturers to improve the efficiency and effectiveness of their protein harvest step.

TechniKrom™ Asahi Kasei Bioprocess, Inc. provides technologically advanced bioprocess equipment, products, and services to the biopharmaceutical, pharmaceutical, veterinary, and nutraceutical industries that permit the lowest cost of production and highest innate product quality, especially in cGMP-regulated environments. We help our clients implement true Manufacturing Science in their facilities to enable their achievement of these goals.

GE Healthcare Process scale hollow fiber cartridges offered by GE Healthcare are provided in eight basic configurations covering a membrane area range of 0.92–28 m² (9.9–300 ft²) depending on the fiber internal diameter.

Spectrum Spectrum single-use CellFlo hollow fiber membranes are specially designed for the gentle and efficient separation of whole cells in microfiltration, bioreactor perfusion, and culture harvest applications. Much like our own circulatory system, CellFlo combines the advantages of tangential flow microfiltration (0.2 and 0.5 μm pores sizes) with larger hollow fiber flow channels (1 mm ID) to provide gentler efficacious microseparations without the risk of cell lysis. Other cell separation technologies have a higher risk of lysis, resulting in ruptured cells and culture harvests contaminated with intracellular macromolecules. Perfectly suited for continuous cell perfusion and bacterial fermentation, CellFlo membranes isolate secreted proteins while eliminating spent media containing metabolic wastes and drawing in fresh nutrient-rich media. Consequently, CellFlo enables cultures to grow to a higher cell density with higher viability providing as much as a tenfold increase in daily production of secreted proteins.

Whether performing a cell perfusion or conducting a simple microfiltration, single-use CellFlo modules can either be autoclaved or purchased irradiated (irr) for quick sterile assembly. Spectrum offers single-use CellFlo membrane modules in the full range of MiniKros and KrosFlo sizes and surface areas for processing volumes ranging from 500 mL to 1000 L. All CellFlo membranes have a 1 mm fiber inner diameter, available in 0.2 and 0.5 μm pore sizes.

MaxCell and Spectrum Laboratories CellFlo® and KrosFlo® module families.

Millipore has developed a derivative hollow fiber system, Pellicon, and how it compares with standard hollow fiber filters, is shown in Table 9.10.

Table 9.10 Hollow Fiber vs. Pellicon 2 Summary

Feature	Hollow Fiber	Pellicon 2
Robustness and reliability	Fibers are prone to stress failure	Very robust
Pressure capability	Low	High
Membrane choices	1. Polyethersulfone	1. Polyethersulfone 2. Regenerated cellulose 3. PVDF (for MICRO filtration)
Flow rate required to operate TFF processes	High flow rate resulting in • High-energy consumption • Large piping • Compromised concentration ratio • Increased demand for floor space	Low flow rate resulting in • Low-energy consumption • Small piping • High concentration ratio • Compact system size
Linear scale-up	Compromised by differences in the length of the flow channel in laboratory vs. process scale cartridges	Identical length of flow channel length in all cartridge sizes to facilitate predictable scaling results
Consistency of retention relative to retentate channel design	Compromised consistency due to • Open flow channels. No internal mixing	High consistency due to • Build-in static mixer efficient internal mixing
Consistency of retention relative to the retentate flow channel length	Compromised consistency due to • Long flow channels	High consistency due to • Very short flow channels

Flat sheets can be arranged in a stack in place of pleated or hollow fiber filter media. The sheets are sealed such that they leave channels open to allow feed to pass through.

Conclusion

From prefiltration to remove large sediments to harvesting bacterial cultures to removing viruses in the final stages of the biological drug manufacturing, filters play one of the most important roles. There are dozens of companies specializing in specific filtration processes and giants dominate the market with their Millipore Pod® system, the Pall Stax® system, or Sartoclear® XL Drums from Sartorius Stedim Biotech. They also keep bringing newer filters, housings, arrangement, and recently in an integrated setup in their single-use factories concept. The user is highly advised to consult the current literature on the suitability of the type of filters used. More often, the suppliers would be more than willing to understand the process and make recommendations. Obviously, cost is a serious concern, but when it is realized that in a cGMP manufacturing environment having qualified a particular filter or a housing for a unit process, it is not easy to switch over to another type of filter or another supplier, the selection of these filter becomes a serious concern.

Controls

> The goal of PAT is to understand and control the manufacturing process, which is consistent with our current drug quality system: *quality cannot be tested into products; it should be built-in or should be by design.*

> **Food and Drug Administration, 2009**

According to the FDA guidance for industry, process analytical technology (PAT) is intended to support innovation and efficiency in pharmaceutical development. PAT is a system for designing, analyzing, and controlling manufacturing through timely measurements (i.e., during processing) of critical quality and performance attributes of raw and in-process materials and processes with the goal of ensuring final product quality. It is important to note that the term analytical in PAT is viewed broadly to include chemical, physical, microbiological, mathematical, and risk analysis conducted in an integrated manner.

To fulfill process requirements, single-use sensors, which are either integrated with the single-use bioreactor or included in the cover and are disposed of with the bioreactor, are required. They provide a continuous signal and allow information about the status of a cell culture to be gathered at any time. The traditional batch analysis, such as HPLC, electrochemistry, and wet chemical analysis in place of single-use sensors, would increase the risk of contamination.

Since the single-use bioreactors are new to the industry, the first attempt to monitor the product in the bioreactor was to use the traditional biosensors used in hard-walled systems to measure bioreactor temperature, dissolved oxygen (DO), pH, conductivity, and osmolality. These probes must first be sterilized (via autoclaving) and then attached to penetration adapter fittings that are welded into bioreactor bags. Not surprisingly, this is a labor-intensive and time-consuming process that has the potential to compromise the integrity and sterility of single-use bioreactor bags and has been largely discarded in favor of truly single-use sensors. Critical process parameters that are often monitored include pressure, pH, dissolved oxygen, conductivity, UV absorbance, flow, and turbidity. The packages that contain the traditional technologies for monitoring these parameters are not usually compatible with or effective when integrated into single-use assemblies for many reasons: cost, cross-contamination, inability to maintain a closed system, and system incompatibility with gamma irradiation.

The practice of integrating bags, tubing, and filters into preassembled, ready-to-use bioprocess solutions is optimized if noninvasive sensing of critical process parameters is part of the package instead of using sensors that may require sterilization and cleaning validation, the core processes that are obviated in the use of single-use bioreactors.

Even though these obstacles do not always preclude the use of traditional measurement technologies, single-use solutions for monitoring process parameters eliminate the need for equipment cleaning and autoclaving small parts, reduce the risk and cost involved with making process connections, and may be more cost effective than tracking and maintaining traditional technologies. For example, a sanitary, autoclavable pressure transducer that is qualified for a certain number of autoclave cycles and requires recalibration may be more expensive to use versus a single-use pressure sensor.

Adoption of single-use sensors requires a keen understanding of their need and utilization. Their suitability would be determined by their material properties, sensor manufacturing, process compatibility, performance requirements, control system integration, compatibility with treatments before use, and regulatory requirements.

Several companies, including Finesse and Fluorometrix (recently acquired by Sartorius Stedim), have created single-use, membrane biosensors that can be added to or directly incorporated (during manufacturing) into single-use bioreactor bags.

There are two options in using single-use sensors; one where the sensors are placed in situ in contact with the liquid, and the other where the external sensors contact the medium either optically (ex situ) or via a sterile (and single-use) sample removal system (on line). Single-use sensors must be sterilizable if they come in contact with media, these must also be cost-effective and reliable. Better designs use inexpensive sensing elements can be located inside a single-use reactor and combined with reusable (and more expensive) analytical equipment outside the reactor. Inexpensive, single-use sensors can also be placed on transistors and placed either in the headspace, inlet, outlet, or into the cultivation broth for liquid-phase analysis (temperature, pH, pO_2). These can also be optical sensors, which allow noninvasive monitoring through a transparent window.

Sampling systems

Continuous sampling from a bioreactor can be accomplished using a sterile filter and a peristaltic pump to obtain cell-free sample and where the dead volume of sample is of concern (smaller bioreactors), microfiltration membranes can be used, which may be placed inside the bioreactor; single-use forms of these are not yet available. Suppliers include TraceBiotech (www.trace.de) and Groton (www.grotonbiosystems.com).

Where removing cells proves cumbersome, the samples may be treated to stop their metabolic activity by freezing or using inactivation chemicals.

One way to solve the sampling problem is to use a pre-sterilized sampling container, including a needleless syringe that can be welded to the sampling module of the bag bioreactor. A sample is pumped into the container; the sampling containers can be removed and the tube heat-sealed. Sartorius and GE use this method. Other fully sampling systems involve connecting to a bioreactor a pre-sterilized Luer connection including a one-way valve to prevent the sample from flowing back into the reactor. The sample is withdrawn from the reactor by a syringe and directed through a sample line into a reservoir. Cellexus Biosystems (Cambridgeshire, UK) and Millipore (Billerica, MA) use this approach.

The Cellexus system is connected to the sample line, and there are up to six sealed sample pouches. The sample from the reservoir can then be pushed into the pouches that are subsequently separated by a mechanical sealer resulting in sealed, sterile samples.

The proprietary Millipore system comprises a port insert that can be fitted to several reactor side ports and a number of flexible conduits that can be opened and closed individually for sampling and are connected to flexible, single-use sampling containers. Sampling is limited to a number of available conduits in each module.

These sampling systems allow aseptic sampling but are limited by the number of samples taken per module and the lack of automation. And while these methods come with good validation data, the risk of contamination cannot be removed since the bioreactor is indeed breached every time a sample is withdrawn. There is a need to develop other methods that would not require contact with media.

Optical sensors

Optical sensors work on the principle of the effect of electromagnetic waves on molecules. It is a totally noninvasive method and can provide continuous results of many parameters at the same time. It is relatively easy to use them through a transparent window in the bioreactors. The detector part of the system can be physically separated allowing utilization of expensive analytical devices allowing optical sensors to be used in situ or online.

Fluorescence sensors can be optimized for measurements of nicotinamide adenine dinucleotide phosphate (NADPH) and used for both biomass estimation and differentiating between aerobic and anaerobic metabolism. The two-dimensional process fluorometery enables simultaneous measurement of several analytes by scanning through a range of excitation and emission wavelengths including proteins, vitamins, coenzymes, biomass, glucose, and metabolites such as ethanol, adenosine-5′-triphosphate (ATP), and pyruvate. Thus, it is possible to use fluorometry to characterize the fermentation process. Generally, a fiber optic light attached to a bioprocessor and shining the light through a glass window in the bioprocessor would work very well. Examples of this are the fluorometers from BioView system (www.delta.dk). The BioView® sensor is a multichannel fluorescence detection system for application in biotechnology, pharmaceutical and chemical industry, food production, and environmental monitoring. It detects specific compounds and the state of microorganisms as well as their chemical environment without interfering with the sample. The BioView system measures fluorescence online directly in the process. An interference with the sample is eliminated. There is no need to take samples for off-line analysis, which saves manpower and reduces the risk of contamination. However, in view of the complexity of spectra of multiple components, high-level resolution programming is required.

Finesse Solutions, LLC (www.finesse.com), a manufacturer of measurement and control solutions for life sciences process applications, offers its SmartBags, which are designed to be plug-and-play bio-processing containers having full measurement capability for at least 21 days.

The SmartBag SensorPak leverages TruFluor pH and dissolved oxygen phase fluorometric technology in a compact assembly that is pre-calibrated using a SmartChip and provides accurate, drift-free, in situ measurements. The combined pH and DO optical reader uses advanced optical components including a large area photodiode that minimizes photodegradation of the active sensing elements. The SensorPak also leverages TruFluor temperature 316L stainless steel thermal window for highly stable readings. The SensorPak is welded into the single-use vessel and eliminates the need for sterile connectors and their associated complications such as leakage and batch contamination. All wetted materials of the SensorPak are USP class VI compliant, and being identical to TruFluor, allow directly measurement comparisons and scale-up from 10 L rocker bags to 2000 L SUBs.

The biosensors manufactured by Fluorometrix are noninvasive, membrane sensors developed using optical fluorometric chemistries that can be directly incorporated into any single-use bioreactor bags. Because the sensors can be manufactured into any type of single-use bag, they are useful for both upstream and downstream applications. Also, these are compatible with the FDA's PAT initiative, which aims to monitor and control biological manufacturing by measuring critical parameters throughout the manufacturing process.

Many metabolic products in a bioreactor can be readily detected by IR spectroscopy, but water absorbed IR beam can only be NIR or SIR for biomass analysis when using in transmission mode. However, attenuated total reflectance spectroscopy (ATRIR) is based on the reflection of light at an interface between two phases with different indices of refraction, and the light beam penetrates into the medium with the lower refraction index in the dimension of one wavelength. Absorption of IR results in a decrease in the intensity of the reflected beam to detect the analyte. Probes for both types of IR spectroscopy are used. Hitec-Zang (www.hitec-zang.de) offers a large range of PAT devices including IR systems.

NIR transmission probes, ATR-IR probes for bioreactors are now commercially available. These are connected through silver halide fibers or radiofrequency connectors.

In addition to IR and fluorescence, optical methods based on photoluminescence, reflection, and absorption are also used. The optical electrodes or "optodes" can

401

be attached using glass fibers leaving the measurement equipment outside of the bioreactor as discussed earlier for fluorescence detectors, allowing the use of these chemosensors in situ or online.

Oxygen sensors work by quenching fluorescence by molecular oxygen; measurement required a fluorescent dye (metal complexes) immobilized and attached to one end of an optical fiber, and the other end of the fiber is interfaced with an excitation light source. The duration and strength of fluorescence depend on the oxygen concentration in the environment around the dye. The emitted fluorescence light is collected and transmitted for reading outside of the bioreactor. These electrodes work better than the traditional platinum probe electrodes to detect oxygen, working in both liquid and gas phase. Examples of oxygen sensors I include PreSens (www.presens.de) noninvasive oxygen sensors that measure the partial pressure of both dissolved and gaseous oxygen. These sensor spots are used for glassware and single uses. The sensor spots are fixed on the inner surface of the glass or transparent plastic material. The oxygen concentration can, therefore, be measured in a noninvasive and nondestructive manner from outside, through the wall of the vessel. Different coatings for different concentration ranges are available. It offers online monitoring of concentration ranges from 1 ppb up to 45 ppm dissolved oxygen, with dependence on flow velocity and measuring oxygen in the gas phase as well; these can be autoclaved.

The Ocean Optics (www.oceanoptics.com) offers the world's first miniature spectrometer with a wide array of sensors for oxygen, pH, and in the gas phase.

The pH sensors work by fluorescence or absorption and for fiber-optic pH measurements, both fluorescence- and absorbance-based pH indicators can be applied. For fluorescence, the most common dyes are 8-hydroxy1,3,6-pyrene trisulfonic acid and fluorescein derivatives, while phenol red and cresol red are used for absorption type measurements. Fluorescent dyes are sensitive to ionic strength limiting their use for broad pH measurement, more than 3 units.

The new transmissive pH probes from Ocean Optics use a proprietary sol–gel formulation infused with a colorimetric pH indicator dye. This material is coated with our exclusive patches to reflect light back through the central read fiber or to transmit light through in order to sense the color change of the patch at a specific wavelength. While typical optical pH sensors are susceptible to drastic changes in performance in various ionic strength solutions, Ocean Optics' sensory layer has been chemically modified (esterified?) to allow accurate sensing in both high and low salinity samples. The transmissive pH probes from Ocean Optics can be used with a desktop system as well as with the Jaz handheld spectrometer suite. The desktop system uses a module is SpectraSuite software that allows for simplified calibration, convenient pH readings, customizable data logging, and comprehensive exportation of data and calibration information.

- Proprietary organically modified sol–gel formulation engineered to maximize immunity to ionic strength sensitivity
- Compatible with some organic solvents (acetone, alcohols, aromatics, etc.)
- Sol–gel material chosen over typical polymer method, allowing for faster response time, versatility in the desired dopants, greater chemical compatibility, flexible coating, and enhanced thermal and optical performance
- Indicator molecule allows high-resolution measurement in biological range (pH 5–9)
- Simplified algorithm takes analytical and baseline wavelengths into account to reduce errors caused by optical shifts

The TruFluor™ (www.finesse.com) dissolved oxygen and temperature sensor is a single-use solution consisting of a single-use sheath, an optical reader, and a transmitter. The single-use sheath can be pre-inserted in a single-use bioreactor bag port and irradiated with the bag in order to both preserve and guarantee the sterile barrier. All wetted materials of the sheath are USP class VI compliant. The optical reader utilizes a LED and a large area photodiode with integrated optical filtering that minimizes photo-degradation of the acting sensing element. The design has been optimized to provide an accurate in situ measurement of dissolved oxygen using phase fluorometric detection in real time. The temperature measurement leverages a 316L stainless steel thermal window embedded in the sheath and provides a highly accurate temperature measurement that can be used as a process variable or for temperature compensation.

Carbon dioxide sensors work on the principle of measure pH of a carbonate buffer embedded in a CO_2 permeable membrane. The reaction time of the sensors is long, and use of quarternary ammonium hydroxide has been made to achieve a faster response. Fluorescence-based sensors are attractive as they facilitate the development of portable and low-cost systems that can be easily deployed outside the laboratory environment. The sensor developed for this work exploits a pH fluorescent dye 1-hydroxypyrene-3,6,8-trisulfonic acid, ion-paired with cetyltrimethylammonium bromide (HPTS-IP), which has been entrapped in a hybrid sol–gel–based matrix derived from n-propyltriethoxysilane along with the liphophilic organic base. The probe design involves the use of dual-LED excitation in order to facilitate ratiometric operation and uses a silicon PIN photodiode. HPTS-IP exhibits two pH-dependent changes in excitation bands, which allows for dual excitation ratiometric detection as an indirect measure of the pCO_2. Such measurements are insensitive to changes in dye concentration, leaching and photobleaching of the fluorophore and instrument fluctuations, unlike unreferenced fluorescence intensity measurements. The performance of the sensor system is characterized by a high degree of repeatability, reversibility, and stability.

The YSI 8500 CO_2 monitor measures dissolved carbon dioxide in bioprocess development applications. Engineered to fit a variety of bioreactors, the unit delivers precise, real-time data that increases an understanding of critical fermentation and cell culture processes. These data can help in gaining insight into cell metabolism, cell culture productivity, and other changes within bioreactors.

An in situ monitor based on the reliable optochemical technology was developed by Tufts University and YSI Incorporated (www.ysilifesciences.com). The technology involves the use of a CO_2 sensor capsule consisting of a small reservoir of bicarbonate buffer covered by a gas permeable silicone membrane. The buffer contains HPTS (hydroxypyrene trisulfonic acid), a pH-sensitive fluorescent dye. CO_2 diffuses through the membrane into the buffer, changing its pH. As the pH changes, the fluorescence of the dye changes. The model 8500 monitor compares the fluorescence of the dye at two different wavelengths to determine the CO_2 concentration of the sample medium. The sensor can be autoclaved multiple times. It will measure dissolved CO_2 over the range of 1%–25%, with an accuracy of 5% of the reading, or 0.2% absolute. Previously, CO_2 was measured either in the exit gases from the fermentation process or by taking a manual sample. The new optical-chemical technology uses a fiber-optic cable transfer light through a stainless-steel probe into a single-use sensor capsule, which contains a pH-sensitive dye. The dissolved carbon dioxide diffuses through a polymer membrane to change the color of the dye, which is then relayed by fiber optic cable back to a rack-mounted monitor that determines and displays the dissolved CO_2 level.

Biomass sensors

Information about the biomass concentration can also be obtained via turbidity sensors. Generally, these sensors are based on the principle of scattered light. Most turbidity sensors have the disadvantage that there is only a linear correlation for low-particle concentrations. But sensors that use backscattering light (180°) also have linear properties for high-particle concentrations. A window that is translucent for the desired wavelength in the IR region is necessary for the use in single-use reactors. The S3 Mini-Remote Futura line of biomass detectors (www.applikonbio.com) make it possible to incorporate sensors inside single-use bioreactors. This system incorporates an ultra lightweight pre-amplifier for connecting to the ABER single-use probe. The main Futura housing can be mounted away from the single-use bioreactor vessel. Cells with intact plasma membranes in a fermenter can be considered to act as tiny capacitors under the influence of an electric field. The nonconducting nature of the plasma membrane allows a buildup of charge. The resulting capacitance can be measured: it is dependent upon the cell type and is directly proportional to the membrane bound volume of these viable cells. The choice of in situ steam sterilizable probes includes a single-use, sterilizable flow through the cell.

Electrochemical sensors

Electrochemical sensors include potentiometric, conductometric, and voltammetric sensors. Thick-and thin-film sensors, as well as chemically sensitive field-effect transistors (ChemFETs), possess potential as potentiometric single-use sensors in bioprocess control because they can be produced inexpensively and in large quantities.

Many pH-sensing systems rely on amperometric methods, but they require constant calibration due to instability or drift. The setups of most amperometic sensors are based on the pH-dependent selectivity of membranes or films on the electrode surface.

While turbidity sensors detect the total amount of biomass concentration, capacitance sensors provide information specifically about the viable cell mass. The electrical properties of cells in an alternating electrical field are generally characterized by an electrical capacitance and conductance. The integrity of the cell membrane exerts a significant influence on the electrical impedance so that only viable cells can be estimated. Biodis Series for monitoring viable biomass in single-use applications is available from Fogale (www.fogalebiotech.com) and Aber (www.aber-instruments.com), the latter has now offered an integrated version with DASGIP (www.dasgip.de). The new Aber Futura Biomass Monitor has been designed so that multiple units can be easily incorporated into bioreactor controllers and SCADA systems.

The sensor CITSens Bio (http://www.c-cit.ch/) can monitor the consumption of glucose and/or the production of L-lactate during cultivation. The CITSens Bio utilizes an enzymatic oxidation process and electron transfer from glucose or lactate to the electrode (anode) via a chemical wiring process, which is catalyzed by an enzyme specific for -D-glucose or L-lactate and a mediator. The sensor function is therefore not affected by oxygen concentration and produces an exceptionally low concentration of side products, such as peroxide. The working principle of this sensor is in contrast to that of a number of well-known alternatives currently on the market, which depend on a sufficient supply of oxygen for their operation as they measure the hydrogen peroxide produced during the bioprocess. The principal feature of the CITSens Bio is a miniaturized, screen-printed electrode comprising a three-electrode system for amperometric detection of the current transmitted to the anode (working, counter, and reference electrode). This three-electrode system ensures a reliable electrical signal with long-term stability. The chemical components, including the enzyme, are deposited in the active field of the working electrode, and the enzyme is cross-linked to form protein and hence is immobilized in this network. The immobilization process itself has an antimicrobial effect. A dialysis membrane is cast over the sensing head to create a barrier between the sensor and the cultivation medium.

Pressure sensors

Another important process parameter that is frequently monitored during bioprocess unit operations like filtration, chromatography, and many others is pressure. Using a traditional stainless steel pressure gauge in conjunction with a single-use experimental setup is possible but has the drawback that the pressure gauge has to be sterilized separately. Furthermore, the connection of the sensor to the previously gamma-radiated single-use assembly can be problematic.

Many bioprocess unit operations are either controlled based on pressure or have significant pressure-related safety issues. Traditional stainless steel reactors are monitored and controlled for pressure, as pressure is used as a means of influencing mass transfer and preventing contamination. In addition, a high-pressure event is a potentially hazardous situation. Single-use bioreactor systems, on the other hand, are frequently not monitored or controlled for pressure because stainless steel pressure transducers are not compatible or cost effective when applied to single-use bioreactors. As a result, a clogged vent filter on a bioreactor can easily rupture bags, spilling the contents of the reactor and exposing the operators to unprocessed bulk.

Another application where pressure monitoring is central to process performance is depth and sterile filtration. A filter's capacity is primarily measured by either flow decay or pressure increases, although adding reusable traditional pressure transducers to a process train defeats the purpose of a single-use process set-up. Depending on the process application, the product contact surface of a traditional device requires either sanitization or moist heat sterilization.

There are traditional devices that are compatible with steam in place, where only the product contact surface is exposed to steam, and even devices that can be placed in an autoclave where the entire device is exposed to steam. Many single-use process components, however, are not compatible with moist heat sterilization temperatures so there may be a requirement for separate sterilization of the stainless steel device and possibly less than optimal connection to a pre-sterilized single-use assembly.

Single-use pressure sensing allows for rapid changeover of product contact parts in both development applications and especially in early-phase clinical manufacture. For example, single-use pressure sensors from PendoTECH were designed to enable pressure measurement with single-use assemblies that have flexible tubing as the fluid path. These single-use pressure sensors are gamma compatible (up to 50 kGy), and the fluid-path materials meet USP Class VI guidelines and are also compliant with EMEA 410 Rev 2 guidelines.

On a single-use bioreactor, a sensor can be installed on a vent line to measure headspace pressure. Even though the sensors are qualified for use up to 75 psi, the core sensor is accurate in the low-pressure range required for a single-use bioreactor.

A better solution with respect to ease of operation and compatibility are single-use pressure sensors, which are now available on the market. The single-use sensors from PendoTECH (www.pendotech.com) can be used with tubing of various sizes (0.25–1 in. in diameter) and can be gamma radiated with tubing and bag assemblies. They are the alternative low-cost solution for use with tubing and bioprocess containers to the existing stainless steel pressure sensors on the market. They are available in caustic resistant polysulfone, so they can be in-line during caustic sanitization processes. The pressure sensors can be integrated into pressure measurement and control with both a PressureMAT™ System (monitor/transmitter) or PendoTECH Process Control System and depending on the number of sensors and process requirements. The data collected by these systems can be output to a PC or another data monitoring device. They also can be integrated into other prequalified third party pumps and monitors (adapters for phone jacks can be made). The pressure sensors are very accurate in the pressure ranges typically used with flexible tubing and single-use process containers and are qualified for use to 75 psi. Applications include multistage depth filtration, TFF/cross-flow filtration, and bioreactor pressure monitoring.

NovaSensor's NPC-100 (www.ge-mcs.com) pressure sensor is specifically designed for use in single-use medical applications. The device is compensated and calibrated per the Association for the Advancement of Medical Instrumentation (AAMI) guidelines for industry acceptability. The sensor integrates a high-performance, pressure sensor dies with temperature compensation circuitry and gel protection in a small, low-cost package.

The SciPres (www.scilog.com) combines pressure-sensing capabilities and the convenience of disposability with easy setup. Each sensor is preprogrammed and barcoded with a unique ID for easy traceability and data documentation when combined with the SciLog SciDoc software. Factory calibration data is also stored on each sensor's chip for out-of-box, plug and play use. The SciPres comes in five different sizes to fit a variety of tubing sizes: Luer, 3/8 in. barb, 1/2 in. barb, 3/4 in. TC (Tri-Clover), and 1.0 in. TC (Tri-Clover).

The SciCon combines temperature sensing capabilities with conductivity sensing capabilities in a compact, single-use, single-use package at a low-price point. Each sensor is pre-programmed and barcoded with a unique ID for easy traceability and data documentation when combined with the SciLog SciDoc software. Factory calibration data are also stored on each sensor's chip for out-of-box, plug and play use. The SciCon comes in five different sizes to fit a variety of tubing sizes: Luer, 3/8 in. barb, 1/2 in. barb, 3/4 in. TC (Tri-Clover), and 1.0 in. TC (Tri-Clover).

Conclusion

The need to monitor the characteristics of a biomass goes across many industries, and most notably in the fermentation industry such as wine-making, the method of online, and in situ monitoring of just about every function that is needed to perform a full PAT work are available. In recent years, there have been significant breakthroughs in the technologies available for monitoring including fluorescence, dye-base pH, and oxygen measurements. While most of the sensors were initially developed for the hard-walled bioreactor industry, the disposable versions of these sensors appear almost every day. The basic principle is that if a sensor is placed inside the bioreactor vessel it should be sterile and preferably single-use; the cost of throwing away a sensor has come down significantly, and it is now possible to monitor just about every function including cell mass, both total and live readily using these single-use sensors. Alternately, many methods are available that work from outside of the bioreactor entirely, particularly those involving fluorescence and optical measurements.

It is anticipated that within the next 5 years, as single-use bioreactors begin to replace large hard-walled systems, much improved systems would become available, especially those consolidating several monitoring functions into one.

Filling and finishing systems

There are two stages where a biological product requires filling in containers for long-term storage. First, it comes at the end of the downstream cycle when a purified solution of a protein drug is ready for filling into containers as a bulk drug that would be shipped out to companies to formulate and finish the product in an end-customer dosage form. The second stage of filling is to formulate the bulk and fill into end-customer dosage form, a syringe, vial, or ampoule.

The bulk product would generally be labeled as a pharmacopoeia product, such as Erythropoietin Concentrated Solution EP, and would thus comply all requirements of the label as required in the pharmacopoeia. There is no need for any formulation additives and the last buffer exchange in the downstream processing had likely already brought it into a formulation that is stable. This stage of filling is conducted as a continuation of downstream processing. Sterile serum bottles are available for packaging the product (e.g., from Thomas Scientific, www.thomassci.com) in size from 25 mL to 9 L. These solutions can alternately be filled in sterile-flexible bags. It would be a prudent step to fill them using a sterile filter as the final stage, perhaps a virus-clearing filter.

The second stage of filling protein solutions is more demanding. The final production step is transferring the new medium from the transfer vessel or bags and into vials for distribution. Traditionally, the final fill operation consisted of stainless steel equipment connected via reusable valves, rigid tubing, and steel pipes. Again, this equipment requires validation and must be subjected to a CIP cycle after each filling cycle is completed. Today, many process engineers are designing this operation with single-use tubing assemblies in place of stainless steel piping to reduce sterilization time and cost.

One example of integrating single-use systems in a final fill operation is for simplifying mobile transfer tanks. These tanks with single-use liners are designed to transfer product from formulation suites to storage areas and ultimately to filling suites. To allow sterile connection to and from these vessels, designers traditionally add three-way valve assemblies to fill and drain ports to facilitate SIP operations. The design of these three-way valves makes it difficult to validate cleaning procedures. Replacing these heavy three-way valve assemblies with single-use tube sets and connectors eliminates cleaning, validation, and maintenance.

Single-use tubing assemblies can either be attached prior to equipment sterilization with single-use SIP connectors (used as either steam access or condensate drainage sites) or steamed separately, just prior to fluid transfer. For vessel outlet, combining a number of single-use components into the transfer line can create a robust system to ensure product safety. For example, outlet transfer lines could incorporate a single-use SIP connector to attach to the sterile holding tank. Then, a through-the-wall fluid transfer system is used to bring a portion of the transfer line into the filling suite. Next, a sterile connector is used to attach the transfer line to a separate portion of the transfer line that has already been steamed onto the filling machine with a single-use SIP connector. Finally, disconnecting the transfer lines using a quick disconnect coupling that has been validated as an aseptic disconnect enables the processor to confidently make an aseptic disconnection from the storage vessel or bag.

The leader in sterile product filling systems remains Bosch Packaging, among others. The following information was developed by Bosch Packaging (www. boschpackaging.com).

Bosch packaging systems

Use of single-use components in product downstream processing and final fill operations is increasing as the technology for performing these steps in a single-use mode also increases. There is a high demand for systems that support single-use purification, formulation, and filling operations. There are several drivers for this. First is the desire to realize increased processing efficiency through the elimination of preparative steps like CIP and SIP for product-contact equipment and parts. For example, pre-sterilized, single-use tubing and bags can be used to replace stainless steel piping and tanks that have to be cleaned and steamed between uses. Second is the reduction of validation efforts related to the product path, in particular, the elimination of cleaning validation. Products that are hard to clean, or are highly potent or toxic, often require dedicated product-contact parts. This is because existing cleaning processes are inconsistent or simply do not work to remove certain products to safe levels. Third is containment of toxic products. Single-use systems can be removed, bagged, and disposed of without breaking connections and exposing the environment to the product. A forth driver is the desire to match existing single-use upstream processes that are available for new products, particularly biopharmaceuticals and other protein-based drugs.

In the past, filling line equipment was commonly dedicated entirely to a single product. However, this approach is no longer economically feasible except for the highest-volume drugs that support nearly continuous filling operations. Most filling lines today support multiproduct operations. Traditional multiproduct operations require validation of the level of product carryover after cleaning operations to ensure subsequent products are not contaminated. Certain product-contact parts are hard to clean to acceptable levels, and are therefore dedicated to specific products.

An alternative to the filling equipment dedication is the use of single-use parts and assemblies. These systems can include components like bulk product bags, capsule filters, silicone tubing, and other plastic fittings and parts. Many of these parts can already be purchased pre-cleaned and pre-sterilized, and double or triple bagged for easy use within clean rooms. However, filling operations are critical enough to require entire single-use systems be assembled and sterilized together rather than having to piece individual components together at the point of use. The dosing system used for filling also affects the characteristics of the single-use components.

The pre-sterilized, single-use concept has already been realized with several off-the-shelf filling systems using peristaltic or gravimetric dosing. The existing systems, however, are designed for low-speed, small batch filling operations. Scale-up of these systems for high-speed filling creates technical obstacles that include a relatively slow dosing speed, lower filling accuracy and precision, and difficulty dosing products with variable temperature and viscosity characteristics.

Peristaltic dosing is ideally suited for single uses, as peristaltic tubing is used for much of the product path. Single-use peristaltic systems are typically comprised of a product hold bag, supply tubing, and a filling needle that are bagged together and sterilized using gamma irradiation. The assembly is removed from the bag and connected to the filling system, which can be as simple as a single peristaltic pump, immediately before use. High-quality peristaltic pumps can be very precise at dosing water-like solutions at slower speeds. However, the tubing directly influences accuracy. The tubing that is located in the pump head changes shape over time due to wear, so accuracy drift is common. Characterizing and compensating for the drift is required. Peristaltic pumps also dose at slower speeds, which means that high-speed peristaltic systems require more pump heads to dose at the same rate as the equivalent piston, time pressure, or rolling diaphragm systems.

Gravimetric dosing uses optical sensors to dose a given volume based on a calculation of the interior volume of a given length of tubing or glass. The entire product path is supplied as a single-use, pre-sterilized assembly. Accuracy and precision of the system with water-like solutions is comparable to other dosing systems. However, dosing speed and accuracy can be directly related to fluid temperature and viscosity. Dosing time is based on the speed at which a liquid will flow through the tubing based on gravity. Thicker solutions flow more slowly and are therefore dosed at a slower rate. Relatively small temperature changes over the course of a filling event can affect product viscosity enough to have a significant effect on the volume filled. Like peristaltic systems, more pump heads are required to dose at the same rate as the equivalent piston, time pressure, or rolling diaphragm systems.

The current single-use, pre-sterilized dosing systems are based on scaling-up technologies designed and used for small-scale filling operations. However, a better approach is to convert an existing high-speed dosing technology to single use. The three most common commercial systems are piston pumps, rolling diaphragm pumps, and time pressure dosing. All three of these systems would require significant technical improvements and modifications to be converted to single-use use.

Piston pumps rely on precise physical tolerance between the pump body and piston to provide dosing accuracy and to ensure the product does not leak during use. Pump bodies and pistons are commonly matched when they are fabricated to ensure they do not gall during use. Existing piston pumps for pharmaceutical dosing can be made using stainless steel or ceramic components. Neither material can be used to make a single-use pump due to the high cost of manufacture. Plastic components are an alternative but cannot be fabricated to the correct tolerances to ensure accuracy. Excessive wear and leaking would also be issues. A catastrophic loss of function will likely result without the use of O-rings, a lubricant or both to separate the moving plastic surfaces. Plastic particles, elastomeric particles in the case of o-ring use, or lubricant will also be shed by the pump and introduced into the product stream.

The rolling diaphragm pump, originally developed by TL Systems (now Bosch Packaging Technology, www.boschpackaging.com), is comprised of a stainless steel pump with a diaphragm. A headpiece and diaphragm make up the liquid chamber. Dosing occurs by actuating a piston that is attached to the diaphragm. It is very similar to piston dosing; only the diaphragm keeps the product from contact with the piston and other internal components. The only stainless steel part in contact with the product is the headpiece. Making this system out of plastic will require the use of O-rings, a lubricant or both to separate the moving piston from the body. Unlike the piston pump, however, these surfaces are separated from the fluid path by the diaphragm, so contamination of product stream will not be an issue. The tolerances for each part are not as critical as with piston pumps, as dose accuracy is related to accurate piston stroke while at the same time maintaining consistent dimensions in the fluid chamber.

Time pressure systems are designed to dispense using a pressurized product supply and timed valve openings. A portion of the product path from the product supply manifold to the filling nozzles is made of elastomeric tubing. This tubing is used in association with an automatic tubing pinch mechanism to create the valve. The use of single-use tubing seems to make the system a good candidate for a single use system. However, this is not the case. These systems often use a small surge tank for product supply, and this tank must be pressurized to 10 psig or more for the system to function. Replacing this tank with a bag would require that the bag be pressurized beyond its normal design pressure. There is currently no good solution for pressurization of a surge bag system for use with time pressure filling.

Ensuring a single-use system dispenses at high speed and at the same time is durable enough for commercial use requires rigorous testing. No dosing system is appropriate for commercial use without proof of accuracy and precision over its operating lifetime. The maximum intended run duration for commercial systems can last for up to a week or more and involve 500,000–1,000,000 dosing cycles per station. This is well beyond the design specification of existing single-use dosing systems.

A limitation of all current single-use pre-sterilized dosing systems is the plastic filling needle. The current plastic needles are not designed for commercial filling operations. Most are too wide to penetrate small containers and/or are too short to perform bottom-up filling. Bottom-up filling, where the filling needle penetrates the container and is drawn out during dosing, is common with high-speed filling to reduce product splash and foaming. The plastic needles are also not shaped to fit correctly within needle holders on common commercial filling systems. Custom fixtures are required to use them on existing machines.

High-speed filling requires needles made to very tight tolerances, particularly the needle diameter, as this has an influence on dosing accuracy and precision. Because high-speed needles travel during and after dispensing, needle drip between doses has to be eliminated. Precise needle opening size and opening shape are also required. Substituting plastic needles with ones made from stainless steel can solve most of these issues but is too expensive for single-use assemblies.

The use of single-use components in downstream processing and final-fill operations is increasing as the technology for performing these steps in a single-use mode advances. This evolution has created significant demand for systems that support single-use purification, formulation, and filling operations.

Similar to the advantages accrued in other uses of single-use systems, the elimination of preparative steps like CIP and SIP for product-contact equipment and parts by using presterilized, single-use tubing and bags can be highly beneficial. The need to reduce validation efforts related to the product path, in particular, the elimination of cleaning validation is an important consideration since many products are hard to clean, or are highly potent or toxic, often require dedicated product-contact parts. This is because existing cleaning processes are inconsistent or simply do not work to remove certain products to safe levels. The problems compound in the filling of biological drugs where even levels of contamination otherwise allowed can have serious effects on the quality of the product. Since it is not possible to predict how various contaminants would affect the three- and four-dimensional structure of protein solutions, the only sure way is to avoid the possibility of any contamination rather than prove that the contact parts are clean.

To meet this standard, in the past, filling-line equipment was commonly dedicated to a single product. This approach is no longer economically feasible. Most filling lines today support multiproduct operations, with the exception of the highest-volume drugs, which support nearly continuous filling operations. However, in the case of protein drugs, a single-machine head-filling multiple products remains very risky.

An alternative to filling-equipment dedication is the use of single-use parts and assemblies. These systems can include components like bulk product bags, capsule filters, silicone tubing, and other plastic fittings and parts. Many of these parts can be purchased precleaned and presterilized as well as double or triple bagged for easy use within cleanrooms. Filling operations are critical enough to require that entire single-use systems be assembled and sterilized together rather than having to piece individual components together at the point of use.

Presterilized, single-use products have been around for a while, with several off-the-shelf filling systems using peristaltic or gravimetric dosing. These existing systems are designed for low-speed, small-batch filling operations. Scale-up of these systems for high-speed filling creates technical obstacles that include a relatively slow dosing speed, lower filling accuracy and precision, and difficulty dosing products with variable temperature and viscosity characteristics.

Peristaltic dosing is ideally suited for single uses. Single-use peristaltic systems are typically composed of a product hold bag, supply tubing, and a filling needle, all of which are bagged together and sterilized using gamma irradiation. The assembly is removed from the bag and connected to the filling system, which can be as simple as a single peristaltic pump, immediately before use.

High-quality peristaltic pumps can precisely dose water-like solutions at slower speeds. The tubing, however, directly influences accuracy; accuracy drift is common as the tubing that is located in the pump head changes shape over time due to wear. Characterizing and compensating for this drift is mandatory. Peristaltic pumps also dose at slower speeds, which means that high-speed peristaltic systems require more pump heads to dose at the same rate as the equivalent piston, time-pressure, or rolling-diaphragm systems.

Gravimetric dosing uses optical sensors to dose at a certain volume based on the calculation of the interior volume of a given length of tubing or glass. The entire product path is supplied as a single-use, presterilized assembly. Accuracy and precision of the system with water-like solutions is comparable to other dosing systems, with speed and accuracy directly related to fluid temperature and viscosity. Dosing time is based on the speed at which a liquid will flow through the tubing based on gravity. Thicker solutions flow more slowly and are therefore dosed at a slower rate.

Relatively small temperature changes over the course of a filling event can affect product viscosity enough to have a significant effect on the volume filled. Like peristaltic systems, more pump heads are required to dose at the same rate as the equivalent piston, time-pressure, or rolling-diaphragm systems.

Current single-use, presterilized dosing systems are based on the scale-up of technologies designed and used for small-scale filling operations. A better approach is to convert an existing high-speed dosing technology to single use. The three most common commercial systems are piston pumps, rolling-diaphragm pumps, and time-pressure dosing. All three of these systems would require significant technical improvements and modifications to be converted to single-use use.

Piston pumps rely on precise physical tolerance between the pump body and piston to provide dosing accuracy and to ensure that product does not leak during use. Plastic components are an alternative, but they cannot be fabricated to the correct tolerances to ensure accuracy. The rolling-diaphragm pump is composed of a stainless steel pump, with a headpiece and a diaphragm making up the liquid chamber. Dosing occurs by actuating a piston that is attached to the diaphragm. A rolling-diaphragm system is similar to a piston-dosing system, the big difference being that the diaphragm keeps the product from contacting the piston and other internal components. The only stainless steel part in contact with the product is

411

the headpiece. Time-pressure systems are designed to dispense using a pressurized product supply and timed valve openings. A portion of the product path from the supply manifold to the filling nozzles is made of elastomeric tubing, which is used in association with an automatic tubing-pinch mechanism to create the valve. The use of single-use tubing seems to make this system a good candidate for single use. This is not the case though. Time-pressure systems often use a small surge tank for product supply, and this tank must be pressurized to 10 psig or more. Replacing this tank with a bag would require that the bag be pressurized beyond its normal design pressure. There is currently no good solution for pressurization of a surge-bag system for use with time-pressure filling.

Ensuring that a single-use system dispenses at high speeds and is also durable enough for commercial use requires rigorous testing. The run duration of commercial systems is typically a week or more, involving 500,000–1,000,000 dosing cycles per station. This is well beyond the design specification of existing single-use dosing systems—substantial improvements are needed.

The plastic filling needle limits all single-use, presterilized dosing systems at this time. Current plastic needles are not designed for commercial filling operations. Most are too wide to penetrate small containers and/or too short to perform bottom-up filling. Bottom-up filling, where the filling needle penetrates the container and is drawn out during dosing, is common with high-speed filling to reduce product splash and foaming. In addition, plastic needles are not shaped to fit correctly within needle holders on common commercial filling systems, requiring custom fixtures to use them on existing machines. High-speed filling requires needles made to tight tolerances, particularly the needle diameter, which influences dosing accuracy and precision. As high-speed needles travel during and after dispensing, needle drip between doses has to be eliminated. Precise needle opening size and opening shape are also required. Substituting plastic needles with ones made from stainless steel can solve most of these issues, but this is too expensive for single-use assemblies.

PDC Aseptic Filling Systems

PDC Aseptic Filling Systems (www.lpsinc.net/pdc) specialize in the design and manufacture of innovative filling solutions used by major Biopharmaceutical companies. Through years of development, testing, and validation by major biopharmaceutical companies, PDC's technology allows for aseptic filling within any clean room classification, at a fraction of fixed stainless steel system cost and manifold filling cost. Further, through the use of proprietary single-use fill system liners, customers have a fully single-use option eliminating the risk of cross-contamination and eliminating much of the cost of cleaning and documentation at all levels of pharmaceutical production.

PDC manufactures aseptic filling systems featuring a unique single-use sterile fill line. Aseptic filling systems are available to fill bottles, bags, or drums from 5 mL up to 1000 L. PDC's single-use filling lines offer single use noncontact aseptic filling that can be custom designed for a variety of biopharmaceutical filling applications.

Other solutions for filling the bulk are available from Sartorius, Pall, and Millipore.

The Integrity™ LevMixer® system is a mobile, flexible mixing system that allows efficient and reproducible single-use mixing of a wide range of volumes in a broad series of applications ranging from buffer preparation to final formulation. The LevMixer system is engineered for use with ATMI single-use mixing bags in cGMP certified cleanrooms without complex instrumentation to control the mixing process. The LevMixer consists of an interchangeable superconducting drive unit and proprietary levitating impeller-based single-use mixing bags fitted into containers on either a portable dolly or a floor-mounted tank. Once properly charged and coupled with the mixing bag, the activation of the motor induces levitation and rotation of the in-bag impeller resulting in effective mixing action inside a hermetically sealed bag. Coupling of the in-bag impeller with the drive motor requires no dynamic seals or shaft penetration inside the bag. The drive motor is enclosed on a portable cart that can be easily disconnected from the bag and reconnected to another mixing bag, allowing mixing in multiple bags of various sizes with a single drive unit. As with all ATMI LifeSciences' single-use mixing systems, the LevMixer utilizes single-use mixing bags made from Integrity TK8 bioprocess film. The product-contacting layer of TK8 film is blow-extruded in-house by ATMI under cleanroom conditions using medical-grade ultra-low-density polyethylene resin. It is then laminated to create a gas barrier film of exceptional cleanliness, strength, and clarity that is ADCF and complies fully with USP Class VI requirements.

Palletank® for LevMixer® is a stainless steel cubical container designed to perfectly fit with the Flexel® Bags for LevMixer with its integrated impeller. It includes a railed port for coupling the mobile LevMixer drive unit with the Flexel Bags for LevMixer and a clamp holder to facilitate powder transfer. The hinged door allows easy installation of the bag system, whereas the front bottom gate facilitates easy tubing installation and access. Windows on lateral and rear sides enable the user to visually control the mixing process. The cubical shape improves the mixing efficiency and offers scalability from 50 to 1000 L.

Pall's Allegro 3D single-use mixers are impellor-based systems for large-volume applications including vaccine and drug formulation, and mixing viscous fluids. Available as 200 L systems for working volumes from 50 to 200 L, Allegro 3D mixers are also ideal for a wide range of liquid–liquid and solid–liquid mixing applications such as compounding, formulation, buffer and media preparation, and pH/conductivity adjustment. Front-loading of the single-use mixer assembly eases set-up. Mixing data are available for a wide range of industry standard applications, demonstrating fast and complete mixing even for preparations requiring high concentrations and large additions of solids.

Summary

There are tremendous advantages to the use of single-use, pre-sterilized dosing systems for commercial filling operations. Increased processing efficiency through the elimination of preparative steps like CIP and SIP, reduction of validation efforts including elimination of cleaning validation, containment of toxic products, and matching existing single-use upstream processes are all compelling arguments for these systems for product filling operations. However, significant technical achievements must be realized before a system can be scaled for high-speed filling operations.

Regulatory compliance

The medical device industry has matured into hundreds of regulatory approvals worldwide and with that has come a keen understanding of biocompatibility, leachability, and safety of the plastic materials used. The single-use systems gaining wide appeal in bioprocessing would have to revert to regulatory opinions on medical devices since there are no current guidelines available from the FDA or EMEA on the evaluation of single-use manufacturing factories for biological drugs.

There is sufficient guidance available on the testing and validation of any measurement device used in cGMP, ranging from calibration to IQ/OQ/PQ to CFR 21 Part 11 compliance when there is a CPU involved.

To date, no major regulatory authority has approved any product manufactured using a single-use bioreactor. It is not because of any risk factors in the use of single-use systems but the inability of the sponsors to file these applications. Generally, a sponsor will file a regulatory marketing authorization application using manufacturing systems that are capable of producing at least 10% of the final batch size that the sponsors plan to make when the drug is approved. The sponsor also wants to make sure that the larger batches would be easily scaled up. When using hard-walled systems, this is relatively easy as the fermenters and bioreactors supplied by large equipment manufacturer offer several sizes, and most of these are readily scalable. When it comes to single-use bioreactor systems, the largest size bioreactor that has become available is 2000 L in size and even that is an expensive offering as most manufacturers have simply emulated the hard-walled systems by placing a liner within them. All other hardware and software remain the same, making it a much more expensive undertaking. Even if the big pharma is willing to accept those costs, the maximum size of 2000 L is too low to convince betting on this system. Recently, a Korean company installed hard-walled animal cell bioreactor greater than 80,000 L. However, this type of sizing is most likely to become obsolete as cell lines become more productive. As an example, a company using a cell line qualified 20 years ago would have to use at least 20 times larger scale than if they were able to replace them with a newer line.

While for drugs like insulin and other bacterial expression drugs the size of production will remain high, drugs made using cell culture are most likely to be made in smaller size bioreactors as the lines enter a productivity of 10 G/L.

Still, there remains a large market for smaller volume drugs, clinical supplies, and more recently, cell therapy that holds great promise for the single-use manufacturing industry to begin to show its promise in terms of approval marketing authorizations.

The requirements for regulatory submission using a single-use system are exactly the same as applicable to hard-walled manufacturing systems. The usual validation, calibration, and operational qualification of the equipment are required. However, that does create a challenge especially when it comes to managing characterization, robustness, and scalability. The fact that the sponsor must submit robust studies to prove that all leachables are identified and that they do not affect the quality of the product if present. Leacheables apply to every component, from the mixing bags to transfer tubes to filters, connectors, and storage devices. The sponsor is also required to demonstrate the scalability of membrane chromatography, flow distribution, and device variability.

There is a belief that membranes are more variable than chromatography resins, and it is the sponsor's responsibility to prove that the variation does not affect the quality through extensive PAT practice.

Finally, the sponsor must demonstrate comparability of the product with the originator product if this is a biosimilar product. This includes lot-to-lot variations as well, and that is where some problems can arise. There will never be two batches of a biological drug that will always be the same; variability is inherent in biological systems, and it is for this reason alone that he regulatory agencies require strict environmental and compliance control.

To understand how the FDA views the single-use or single-use technology, a review of the Manual for Biologicals Compliance Program Guidance Manual, Chapter 45 Biological Drug Products; Inspection of Biological Drug Products (CBER) 7345.848 is required.

Implementation Date: October 1, 2010 (www.fda.gov/cber/cpg/7345848.htm) has one entry for "single-use," and none for "single-use" or "plastic." The entry on "single-use" states as follows:

> **(8c) Filtration** There are various types of filtration methods, such as diafiltration, ultrafiltration and microfiltration that may be used in the purification of vaccine products. Some of the filters used may be single-use, and some may be multi-use. The filters are usually placed in a filter housing apparatus. The criteria used for the evaluation of the column purification should also be applied to the filter housings and the multi-use filters.

The concerns about leachability or extractability remain to be challenged in filed applications, and as with any new event, the industry is in a waiting mode to see how the FDA or EMEA react to these submissions.

Environmental concerns

The carbon footprint of single-use technologies is larger than the reusable systems producing more solid waste; however, this must be studied in the light of the overall impact and not just in isolation.

The solid waste in single-use systems is mainly used plastic components, from bioreactor bags to connectors to filters. Note that even the stainless steel industry produces substantial solid waste and also as plastic waste. The waste is of two types; one that can be folded and readily compressed and the other that is hard to compress such as filter capsules, cartridges, etc. It is important to understand that the size of solid waste produced in a single-use system reduces as a percentage of total waste as the batch sizes become larger; this is because some of the basic components are not related to the size of the batch. A rule of thumb to observe is that the size of solid waste is about 10%–12% of the total batch size, so a batch of 5000 L would produce a waste of about 500 kg can be expected.

Disposal of solid waste from manufacturing can be a cumbersome task if the types of waste are different in terms of biosafety decontamination requirement. It is important that the components to be discarded be identified with the hazard before they are put into use.

Biosafety

How a waste is handled depends to a large degree on its biosafety status. Since all single-use components that come in contact with a GMO may have to be equally treated depending on their biosafety status, a good understanding of the NIH Guidelines for Research Involving Recombinant DNA Molecules should be reviewed; the most recent version of which was issued in January 2011.

Since the bioprocessing industry is concerned mainly with two types of host cells, animal tissue like CHO cells and bacterial organism like *E. coli*, the discussion

Table 9.11 NIH Classification of Hazards

Risk Group 1 (RG1)	Agents that are not associated with disease in healthy adult humans
Risk Group 2 (RG2)	Agents that are associated with human disease that is rarely serious and for which preventive or therapeutic interventions are *often* available
Risk Group 3 (RG3)	Agents that are associated with serious or lethal human disease for which preventive or therapeutic interventions *may be* available (high individual risk but low community risk)
Risk Group 4 (RG4)	Agents that are likely to cause serious or lethal human disease for which preventive or therapeutic interventions are *not usually* available (high individual risk and high community risk)

her pertains to both of these selections. The NIH classifies the hazard into four categories (Table 9.11):

Most of the GMOs that are used would fall in Risk Group 1 or 2. There is also an exempt group that is provided in Appendix C of the NIH Guidelines.

Appendix C-VIII-E. i.e., the total of all genomes within a Family shall not exceed one-half of the genome.

Appendix C-I. (Recombinant DNA in Tissue Culture) states: Recombinant DNA molecules containing less than one-half of any eukaryotic viral genome (all viruses from a single family being considered identical -- see Appendix C-VIII-E, *Footnotes and References of Appendix C*), that are propagated and maintained in cells in tissue culture are exempt from these *NIH Guidelines* with the exceptions listed in Appendix C-I-A.

Appendix C-I-A. Exceptions

The following categories are not exempt from the NIH Guidelines: (i) experiments described in Section III-A which require Institutional Biosafety Committee approval, RAC review, and NIH Director approval before initiation, (ii) experiments described in Section III-B which require NIH/OBA and Institutional Biosafety Committee approval before initiation, (iii) experiments involving DNA from Risk Groups 3, 4, or restricted organisms (see Appendix B, Classification of Human Etiologic Agents on the Basis of Hazard, and Sections V-G and V-L, Footnotes and References of Sections I through IV) or cells known to be infected with these agents, (iv) experiments involving the deliberate introduction of genes coding for the biosynthesis of molecules that are toxic to vertebrates (see Appendix F, Containment Conditions for Cloning of Genes Coding for the Biosynthesis of Molecules Toxic for Vertebrates), and (v) whole plants regenerated from plant cells and tissue cultures are covered by the exemption provided they remain axenic cultures even though they differentiate into embryonic tissue and regenerate into plantlets.

Other exemptions may also apply to bacteria and yeast but with the following limitations:

Appendix C-II. *Escherichia coli* K-12 Host-Vector Systems

Experiments which use Escherichia coli K-12 host-vector systems, with the exception of those experiments listed in Appendix C-II-A, are exempt from the NIH Guidelines provided that: (i) the Escherichia coli host does not contain conjugation proficient plasmids or generalized transducing phages; or (ii) lambda or lambdoid or Ff bacteriophages or non-conjugative plasmids (see Appendix C-VIII. Footnotes and References of Appendix C, Footnotes and References of Appendix C) shall be used as vectors. However, experiments involving the insertion into Escherichia coli K-12 of DNA from prokaryotes that exchange genetic information (see Appendix C-VIII. Footnotes and References of Appendix C, Footnotes and References of Appendix C) with Escherichia coli may be performed with any Escherichia coli K-12 vector (e.g., conjugative plasmid). When a non-conjugative vector is used, the Escherichia coli K-12 host may contain conjugation-proficient plasmids either autonomous or integrated or generalized transducing phages. For these exempt laboratory experiments, Biosafety Level (BL) 1 physical containment conditions are recommended. For large-scale fermentation experiments, the appropriate physical containment conditions need be no greater than those for the host organism unmodified by recombinant DNA techniques; the Institutional Biosafety Committee can specify higher containment if deemed necessary.

Appendix C-II-A. Exceptions

The following categories are not exempt from the NIH Guidelines: (i) experiments described in Section III-A which require Institutional Biosafety Committee approval, RAC review, and NIH Director approval before initiation, (ii) experiments described in Section III-B which require NIH/OBA and Institutional Biosafety Committee approval before initiation, (iii) experiments involving DNA from Risk Groups 3, 4, or restricted organisms (see Appendix B, Classification of Human Etiologic Agents on the Basis of Hazard, and Sections V-G and V-L, Footnotes and References of Sections I through IV) or cells known to be infected with these agents may be conducted under containment conditions specified in Section III-D-2 with prior Institutional Biosafety Committee review and approval, (iv) large-scale experiments (e.g., more than 10 liters of culture), and (v) experiments involving the cloning of toxin molecule genes coding for the biosynthesis of molecules toxic for vertebrates (see Appendix F, Containment Conditions for Cloning of Genes Coding for the Biosynthesis of Molecules Toxic for Vertebrates).

Appendix C-III. Saccharomyces Host-Vector Systems

Experiments involving Saccharomyces cerevisiae and Saccharomyces uvarum host-vector systems, with the exception of experiments listed in Appendix C-III-A, are exempt from the NIH Guidelines. For these exempt experiments, BL1 physical containment is recommended. For large-scale fermentation experiments, the appropriate physical containment conditions need be no greater than those for the host organism unmodified by recombinant DNA techniques; the Institutional Biosafety Committee can specify higher containment if deemed necessary.

Appendix C-III-A. Exceptions

The following categories are not exempt from the NIH Guidelines: (i) experiments described in Section III-A which require Institutional Biosafety Committee approval, RAC review, and NIH Director approval before initiation, (ii) experiments described in Section III-B which require NIH/OBA and Institutional Biosafety Committee approval before initiation, (iii) experiments involving DNA from Risk Groups 3, 4, or restricted organisms (see Appendix B, Classification of Human Etiologic Agents on the Basis of Hazard, and Sections V-G and V-L, Footnotes and References of Sections I through IV) or cells known to be infected with these agents may be conducted under containment conditions specified in Section III-D-2 with prior Institutional Biosafety Committee review and approval, (iv) large-scale experiments (e.g., more than 10 liters of culture), and (v) experiments involving the deliberate cloning of genes coding for the biosynthesis of molecules toxic for vertebrates (see Appendix F, Containment Conditions for Cloning of Genes Coding for the Biosynthesis of Molecules Toxic for Vertebrates).

Appendix C-IV. Bacillus subtilis or Bacillus licheniformis Host-Vector Systems

Any asporogenic Bacillus subtilis or asporogenic Bacillus licheniformis strain which does not revert to a spore-former with a frequency greater than $10-7$ may be used for cloning DNA with the exception of those experiments listed in Appendix C-IV-A, Exceptions. For these exempt laboratory experiments, BL1 physical containment conditions are recommended. For large-scale fermentation experiments, the appropriate physical containment conditions need be no greater than those for the host organism unmodified by recombinant DNA techniques; the Institutional Biosafety Committee can specify higher containment if it deems necessary.

Appendix C-IV-A. Exceptions

The following categories are not exempt from the NIH Guidelines: (i) experiments described in Section III-A which require Institutional Biosafety Committee approval, RAC review, and NIH Director approval before initiation, (ii) experiments described in Section III-B which require NIH/OBA and Institutional Biosafety Committee approval before initiation, (iii) experiments involving DNA from Risk Groups 3, 4, or restricted organisms (see Appendix B, Classification of Human Etiologic Agents on the Basis of Hazard, and Sections V-G and V-L, Footnotes and References of Sections I through IV) or cells known to be infected with these agents may be conducted under containment conditions specified in Section III-D-2 with prior Institutional Biosafety Committee review and approval, (iv) large-scale experiments (e.g., more than 10 liters of culture), and (v) experiments involving the deliberate cloning of genes coding for the biosynthesis of molecules toxic for vertebrates (see Appendix F, Containment Conditions for Cloning of Genes Coding for the Biosynthesis of Molecules Toxic for Vertebrates).

In addition to those exemptions, the NIH distinguishes laboratory use from Good Large-Scale Practice (GSLP), the latter applies to manufacturers and makes exceptions from various compliance required in a laboratory (Table 9.12).

In summary, in the United States, mammalian cells used for the production of recombinant proteins normally require BL1 level in laboratory and a GLSP level at large-scale commercial production, which means there are special containment or decontamination requirements and single-use bioreactors can be discarded along with other solid waste. When using bacteria, this may not be the case, and usual procedures for bio-decontamination would apply unless they fall under IIIC exemption category. Since most of the future products are likely to be produced in CHO cells like the MAbs, it makes the cost of disposing of disposal components easily affordable.

However, discarding in this case would be either incinerating them or taking them to a land as they are less likely to be recycled, according to the general consensus of the industry, though I do not see any reason why they could not. The recycling processes are extremely invasive and should remove any contaminants.

For those components are readily recyclable, these are classified into following categories:

Grade A = recyclable plastic. Pure fractions of identified plastics Examples: preparation bags, cell culture bags, hold bags, cartridge bodies, single tubing, tank liners, packaging material. Currently, grade A wastes constitute a small percentage, less than 10% of a complete single-use production chain. If designed appropriately, single components facilitate source separation of different grade A material batches (PE. Pp, PS). Only grade A wastes are suitable for material recycling.

Grade B = mixed fractions of different plastics or multilayer films comprising polymers or thermosets. Examples: most connectors, transfer systems, manifolds, tripolymer film bags, complete filtration cartridges, most bag bioreactors. Up to 95% of the single-use components of a single-use biomanufacturing chain is grade B material. This fraction is suitable for energy recovery but not for material recycling.

Grade C = fractions with a significant amount (>5%) of nonplastic material such as glass, ceramics, metals, and electronic components. Examples: cell culture bags with sealed-in sensors, pumps and pump heads, centrifuge cartridges, mixing systems, filtration cells, DSP units, single-use sensors. Grade C waste usually constitutes only a minor but increasing fraction generally less than 20% of total SUS waste. If recycled or used for fuel production, grade C waste requires pretreatment (fractionating of plastics. metal removal). It is mandatory to separate electronics and sensors from bulk wastes in the EC and in Switzerland to facilitate collection and recycling of this material.

Table 9.12 Comparison of Good Large-Scale Practice (GLSP) and Biosafety Level (BL)—Large-Scale (LS) Practice

Criterion [See Appendix K-VI-B, *Footnotes of Appendix K of the the NIH Guidelines*]	GLSP	BL1-LS	BL2-LS	BL3-LS
1. Formulate and implement institutional codes of practice for the safety of personnel and adequate control of hygiene and safety measures.	K-II-A	G-I		
2. Provide adequate written instructions and training of personnel to keep the work place clean and tidy and to keep exposure to biological, chemical, or physical agents at a level that does not adversely affect health and safety of employees.	K-II-B	G-I		
3. Provide changing and hand washing facilities, as well as protective clothing, appropriate to the risk, to be worn during work.	K-II-C	G-II-A-1-h	G-II-B-2-f	G-II-C-2-i
4. Prohibit eating, drinking, smoking, mouth pipetting, and applying cosmetics in the work place	K-II-C	G-II-A-1-d G-II-A-1-e	G-II-B-1-d G-II-B-1-e	G-II-C-1-c G-II-C-1-d
5. Internal accident reporting	K-II-G	K-III-A	K-IV-A	K-V-A
6. Medical surveillance	NR	NR		
7. Viable organisms should be handled in a system that physically separates the process from the external environment (closed system or other primary containment)	NR	K-III-B	K-IV-B	K-V-B
8. Culture fluids not removed from a system until organisms are inactivated	NR	K-III-C	K-IV-C	K-V-C
9. Inactivation of waste solutions and materials with respect to their biohazard potential	K-II-E	K-III-C	K-IV-C	K-V-C
10. Control of aerosols by engineering or procedural controls to prevent or minimize release of organisms during sampling from a system, addition of materials to a system, transfer of cultivated cells, and removal of material, products, and effluent from a system	Minimize *Procedure* K-II-F	Minimize *Engineer* K-III-B K-III-D	Prevent *Engineer* K-IV-B K-IV-D	Prevent *Engineer* K-V-B K-V-D
11. Treatment of exhaust gases from a closed system to minimize or prevent the release of viable organisms	NR	Minimize K-III-E	Prevent K-IV-E	Prevent K-V-E
12. A closed system that has contained viable organisms not to be opened until sterilized by a validated procedure	NR	K-III-F	K-IV-F	K-V-F
13. Closed system to be maintained at as low pressure as possible to maintain the integrity of containment features	NR	NR	NR	K-V-G
14. Rotating seals and other penetrations into a closed system designed to prevent or minimize leakage	NR	NR	Prevent K-IV-G	Prevent K-V-H
15. The closed system shall incorporate monitoring or sensing devices to monitor the integrity of the containment	NR	NR	K-IV-H	K-V-I
16. Validated integrity testing of the closed containment system	NR	NR	K-IV-I	K-V-J
17. Closed system to be permanently identified for record keeping purposes	NR	NR	K-IV-J	K-V-K
18. Universal biosafety sign to be posted on each closed system	NR	NR	K-IV-K	K-V-L
19. Emergency plans required for handling large losses of cultures	K-II-G	K-III-G	K-IV-L	K-V-M
20. Access to the work place	NR	G-II-A-1-a	G-II-B-1-a	K-V-N
21. Requirements for controlled access area	NR	NR	NR	K-V-N&O

Notes: NR, not required. Appendix K-VI-B of the NIH Guidelines. The criteria in this grid address only the biological hazards associated with organisms containing recombinant DNA. Other hazards accompanying the large-scale cultivation of such organisms (e.g., toxic properties of products; physical, mechanical, and chemical aspects of downstream processing) are not addressed and shall be considered separately, albeit in conjunction with this grid.

Liquid waste

Single-use factories produce a very small quantity of liquid waste, except for used media, eluant from TFF and downstream columns. The combined liquid wastes for complete manufacturing chains typically contain only 2%–5% loading of CIP agents (caustic, acids) compared with SS systems.

Incineration

According to the US Environmental Protection Agency, "incineration is a widely-accepted waste treatment option with many benefits." Combustion reduces the volume of waste that must be disposed of in landfills, and can reduce the toxicity of waste. Incineration is a method of disposal that is used in many countries, and some companies incinerate as part of their standard disposal policy. In the European Union, a number of directives specifically address the issue of waste incineration and disposal:

- Directive 2000/76/EC of the European Parliament and the Council on the incineration of waste (7).
- Directive 94/67/EC of the Council of the European Communities (8) excludes incinerators for infectious clinical waste unless rendered hazardous according to Directive 91/689/EEC of the Council of the European Communities on hazardous waste (9).

In some cases, incineration also can result in significant energy recovery as discussed later.

Cogeneration is a process in which a facility uses its waste energy to produce heat or electricity. Cogeneration is considered more environmentally friendly than exhausting incinerator heat and emissions directly up a smokestack.

One bioprocess company sends all its waste to a facility that incinerates and uses it to generate electricity for a major U.S. city. Another company uses a waste heat boiler to make low-pressure steam. Although there are wide variations, the heat value of mixed plastics waste is estimated to be about 15,000–20,000 BTUs/lb (34,890–46,520 kJ/kg), which compares favorably to coal at 9,000–12,000 BTUs/lb (20,934–27,912 kJ/kg).

Cogeneration is more widely applied in Europe and Asia than in the United States. In the United States, this process is being installed increasingly at universities, hospitals, and housing complexes for which boilers and chillers can serve multiple large buildings. In the European Union, the European Parliament and Council Directive 94/62/EC on packaging and packaging waste, Article 6 addresses energy recovery by incineration, and Article 10 addresses standardization.

In addition, the standard EN 13431 Packaging—Requirements for Packaging Recoverable in the Form of Energy Recovery, Including Specification of Minimum Inferior Calorific Value specifies requirements for packaging to be classified as recoverable in the form of energy and sets out procedures for assessing conformity with those requirements. The scope is limited to factors under a supplier's control.

Pyrolysis

Pyrolysis is a method for converting oil from plastics such as PE, PP, and polystyrene (PS) that can be used as fuel for internal combustion engines, generators, boilers, and industrial burners. Plastics are separated into oil, gas, and char residue by being heated in a pyrolysis chamber. Gas flowing through a catalytic converter is converted into distillate fraction by the catalytic cracking process (enzymatically breaking the complex molecules down into simpler ones). The distillate is cooled as it passes through a condenser and then is collected in a recovery tank. From the recovery tank, the product is run through a centrifuge to remove contaminates. The clean distillate then is pumped to a storage tank.

About 950 mL of oil can be recovered from 1 kg of certain types of plastics. A comparison of the distillate produced by pyrolysis and regular diesel shows the good similarity between the fuels, with the advantage that distillate from pyrolysis burns cleaner.

In-house incineration has, however, not been widely adopted as a means of mini-mizing solid waste in biopharmaceutical manufacture. With rising disposal costs, this may change rapidly; in most instances, this would be sourced out.

Land filling of plastic wastes is expected to decrease in importance as a disposal option for solid wastes. The supposed advantages of landfills such as low operating and capital costs, and high local availability and energy production may no longer realistically apply, if indeed they ever did. Current land filling practices include both direct disposal of nonhazardous waste and disposal of hazardous wastes after pretreatment.

Grind and autoclave

All materials that have been in contact with biopharmaceutical components or with bioagents must be regarded as hazardous. Typical pretreatment for hazard-ous wastes includes grinding and autoclaving, as is common practice with hospital waste. Some items are pretreated and shredded before landfilling. This option is appealing because it may be accepted as safe in some cases and reduces landfill volume compared with the unshredded product. Additional discussions are ongoing regarding use of other hospital waste treatments such as autoclaving, thereby mak-ing a single-use system or component suitable for disposal in a standard municipal waste incinerator or landfill (if allowed). Some companies dispose of their used components or systems into a grinder–autoclave currently used in many hospitals, the National Cancer Institute, and the National Institutes of Health, among others.

This combination of mechanical and physical pretreatment significantly reduces the amount of waste for disposal, which can be reduced still further if it is then mechanically compacted. Furthermore, high-temperature pretreatment at, for example, 75°C, will almost completely inactivate biological contamination and destroy most (but not all) pharmaceutical contaminants as well. Higher tem-perature treatment up to 130°C, or gamma-ray irradiation, can ensure complete destruction of all temperature-resistant contaminants. One potential drawback of autoclaving is the possible production of leachable by-products from the plastics or biopharmaceutical components, which can also result from pre-decontamina-tion with chlorine dioxide leading to a higher risk of soil and water contamination through landfill leakage. The use of chlorine as a disinfectant is not regarded as environmentally sound because it can lead to undesirable atmospheric emissions. Wastes that are strongly acidic or caustic through contamination with pH control or CIP agents must be neutralized before land filling.

Untreated, and ground and autoclaved (or otherwise decontaminated) waste exhib-its high long-term stability in landfills and is not susceptible to biodegradation. As a consequence, no methane is produced in this fraction of land-filled waste. As energy is required for waste pretreatment and transport, the overall energy balance of waste disposed of in a landfill is negative.

Landfill

Some companies choose landfilling. Its potential as an option varies based on the municipality and regional regulations as well as on a product's use before disposal.

Untreated: One industry player landfills an untreated component because its system does not require prior treatment. Another company whose products do not require landfill pretreatment uses biohazard bags before disposal in the solid waste trash.

Treatment

Based on the application, some companies decontaminate with a dose of chlorine dioxide or other deactivator and then dispose of the item in a landfill. This option is more expensive than disposing of an untreated component because it requires extra steps before landfilling. It does, however, allow for the product to be landfilled after use without other cleaning and decontamination steps.

All materials used in the manufacture of single uses contain extractables and leachables. In landfills, where both aerobic (oxidative) and anaerobic (reductive) conditions occur, both of these types of products will be found in the vicinity of deposited waste material. Leachables, by their definition, will be released from plastics under normal landfill conditions. These may also have biohazardous substances associated with them if the waste has not been presterilized and hence may cause an environmental risk. Extractables may also be released from landfilled plastics over extended periods as aggressive chemical conditions such as acidic, caustic, or even dilute solvent microclimates may be established. Furthermore, high-temperature pretreatment (presterilization) has the potential to convert extractables into leachables. The assessment of the risk associated with extractables and leachables has so far (by definition) been focused on the product and the associated health risks and not on potential environmental risks.

Overall environmental impact

It is generally believed that the environmental impact of conventional stainless steel systems is lower than the single-use systems, and thus the induction of single-use technology carries a negative connotation, particularly in the United States, where 5% of the world's population produces 95% of world garbage.

There is a dire need to apprise the users that it is the overall impact on the environment that matters. Several comprehensive studies have confirmed that a fully single-use biopharmaceutical factory can be environmentally advantageous compared with a conventional stainless steel biomanufacturing for following reasons:

- Single-use systems require 1/10th the water to process an equivalent amount of product—water is the most precious resource.
- The energy footprint of stainless steel systems far outweighs the energy needed to incinerate plastic waste.
- Despite their long life, the disposal of stainless steel produces a much higher impact since it is not possible to incinerate it.
- Single-use system factors run with much smaller energy requirement and thus add less to the carbon footprint.
- Single-use system factories are less labor intensive mainly because of elimination of the SIP/CIP and testing for validation.

Summary

One of the greatest impediments to the acceptance of single-use systems has been their image as contributor to the carbon footprint and adversely affecting the environment even in a society that is perhaps the most waste producer in the world, the United States. However, most of these perceptions are wrong and misled by the defenders of the stainless steel industry, which should begin to feel the shudder of these plastic parts. Would there be a time soon when we would have biodegradable bioreactors, perhaps; would there be a time when we would have a nonleaching bioreactors and other plastic component, perhaps sooner. However, until such changes come about, the user should realize that a lot of work goes on developing

Table 9.13 Overview of Comparisons of Various Disposal Options

Option	Advantages	Disadvantages
Landfill, untreated	Lowest operating cost, no capital cost	No, an option for hazardous waste; perceived as environmentally unfriendly
Landfill, treated	Inexpensive, no capital cost	Perceived as environmentally unfriendly
Grind, autoclave, and landfill	Generally accepted as safe, reduces landfill volume	Significant capital cost requires extra handling
Recycling	Environmentally appealing	Impractical for mixed materials
Incinerate	Generally accepted as safe	May be legally restricted and costly
Incinerate with generation of steam or electricity (cogeneration)	Most environmentally benign, some return on investment	May be legally restricted, and presents the highest capital cost
Pyrolysis	Produces usable pure diesel fuel; fuel produced burns more cleanly than produced from refinery	New technology—few options available; subpar efficiency

these components, and equipment suppliers are not likely to switch over to different materials from the ones they have had a lot of experience working with. Design and material changes are difficult to make, as it requires a tremendous amount of validation work to assure that a component would work as it is supposed to every time it is used. Given next is an overview of comparisons of various disposal options (Table 9.13).

The environmental hazard threat from single-use components is unfounded; more so, when we examine how the world is looking at even the very core of the environment impact theory. A poll by the Pew Center For The People And The Press released November 18, 2010, found that 53% of Republicans believe that there is no evidence that the earth is warming. Seventy percent of Tea Party Republicans say that global warming is bull.

Bibliography

AAMI TIR17. Radiation sterilization—Material qualification, 2008. www.aami.org; http://webstore. ansi.org/ansidocstore/product.asp?sku=AAMI+TIR17%3A1997 (accessed September 15, 2015).

AAMI TIR33. Sterilization of health care products—Radiation, substantiation of a selected sterilization dose–method VDmax, 2005. http://www.aami.org; http://www.webstore.ansi.org/ansidocstore/product.asp?sku=AAMI+TIR33%3A2005 (accessed September 15, 2015).

Abdullah G. Making the most of a powerful nuisance. *Penn State Agric Mag* 2003;17–19. http://www.aginfo.psu.edu/psa/f2003/nuisance.html (accessed September 15, 2015).

AC Engineering. Disposable pumps—Product description, 2010. http://www.acengineering.co.il/index.aspx?id=2152 (accessed September 15, 2015).

Adam E, Sarrazin S, Landolfi C, Motte V, Lortat-Jacob H, Vassalle P, Delehedde M. Efficient long-term and high-yielded production of a recombinant proteoglycan in eukaryotic HEK293 cells using a membrane-based bioreactor. *Biochem Biophys Res Commun* 2008;369: 297–302.

Advanced Scientifics. Three60 sampling system—Product description, 2010. http://www.asi360.com (accessed September 15, 2015).

Agalloco J, Akers J. Sterile product manufacture. In: *Pharmaceutical Manufacturing Handbook-Production and Processes*, Gad SC, ed. John Wiley & Sons, Hoboken, NJ, 2008, pp. 99–136.

Agalloco J, Akers J, Madsen R. Choosing technologies for aseptic filling: "Back to the future, forward to the past?" *Pharm Eng* 2007;27:8–16.

Albis Technologies AG. 2010. Zurich, Switzerland.

Albrecht W, Fuchs H, Kittelmann W, eds. *Nonwoven Fabrics*. Wiley-VCH, Weinheim, Germany, 2002.

Aldington S, Bonnerjea J. Scale-up of monoclonal antibody purification processes. *J Chromatogr B* 2007;848:64–78.

Altaras GM, Eklund C, Ranucci C, Maheswari G. Quantitation of lipids with polymer surfaces in cell culture. *Biotechnol Bioeng* 2007;96:999–1007.

American Chemistry Council. PlasticsResource.com: Information on plastics and the environment. http://www.plasticsresource.com/s_plasticsresource/index.asp (accessed September 15, 2015).

American Chemistry Council. Resin identification codes. http://www.americanchemistry.com/s_plastics/bin.asp?CID=1102&DID=4645&DOC=FILE (accessed September 15, 2015).

Anderlei T, Cesana C, De Jesus M, Kühner M, Wurm F. Shaken bioreactors provide culture alternative. *Gen Eng Biotechnol News* 2009;29:44. http://www.genengnews.com/issues/articleindex.aspx (accessed September 15, 2015).

Anders KD, Akhnoukh R, Scheper T, Kretzmer G. Culture fluorescence measurements for the monitoring and characterization of insect cell cultivation in bioreactors. *Chemie Ingenieur Technik* 1992;64(6):572–573.

Anders KD, Wehnert G, Thordsen O, Scheper T, Rehr B, Sahm H. Biotechnological applications of fiber-optic sensing—Multiple uses of a fiber-optic fluorometer. *Sens Actuat B Chem* 1993;11(1–3):395–403.

Anderson KP, Lie YS, Low MA, Williams SR, Fennie EH, Ngyen TP, Wurm FM. Presence and transcription of intracisternal A-particle related sequences in CHO cells. *J Virol* 1990;64:2021–2032.

Anicetti V. Biopharmaceutical processes: A glance into the 21st century. *BioProcess Int* 2009;7(4):S4–S11.

ANSI/AAMI ST67. Sterilization of health care products—Requirements for products labeled "Sterile." http://www.ansi.org (accessed September 15, 2015).

ANSI/AAMI/ISO 11137-1. Sterilization of health care products—Radiation, Part 1: Requirements for development, validation and routine control of a sterilization process for medical devices, 2006. http://www.iso.org (accessed September 15, 2015).

ANSI/AAMI/ISO 11137-2. Sterilization of health care products—Radiation, Part 2: Establishing the sterilization dose, 2006. http://www.iso.org (accessed September 15, 2015).

ANSI/AAMI/ISO 11137-3. Sterilization of health care products—Radiation, Part 3: Guidance on dosimetric aspects, 2006. http://www.iso.org (accessed September 15, 2015).

ANSI/AAMI/ISO 11737-1. Sterilization of medical devices—Microbiological methods, Part 1: Determination of a population of microorganisms on products, 2006. http://www.iso.org (accessed September 15, 2015).

ANSI/AAMI/ISO 11737-2. Sterilization of medical devices—Microbiological methods, Part 2: Tests of sterility performed in the validation of a sterilization process (first edition), 1998. http://www.iso.org (accessed September 15, 2015).

Arnold SA, Gaensakoo R, Harvey LM, McNeil B. Use of at-line and in-situ near-infrared spectroscopy to monitor biomass in an industrial fed-batch *Escherichia coli* process. *Biotechnol Bioeng* 2002;80(4):405–413.

Article 16 of and Annex II to Council Directive 1999/31/EC. Criteria and procedures for the acceptance of waste at landfills. European Union, 1999. http://eur-lex.europa.eu/legal-content/EN/TXT/?qid=1443594664940&uri=CELEX:32003D0033 (accessed September 15, 2015).

Arunakumari A. Integrating high titer cell culture processes with highly efficient purification processes for the manufacturing of human monoclonal antibodies. *IBC Conference*, San Francisco, CA, November 2006.

Arunakumari A, Wang J, Ferreira G. Improved downstream process design for human monoclonal antibody production. *BioPharm Int* 2007;20:36–40.

Arunakumari A, Wang JM, Ferreira G. Alternatives to protein A: Improved downstream process design for humanly monoclonal antibody production. *BioPharm Int* 2007;2:36–40.

ATMI Life Sciences. Mixing and resuspension of high powder loads using a magnetic mixer. Application note, 2008a. http://www.pall.com/pdfs/Biopharmaceuticals/BPI-Mixers-Rawlings.pdf (accessed September 15, 2015).

ATMI Life Sciences. Mixing a diatomaceous earth slurry using the Jet-Drive mixer. Application note, 2008b. Available from ATMI Life Sciences. www.pall.com.

ATMI Life Sciences. Mixing a diatomaceous earth slurry using a Pad-Drive mixer. Application note, 2008c. Available from ATMI Life Sciences. www.pall.com.

ATMI Life Sciences. Mixing a diatomaceous earth slurry using the WandMixer. Application note, 2008d. Available from ATMI Life Sciences. www.pall.com.

ATMI Life Sciences. Mixing of high powder loads using a LevMixer. Application note, 2008e. Available from ATMI Life Sciences. www.pall.com.

ATMI Life Sciences. Particle generation in the Jet-Drive mixer. Application note, 2008f. Available from ATMI Life Sciences. www.pall.com.

ATMI Life Sciences. Particle generation in the Magnetic Mixer. Application note, 2008g. Available from ATMI Life Sciences. www.pall.com.

ATMI Products. Catalog, 2009. www.pall.com.

ATV. Abwasser aus gentechnischen Produktionsanlagen und vergleichbaren Einrichtungen. DWA, Hennef, Germany, 1996.

ATV. Abwasser aus Krankenhaäusern und anderen medizinischen Einrichtungen. DWA, Hennef, Germany, 2001.

Aunins JB, Bibila TA, Gatchalian S et al. Reactor development for the hepatitis A vaccine VAQTA. In: *Animal Cell Technology: From Vaccine to Genetic Medicine*, Carrondo MJT, Griffiths B, Moreira JLP, eds. Kluwer, Dordrecht, the Netherlands, 1997, pp. 175–183.

Aunins JG et al. Reactor development for the hepatitis A vaccine VAQTA. In: *Animal Cell Technology: From Vaccine to Genetic Medicine*, Carrondo MJT, Griffiths B, Moreira JLP, eds. Kluwer, Dordrecht, the Netherlands, 1997, pp. 175–183.

Bail P, Crawford B, Lindström K. 21st century vaccine manufacturing. *BioProcess Int* 2009;4:18–28.

Bandrup J. Verfahrenswege der Kunststoffverwertung aus ökonomischer und ökologischer Sicht. In: *Bio- und Restabfallbehandlung*, Wiemer K, Kern M, eds. Witzenhausen Institute, Witzenhausen, Germany, 1999.

Barbaroux M, Sette A. Properties of materials used in single-use flexible containers: Requirements and analysis, 2006. http://biopharminternational.findpharma.com/biopharm/article/articleDetail.jsp?id=423541&sk=&date=&pageID=7 (accessed September 15, 2015).

Barnoon B, Bader B. Lifecycle cost analysis for single-use systems. *BioPharm Int* 2008.

Bartolome AJ, Ulber R, Scheper T, Sagi E, Belkin S. Genotoxicity monitoring using a 2D-spectroscopic GFP whole cell biosensing system. *Sens Actuat B Chem* 2003;89(1–2):27–32.

Bean B, Matthews T, Daniel N, Ward S, Wolk B. Guided wave radar at Genentech: A novel technique for noninvasive volume measurement in disposable bioprocess bags: GWR may be a cheaper, more practical alternative to traditional methods, 2008. http://www.pharma-manufacturing.com/articles/2008/185.html (accessed September 15, 2015).

Becker T, Mitzscherling M, Delgado A. Hybrid data model for the improvement of an ultrasonic-based gravity measurement system. *Food Control* 2002;13(4–5):223–233.

Beeksma LA, Kompier R. Cell growth and virus propagation in the costar cell cube system. In: *Animal Cell Technology: Developments towards the 21st Century*, Beuvery EC, Griffiths JB, Zeijlemaker WP, eds. Kluwer, Dordrecht, the Netherlands, 1995, pp. 661–663.

Behme S. *Manufacturing of Pharmaceutical Proteins*. Wiley-VCH Verlag GmbH & Co. KGaA, Weinheim, Germany, 2009a.

Behme S. Production facilities. In: *Manufacturing of Pharmaceutical Proteins*, Behme S, ed. Wiley VCH, Weinheim, Germany, 2009b, pp. 227–275.

Bender J, Wolk B. Putting a spin on CHO harvest. Centrifuge technology development. *ACS Meeting*, Boston, MA, 1998.

Bennan J et al. Evaluation of extractables from product-contact surfaces. *BioPharm Int* 2002;15(12):S22–S34.

Bentebibel S, Moyano E, Palazón J, Cusidó RM, Bonfill M, Eibl R, Piñol MT. Effects of immobilization by entrapment in alginate and scale-up on paclitaxel and baccatin III production in cell suspension cultures of *Taxus baccata*. *Biotechnol Bioeng* 2005;89:647–655.

Bergveld P. Development of an ion-sensitive solid-state device for neurophysiological measurements. *IEEE Trans Biomed Eng* 1970;19:70–71.

Bergveld P. Thirty years of ISFETOLOGY: What happened in the past 30 years and what may happen in the next 30 years. *Sens Actuat B Chem* 2003;88(1):1–20.

Bernard F et al. Disposable pH sensors. *BioProcess Int* 2009;7(1): S32–S36.

Bestwick D, Colton R. Extractables and leachables from single-use disposables. *BioProcess Int* 2009;7(1):S88–S94.

Bioengineering. Aerosol-free Hi containment sampling system. Bioengineering-Culture in Hygienic Design, 2001a, pp. 37–40.

Bioengineering. Aseptic connection of two sterile spaces. Bioengineering-Culture in Hygienic Design, 2001b, pp. 21–36.

BioProcess Systems Alliance (BPSA) Disposal Subcommittee. Guide to disposal of single-use bioprocess systems. *BioProcess Int* 2008;6(5):S24–S27.

BioWorld Snapshots. Biotechnology products on the markets since 1982. *BioWorld Today*, Atlanta, GA, 2009.

Boehl D, Solle D, Hitzmann B, Scheper T. Chemometric modelling with two-dimensional fluorescence data for *Claviceps purpurea* bioprocess characterization. *J Biotechnol* 2003;105(1–2):179–188.

Boehm J, Bushnell B. Providing sterility assurance between stainless steel and single-use systems. *BioProcess Int* 2007;5(4):S66–S71.

Booth A. *Radiation Sterilisation, Validation and Routine Operations Handbook.* Davis Healthcare International Publishing, River Grove, IL, 2007.

Bosch Packaging Technology. New Prevas disposable dosing system for filling liquid pharmaceuticals reduces validation and production costs, 2008. http://www.pharmaceuticalonline.com/article.mvc/New-Prevas-Disposable-Dosing-System-For-Filli-0001?VNETCOOKIE=NO (accessed September 15, 2015).

Boss J. Evaluation of the homogeneity degree of a mixture. *Bulk Solids Handling* 1986;6(6):1207–1215.

Brecht R. Disposable bioreactors—Maturation into pharmaceutical glycoprotein manufacturing. In: *Disposable Bioreactors,* Series: *Advances in Biochemical Engineering/Biotechnology,* Vol. 115, Eibl D, Eibl R, eds. Springer, Berlin, Germany, 2009, pp. 1–31.

British Standards Online. EN 13431 Packaging—Requirements for packaging recoverable in the form of energy recovery, including specification of minimum inferior calorific value, 2007. http://www.bsonline. si-global.com.

Brorson K. CDER/FDA. Virus filter validation and performance. *Recovery of Biological Products XII,* Phoenix, AZ, 2006.

Brorson K, Krejci S, Lee K, Hamilton E, Stone K, Xu Y. Bracketed generic inactivation of rodent retroviruses by low pH treatment for monoclonal antibodies and recombinant proteins. *Biotechnol Bioeng* 2003;82:321–329.

Brough H, Antoniou C, Carter J, Jakubik J, Xu Y, Lutz H. Performance of a novel Viresolve NFR virus filter. *Biotechnol Prog* 2002;18:782–795.

Brown LF, Mason JL. Disposable PVDF ultrasonic transducers for nondestructive testing applications. *IEEE Trans Ultrason Ferroelectr Freq Control* 1996;43(4):560–568.

Bruce MP, Boyd V, Duch C, White JR. Dialysis-based bioreactor systems for the production of monoclonal antibodies—Alternatives to ascites production in mice. *J Immunol Methods* 2002;264:59–68.

Büchs J, Maier U, Milbradt C, Zoels B. Power consumption in shaking flasks on rotary shaking machines: I. Power consumption measurements in unbaffled flask at low viscosity. *Biotechnol Bioeng* 2000;68:589–593.

Burrill GS. Biotech IPOs completed 1996–2009. Burrill & Company, San Francisco, CA, 2010.

Bush L. Disposal of disposables. *BioPharm Int* 2008;7:12.

Cabatingan M. Impact of virus stock quality on virus filter validation: A case study. *BioProcess Int* 2005;1:39–43.

Cai K, Gierman TM, Hotta J, Stenland CJ, Lee DC, Pifat DY, Petteway SR. Ensuring the biologic safety of plasma-derived therapeutic proteins. *BioDrugs* 2005;19:79–96.

Cappia J-M, Holman NBT. Integrating single-use disposable processes into critical aseptic processing operations. *BioProcess Int* 2004;2(4). http://www.bioprocessintl.com/multimedia/archive/00078/0209su12_78577a.pdf (accessed September 15, 2015).

Cardona M, Allen B. Incorporating single-use systems in biopharmaceutical manufacturing. *BioProcess Int* 2006;4(4, Suppl.):10–14.

Castellarnau M, Zine N, Bausells J, Madrid C, Juárez A, Samitier J, Errachid A. Integrated cell positioning and cell-based ISFET biosensors. *Sens Actuat B Chem* 2007;120(2):615–620.

Castillo J, Vanhamel S. Cultivating anchorage-dependent cells. *Genet Eng Biotechnol News* 2007;16:40–41.

Cellexus Biosystems. Aseptic sampling from bioreactors, 2006. http://www.proteigene.com/pdf/Cellexussampler.pdf (accessed September 15, 2015).

Center for Drug Evaluation and Research. Guidance for industry: Container closure systems for packaging human drugs and biologics—Chemistry, manufacturing, and controls documentation. U.S. Food and Drug Administration, Rockville, MD, 1999. http://www.fda.gov/cder/guidance/1714fnl.htm (accessed September 15, 2015).

Charles I, Lee J, Dasarathy Y. Single-use technologies—A contract biomanufacturer's perspective. *BioPharm Int* 2007;20(Suppl.):31–36.

Cheng LJ, Li JM, Chen J, Ge YH, Yu ZR, Han DS, Zhou ZM, Sha JH. NYD-SP16, a novel gene associated with spermatogenesis of human testis. *Biol Reprod* 2003;68(1):190–198.

Chmiel H. Bioreaktoren. In: *Bioprozesstechnik*, Chmiel H, ed. Elsevier, München, Germany, 2006, pp. 195–215.

CHMP/CVMP. Guideline on plastic immediate packaging materials. European Medicines Evaluation Agency, London, U.K., 2005. http://www.emea.europa.eu/pdfs/human/qwp/435903en.pdf (accessed September 15, 2015).

Chovelon JM, Fombon JJ, Clechet P, Jaffrezic-Renault N, Martelet C, Nyamsi A, Cros Y. Sensitization of dielectric surfaces by chemical grafting: Application to pH ISFETs and REFETs. *Sens Actuat B Chem* 1992;8(3):221–225.

Clutterbuck A, Kenworthy J, Lidell J. Endotoxin reduction using disposable membrane adsorption technology in cGMP manufacturing. *BioPharm Int* 2007;20:24–31.

Code of Federal Regulations. Title 21 CFR 211. Current good manufacturing practice for finished pharmaceuticals. National Archives and Records Administration, Office of the Federal Register of the USA, 2009a.

Code of Federal Regulations. Title 21, part 210. Current good manufacturing practice for finished pharmaceuticals. National Archives and Records Administration. Office of the Federal Register of the USA, 2009b.

Colder Products. Steam-Thru connections—Product description, 2010. http://www.colder.com/Products/SteamThruConnections/tabid/740/Default.aspx (accessed September 15, 2015).

Cole G. *Pharmaceutical Production Facilities: Design and Applications.* Pharmaceutical Science Series. Taylor & Francis/Informa Health Care, London, U.K., 1998.

Cole Parmer. Silicon tubing—Product description, 2010. http://www.coleparmer.com/Catalog/product_view.asp?sku=9610500 (accessed September 15, 2015).

Colton R. Recommendations for extractables and leachables testing—Part 1. *BioProcess Int* 2007;11:36–44.

Colton R. The extractables and leachables subcommittee of the bio-process systems alliance. Recommendations for extractables and leachables testing, part 1: Introduction, regulatory issues and risk assessment. *BioProcess Int* 2007;12:36–44.

Colton R. Recommendations for extractables and leachables testing—Part 2. *BioProcess Int* 2008a;1:44–52.

Colton R. The extractables and leachables subcommittee of the bio-process systems alliance. Recommendations for extractables and leachables testing, part 2: Executing a program. *BioProcess Int* 2008b;1:44–52.

Committee for Proprietary Medicinal Products (CPMP). Mark for guidance on virus validation studies: The design, contribution and interpretation of studies validating the inactivation and removal of viruses. The European Agency for the Evaluation of Medicinal Products, London, U.K. CPMP/BWP/268/95, 1996.

Committee for Proprietary Medicinal Products (CPMP). Annex to note for guidance on development pharmaceutics: Decision trees for the selection of sterilisation methods. The European Agency for the Evaluation of Medicinal Products, London, U.K., 2000.

CPMP. Mark for guidance on quality of biotechnology products, viral safety evaluation of biotechnology products derived from cell lines of humanly or animal origin. CPMP/I/295/9S. London, U.K., 1997.

CPT Consolidated Polymer Technologie, Inc. Material comparison. Brochure, 2002. http://www.stiflow.com/CFIex-Tubing-Material-Comparison.pdf (accessed September 15, 2015).

Croughan G. 2010. Beyond just high titers: The future of cell line engineering. *IBC Cell Line Development and Engineering Conference*, San Francisco, CA, 2010.

Curling J, Gottschal U. Process chromatography: Five decades of innovation. *BioPharm Int* 2007;21:70–94.

Curtis S, Lee K, Blank GS, Brorsen K, Xu Y. Generic/matrix evaluation of SV40 clearance by anion exchange chromatography in flow-through mode. *Biotechnol Bioeng* 2003;84:179–186.

Curtis WR. Achieving economic feasibility for moderate-value food and flavour additives. In: *Plant Cell and Tissue Culture for the Production of Food Ingredients*, Fu T, Singh G, Curtis WR, eds. Kluwer Academic, New York, 1999, pp. 225–236.

Curtis WR. Growing cells in a reservoir formed of a flexible sterile plastic liner. United States Patent 6709862B2, 2004.

D'Aquino R. Bioprocessing systems go disposable. *Chem Eng Prog* 2006;102(5):8–11.

Dancette OP, Taboureau JL, Tournier E, Charcosset C, Blond P. Purification of immunoglobulins G by protein A/G affinity membrane chromatography. *J Chromatogr B* 1999;723:61–68.

Danckwerts PV. Continuous flow systems: Distribution of residence times. *Chem Eng Sci* 1953;2:1–13.

Davis JM. Hollow fibre cell culture. In: *Animal Cell Biotechnology: Methods and Protocols,* Series: *Methods in Biotechnology*, Vol. 24, Pörtner R, ed. Humana Press, Totowa, NJ, 2007a, pp. 337–352.

427

Davis JM. Systems for cell culture scale-up. In: *Medicines from Animal Cell Culture*, Stacey G, Davis JM, eds. John Wiley & Sons, Chichester, U.K., 2007b, pp. 145–171.

Davis RM, Taylor G. The mechanics of large bubbles rising through liquids in tubes. *Proc R Soc Lond A* 1950;200:375–392.

De Jesus MJ, Girard P, Bourgeois M, Baumgartner G, Kacko B, Amstutz H, Wurm FM. TubeSpin satellites: A fast track approach for process development with animal cells using shaking technology. *Biochem Eng J* 2004;17:217–223.

De Jesus MJ, Wurm FM. Medium and process optimization for high yield, high density suspension cultures: From low throughput spinner flasks to high throughput millilitre reactors. *BioProcess Int* 2009;7(Suppl. 1):12–17.

De Wilde D, Noack U, Kahlert W, Barbaroux M, Greller G. Bridging the gap from reusable to single-use manufacturing with stirred, single-use bioreactors. *BioProcess Int* 2009;7(Suppl. 4):36–41.

DeGrazio FL. 2006. The importance of leachables and extractables testing for a successful product launch. http://pharmtech.fmdpharma.com/pharmtech/Article/The-Importance-of-Leachables-and-Extractables-Test/ArticleStandard/Article/detail/482447 (accessed September 15, 2015).

Denton A, Jones C, Tarrach K. Integration of large-scale chromatography with nanofiltration for an ovine polyclonal product. *Pharm Technol* 2009;1:62–70.

DePalma A. Options for anchorage-dependent cell culture. *Genet Eng Biotechnol News* 2002;22:58–62.

DePalma A. Bioprocessing on the way to total-disposability manufacture. *Genet Eng Biotechnol News* 2004;24(3):40–41.

DePalma A. Liquid mixing: Solid challenges, 2005. http://www.pharmamanufacturing.com/articles/2005/297.html (accessed September 15, 2015).

DePalma A. Bright sky for single-use bioprocess products. *Genet Eng Biotechnol News* 2006;26(3):50–57.

Desai MA, Rayner M, Burns M, Bermingham D. Application of chromatography in the downstream processing of biomolecules. *Methods Biotechnol* 2000;9:73–94.

DeWilde D, Noack U, Kahlert W, Barbaroux M, Greller G. Bridging the gap from reusable to single-use manufacturing with stirred, single-use bioreactors. *BioProcess Int* 2009;7(Suppl. 4):36–42.

DiBlasi K, Jornitz MW, Gottschalk U, Priebe PM. Disposable biopharmaceutical processes—Myth or reality? *BioPharm Int* 2006;11:2–10.

Diehl T. Application of membrane chromatography in the purification of humanly monoclonal antibodies. *Downstream Technology Forum*, King of Prussia, PA, 2006.

DiMasi JA, Grabowski HG. The cost of biopharmaceutical R&D: Is biotech different? *Manage Decis Econ* 2007;28:469–479.

DIN EN 556. Sterilization of medical devices—Requirements for medical devices to be designated "Sterile." http://webstore.ansi.org.

DIN ISO 15378. Primärverpackungen für Arzneimittel—Besondere Anforderungen für die Anwendung von ISO 9001:2000 entsprechend der Guten Herstellungspraxis (GMP) (ISO 15378:2006), October 2007, DIN-Norm.

Ding W, Martin J. Implementing single-use technology in biopharmaceutical manufacturing: An approach to extractables/leachables studies, part one—Connections and filters. *BioProcess Int* 2008;6(9):34–42.

Directive 2000/76/EC of the European Parliament and of the Council on the Incineration of Waste. *Off J Eur Communities* 2000;332:91–111. http://www.eur-lex.europa.eu/LexUriServ/LexUriServ.do?uri=OJ:L:2000:332:0091:0111:EN:PD (accessed September 15, 2015).

Dremel BAA, Schmid RD. Optical sensors for bio-process control. *Chemie Ingenieur Technik* 1992;64(6):510–517.

Drugmand JC, Havelange N, Debras F, Collignon F, Mathieu E, Castillo J. Human and animal vaccine production in a new disposable fixed-bed bioreactor, 2009. http://www.artelis.be./uploads/pdf.POSTER%20ARTEFIX%20BD.pdf (accessed September 15, 2015).

Ducos JP, Terrier B, Courtois D. Disposable bioreactors for plant micropropagation and cell cultures. In: *Disposable Bioreactors,* Series: *Advances in Biochemical Engineering/Biotechnology*, Vol. 115, EiblD, Eibl R, eds. Springer, Berlin, Germany, 2009, pp. 89–115.

Ducos JP, Terrier B, Courtois D, Pétiard V. Improvement of plastic-based disposable bioreactors for plant science needs. *Phytochem Rev* 2008;7:607–613.

Edelmann W, Arnet M, Schwarzenbach HU, Stutz E. *Kunststoffverwertung im Kanton Zug.* ZEBA, Zug, Switzerland, 2004.

Eibl R, Eibl D. Bioreactors for plant cell and tissue cultures. In: *Plant Biotechnology and Transgenic Plants*, Oksman-Caldentey KM, Barz WH, eds. Marcel Dekker, New York, 2002, pp. 163–199.

Eibl R, Eibl D. Design and use of the wave bioreactor for plant cell culture. In: *Plant Tissue Culture Engineering*, Series: *Focus on Biotechnology*, Vol. 6, Dutta Gupta S, Ibaraki Y, eds. Springer, Dordrecht, the Netherlands, 2006, pp. 203–227.

Eibl R, Eibl D. Disposable bioreactors for cell culture-based bioprocessing. *ACHEMA Worldwide News* 2007a;2:8–10.

Eibl R, Eibl D. Disposable bioreactors for inoculum production and protein expression. In: *Animal Cell Biotechnology: Methods and Protocols*, Series: *Methods in Biotechnology*, Vol. 24, Pörtner R, ed. Humana Press, Totowa, NJ, 2007b, pp. 321–335.

Eibl R, Eibl D. Application of disposable bag bioreactors in tissue engineering and for the production of therapeutic agents. In: *Bioreactor Systems for Tissue Engineering*, Vol. 112, Kasper C, van Griensven M, Pörtner R, eds. Springer, Berlin, Germany, 2008a, pp. 183–207.

Eibl R, Eibl D. Bioreactors for mammalian cells: General overview. In: *Cell and Tissue Reaction Engineering*, Eibl R, Eibl D, Pörtner R, Catapano G, Czermak P, eds. Springer, Heidelberg, Germany, 2008b, pp. 55–82.

Eibl R, Eibl D. Design of bioreactors suitable for plant cell and tissue cultures. *Phytochem Rev* 2008c;7:593–598.

Eibl R, Eibl D. Disposable bioreactors in cell culture-based upstream processing. *BioProcess Int* 2009a;7(Suppl. 1):18–23.

Eibl D, Eibl R (eds.). *Disposable Bioreactors*, Series: *Advances in Biochemical Engineering/Biotechnology*, Vol. 115. Springer, Berlin, Germany, 2009b.

Eibl R et al. Disposable bioreactors: The current state-of-the-art and recommended applications in biotechnology. *Appl Microbiol Biotechnol* 2010;86(1):41–49.

Eibl R, Rutschmann K, Lisica L, Eibl D. Kosten reduzieren durch Einwegbioreaktoren? *BioWorld* 2003;5:22–23.

Eibl R, Werner S, Eibl D. Bag bioreactor based on wave-induced motion: Characteristics and applications. In: *Disposable Bioreactors*, Series: *Advances in Biochemical Engineering/Biotechnology*, Vol. 115, Eibl D, Eibl R, eds. Springer, Berlin, Germany, 2009a, pp. 55–87.

Eibl R, Werner S, Eibl D. Disposable bioreactors for plant liquid cultures at litre-scale: Review. *Eng Life Sci* 2009b;9:156–164.

EMEA/CPMP. Mark for guidance on virus validation studies: The design, contribution and interpretation of studies validating the inactivation and removal of viruses, 2008. http://www.emea.europa.eu/pdfs/human/bwp/026895en.pdf (accessed September 15, 2015).

EPA. Guide for industrial waste management, 2009. www.epa.gov/epawaste/nonhaz/industrial/guide (accessed September 15, 2015).

EPA Computational Toxicology Program. Distributed Structure-Searchable Toxicity (DSSTox) database network, 2007. U.S. Environmental Protection Agency, Washington, DC. http://www.epa.gov/ncct/dsstox (accessed September 15, 2015).

Equipment Construction. Code of Federal Regulations, Food and Drugs Title 21, Part 211.65, 2006. U.S. Government Printing Office, Washington, DC. http://www.accessdata.fda.gov/scripts/cdrh/cfdocs/cfcfr/CFRSearch.cfm?CFRPart=211 (accessed September 15, 2015).

Ernst & Young. Beyond Borders Global Biotechnology Report. Ernst & Young, Zurich, Switzerland/Boston, MA, 2007.

Etzel M, Gadam S, eds. *Process Scale Bioseparations for the Biopharmaceutical Industry*. Taylor & Francis, Boca Raton, FL, 1998.

Etzel M, Riordan W. Membrane chromatography: Analysis of breakthrough curves and viral clearance. In: *Process Scale Bioseparations for the Biopharmaceutical Industry*, Shukla A, Etzel M, Gadam S, eds. Taylor & Francis, Boca Raton, FL, 2006, pp. 277–296.

EU. Directive 1999/31/EC; criteria and procedures for the acceptance of waste to landfills. *Off J Eur Union* 1999;182: 1–19.

EU. Directive 2000/76/EC; incineration of waste. *Off J Eur Union* 2000;332:91–111.

EU. Directive 2002/96/EC; waste electrical and electronic equipment. *Off J Eur Union* 2003;37:24–38.

EUDRALEX. Volume 4: Good manufacturing practices, medicinal products for human and veterinary use. European Commission, Brussels, Belgium, 1998. http://www.ec.europa.eu/enterprise/pharmaceuticals/eudralex/homev4.htm (accessed September 15, 2015).

European Commission. Eudralex Vol. 4, Annex 2: Manufacture of biological medicinal products for human use. European Commission, Brussels, Belgium, 1992a.

European Commission. Eudralex Vol. 4, Annex 12: Use of ionising radiation in the manufacture of medicinal products. European Commission, Brussels, Belgium, 1992b.

European Commission. Eudralex Vol. 4, Annex 14: Manufacture of products derived form human blood or human plasma. European Commission, Brussels, Belgium, 1992c.

European Commission. Eudralex Vol. 4, Annex 15: Qualification and validation. European Commission, Brussels, Belgium, 2001.

European Commission. Eudralex Vol. 4, EU Guidelines to Good Manufacturing Practice, part II: Basic requirements for active substances used as starting materials. European Commission, Brussels, Belgium, 2005.

European Commission. Eudralex Vol. 4, Annex 1: Manufacture of sterile medicinal products. European Commission, Brussels, Belgium, 2008a.

European Commission. Eudralex Vol. 4, EU Guidelines to Good Manufacturing Practice, part I: Basic requirements for medicinal products. European Commission, Brussels, Belgium, 2008b.

European Commission (Enterprise Directorate General). EMEA guideline on virus safety evaluation of biotechnological investigational medicinal products. London, U.K., 2006.

European Commission Enterprise Directorate General. Working party on control of medicines and inspections. EU Guide to Good Manufacturing Practice, Vol. 4, Annex 15, July 2001, Cleaning Validation.

European Directorate for the Quality of Medicines & Healthcare. European pharmacopeia, Vol. 6.5, 6th edn. Council of Europe, Strasbourg, France, 2009.

European Medicines Evaluation Agency (EMEA). Guideline on plastic immediate packaging materials. The European Agency for the Evaluation of Medicinal Products, London, U.K. CHMP/CVMP 205/04, 2005.

European Parliament and the Council of the European Union on Packaging and Packaging Waste. European Parliament and Council Directive 94/62/EC on Packaging and Packaging Waste, 1994. http://www.eur-lex.europa.eu/LexUriServ/LexUriServ.do?uri=CELEX:31994L0062: EN:HTML (accessed September 15, 2015).

Extractables and Leachables Subcommittee of the Bio-Process Systems Alliance. Recommendations for extractables and leachables testing. *BioProcess Int* 2008;6(5): S28–S39.

Fahrner RL, Iyver HV, Blank GS. The optimal flow rate and column length for maximum production rate of protein A chromatography. *Bioprocess Eng* 1999;21:287–292.

Fahrner RL, Knudsen HL, Basey CD, Galan W, Feuerhelm D. Industrial purification of pharmaceutical antibodies: Development, operation, and validation of chromatography processes. *Biotechnol Genet Eng Rev* 2001;18:301–327.

Falch FA, Heden CG. Disposable shaker flasks. *Biotechnol Bioeng* 1963;5:211–220.

Falkenberg FW. Production of monoclonal antibodies in the miniPerm bioreactor: Comparison with other hybridoma culture methods. *Res Immunol* 1998;6:560–570.

Farid SS. Process economics of industrial monoclonal use systems, part 2. *BioProcess Int* 2007;(Suppl. 3):51–56.

Farid SS. Process economic drivers in industrial monoclonal antibody manufacture. In: *Process Scale Purification of Antibodies*, Gottschalk U, ed. Wiley, New York, 2009, pp. 239–262.

Farid SS, Washbrook J, Titchener-Hooker NJ. Decision-support tool for assessing biomanufacturing strategies under uncertainty: Stainless steel versus disposable equipment for clinical trial material preparation. *Biotechnol Prog* 2005;21(2):486–497.

Farshid M, Taffs RE, Scott D, Asher DM, Brorson K. The clearance of viruses and transmissible spongiform encephalopathy agents from biologicals. *Curr Opin Biotechnol* 2005;16:561–567.

FDA. Guideline on sterile drug products produced by aseptic processing. FDA, Rockville, MD, 1987.

FDA. Equipment cleaning and maintenance. Code of Federal Regulations (CFR) Part 211.67 Title 21 Rev. FDA, Rockville, MD, May 25, 2004.

Fenge C, Lüllau E. Cell culture bioreactors. In: *Cell Culture Technology for Pharmaceutical and Cell-Based Therapies*, Ozturk SS, Hu WS, eds. CRC, New York, 2006, pp. 155–224.

Fermentation—History—Ferments, living, cell, debate, result, and cells. *BioProcess Intl 2015*, Boston, MA. http://science.jrank.org/pages/2675/Fermentation-History. html#ixzz1IeEDOutS (accessed September 15, 2015).

Fichtner S, Giese U, Phal I, Reif OW. Determination of "extractables" on polymer materials by means of HPLC-MS. *PDA J Pharm Sci Technol* 2006;60:291–301.

Forster R, Ishikawa M. The methodologies for impact assessment of plastic waste management options—How to handle economic and ecological impacts? *Proceedings R99*, Geneva, Switzerland, 1999.

Foulon A et al. Using disposables in an antibody production process. *BioProcess Int* 2008;6(6): 12–18.

Foulon A, Trach F, Pralong A, Proctor M, Lim J. Using disposables in an antibody production process: A cost-effectiveness study of technology transfer between two production sites. *BioProcess Int* 2008;6(Suppl. 3):12–18.

Fox S. Disposable processing: The impact of disposable bioreactors on the CMO industry. *Contract Pharma* 2005;6:62–74.

Fraud N, Kuczewski M, Zarbis-Papastoitsis G, Hirai M. Hydrophobic membrane adsorbers for large-scale downstream processing. *BioPharm Int* 2009;23:24–27.

Fries S, Glazomitsky K, Woods A, Forrest G, Hsu A, Olewinski R, Robinson D, Chartrain M. Evaluation of disposable bioreactors. *BioProcess Int* 2005;3(Suppl. 6):36–44.

Fuller M, Pora H. Introducing disposable systems into biomanufacturing: A CMO case study. *BioProcess Int* 2008;6(10):30–36.

Galliher P. Achieving high-efficiency production with microbial technology in a single-use bioreactor platform. *BioProcess Int* 2008;11:60–65.

Ganzlin M, Marose S, Lu X, Hitzmann B, Scheper T, Rinas U. In situ multi-wavelength fluorescence spectroscopy as effective tool to simultaneously monitor spore germination, metabolic activity and quantitative protein production in recombinant *Aspergillus niger* fed-batch cultures. *J Biotechnol* 2007;132(4):461–468.

GE Healthcare. CHO cell supernatant concentration with Kvick Lab cassettes. Application note 11-0013-62, 2006.

GE Healthcare. Purification of a monoclonal antibody using ReadyToProcess columns. Application note 28-9198-56 AA, 2007.

GE Healthcare. Rapid production of clinical grade T lymphocytes in the Wave Bioreactor, 2008. http://www.5.gelifesciences.com/aptrix/upp0091.nsf./Content/F7AD616DACC22171C125747400812B51/$file/28933149AA.pdf (accessed September 15, 2015).

GE Healthcare. A flexible antibody purification process based on ReadyToProcess products. Application note 28-9403-48 AA, 2009a.

GE Healthcare. High-throughput screening and optimization of a multi modal polishing step in a monoclonal antibody purification process. Application note 28-9509-60, 2009b.

GE Healthcare. High-throughput screening and optimization of a Protein A capture step in a monoclonal antibody purification process. Application note 28-9468-58, 2009c.

GE Healthcare. Scale-up of a downstream monoclonal antibody purification process using HiScreen and AxiChrom columns. Application note 28-9409-49, 2009d.

GE Healthcare. Hot lips tube sealer—Product description, 2010a. http://www.gelifesciences.com/aptrix/upp01077.nsf/Content/wave_bioreactor_home~wave_fluid_transfer~hot_lips_tube_sealer (accessed September 15, 2015).

GE Healthcare. ReadyMate DAC—Product description, 2010b. http://www5.gelifesciences.com/aptrix/upp00919.nsf/Content/692F8252BA8B1477C125763C00827AEI/$file/28937902+AC+.pdf (accessed September 15, 2015).

GE Healthcare. Sterile tube fuser—Technical information, 2010c. http://www5.gelifesciences.com/aptrix/upp01077.nsf/Content/Products?OpenDocument&parentid=986919&moduleid=167710&zone= (accessed September 15, 2015).

GE Healthcare. Wave mixer concept, 2010d. http://wwwI.gelifesciences.com/aptrix/upp01077.nsf/Content/wave_bioreactor_home~how_it_works~how_it_works_wave_mixer (accessed September 15, 2015).

GE. NPC-100 pressure sensor, 2009. http://www.gesensing.com/products/npc_100_series.htm?bc=bc_indust+bc_med_fluid (accessed September 15, 2015).

Gebauer A, Scheper T, Schugerl K. Penicillin acylase production by *E. coli*. *Bioprocess Eng* 1987;2(2):55–58.

Gebauer K, Thommes J, Kula M. Plasma protein fractionation with advanced membrane adsorbents. *Biotechnol Bioeng* 1997a;54:181–189.

Gebauer KH, Thommes J, Kula M. Breakthrough performance of high capacity membrane adsorbers in protein chromatography. *Chem Eng Sci* 1997b;52:405–419.

Genzel Y, Olmer RM, Schaefer B, Reichl U. Wave microcarrier cultivation of MDCK cells for influenza virus production in serum containing and serum-free media. *Vaccine* 2006;24:6074–6087.

Georgiev MI, Weber J, Maciuk A. Bioprocessing of plant cell cultures for mass propagation of targeted compounds. *Appl Microbiol Biotechnol* 2009;83:809–823.

Ghosh R. Protein separation using membrane chromatography: Opportunities and challenges. *Chromatogr A* 2004;952:13–27.

Girard LS, Fabis MJ, Bastin M, Courtois D, Pétiard V, Koprowski H. Expression of a human anti-rabies virus monoclonal antibody in tobacco cell culture. *BBRC* 2006;345:602–607.

Glaser V. Disposable bioreactors become standard fare. *Genet Eng Biotechnol News* 2005;25(14):80–81.

Glaser V. Bioreactor and fermentor trends. *Genet Eng Biotechnol News* 2009. http://www.genengnews.com/issues/item.aspx (accessed September 15, 2015).

Gold LS. Carcinogenic potency database (CPDB). University of California, Berkeley, CA, 2007. http://www.potency.berkeley.edu/cpdb.html.

Goldstein A, Loesch J, Mazzarella K, Matthews T, Luchsinger G, Javier DS. Freeze bulk bags: A case study in disposables implementation: Genentech's evaluation of single-use technologies for bulk freeze-thaw, storage, and transportation, 2009. http://biopharminternational.findpharma.com/biopharm/Disposables+Articles/Freeze-Bulk-Bags-A-Case-Study-in-Disposables-Imple/ArticleStandard/Article/detail/637583 (accessed September 15, 2015).

Gorter A, van de Griend RJ, van Eendenburg JD, Haasnot WH, Fleuren GJ. Production of bi-specific monoclonal antibodies in a hollow-fibre bioreactor. *J Immunol Methods* 1993;161:145–150.

Gottschaik U. Downstream processing of monoclonal antibodies: From high dilution to high purity. *BioPharm Int* 2005a;19:42–58.

Gottschalk U. New and unknown challenges facing biomanufacturing. *BioPharm Int* 2005b;19:24–28.

Gottschalk U. Thirty years of monoclonal antibodies: A long way to pharmaceutical and commercial success modern pharmaceuticals. Wiley-VCH Verlag GmbH & Co. KGaA, Weinheim, Germany, 2005c.

Gottschalk U. The renaissance of protein purification. *BioPharm Int* 2006;6(Suppl. 6):8–9.

Gottschalk U. New standards in virus and contaminant clearance. *BioPharm Int* 2007a;10:5.

Gottschalk U. The renaissance of protein purification. *BioPharm Int* 2007b;10:41–42.

Gottschaik U. Bioseparation in antibody manufacturing: The good, the bad and the ugly. *Biotechnol Prog* 2008;24:496–503.

Gottschalk U. Disposables in downstream processing. In: *Disposable Bioreactors*, Series: *Advances in Biochemical Engineering/Biotechnology*, Vol. 115, Eibl R, Eibl D, eds. Springer, Berlin, Germany, 2009, pp. 172–183.

Greb E. The debate over preuse filter-integrity testing. PharmTech, 2009. http://pharmtech.findpharma.com/pharmtech/Article/The-Debate-over-Preuse-Filter-Integrity-Testing/Artic!eStandard/Article/detail/612612 (accessed September 15, 2015).

Groton Biosystems. 24/7 online reactor autosampling, 2009. http://www.grotonbiosystems.com/pressroom/press_docs/MKT-046_ProductBulletin_SampleProbe.pdf (accessed September 15, 2015).

Guide for Industrial Waste Management. U.S. Environmental Protection Agency, Washington, DC, 2008. http://www.epa.gov/epawaste/nonhaz/industrial/guide/index.htm (accessed September 15, 2015).

Gupta A, Rao G. A study of oxygen transfer in shake flasks using a non-invasive oxygen sensor. *Biotechnol Bioeng* 2003;84:351–358.

Guth U, Gerlach F, Decker M, Oelßner W, Vonau W. Solid-state reference electrodes for potentiometric sensors. *J Solid State Electrochem* 2009;13(1):27–39.

Hagedorn A, Levadoux W, Groleau D, Tartakovsky B. Evaluation of multiwavelength culture fluorescence for monitoring the aroma compound 4-hydroxy-2(or 5)-ethyl-5(or 2)-methyl-3(2H)-furanone (HEMF) production. *Biotechnol Prog* 2004;20(1):361–367.

Haldankar R, Li D, Saremi Z, Baikalov C, Deshpande R. Serum-free suspension large-scale transient transfection of CHO cells in Wave bioreactors. *Mol Biotechnol* 2006;34:191–199.

Hami LS, Chana H, Yuan V, Craig S. Comparison of a static process and a bioreactor-based process for the GMP manufacture of autologous Xcellerated T cells for clinical trials. *BioProcess J* 200;32:1–10.

Hami LS, Green C, Leshinsky N, Markham E, Miller K, Craig S. GMP production of Xcellerated T cells for the treatment of patients with CLL. *Cytotherapy* 2004;6:554–562.

Hamid Mollah A. Cleaning validation for bio-pharmaceutical manufacturing at Genentech. *BioPharm Int* 2008;2:36–41.

Hantelmann K, Kollecker A, Hull D, Hitzmann B, Scheper T. Two-dimensional fluorescence spectroscopy: A novel approach for controlling fed-batch cultivations. *J Biotechnol* 2006;121(3):410–417.

Harris M. The billion-plus blockbusters: The top 25 Biotech drugs. *BioWorld* 2009. http://www.bioworld.com/servlet/com.accumedia.web.Dispatcher?next=bioWorldHeadlines_article&forceid=51907 (accessed September 15, 2015).

Haughney H, Hutchinson J. A disposable option for bovine serum filtration and packaging. *BioProcess Int Suppl* 2004;4(9):2–5.

Haughney H, Hutchinson J. Single-use systems reduce production timelines. *Gen Eng News* 2004;24(8). www.hyclone.com/pdf/bp_single_use24gen8.pdf (accessed September 15, 2015).

Heath C, Kiss R. Cell culture process development: Advances in process engineering. *Biotechnol Prog* 2007;23:46–51.

Heinzle E, Biwer AP, Cooney CL. Monoclonal antibodies. In: *Development of Sustainable Bioprocesses-Modelling and Assessment*, Heinzle E, Biwer AP, Cooney CL, eds. John Wiley & Sons, Chichester, U.K., 2006, pp. 241–260.

Heller A, Feldmann B. Electrochemical glucose sensors and their applications in diabetes management. *Chem Rev* 2008;108:2482–2505.

Hemmerich KJ. Polymer materials selection for radiation-sterilized products. *MDDI* 2000. http://www.devicelink.com/mddi/archive/00/02/006.html (accessed September 15, 2015).

Henning B, Rautenberg J. Process monitoring using ultrasonic sensor systems. *Ultrasonics* 2006;44:e1395–e1399.

Henzler HJ. Particle stress in bioreactors. In: *Influence of Stress on Cell Growth and Product Formation*, Series: *Advances in Biochemical Engineering/Biotechnology*, Vol. 67, Schügerl K, Kretzmer G, eds. Springer, Berlin, Germany, 2000, pp. 38–82.

Hess S, Baier U, Lettenbauer C, Hafner D. A new application for the wave bioreactor 20: Cultivation of Erynia neoaphidis, a mycel producing fungus. *IOBC Meeting "Insect Pathogens and Insect Parasitic Nematodes"*, Birmingham, U.K., 2002.

Higuchi A, Kyokon M, Murayama S, Yokogi M, Hirasaki T, Manabe SI. Effect of aggregated protein sizes on the flux of protein solution through microporous membranes. *J Membr Sci* 2004;236:137–144.

Hilmer J-M, Scheper T. A new version of an in situ sampling system for bioprocess analysis. *Acta Biotechnol* 1996;16(2–3):185–192.

Hirschy O, Schmid T, Grunder JM, Andermatt M, Bollhalder F, Sievers M. Wave reactor and the liquid culture of the entomopathogenic nematode *Steinerma feltiae*. In: *Developments in Entomopathogenic Nematode/Bacterial Research*, Griffin CT, Burnell AM, Downes MJ, Mulder R, eds. DG XII, COST 819. Luxembourg, 2001.

Hitchcock T. Production of recombinant whole-cell vaccines with disposable manufacturing systems. *BioProcess Int* 2009;5:36–43.

Hitzmann B, Broxtermann O, Cha YL, Sobieh O, Stark E, Scheper T. The control of glucose concentration during yeast fed-batch cultivation using a fast measurement complemented by an extended Kalman filter. *Bioprocess Eng* 2000;23(4):337–341.

Hopkinson J. Hollow fibre cell culture systems for economical cell-product manufacturing. *BioTechnology* 1985;3:225–230.

Horowitz B, Lazo A, Grossberg H, Page G, Lippin A, Swan G. Virus inactivation by solvent/detergent treatment and the manufacture of SD-plasma. *Vox Sang* 1998;74(Suppl. I):203–206.

Hou K, Mandaro R. Bioseparation by ion exchange cartridge chromatography. *BioTechniques* 1986;4:358–366.

Houtzager E, van der Linden R, de Roo G, Huurman S, Priem P, Sijmons PC. Linear scale-up of cell cultures. The next level in disposable bioreactor design. *BioProcess Int* 2005;6:60–66.

Hsiao TY, Bacani FT, Carvalho EB, Curtis WR. Development of a low capital investment reactor system: Application for plant cell suspension culture. *Biotechnol Prog* 1999;15:114–122.

Hübl D, Böhm HP. Arzneimittelrückstände im Wasser und Abwasser. I. *Magdeburger Workshop*, Magdeburg, Germany, 2007.

Hundt B, Best C, Schlawin N, Kassner H, Genzel Y, Reichl U. Establishment of a mink enteritis vaccine production process in stirred-tank reactor and Wave®Bioreactor microcarrier cultures in 1–10 L scale. *Vaccine* 2007;25:3987–3995.

Hynetics Co. Description of the HyNetics disposable mixing system, 2010. http://www.hynetics.com/default/WhatWeOffer.htm (accessed September 15, 2015).

ICH Q7. Good manufacturing practice guide for active pharmaceutical ingredients. *Fed Register* 2000;66(186):49028–49029. http://www.ich.org/LOB/media/MEDIA433.pdf (accessed September 15, 2015).

Immelmann A, Kellings K, Stamm O, Tarrach K. Validation and quality procedures for virus and prion removal in biopharmaceuticals. *BioProcess Int* 2005;3:38–45.

Incardona NL, Tuech JK, Murti G. Irreversible binding of phage phi XI74 to cell-bound lipopolysaccharide receptors and release of virus-receptor complexes. *Biochemistry* 1985;24: 6439–6446.

Innovative Pharmaceutical Development, Manufacturing, and Quality.

International Conference on Harmonization. Q5B: Quality of biotechnological products: Analysis of the expression construct in cells used for production of r-DNA derived protein products, 1995a. Geneva, Switzerland.

International Conference on Harmonization. Q5C: Quality of biotechnological products: Stability testing of biotechnological/biological products, 1995b. Geneva, Switzerland.

International Conference on Harmonization. ICH harmonized tripartite guideline characterisation of cell substrates used for production of biotechnological/biological products; Q5D Current step; July 16, 1997.

International Conference on Harmonization. S6: Preclinical safety evaluation of biotechnology-derived pharmaceuticals, 1997. Geneva, Switzerland.

International Conference on Harmonization. Q5A: Viral safety evaluation of biotechnology products derived from cell lines of human or animal origin Q5A, 1999. Geneva, Switzerland.

International Conference on Harmonization. Q7: Good manufacturing practice guide for active pharmaceutical ingredients, 2000. Geneva, Switzerland.

International Conference on Harmonization. Q5E: Comparability of biotechnological/biological products subject to changes in their manufacturing process, 2002. Geneva, Switzerland.

International Conference on Harmonization. Q9: Quality risk management, 2005. Geneva, Switzerland.

International Conference on Harmonization. Q3A: Impurities in new drug substances, 2006a. Geneva, Switzerland.

International Conference on Harmonization. Q3B: Impurities in new drug products, 2006b. Geneva, Switzerland.

International Conference on Harmonization. Q3C: Impurities: Guideline for residual solvents, 2007. Geneva, Switzerland.

International Conference on Harmonization. Q10: Pharmaceutical quality management system, 2008a. Geneva, Switzerland.

International Conference on Harmonization. Q8: Pharmaceutical development, 2008b. Geneva, Switzerland.

International Organization for Standardization. Sterilization of health care products—Radiation. ISO 11137, ISO's Central Secretariat, Geneva, Switzerland, 2006a.

International Organization for Standardization. Sterilization of health care products—Biological indicators. ISO 11138, ISO's Central Secretariat, Geneva, Switzerland, 2006b.

International Organization for Standardization. Sterilization of health care products—Vocabulary. ISO 11139, ISO's Central Secretariat, Geneva, Switzerland, 2006c.

International Organization for Standardization. Sterilization of health care products—Ethylene oxide. ISO 11135, ISO's Central Secretariat, Geneva, Switzerland, 2008.

International Organization for Standardization. Sterilization of health care products—General requirements for characterization of sterilizing agent and the development, validation and routine control of sterilization process for medical devices. ISO 14937, ISO's Central Secretariat, Geneva, Switzerland, 2009a.

International Organization for Standardization. Sterilization of health care products—biological indicators—Guidance for the selection, use and interpretation of results. ISO 14161, ISO's Central Secretariat, Geneva, Switzerland, 2009b.

Internationally Conference on Harmonization. Q5A: Viral safety evaluation of biotechnology product derived from cell lines of humanly or animal origin. Geneva, Switzerland, 1998.

Ireland T, Lutz H, Siwak M, Bolton G. Viral filtration of plasma-derived humanly IgG:A case study using viresolve NFR, *BioPharm Int* 2004;11:38–44.

Isberg EA. Advanced aseptic processing: RABS and isolator operations. *Pharm Eng* 2007;27:18–21.

Ishihara T, Kadoya T. Accelerated purification process development of monoclonal antibodies for shortening time to clinic design and case study of chromatography processes. *J Chromatogr A* 2007;176:149–156.

Iwaya M, Iron Mountain S, Bartok K, Denhardt G. Mechanism of replication of single stranded PhiXI74 DNA. VII. Circularization of the progeny viral strand. *J Virol* 1973;12:808–818.

Jablonski-Lorin C, Mellio V, Hungerbühler E. Stereoselective bioreduction to a chiral building block on a kilogram scale. *Chimia* 2003;57:574–576.

Jagschies G. Flexible manufacturing: Driving monoclonal antibody process economics. *GIT BIOprocess* 2009;1:30–31.

Jagschies G, O'Hara A. Debunking downstream bottleneck myth. *Genet Eng News* 2007;27:3.

Jain E, Kumar A. Upstream processes in antibody production: Evaluation of critical parameters. *Biotechnol Adv* 2008;26:46–72.

Jenke D. An extractables/leachables strategy facilitated by collaboration between drug product vendors and plastic material/system suppliers. *PDA J Pharm Sci Technol* 2007;61:17–23.

Jenke D. Evaluation of the chemical compatibility of plastic contact materials and pharmaceutical products; safety consider-ations related to extractables and leachables. *J Pharm Sci* 2007;96:2566–2581.

Jenke D, Story J, Lalani R. Extractables/leachables from plastic tubing used in product manufacturing. *Int J Pharm* 2006;315:75–92.

Jia Q, Li H, Hui M, Hui N, Joudi A, Rishton G, Bao L, Shi M, Zhang X, Luanfeng L, Xu J, Leng G. A bioreactor system based on a novel oxygen transfer method. *BioProcess Int* 2008;6:66–78.

Joeris K, Frerichs JG, Konstantinov K, Scheper T. In-situ microscopy: Online process monitoring of mammalian cell cultures. *Cytotechnology* 2002;38(1–2):129–134.

Joeris K, Scheper T. Visualizing transport processes at liquid-liquid interfaces—The application of laser-induced fluorescence. *J Colloid Interface Sci* 2003;267(2):369–376.

Jones C, Dent NA. Integration of generous scale chromatography with nanofiltration for in ovine polyclonal product. *Second European Downstream Technology Forum*, Goettingen, Germany, 2007.

Jones SCB, Smith MP. Evaluation of an alkali stable Protein A matrix versus Protein A sepharose fast flow and considerations on process scale-up to 20,000 L. *Third International Symposium on Downstream Processing of Genetically Engineered Antibodies and Related Molecules*, Nice, France, October 2004.

Joo S, Brown RB. Chemical sensors with integrated electronics. *Chem Rev* 2008;108(2):638–651.

Jornitz MW, Meltzer TH. *Filtration Handbook Liquids*. PDA, DHI LLC, River Grove, IL, 2004.

Jornitz MW, Meltzer TH. *Pharmaceutical Filtration—The Management of Organism Removal*. PDA, DHI LLC, River Grove, IL, 2006.

Jornitz MW, Meltzer TH, eds. Filtration and purification in the biopharmaceutical industry. In: *Drugs and Pharmaceutical Sciences*, 2nd ed., Vol. 174, Swarbrick J, ed. Informa Healthcare, New York, 2008.

Kallenbach NO, Cornelius PA, Negus D, Montgomerie D, Englander S. Inactivation of viruses by ultraviolet light. *Curr Stud Hematol Blood Transfus* 1989;56:70–82.

Karlsson E, Ryden L, Brewer J. Ion exchange chromatography. In: *Protein Purification: Principles, High-Resolution Methods and Applications*, Janson JC, Ryden L, eds. VCH, New York, 1989, pp. 107–115.

Kato Y, Peter CP, Akgün A, Büchs J. Power consumption and heat transfer resistance in large rotary shaking vessels. *Biochem Eng J* 2004;21:83–91.

Kelley B. Industrialization of mAb production technology. The bioprocessing industry at a crossroads. *MAbs* 2009;1(5):443–452.

Kelley K. Very large scale monoclonal antibody purification: The case for conventional unit operations. *Biotechnol Prog* 2007;23:995–1008.

Kemeny MM, Cooke V, Melester TS, Halperin IC, Burchell AR, Yee JP, Mills CB. Splenectomy in patients with AIDS and AIDS-related complex. *AIDS* 1993;7(8):1063–1067.

Kemplen R, Preissmann A, Berthold W. Assessment of a disc stack centrifuge for use in mammalian cell separation. *Biotechnol Bioeng* 1995;46:132–138.

Kermis HR, Kostov Y, Rao G. Rapid method for the preparation of a robust optical pH sensor. *Analyst* 2003;128(9):1181–1186.

Khanna V. ISFET (ion-sensitive field-effect transistor)-based enzymatic biosensors for clinical diagnostics and their signal conditioning instrumentation. *IETE J Res* 2008;54(3): 193–202.

Khanna VK, Ahmad S, Jain YK, Jayalakshmi M, Vanaja S, Madhavendra SS, Manorama SV, Kantam ML. Development of potassium-selective ion-sensitive field-effect transistor (ISFET) by depositing ionophoric crown ether membrane on the gate dielectric, and its application to the determination of K^+-ion concentrations in blood serum. *IJEMS* 2007;14(2):112–118.

Khanna VK, Kumar A, Jain YK, Ahmad S. Design and development of a novel high-transconductance pH-ISFET (ion-sensitive field-effect transistor)-based glucose biosensor. *Int J Electron* 2006;93(2):81–96.

Kinney SD, Phillips CW, Lin KJ. Thermoplastic tubing, welders and sealers. *BioProcess Int* 2007;5(4):52–61.

Kleenpak data sheet biopharm_34125. Pall Corporation, 2009.

Klimant I, Kuhl M, Glud RN, Hoist G. Optical measurement of oxygen and temperature in microscale: Strategies and biological applications. *Sens Actuat B Chem* 1997;38(1–3):29–37.

Knazek RA, Gullino PM, Kohler PO, Dedrick RL. Ceil culture on artificial capillaries: An approach to tissue growth in vitro. *Science* 1972;178:65–67.

Knevelman C, Hearle DC, Osman JJ, Khan M, Dean M, Smith M, Aiyedebinu Cheung K. Characterization and operation of a disposable bioreactor as a replacement for conventional steam-in-place inoculum bioreactors for mammalian cell culture processes, 2002. http://www.5.gelifesciences.com/aptrix/upp01077.nsf/Content/wave_bioreactor_home-wave_literature_WindowsInternetExplorer.

Knudsen HL, Fahrner RL, Xu Y, Morling LA, Blank GS. Membrane ion-exchange chromatography for process-scale antibody purification. *J Chromatogr A* 2001;907:145–154.

Knuttel T, Hartmann T, Meyer H, Scheper T. On-line monitoring of a quasi-enantiomeric reaction with two coumarin substrates via 2D-fluorescence spectroscopy. *Enzyme Microb Technol* 2001;29(2–3):150–159.

435

Knuttel T, Meyer H, Scheper T. Application of 2D-fluorescence spectroscopy for on-line monitoring of pseudoenantiomeric transformations in supercritical carbon dioxide systems. *Anal Chem* 2005;77(19):6184–6189.

Knuttel T, Meyer H, Scheper T. The application of two-dimensional fluorescence spectroscopy for the on-line evaluation of modified enzymatic enantioselectivities in organic solvents by forming substrate salts. *Enzyme Microb Technol* 2006;39(4):607–611.

Kohls O, Scheper T. Setup of a fiber optical oxygen multisensor-system and its applications in biotechnology. *Sens Actuat B Chem* 2000;70(1–3):121–130.

Koneke R, Comte A, Jurgens H, Kohls O, Lam H, Scheper T. Faseroptische Sauerstoffsensoren für Biotechnologie, Umwelt- und Lebensmitteltechnik. *Chemie Ingenieur Technik* 1998;70(12):1611–1617.

Kornmann H, Valentinotti S, Duboc P, Marison I, von Stockar U. Monitoring and control of Gluconacetobacter xylinus fed-batch cultures using in situ MID-IR spectroscopy. *J Biotechnol* 2004;113:231–245.

Kranjac D. Validation of bioreactors: Disposable vs. reusable. *BioProcess Int Industry Yearbook* 2004;86.

Kraume M, ed. *Mischen und Rühren-Grundlagen und moderne Verfahren.* Wiley-VCH Verlag GmbH & Co. KGaA, Weinheim, Germany, 2003.

Kresta SM, Brodkey RS. Turbulence in mixing applications. In: *Handbook of Industrial Mixing-Science and Practice*, Paul EL, Atiemo-Obeng VA, Kresta SM, eds. John Wiley & Sons, Hoboken, NJ, 2004, pp. 19–88.

Kubota N, Konno Y, Miura S, Saito K, Sugita K, Watanabe K, Sugo T. Comparison of two convection-aided protein adsorption methods using porous membranes and perfusion beads. *Biotechnol Prog* 1996;12:729–876.

Kullick T, Bock U, Schubert J, Scheper T, Schugerl K. Application of enzyme-field effect transistor sensor arrays as detectors in a flow-injection analysis system for simultaneous monitoring of medium components. Part II. Monitoring of cultivation processes. *Anal Chim Acta* 1995;300(1–3):25–31.

Kullick T, Quack R, Röhrkasten C, Pekeler T, Scheper T, Schügerl K. Pbs-field-effect-transistor for heavy metal concentration monitoring. *Chem Eng Technol* 1995;18(4):225–228.

Kybal J, Sikyta B. A device for cultivation of plant and animal cells. *Biotechnol Lett* 1985;7:467–470.

Kybal J, Vlcek V. A simple device for stationary cultivation of microorganisms. *Biotechnol Bioeng* 1976;18:1713–1718.

Lam P, Sane S. Design and testing of a prototype large-scale bag freeze-thaw system: The development of a large-scale bag freeze-thaw system will have many benefits for the biopharmaceutical industry, 2007. http://biopharminternational.findpharma.com/biopharm/Disposables/Design-and-Testing-of-a-Prototype-Large-Scale-Bag-/ArticleStandard/Article/detail/473322 (accessed September 15, 2015).

Lamproye A. Viral clearance by nanofiltration, strategies for successful validation studies. *First European Downstream Forum*, Goettingen, Germany, 2006.

Landon R, Baloda S. Disposable technology: Validation of a novel disposable connector for sterile fluid transfer. *BioProcess Int Industry Yearbook* 2005;88.

Langer E. Fifth annual report and survey of biopharmaceutical manufacturing capacity and production. BioPlan Associates Inc., Rockville, MD, 2008.

Langer E. Sixth annual report and survey of biopharmaceutical manufacturing capacity and production. BioPlan Associates, Rockville, MD, 2009.

Langer ES. Trends in single-use bioproduction. *BioProcess* 2009.

Langer ES, Price BJ. Biopharmaceutical disposables as a disruptive future technology. *BioPharm Int* 2007;20(6):48–56.

Langer ES, Ranck J. The ROI case: Economic justification for disposables in biopharmaceutical manufacturing. *BioProcess Int* 2005;27(Suppl. 1):46–50.

Lawrence NS, Pagels M, Hackett SFJ, McCormack S, Meredith A, Jones TGJ, Wildgoose GG, Compton RG, Jiang L. Triple component carbon epoxy pH probe. *Electroanalysis* 2007;19(4):424–428.

Lee SK, Suh EK, Cho NK, Park HD, Uneus L, Spetz AL. Comparison study of ohmic contacts to 4H-silicon carbide in oxidizing ambient for harsh environment gas sensor applications. *Solid State Electron* 2005;49(8):1297–1301.

Lee SY, Kim SH, Kim VN, Hwang JH, Jin M, Lee J, Kim S. Heterologous gene expression in avian cells: Potential as a producer of recombinant proteins. *J Biomed Sci* 1999;6:8–17.

Lehmann J, Heidemann R, Riese U, Lütkemeyer D, Büntemeyer H. Der Superspinner—Ein Brutschrankfermenter für die Massenkultur tierischer Zellen. *BioEngineering* 1992;5/6:112–117.

Leventis HC, Streeter I, Wildgoose GG, Lawrence NS, Jiang L, Jones TGJ, Compton RG. Derivatised carbon powder electrodes: Reagentless pH sensors. *Talanta* 2004;63(4): 1039–1051.

Levine B. Making waves in cell therapy: The Wave bioreactor for the generation of adherent and non-adherent cells for clinical use, 2007. http://www.wavebiotech.com/pdf/literature/ISCT_2007_Levine_Final.pdf (accessed September 15, 2015).

Lewcook A. Disposables crucial in future of patient-specific meds, 2007. http://www.in-pharmatechnologist.com/Processing/Disposables-crucial-in-future-of-patient-specific-meds (accessed September 15, 2015).

Li CY, Zhang XB, Han ZX, Akermark B, Sun LC, Shen GL, Yu RQ. A wide pH range optical sensing system based on a sol-gel encapsulated amino-functionalised corrole. *Analyst* 2006;131(3):388–393.

Lim JAC, Sinclair A. Process economy of disposable manufacturing: process models to minimize upfront investment. *Am Pharm Rev* 2007;10:114–121.

Lim JAC, Sinclair A, Kim DS, Gottschalk U. Economic benefits of single-use membrane chromatography in polishing. A cost of goods model. *BioProcess Int* 2007;5:60–64.

Limke T. Comparability between the Mobius CellReady 3 L bioreactor and 3 L glass bioreactors. *BioProcess Int* 2009;7:122–123.

Lindemann C, Marose S, Scheper T, Nielsen HO, Reardon KF. Fluorescence techniques for bioprocess monitoring. In: *Encyclopedia of Bioprocess Technology: Fermentation, Biocatalysis, and Bioseparation*, Flickinger MC, Drew SW, eds. John Wiley & Sons, Inc., 1999.

Liu CM, Hong LN. Development of a shaking bio-reactor system for animal cell cultures. *Biochem Eng J* 2001;2:121–125.

Liu CZ, Towler MJ, Medrano G, Cramer CL, Weathers PJ. Production of mouse interleukin-12 is greater in tobacco hairy roots grown in a mist reactor than in an airlift reactor. *Biotechnol Bioeng* 2009;102:1074–1086.

Liu P. Strategies for optimizing today's increasing disposable processing environments. *BioProcess Int* 2005a;9:10–15.

Liu P. Strategies for optimizing today's increasingly disposable processing environments. *BioProcess Int* 2005b;3(6, Suppl.):10–15.

Lloyd-Evans P, Phillips DA, Wright, ACC, Williams RK. Disposable process for cGMP manufacture of plasmid DNA. *BioPharm Int* 2007;1:18–24.

Lok M, Blumenblat S. Critical design aspects of single-use systems: Some points to consider for successful implementation. *BioProcess Int Suppl* 2007;5(S):28–31.

Lonza. CELL-tainer single-use bioreactors. Brochure. Lonza Walkersville, Walkersville, MD, 2008.

Lorenz CM, Wolk BM, Quan CP, Alcal EW, Closely M, McDonald DJ, Matthew TC. The effect of low intensity ultraviolet C light on monoclonal antibodies. *Biotechnol Prog* 2009;25:476–482.

Lorenzelli L, Margesin B, Martinoia S, Tedesco MT, Valle M. Bioelectrochemical signal monitoring of in-vitro cultured cells by means of an automated microsystem based on solid state sensor-array. *Biosens Bioelectron* 2003;18(5–6):621–626.

Lu JZ, Rosenzweig Z. Nanoscale fluorescent sensors for intracellular analysis. *Fresenius J Anal Chem* 2000;366(6–7):569–575.

Lute P, Bailey M, Combs J, Sukumar M, Brorson K. Phage passage after extensive processing in small-virus retentive filters. *Biotechnol Appl Biochem* 2007;47:141–151.

Lytle C, Sagripanti J. Predicted inactivation of viruses of relevance to biodefense by solar radiation. *Virology* 2005;79:14244–14252.

Maa YF, Hsu CC. Performance of sonication and microfluidization for liquid-liquid emulsification. *Pharm Dev Technol* 1999;4(2):233–240.

Mach CJ, Riedman D. Reducing microbial contamination risk in biotherapeutic manufacturing: Validation of sterile connections. *BioProcess Int* 2008;6(8):20–26.

Maharbiz MM, Holtz WJ, Howe RT, Keasling JD. Microbioreactor arrays with parametric control for high-throughput experimentation. *Biotechnol Bioeng* 2004;86(4):485–490.

Maier U, Büchs J. Characterization of the gas-liquid mass transfer in shaking bioreactors. *Biochem Eng J* 2001;7:99–106.

Maier U, Losen M, Büchs J. Advances in understanding and modeling the gas–liquid mass transfer in shake flasks. *Biochem Eng J* 2003;17:155–167.

Mardirosian D, Guertin P, Crowell J, Yetz-Aldape J, Hall M, Hodge G, Jonnalagadda K, Holmgren A, Galliher P. Scaling up a CHO-produced hormone-protein fusion product. *BioProcess Int* 2009;7(Suppl. 4):30–35.

Margesin R, Schneider M, Schinner F. *Praxis der biotechnologischen Abluftreinigung*. Springer, Berlin, Germany, 1995.

437

Marks DM. Equipment design considerations for large scale cell culture. *Cytotechnology* 2003;42:21–33.

Marose S, Lindemann C, Scheper T. Two-dimensional fluorescence spectroscopy: A new tool for on-line bioprocess monitoring. *Biotechnol Prog* 1998;14(1):63–74.

Marose S, Lindemann C, Ulber R, Scheper T. Optical sensor systems for bioprocess monitoring. *Trends Biotechnol* 1999;17(1):30–34.

Martin J. Case study: Orthogonal membrane technologies for virus and DNA clearance. *PDA Internationally Congress*, Rome, Italy, 2004.

Marx U. Membrane-based cell culture technologies: A scientifically economically satisfactory alternative malignant ascites production for monoclonal antibodies. *Res Immunol* 1998;6:557–559.

Maslowski J. 2007. http://www.asepticfilling.com/PDC-disposable.htm (accessed September 15, 2015).

Masser D. Advanced scientifics' single use systems. *BioProcess Int Industry Yearbook* 2008;134.

Matthews T et al. An integrated approach to buffer dilution and storage. Pharmaceutical Manufacturing, 2009.

Mauter M. Environmental life-cycle assessment of disposable bioreactors. *BioProcess Int* 2009a;4:18–29.

Mauter M. Environmental life-cycle comparison of conventional and disposable reactors. GE Healthcare, 2009b. http://www.bioprocessintl.com/manufacturing/supply-chain/environmental-life-cycle-assessment-of-disposable-bioreactors-184138/ (accessed September 15, 2015).

Mazarevica G, Diewok J, Baena JR, Rosenberg E, Lendl B. On-line fermentation monitoring by mid-infrared spectroscopy. *Appl Spectrosc* 2004;58(7):804–810.

McArdle J. Report of the workshop on monoclonal antibodies. *ATM* 2004;32(Suppl. 1):119–122.

McDonald KA, Hong LM, Trombly DM, Xie Q, Jackman AP. Production of human α-l-antitrypsin from transgenic rice cell culture in a membrane bioreactor. *Biotechnol Prog* 2005;21:728–734.

Meissner. SALTUS—The disposable mixing and bio-reactor system. Brochure 2009.

Menter FR. Zonal two equation k-ω turbulence models for aerodynamic flows. AIAA Paper 93-2906, 1993.

Menzel C, Lerch T, Scheper T, Schugerl K. Development of biosensors based on an electrolyte isolator semiconductor (EIS)-capacitor structure and their application for process monitoring. Part I. Development of the biosensors and their characterization. *Anal Chim Acta* 1995;317(1–3):259–264.

Meyeroltmanns F, Schmitz J, Nazlee M. Disposable bioprocess components and single-use concepts for optimized process economy in biopharmaceutical production. *BioProcess Int* 2005;3:60–66.

Migita S, Ozasa K, Tanaka T, Haruyama T. Enzyme-based field-effect transistor for adenosine triphosphate (ATP) sensing. *Anal Sci* 2007;23(1):45–48.

Mikola M, Seto J, Amanullah A. Evaluation of a novel Wave Bioreactor cellbag for aerobic yeast cultivation. *Bioprocess Biosyst Eng* 2007;30:231–241.

Millipore Corp. Lit. No. DS1002 EN00, 2005. http://www.millipore.com/catalogue/module/c9423 (accessed September 15, 2015).

Millipore. Mobius MIX500 disposable mixing system characterization. Brochure. Millipore Corporation, Billerica, MA, 2008.

Millipore. Datasheet Mobius® CellReady 3 L Bioreactor, 2009a. http://www.millipore.com/publications.nsf/a73664f9f981af8c852569b9005b4eee/228eeedd2285 ebe1852575de00570375/$FILE/DS26770000.pdf (accessed September 15, 2015).

Millipore. Mobius disposable mixing systems. Brochure. Millipore Corporation, Billerica, MA, 2009b.

Millipore. Lynx S2S—Product description, 2010a. http://www.millipore.com/catalogue/module/c9502 (accessed September 15, 2015).

Millipore. Lynx ST Connectors—Data sheet, 2010b. http://www.millipore.com/publications.nsf/a73664f9f981af8c852569b9005b4eee/402ffe097a6bIdca85256d510043ba64/$FILE/DS1750EN00.pdf (accessed September 15, 2015).

Millipore. NovaSeptum sampling system—Product description, 2010c. http://www.millipore.com/catalogue/module/c10713 (accessed September 15, 2015).

Mills A, Chang Q, Mcmurray N. Equilibrium studies on colorimetric plastic film sensors for carbon dioxide. *Anal Chem* 1992;64(13):1383–1389.

Mitchell DL. The relative cytotoxicity of (6-4): Photoproducts and cyclobutane dimers in mammalian cells. *Photochem Photobiol* 1988;48:51–57.

438

Mitchell JP, Court J, Mason MD, Tabi Z, Clayton A. Increased exosome production from tumor cell cultures using the Integra CELLine culture system. *J Immunol Methods* 2008;335:98–105.

Mohri S, Shimizu J, Goda N, Miyasaka T, Fujita A, Nakamura M, Kajiya F. Measurements of CO_2, lactic acid and sodium bicarbonate secreted by cultured cells using a flow-through type pH/CO_2 sensor system based on ISFET. *Sens Actuat B Chem* 2006;115(1): 519–525.

Monge M. Single-use bag manifolds: Applications. World Pharmaceutical Developments, 2002, pp. 85–86.

Monge M. Successful project management for implementing single-use bioprocessing systems. *BioPharm Int* 2006. http://www.biopharminternational.findpharma.com/successful-project-management-implementing-single-use-bioprocessing-systems (accessed September 15, 2015).

Monge M. Disposables: Process economics—Selection, supply chain, and purchasing strategies. *BioProcess International Conference*, Anaheim, CA, 2008.

Mora J, Sinclair A, Delmdahl N, Gottschalk U. Disposable membrane chromatography. Performance analysis and economic cost model. *BioProcess Int* 2006;4(Suppl. 4):38–43.

Morrow K. Disposable bioreactors gaining favor. *Genet Eng Biotechnol News* 2006;26(12):42–45.

Mukherjee J, Lindemann C, Scheper T. Fluorescence monitoring during cultivation of *Enterobacter aerogenes* at different oxygen levels. *Appl Microbiol Biotechnol* 1999;52(4):489–494.

Muller C, Hitzmann B, Schubert F, Scheper T. Optical chemo- and biosensors for use in clinical applications. *Sens Actuat B Chem* 1997;40(1):71–77.

Muller N, Girard P, Hacker D, Jordan M, Wurm FM. Orbital shaker technology for the cultivation of mammalian cells in suspension. *Biotechnol Bioeng* 2004;89:400–406.

Muller W, Anders KD, Scheper T. Culture fluorescence measurements on immobilized yeast. *Chemie Ingenieur Technik* 1989;61(7):564–565.

Muller W, Wehnert G, Scheper T. Fluorescence monitoring of immobilized microorganisms in cultures. *Anal Chim Acta* 1988;213(1–2):47–53.

Munkholm C, Walt DR, Milanovich FP. A fiberoptic sensor for CO_2 measurement. *Talanta* 1988;35(2):109–112.

Murphy T, Schmidt J. Animal-derived agents in disposable systems. *Gen Eng News* 25(14). http://www.genengnews.com/articles/chtitem.aspx?tid=1090&chid=3 (accessed September 15, 2015).

Nagel A, Koch S, Valley U, Emmrich F, Marx U. Membrane-based cell culture systems—An alternative to in vivo production of monoclonal antibodies. *Dev Biol Stand* 1999;101: 57–64.

Nauman EB. Residence time distributions. In: *Handbook of Industrial Mixing-Science and Practice*, Paul EL, Atiemo-Obeng VA, Kresta SM, eds. John Wiley & Sons, Hoboken, NJ, 2004, pp. 1–18.

Negrete A, Kotin RM. Production of recombinant adeno-associated vectors using two bioreactor configurations at different scales. *J Virol* 2007;145:155–161.

Negrete A, Kotin RM. Large-scale production of recombinant adeno-associated viral vectors. In: *Gene Therapy Protocols*: Vol. I, Series: *Methods in Molecular Biology*, Vol. 433, Le Doux JM, ed. Humana Press, Totowa, NJ, 2008, pp. 79–96.

Nelson K, Bielicki J, Anson DS. Immobilization and characterization of a cell line exhibiting a severe multiple sulphatase deficiency phenotype. *Biochem J* 1997;326:125–130.

Nguyen HJ. Development of a membrane adsorber direct capture step: Process comparison of two commercial membranes. Presentation given at *Sartorius Stedim Biotech Downstream Technology Forum*, Berkeley, CA, October 1, 2009.

Nicklin DJ, Wilkes JO, Davidson JF. Two-phase flow in vertical tubes. *Trans Inst Chem Engrs* 1962;40:61–68.

Nienow AW. Reactor engineering in large scale animal cell culture. *Cytotechnology* 2006;50:9–33.

Norling L, Lute S, Emery R, Khuu W, Voisard M, Xu Y, Chen Q, Blan G, Brorson K. Impact of multiple reuse of anion exchange chromatography media on virus removal. *J Chromatogr A* 2005;1069:79–89.

Norris BJ, Gramer MJ, Hirschek MD. Growth of cell lines in bioreactors. In: *Basic Methods in Antibody Production and Characterization*, Howard GC, Bethell DR, eds. CRC Press, Boca Raton, FL, 2000, pp. 87–104.

Norwood D, Bail D, Blanchard J et al. PQRI Leachables and Extractables Working Group. Safety thresholds and best practices for extractables and leachables in orally inhaled and nasal drug products. Product Quality Research Institute, Arlington VA, 2006.

Norwood D, Paskiet D, Ruberto M et al. Best practices for extractables and leachables in orally inhaled and nasal drug products: An overview of the PQRI recommendations. *Pharm Res* 2007;4:727–739.

Norwood DL et al. Analysis of polycyclic aromatic hydrocarbons in metered dose inhaler drug formulations by isotope dilution gas chromatography/mass spectrometry. *J Pharm Biomed Anal* 1995;13(293):293–304.

Nova Biomedical Corporation. BioProfile, automated chemistry analyzer. Waltham, MA. http://www.novabiomedical.com (accessed September 15, 2015).

Novais JL, Titchener-Hooker NJ, Hoare M. Economic comparison between conventional and disposables-based technology for the production of biopharmaceuticals. *Biotechnol Bioeng* 2001;75(2):143–153.

Oashi R, Singh V, Hamel JFP. Perfusion culture in disposable bioreactors. *GEN* 2001;21(40):78.

Odian G. *Principles of Polymerization*, 2nd edn. Wiley-Interscience Publication, New York, 1981.

Oelßner W, Zosel J, Guth U, Pechstein T, Babel W, Connery JG, Demuth C, Grote Gansey M, Verburg JB. Encapsulation of ISFET sensor chips. *Sens Actuat B Chem* 2005;105(1):104–117.

Ohlrogge K, Eberts K, eds. *Membranen*. Wiley-VCH, Weinheim, Germany, 2006.

Okonkowski J, Balasubramanian U, Seamans C, Fischrogen Z, Zhang J, Lachs P, Robinson D, Chartrain M. Cholesterol delivery to NS0 cells: Challenges and solutions in disposable linear low-density polyethylene-based bioreactors. *J Biosci Bioeng* 2007;103:50–59.

Öncül AA, Kalmbach A, Genzel Y, Reichl U, Thèvenin D. Numerische und experimentelle Untersuchung der Fliessbedingungen in Wave-Bioreaktoren. *CIT* 2009;81:1241.

Outlook. 2008.© Tufts University, Outlook 2008 report. Tufts Center for the Study of Drug Development, Boston, MA.

Ozmotech Pty Ltd (Australia). Ozmotech Energy Technology. www.ozmoenergy.com/technology (accessed September 15, 2015).

Ozturk SS. Comparison of product quality: Disposable and stainless steel bioreactor. *BioProduction 2007*, Berlin, Germany, 2007.

Pahl M. Mischtechnik, Aufgaben und Bedeutung. In: *Mischen und Rühren-Grundlagen und moderne Verfahren*, Kraume M, ed. Wiley-VCH Verlag GmbH & Co. KGaA, Weinheim, Germany, 2003, pp. 1–19.

Palazón J, Mallol A, Eibl R, Lettenbauer C, Cusidó RM, Piñol MT. Growth and ginsenoside production in hairy root cultures of *Panax* ginseng using a novel bioreactor. *Planta Med* 2003;69:344–349.

Pall. Kleenpak sterile connectors—Product description, 2010. http://www.pall.com/variants/pdf/pdf/biopharm_34125.pdf (accessed September 15, 2015).

Pandurangappata M, Lawrence NS, Jiang L, Jones TG, Compton RG. Physical adsorption of *N*,*N*′-diphenyl-*p*-phenylenediamine onto carbon particles: Application to the detection of sulfide. *Analyst* 2003;128(5):473–479.

Pang J, Blanc T, Brown J, Labrenz S, Villalobos A, Depaolis A, Gunturi S, Grossman S, Lisi P, Heavner GA. Recognition and identification of UV absorbing leachables in EPREX® pre-filled syringes: An unexpected occurrence at a formulation-component interference. *PDA J Pharm Sci Technol* 2007;61:423–432.

Parenteral Drug Association. Technical report no. 1. *J Pharm Sci Technol* 2007;61(Suppl.):S-1.

Parkinson S. User's guide thermo scientific hyclone single-use mixer. User guide, UG004 Rev4. Thermo Scientific User Guide, 2009.

Paschedag AR. *CFD in der Verfahrenstechnik*. Wiley-VCH, Weinheim, Germany, 2004.

Patterson BJ. A closer look at automated in-line dilution, 2009. http://pharmtech.findpharma.com/pharmtech/article/articleDetail.jsp?id=632935&sk=&date=&pageID=4 (accessed September 15, 2015).

Paul EL, Atiemo-Obeng VA, Kresta SM. Introduction. In: *Handbook of Industrial Mixing-Science and Practice*, Paul EL, Atiemo-Obeng VA, Kresta SM, eds. John Wiley & Sons, Hoboken, NJ, 2004, pp. xxxiii–lviii.

PDC Aseptic Filling Systems. Thermoelectric tube sealer—Technical data sheet, 2010. http://www.asepticfilling.com/Thermoelectric%20Tube%20Sealer%20Data%20Sheet%20Reviewed%20Rev%205.pdf (accessed September 15, 2015).

Peacock L, Auton KA. Comparing shaker flasks with a single-use bioreactor for growing yeast seed cultures. *BioProcess Int* 2008;6:54–57.

Pekeler T, Lindemann C, Scheper T, Hitzmann B. Prediction of bioprocess parameters from two-dimensional fluorescence spectra. *Chemie Ingenieur Technik* 1998;70(12):1610–1611.

PendoTECH. Process scale single use pressure sensors, 2009. http://www.pendotech.com/products/disposable_pressure_sensors/disposable_pressure_sensors.htm (accessed September 15, 2015).

Peter CP, Suzuki Y, Büchs J. Hydromechanical stress in shake flask: Correlation for the maximum local energy dissipation rate. *Biotechnol Bioeng* 2006;93:1164–1176.

Pharmaceutical Inspection Co-operation Scheme. Aide memoire: Inspection of biotechnology manufacturers. PI 024-2. PIC/S Secretariat, Geneva, Switzerland, 2007a.

Pharmaceutical Inspection Co-operation Scheme. Isolators used for aseptic processing and sterility testing. PI 014-3. PIC/S Secretariat, Geneva, Switzerland, 2007b.

Pharmaceutical Inspection Co-operation Scheme. Recommendations on validation master plan, installation and operational, nonsterile process validation, cleaning validation. PI 006-3. PIC/S Secretariat, Geneva, Switzerland, 2007c.

Pharmaceutical Inspection Co-operation Scheme. Recommendation on sterility testing. PI 012-3. PIC/S Secretariat, Geneva, Switzerland, 2007d.

Pharmaceutical Inspection Co-operation Scheme. Recommendations on the validation of aseptic processes. PI 007-5. PIC/S Secretariat, Geneva, Switzerland, 2009.

Phillips M, Cormier J, Ferrence J, Dowd C, Kiss R, Lutz, H. Carter J. Performance of a membrane adsorber for trace impurity removal in biotechnology manufacturing. *J Chromatogr A* 2005;1078:74–82.

Pierce LN, Shabram PW. Scalability of a disposable bioreactor from 25 L–500 L run in perfusion mode with a CHO cell-based cell line: A tech review. *BioProcess J* 2004;4:51–56.

Pinto F. Effect of experimental parameters on plastics pyrolysis reactions. *Proceedings R99*, Geneva, Switzerland, 1999.

Pora H. Increasing bioprocessing efficiency, single use technologies. *Pharm Technol Eur* 2006;18:24–29.

Pora H. The case for disposable manufacturing equipment to accelerate vaccine development. *BioPharm Int* 2006;19(6).

Pora H, Rawlings B. A user's checklist for introducing single-use components into process systems. *BioProcess Int* 2009a;4:9–16.

Pora H, Rawlings B. Managing solid waste from single-use systems in biopharmaceutical manufacturing. *BioProcess Int* 2009b;7(1):18–25.

Potera C. Firm on quest to improve biomanufacturing. *Genet Eng Biotechnol News* 2009;7:20–21.

PQRI Leachables and Extractables Working Group. Safety thresholds and best practices for extractables and leachables in orally inhaled and nasal drug products. Product Quality Research Institute, Arlington, VA, 2006. http://www.pqri.org/pdfs/LE_Recommendations_to_FDA_09-29-06.pdf (accessed September 15, 2015).

Prashad M, Tarrach K. Depth filtration aspects for the clarification of CHO cell derived biopharmaceutical feed streams. *FISE* 2006;9:28–30.

Press/BioPlan Associates, Scale biomanufacturing and scale up production. pp. 1–28.

Proulx SP, Furey JF. Disposable, pre-sterilized fluid receptacle sampling device. 7293477 B2, Millipore Corporation, Billerica, MA, 2007.

Purefit data sheet FLS-3309A-1.5M-1008-SGCS. Saint Gobain Performance Plastics 2008.

Quattroflow. Catalogue, 2010. http://www.quattroflow.com/055c079b5a0c11315/index.html (accessed September 15, 2015).

Rafa B, Panofen F. Repräsentativ und kontaminations-frei. *P&A Biotech* 2009;1:37.

Ransohoff T. Disposable chromatography: Current capabilities and future possibilities. BPD North Carolina Biotechnology Center, Triangle Park, NC, November 18, 2004.

Rao G, Moreira A, Brorson K. Disposable bioprocessing: The future has arrived. *Biotechnol Bioeng* 2009;102(2):348–356.

Rathore N, Rajan RS. Current perspectives on stability of protein drug products during formulation, fill and finish operations. *Biotechnol Prog* 2008;24(3):504–514.

Rauth AM. The physical state of viral nucleic acid and the sensitivity of viruses to ultraviolet light. *Biophys J* 1965;5:257–273.

Raval K, Liu CM, Büchs J. Large-scale disposable shaking bioreactors. *BioProcess Int* 2006;1:46–49.

Ravise A et al. Hybrid and disposable facilities for manufacturing of biopharmaceuticals: Pros and cons. *Adv Biochem Eng Biotechnol* 2009;115:185–219.

Rawlings B, Pora H. Environmental impact of single-use and reusable bioprocess systems. *BioProcess Int* 2009;2:18–25.

Ray S, Tarrach K. Virus clearance strategy using a three-tier orthogonal technology platform. *BioPharm Int* 2008;22:50–58.

Reardon KF, Scheper T, Anders KD, Muller W, Buckmann AF. Novel applications of fluorescence sensors. *11th Symposium on Biotechnology for Fuels and Chemicals*, Springs, CO, 1989.

Reardon KF, Scheper T, Bailey JE. In situ fluorescence monitoring of immobilized *Clostridium acetobutylicum*. *Biotechnol Lett* 1986;8(11):817–822.

Reardon KF, Scheper TH, Bailey JE. Use of a fluorescence sensor for measurement of NAD(P) H-dependent culture fluorescence of immobilized cell systems. *Chemie Ingenieur Technik* 1987;59(7):600–601.

Reif OW, Solkner P, Rupp J. Analysis and evaluation of filter cartridge extractables for validation of pharmaceutical downstream processing. *J Pharm Sci Technol* 1996;50:399–407.

Rhee JI, Lee KI, Kim CK, Yim YS, Chung SW, Wei JQ, Bellgardt KH. Classification of two-dimensional fluorescence spectra using self-organizing maps. *Biochem Eng J* 2005;22(2):135–144.

Ries C. The process engineering characteristics of the Thermo Fisher Scientific Single-Use Bioreactor 50 L: Determination of mixing time, power input and kLa values. Scientific report [unpublished]. Zurich University of Applied Sciences, Switzerland, 2008.

Ries C, John C, Eibl R. Einwegbioreaktoren für die Prozessentwicklung mit Insektenzellen. *Bioforum* 2009;3:11–13.

Rios M. Process considerations for cell-based influenza vaccines. *Pharm Technol* 2006;4:1–6.

Ritala A, Wahlström EH, Holkeri H, Hafren A, Mäkelainen K, Baez J, Mäkinen K, Nuutila AM. Production of a recombinant industrial protein using barley cell cultures. *Protein Expr Purif* 2008;59:274–281.

Röll M. Thermal welding for sterile connections. *Genet Eng Biotechnol News* 2006;26:64.

Rombach C. Single-use benefits and hopes. *BioProcess Int* 2005;3(2):88.

Roy J. Pharmaceutical impurities—A mini-review. *AAPS Pharm Sci Technol* 2002;3:1–6.

Royce J. et al. Guidelines for selecting normal flow filters. Pharmaceutical Technology, 2008. http://www.pharmtech.com/guidelines-selecting-normal-flow-filters (accessed September 15, 2015).

Roychoudhury P, Harvey LM, McNeil B. At-line monitoring of ammonium, glucose, methyl oleate and biomass in a complex antibiotic fermentation process using attenuated total reflectance-mid-infrared (ATR-MIR) spectroscopy. *Anal Chim Acta* 2006;561:218–224.

Rudolph G, Bruckerhoff T, Bluma A, Korb G, Scheper T. Optical inline measurement procedure for cell count and cell size determination in bioprocess technology. *Chemie Ingenieur Technik* 2007;79(1–2):42–51.

Ryder M, Fisher S, Hamilton G, Hamilton M, James G. Bacterial transfer through needlefree connectors: Comparison of nine different devices, 2007. Poster. http://www.icumed.com/Docs-Clave/Ryder%20SHEA%202007%20Poster.pdf (accessed September 15, 2015).

Safety thresholds and best practices for extractables and leachables in orally inhaled and nasal drug products. Product Quality Research Institute, Arlington, VA, 2006.

Sagi E, Hever N, Rosen R, Bartolome AJ, Premkumar JR, Ulber R, Lev O, Scheper T, Belkin S. Fluorescence and bioluminescence reporter functions in genetically modified bacterial sensor strains. *Sens Actuat B Chem* 2003;90(1–3):2–8.

Sandstrom C, Schmidt B. Facility-design considerations for the use of disposable bags. *BioProcess Int* 2005;(Suppl. 4):56–60.

Sartorius Stedim Biotech. Sacova valve—Technical information, 2000. http://sartorius.or.kr/B_Braun_Biotech/Fermenters_and_Bioreactors/pdf/TI_SACOVAe_02-00.pdf. (accessed September 15, 2015)

Sartorius Stedim Biotech. Flexel 3D LevMix system for Palletank. Brochure. Sartorius Stedim Biotech, Göttingen, Germany, 2009a.

Sartorius Stedim Biotech. Flexel 3D system for recirculation mixing. Brochure. Sartorius Stedim Biotech, Göttingen, Germany, 2009b.

Sartorius Stedim Biotech. High cell density bioreactor offload clarification using Sartoclear® P Depth Filters. Application Note 5. Sartorius Stedim Biotech, Göttingen, Germany, 2008.

Sartorius Stedim Biotech. White Paper: Evolving toward single-use bioprocessing: From solutions to holistic value creation, 2009c. http://www.sartorius.us/us/products/bioprocess/aseptic-transfer-systems/biosafe-aseptic-transfer-single-use-bag/ (accessed September 15, 2015).

Sartorius Stedim Biotech. Aseptic transfer system—Definition of the technology, 2010a. Sartorius Stedim Biotech. Opta SFT-I—Product description, 2010b. http://www.sartorius.com/fileadmin/sartorius_pdf/alle/biotech/Data_Opta_SFT-1_SLO2000-e.pdf (accessed September 15, 2015).

Sato K, Yoshida Y, Hirahara T, Ohba T. On-line measurement of intracellular ATP of *Saccharomyces cerevisiae* and pyruvate during sake mashing. *J Biosci Bioeng* 2000;90(3):294–301.

Schears G, Schultz SE, Creed J, Greeley WJ, Wilson DF, Pastuszko A. Effect of perfusion flow rate on tissue oxygenation in newborn piglets during cardiopulmonary bypass. *Ann Thorac Surg* 2003;75(2):560–565.

Scheper T. Biosensors for process monitoring. *J Ind Microbiol* 1992;9(3–4):163–172.

Scheper T, Brandes W, Grau C et al. Applications of biosensor systems for bioprocess monitoring. *Third International Symposium on Analytical Methods in Biotechnology (Anabiotec 90)*, San Francisco, CA, 1990.

Scheper T, Brandes W, Maschke H, Ploetz F, Mueller C. Two FIA-based biosensor systems studied for bio-process monitoring. *J Biotechnol* 1993;31(3):345–356.

Scheper T, Buckmann AF. A fiber optic biosensor based on fluorometric detection using confined macromolecular nicotinamide adenine-dinucleotide derivatives. *Biosens Bioelectron* 1990;5(2):125–135.

Scheper T, Gebauer A, Kuhlmann W, Meyer HD, Schugerl K. Dechema-Monographien 1984a;95.

Scheper T, Gebauer A, Sauerbrei A, Niehoff A, Schugerl K. Measurement of biological parameters during fermentation processes. *Anal Chim Acta* 1984;163;111–118.

Scheper T, Gebauer A, Schugerl K. Monitoring of NADH-dependent culture fluorescence during the cultivation *Escherichia coli*. *Chem Eng J* 1987;34(1):B7–B12.

Scheper T, Hitzmann B, Stark E, Ulber R, Faurie R, Sosnitza P, Reardon KF. Bioanalytics: Detailed insight into bioprocesses. *Anal Chim Acta* 1999;400:121–134.

Scheper T, Jornitz MW (eds.). *Sterile Filtration: Advances in Biochemical Engineering/Biotechnology*. Springer, Berlin, Germany, 2006.

Scheper T, Lorenz T, Schmidt W, Schugerl K. Online measurement of culture fluorescence for process monitoring and control of biotechnological processes. *Ann N Y Acad Sci* 1987;506:431–445.

Scheper T, Schugerl K. Bioreactor characterization by in situ fluorometry. *Chemie Ingenieur Technik* 1986a;58(5):433–433.

Scheper T, Schugerl K. Characterization of bioreactors by in situ fluorometry. *J Biotechnol* 1986b;3(4):221–229.

Scheper T, Schugerl K. Culture fluorescence studies on aerobic continuous cultures of *Saccharomyces cerevisiae*. *Appl Microbiol Biotechnol* 1986c;23(6):440–444.

Scheper TH, Hilmer JM, Lammers F, Muller C, Reinecke M. Biosensors in bioprocess monitoring. *J Chromatogr A* 1996;725(1):3–12.

Schmidt S, Kauling J. Process and laboratory scale UV inactivation of viruses and bacteria using to innovative coiled tube reactor. *Chem Eng Technol* 2007;30:945–950.

Schmidt S, Mora J, Dolan S, Kauling J. An integrated concept for robustly and efficient virus clearance and contaminant removal in biotech processes. *BioProcess Int* 2005;8:26–31.

Schneditz D, Kenner T, Heimel H, Stabinger H. A sound-speed sensor for the measurement of total protein-concentration in disposable, blood-perfused tubes. *J Acoust Soc Am* 1989;86(6):2073–2080.

Schöning MJ, Brinkmann D, Rolka D, Demuth C, Poghossian A. CIP (cleaning-in-place) suitable non-glass pH sensor based on a Ta_2O_5-gate EIS structure. *Sens Actuat B Chem* 2005;111–112:423–429.

Schreyer HB, Miller SE, Rodgers S. Application note: High-throughput process development. *Genet Eng Biotechnol News* 2007;27(17). http://www.genengnews.com/issues.com/issues/item.aspx?issue_id=78 (accessed September 15, 2015).

Schugerl K. Progress in monitoring, modeling and control of bioprocesses during the last 20 years. *J Biotechnol* 2001;85(2):149–173.

Schugerl K, Bellgardt KH, Kretzmer G, Hitzmann B, Scheper T. In-situ and online monitoring and control of biotechnological processes. *Chemie Ingenieur Technik* 1993;65(12):1447–1456.

Schugerl K, Lindemann C, Marose S, Scheper T. *Bioprocess Engineering Course*, 1998, p. 400.

Schugerl K, Lorenz T, Lubbert A, Niehoff J, Scheper T, Schmidt W. Pros and cons—On-line versus off-line analysis of fermentations. *Trends Biotechnol* 1986;4(1):11–15.

Schugerl K, Lubbert A, Scheper T. Online measurement of bioreactor performance. *Chemie Ingenieur Technik* 1987;59(9):701–714.

Schwan S, Fritzsche M, Cismak A, Heilmann A, Spohn U. In vitro investigation of the geometric contraction behavior of chemo-mechanical P-protein aggregates (forisomes). *Biophys Chem* 2007;125(2–3):444–452.

Schwander E, Rasmusen H. Scalable, controlled growth of adherent cells in a disposable, multilayer format. *Genet Eng Biotechnol News* 2005;25:29.

Scientific HyClone BPCs® Products and Capabilities. Catalog, 2008/2009.

SciLog. Disposable, pre-calibrated SciCon conductivity sensors, 2009a. www.scilog.com/sensor/conductivity.php (accessed September 15, 2015).

SciLog. Pressure sensors & monitors, 2009b. http://www.scilog.com/sensor/pressure.php (accessed September 15, 2015).

Scott C. 'Single-use' doesn't necessarily mean 'disposable.' *BioProcess Int* 2007a;5(4, Suppl.):4.

Scott C. Disposables qualification and process validation. *BioProcess Int* 2007b;5:24–27.

Scott C. Single-use bioreactors: A brief review of current technology. *BioProcess Int Suppl* 2007c;5(5):44–51.

Scott C. Biotech leads a revolution in vaccine manufacturing. *BioProcess Int* 2008;6(Suppl. 6):12–18.

Scott LE, Aggett H, Glencross DK. Manufacture of pure antibodies by heterogeneous culture without downstream purification. *Biotechniques* 2001;31:666–668.

SEBRA. Aseptic sterile welder—Product description, 2010a. http://www.sebra.com/BCP-3960.html (accessed September 15, 2015).

SEBRA. Tube sealer—Product descriptions, 2010b. http://www.sebra.com/BCP-biopharmaceutical.html (accessed September 15, 2015).

Selvanayagam ZE, Neuzil P, Gopalakrishnakone P, Sridhar U, Singh M, Ho LC. An ISFET-based immunosensor for the detection of [beta]-Bungarotoxin. *Biosens Bioelectron* 2002;17(9):821–826.

Severinghaus JW, Bradley AF. Electrodes for blood pO_2 and pCO_2 determination. *J Appl Physiol* 1958;13(3):515–520.

Sevilla F, Kullick T, Scheper T. A bio-FET sensor for lactose based on co-immobilized β-galactosidase/glucose dehydrogenase. *Biosens Bioelectron* 1994;9(4–5):275–281.

Sharma B, Bader F, Templeman T, Lisi B, Ryan M, Heavner GA. Technical investigations into the cause of the increased incidence of antibody-mediated pure red cell aplasia associated with EPREX®. *Eur J Hosp Pharm* 2004;5:86–91.

Shire SJ. Formulation and manufacturability of biologics. *Curr Opin Biotechnol* 2009;6:708–714.

Shukla A, Hubbard B. Tressel T, Guhan S, Low D. Downstream processing of monoclonal antibodies—Application of platform approaches. *J Chromatogr B* 2007;828:28–39.

Simonet J, Gantzer C. Inactivation of poliovirus I and F-specific RNA phages and degradation of their genomes by UV irradiation at 254 nanometers. *Appl Environ Microbiol* 2006;72:7671–7677.

Simonis A, Dawgul M, Luth H, Schöning MJ. Miniaturised reference electrodes for field-effect sensors compatible to silicon chip technology. *Electrochim Acta* 2005;51(5):930–937.

Simonis A, Luth H, Wang J, Schöning MJ. New concepts of miniaturised reference electrodes in silicon technology for potentiometric sensor systems. *Sens Actuat B Chem* 2004;103(1–2):429–435.

Sinclair A. Biological products manufacturing: Cost challenges and opportunities now and in the future. *BPI*, Raleigh, NC, October 12, 2009.

Sinclair A. How geography affects the cost of bio-manufacturing. *BioProcess Int* 2010;516–519.

Sinclair A et al. The environmental impact of disposable technologies. *BioPharm Int* 2008a. http://www.biopharminternational.findpharma.com/biopharm/Disposables+Articles/The-Environmental-Impact-of-Disposable-technologie/ArticleStandard/Article/detail/566014 (accessed September 15, 2015).

Sinclair A et al. The environmental impact of disposables technologies. *BioPharm Int (Suppl.)* 2008b. http://www.biopharminternational.findpharma.com/biopharm/article/articleDetail.jsp?id=566014&pageID=1&sk=&date= (accessed September 15, 2015).

Sinclair A et al. The environmental impact of disposable technologies. The Biopharm international guide, November 2008c; Base of the analysis: Typical mAb process at 3 × 2000 L scale.

Sinclair A, Leveen L, Monge M, Lim J, Cox S. The environmental impact of disposable technologies—Can disposables reduce your facility's environmental. *BioPharm Int* 2008c;11: 1–11.

Sinclair A, Monge M. Quantitative economic evaluation of single use disposables in bioprocessing. *Pharm Eng* 2002;22:20–34.

Sinclair A, Monge M. Biomanufacturing for the 21st century: Designing a concept facility based on single-use systems. *BioProcess Int* 2004;2:26–31.

Sinclair A, Monge M. Concept facility based on single antibody manufacture. *J Chromatogr B* 2005a;8–18:848.

Sinclair A, Monge M. Concept facility based on single-use systems: Part 2. *BioProcess Int* 2005b;3(9):S51–S55.

Sinclair A, Monge M. Disposables cost contributions: A sensitivity analysis. *BioPharm Int* 2009a;22(4):14–18.

Sinclair A, Monge M. Evaluating disposable mixing systems. *Biopharm Int* 2009b;22(2):24–29.

Singh SK, Kolhe P, Wang W, Nema S. Large-scale freezing of biologics: A practitioner's review, part 1: Fundamental aspects. *BioProcess Int* 2009a;7(9):32–44.

Singh SK, Kolhe P, Wang W, Nema S. Urge-scale freezing of biologics: A practitioner's review, part 2: Practical advice. *BioProcess Int* 2009b;7(10):34–42.

Singh V. Disposable bioreactor for cell culture using wave-induced motion. *Cytotechnology* 1999;30:149–158.

Singh V. Bioprocessing tutorial: Mixing in large disposable containers. *Genet Eng Biotechnol News* 2004;24(3):42–43.

Singh, V. *The Wave Bioreactor Story*. Wave Biotech LLC, Somerset, U.K., 2005.

Slivac I, Srček VG, Radoševic K, Kmetič I, Kniewald Z. Aujeszky's disease virus production in disposable bioreactors. *J Biosci* 2006;3:363–368.

Smiley D, Rhee M, Ziemer D, Gallina D, Phillips LS, Kolm P, Umpierrez GE. Lack of follow-up care: A major obstacle to optimal care in Latinos with diabetes. *Diabetes* 2005;54:A588–A589.

Smith M. An evaluation of protein A and non-protein A methods for the recovery of mono-clonal antibodies and considerations for process scale-up. Presented at the *Scaling-up of Biopharmaceutical Products*, The Grand, Amsterdam, the Netherlands, January 2004.

Smith M. Strategies for the purification of high titer, high volume mammalian cell culture batches. *BioProcess International European Conference and Exhibition*, Berlin, Germany, April 2005.

Sofer G, Lister DC. Inactivation methods grouped by virus: Virus inactivation in the 1990s and into the 21st century. *BioPharm Int* 2003;6:37–42.

Solle D, Geissler D, Stark E, Scheper T, Hitzmann B. Chemometric modelling based on 2D-fluorescence spectra without a calibration measurement. *Bioinformatics* 2003;19(2):173–177.

Song K-S, Zhang G-J, Nakamura Y, Furukawa K, Hiraki T, Yang J-H, Funatsu T, Ohdomari I, Kawarada H. Label-free DNA sensors using ultrasensitive diamond field-effect transistors in solution. *Phys Rev E: Stat Nonlin Soft Matter Phys* 2006;74(4):041919.

Stärk E, Hitzmann B, Schügerl K, Scheper T, Fuchs C, Köster D, Märkl H. In-situ-fluorescence-probes: A useful tool for non-invasive bioprocess monitoring. *Adv Biochem Eng Biotechnol* 2002;74:21–38.

Stein A, Kiesewetter A. Cation exchange chromatography in antibody purification: pH screening for optimized binding and HCP removal. *J Chromatogr B* 2007;848:151–158.

Sterilizing filtration of liquids. PDA technical report no. 26. *PDA J Pharmaceut Sci Technol* 1998;52:S1.

Stone TE, Goel V, Leszczak J. Methodology for analysis of extractables: A model stream approach. *Pharm Technol* 1994;18:116–121.

Streeter I, Leventis HC, Wildgoose GG, Pandurangappa M, Lawrence NS, Jiang L, Jones TGJ, Compton RG. A sensitive reagentless pH probe with a ca. 120 mV/pH unit response. *J Solid State Electrochem* 2004;8(10):718–721.

Suhr H, Wehnert G, Schneider K, Bittner C, Scholz T, Geissler P, Jahne B, Scheper T. In-situ microscopy for online characterization of cell-populations in bioreactors, including cell-concentration measurements by depth from focus. *Biotechnol Bioeng* 1995;47(1):106–116.

Sukumar M, Brorson K. Phage passage after extensive processing in small-virus retentive filters. *Biotechnol Appl Biochem* 2007;47:141–152.

Surribas A, Geissler D, Gierse A, Scheper T, Hitzmann B, Montesinos JL, Valero F. State vari-ables monitoring by in situ multi-wavelength fluorescence spectroscopy in heterologous protein production by *Pichia pastoris*. *J Biotechnol* 2006;124(2):412–419.

Szalai ES, Alvarez MM, Muzzio FJ. Laminar mixing: A dynamic systems approach. In: *Handbook of Industrial Mixing-Science and Practice*, Paul EL, Atiemo-Obeng VA, Kresta SM, eds. John Wiley & Sons, Hoboken, NJ, 2004, pp. 89–144.

Tallentire A. The spectrum of microbial radiation sensitivity. *Radiat Phys Chem* 1980;15:83–89.

Tamachi T, Maezawa Y, Ikeda K et al. IL-25 enhances allergic airway inflammation by amplify-ing a TH2-cell dependent pathway in mice. *J Allergy Clin Immunol* 2006;118:606–614.

Tan WH, Shi ZY, Kopelman R. Development of submicron chemical fiber optic sensors. *Anal Chem* 1992;64(23):2985–2990.

Tarrach K. Integrative strategies for viral clearance. *Fourth Annual Biological Production Forum*, Edinburgh, U.K., 2005.

Tarrach K. Process economy of disposable chromatog-raphy in antibody manufacturing: Development and production of antibodies, vaccines, and gene vectors. *WilBio's Bioprocess Technology*, Amsterdam, the Netherlands, 2007a.

Tarrach K. Virus filter positioning in the purification process of cell culture intermediates and flow decay aspects associated with small non-enveloped virus retention. *Bioprocess Internationally European Conference and Exhibition*, Paris, France, 2007b.

Tarrach K, Meyer A, Dathe JE, Sunn H. The effect of flux decay on a 20 nm nanofilter for virus retention. *BioPharm Int* 2007;4:58–63.

Tartakovsky B, Sheintuch M, Hilmer JM, Scheper T. Application of scanning fluorometry for monitoring of a fermentation process. *Biotechnol Prog* 1996;12(1):126–131.

Tartakovsky B, Sheintuch M, Hilmer JM, Scheper T. Modelling of *E. coli* fermentations: Comparison of multicompartment and variable structure models. *Bioprocess Eng* 1997;16(6):323–329.

Taylor I. The CellMaker plus single-use bioreactor: A new bioreactor capable of culturing bacteria, yeast, insect and mammalian cells. *Biotechnica*, Hannover, Germany, 2007.

Teixeira AP, Portugal CAM, Carinhas N, Dias JML, Crespo JP, Alves PM, Carrondo MJT, Oliveira R. In situ 2D fluorometry and chemometric monitoring of mammalian cell cultures. *Biotechnol Bioeng* 2009;102(4):1098–1106.

Terrier B, Courtois C, Hénault N, Cuvier A, Bastin M, Aknin A, Dubreuil J, Pétiard V. Two new disposable bioreactors for plant cell cultures: The wave & undertow bioreactor and the slug bubble bioreactor. *Biotechnol Bioeng* 2007;96:914–923.

Terumo. Sterile tubing welders—Website, 2010a. http://www.terumotransfusion.com/ProductCategory.aspx?categoryId=6 (accessed September 15, 2015).

Terumo. Teruseal tube sealer—Product description, 2010b. http://www.terumotransfusion.com/ProductDetails.aspx?categoryId=5 (accessed September 15, 2015).

The Irradiation and Sterilization Subcommittee of the Bio-Process Systems Alliance. Guide to irradiation and sterilization validation of single-use bioprocess systems, part 2. *BioProcess Int* 2007;5(10):60–70.

The United States Pharmacopeia. USP 31, NF 26. Rockville, IN, 2008.

Thermo Fisher Scientific. Thermo scientific nalgene bioprocess bag management system, 2009. http://www.nalgenelabware.com/features/featureDetail.asp?featureID=70 (accessed September 15, 2015).

Thoemmes J, Kula M. Membrane chromatography: An integrative concept in the downstream processing of proteins. *Biotechnol Prog* 1995;11:357–367.

Thommes J, Etzel M. Alternatives to chromatographic separations. *Biotechnol Prog* 2007;23:42–45.

Thordsen O, Lee SJ, Degelau A, Scheper T, Loos H, Rehr B, Sahm H. A model system for a fluorometric biosensor using permeabilized *Zymomonas mobilis* or enzymes with protein confined dinucleotides. *Biotechnol Bioeng* 1993;42(3):387–393.

Tollnik C. 2009. Einsatz von Disposables in der Praxis—ein Erfahrungsbericht zu Design und Betrieb einer Pilotanlage für klinische Wirkstoffproduktionen. 2. Konferenz Einsatz von Single-Use-Disposables (Concept Heidelberg). Mannheim, Germany.

Trebak M, Chong JM, Herlyn D, Speicher DW. Efficient laboratory-scale production of monoclonal antibodies using membrane-based high-density cell culture technology. *J Immunol Methods* 1999;230:59–70.

Trevisan MG, Poppi RJ. Direct determination of ephedrine intermediate in a biotransformation reaction using infrared spectroscopy and PLS. *Talanta* 2008;75:1021–1027.

Tservistas M, Koneke R, Comte A, Scheper T. Oxygen monitoring in supercritical carbon dioxide using a fibre optic sensor. *Enzyme Microb Technol* 2001;28(7–8):637–641.

Tutorial. High-yield single-use cell culture systems. *Genet Engineering and Biotechnology News*, 2005. http://www.genengnews.com/articles/chtitem.aspx?tid=1093&chid=3 (accessed September 15, 2015).

Ulber R, Frerichs JG, Beutei S. Optical sensor systems for bioprocess monitoring. *Anal Bioanal Chem* 2003;376(3):342–348.

Ulber R, Hitzmann B, Scheper T. Innovative bio-process analysis—New approaches to understanding biotechnological processes. *Chemie Ingenieur Technik* 2001;73(1–2):19–26.

Ulber R, Protsch C, Solle D, Hitzmann B, Willke B, Faurie R, Scheper T. Use of bioanalytical systems for the improvement of industrial tryptophan production. *Chem Eng Technol* 2001;24(7):15–17.

Ulber R, Scheper T. Enzyme biosensors based on ISFETs. In: *Enzyme and Microbial Biosensors*, Mulchandani A, Rogers KR, eds. Springer, Berlin, Germany, 1998, pp. 35–50.

U.S. Environmental Protection Agency. Combustion. http://www.epa.gov/epaoswer/hazwaste/combust.htm (accessed September 15, 2015) (links to Solid Waste Combustion/Incineration, and Hazardous Waste Combustion).

U.S. Department of Health and Human Services. Guidance for industry: Guideline for validation of Limulus amebocyte lysate test as an end-product endotoxin test for human and animal parenteral drugs, biological products and medicinal devices. Food and Drug Administration, Rockville, IL, 1987.

U.S. Department of Health and Human Services. Guidance for industry: Manufacturing, processing or holding active pharmaceutical ingredients. Food and Drug Administration, Rockville, IL, 1998.

U.S. Department of Health and Human Services. Guidance for industry: Container closure systems for packaging human drugs and biologies—Chemistry, manufacturing and controls documentation. Food and Drug Administration, Rockville, IL, 1999.

U.S. Department of Health and Human Services. Guidance for industry: Sterile drug products produced by aseptic processing—Current good manufacturing practice. Food and Drug Administration, Rockville, IL, 2004a.

U.S. Department of Health and Human Services. Guidance for industry: PAT process analytical technology—A framework for innovative pharmaceutical manufacturing and quality assurance. Food and Drug Administration, Rockville, IL, 2004b.

U.S. Department of Health and Human Services. Guidance for industry: Investigating out of specification (OOS) test results for pharmaceutical production. Food and Drug Administration, Rockville, IL, 2006.

U.S. Department of Health and Human Services. Chapter 45: Biological drug products, inspection of biological drug products (CBER). In: *Compliance Program Guidance Manual.* Food and Drug Administration, Rockville, IL, 2008a.

U.S. Department of Health and Human Services. Guidance for industry: Process validation: General principles and practices. Food and Drug Administration, Rockville, IL, 2008b.

U.S. Department of Health and Human Services, Food and Drug Administration, Center for Drug Evaluation and Research (CDER), Center for Veterinary Medicine (CVM), Office of Regulatory Affairs (ORA), Pharmaceutical CGMPs. Guidance for industry PAT: A framework for innovative pharmaceutical development, manufacturing, and quality assurance, 2004 http://www.fda.gov/downloads/Drugs/Guidances/ucm070305.pdf (accessed September 15, 2015).

U.S. Pharmacopeia (USP). <1211>, Sterilization and sterility assurance. http://www.usp.org; www.uspnf.com/uspnf/login (accessed September 15, 2015).

USP. Chapter <87> Biological reactivity tests, in vitro, USP 30. United States Pharmacopeial Convention, Rockville, MD, 2007a.

USP. Chapter <88> Biological reactivity tests, in vivo, USP 30. United States Pharmacopeial Convention, Rockville, MD, 2007b.

Uttamlal M, Walt DR. A fiber-optic carbon dioxide sensor for fermentation monitoring. *Biotechnology* 1995;13(6):597–601.

Valax P, Charbaut E, Dathe JE, Tarrach K, Lamproye A, Broly H. Robustness of parvovirus-retentive membranes and implications for virus clearance validation requirements. *BioProcess Int* 2009;7(Suppl. 1):56–62.

Valentine P. Implementation of a single-use stirred bioreactor at pilot and GMP manufacturing scale for mammalian cell culture. *ESACT 2009 Meeting*, Dublin, Ireland, 2009.

Van den Vlekkert HH, de Rooij NF, van den Berg A, Grisel A. Multi-ion sensing system based on glass-encapsulated pH-ISFETs and a pseudo-REFET. *Sens Actuat B Chem* 1990;1(1–6):395–400.

Van Tienhoven EAE, Korbee D, Schipper L, Verharen HW, De Jong WH. In vitro and in vivo (cyto) toxicity assays using PVC and LDPE as model materials. *J Biomed Mater Res* 2006;A78:175–182.

Verjans B, Thilly J, Hennig H, Vandecasserie C. Qualification results of a new system for rapid transfer of sterile liquid through a containment wall. *Pharm Technol* 2007;31:184–195.

Vogel J. Fast capture of biopharmaceuticals from continuous cell culture. *Comprehensively Chromatography Conference*, Emeryville, CA, 2005.

Vogt R, Paust T. Disposable factory of tailor made integration of single-use systems. *BioProcess Int* 2008;7(Suppl. 1):72–77.

Walker J, Sanderson K, Norris L, Stager A, Bellafiore L. Buffer blending for cGMP bioprocessing in the 21st century. Asahi Kasei Bioprocess, Evanston, IL, 2005. Poster reprint; www.technikrom.com/pdf/BufferblendingTechniKromL.pdf (accessed September 15, 2015).

Walsh G. Engineering biopharmaceuticals. *BioPharm Int* 2007;20:64–68.

Walter JK. Strategies and considerations for advanced economy in downstream processing of biopharmaceutical proteins. In: *Bioseparation and Bioprocessing. Processing, Quality and Characterization, Economics, Safety and Hygiene*, Subramanian G, ed. Wiley-VCH, Weinheim, 1998, pp. 447–460.

Walter JK, Nothelfer F, Werz W. Validation of viral safety for pharmaceutical proteins. In: *Biodissolution and Bioprocessing*, Vol. I, Subramanian G, ed. Wiley-VCH, Weinheim, Germany, 1998, pp. 465–496.

Watler PK. Solving 21st century manufacturing challenges with single-use technologies. Presentation given at *the Bio-Process Systems Alliance Meeting*, La Jolla, CA, 2009. http://www.BPSalliance.org (accessed September 15, 2015).

Wang E. Cryopreservation, storage and transportation of biological process intermediates. *BioProcess Int Ind Yearbook* 2006:78–79.

Wang J, Moult A, Chao SF, Remington K, Treckmann R, Emperor K, Pifat D, Hotta J. Virus inactivation and protein recovery in a novel ultraviolet C reactor. *Vox Song* 2004;86:230–238.

Weathers PJ, Towler MJ, Xu J. Bench to batch: Advances in plant cell culture for producing useful products. *Appl Microbiol Biotechnol* 2010;85:1339–1351.

Weber W, Weber E, Geisse S, Memmert K. Optimisation of protein expression and establishment of the wave bioreactor for baculovirus/insect cell culture. *Cytotechnology* 2002;38: 77–85.

Wehnert G, Anders KD, Bittner C, Kammeyer R, Hubner U, Niellsen J, Scheper T. Combined fluorescence scattered-light detector and its use in process monitoring in biotechnology. *1989 Annual Meeting of Process Engineers*, Berlin, Germany, 1989.

Wei J, Yang H, Sun H, Lin Z, Xia S. A full CMOS integration including ISFET microsensors and interface circuit for biochemical applications. *Rare Met Mater Eng* 2006;35(3):443–446.

Weidner J, Jimenez F. Scale-up case study for long term storage of a process intermediate in bags. *Am Pharm Rev* 2008. http://www.americanpharmaceuticalreview.com/ViewArticle.aspx?ContentID=3486 (accessed September 15, 2015).

Weigl BH, Wolfbeis OS. Sensitivity studies on optical carbon dioxide sensors based on ion pairing. *Sens Actuat B Chem* 1995;28(2):151–156.

Weitzmann KH. The use of model solvents for evaluating extractables from filters used to process pharmaceutical products. *Pharm Technol* 1997;21:73–79.

Wells B. Guide to disposal of single-use bioprocess systems. *BioProcess Int* 2007;11:22–28.

Wells B et al. Guide to disposal of single-use bioprocess systems. *BioProcess Int* 2007;6(5):S24–S27.

Wendt D. BioTrends: Disposable processing systems: How suppliers are meeting today's biotech challenges form fluid handling to filtration. *BioPharm Int* 2003;15:18–22.

Werner S, Nägeli M. Good vibrations. *BioTechnology* 2007;3:22–24.

WHO Expert Committee on Specifications for Pharmaceutical Preparations. Good manufacturing practices for pharmaceutical products: Main principles. Technical Report Series No. 908, Annex 4. World Health Organization, Geneva, Switzerland, 2003.

Williamson C, Fitzgerald R, Shukla AA. Strategies for implementing a BPC in commercial biologics manufacturing. *BioProcess Int* 2009;7(10):24–33.

Wilson JS. A fully disposable monoclonal antibody manufacturing train. *BioProcess Int* 2006;4(Suppl. 4):34–36.

Wolfbeis OS. Materials for fluorescence-based optical chemical sensors. *J Mater Chem* 2005;15(27–28):2657–2669.

Wong R. Disposable assemblies in biopharmaceutical production: Design, implementation and troubleshooting. *BioProcess Int Suppl* 2004;4(9):36–38.

Wu Y, Ahmed A, Waghmare R, Genest P, Issacson S, Krishnan M, Kahn DW. Validation of adventitious virus removal by virus filtration: A novel procedure for monoclonal antibody of process. *BioProcess Int* 2008;5:54–59.

Wurm FM. Novel technologies for rapid and low cost provisioning of antibodies and process details in mammalian cell culture-based biomanufacturing. *BioProduction 2007*, Berlin, Germany, 2007.

Xu J, Zhao W, Luo X, Chen H. A sensitive biosensor for lactate based on layer-by-layer assembling MnO_2 nanoparticles and lactate oxidase on ion-sensitive field-effect transistors. *Chem Commun* 2005;14;(6):792–794.

Yates DE, Levine S, Healy TW. Site-binding model of the electrical double layer at the oxide/water interface. *J Chem Soc Faraday Trans* 1974;170:1807–1818.

YSI Life Sciences. YSI STAT 2300, YSI 2700 SELECT, 2010. http://www.ysilifesciences.com (accessed September 15, 2015).

Zambaux JP. How synergy answers the biotech industry needs. *BioProduction 2007*, Berlin, Germany, 2007.

Zambeaux JP, Vanhamel S, Bosco F, Castillo J. Disposable bioreactor. Patent EP 1961606A2, 2007.

Zandbergen JE, Monge M. Disposable technologies for aseptic filling. *BioProcess Int* 2006;6:48–51.

Zeta. FreezeContainer®, 2009. http://www.zeta.com/DE/Produkte/Freeze-Thaw-Systeme/. (accessed September 15, 2015)

Zhang R, Bouamama T, Tabur P, Zapata G, Gottschalk U, Mora J, Reif O. Viral clearance feasibility study with Sartobind Q membrane adsorber for humanly antibody purification. *IBC Third European Event BioProduction 2004*, Munich, Germany, 2004.

Zhang H, Williams-Dalson W, Keshavarz-Moore E, Shamlou PA. Computational-fluid-dynamics (CFD) analysis of mixing and gas-liquid mass transfer in shake flasks. *Biotechnol Appl Biochem* 2005;41:1–8.

Zhang X, Bürki CA, Stettler M et al. Efficient oxygen transfers by surface aeration in shaken cylindrical containers for mammalian cell cultivation at volumetric scales up to 1000 L. *Biochem Eng J* 2009;45:41–47.

Zhang X, Stettler M, De Sanctis D, Perrone M, Parolini N, Discacciati M, De Jesus M, Hacker D, Quarteroni A, Wurm F. Use of orbital shaken disposable bioreactors for mammalian cell cultures from the mL scale to the 1,000 L scale. In: *Disposable Bioreactors, Series: Advances in Biochemical Engineering/Biotechnology*, Vol. 115, Eibl D, Eibl R, eds. Springer, Berlin, Germany, 2009, pp. 33–53.

Zheng R. The game changer. *BioProcess Int* 2010;8(4):S4–S9.

Zhou J, Dehghani H. Development of viral clearance strategies for large-scale monoclonal antibody production. In: *Advances in Large Scale Biomanufacturing and Scale-Up Production*, 2nd edn., Langer ES, ed. ASM, Rockville, MD, 2007, pp. 558–586.

Zhou JX, Tressel T. Basic concepts in Q membrane chromatography for generous scale antibody production. *Biotechnol Prog* 2006;22:341–349.

Zhou JX, Solamo F, Hong T, Shearer M. Tressel T. Viral clearance using disposable systems in monoclonal antibody commercial downstream processing. *Biotechnol Bioeng* 2008;100:488–496.

Zhou JX, Tressel T, Gottschalk U et al. New Q membrane scale-down model for process-scale antibody purification. *J Chromatogr A* 2006;1134:66–73.

Zhou JX, Tressel T, Guhan S. Disposable chromatography: Single-use membrane chromatography as a polishing option during antibody purification is gaining momentum. *BioPharm Int* 2007;1(Suppl. 1):26–35.

Zhou JX, Tressel T, Yang X, Seewoester T. Implementation of advanced technologies in commercial monoclonal antibody production. *Biotechnology* 2008;3:1185–1200.

Ziv M. Organogenic plant regeneration in bioreactors. In: *Plant Biotechnology and In Vitro Biology in the 21st Century*, AAltmann, M Ziv, SIzhar, eds. Kluwer, Dordrecht, The Netherlands, 1999, pp. 673–676.

Ziv M. Bioreactor technology for plant micropropagation. *Hort Rev* 2000;24:1–30.

Ziv M. Simple bioreactors for mass propagation of plants. *Plant Cell Tissue Organ Cult* 2005;81:277–285.

Ziv M, Ronen G, Raviv M. Proliferation of meristematic clusters in disposable pre-sterilized plastic biocontainers for the large-scale propagation of plants. *In Vitro Cell Dev Biol Plant* 1998;34:152–158.

Zlokarnik M. *Stirring-Theory and Practice*. Wiley-VCH Verlag GmbH & Co. KGaA, Weinheim, Germany, 2001.

Zlokarnik M. *Scale-Up in Chemical Engineering*. Wiley-VCH Verlag GmbH & Co. KGaA, Weinheim, Germany, 2006.

Zweifel H. *Plastic Additives Handbook*, 5th edn. Hanser Publishers, Munich, Germany, 2001.

Chapter 10 Commercial manufacturing overview

Introduction

Commercial manufacturing of biosimilar products requires one special consideration—cost of production since the goal is to compete in the market on cost. As a result, we have provided in this book many options available to biosimilar product manufacturers, including the use of single-use technologies.

The overall process of manufacturing remains similar to what the originators use, except perhaps for the scale of production. Since the biosimilar product manufactures are likely to dedicate the facility to the production of single products, several interesting possibilities arise in designing the facilities. A few examples are provided. Also described here are specific components, equipment, and systems used for these manufacturing processes.

Media

Media must be carefully selected to provide the proper rate of growth and the essential nutrients for the organisms producing the desired product. Raw materials should not contain any undesirable and toxic components that may be carried through the cell culture, fermentation, and the purification process to the finished product. Water is an important component of the media, and the quality of the water will depend on the recombinant system used, the phase of manufacture, and intended use of the product. Raw materials considered to be similar when supplied by a different vendor should meet the acceptance criteria before use. In addition, a small-scale pilot run followed by a full-scale production run is recommended when raw materials from a different vendor are used to assure that growth parameters, yield, and final product purification remain the same.

Most mammalian cell cultures require serum for growth. Frequently, serum is a source of contamination by adventitious organisms, especially mycoplasma, and firms must take precautions to assure sterility of the serum. There is an additional concern that bovine serum may be contaminated with bovine spongiform encephalopathy (BSE) agent. Because there is no sensitive *in vitro* assay to detect the presence of this agent, it is essential that the manufacturers know the source of the serum and request certification that the serum does not come from areas where BSE is endemic. Other potential sources of BSE may be proteases and other enzymes derived from bovine sources. Biological product manufacturers have been requested to determine the origin of these materials used in manufacturing.

The media used must be sterilized generally by sterilizing in place (SIP) or by using a continuous sterilizing system (CSS) process. Any nutrients or chemicals added beyond this point must be sterile. Air lines must include sterile filters. The following checklist, though not totally inclusive, should be used frequently:

- Confirm the compliance of the source of serum.
- Confirm that the sterilization cycle has been properly validated to ensure that the media will be sterile.
- Verify that all raw materials have been tested by quality control. Determine the origin of all bovine material.
- Document instances where the media failed to meet all specifications.
- Verify that expired raw materials have not been used in manufacture.
- Check that media and other additives have been properly stored.

Culture growth

Cell cultures are run in batch, fed batch, or continuous mode depending on the expression system used. Continuous systems may take weeks to complete, and several harvest pools often result, making it necessary to clearly define the batch strategy. Bioreactor inoculation, transfer, and harvesting operations must be done using validated aseptic techniques. Additions or withdrawals from industrial bioreactors are generally done through steam-sterilized lines and steam-lock assemblies. Steam may be left on in situations for which the heating of the line or bioreactor vessel wall would not be harmful to the culture (Table 10.1).

It is important for a bioreactor system to be closely monitored and tightly controlled to achieve the proper and efficient expression of the desired product. The parameters for the fermentation process must be specified and monitored. These may include growth rate, pH, waste byproduct level, viscosity, addition of chemicals, density, mixing, aeration, foaming, etc. Other factors that may affect the finished product include shear forces, process-generated heat, and effectiveness of seals and gaskets.

Many growth parameters can influence protein production. Some of these factors may affect deamidation, isopeptide formation, or host cell proteolytic processing. Although nutrient-deficient media are used as a selection mechanism in certain cases, media deficient in certain amino acids may cause substitutions. For example, when *Escherichia coli* is starved of methionine and/or leucine while growing, the organism will synthesize norleucine and incorporate it in a position normally occupied by methionine, yielding an analog of the wild-type protein. The presence of these closely related products will be difficult to separate chromatographically; this may have implications both for the application of release specifications and for the effectiveness of the product purification process.

Computer programs used to control the course of fermentation, data logging, and data reduction and analysis should be validated. (See text under Validation.)

Bioreactor systems designed for recombinant microorganisms require not only that a pure culture is maintained, but also that the culture be contained within the systems. The containment can be achieved by the proper choice of a host–vector system that is less capable of surviving outside a laboratory environment and by physical means when this is considered necessary. The NIH Guidelines described under Environment Control describe the details further.

Process overview

Table 10.1 Process Overview for Different Expression Systems

General	Cell culture	Batch, fed batch, or perfusion; 3–40 days.
	Harvest	Centrifugation/filtration; omit if using expanded bed.
	Capture	Chromatographic unit operation is typically based on affinity or IEC. Remove major host- and process-related impurities and water.
	Variable unit operation	The unit operation may be refolding (if *E. coli* is used) or virus inactivation if insect cells, mammalian cells, or transgenic animals are used.
	Intermediate purification	Chromatographic unit operation typically based on HIC, IEC, or HAC stepwise gradient technology used to remove host and process-related impurities.
	Variable unit operation	This unit operation may include tag removal (if *E. coli* is used) or virus removal by filtration if insect cells, mammalian cells, or transgenic animals are used.
	Polishing	Chromatographic unit operation typically based on HP-IEDC or HP-RPC stepwise/linear gradient technology used to remove product-related impurities.
	Variable unit operation	SEC or ultra-filtration to assure proper drug substance formulation.
	Drug substance	The conversion of the drug substance to drug product typically include change of buffer, precipitation, or crystallization.
	Formulation	Batch manufacturing often including stabilizers such as albumin.
	Finished drug product	Filling in appropriate containers such as vials or prefilled syringes.
E. coli (Gram negative)	Fermentation	Expression of N-terminally extended target protein to overcome formation of Met-protein.
	Harvest	Harvest of cells by centrifugation prior to cell disruption.
	Cell disruption	Disruption with French press or like; wash out inclusion bodies.
	Extraction	Extraction under reducing and denaturing conditions (e.g., 0.1 M cystein, 7 M urea pH 8.5).
	Capture	Purification under reducing and denaturing conditions (e.g., IEC or IMAC if protein is His-tagged)
	Renaturation	Controlled folding of the target protein using hollow fiber, SEC, dilution, or buffer exchange
	Intermediate purification	Purification of the folded target protein (e.g., IEC, HIC, HAC)
	Enzyme cleavage	Cleavage of the N-terminal extension with exo- or endo- proteases
	Polishing 1	Purification of the target protein (e.g., HP-IEC, HP-RPC)
	Polishing 2	Purification of the target protein by SEC (not always included)
	Drug substance	The purified bulk product.
	Formulation	Reformulation of the drug substance preparing for administration to humans.
	Drug product	The final product.
Gram positive bacteria	Fermentation	Expression of the target protein to the periplasmatic room or medium.
	Harvest	Harvest of cells by centrifugation prior to cell disruption. This step may be bypassed by means of expanded bed technology.
	Capture	Purification of the target protein from the supernatant.
	Intermediate purification	Purification of the target protein (e.g., IEC, HIC, HAC).
	Polishing 1	Purification of the target protein (e.g., HP-IEC, HP-RPC).
	Polishing 2	Purification of the target protein by SEC (not always included).
	Drug substance	The purified bulk product.
	Formulation	Reformulation of the drug substance preparing for administration to humans.
	Drug substance	The final product.
Yeast	Fermentation	Expression the target protein to the medium.
	Harvest	Harvest of cells by centrifugation prior to cell disruption. This step may be bypassed by means of expanded bed technology.
	Capture	Purification of the target protein from the supernatant or by expanded bed technology.
	Intermediate purification	Purification of the target protein (e.g., IEC, HIC, HAC).
	Polishing 1	Purification of the target protein (e.g., HP-IEC, HP-RPC).

(*Continued*)

453

Table 10.1 (*Continued*) Process Overview for Different Expression Systems

	Polishing 2	Purification of the target protein by SEC (not always included).
	Drug substance	The purified bulk product.
	Formulation	Reformulation of the drug substance preparing for administration to humans.
	Drug product	The final product.
Insect cells	Cell culture	Expression of the target protein to the medium.
	Harvest	Harvest of cells by centrifugation prior to cell disruption. This step may be bypassed by means of expanded bed technology.
	Capture	Purification of the target protein from the supernatant or by expanded bed technology.
	Virus inactivation	Inactivation by means of low pH, high temperature, detergents, etc.
	Intermediate purification	Purification of the target protein (e.g., IEC, HIC, HAC).
	Virus filtration	Nano-filtration.
	Polishing	Purification of the target protein (e.g., HP-IEC, HP-RPC, SEC).
	Drug substance	The purified bulk product.
	Formulation	Reformulation of the drug substance preparing for administration to humans.
	Drug product	The final product.
Mammalian cells	Cell culture	Expression of the target protein to the medium.
	Harvest	Harvest of cells by centrifugation prior to cell disruption: This step may be bypassed by means of expanded bed technology.
	Capture	Purification of the target protein from the supernatant or by expanded bed technology.
	Virus inactivation	Inactivation by means of low pH, high temperature, detergents, etc.
	Intermediate purification	Purification of the target protein (e.g., IEC, HIC, HAC).
	Virus filtration	Nano-filtration.
	Polishing	Purification of the target protein (e.g., HP-IEC, HP-RPC, SEC).
	Drug substance	The purified bulk product.
	Formulation	Reformulation of the drug substance preparing for administration to humans.
	Drug product	The final product.
Transgenic animals	Raw milk	Milking of animals according to good agricultural practices.
	Skim milk	Centrifuged raw milk with low fat content.
	Capture	Purification of the target protein from the skim milk.
	Virus inactivation	Inactivation by means of low pH, high temperature, detergents, etc.
	Intermediary purification	Purification of the target protein (e.g., IEC, HIC, HAC).
	Virus filtration	Nano-filtration.
	Polishing	Purification of the target protein (e.g., HP-IEC, HP-RPC, SEC).
	Drug substance	The purified bulk product.
	Formulation	Re-formulation of the drug substance preparing for administration to humans.
	Drug product	The final product.

Process maturity

As soon as a project enters into the development phase, the time of delivery of material for preclinical and clinical trials becomes a major issue. The amounts needed far exceed what can be produced in a development laboratory, and material for clinical trials must be produced according to cGMP. The process designers face a dilemma: how much should process design be compromised in order to produce the material needed in the shortest possible time? An immature process will lead to process redesign late in the project, resulting in tedious repeats of analytical and biological testing. A too long process development period will—on the other hand—not be accepted by upper management due to tight project time constraints. It is important to understand the depth of the dilemma. Imagine a mammalian pilot scale process being used to produce material for initial drug substance/product stability studies, references, tox studies, and virus validation studies. A major

Table 10.2 Maturity of Process

Maturity Criteria	Comments
Cell line	QA release of the MCB required prior to transferring cells into the cGMP facility.
Process design	Design should be robust with provisions for removal of host, process, and product-related impurities. Where refolding is involved, include a well-defined renaturation step; where virus contamination can be an issue, integrate virus removal and inactivation steps with due consideration for denaturation of protein.
Raw materials	A list of raw materials should be provided; do not use materials not qualified to be used in cGMP manufacturing (see DMF discussion).
Intermediary compounds	Data from three small-scale batches should be provided; critical parameters and their interaction should be defined. Holding times of relevance to large-scale operations should be provided.
Drug substance	Short-term stability of the intermediary compounds to be documented.
Drug product (DP)	Formulation to be flexible enough to change composition of the bulk; short-term stability documented.
In-process control	Parameters stated in intervals (validated and proven acceptable ranges) and monitored; analytical method description plus data for in-process analyses to be provided.
Specifications	Acceptance criteria for DS and DSP provided where possible.
Quality control	DS and DSP quality control plan defined and analytical method description plus typical data provided.

process change will raise doubt of all the data obtained with the original process, and although much can be accomplished with comparability studies, stability and virus validation studies have to be repeated severely delaying the project. Project management must also consider other issues such as process economy, which does not compromise safety but can result in a no-go decision.

Obviously, there is a need to define the process maturity level before scale-up or in other words to define a priori process maturity criteria. Ideally, criteria listed in the following table should be accomplished before tech transfer, but the real world is not perfect and certain tasks may remain unfinished at scale-up with the risk of major adjustments and perhaps process redesign at a later stage. Some companies do plan with a process redesign between phase 2 and 3 as a compromise between time to market and extra development costs. Maturity criteria for technology transfer and scale-up are described in Table 10.2.

Process optimization

Once the process has been scaled up, the process design is locked, but there is room for optimization within the parameter intervals stated (proven acceptable range). The optimization process involves adjustment of the process in order to improve yields, process economy, process robustness, column lifetime, use of raw materials, and labor savings.

Tech transfer and documentation

Several factors should be considered before transferring the process from the development laboratory to the pilot facility. Culturally, there are huge differences between the two areas. The development staff tends to concentrate on the protein chemistry striving for maximal resolution and scientifically elegant solutions while the pilot staff tends to focus on engineering issues and optimal use of the equipment. It is therefore important to assure excellent communication between the two areas and to stress the importance of "design in." The size of the tech transfer package very much reflects the quality and maturity of the process and its associated activities. What follows are listed issues of relevance for the tech transfer between the development laboratory and the pilot facility.

The technology transfer package for recombinant protein manufacturing is a comprehensive document that addresses various issues in sufficient detail to allow the transferee to replicate the process of each transfer package.

> *Overview*: Process rationale and brief strategy of production.
> *Molecule*: Physical and chemical properties of target protein.
> *Cell line*: History of cell line; MCB vials, WCB vials, WCB release documentation; safety profile; reworking and propagation documentation.
> *Process*: List of raw materials and equipment; hazardous element identification; development history documentation; description of unit operations; manufacturing process protocol; small batch data; critical parameters; related impurities description: host, process and product-related; process flow sheet; major impurity removal steps; batch production plan.
> *Fill and finish*: Master production document; batch record; formulation deviation parameters; development history document; compliance record.
> *Storage*: Conditions for all in-process materials, drug substance, and drug product.
> *Shipping*: Conditions and instructions, packaging specifications for both drug substance and drug product.
> *In-process control*: Required tests and when needed; parameter intervals (tolerance and specifications) for pH, conductivity, redox potential, protein concentration, temperature, holding time, load, transmembrane pressure, linear flow, etc.
> *Specifications*: Acceptance criteria for appearance, identity, biological activity, purity and quantity; pharmacopeial specifications where applicable.
> *Analytical methods*: Description, typical data, method qualification level, methods transfer protocols, and reference material.
> *Stability*: Real-time data and extrapolated data on both drug substance and drug product.
> *Virus validation*: Documentation and protocol; compliance report.
> *Master validation plan*: Update.
> *Training*: Employee training protocol, end of training objectives, and ongoing support.

Validation

The economy of the manufacturing process (facility depreciation, raw material costs, quality control, and quality assurance) must match the expected results. As a rule of thumb, the expenses on manufacture should not exceed 15% of the price per dose (vial). Some of the factors influencing the process economy are manpower, chromatographic media, filters and membranes, buffers, other raw materials, number and type of in-process control analyses, and the overall yield obtained. Additional cost comes from complying with environmental requirements. Where organic solvents are used, their disposal is further costly. Waste disposal, particularly the requirement that all material in contact with the genetically modified cells, must be properly sterilized and disposed, and adds considerable overheads to the overall production costs.

Typically, manufacturers develop purification processes on a small scale and determine the effectiveness of the particular processing step. When scale-up is performed, allowances must be made for several differences when compared with the laboratory-scale operation. Longer processing times can affect product quality adversely since the product is exposed to conditions of buffer and temperature for longer periods. Product stability, under purification conditions, must be carefully defined. It is important to define the limitations and effectiveness of the particular step. Process validation on the production size batch should then compare the effect of scale-up. Whereas the data on the small scale can help in the validation, it is important that validation be performed on the production size batches. There are specific situations such as where columns are regenerated to allow repeated use; this requires proper validation procedures performed and the process periodically monitored for chemical and microbial contamination.

Where batches are rejected, it is important to identify the specific manufacturing and control systems that resulted in the failure and thus appropriated action plan brought in place to prevent reoccurrence of the mistake.

Scale-up

The downstream process is designed such (in a linear manner) where most parameters are fixed and only column diameter in chromatographic unit operations, and the membrane area in filtration operations. The perpetual conflict between the pilot scale mentality and the commercial sale requirements continue to be the greatest impediment—the human factor. The initiative at the US FDA in Process and Analytical Technology (PAT) Initiative (http://www.fda.gov/cder/OPS/PAT.htm) discussed in detail elsewhere offers an ideal solution adopting the concept of "design in," where issues related to process safety, robustness, cGMP compliance, facility constraints, economy, and time to market is build into the process at small-scale development. The process is tested in small scale before transfer against predefined process maturity factors, such as parameter intervals, in-process control points, critical parameters and their interactions, robustness, yield, documentation level, raw material qualification, and maturity of analytical procedures. This concept challenges the classic stepwise scale-up (a short-cut approach to accommodate unrealistic timelines), where cell culture and initial purification steps (e.g., capture) are scaled up before the process is developed in small scale. The stepwise scale-up procedure creates much confusion, is time consuming, and invites late process redesign. It is also common to outsource this stage of work, particularly if the company plans to outsource manufacturing as well. Transferring an immature process into the pilot environment may result in delayed phase 1–2 clinical material manufacture as process changes may raise doubts regarding the stability, virus validation, and preclinical data. Any signs of an immature process should therefore be dealt with immediately. Immature process indicators include

- Yields are low or vary from batch to batch
- Target protein instability
- Changing impurity profiles
- UV diagrams are not super imposable
- Too many batches are discarded or do not meet specifications
- Manufacture is terminated during processing
- Reworking of unit operations is necessary
- The process is frequently being redesigned

Specific scale-up issues

The optimal scale-up and the choice of the scale is influenced by several factors in an integrated model. Given next is a cost estimation model.

Cost calculations

The costs of producing a batch of recombinant-derived product can be calculated from the input figures listed in the following table. It is recommended to use an excel spreadsheet for cost calculations including specific costs not mentioned in the table.

The cost contributions at various steps are given later. Opportunities should be recognized where costs can be cut, particularly in deciding the magnitude of scale-up. The relative contribution of each of the step should be the first criterion of selection on which step to be reworked to reduce costs. The relative costs should be considered in terms of long-term impact. For example, a 1% recurring cost can add a substantial savings if reduced.

Upstream cost = (B1*B2)/12 + B3*B4 + B5 (asterisk is used for multiplier)
Downstream = (C1*C2)/12 = (B3*A1*C5*C3)/(C4*C6*1000) + (C7*C8) + C9 + C10 C11
Fill and Pack = D1*D2* + D3
Total cost/batch = Upstream + Downstream + Fill and pack + E1*E2 + F1*F2 + G1*G2 + H1
Yield/batch = (B3*A1*A3)/100,000
Cost/G = Total cost/Yield (per batch)
Cost per vial = Cost per batch/D1
Cost per dose = (Cost per batch/D1)*A6

Based on this model, we can study what components of the entire process can be altered to achieve the best possible production costs.

- *Quantity required.* The cost is inversely proportional to quantity produced, ranging from about $85/mg for insulin to $1000+/mg for some cytokines. For expensive products, it may be worthwhile not to look at production cost optimization as timely entry may be more relevant.
- Batch is defined as the *batch* or fed batch cell culture volume processed in downstream as a combined amount of material resulting in the drug substance. In continuous cell cultures or by transgenic animal technology, the harvest or milk may be pooled in subfractions making it more difficult to define the batch and the batch scheme must provide information about the flow, the harvest procedures, pools, analytical in-process control programs, and intermediary compounds. Also included here are details if several columns are used or where parallel processing or splitting of processing is envisioned. The fact that inclusion bodies can be combined from several fermentation batches and then processed together requires clear identification of starting sub-batches.
- Batch variations document specifies variability and lack of reproducibility; this may be due to scale-up procedure due to formation of concentration gradients in large reactors or containers.
- Buffer preparation at large scale that cannot be handled require automation and robotics management creating an entirely different set of validation requirements and tools including computer validation (see Section 211.68 (a, b) of FDA CGMP for validation of automated systems, mechanized racks, and computers).

- Column life when prolonged reduces cost significantly; remember that almost 70% of total production cost goes into downstream processing, mainly in the cost of chromatographic media. Measure taken to prolong column life include longer usage, recharging, etc., must be properly validated and documents.
- Environmental issues relate to disposal of large waste; for example, use of ammonium sulfate will severely affect the environment in large-scale operations.
- Equipment interaction can determine the choice of chemicals used; sodium chloride is a corrosive agent to stainless steel, whereas sodium acetate is not. These issues should be addressed as early as possible in-process development.
- Facility costs can be very high because of specialize area requirements, specialized manpower required, environment controls, and waste disposal needs, etc. A prospective biogeneric marketer would be wise to look into outsourcing manufacturing, especially if several products are involved that may require separate processing suites. One of the control areas that is often not given full budgeting is the monitoring of environment; a 5000 ft^2 facility may cost upwards of $2 million per year only to comply with the monitoring standards. As the regulatory environment is still evolving, the area requirements are likely to change that may cost substantial redesigning, another reason to outsource manufacturing until such time that the market is firmly established. The price per gram drug substance is significantly reduced by linking process design, scale-up factor, and batch logistics to the facility design and thereby reducing the occupancy time. This requires several levels of set up, one for transfer of technology to pilot scale, from pilot scale to first-stage manufacturing and from first-stage manufacturing to full-scale manufacturing.
- Validation of biological processes is an expensive exercise that continues throughout the commercial manufacturing operations. Typically, validation steps are initiated when all separation and purification steps are described in detail and presented with flowcharts. Adequate descriptions and specifications should be provided for all equipment, columns, reagents, buffers, and expected yields. The FDA defines process validation in the May 1987 "Guideline on General Principles of Process Validation" as: "validation—establishing documented evidence which provides a high degree of assurance that a specific process will consistently produce a product meeting its predetermined specifications and quality attributes." As a result, there is a need to establish comprehensive documentary proof to justify the process and demonstrate that the process works consistently. Validation reports for the various key processes would be dependent on the process involved; for example, if an ion-exchange column is used to remove endotoxins, there should be data documenting that this process is consistently effective as done by determining endotoxin levels before and after processing. It is important to monitor the process before, during, and after to determine the efficiency of each key purification step. One method commonly used to demonstrate validation is to "spike" the preparation with a known amount of a contaminant and then demonstrate its absence.
- Harvesting can be programmed to store inclusion bodies for longer time (even 2 years) and most large-scale operations should validate this storage step.

- Holding times can be long and add cost in commercial production; these are often not considered in developing processes; it is advisable that in the initial phases, realistic times should be validated. This aspect is related more to logistics than to science. It takes much longer to empty a 4000 L tank than it does to dump a 2 L flask. Often, the practical considerations of shift-change (if the process requires more than 8 h) is necessary in designing the process.

- Cleaning, sanitization and storage of columns, equipment and utensils is an integrated part of the manufacturing program. Whereas liberties are routinely taken in small-scale production, these issues can add substantial costs in a poorly designed process and facility but also raise contamination risks that may not be acceptable by the regulatory authorities.

- Analytical assays to test quality may soon become less expensive, if manufacturers are able to avoid them in processes that are well developed. This is particularly important as large-scale manufacturing require larger testing protocols. Recently, the FDA released a draft on Process Analytical Technology (PAT) at http://www.fda.gov/cder/guidance/5815dft.htm). This extended approach, taken during the development phase, not only includes traditional analytical testing, but also real-time monitoring of the process parameters and responses. A better monitoring of parameters and responses may result in a reduced analytical in-process control program.

- Labor-intensive processes are expensive and costs may be reduced by reducing the number of unit operations and by automation. Each unit operation adds to cost and reduces yield. However, there is a limit to how the number of process steps due to requirement for effective impurity removal and virus inactivation if insect cells, mammalian cells, or transgenic animals have been used for protein expression. There is an extensive effort to automate systems as generic companies, which have reasons to adopt more cost-effective systems, enter the market. From robotics to continuous processing systems to wave bioreactors are all efforts to reduce human resource cost. A prospective manufacturer must look at all available and soon to be available alternates in designing the process. However, the US FDA requirements for validating automated process and computer systems should be implemented as early as possible.

- Monitoring through online data collection systems are needed to control and adjust parameters critical to processes. This creates a problem because small-scale instruments may not have the same level of monitoring potential as do the large ones (which can be ordered with custom features).

- Process design depends on the nature of the target protein, the expression system used, and the demand for safety, robustness, and compliance with cGMP. In order to produce a safe product, qualified raw materials must be used and host, process, and product-related impurities removed. Although variations in process design may occur, certain general rules apply such as use of at least three chromatographic principles, virus inactivation, and filtration in process based on insect cells, mammalian cells, or transgenic animals. Target protein stability must be documented throughout processing and upon storage. Process design is a small-scale activity and in principle no process should be transferred for scale-up before it is reasonably tested in small scale. The major process maturity criteria are linked to specifications, robustness, and cGMP compliance. Although drug substance and drug product specifications and their accept criteria are not fully defined at this early stage, data from at least three small-scale batches must be provided and tested against available specifications and accepts criteria. The process must be robust and to some extent provide the expected outcome. Knowledge of critical parameters and their interactions is valuable information prior to scale-up. The process should be tested with respect to cGMP to make sure that the process complies with the facility design and manufacturing procedures. Difficulties should be expected if the expression yield is low, the protein unstable, the purification yield is low, or the process comprises too many steps.

- Process economy determines the scale-up factor adopted as there is always a certain maximum capacity to which a process can be scaled with economic advantage. A good example of this would be the manufacturing of insulin that is produced in quantities of tons rather than grams. If a scale-up process moves from 100 L fermenter to 30,000 L, the cost is not necessarily proportional to the capacity and may render the project commercially not feasible. Also, when outsourcing, it is always a good idea to adjust the process to available bioreactor size as not all CMOs carry all sizes of bioreactors. When processing bacterial cultures, it may be worthwhile to take advantage of pooling inclusion bodies when only smaller size fermenters are available. However, as the batch size decreases, the costs for analytical assays increases, so batch size and analytical programs should be carefully balanced.

- Quality control testing is required for both drug substance and drug product; reduced test programs should be reconsidered upon scale-up. Some test analyses introduced in an early development phase may be skipped from the batch release program as combined data have confirmed process robustness. The rationale for removing a given analytical assay should be given. The PAT Initiative at FDA may reduce the testing of products.

- Quality assurance systems should assure that the process scale-up does not alter process safety, robustness, or compliance with regulatory demands. If the process is redesigned during scale-up, the validity of preclinical data should be considered. Robust systems introduce strict control of unit operation parameters and define parameters in intervals rather than set points. The parameter intervals are tested and justified in small scale (proven acceptable range), making room to adjust the intervals according to large-scale needs. In the linear scale concept, these parameter intervals are kept constant upon scale-up. Other factors, such reactor volumes, sample loads, and column diameters, are increased—all in a linear fashion.

- Cell cultures offer most opportunities and most problems in scale-up. For example, large-scale animal cell cultures are fundamentally different from conventional microbial fermentation due to the fragility of mammalian cells. The cells are easily damaged by mechanical stress, making it impossible to use conditions of high aeration and agitation; this includes the use of the newly introduced wave bioreactors. Fortunately, animal cells grow slowly and at less cell densities and therefore do not require the high oxygen inputs typical of microbial cultures. Cell culture scale-up often results in changes in the cell culture supernatant composition, which may affect the downstream process. However, except for reactor volume, other parameters like culture medium, pH, temperature, redox potential, osmolality, agitation rate, flow rate, ammonia, glucose, glutamine, lactate

461

concentrations, pCO_2, and OUR remain constant within the prescribed interval limits.

- Precipitation step scale-up involves only change in the amount of sample and the volume of reagent as all other parameters like sample pH, conductivity, temperature, concentration, redox potential, holding time, reagent concentration, and precipitation time remain constant within interval limits. The procedure of precipitation also remains identical.

- Chromatography scale-up produces more problems than any other operation in the process. Larger equipment may cause extra-column zone broadening due to different lengths and diameters of outlet pipes, valves, monitor cells, etc. An increase in column diameter may result in decreased flow rate due to a reduction in supportive wall forces (at constant pressure drop). For example, a decrease in flow of 30%–45% is observed for a column packed with Sepharose 6 FF when the diameter is increased from 2.6 to 10 cm. Prolonged sample holding times on column may result in precipitation of material, resulting in clotting of pipes, valves, or chromatographic columns. Parameters that are changed proportionally (linear scale-up) include sample volume, sample load, column diameter, column area, and column volume, flow rate; residence time remains constant (an alternate method would keep residence time constant allow for variations in both column area and height). Parameters that remain constant within the interval limits include sample pH, conductivity, temperature, concentration, redox potential, holding time, bed height, residence time, linear flow rate, binding capacity, back pressure, buffers, equilibration procedure, wash procedure, elution procedure, CIP procedure. Gradients should not be changed linearly but stepwise.

- Filtration step scale-up does not change the intervals limits for pH, redox potential, temperature, concentration, conductivity, holding time, membrane type, transmembrane pressure, retentate pressure, feed pressure, cross-flow velocity, filtrate velocity, C wall, flux, and CIP procedures. Parameters that are increased linearly include sample volume and membrane area.

- Equipment change is the most significant aspect of scale-up from laboratory scale to production scale. Not all equipment is available in scalable type. This consideration should be the prime deciding factor in laboratory scale development. Manufacturers like New Brunswik, Amersham, Pall, etc., offer a broad line of products in terms of capacity and are always preferred to single-source suppliers even if the initial cost is higher to use this equipment. Large-scale hardware often has a different design, for example, pump design may change from high precision piston or displacement pumps to rotary, diaphragm, or peristaltic pumps. Similarly, low-volume multiport valves are replaced by simple one-way valves, which combined with large-scale tubing, may expand volumes of equipment accessories substantially and lead to extra dispersion of the target protein molecules. Large-scale equipment is often built of stainless steel and does not withstand high concentrations of sodium chloride. The scale-up issues to consider in relation to large-scale equipment include differences in chromatographic column physics of movement, the choice and placement of tubing, valve and reservoir, chemical resistance of construction material, the choice of CIP and SIP, etc. (Table 10.3).

Table 10.3 Input Figures for Cost Calculations

Category	Issue	Index	Comments
General	Expression level	A1	Amount in g/L expressed
	# of batches	A2	
	Process yield	A3	% of purified protein (drug product)
	Dose	A4	mg/dose
	Pack size	A5	mg/vial
	Vials needed	A6	# of vials/dose
Upstream	Facility	B1	Yearly cost ($) of using upstream component of cGMP facility (including maintenance and manpower)
	Utilization	B2	#Months the upstream component is used for given project
	Culture volume	B3	Volume in liters in a given batch
	Media cost	B4	Price in $/L of culture media
	Utensils	B5	Price in $ for utensils used (e.g., filters, bags, etc.)
Downstream	Facility	C1	Yearly cost ($) for using downstream component of the cGMP facility (including maintenance and manpower)
	Utilization	C2	#Months the downstream component is used for a given project
	Chromatography steps	C3	Number of chromatography steps
	Binding capacity	C4	Average binding capacity in mg/mL
	Media cost	C5	Chromatography media cost in $/L
	Buffer volume	C7	Total consumption in L (on an average 15 column volumes are used/step)
	Buffer cost	C8	$/L
	Utensils	C9	Cost in $ for components used (filters, membranes, bags, etc.)
	Raw materials	C10	Cost in $ for expensive reagents, enzymes, etc.
	Formulation	C11	Cost in $ for formulation of drug substance
Fill and pack	Number of vials	D1	Vials/batch
	Price	D2	Price/vial
	Shipping cost	D3	
In-process control	# of analyses	E1	Total number per batch
	Cost of analysis	E2	Average in $/IPC analysis
DS quality control	# of analyses	F1	Total number of drugs substance quality analysis per batch
	Cost	F2	$/ analysis
DP quality control	# of analyses	G1	Total number per batch of drug product
	Cost	G2	$/analysis
QA release	Cost	H1	$/batch

Specific economy issues

Economy considerations start already with the choice of expression system, culture conditions, and demand for process robustness, hence this is part of the "design-in" strategy recommended. A rule of thumb tells that manufacturing comprising upstream, downstream, formulation, fill and pack, quality control, and documentation should add up to 15% of the vial price, but no more. Since a cGMP facility is expensive, outsourcing is highly recommended for new entrants to biogeneric field. This advice is not merely a cost-saving measure but also offers only logistic solutions. For example, where a transgenic animal is involved, few pharmaceutical manufacturers would know how to handle farming of animals and comply with good cattle-raising practices. The costs of upstream, downstream, and quality control are closely interrelated not only by variations in process design but also in respect to batch sizes versus in-process control and analytical control programs. A robust process will allow for large batches, thus reducing the demand for manpower and the number of samples to be analyzed.

Process design should be distinguished from process optimization, where factors such as labor, automation, lean management, column lifetime, reuse of utensils, and batch planning affect the process economy. The latter issues are dealt with at a later development stage, typically when the process design has been locked.

- Expression system selection and eventually cell line takes place early in the process and is the most important decision. The choice depends on the nature of the target protein (glycosylation, phosphorylation, acylation, size, etc.), expected expression levels, expression system development time, risk of batch failure, safety considerations, amount needed, and regulatory record.
  - The target proteins without posttranslational modifications can be expressed in all expression system (bacteria, yeast, insect cells, mammalian cells, transgenic animals, or transgenic plants); bacterial system is historically preferred for cost considerations but better expression yields have been obtained with yeast, mammalian cell cultures, and the introduction of transgenic animals and plants over the past 10 years challenge the bacterial systems, where intracellular expression of Met-protein and in vitro folding increases the complexity of the downstream process. Posttranslational modified proteins or more complex proteins are expressed in insect cells, mammalian cells, transgenic animals, or plants. Microbial systems cannot be used.
  - Expression level varies with the host organism used and the nature of the target protein. Typical expression levels vary widely. In most hosts, these range at less than 1 g/L; *E. coli* generally gives a better yield of 1–4 g/L while mammalian cells when used for antibodies generally produce much higher levels; transgenic animals provide the highest yield of 5–40 g/L; the yield in transgenic plants is uncertain and not widely available for evaluation. The nature and quality of the expressed protein influences the purification yield. Although expression levels of *E. coli* usually are high, expression of N-terminal extended target protein and the need for in vitro refolding significantly influences the overall process yield. Further, stressed cells tend to express less stable protein, resulting in great losses during purification or production of drug substance/product with shortened lifetime.
  - Expression system development time vary widely between various expression systems. However, for most hosts, 4–6 months are required to develop the system. Once developed, the time to target protein expression depends on the bioreactor system deployed but generally range from a few days, for example, 5 days for *E. coli* and other bacteria, 2 weeks for yeast, 4 weeks for inset cells, and 2–16 weeks for mammalian cells. The longer development time for transgenic goats and cows (18–24 months) is partly compensated for by the relative high expression levels and the fast access to target protein, once the system has been developed.
  - Risk of batch failure is mainly due to infections; large-scale mammalian cell cultures, having long cell expansion times including several bioreactors and running over long time intervals, are associated with higher risk factors than other expression systems. Due to the high cell culture media cost, the economic loss can be substantial.
  - Safety considerations add to cost significantly. Insect, cells, mammalian cells, and transgenic animals can be infected with viruses. Costly virus testing, virus reduction unit operations, and validation programs are needed to assure product safety. The potential prion infection risk

of sheep, goats, and cows is being debated, emphasizing the need for controlled herds.

- Amount needed. A 1000 L bioreactor with an expression level of 1 g/L produces 1 kg of target protein per reactor volume. A batch or fed-batch culture typically runs for 7 days offering a productivity of 100 g/day. A 1000 L perfusion bioreactor with a 2 × flow per 24 h and with an expression level of 1 g/L produces 200 g/day. A transgenic cow expressing 20 g/L milk produces 400 g/day assuming a volume of 20 L milk per day. In terms of output, the transgenic cow is a far more efficient expression system than the cell culture–based systems and probably less risky. Animal-based expression systems should therefore seriously be considered for large-scale operations.
- Regulatory record of insect cells, transgenic animals, and plants is poor. This is an important consideration when selecting a system to assure that favorable regulatory review is forthcoming.

- Raw materials are very different in price between culture media used for microbial, insect, and mammalian cell cultures, and commercial scale fermenter media, the former being the most expensive per liter. Unfortunately, mammalian cell cultures usually offer relative low expression levels compared to microbial systems, making expenses to culture media a major cost contributor. For an expression system yielding 1000 g/L with 40% yield, the contribution of the cost of media is about $25/g of protein; when the expression level drops to 10 g/L, the cost contribution of media rises to $2500/g of protein. Obviously, low mammalian cell expression levels adversely affect the process economy.
- Cell growth is normally achieved in batch, fed-batch, or continuous mode, the latter mainly being used for mammalian cell bioreactors up to a volume of maximum 300 L, at present. A bioreactor run in the batch mode has a fixed working volume and the cell culture is grown to a defined cell density before harvest. The harvest volume defines the batch and the yield is the expression level x harvest volume. For example, a bioreactor of 100 L working volume with an expression level 300 mg/L results in a batch yield of 30 g. The productivity is thus 30 g/week or 0.3 g/L culture medium, assuming it takes 1 week to complete the batch. A bioreactor run in fed-batch mode may result in a final volume of 130 L equivalent to a productivity of 39 g/week or 0.3 g/L culture medium assuming identical culture time and expression level. A 100 L working volume bioreactor run in continuous mode at a flow of 2 reactor volumes per day and an expression level of 100 mg/L produces $100 \times 2 \times 7 \times 100 = 140$ g/week or $140/1400 = 0.1$ g/L culture medium or 1/3 of the batch culture. The increased overall productivity may compensate for the decreased productivity per L culture medium.
- In-process control is less expensive for batch and fed-batch systems compared to continuous cultures, but the overall cost may not be too different and is worth considering in light of the PAT Initiative of FDA. Milking procedures may result in production of small volume bags increasing the cost for analytical control programs. The insect and mammalian cell end-of-production test comprising sterility, fungi, mycoplasma, and virus testing should be included in the cost calculations.
- Yield is the composite of each purification step's output. A downstream process comprising 10 unit operations with an average recovery of 95% will result in an overall yield of 57%, which is acceptable. However, an average recovery of 80% will result in a total yield of 11%, which in most circumstances is not acceptable. In several cases, more than 10 unit

operations are needed to guarantee a safe product, making it fair to conclude that one should aim for more than 95% recovery in most, if not all, unit operations. Some of the most recent trends to alter the molecular structure, for example, pegylation have met with lower yields. Because of the mathematical nature of proportional reduction, small changes in step yields result in dramatic changes in the total yield. For example, a step yield of 95% where 15 steps are involved gives 46% total yield; the same 15 steps in 75% step yield would give only 1% of the total yield. In most instances, 5–10 steps are minimally involved; at 10 steps, therefore, the total yield ranges from 60% to 6% in 95% to 75% step yield transition.

- Batch size is decided on a variety of consideration, not all of them have economic optimization. Large batches require automated facilities, but results in relatively small number of samples to analyze.
- Chromatographic media costs are a minor fraction of the entire manufacturing costs, and it may be wise to select media from other criteria than price per liter. Service, trouble shooting, linking media, column and equipment to the same supplier, and regulatory support files may be far more important aspects as the number of failed batches can be reduced by such actions. Suppliers like Amersham or others with wide range of offering and validation should be the first choice as media vendors. The major factors influencing the economy of chromatographic unit operations are media cost, binding capacity, recovery, column lifetime, linear flow, and shelf life.
- Utensils are bags, filters, or any other equipment exchanged at regular intervals. It has become common practice to use bags, filters, tubes, etc., only once in order to reduce cost and time to cleaning in place procedures.
- Number of steps in a process directly affect cost and yield; however, reducing the number of steps process can affect robustness and safety and even later costs in additional testing as may be required by the regulatory authorities. For example, a choice may have to be made whether to add an additional adventitious agent removal process of validating the system.
- In-process control testing should be minimal, and this is only possible with a well-defined system, which may incur higher costs initially. The trend is to reduce the number of in-process analytical methods and to expand on monitoring of parameters and responses, thereby keeping strict control of the process in real time. The number of samples to analyze is inversely related to the batch size; large batches reduce the cost for in-process control.
- Formulation, fill, and pack operations of the API can also be subject to cost reduction depending on the formulation. Obviously, a lyophilized product will cost more, and if a ready-to-inject formula can be developed, that should be a better choice. The components such as syringes, pen systems, etc., added at this stage add substantially to the cost of the product.
- In-process and quality control testing is extensive (see USP, EP, or BP) and expensive because of the nature of tests involved, notwithstanding other requirements of validation common to all types of testing. The assays used for in-process control can be justified as they provide essential process information during development or the assays that are used to monitor a given outcome important for in-process control in manufacturing processes. Thus, reducing cost would mean reducing the number of in-process analyses without compromising process control. It is common practice to revise the program as more and more data become available and the process matures. Another way to reduce costs is to use the PAT approach recently suggested by the FDA (http://www.fda.gov/cder/guidance/5815dft.htm). PAT is considered to be a system for designing,

analyzing, and controlling manufacturing through time measurements of critical quality and performance attributes of raw and in-process materials and processes with the goal of ensuring final product quality. The goal of PAT is to understand and control the manufacturing process as quality cannot be tested into products, but should be built-in. This approach can be taken during the development phase extending the in-process control program to not only include traditional analytical testing, but also real-time monitoring of the process parameters and responses. A desired goal of the PAT framework is to design and develop processes that can consistently ensure a predefined quality at the end of the manufacturing process. Such procedures would be consistent with the basic tenet of quality by design and could reduce the risks to quality and regulatory concerns while improving efficiency. A third way to reduce costs is to lower the price per sample by assuring a continuous flow of samples, thereby reducing time spent on method set up and calibration.

- Quality control testing is performed on both, the drug substance (DS) and the drug product (DP) (see BP/EP/USP). The testing typically comprises from 10 to 15 different analytical methods with an average price between $500 and $3000/sample. If outside laboratories are inducted to provide additional testing, the cost can skyrocket. For example, NIBSC would typically charge about $20,000 to test one sample; animal assays and viral assays can be extremely expensive when outsourced; yet, outsourcing is still the preferred way of doing these assays to obviate the large cost of maintaining animal houses or viral containment systems and validating the methods.

Process materials

The quality of water should depend on the intended use of the finished product. For example, CBER requires water for injection (WFI) quality for process water. On the other hand, for in vitro diagnostics, purified water may suffice. For drugs, the quality of water required depends on the process. Also, because processing usually occurs cold or at room temperature, the self-sanitization of a hot WFI system at 75°C–80°C is lost.

For economic reasons, many of the biotech companies manufacture WFI by reverse osmosis rather than by distillation, which may result in contaminated systems because of the nature of processing equipment that is often difficult to sanitize. Any threads or drops in a cold system provide an area where microorganisms can lodge and multiply. Some of the systems employ a terminal sterilizing filter. However, the primary concern is endotoxins, and the terminal filter may merely serve to mask the true quality of the WFI used. The limitations of relying on a 0.1 mL sample of WFI for endotoxins from a system should also be recognized. The system should be designed to deliver high-purity water, with the sample merely serving to assure that it is operating adequately. As with other WFI systems, if cold WFI water is needed, point-of-use heat exchangers can be used.

Buffers can be manufactured as sterile, nonpyrogenic solutions and stored in sterile containers. Some of the smaller facilities have purchased commercial sterile, non-pyrogenic buffer solutions.

The production and/or storage of nonsterile water that may be of reagent grade or used as a buffer should be evaluated from both a stability and microbiological aspect.

WFI systems for BDP are the same as WFI systems for other regulated products. As with other heat-sensitive products, cold WFI is used for formulation. Cold systems are prone to contamination. The cold WFI should be monitored both for endotoxins and microorganisms.

Environment control

Microbiological quality of the environment during various processing is very important, particularly as the process continues downstream, more intensive control and monitoring is recommended. The environment and areas used for the isolation of the BDP should also be controlled to minimize microbiological and other foreign contaminants. The typical isolation of BDP should be of the same control as the environment used for the formulation of the solution prior to sterilization and filling.

NIH guidelines for handling DNA material

The recombinant technology is associated with a number of safety issues related to the expression system used, cell banking, fermentation and cell cultures, raw materials used, downstream processing, and unintended introduction of adventitious agents (bacteria, viruses, mycoplasma, prions). One of the major purposes of the purification process is to provide a rational design assuring the removal of the said adventitious agents and other harmful impurities. A rule of thumb says that at least three different chromatographic principles should be used in a biopharmaceutical downstream process. If insect cells, mammalian cells, or transgenic animals have been used, a virus inactivation step and an active virus filtration step should be considered. The adventitious agents and their relation to the expression system used is better appreciated realizing that endotoxins, nucleic acids, bioburden, viruses, and prions are an issue in all systems of expression except that viruses and prions are not an issue in microbial systems; prions are also not an issue in other systems except in transgenic animals. The use of raw materials should be carefully investigated. Only raw materials suited for biopharmaceutical processing should be accepted, based on solid documentation on safety and quality.

The purpose of the *NIH guidelines* is to specify practices for constructing and handling: (1) recombinant deoxyribonucleic acid (DNA) molecules, and (2) organisms and viruses containing recombinant DNA molecules. Any recombinant DNA experiment, which according to the *NIH guidelines* requires approval by NIH, must be submitted to NIH or to another Federal agency that has jurisdiction for review and approval. Once approvals, or other applicable clearances, have been obtained from a Federal agency other than NIH (whether the experiment is referred to that agency by NIH or sent directly there by the submitter), the experiment may proceed without the necessity for NIH review or approval. For experiments involving the deliberate transfer of recombinant DNA, or DNA or RNA derived from recombinant DNA, into human research participants (human gene transfer), no research participant shall be enrolled until the RAC review process has been completed.

In the context of the *NIH guidelines*, recombinant DNA molecules are defined as either (1) molecules that are constructed outside living cells by joining natural or synthetic DNA segments to DNA molecules that can replicate in a living cell, or (2) molecules that result from the replication of those described in (1). Synthetic DNA segments that are likely to yield a potentially harmful polynucleotide or polypeptide (e.g., a toxin or a pharmacologically active agent) are considered as equivalent to their natural DNA counterpart. If the synthetic DNA segment is not expressed in vivo as a biologically active polynucleotide or polypeptide product, it is exempt from the *NIH guidelines*.

Genomic DNA of plants and bacteria that have acquired a transposable element, even if the latter was donated from a recombinant vector no longer present, are not subject to the *NIH guidelines* unless the transposon itself contains recombinant DNA.

Risk assessment is ultimately a subjective process. The investigator must make an initial risk assessment based on the risk group (RG) of an agent. Agents are classified into four RGs according to their relative pathogenicity for healthy adult humans by the following criteria: (1) risk group 1 (RG1) agents are not associated with disease in healthy adult humans. (2) Risk group 2 (RG2) agents are associated with human disease, which is rarely serious and for which preventive or therapeutic interventions are *often* available. (3) Risk group 3 (RG3) agents are associated with serious or lethal human disease for which preventive or therapeutic interventions *may be* available. (4) Risk group 4 (RG4) agents are likely to cause serious or lethal human disease for which preventive or therapeutic interventions are *not usually* available.

Considerable information already exists about the design of physical containment facilities and selection of laboratory procedures applicable to organisms carrying recombinant DNA. The existing programs rely upon mechanisms that can be divided into two categories: (1) a set of standard practices that are generally used in microbiological laboratories; and (2) special procedures, equipment, and laboratory installations that provide physical barriers that are applied in varying degrees according to the estimated biohazard.

There are six categories of experiments defined in the NIH guidelines involving recombinant DNA that require the following:

1. Institutional Biosafety Committee (IBC) approval, RAC review, and NIH Director approval before initiation. Experiments considered as *major actions:* The deliberate transfer of a drug resistance trait to microorganisms that are not known to acquire the trait naturally, if such acquisition could compromise the use of the drug to control disease agents in humans, veterinary medicine, or agriculture, will be reviewed by RAC.

2. NIH/OBA and Institutional Biosafety Committee approval before initiation. Experiments involving the cloning of toxin molecules with LD50 of less than 100 nanograms per kilogram body weight. Deliberate formation of recombinant DNA containing genes for the biosynthesis of toxin molecules lethal for vertebrates at an LD50 of less than 100 ng/kg body weight (e.g., microbial toxins such as the botulinum toxins, tetanus toxin, diphtheria toxin, and *Shigella dysenteriae* neurotoxin). Specific approval has been given for the cloning in *Escherichia coli* K-12 of DNA containing genes coding for the biosynthesis of toxic molecules that are lethal to vertebrates at 100 ng to 100 µg/kg body weight.

3. Institutional Biosafety Committee and Institutional Review Board approvals and RAC review before research participant enrollment. Experiments involving the deliberate transfer of recombinant DNA, or DNA or RNA derived from recombinant DNA, into one or more human research participants. For an experiment involving the deliberate transfer of recombinant DNA, or DNA or RNA derived from recombinant DNA, into human research participants (human gene transfer), no research participant shall be enrolled until the RAC review process has been completed. In its evaluation of human gene transfer proposals, the RAC will consider whether

a proposed human gene transfer experiment presents characteristics that warrant public RAC review and discussion. The process of public RAC review and discussion is intended to foster the safe and ethical conduct of human gene transfer experiments. Public review and discussion of a human gene transfer experiment (and access to relevant information) also serves to inform the public about the technical aspects of the proposal, meaning and significance of the research, and any significant safety, social, and ethical implications of the research.

4. Institutional Biosafety Committee approval before initiation. Prior to the initiation of an experiment that falls into this category, the Principal Investigator must submit a registration document to the Institutional Biosafety Committee, which contains the following information: (1) the source(s) of DNA; (2) the nature of the inserted DNA sequences; (3) the host(s) and vector(s) to be used; (4) if an attempt will be made to obtain expression of a foreign gene, and if so, indicate the protein that will be produced; and (5) the containment conditions that will be implemented as specified in the *NIH guidelines.*

   a. *Experiments using RG2, RG3, RG4, or restricted agents as host–vector systems.* Experiments involving the introduction of recombinant DNA into RG2 agents will usually be conducted at Biosafety Level (BL) 2 containment. Experiments with such agents will usually be conducted with whole animals at BL2 or BL2-N (Animals) containment.

   b. *Experiments involving the use of infectious DNA or RNA viruses or defective DNA or RNA viruses in the presence of helper virus in tissue culture systems caution:* Special care should be used in the evaluation of containment levels for experiments that are likely to either enhance the pathogenicity (e.g., insertion of a host oncogene) or to extend the host range (e.g., introduction of novel control elements) of viral vectors under conditions that permit a productive infection. In such cases, serious consideration should be given to increasing physical containment by at least one level.

   c. *Experiments involving whole animals.* This section covers experiments involving whole animals in which the animal's genome has been altered by stable introduction of recombinant DNA, or DNA derived therefrom, into the germ-line (transgenic animals) and experiments involving viable recombinant DNA-modified microorganisms tested on whole animals. For the latter, other than viruses that are only vertically transmitted, the experiments may *not* be conducted at BL1-N containment. A minimum containment of BL2 or BL2-N is required.

   d. *Experiments involving whole plants.* Experiments to genetically engineer plants by recombinant DNA methods, to use such plants for other experimental purposes (e.g., response to stress), to propagate such plants, or to use plants together with microorganisms or insects containing recombinant DNA.

   e. *Experiments involving more than 10 L of culture.* The appropriate containment will be decided by the Institutional Biosafety Committee. Where appropriate, Appendix K, *Physical containment for large-scale uses of organisms containing recombinant DNA molecules,* shall be used. Appendix K describes containment conditions Good Large Scale Practice through BL3-Large Scale.

5. Institutional Biosafety Committee notification simultaneous with initiation. Examples include experiments in which all components derived

from nonpathogenic prokaryotes and nonpathogenic lower eukaryotes and may be conducted at BL1 containment.

a. Experiments involving the formation of recombinant DNA molecules containing no more than two-thirds of the genome of any eukaryotic virus. Recombinant DNA molecules containing no more than two-thirds of the genome of any eukaryotic virus (all viruses from a single family being considered identical may be propagated and maintained in cells in tissue culture using BL1 containment). For such experiments, it must be demonstrated that the cells lack helper virus for the specific families of defective viruses being used. The DNA may contain fragments of the genome of viruses from more than one family but each fragment shall be less than two-thirds of a genome.

b. Experiments involving whole plants

c. Experiments involving transgenic rodents

6. Are exempt from the *NIH guidelines.* The following recombinant DNA molecules are exempt from the *NIH guidelines* and registration with the Institutional Biosafety Committee is not required:

a. Those that are not in organisms or viruses.

b. Those that consist entirely of DNA segments from a single nonchromosomal or viral DNA source, though one or more of the segments may be a synthetic equivalent.

c. Those that consist entirely of DNA from a prokaryotic host including its indigenous plasmids or viruses when propagated only in that host (or a closely related strain of the same species), or when transferred to another host by well-established physiological means.

d. Those that consist entirely of DNA from an eukaryotic host including its chloroplasts, mitochondria, or plasmids (but excluding viruses) when propagated only in that host (or a closely related strain of the same species).

e. Those that consist entirely of DNA segments from different species that exchange DNA by known physiological processes, though one or more of the segments may be a synthetic equivalent. A list of such exchangers will be prepared and periodically revised by the NIH Director with advice of the RAC after appropriate notice and opportunity for public comment.

f. Those that do not present a significant risk to health or the environment as determined by the NIH Director, with the advice of the RAC, and following appropriate notice and opportunity for public comment.

Biosafety levels

There are four biosafety levels described. These biosafety levels consist of combinations of laboratory practices and techniques, safety equipment, and laboratory facilities appropriate for the operations performed and are based on the potential hazards imposed by the agents used and for the laboratory function and activity. Biosafety Level 4 provides the most stringent containment conditions, Biosafety Level 1 the least stringent. Experiments involving recombinant DNA lend themselves to a third containment mechanism, namely, the application of highly specific biological barriers. Natural barriers exist that limit either: (1) the infectivity of a vector or vehicle (plasmid or virus) for specific hosts, or (2) its dissemination and survival in the environment. Vectors, which provide the means for recombinant DNA and/or host cell replication, can be genetically designed to decrease, by many orders of magnitude, the probability of dissemination of recombinant DNA outside the laboratory.

Since these three means of containment are complementary, different levels of containment can be established that apply various combinations of the physical and biological barriers along with a constant use of standard practices. Categories of containment are considered separately in order that such combinations can be conveniently expressed in the *NIH guidelines*.

Physical containment conditions within laboratories may not always be appropriate for all organisms because of their physical size, the number of organisms needed for an experiment, or the particular growth requirements of the organism. Likewise, biological containment for microorganisms may not be appropriate for all organisms, particularly higher eukaryotic organisms. However, significant information exists about the design of research facilities and experimental procedures that are applicable to organisms containing recombinant DNA that is either integrated into the genome or into microorganisms associated with the higher organism as a symbiont, pathogen, or other relationship. This information describes facilities for physical containment of organisms used in nontraditional laboratory settings and special practices for limiting or excluding the unwanted establishment, transfer of genetic information, and dissemination of organisms beyond the intended location, based on both physical and biological containment principles. Research conducted in accordance with these conditions effectively confines the organism.

Revision of Appendix K of the NIH guidelines revised in April 2002 (http://www4.od.nih.gov/oba/rac/guidelines/guidelines.html) reflects a formalization of suitable containment practices and facilities for the conduct of large-scale experiments involving recombinant DNA-derived industrial microorganisms. Appendix K replaces portions of Appendix G when quantities in excess of 10 L of culture are involved in research or production. For large-scale research or production, four physical containment levels are established: GLSP, BL1-LS, BL2-LS, and BL3-LS.

- *GLSP*: (good large-scale practice) Level of physical containment is recommended for large-scale research of production involving viable, nonpathogenic and nontoxigenic recombinant strains derived from host organisms that have an extended history or safe large-scale use. The GLSP level of physical containment is recommended for organisms such as those that have built-in environmental limitations that permit optimum growth in the large-scale setting but limited survival without adverse consequences in the environment.
- *BL1-LS*: (biosafety level 1-large scale) Level of physical containment is recommended for large-scale research or production of viable organisms containing recombinant DNA molecules that require BL1 containment at the laboratory scale.
- *BL2-LS*: Level of physical containment is required for large-scale research or production of viable organisms containing recombinant DNA molecules that require BL2 containment at the laboratory scale.
- *BL3-LS*: Level of physical containment is required for large-scale research or production of viable organisms containing recombinant DNA molecules that require BL3 containment at the laboratory scale.
- *BL4-LS*: No provisions are made at this time for large-scale research or production of viable organisms containing recombinant DNA molecules that require BL4 containment at the laboratory scale.

There should be no adventitious organisms in the system during cell growth. Contaminating organisms in the bioreactor may adversely affect both the product yield and the ability of the downstream process to correctly separate and purify the desired protein. The presence or effects of contaminating organisms in the bioreactor can be detected in a number of ways—growth rate, culture purity, bacteriophage assay, and fatty acid profile.

To assure compliance with the NIH guidelines:

- Verify that there are written procedures to assure absence of adventitious agents and criteria established to reject contaminated runs.
- Maintain cell growth records and verify that the production run parameters are consistent with the established pattern.
- Establish written procedures to determine what investigations and corrective actions will be performed in the event that growth parameters exceed established limits.
- Assure proper aseptic techniques during cell culture techniques and appropriate in-process controls in their processing.

The FDA is responsible under the National Environmental Policy Act (NEPA) for ascertaining the environmental impact that may occur due to the manufacture, use, and disposal of FDA-regulated products. The FDA makes sure that the product sponsor is conducting investigations safely. Typically, a product sponsor describes environmental control measures in environmental assessments (EAs) that are part of the product application. When the product is approved, the EA is released to the public. Of particular importance are the NIH guidelines for Recombinant DNA Research and particularly the Appendix K (2002), regarding the establishment of guidelines for the level of containment appropriate to Good Industrial Large-Scale Practices. It must be assured that the equipment and controls described in the EA as part of the biocontainment and waste processing systems are validated to operate to the standards; the equipment is in place, is operating, and is properly maintained. Such equipment may include, for example, HEPA filters, spill collection tanks with heat or hypochlorite treatment, and diking around bioreactors and associated drains. SOPs should be established for the cleanup of spills, for actions to be taken in the case of accidental exposure of personnel, for opening and closing of vessels, for sampling and sample handling, and for other procedures that involve breaching containment or where exposure to living cells may occur.

Good manufacturing controls of active pharmaceutical ingredients

The Good Manufacturing Practices prescribed for the API depend to a great degree on the type of API manufactured; for example, shown next is the increasing compliance requirement in various API manufacturing types (Figure 10.1).

Manufacturing systems and layout

Each unit operation will require careful adjustment of the sample parameters (pH, conductivity, protein concentration, redox potential, load). In some cases, desalting, ultra-filtration, or addition of co-solvents, are needed in order to assure proper conditions. One should be aware that ion exchange generally requires samples of low ionic strength, while hydrophobic interaction chromatography is generally used for samples of high ionic strength. Such observations can be used in the design of the process, thus reducing the number of unit operations. The particle size of the media will decrease from capture to polishing to accommodate the need for resolution by decreasing zone spreading on the column. The manufacturing layout will depend on the unit operations included in the manufacturing system.

Type of manufacturing	Application of cGMP to steps (shown in gray) used in this type of manufacturing				
Chemical manufacturing	Production of the API starting material	Introduction of the API starting material into process	Production of intermediate(s)	Isolation and purification	Physical processing, and packaging
API derived from animal sources	Collection of organ, fluid, or tissue	Cutting, mixing, and/or initial processing	Introduction of the API starting material into process	Isolation and purification	Physical processing, and packaging
API extracted from plant sources	Collection of plant	Cutting and initial extraction(s)	Introduction of the API starting material into process	Isolation and purification	Physical processing, and packaging
Herbal extracts used as API	Collection of plants	Cutting and initial extraction		Further extraction	Physical processing, and packaging
API consisting of comminuted or powdered herbs	Collection of plants and/or cultivation and harvesting	Cutting/comminuting			Physical processing, and packaging
Biotechnology: fermentation/cell culture	Establishment of master cell bank and working cell bank	Maintenance of working cell bank	Cell culture and/or fermentation	Isolation and purification	Physical processing, and packaging
"Classical" Fermentation to produce an API	Establishment of cell bank	Maintenance of the cell bank	Introduction of the cells into fermentation	Isolation and purification	Physical processing, and packaging

Figure 10.1 Application of the cGMP guidance in API manufacturing.

A biopharmaceutical manufacturing process must fulfill the criteria of consistency and robustness, meaning that the outcome of each of the unit operations and the entire process shall be the same from lot to lot. Not only shall the acceptance criteria of the drug substance specification program be met, but also in-process acceptance criteria must be specified and met. Each parameter of each of the unit operations should be defined as a proven acceptable range within which the operation has to take place. It is today common practice to qualify and validate the downstream process unit operations by means of statistical factorial design analysis. Small variations in handling procedures or in parameter set points must not influence the outcome of the manufacturing process. Most proteins are unstable in aqueous solutions and the demand for robustness is not easily obtained. Procedures that are easy to carry out in a laboratory scale (pH adjustment, chromatographic gradients, fraction collection) are technically complicated in large scale. Much can be gained if the process designers think ahead by, for instance, defining broad parameter intervals.

Where multiple products manufacturing is envisioned, the manufacturer is faced with the dilemma of design parameters that would comply with the regulatory requirements internationally. Given the cost of establishing such facilities, global compliance, rather than a regional approval, is recommended. It is well established that whereas the US FDA follows certain strict requirements, the EU requirements of cGMP compliance often exceed the US FDA requirements. Whereas the US FDA has moved therapeutic proteins from the CBER to CDER, these remain pretty much covered under the same regulations as are the biological products, including the timely visits and approvals by the agency, of the manufacturing facility. BLA 357 provides details of this inspection schedule. The fundamental question whether a facility can be used to manufacture a multitude of molecules needs examination. If we broadly classify the processes involved based on the type of fermenter required, the decision can be relatively straightforward. Bacterial and yeast fermentation

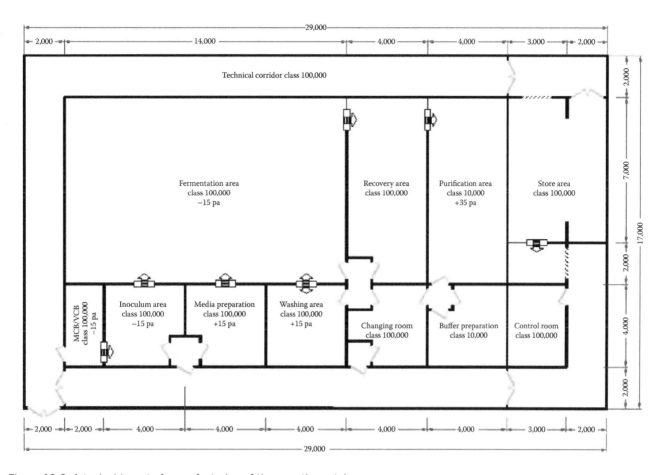

Figure 10.2 A typical layout of manufacturing of therapeutic proteins.

requires a faster agitating fermenter requiring at least two sets of line of production, one for mammalian cells (which require slow and more gentle stirring) and the other types that require relatively more agile stirring. Many newer systems now offer an opportunity for a closed system of transfer of culture to the larger fermenting vessels reducing the environment definition, say from 100,000 area classification to pharmaceutical grade (unclassified) conditions. With computerized CIP and SIP available on most equipment, such systems are highly recommended. However, where closed systems are not utilized, the rooms where fermentation is installed should be at least of 100,000 class. In all areas where the product is exposed such as in the cell culture transfer and downstream processing purification, the rooms must be class 10,000 with separation equipment placed under class 1000 laminar flow curtains. Most regulatory authorities will not allow GMCs (genetically modified cells) to be let out in the environment. This requires an elaborate setup to sterilize the media prior to its discharge. A word of caution applies here where bacterial inclusion bodies are involved. Whereas most of the cells are ground, some remain, requiring an equally intensive treatment of media.

Figure 10.2 shows a typical layout for a biotechnology manufacturing unit. The manufacturing layout of biological products is determined by two major factors: the size of production and the type of production; in most cases, the conditions of containment described earlier and the need to process products under clean room conditions are similar to the processing of sterile products otherwise. As a rule of thumb, the environment should be comparable to preparation room environment for sterile products. There are four major types of work performed in a biological product manufacturing and given later are the requirements for each of these phases:

1. *Master Cell Bank and Working Cell Bank.* A dedicated room is to be made available for each GMC; this room includes a cold storage systems (often a liquid nitrogen system) and a cold (−70°C) cabinet. The cells in the MCB are used to create WCB and both of these are kept under high security. It is recommended that this be a vaulted area with class 100,000 environment; generally a 100–200 ft^2 area would suffice for this purpose. Some manufacturers divide the room into two, one for MCB and one for WCB with restricted entrance to each. In some designs, a direct transfer from WCB to the inoculum/culture room (see later text) is allowed through a transfer window under negative pressure. However, this practice is questioned on the basis of need to maintain a certain area classification; as a result, the MCB/WCB is to be considered as supply center at the time of use, processed through materials dispensing. In all instances, a duplicate MCB shall be maintained off site from the immediate manufacturing area. These rooms should have a backup supply of electricity to assure no power breakdown losses along with alarms to record temperature variations in the cabinets storing the GMCs; an electronic recording device that would transmit the temperature at which the GMCs are stored should be installed. The room should also be equipped with an automated system of pressure differential; the room to be maintained negative with respect to the corridor. This room is dedicated to each GMC; the reason being to avoid mixing up of cultures, access restrictions to different personnel, and the storage requirements.

2. *Inoculum room.* This is the first room where the culture tubes are opened for the purpose of making WCB, or for making the inoculum for fermentation should be a class 10,000 room. If the fermentation system used is a closed inline system, then this room will also have a 4–8 L fermenter to make the starter culture; the culture will then be directly transferred to larger size fermenters. Where roller bottles are used, this room will serve as the staging room to prepare the culture for inoculation into the bottles, which would be done in another room because of the size of operation involved. The culture is handled under a biosafety laminar flow hood; (biosafety hoods prevent exposure of operator). The room should have a 10,000 classification with class 100 under the laminar flow hood. Generally, this room will be connected to the fermentation area, the recovery area, and the roller bottle preparation area, preferably through a negative pressure passing carousel. The room should be of smallest size possible (100–200 ft^2).

3. *Fermentation room.* This is generally the largest room of the facility or may comprise a series of rooms depending on the size of production. Where larger fermenters such as 25,000 L sizes are involved, this may take a three-stage fermenting. A facility of this size would likely be a 20,000+ ft^2 facility that may comprise several floors to accommodate large fermenters. However, for many therapeutic proteins of low dosing, fermenters of 500 L should be sufficient and these can be accommodated on a single-floor basis. The area classification for this room is 100,000 unless a complete closed system is used (which is recommended) wherein general pharmaceutical grade classification (unclassified) may be used.

4. *Roller bottle room.* Where mammalian cell cultures are used in roller bottles, the fermentation room (given earlier) is replaced with two rooms, one for staging the roller bottles and the other a 37°C room to roll the

bottles. The classification of both rooms is 100,000 with bottles being opened under a class 100 biosafety laminar flow hood. There are several issues involved in using roller bottles, the most important being the cost and time constraints in processing large number of bottles. Newer systems offer robotic controls and also recently a multiple bottle automated continuously flowing system; these will be discussed later in the manufacturing details. Where large quantities are involved, manufacturers will be advised to develop a fermentation-based system; much progress has been made in this area and it would be discussed later as well. The size of roller bottle room will depend on the number of rolling racks involved and the size of the staging room will depend on the type of process involved (robotics, manual, continuous flow, etc.). For a medium-scale operation, the 37°C room is about 150 ft^2 and the staging area about 400 ft^2.

5. *Recovery room.* The product of fermentation (either from fermenters or from roller bottles) is brought to this room for the first stage of processing. Where inclusion bodies are involved (such as in the use of bacterial cultures), this room will be cold centrifuge and cell disrupters; for the mammalian culture systems where the protein is secreted, this will be the first stage of reduction of volume through filtration. This room is also used to store the product where refolding is involved at 2°C –8°C environment generally (provided by walk-in refrigerators). This area classification of this room remains 100,000. A room of about 500 ft^2 is required for this purpose.

6. *Downstream processing room.* This is the second 10,000 classification room with large square footage under laminar flow hoods for the purification process. It is noteworthy that dedicated contact equipments (columns, vessels, etc.) are required for each product. A room of about 500 ft^2 is required as a minimum; the size of room will depend on the volume of production and the steps involved; in some instances, this may take more than 5000 ft^2 where large-scale filtration equipment is involved and may comprise several floors. Some manufacturers do their downstream processing in different facilities; in such instance, there should be proper SOPs describing the packaging and transportation of the fermentation product.

7. *Media and buffer preparation rooms.* Each process requires a specific buffer and media; depending on the size of operation, the quantity of these liquids can be substantial. For example, where using large fermenters (1000 L or up), it may be advisable to switch to closed systems of media preparation and transfer to fermenters; however, for most medium- and small-scale operations, a large media preparation room is required (500 ft^2) with 10,000 classification and work under class 100 hood; the media prepared is then transferred to storage area and issued a specification code; the same applies to buffers used in downstream processing. Buffers should be prepared in a separate room and transported in closed containers to the storage area prior to dispensing.

8. *Storage rooms.* Incoming material is stored in special environment controlled areas; large refrigerated space is required for many components including media and buffer. This room is also a 100,000 classification room of about 1,000 ft^2. A part of the room is dedicated to staging of supplies at the time of batch issue for production wherein material will be gathered from the WCB, media, and buffer rooms.

9. *Finished product storage room.* This is a relatively smaller room, about 100 ft^2 where concentrate is stored at refrigerated temperature. The classification remains 100,000.

Cleaning procedures

Validation of the cleaning procedures for the processing of equipment, including columns, should be carried out. This is especially critical for a multiproduct facility. The manufacturer should have determined the degree of effectiveness of the cleaning procedure for each BDP or intermediate used in that particular piece of equipment. Validation data should verify that the cleaning process will reduce the specific residues to an acceptable level. However, it may not be possible to remove absolutely every trace of material, even with a reasonable number of cleaning cycles. The permissible residue level, generally expressed in parts per million (ppm), should be justified by the manufacturer. Cleaning should remove endotoxins, bacteria, toxic elements, and contaminating proteins, while not adversely affecting the performance of the column. There should be established a written equipment cleaning procedure that provides details of what should be done and the materials to be utilized. Some manufacturers list the specific solvent for each BDP and intermediate. For stationary vessels, often clean-in-place (CIP) is used and in these instances necessary diagrams should be drawn to identify specific parts (e.g., valves, etc.) that are part of the cleaning protocol.

After cleaning, there should be some routine testing to assure that the surface has been cleaned to the validated level. One common method is the analysis of the final rinse water or solvent for the presence of the cleaning agents last used in that piece of equipment. There should always be direct determination of the residual substance.

The efficiency of the cleaning system would depend to a large degree on the robustness of the analytical system used to characterize the cleaning end points. The sensitivity of modern analytical apparatus has lowered some detection thresholds below parts per million (ppm), even down to parts per billion (ppb). The residue limits established for each piece of apparatus should be practical, achievable, and verifiable. There should be a rationale for establishing certain levels that must be documented to prove their scientific merit. Another factor to consider is the possible nonuniform distribution of the residue on a piece of equipment. The actual average residue concentration may be more than the level detected.

Processing and filling

The products of biotechnology are proteins and peptides that are relatively unstable molecules compared to most organic pharmaceuticals. Most biotechnology processes involve the transfer of proteins from one stabilizing or solubilizing buffer to another during the purification process. Ultimately, the protein is exchanged into its final solution dosage form where long-term stability is achieved. In addition, these products often require lyophilization to achieve long-term stability because of the potential for degradation by a variety of mechanisms, including deamidation, aggregation, oxidation, and possible proteolysis by trace levels of host cell proteases. The final dosage form of the protein usually contains stabilizing compounds that result in the optimal pH and solution conditions necessary for long-term product stability and/or the desired properties for administration of the product (tonicity). These compounds include proteins, polyhydric alcohols, amino acids, carbohydrates, bulking agents, inorganic salts, and nonionic surfactants. In addition, these excipients may be required for stable lyophilized cake formation. There are special requirements for lyophilized products, such as the control of moisture levels, that generally are defined in the individual USP monograph and that may be important to product stability. Significantly, the assessment of protein stability usually requires the use of multiple analytical methods, each of which may be used to assess a specific mode of protein degradation. The use of accelerated stability studies to predict the shelf life of protein formulations is often complicated by the effects of temperature on protein conformation, resulting in non-Arrhenius behavior. Thus, reliance on real-time, recommended storage condition stability studies is often required for establishing the expiration dating of biotechnology-derived products.

Most BDP cannot be terminally sterilized and must be manufactured by aseptic processing. The presence of process-related contaminants in a product or device is chiefly a safety issue. The sources of contaminants are primarily the cell substrate (DNA, host cell proteins, and other cellular constituents, viruses), the media (proteins, sera, and additives), and the purification process (process-related chemicals and product-related impurities).

Because of stability considerations, most BDP are either refrigerated or lyophilized. Low temperatures and low moisture content are also deterrents to microbiological proliferation. For the validation of aseptic processing of the nonpreserved single-dose biopharmaceutical (that is aseptically filled) stored at room temperature as a solution, the limitations of 0.1% media fill contamination rate should be recognized.

Media fill data and validation of the aseptic manufacturing process should be well documented. Some BDP may not be very stable and may require gentle mixing and processing. Whereas double filtrations are relatively common for aseptically filled parenterals, single filtration at low pressures is usually performed for BDP. It is for this reason that manufacturing directions should be specific, with maximum filtration pressures given.

The environment and accessibility for the batching of the nonsterile BDP should be controlled. Because many of these products lack preservatives, inherent bacteriostatic, or fungistatic activity, bioburden before sterilization should be low and this bioburden should be determined prior to sterilization of these bulk solutions and before filling. Obviously, the batching or compounding of these bulk solutions should be controlled in order to prevent any potential increase in microbiological levels that may occur up to the time that the bulk solutions are filtered (sterilized). One concern with any microbiological level is the possible increase in endotoxins that may develop. Good practice for the compounding of these products would also include batching in a controlled environment and in sealed tanks, particularly if the solution is to be stored prior to sterilization. Good practice would also include limitations on the length of manufacturing time between formulation and sterilization.

In-process testing is an essential part of quality control and ensures that the actual, real-time performance of an operation is acceptable. Examples of in-process controls are stream parameters, chromatography profiles, protein species and protein concentrations, bioactivity, bioburden, and endotoxin levels. This set of in-process controls and the selection of acceptance criteria require coordination with the results from the validation program.

The filling of BDP into ampoules or vials presents many of the same problems as with the processing of conventional products. In established companies, these issues are addressed using adequate documentation; however, for the new BDP facility, attempting to develop and prove clinical effectiveness and safety along with validation of sterile operations, equipment and systems, can be a lengthy process, particularly if requirements are not clearly understood.

The batch size, at least when initially produced, likely will be small. Because of the small batch size, filling lines may not be as automated as for other products typically filled in larger quantities. Thus, there is more involvement of people filling these products, particularly at some of the smaller, newer companies. This can bring quality inconsistencies. Problems during filling include inadequate attire; deficient environmental monitoring programs; hand-stoppering of vials, particularly those that are to be lyophilized; and failure to validate some of the basic sterilization processes. Because of the active involvement of people in filling and aseptic manipulations, the number of persons involved in these operations should be minimized, and an environmental program should include an evaluation of microbiological samples taken from people working in aseptic processing areas.

Another concern about product stability is the use of inert gas to displace oxygen during both the processing and filling of the solution. As with other products that may be sensitive to oxidation, limits for dissolved oxygen levels for the solution should be established. Likewise, validation of the filling operation should include parameters such as line speed and location of filling syringes with respect to closure, to assure minimal exposure to air (oxygen) for oxygen-sensitive products. In the absence of inert gas displacement, the manufacturer should be able to demonstrate that the product is not affected by oxygen.

Typically, vials to be lyophilized are partially stoppered by machine. Where an operator places stopper manually, serious problems can arise. Another major concern with the filling operation of a lyophilized product is assurance of fill volumes. Obviously, a low fill would represent a subpotency in the vial. Unlike a powder or liquid fill, a low fill would not be readily apparent after lyophilization, particularly for a product where the active ingredient may be only a milligram. Because of the clinical significance, subpotency in a vial potentially can be a very serious situation, clinically.

Laboratory testing

The following tests may be applicable to component, in process, bulk and/or final product testing. The tests that are needed will depend on the process and the intended use of the product.

- *Quality*:
  - Color/appearance/clarity
  - Particulate analysis
  - pH determination
  - Moisture content
  - Host cell DNA
- *Identity*: A single test for identity may not be sufficient. Confirmation is needed that the methods employed are validated. A comparison of the product to the reference preparation in a suitable bioassay will provide additional evidence relating to the identity and potency of the product.
  - Peptide mapping (reduced/nonreduced)
  - Gel electrophoresis
  - SDS PAGE
  - Isoelectric focusing (IEF)
  - Immunoelectrophoresis
  - Two-Dimensional electrophoresis
  - Capillary electrophoresis
  - HPLC (chromographic retention)

- Immunosassay
- ELISA
- Western blot
- Radioimmunoassay
- Amino acid analysis
- Amino acid sequencing
- Mass spectroscopy
- Molecular weight (SDS PAGE)
- Carbohydrate composition analysis (glycosylation)
- *Protein concentration/content*:
  - Tests that may be encountered:
  - Protein quantitations
  - Lowry
  - Biuret method
  - UV spectrophotometry
  - HPLC
  - Amino acid analysis
  - Partial sequence analysis
- *Purity*: "Purity" means relative freedom from extraneous matter in the finished product, whether or not harmful to the recipient or deleterious to the product. Purity includes, but is not limited to, relative freedom from residual moisture or other volatile substances and pyrogenic substances. Protein impurities are the most common contaminants. These may arise from the fermentation process, media, or the host organism. Endogenous retroviruses may be present in hybridomas used for monoclonal antibody production. Specific testing for these constituents is imperative in in vivo products. Removal of extraneous antigenic proteins is essential to assure the safety and the effectiveness of the product.
  - Tests for protein impurities:
    - Electrophoresis
      - SDS PAGE
      - IEF
    - Dimensional electrophoresis
    - Peptide mapping
    - Multiantigen ELISA
    - HPLC size exclusion HPLC reverse-phase HPLC
  - Tests for nonprotein impurities:
    - Endotoxin testing
    - U.S.P. Rabbit pyrogen test
    - Limulus amebocyte lysate (LAL) E
    - Endogenous pyrogen assay
- *Pyrogen contamination*: Pyrogenicity testing should be conducted by injection of rabbits with the final product or by the LAL assay. The same criteria used for acceptance of the natural product should be used for the biotech product. The presence of endotoxins in some in vitro diagnostic products may interfere with the performance of the device. Also, it is essential that in vivo products be tested for pyrogens. Certain biological pharmaceuticals are pyrogenic in humans despite having passed the LAL test and the rabbit pyrogen test. This phenomenon may be due to materials that appear to be pyrogenic only in humans. To attempt to predict whether human subjects will experience a pyrogenic response, an endogenous pyrogen assay is used. Human blood mononuclear cells are cultured in vitro with the final product, and the cell culture fluid is injected into

481

rabbits. A fever in the rabbits indicates the product contains a substance that may be pyrogenic in humans.
- U.S.P. Rabbit Pyrogen Test
- LAL
- Assay endogenous pyrogen assay
- *Viral contamination*: Tests for viral contamination should be appropriate to the cell substrate and culture conditions employed. Absence of detectable adventitious viruses contaminating the final product should be demonstrated.
  - Cytopathic effect in several cell types
  - Hemadsorption embryonated egg testing
  - Polymerase chain reaction (PCR)
  - Viral antigen and antibody immunoassay
  - Mouse antibody production (MAP)
- *Nucleic acid contamination*: Concern about nucleic acid impurities arises from the possibility of cellular transformation events in a recipient. Removal of nucleic acid at each step in the purification process may be demonstrated in pilot experiments by examining the extent of elimination of added host cell DNA. Such an analysis would provide the theoretical extent of the removal of nucleic acid during purification. Direct analyses of nucleic acid in several production lots of the final product should be performed by hybridization analysis of immobilized contaminating nucleic acid utilizing appropriate probes, such as nick-translated host cell and vector DNA. Theoretical concerns regarding transforming DNA derived from the cell substrate will be minimized by the general reduction of contaminating nucleic acid.
  - DNA hybridization (dot blot)
  - Polymerase chain reaction (PCR)
- *Protein contamination*:
  - SDS PAGE
  - PLC
  - IEF
- *Foreign protein contamination*:
  - Immunoassays
  - Radioimmunoassays
  - ELISA
  - Western blot
  - SDS page
  - Two-dimensional electrophoresis
- *Microbial contamination*: Appropriate tests should be conducted for microbial contamination that demonstrates the absence of detectable bacteria (aerobes and anaerobes), fungi, yeast, and mycoplasma, when applicable.
  - U.S.P. sterility test
  - Heterotrophic plate count and total yeasts and molds
  - Total plate count
  - Mycoplasma test
  - LAL/pyrogen
- *Chemical contaminants*: Other sources of contamination must be considered, for example, allergens, petroleum oils, residual solvents, cleaning materials, column leachable materials, etc.
- *Potency (activity)*: "Potency" is interpreted to mean the specific ability or capacity of the product, as indicated by appropriate laboratory tests or by adequately controlled clinical data obtained through the administration of the product in the manner intended, to produce a given result.

Tests for potency should consist of either in vitro or in vivo tests, or both, which have been specifically designed for each product so as to indicate its potency. A reference preparation for biological activity should be established and used to determine the bioactivity of the final product. Where applicable, in-house biological potency standards should be cross-referenced against international (World Health Organization [WHO], National Institute of Biological Standards and Control [NIBSC]) or national (National Institutes of Health [NIH], National Cancer Institute [NCI], Food and Drug Administration [FDA]) reference standard preparations, or USP standards. Validated method of potency determination include

- Whole animal bioassays
- Cell culture bioassays
- Biochemical/biophysical assays
- Receptor-based immunoassays
- Potency limits
    - Identification of agents that may adversely affect potency
    - Evaluation of functional activity and antigen/antibody specificity
    - Various immunodiffusion methods (single/double)
    - Immunoblotting/radio- or enzyme-linked Immunoassays
    - HPLC-validated to correlate certain peaks to biological activity
- *Stability*: "Stability" is the capacity of a product to remain within specifications established to ensure its identity, strength, quality, purity, safety, and effectiveness as a function of time. Studies to support the proposed dating period should be performed on the final product. Real-time stability data would be essential to support the proposed dating period. Testing might include stability of potency, pH, clarity, color, particulates, physiochemical stability, moisture, and preservatives. Accelerated stability testing data may be used as supportive data. Accelerated testing or stress tests are studies designed to increase the ratio of chemical or physical degradation of a substance or product by using exaggerated storage conditions. The purpose is to determine kinetic parameters to predict the tentative expiration dating period. Stress testing of the product is frequently used to identify potential problems that may be encountered during storage and transportation and to provide an estimate of the expiration dating period. This should include a study of the effects of temperature fluctuations as appropriate for shipping and storage conditions. These tests should establish a valid dating period under realistic field conditions with the containers and closures intended for the marketed product. Some relatively fragile biotechnically derived proteins may require gentle mixing and processing and only a single filtration at low pressure. The manufacturing directions must be specific with maximum filtration pressures given in order to maintain stability in the final product. Products containing preservatives to control microbial contamination should have the preservative content monitored. This can be accomplished by performing microbial challenge tests (i.e., U.S.P. antimicrobial preservative effectiveness test) or by performing chemical assays for the preservative. Areas that should be addressed are
    - Effective monitoring of the stability test environment (i.e., light, temperature, humidity, residual moisture)
    - Container/closure system used for bulk storage (i.e., extractables, chemical modification of protein, change in stopper formulations that may change extractable profile)
    - Identify materials that would cause product instability and test for presence of aggregation, denaturation, fragmentation, deamination, photolysis, and oxidation

- *Tests to determine aggregates or degradation products*:
  - SDS PAGE
  - IEF
  - HPLC
  - Ion exchange chromatography
  - Gel filtration
  - Peptide mapping
  - Spectrophotometric methods
  - Potency assays
  - Performance testing
  - Two-dimensional electrophoresis
- *Batch to batch consistency*: The basic criterion for determining that a manufacturer is producing a standardized and reliable product is the demonstration of lot-to-lot consistency with respect to certain predetermined release specifications.
  - *Uniformity*: Identity, purity, functional activity.
  - *Stability*: Acceptable performance during shelf life, precision, sensitivity, specificity. Like other small molecules, protein drugs are subject to demonstration of stability (providing a predetermined minimum potency) to the time of use and, in addition, a safety profile since the degradation products of protein drugs can be immunogenic, compared to small molecules where the concern is mainly creation of toxic molecules. The stability studies for therapeutic proteins are conducted at three levels: preformulation, formulation development, and formal GMP studies. The preformulation studies determine basic stability properties of bulk protein or peptide and the accelerated studies at this stage are primarily intended to establish stability indicating assays and other analytic methods. The formulation development studies are intended for the candidate formulation and encompass large studies that evaluate the effects of excipients, container/closure systems, and where lyophilized, a study of myriad factors that can alter the characteristics of products. The data generated in the formulation development studies are used to select final formulation and to design the studies that follow: formal GMP studies. The formal stability studies are used to support clinical use, IND, and then all the way through a Biological License Application (submitted to CDER; effective June 2003, therapeutic proteins are now handled by CDER). When preparing supplies for the clinical use, it is important to know that there is no need to demonstrate shelf life for the commercial dosage form and only stability demonstration is required during the testing phase, such as 6 months; many manufacturers used frozen product to assure adequate stability; this may create a logistics problem of assuring that the clinical sites can store the produce frozen. Obviously, products that may be adversely affected by freezing will not be subjected to this method of reducing the clinical startup time. Also, at this stage of initial clinical testing under an IND, the test methods need not be fully validated or having demonstrated robustness; as long as reproducibility and repeatability is demonstrated, this should be acceptable to the FDA. The formal GMP studies monitor commercial lots and clear ICH guidelines are available to follow the protocols of these studies (see ICH, 1996).

Laboratory controls

A quality control program for the drug substance and drug product must be defined and acceptance criteria set for each analysis. The setting of acceptance criteria is an ongoing activity throughout development (and scale-up) as more and more data become available. A batch is released provided all analytical results are within the specified ranges. A high acceptance rate should be expected if the process is robust and in compliance, implying that both the regulatory authorities and the manufacturer often share identical views. The process designer is advised to carefully consider the aforementioned issues when designing the purification process (the design in principle). Much of the design can be carried out before entering the laboratory due to the restrictions governing biopharmaceutical processing. The predesign phase is called process modeling, thus preceding the experimental design phase for optimization and testing, which takes place in the laboratory.

In general, quality control systems for biotechnology-derived products are very similar to those quality control systems routinely employed for traditional pharmaceutical products in such areas as raw material testing and release, manufacturing and process control documentation, and aseptic processing. Biotechnology-derived products' quality control systems incorporate some of the same philosophies applied to the analysis of low molecular weight pharmaceutical products. These include the use of chemical reference standards and validated methods to evaluate a broad spectrum of known and/or potential product impurities and potential breakdown products. The quality control systems for biotechnology-derived products are generally analogous to those established for traditional biologicals with respect to determining product sterility, product safety in experimental animals, and product potency. The fundamental difference between quality control systems for biotechnology-derived products and traditional pharmaceuticals is in the types of methods that are used to determine product identity, consistency, purity, and impurity profiling. Furthermore, in biotechnology quality control, it is frequently necessary to use a combination of final product and validated in-process testing and process validation to ensure the removal of undesired real or potential impurities to the levels suggested by regulatory agencies. Biotechnology-derived products generally require a detailed characterization of the production organism (cell), a complete assessment of the means of cell growth/propagation, and explicit analysis of the final product recovery process.

The complexity of the quality control systems for biotechnology-derived products is related to both the size and structural characteristics of the product and manufacturing process. In general, the quality control systems required for products produced in prokaryotic cells are less complex than the systems required for products produced in eukaryotic cells. The quality control systems for prokaryotic production organisms usually entail documentation of the origin of the producer strain and encompass traditional testing for adventitious organisms, karyology, phenotyping, and antibiotic resistance. In addition, newer techniques such as DNA restriction mapping, DNA sequence analysis, and routine monitoring that may include measurement of mRNA and/or plasmid DNA levels may be useful. The quality control of the master cell bank and working cell bank for eukaryotic production organisms generally includes testing for adventitious organisms, karyology, identity, and stability monitoring. All eukaryotic cell lines (except yeast) are generally tested for the presence of retroviruses, retroviral activity markers, and tumorigenicity, although many of these tests may be of limited value.

The laboratory controls are similar to what is expected in normal cGMP/GLP compliance for all pharmaceutical products with special consideration given to unique materials and their handling:

- *Training*: Laboratory personnel should be adequately trained for the jobs they are performing.
- *Equipment maintenance/calibration/monitoring*: Documentation and scheduling for maintenance, calibration, and monitoring of laboratory equipment involved in the measurement, testing and storage of raw materials, product, samples, and reference reagents.
- *Validation*: All laboratory methods should be validated with the equipment and reagents specified in the test methods. Changes in vendor and/ or specifications of major equipment/reagents would require revalidation. Raw data should support validation parameters in submitted applications.
- *Standard/reference material*: Reference standards should be well characterized and documented, properly stored, secured, and utilized during testing.
- *Storage of labile components*: Laboratory cultures and reagents, such as enzymes, antibodies, test reagents, etc., may degrade if not held under proper storage conditions.
- *Laboratory SOPs*: Procedures should be written, applicable and followed. Quality control samples should be properly segregated and stored.

Documentation

The development program comprises a variety of activities comprising project planning, cell banking, process development, development of analytical procedures, scale-up, manufacture, stability studies, preparation of reference materials, and quality assurance. The work will ultimately lead to a process for the manufacture and control of the licensed product. In order to obtain a license, extensive documentation must be provided (new drug application, biological license application). However, much of the work carried out during development and scale-up is not included in the said applications, and it is up to the project owner to provide the development documentation upon inspection from regulatory authorities. The statement "if it is not documented, it has not been carried out" should be taken rigorously.

A major part of the documentation required, can be planned (e.g., cell banking reports, unit operation descriptions, development report, analytical method descriptions, batch records). Other reports (e.g., summary reports) are written along with the experimental work. Although such reports are not providing a description of the final work, they are very useful for informing coming users about the rationale for decision-making. It is therefore recommended to include summary reports in the tech transfer package.

The drug development program produces hundreds or even thousands of documents written by different people from different departments and often from different companies. Any of these documents may be needed at a later stage and it is necessary to set up an efficient documentation system to assure document tracking. An important part of the tracking procedure is to ensure efficient authentication of the document comprising information of author, date, version, company, facility, etc.

Technical package

The documents listed earlier serve as the basis for the tech transfer between laboratory scale, pilot scale-up, and non-GMP manufacture and cGMP manufacture.

- According to the Common Technical Document, ICH (www.nihs.go.jp/dig/ich/m4index-e.html) information about the nomenclature and the chemical/physical properties of the target protein should be given.
- The rationale and strategy for the process design should be documented.
- The process development work should be documented in laboratory notebooks, summary reports, unit operation descriptions, a development report, and a process protocol. Batch records should be included for test batches used for process evaluation in small scale.
- It is common practice to include a list of raw materials together with raw material information and qualification documentation.
- The manufacturer must ensure the quality and safety of the drug substance by a series of test procedures with established acceptance criteria to which the drug substance must conform. An important part of the total control strategy is to provide analytical batch data and compare these with the specified acceptance criteria. A complete set of acceptance criteria cannot be disclosed a priori as some acceptance criteria are established and justified during development and production of preclinical and clinical test batches. Acceptance criteria are linked to the total analytical program established as part of the quality control and drug substance characterization program. As more and more data are collected during development and early phase pilot production the acceptance criteria may be adjusted or changed, thus making it difficult to establish defined batch release procedures in the development phase.
- Updated unit operation documents, raw material list and process protocol should be included together with scale-up summary reports and batch records.
- Data collected during development, scale-up, and manufacture comprise a valuable repository of information, which can be used for trouble-shooting, comparability, and equivalence studies. Control should be exerted on all levels and should constitute an integrated part of the total quality assurance. Each analytical method used in development and for in-process control, drug substance, and drug product release should be described in an individual report including relevant analytical data. The report will gradually develop into a validation document for the method of choice.
- Short- and long-term stability studies for intermediary compounds, drug substance, and drug product should be documented.
- Reference materials and standards are usually extensively characterized. The process protocol, analytical methods used, and the analytical data must be provided in an accompanying report.

Bibliography

Asenjo JA, Patrick I. Large-scale protein purification. In: *Protein Purification Applications. A Practical Approach.* Harris ELV, Angal S, eds. IRL Press, Oxford, U.K., 1990, pp. 1–28.

Bödeker BGD, Newcomb R, Yuan P, Braufman A, Kelsey W. Production of recombinant factor VIII from perfusion cultures: I. Large-scale fermentation. In: *Animal Cell Technology. Products of Today, Prospects for Tomorrow.* Spier RE, Griffiths JB, Berthold W, eds. Butterworth Heinemann, Oxford, U.K., 1994, pp. 580–583.

Bylund F, Castan A, Mikkola R, Veide A, Larsson G. Influence of scale-up on the quality of recombinant human growth hormone. *Biotechnol Bioeng* 2000;69(2):119–128.

Davis CJA. Large-scale chromatography: Design and operation. In: *Bioseparation and Bioprocessing. Vol I. Biochromatography, Membrane Separations, Modeling, Validation.* Subramanian G, ed. Wiley-VCH, Weinheim, Germany, 1998, pp. 125–143.

Gerstner JA. Economics of displacement chromatography—A case study: Purification of oligonucleotides. *BioPharm* 1996;9(1):30–35.

ICH. Stability guidelines for biologics. *Fed Reg.* July 1996;61(133):36466–36469.

Janson J-C. Scaling up of affinity chromatography, technological and economical aspects. In: *Affinity Chromatography and Related Techniques.* Gribnau TCJ, ed. Elsevier, Amsterdam, the Netherlands, 1982, pp. 503–512.

King LA, Possee RD. Scaling up the production of recombinant protein in insect cells; laboratory bench level. In: *The Baculovirus Expression System. A Laboratory Guide.* Chapman & Hall, London, U.K., 1992, pp. 171–179.

Munshi CB, Fryxell KB, Lee HC, Branton WD. Large-scale production of human CD38 in yeast by fermentation. *Methods Enzymol* 1997;280:318–330.

Rathore AS, Latham P, Levine H, Curling J, Kaltenbrunner O. Costing issues in the production of biopharmaceuticals. *BioPharm International* 2004;17(2):46–55.

Sadana A, Beelaram AM. Efficiency and economics of bioseparation: Some case studies. *Bioseparation* 1994;4(4):221–235.

Sofer G, Hagel L. Economy. In: *Handbook of Process Chromatography: A Guide to Optimization, Scale-up and Validation.* Academic Press, San Diego, CA, 1997a, pp. 227–243.

Sofer G, Hagel L. *Handbook of Process Chromatography: A Guide to Optimization, Scale-up and Validation.* Academic Press, San Diego, CA, 1997b.

Sofer GK, Nystrom LE. Economics. In: *Process Chromatography. A Practical Guide.* Sofer GK, Nystrom LE, eds. Academic Press, London, U.K., 1989a, pp. 107–116.

Sofer GK, Nystrom LE. *Process Chromatography. A Practical Guide.* Academic Press, London, U.K., 1989b.

Walter JK. Strategies and considerations for advanced economy in downstream processing of biopharmaceutical proteins. In: *Bioseparation and Bioprocessing. Vol II. Processing, Quality and Characterization, Economics, Safety and Hygiene.* Subramanian G, ed. Wiley-VCH 1998, pp. 447–460.

Walter JK, Werz W, Berthold W. Virus removal and inactivation—Concept and data for process validation of downstream processing. *Biotech Forum Europe* 1992;9:560–564.

Walter JK, Werz W, Berthold W. Process scale considerations in evaluation studies and scale-up. *Dev Biol Stand* 1996;88:99–108.

Warner TN, Nochumsen S. Rethinking the economics of chromatography. New technologies and hidden costs. *BioPharm Int* 2003;16(1):58–60.

Wheelwright SM. *Protein Purification: Design and Scale-up of Downstream Processing.* John Wiley & Sons, New York, 1993.

Yamamoto S, Nomura M, Sano Y. Resolution of proteins in linear gradient elution ion-exchange and hydrophobic interaction chromatography. *J Chromatogr* 1987;409:101–110.

Chapter 11 Outsourcing considerations

Background

Growing by approximately 15%–18% every year, biotechnology (encompassing biopharmaceuticals and vaccines) is the fastest growing sector in the industry. The growth is being driven by multiple factors including opportunities for expansion into China, India, and other emerging markets. Global biopharma revenues for 2010 are estimated at more than $100 billion, approximately 14% of total industry revenues. More than 30 biotech products have now reached blockbuster status (annual revenues of $1 billion or greater). On the R&D front, some 12,000 large molecules are in preclinical discovery or clinical trials, offering promise for patients in nearly 150 disease areas. It is estimated that biotechnology will make up as much as one-third of future new drug approvals. In addition, some industry projections show that, by 2014, 7 of the top 10 drugs globally will be biotech drugs. However, pressured to develop and deliver new drugs that are accessible, affordable, and revenue enhancing, these biotechnology companies are now trending away from solutions that involve "stainless steel" (capital investment in manufacturing capabilities and capacity) and toward those more rooted in "science" (biotechnology process development, to be precise)—with important outsourcing implications for CMOs and biotech drug sponsors alike.

The novelty of biosimilar products, the financial challenges and rewards associated with them, and the scientific and regulatory difficulties anticipated have provided a fertile ground for outsourcing resulting in many alliances in this field. A list of major biosimilar players and their alliances is as follows:

- Teva/Lonza (dissolved September 2013)
- Actavis/Amgen/Synthon
- Mylan/Biocon
- Hospira/Celltrion
- Merck (MSD)/Hanwha/Others
- Merck/Samsung Bioepis
- Samsung/Quintiles Transnational/Biogen
- Baxter/Momenta
- Lilly/Boehringer
- Dr. Reddy's/Merck-Serono
- Sandoz—No alliance
- Therapeutic Proteins International/Amneal

In several instances, the developer licenses out the product to a prospective distributor who may otherwise have an established system of distribution. Nonetheless, none of these alliances have produced any results so far. However, the business model used to manufacture biosimilars is currently undergoing a few changes.

Value of outsourcing

The concept of outsourcing dates back to about 10,000 years ago when the Ice Age glaciers receded and humans came out of their caves and got to know their fellow cave dwellers and realized that the efficiency (and therefore survival) depends on sharing and exploiting their varying types of expertise. Much more recently, vertical integration of businesses over the past 50 years has produced giant industries that ultimately proved too inefficient to survive as globalization caught on. Competitive forces determined what each company's core activity was and what its supporting or ancillary functions were.

Today, a "virtual company" represents an ultimate example of the benefits of outsourcing. A new expertise of professionals who can manage outsourcing has developed, and most companies today consider it an essential core competency. Those without extensive experience in outsourcing soon learn that there are no perfect outsourcing solutions, just choices that are more efficient, and experience leads to more predictable systems.

Outsourcing is a well-discussed topic. A large volume of literature is available addressing various issues that relate to outsourcing difficulties as they appear in different industries. Journal articles have even been written on the history of outsourcing.

The literature is full of do's and don'ts of advice from the gurus of the industry, the CSO (contract service organization) sector, consultants, and economic experts. But no advice is more valuable than what we learn from real experience.

The concept of a virtual company is not new, however often those who are perpetually enamored with the smell of brick and mortar may disdain it. What is new is the realization that all the sensitive issues that have historically been responsible for too much vertical integration can be resolved with good planning. While most companies currently rely on outsourcing to some degree, outsourcing biotech manufacturing should be a strategic decision, one that aligns with a company's overall business strategy. The strategic outsourcing approach used will depend largely on which aspects of the business are to be optimized and what challenges need to be addressed. Historical drivers such as product cost, capital avoidance, demand variability, and internal versus external core competencies for outsourcing in the pharma industry hold true for biotech as well.

Factors to consider in outsourcing

A closer look at the important trends, issues, and challenges impacting the industry now can provide clues to the future state of biotechnology outsourcing. Industry factors worth noting include the following:

- The draw of globalization and related challenges to manage a supply chain of highly sensitive products
- Pressing needs to deliver more affordable, accessible products and therapies
- Significant advances in technology and innovation in bioprocessing that can readily make an in-house facility obsolete
- Competition from other companies creating an inevitable need for first to market
- Complexity and critical specialization of biopharmaceutical manufacturing, testing and clinical trials

The driving forces for outsourcing will also include the following:

- *Capital expenditures*: Since there are years involved before a biosimilar product is launched, having a manufacturing facility on hand can create heavy financial burden.
- *Unique technologies*: Flexibility of exploiting different technologies that are molecule-dependent, such as cytokine versus mammalian cells, can be better addressed when outsourced.
- *Flexible capacity*: Despite the projections made, it is not always possible to accurately judge both the timing and volume of capacity required for launch for first few years.
- *Commercial strategy*: Marketing and selling of biosimilars creates a new model that is currently not existing—those who are already in the field may reduce the time to commercialize—this model has worked nicely for several alliances as described earlier.
- *Biosimilars growth*: A decision can be made by the developer to focus on specific parts of development such as clinical pharmacology lots that may require more extensive analytical similarity testing and leave out other parts such as fill and finish to CMOs.

Key requirements

Unlike small molecule products where over the decades a clear understanding has developed as to what constituted cGMP compliance, risk factors, and issues like cross-contamination, many of these issues are yet to be decided for biosimilar products when outsourced. The burden on CMOs is therefore very different in the biotechnology field:

- *Quality and continuous improvement*: These are cGMP issues and a BLA license brings different challenges than drug authorization; the issues relating to quality systems, some of which can be product-specific must be in place. The CMO must be able to guarantee continuous improvements and compliance with CBE compliance.
- *GMP management and remediation*: Manufacturing multiple products as most CMOs will do brings a bigger challenge of cross-contamination and this may limit the availability of CMOs, for example, those who have dedicated manufacturing suites for mammalian cells. This may not be an issue in the development phase but the regulatory agencies will not allow multiple product facilities to produce biosimilars, if this can affect their quality. The quality agreements for these products are difficult to draw and even more difficult to maintain.
- *Cost*: Since biosimilars are predicted to have a lower cost, the CMOs may have to adopt systems that allow this competition to thrive; the biosimilar developer may find itself with little choice but to do it on its own, if cost advantages are not presented.
- *Supply assurance*: Since it is not always possible to make accurate predictions and since the CMO must book its facility ahead of time, a discord always exists in assuring supply—this may be a serious drawback as the biosimilar developer begins to work with distributors who would want assurance of short-notice change in the supply order.
- *Technology platforms*: CMOs specialize in technology they can offer, such as disposable vs. fixed wall bioreactors, the size of lots (determined often by downstream processing), and many other constraints. A drawback therefore arises if the developer decided to change the technology platform to stay competitive. While comparability protocols will be

available, it will inevitably require significant investment by the CMOs, for which they may not be ready.

- *IP strategy*: The biosimilar developer may have developed its own IP that will not be wise to share with some CMOs, especially those that fall outside of immediate legal jurisdictions; alternately, the CMO may use its own IP that the biosimilar may not be able to take with him if the choice of CMO is changed.
- *Partnership mentality*: The best models will involve some type of partnership where both parties have equal stakes.

What to outsource?

The level of outsourcing depends how much a company is willing to take risk and a great deal on in-house preparedness. It can take place at any of the following manufacturing process steps for recombinant biosimilar drugs:

- Molecular biology (gene construct)
- Fermentation development (upstream)
- Purification (downstream)
- Protein characterization and analysis
- Preclinical testing
- Regulatory filing
- Clinical supplies (CGMP)
- Full-scale manufacturing

When deciding the nature and extent of outsourcing, companies have several questions to answer:

- Which contract manufacturer should we partner with?
- Which research organizations should we partner with?
- How do we negotiate the contracts?
- What are the liability issues?
- What will it cost?
- What is the timeline?

The first thing a biosimilar developer would do is to establish a project management team, identify single contact for each type of outsourcing within the company and prepare financial projections of what would be considered an acceptable cost of outsourcing, both in terms of time and money.

Since the interaction between a contract organization and the developer will ultimately be a technical one, several subject matter experts are engaged on both sides. A team building with CO starts with assuring compatibility between the leads of the two companies. More often projects have fallen because the two leads are not compatible for reasons entirely other than the mechanical or scientific. An expected qualification on both sides is to be able to understand the requirements and limitations of each, ahead of the discussion. For this reason, the CO may want to spell out upfront the limits of any expectations and the developer to be succinct in describing their minimal expectation, such as a timeframe to have the clinical material ready.

A new trend, which has a lot of weight is to outsource the outsourcing team, if one can be found technically qualified enough for the task. Many smaller companies may find this to be a viable choice instead of adding headcounts that may not prove compatible at a later stage—an outsourced outsourcing removes that constraint.

The universe of CSOs is expanding fast, with some offering integrated functions, of which many are highly specialized. Adding them all to this list would make

Table 11.1 Potential Suppliers of Recombinant Cell Lines

www.antitope.com	www.lifetechnologies.com
www.mdpi.com	www.genscript.com
www.precisionantibody.com	www.lfbbiomanufacturing.com
www.selexis.com	www.biomedcentral.com
www.fishersci.com	www.sbproteinexpression.com
www.creative-biolabs.com	www.cobrabio.com
www.innoprot.com	www.astratebio.com
www.lonza.com	www.infoscience.epfl.ch
www.invivo.de	www.ibclifesciences.com
www.sigmaaldrich.com	www.speedbiosystems.com
www.cobrabio.com	www.crcnetbase.com
www.bpscience.com	www.catalent.com
www.alphalyze.com	www.kar.kent.ac.uk
www.xtalks.com	www.biomedcentral.com
www.sydlabs.com/	www.pangen.com
www.protalix.com	www.amsbio.com
www.assaydepot.com	

it unwieldy, so our intent is to list only companies that can be of general use to many potential clients. Listing a company here does not constitute an endorsement, although many need no endorsement anyway.

Cell line suppliers

Once the financial aspects are worked out, the developer needs to ascertain the risk to product value if it is outsourced given the above constraints and also how the speed to market is impacted. For new molecules, this is of lesser importance but when the timeline for submission of 351(k) packages is tight, choosing a CMO can be extremely difficult. This requires a storyboard and a clear list of expectations, all of which can change the first cost assumptions.

There are several levels of outsourcing possibilities. The first step is the creation and validation of a recombinant cell line. In all likelihood, a new developer of biosimilars will not be equipped with in-house expertise to develop these lines. Whereas the developer is advised not to make any novel selections of cell lines, there may be situations where an alternate cell line to the originator process may be more desirable. Fortunately, this is the easiest part of outsourcing, with least constraints and complexities. Table 11.1 lists a few suppliers of recombinant cell lines. Many of these suppliers can also provide small-scale 5–25 L size working process as well.

A supply agreement must include, among other things, an IP clearance, data on stability of cell lines over several generations, and must precede by providing samples of the product for initial analytical similarity testing. This is particularly important for monoclonal antibodies where difference in the glycan patterns may make a cell line totally useless.

CMOs

In all likelihood, the developer, unless it is a virtual operation, would have the ability to judge the quality of expression in small lots; however, this is also the time when the developer should start engaging biopharmaceutical manufacturers to contract out since at this time the developer knows the overall nature of the process. Table 11.2 provides a list of a few companies that can provide this function.

Table 11.2 Biopharmaceutical CMOs

Company	URL
3P Biopharmaceuticals	www.3pbio.com
AbbVie Contract Manufacturing	www.abbviecontractmfg.com
ABL, Inc.	www.ablinc.com
Affinity Life Sciences, Inc.	www.affinitylifesciences.com
Alpha Biologics	www.alphabiologics.com
Ajinomoto Althea, Inc.	www.altheatech.com
AMRI	www.amriglobal.com
Apotex Fermentation, Inc.	www.apoferm.com
Aptuit, Inc.	www.aptuit.com
Asahi Glass Company Ltd./Aspex Division	www.agc.com/english/aspex/
AutekBio, Inc.	www.autekbio.com
Avid Bioservices, Inc.	www.avidbio.com
Bio Base Europe Pilot Plant	www.bbeu.org
Biocon Ltd.	www.biocon.com
Bio Elpida	www.bio-elpida.com
Biological Process Development Facility	www.bpdf.unl.edu
Biomay AG	www.biomay.com
BIOMEVA GmbH	www.biomeva.com
Biopharmaceutical Development Program, SAIC-Frederick, Inc.	web.ncifcrf.gov/research/bdp
BioReliance Corporation/SAFC	www.bioreliance.com
Biotechpharma UAB	www.biotechpharma.lt
BioVectra, Inc.	www.biovectra.com
Biovian Ltd.	www.biovian.com
BioVolutions	www.biovolutions.com
Boehringer Ingelheim GmbH	www.BioXcellence.com
Callaghan Innovation	www.callaghaninnovation.govt.nz
Cambrex Corp.	www.cambrex.com
Catalent Pharma Solutions	www.catalent.com
Cellca GmbH	www.cellca.de
Celonic AG	www.celonic.de
CMC Biologics	www.cmcbio.com
Cobra Biologics	www.cobrabio.com
Cook Pharmica LLC	www.cookpharmica.com
Cytovance Biologics, Inc.	www.cytovance.com
Development Center for Biotechnology	www.dcb.org.tw
DSM Biologics	www.dsmpharmaceuticals.com
Eurogentec Biologics	biologics.eurogentec.com
Florida Biologix®	www.floridabiologix.ufl.com
Fujifilm Diosynth Biotechnologies	www.fujifilmdiosynth.com
Gala Biotech/A Catalent Pharma Solutions Company	www.gala.com
Galilaeus Oy Ltd.	www.galilaeus.fi
Gallus BioPharmaceuticals	www.gallusbiopharma.com
Genhelix	www.genhelix.com
GlaxoSmithKline Biopharmaceuticals	www.gsk.com/biopharm
Glycotope Biotechnology	www.glycotope-bt.com
Goodwin Biotechnology, Inc.	www.goodwinbio.com

(Continued)

Table 11.2 (*Continued*) Biopharmaceutical CMOs

Company	URL
Hong Kong Institute of Biotechnology Ltd.	www.hkib.org.hk
IDT Biologika GmbH	www.idt-biologika.de
Inno Biologics Sdn. Bhd.	www.innobiologics.com
Innovent Biologics, Inc.	www.innoventbio.com
Intas Pharmaceuticals Ltd.	www.intaspharma.com
KBI BioPharma, Inc.	www.kbibiopharma.com
Kemwell Biopharma Pvt. Ltd.	www.kemwellpharma.com
Lentigen Corp./A Miltenyi Biotec Company	www.lentigen.com
LFB Biomanufacturing	www.LFBbiomanufacturing.com
Lonza, Inc.	www.lonza.com
Maine Biotechnology Services, Inc.	www.mainebiotechnology.com
MBI International	www.mbi.org
Medicago	www.medicago.com
Meridian Life Science, Inc.	www.meridianlifescience.com
National Research Council Canada	www.nrc-cnrc.gc.ca
Novasep	www.novasep.com
Omnia Biologics	www.omniabiologics.com
PacificGMP	www.pacificgmp.com
Paragon Bioservices	www.paragonbioservices.com
Patheon	www.patheon.com
PharmedArtis GmbH	www.pharmedartis.de
Polymun Scientific	www.polymun.com
ProBioGen AG	www.probiogen.de
Progenitor Cell Therapy LLC	www.progenitorcelltherapy.com
Promab Biotechnoligies, Inc.	www.promab.com
Provantage Biodevelopment Center	www.emdmillipore.com/provantage
Public Health England	www.hpa.gov.uk
PX' Therapeutics	www.px-therapeutics.com
Reliance Life Sciences	www.rellife.com
Rentschler Biotechnologie GmbH	www.rentschler.de
Richter-HELM Biotec GmbH and Company	www.richter-helm.eu
Rutgers University Cell and Cell Products Fermentation Facility	waksman.rutgers.edu/ferment
SAFC Corporation	www.sigma-aldrich.com/safc
Samsung Biologics	www.samsungbiologics.com
Sandoz GmbH/A Novartis Company	www.sandoz.com
Shasun	www.shasun.com
SRC BioManufacturing	www2.src.sk.ca/Industries/Ag-and-BioTech/ Pages/Contract-Biomanufacturing.aspx
Statens Serum Institut	www.ssi.dk
SynCo Bio Partners	www.syncobiopartners.com
Syngene International Ltd.	www.syngeneintl.com
Therapure Biopharma, Inc.	www.therapurebio.com/CDMO
Toyobo Biologics, Inc.	www.toyobobiologics.com/index_e.html
TranXenoGen, Inc.	www.tranxenogen.com
University of Alabama Fermentation Facility	fermentation.bmg.uab.edu
University of Iowa Center for Biocatalysis and Bioprocessing	www.uiowa.edu/~biocat/

(*Continued*)

Table 11.2 (*Continued*) Biopharmaceutical CMOs

Company	URL
University of Iowa Center Pharmaceuticals	uip.pharmacy.uiowa.edu
Vibalogics	www.vibalogics.com
Vista Biologicals Corp.	www.vistabiologicals.com
Wacker Biotech GmbH	www.wacker.com/biologics
Waisman Biomanufacturing	www.gmpbio.org
Wockhardt Ltd.	www.wockhardt-knowhow.com
WuXi Apptec, Inc.	www.wuxiapptec.com

Table 11.3 List of Companies Providing Monoclonal Antibody Manufacturing Support

Abeome Corporation	AbSea Biotechnology, Inc.	Acorda Therapeutics, Inc.
Advanced Biotechnologies	Allele Biotechnology	Amsbio
Antibody Research	Aragen Bioscience	Assay Biotechnology Company
Bio Basic	Bio X Cell	Biogen Idec
BioProcessing, Inc.	Bio-World	Cambridge Research Biochemicals
Cell Biolabs, Inc.	Cell Sciences	Cell Signaling Technology, Inc.
Cellerant Therapeutics, Inc.	Centocor	Charles River Laboratories International, Inc.
Creative Biolabs	Denovo Biotechnology	EastCoast Bio
Epigentek	Epitomics	EtendX, Inc.
FUJIFILM Diosynth	GeneTex, Inc.	Glycotope Group
Groton Biosystems	Icosagen Group	Irvine Pharmaceutical Services
Lampire	Leinco Technologies	Life Technologies
LifeTein LLC	Lofstrand Labs Ltd.	Meridian Life Science, Inc.
Molecular Diagnostic Services (MDS)	NEO Bioscience	Open Monoclonal Technology
OriGene Technologies, Inc.	Paris Biotech (P.A.R.I.S.)	ProMab Biotechnologies, Inc.
ProSci Incorporated	PX Therapeutics	QED Bioscience
Santa Cruz Biotechnology, Inc.	SBH Sciences	Somru Bioscience
SouthernBiotech	StemCell Technologies, Inc.	TGA Sciences, Inc.
Thermo Scientific	Trevigen Inc.	United Biosystems
Vrisko	Wincon TheraCells Biotechnologies Co., Ltd.	Xoma
Zyagen		

There are some companies that specialize in monoclonal antibody services and these are listed in Table 11.3.

Biopharmaceutical support services

While many CMOs listed earlier have a good breadth of services they can offer, it is often inevitable to establish relationship with other contract organizations that are more focused on support, which may include cell line characterization, product similarity testing, development of analytical methods, testing, and release of biosimilar products (Table 11.4).

Table 11.4 Biopharmaceutical Support Providers

Company Name	URL
AAIPharma/Cambridge Major Laboratories	www.aaipharma.com
ABC Laboratories	www.abclabs.com
ABL, Inc.	www.ablinc.com
Affinity Life Sciences, Inc.	www.affinitylifesciences.com
Ajinomoto Aminoscience LLC	www.ajiaminoscience.com
AMRI	www.amriglobal.com
Apotex Fermentation, Inc.	www.apoferm.com
Aptuit, Inc.	www.aptuit.com
Bilcare Global Clinical Supplies	www.bilcaregcs.com
Bio-Concept Laboratories, Inc.	www.bioconcept.com
Biomay AG	www.biomay.com
BioOutsource Ltd.	www.biooutsource.com
BioReliance Corporation	www.bioreliance.com
Bio-Technical Resources	www.biotechresources.com
Charles River Laboratories	www.criver.com
Cytovance Biologics, Inc.	www.cytovance.com
Dalton Pharma Services	www.dalton.com
DPT Laboratories Ltd.	www.dptlabs.com
Eurofins Lancaster Laboratories	www.EurofinsLancasterLabs.com
Fujifilm Diosynth Biotechnologies	www.fujifilmdiosynth.com
Galilaeus	www.galilaeus.fi
Gallus BioPharmaceuticals	www.gallusbiopharma.com
IDBS	www.idbs.com
Irvine Pharmaceutical Services	www.irvinepharma.com
KBI BioPharma, Inc.	www.kbibiopharma.com
Lampire Biological Laboratories, Inc.	www.lampire.com
Leinco Technologies, Inc.	www.leinco.com
Life Technologies, Inc.	www.lifetechnologies.com
Maine Biotechnology Services, Inc.	www.mainebiotechnology.com
Medicago	www.medicago.com
Meridian Life Science, Inc.	www.meridianlifescience.com
Miltenyi Biotec GmbH	www.miltenyibiotec.com
Novoprotein	www.novoprotein.com
Nvigen, Inc.	www.ngiven.com
Onyx Scientific Ltd.	www.onyx-scientific.com
Pall Life Sciences	www.pall.com/services
Patheon, Inc.	www.patheon.com
PII—Pharmaceuticals International, Inc.	www.pharm-int.com
Pilgrim Quality Solutions	www.pilgrimquality.com
Progenitor Cell Therapy LLC	www.progenitorcelltherapy.com
ProMetic Life Sciences, Inc.	www.prometic.com
Puresyn, Inc.	www.puresyn.com
Rockland Immunochemcials, Inc.	www.rockland-inc.com
SAFC Corporation	www.sigma-aldrich.com/safc
Sartorius	www.sartorius.com
Scientific Protein Laboratories	www.spl-pharma.com
Selexis SA	www.selexis.com

(Continued)

Table 11.4 (*Continued*) Biopharmaceutical Support Providers

Company Name	URL
Synexa Life Sciences	www.synexagroup.com
Taconic Biotechnology	www.taconic.com
UniTargetingResearch AS	www.unitargeting.com
Vista Biologicals Corp.	www.vistabiologicals.com
WuXi Apptec, Inc.	www.wuxiapptec.com

Analytical support providers

Protein science requires highly specialized training and equipment to implement and whereas many CMOs and CSOs provide a variety of services including analytical support, it is always a good idea to retain highly specialized companies that have already done the assay development. Table 11.5 lists some of these companies.

Contract research organizations

Development of biosimilars may entail several stages of research including the device component development, altering the manufacturing process, formulation composition, conducting animal toxicology, animal PK/PD and human PK/PD studies. In all likelihood, an outsourced operation will include several contract research organizations (CROs); in some instances, these may double up as CMOs or analytical services. Table 11.6 lists a few known CROs in this field.

Table 11.5 List of Companies Providing Assay Development Services

Absorption Systems	Accelero Bioanalytics	AddexBio
Affinity Life Sciences, Inc.	Allele Biotechnology	Amsbio
Antibody Research	Antitope	Aragen Bioscience
Arvys Proteins, Inc.	Assay Depot	Aushon Biosystems
Biaffin	Bio Basic	BioReliance
Bio-World	Brighter Ideas, Inc.	Bucher Biotech
Cambridge Biomedical	Cambridge Peptides	Cambridge Research Biochemicals
CARBOGEN AMCIS	Celerion	Cell Biolabs, Inc.
Cell Technology, Inc.	Charles River Laboratories International, Inc.	Clongen Laboratories LLC
Creative Biolabs	DiscoveryBioMed, Inc.	Enzo Biochem
Epitomics	Eurofins Scientific	Evotec
Frontier BioSciences, Inc.	GE Healthcare	GeneTex, Inc.
Genpharmtox.com	Glycotope Group	Hardy Diagnostics
Irvine Pharmaceutical Services	Kronos Science	Leinco Technologies
Life Technologies	Lonza Group Ltd.	Medpace, Inc.
Meridian Life Science, Inc.	Molecular Devices LLC	MRM Proteomics, Inc.
NEO Bioscience	Novatia LLC	PPD
ProMab Biotechnologies, Inc.	Promega Corporation.	ProtTech Analytical LLC
SGS Life Science Services— Clinical Research	Shanghai Medicilon Inc.	Smithers Pharma Services LLC
StemCell Technologies, Inc.	Synelvia	Tecan
TGA Sciences, Inc.	Warnex, Inc.	Washington Biotechnology, Inc.
Zelle Biotechnology	Zinsser NA	

Product development services

While the bulk of the technology-driven activities involve the upstream and downstream optimization, formulation and product development activities go side by side. This is more important in those instances where the developer may have limitations in copying the originator formula for IP reasons. A large number of formulation and product development studies are generally required to establish these designs and details. Table 11.7 provides a listing of some of these product development service providers.

Fill and finish providers

There are wide choices available for finishing clinical lots as well as commercial lots. Since the fill and finish steps do not require any special considerations except the issues of cross-contamination and degradation of product during the filling, the developer may find it convenient to quality multiple fill and finish facilities. Table 11.8 provides a list of some prospective fill and finish facilities specializing in biological products.

Legal matters

Engaging COs creates an inevitable legal burden; there are issues of delivery, liability, loss of product, lot failure, accountability, secrecy, regulatory support, regulatory reporting, and many other aspects of business relationship that need

Table 11.6 List of Contract Research Organizations

Advanced Clinical Research Services, Inc.	Advion BioSciences, Inc.	Affinity Life Sciences, Inc.
Allphase Clinical Research	Alphora Research, Inc.	Altogen Labs
America Pharma Source	Amsbio	ANALIZA, Inc.
ANTAEA Medical Services Ltd.	Arvys Proteins, Inc.	ASKA Research
Assay Depot	Assay Depot, Inc.	Averion, Inc.
Celerion	Certus International	Charles River Laboratories International, Inc.
Concentrics Research	CoreLab Partners, Inc.	Curo Clinical Data Management
DiscoveryBioMed, Inc.	Dolby Research	Dow Pharmaceutical Sciences
EDETEK	Epistem	EtendX, Inc.
ethica CRO, Inc.	Eurotrials, Scientific Consultants	Glycotope Group
HH BioTechnologies Pvt. Ltd.	ICON Medical Imaging	IIT Research Institute
INC Research	INDOOR Biotechnologies, Inc.	Irvine Pharmaceutical Services
ISIS Services LLC	Lambda Therapeutic Research	LBBM Research
Medelis, Inc.	Medpace, Inc.	MRM Proteomics, Inc.
NERI	Octalsoft	PAREXEL International
PharmaCore	PPD	PRC Clinical
Rho	SGS Life Science Services— Clinical Research	Smithers Pharma Services LLC
Sygnature	Synelvia	Synteract
TGA Sciences, Inc.	The KPC Group	Trio Clinical Research
Veristat	Warnex, Inc.	Wincon TheraCells Biotechnologies Co., Ltd.
Wockhardt		

Table 11.7 List of Companies Offering Product Development Services

	Accelsiors	AddexBio
Allphase Clinical Research	America Pharma Source	ANALIZA, Inc.
Angel Biotechnology	Assay Depot	Beckloff Associates (Cardinal Health)
Biaffin	Biotage	Biotech Advisor
Brighter Ideas, Inc.	Cambridge Biomedical	Cambridge Major Laboratories, Inc.
Cambridge Peptides	CARBOGEN AMCIS	Celerion
Certus International	Charles River Laboratories International, Inc.	ChemoGenics BioPharma
ClinAudits LLC	Concentrics Research	CTI Clinical Trial and Consulting Services
deCODE chemistry	Deloitte Recap LLC	DynPort Vaccine Company LLC
Elan Corporation	Endpoint Research	Epistem.
Eurofins Scientific	Frontier BioSciences, Inc.	FUJIFILM Diosynth
HTD Biosystems, Inc.	Iris Pharma	Irvine Pharmaceutical Services
LabVantage Solutions	Medpace Inc	Microbiology & Quality Associates
Optimus Pharma Consulting	Pathology Experts LLC	PharmaNet Development Group, Inc.
Pharmatek	PPD	Quality & Compliance Consulting, Inc.
SEPS Pharma	Society for Laboratory Automation and Screening	Solvias
Synelvia	Synteract	Technology Catalysts International (TCI)
The KPC Group	UPM Pharmaceuticals, Inc.	Washington Biotechnology, Inc.
Wincon TheraCells Biotechnologies Co., Ltd.	Xoma	

to be made clear before any work begins. In all likelihood, these conditions will be tested and often the agreements will be revised to take into account unanticipated events. This requires a highly specialized team of lawyers, not just any contract lawyers but a team that is technically qualified to understand the nuances of biological development and manufacturing. In most instances the COs will offer their versions of contracts and these form a good basis to start with.

Here are some examples of clauses that need to be specified:

- *Cost over runs*—determine whether they are a fault of the sponsor company, a lack of understanding by the CMO, or errors by the CMO in such stages as technology transfer—and to determine who "picks up the tab." This can only be understood by lawyers who understand the filing process.
- *Authority assignment*—who is authorized to sign-off on the product lot. This could be released by the CO or the developer.
- *Secrecy*—a lot depends on being able to beat the development cycle and for this reason the CO must assure complete secrecy; generally, this is a major issue—more significant is how we would evaluate damages if advertent or inadvertent disclosures are made by the CO.

Table 11.8 Biopharmaceutical Fill and Finish Contractors

Company	URL
AAIPharma	www.aaipharma.com
ABL, Inc.	www.ablinc.com
Aesica Pharmaceuticals Ltd.	www.aesica-pharma.co.uk
Affinity Life Sciences, Inc.	www.affinitylifesciences.com
Alkermes	www.alkermes.com/contract
Althea Technologies, Inc.	www.altheatech.com
AMP a Siegfried Company	www.siegfried-amp.com
AMRI	www.amriglobal.com
Aptuit, Inc.	www.aptuit.com
Baxter BioPharma Solutions	www.baxterbiopharmasolutions.com
Bio-Concept Laboratories, Inc.	www.bioconcept.com
BioConvergence	www.bioc.us
Biotechpharma UAB	www.biotechpharma.lt
Biovian Oy	www.biovian.com
Boehringer Ingelheim Pharma GmbH & Co. KG	www.BioXcellence.com
Catalent Pharma Solutions	www.catalent.com
Cenexi Laboratoires Thissen SA	www.cenexi.com
Cobra Biologics Ltd.	www.cobrabio.com
Coldstream Laboratories, Inc.	www.coldstreamlabs.com
Cook Pharmica	www.cookpharmica.com
Cytovance Biologics, Inc.	www.cytovance.com
Dalton Pharma Services	www.dalton.com
DPT Laboratories Ltd.	www.dptlabs.com
Dr. Reddy's Laboratories	www.drreddys-cps.com
DSM Pharmaceuticals, Inc.	www.dsmpharmaceuticals.com
Emergent BioSolutions	www.emergentcontractmanufacturing.com/ commercial_manufacturing
Florida Biologix	www.floridabiologix.ufl.com
Fresenius Kabi Product Partnering	www.freseniuskabi-productpartnering.com
Gallus BioPharmaceuticals	www.gallusbiopharma.com
Glycotope Biotechnology	www.glycotope.com
Goodwin Biotechnology, Inc.	www.goodwinbio.com
Grand River Aseptic Manufacturing	www.grandriverasepticmfg.com
Hospira One 2 One	www.one2onecmo.com
IDT Biologika GmbH	www.idt-biologika.com
Intas Pharmaceuticals Ltd.	www.intaspharma.com
Irvine Pharmaceutical Services	www.irvinepharma.com
JHP Pharmaceuticals	www.parsterileproducts.com/ contract-manufacturing
Jubilant HollisterStier Contract Manufacturing & Services	www.jublhs.com
Kemwell Biopharma Pvt. Ltd.	www.kemwellbiopharma.com
LSNE Contract Manufacturing	www.lyophilization.com
Medicago	www.medicago.com
Meridian Life Science	www.meridianlifescience.com
Novasep	www.novasep.com
Oakwood Laboratories, LLC	www.oakwoodlabs.com
Omnia Biologics, Inc.	www.omniabiologics.com

(*Continued*)

Table 11.8 (*Continued*) Biopharmaceutical Fill and Finish Contractors

Company	URL
OSO BioPharmaceuticals Manufacturing, LLC	www.osobio.com
PacificGMP	www.pacificgmp.com
Paragon Bioservices, Inc.	www.paragonbioservices.com
Patheon, Inc.	www.patheon.com
Pharmalucence	www.pharmalucence.com
Pharmaniaga LifeScience SDN BHD	www.pharmaniaga-lifescience.com
Pii—Pharmaceutics International, Inc.	www.pharm-int.com
Piramal Healthcare	www.piramalpharmasolutions.com
Quality BioResources, Inc.	http://qualbio.com
Recipharm AB	www.recipharm.com
Reliance Life Sciences	www.rellife.com
Rentschler Biotechnologie GmbH	www.rentschler.de
Richter-HELM Biotec GmbH and Company	www.richter-helm.eu
ROVI Contract Manufacturing	www.rovicm.es
SAFC Corporation	www.safcglobal.com
Samsung Biologics	www.samsungbiologics.com
Sandoz GmbH/A Novartis Company	www.sandoz.com
SCM Pharma	www.scmpharma.com
Statens Serum Institut	www.ssi.dk
Symbiosis Pharmaceutical Services Ltd.	www.symbiosis-pharma.com
SynCo Bio Partners	www.syncobiopartners.com
Sypharma Pty Ltd.	www.sypharma.com.au
Syrinx Pharmaceuticals P/L	www.syrinxpharmaceuticals.com
Therapure Biopharma, Inc.	www.therapurebio.com
University of Iowa Pharmaceuticals	www.pharmacy.uiowa.edu
Vetter Pharma International GmbH	www.vetter-pharma.com
Vibalogics	www.vibalogics.com
Waisman Biomanufacturing	www.gmpbio.org
Wockhardt Ltd.	www.wockhardt-knowhow.com
WuXi AppTec Co. Ltd.	www.wuxiapptec.com

- *Material transfer agreements*—there will be several stages, back and forth; how would the material, both the product of equipment or supplies be treated is important to define upfront.
- *Legal jurisdiction*—this is a sticky issue if a non-U.S. based CO is used by a U.S. firm; understanding how the contract will be enforced should be concluded ahead of signing the agreement. In many instances, the developer may find itself totally frustrated and not being able to work through the legal system of other countries where delays and graft are common.
- *IP*—when transferring IP, the consideration of secrecy go beyond ordinary clauses; in most cases the risk of IP transfer to a country where the IP laws are not enforced means that the developer should be prepared to lose IP and weight that out against any financial gains possible in that relationship.

- *Electronic data storage and rooms*—many Cos will provide electronic data storage and access in data rooms. Recently, highly sophisticated and robust systems such as those offered by Pro Softnet Corporation of Woodland Hills, CA (www.ibackup.com) serve several purposes. A more sophisticated method may require using providers such as Master Control (www.mastercontrol.com) or www.box.com. They offer time-stamping by a third party, for one, and with costs of such services dropping, redundant time histories can be maintained for posterity. These servers are more reliable than company in-house systems. The data may be encrypted using such widely available protocols as the PGP encryption (Pretty Good Privacy, www.pgp.com) for all documents after their conversion to PDF; PDF is not a proprietary file system, even though most associate it with Adobe Acrobat, which as the originator is probably the best PDF system.
- *Partnerships*—the ultimate tool to make a relationship succeed; several CROs have begun offering this to clients and it makes sense to make them gain benefits so they will more likely be encouraged to protect the interest of the clients.
- *Quality monitoring*—a detailed map of how immediate access to quality parameters will be available to developer is important to make sure that the product is continuously developed and manufactured under the cGMP. The developer may start this exercise by buying out the 483s issued to the CO in the past to develop an understanding of the quality culture of the CMO. Any sign of impropriety, albeit superficially insignificant should be treated as a red flag and generally would be a reason to drop the CMO immediately. The contract should allow periodic audits by the developers of its assigned outsourced contractor to conduct these audits.
- *Conflicts*—always arise as launch delays, cost escalation, regulatory fall outs come into play. How would the conflicts be resolved is important, such as through arbitration; going to court is the last thing the developer would like to do and the CO would like to see happen. So there is inventive on both sides to resolve conflict amicably.

Managing outsourcing

Managing outsourcing effectively requires building a team that will be fully conversant with the technological aspects of outsourcing. Table 11.9 lists some of the key checklist items for this team to consider.

Outlook

For the past few years, demand for biopharmaceutical contract manufacturing services has continued to grow, due to new drug commercialization, higher funding rates of biotechnology companies, and expanding service offerings for novel therapeutics as well as biosimilars. As business conditions continue to improve, the market of COs should reach over $3 billion this year. Whereas most of this growth has come in the novel therapeutic area, some companies have begun to see significant increase in the biosimilar market as well.

Table 11.9 Key Checklist Items for Outsourcing

Management, legal	Project plan, batch plans, budgets, worksheets, communication, Gantt chart, personnel qualifications, facility layout, GMP certification, confidentiality disclosure agreement, material transfer forms, research agreement, noncompete agreement
Cell work	Cell line optimization, master cell bank (sterility, characterization, storage, and stability), working cell bank (sterility, characterization, storage, and stability)
Upstream	Media, conditions, holding times, etc.; process scale, critical development parameters and interactions, production of three consecutive batches at small-scale to meet specifications
Process development	Communication of process rationale and strategy, process scale, process design aiming for CGMP production, three consecutive small-scale batches meet specifications, critical parameters and interactions, protein stability during processing
Non-CGMP	Amount to be produced, batch size, and time schedules production of drug substance for preclinical testing
Formulation development	Drug product development
Reference materials	Amount of reference standard to be produced, conditions for storage
In-process controls	Protein concentration assay, one-dimensional SDS-PAGE, high-performance ion-exchange and reversed-phase chromatography, size-exclusion chromatography, and tests of specific biological activity
Quality control	Protein concentration assays; one-dimensional SDS-PAGE; high-performance ion-exchange and reversed-phase chromatography; size-exclusion chromatography; and tests for endotoxins, host-cell proteins, DNA, bioburden, viruses, leachables, contaminants, carbohydrate patterns, specific biological and immunological activity, toxicity, and appearance
Drug-substance characterization	Mass spectrometry, amino-acid sequencing and analysis, peptide mapping, two-dimensional electrophoresis, ultraviolet scanning, infrared spectroscopy, optical rotation, fluorescence, circular dichroism, nuclear magnetic resonance, diffraction patterns, Fourier-transform infrared spectroscopy, and carbohydrate structure
Drug-substance stability studies	3-, 6-, 12-, and 24-month stability studies including high-performance ion-exchange and reversed-phase chromatography, size-exclusion chromatography, and biological activity
Drug-product stability studies	3-, 6-, 12-, and 24-month stability studies including high-performance ion-exchange and reversed-phase chromatography, size-exclusion chromatography, and biological activity
CGMP production of product for Phase 1 and 2	Specifications, amount of drug product to be made, batch-release drug procedures, list of raw materials, filling and packing, documentation, and data collection and storage
Documentation	Laboratory notebook system, cell work reports, fermentation reports, downstream processing (strategy and rationale report), summary reports, unit-operation reports, analytical reports (QC), drug-substance characterization reports, reference materials report, development report, batch records, manufacturing protocols, list of raw materials, change-control and deviation reports, facility documentation, equipment documentation, and utility documentation
Technology transfer	Discovery to development, development to pilot production, technology transfer packages and plan

Source: Niazi, S. and Flynn III, T., *Bioprocess Int.*, 10–16, September 2006.

Bibliography

Alper J. California's biogoddess. *Nat Biotechnol*. 2003 July;21(Suppl.):BE13–BE14.

Bienias T. Supplying the health-care explosion. *Med Dev Technol*. 2001;12(10):14–15.

Biopharmaceutical Contract Manufacturing 2005. Improved processes and new capacity for pipeline to commercial production—The comprehensive report. HighTech Business Decisions, Moraga, CA, 2005, www.hightechdecisions.com.

Broderson HR. Virtual reality: The promise and pitfalls of going virtual. *Nat Biotechnol*. 2005;23(10):1205–1207.

Burrill GS. The future of biotech outsourcing: Maximizing potential. *BioProcess Int*. 2003;1 (Suppl. 9):50–53.

Cavalla D. The extended pharmaceutical enterprise. *Drug Discov Today* 2003;8(6):267–274.

Coile Jr. RC. Physician executives straddle the digital divide. *Physician Exec*. 2001;27(2):12–19.

Demir SS, Hussein A. Brokering knowledge in biosciences with innocentive. *IEEE Eng Med Biol Mag*. 2003;22(4):26–27.

France offers good prospects. *Med Device Technol*. 2005;16(4):48.

Gura T. Careers: Joining a trend, scientists increasingly say "Call My Agent." *Science*. 2004;303(5656): 303–305.

Hecker SJ, Preston C, Foote M. Production of high-quality marketing applications: Strategies for biotechnology companies working with contract research organizations. *Biotechnol Annu Rev*. 2003;9:269–277.

Hogan J. Create products by going virtual. *Med Device Technol*. 2005;16(9):38–39.

Mills E. Virtual recruitment. *Nat Biotechnol*. 2002;20(8):853.

Mintz CS. Developing and managing successful bio-outsourcing relationships, Report #9210. D&MD Publications, Westborough, MA, March 2006, pp. 1–4.

Mullin R. Wyeth tears a page from the book of DuPont. *Chem Eng News* 2004;82(7):26–27.

Niazi S, Flynn III, T. A practical model for outsourced biomanufacturing. *Bioprocess Int*., 2006 September;10–16.

Niazi SK. *Handbook of Biogeneric Therapeutic Proteins: Regulatory, Manufacturing, Testing, and Patent Issues*. Taylor & Francis (CRC Press), Boca Raton, FL, 2005.

Pissarra PN. Changes in the business of culture. *Nat Biotechnol*. 2004;22(11):1355–1356.

Smith A. Life beyond the walls. *Nature*. 2002;415(6872):7.

Steffy CP. Chapter one: The birth of outsourcing—A rough beginning leaves few survivors; Chapter two: The evolution of outsourcing—An industry reinvents itself. *BioProcess Int*. 2003;1(Suppl. 9):6–20.

Strauss L, Geimer H. Outsourcing manufacturing. *Med Device Technol*. 2001;12(10):10–13.

Thiel KA. Biomanufacturing, from bust to boom... to bubble? *Nat Biotechnol*. 2004;22(11): 1365–1372.

Valdes S, McGuire P. Contract research organizations (CROs) may be the next trend in clinical trials liability. *J Biolaw Bus*. 2004;7(3):11–15.

Van Eldik LJ, Koppal T, Watterson DM. Barriers to Alzheimer disease drug discovery and development in academia. *Alzheimer Dis Assoc Disord*. 2002;16(Suppl. 1):S18–S28.

Appendix: Glossary of terms

7-AAD: 7-Aminoactinomycin D

AATB: American Association of Tissue Banks

AAV: Adeno-associated virus

Absorption: Removing a particular antibody or antigen from a sample (e.g., from serum) by adding the corresponding antigen or antibody to that sample.

Acceptance criteria: Numerical limits, ranges, or other suitable measures for acceptance of the results of analytical procedures in which the drug substance or drug product or materials at other stages of manufacture should meet.

Action limit: An internal (in-house) value used to assess the consistency of the process at less critical steps. Generally, the action limits are tighter than the specification limits; however, to comply with current good manufacturing practice and good laboratory practice requirements, action limits must be listed in the testing standard operating procedure and an action plan listed for values going beyond action limits (the corrective action), though they remain within the specification.

Active substance: Active ingredient or molecule that goes into a specific medicine and that provides this medicine with properties for treating or preventing one or several specific disease(s).

ADA: Aminodeaminase

Additive: A substance added to another in relatively small amounts to effect a desired change in properties.

Adsorption: The binding of molecules to a surface as a result of a chemical or physioelectric interaction between the membrane surface or the chromatographic resin and the molecule. Nonspecific adherence of substances in solution or suspension to cells or other particulate matter.

Adventitious agents: Acquired, sporadic, accidental contaminants. Examples include bacteria, yeast, mold, mycoplasma, or viruses that can potentially contaminate prokaryote or eukaryote cells used in production. Potential sources of adventitious organisms include the serum used in cell culture media, persistently or latently infected cells, or the environment.

Adventitious virus: Unintentionally introduced contaminant virus. See virus.

Adverse reaction: A response to a medicinal product that is noxious and unintended.

Aerobe: An aerobic organism is one that grows in the presence of oxygen. A strict aerobe grows only under such a condition.

Affinity chromatography (AC): A chromatography separation method based on a chemical interaction specific to the target species. In AC, a biospecific adsorbent is prepared by coupling a specific ligand (such as a protein, peptide, or nickel) for the molecule of interest to a solid support. The inherent high specificity of ligand/target interactions makes AC particularly suitable for the capture stage of downstream processing. As a result, the types of affinity methods are biosorption-site recognition (e.g., monoclonal antibody, protein A), hydrophobic interaction (contacts between nonpolar regions in aqueous solutions), dye–ligand specific binding of macromolecules to triazine and triphenylmethane dyes, metal chelate–matrix bound chelate complexes with target molecule by exchanging low-molecular-weight metal bound ligands, and covalent-disulfide bonding reversible under mild conditions.

Affinity (see affinity chromatography): The thermodynamic quantity defining the energy interaction or binding of two molecules, usually that of antibody with its corresponding antigenic determinant.

Agarose: High-molecular-weight polysaccharide used as a separation medium in bead form for biochromatography.

Aggregate: A clustered mass of individual cells—solid, fluffy, or pelletized—that can clog the pores of filters or other fermentation apparatus.

Agrobacterium tumefaciens: A bacterial plant pathogen, commonly found in soil, which contains a plasmid used to introduce desired sections of DNA into plants.

Air diffusion rate: The rate at which air diffuses through the wetted pores of membrane at a given differential pressure. Measuring the air diffusion rate is a method used to check the integrity of a membrane filter.

Alpha (α helix) and beta (β strand): In a protein, certain domains may form specific structures such as α helix and β strand, which constitute the secondary structure of the protein. An α helix has the following features: every 3.6 residues make one turn, the distance between two turns is 0.54 nm, and the C=O (or N–H) of one turn is hydrogen bonded to N–H (or C=O) of the neighboring turn. In a β strand, the torsion angle α in the backbone is about 120°.

Amino acid composition analysis: Used to determine the amino acid composition and/or the protein quantity. A two-step process involving a complete hydrolysis (chemical or enzymatic) of the protein into its component amino acids followed by chromatographic separation and quantitation via high-performance liquid chromatography. The complete amino acid composition of the peptide or protein should include accurate values for methionine, cysteine, and tryptophan. The amino acid composition presented should be the average of at least three separate hydrolysates of each lot number. Integral values for those amino acid residues generally found in low quantities, such as tryptophan and/or methionine, could be obtained and used to support arguments of purity.

Amino acid sequencing: A partial sequencing (8–15 residues) of amino acids within a protein or polypeptide by either amino-terminal or carboxy-terminal sequencing. This method is used to obtain information about the primary structure of the protein, its homogeneity, and the presence or absence of polypeptide cleavages. The sequence data determined by high-performance liquid chromatography (HPLC) analysis is presented in tabular form and should include the total yield for every amino acid at each sequential cleavage cycle. Full sequence is often done by sequencing the peptide fragments isolated from HPLC fractionation. Amino acid sequencing is a required test in British Pharmacopeia/European Pharmacopeia (BP/EP) for listed monographs of therapeutic proteins.

Amino acids: An amino acid is defined as the molecule containing an amino group (NH_2), a carboxyl group (COOH), and an R group. It has the following general formula: $R–CH(NH_2)–COOH$. The R group differs among various amino acids. In a protein, the R group is also called a side chain. There are over 300 naturally occurring amino acids on earth, but the number of different amino acids in proteins is only 20.

Anaerobe: An anaerobic organism grows in the absence of air or oxygen. Some anaerobic organisms are killed by brief exposure to oxygen, whereas oxygen may just retard or stop the growth of others.

Anaphylaxis: An acute and severe allergic reaction in humans.

Anemia: Low red blood cell count.

Anion exchange: The ion-exchange procedure used for the separation of anions. The tetra-alkyl-ammonium group is a typical strong anion-exchange functional group.

Antibiotic resistance: The gene coding for antibiotic resistance is needed for identifying transformants and to ensure antibiotic selective pressure, that is, only cells that harbor an expression vector will divide, thus preventing plasmid loss. Genes conferring ampicillin, tetracycline, or kanamycin resistance are commonly used in expression vectors. Ampicillin resistance is mostly used only on a laboratory scale because the lactamase, which confers the resistance, degrades ampicillin, and thus the selective pressure is lost after a few generations of cell growth. Furthermore, ampicillin has been thought to be potentially allergenic and is therefore usually not the antibiotic of choice in the production of biotherapeutics intended for human use. The U.S. Food and Drug Administration and other regulatory authorities discourage the use of β-lactam antibiotics

Antibody (see also immunoglobulin): A protein molecule having a characteristic structure consisting of two types of peptide chains: heavy (H) and light (L). Antibodies contain areas (binding sites) that specifically fit to and can bind to its corresponding determinant site on an antigen, which has induced the production of that antibody by the B lymphocytes and plasma cells in a living species.

Antifoam agent: A chemical added to the fermentation broth to reduce surface tension and counteract the foaming that can be caused by mixing, sparging, or stirring.

Antigen: A foreign protein or carbohydrate, which, when introduced into an organism, activates specific receptors on the surface immunocompetent T and B lymphocytes. After interaction between antigen and receptors, there usually will be induction of an immune response, that is, production of antibodies capable of reacting specifically with determinant sites on the antigen.

Antigen determinant: The specific part of a structure of an antigen that will induce an immune response, that is, it will fit to the receptors on T and B lymphocytes and will also be able to react with the antibodies produced.

Antigenicity: Antigenicity is the capacity of a substance to function as an antigen—to trigger an immune response. Any agent, often a large molecule that stimulates production of an antibody that will react specifically with it. Each antigen may contain more than one site capable of binding to a particular antibody. An immunogen can cause the production of a number of antibodies with different specificities. Antigenicity of therapeutic proteins is one of the major issues in the comparability of generic therapeutic proteins.

Antiserum: Blood serum that contains antibodies against a particular antigen or immunogen. This frequently means serum from an animal that has been inoculated with the antigen.

Apoptosis: Apoptosis describes the molecular and morphological changes that characterize controlled cellular self-destruction, often called "programmed cell death."

AR: Annual report

Artificial chromosome: Synthesized DNA in chromosomal form for use as an expression vector.

Ascites: Liquid accumulations in the peritoneal cavity. Monoclonal antibodies can be purified from the ascites of mice that carry a transplanted hybridoma.

Aseptic: Sterile, free from bacteria, viruses, and contaminants such as foreign DNA.

Association constant: A reaction between antibody and its determinant that comprises a measure of affinity. The constant is quantitated by mass action law rate constants for association and for dissociation.

Asymmetric membrane: A membrane that is made such that the pore size increases through the membrane matrix.

Asymmetry: See asymmetry factor.

Asymmetry factor: Factor describing the shape of a chromatographic peak. Theory assumes a Gaussian shape and that peaks are symmetrical. The peak asymmetry factor is the ratio (at 10% of the peak height) of the distance between the peak apex and the backside of the chromatographic curve to the distance between the peak apex and the front side of the chromatographic curve. A value >1 is a tailing peak, while a value <1 is a fronting peak.

Attenuated: Weakened (attenuated) viruses often used as vaccines; they can no longer produce disease but still stimulate a strong immune response similar to the natural virus. Examples include oral polio, measles, mumps, and rubella vaccines.

Autoclave, autoclavability: An autoclave is a device that uses saturated steam at a specified pressure over time to kill microorganisms and thus achieve sanitization or sterilization. Because many materials change properties when exposed to moisture, heat, and pressure, products destined for this process must be specially engineered for autoclavability.

Autoradiography: Detection of radioactively labeled molecules on x-ray film.

Avidity: The total binding strength between all available binding sites of an antibody molecule and the corresponding determinants present on antigen.

Bacillus subtilis: A microbe commonly found in soil and vegetation, normally considered nonpathogenic, sometimes used in recombinant microbial fermentation.

Back flushing, backwash: Back flushing is used to elute strongly held compounds at the head of the column by reversing the flow direction of the mobile phase through a chromatographic column.

Bacterial expression: The most common microbial source for recombinant protein production is *Escherichia coli* because of its well-understood genetics. *Bacillus subtilis* and its relatives have been used, mainly because of their greater tendency to secrete proteins into their environment. Various *Streptomyces* species are under study in recombinant fermentation, but so far they have demonstrated low expression levels. *Pseudomonas fluorescens* may have greater potential. But *E. coli* remains one of the most attractive because of its ability to grow rapidly and at high density on inexpensive substrates, its well-characterized genetics, and the availability of an increasingly large number of cloning vectors and mutant host strains. Bacterial genes are contained in a circular genome and on small circular pieces of extragenomic double-stranded DNA elements called plasmids in their nucleus-free cells. Self-replicating plasmids contain regulatory regions (promoter regions and origins of replication) that make them ideal candidates for use in genetic engineering. They can be manipulated using restriction enzymes, cloning vectors (such as bacteriophage viruses), and relatively simple procedures. Certain gene segments with the ability to promote (promoters), direct, or terminate transcription of the foreign DNA are often involved as well.

Bacteriophage: A virus that infects bacteria, sometimes used as a vector. The lambda bacteriophage is frequently used as a vector in recombinant gene experiments. Examples include dsDNA phages with contractile tails, such as T4; dsDNA phages with long flexible tails, such as λ, which are most commonly used in DNA cloning; dsDNA phages with stubby tails, such as p22; ssDNA phages, such as phi X 174; and ssRNA phages, such as MS2.

Baculovirus: A class of insect virus used as vectors for recombinant protein expression in insects; baculovirus is noninfective to humans.

Base pair: Two bases on different strands of nucleic acid that join together. In DNA, cytosine (C) always pairs with guanine (G), and adenine (A) always links to thymine (T). In RNA molecules, adenine joins to uracil (U).

Baseline: The more or less constant signal observed when only the mobile phase passes through the detector. This steady baseline portion of the chromatogram is the reference from which quantitative measurements can be made.

Batch culture: Large-scale cell culture in which cell inoculum is cultured to a maximum density in a tank or airlift fermenter, harvested, and processed as a batch.

Batch fermentation mode: The most commonly used type of fermentation, in which microbes are added to a sterile nutrient broth and allowed to ferment without the addition of further nutrients (except oxygen). A batch fermentation is a closed system where microbes are added to a sterilized nutrient solution in the fermenter and then allowed to incubate. Nothing more is added except oxygen (most microorganisms used in biotechnology are aerobic species), an antifoam agent (to prevent bubble formation), and acid or base to control the solution pH. Consequently, the mixture changes constantly as a result of cellular metabolism, with waste products accumulating and biomass increasing over time. Four growth phases follow inoculation: a lag phase (microorganisms physicochemical equilibrating with their environment), a log phase (cells have adapted to their new surroundings and begin doubling their number logarithmically), a stationary phase (available food has been used up and growth slows or stops; this is when metabolites or recombinant proteins are best harvested), and a death phase (their energy depleted, the cells die off). Doubling times during the log phase vary according to the size and complexity of the microbe: bacteria double in less than an hour, yeasts in 1 or 2 h, filamentous fungi in 2–6 h. In fed-batch fermentation, nutrients are added in increments as growth progresses. Certain ingredients are present at inoculation, and they continue to be added in small doses throughout production.

Bed volume: $V = (\pi \times r^2) \times h$, where r is the inner radius of the column tube and h is the height of the column tube.

Beta ratio: A standard method of rating a filter's ability to remove particles. Beta $(x) = \#$ particles $>$ size (x) upstream$/\#$ particles $>$ size (x) downstream, where x is the particle size in microns.

Binding: The process by which some components in a feed solution adhere to the membrane. Binding can be desirable in some instances, but often, as in the case of protein binding during sterile filtration, can result in a loss of valuable product.

Binding capacity: The binding capacity describes the actual amount of a sample that will bind to the medium packed in a column under defined conditions.

511

It is determined by saturating a gel with sample and then measuring the amount of sample that binds. Parameters like pH, ionic strength, the counterion and the sample all influence the available capacity. Available capacity will change depending on the experimental conditions. It is essential to take these conditions into consideration when comparing the available capacities of different chromatography media.

Binding site: The part of the antibody molecule that will specifically bind antigen.

Bioactivity: A protein's ability to function correctly after it has been delivered to the active site of the body (in vivo).

Bioactivity: The level of specific activity or potency as determined by animal model, cell culture, or in vitro biochemical assay.

Bioavailability: Measure of the true rate and the total amount of drug that reaches the target tissue after administration.

Biologic: A therapeutic agent derived from living things.

Biological activity: The specific ability or capacity of the product to achieve a defined biological effect. Potency is the quantitative measure of the biological activity.

Biological containment: Characteristics of an organism that limit its survival and/or multiplication in the environment.

Biological response modifier: Generic term for hormones, neuroactive compounds, and immunoreactive compounds that act at the cellular level; many are possible candidates for biotechnological production.

Biopharmaceutical: A therapeutic product created through the genetic manipulation of living things, including (but not limited to) proteins and monoclonal antibodies, peptides, and other molecules that are not chemically synthesized, along with gene therapies, cell therapies, and engineered tissues.

Biopharmaceuticals: Medicines made, or derived, from living organisms using biotechnology.

Bioprocessing: Using organisms or biologically derived macromolecules to carry out enzymatic reactions or to manufacture products.

Bioreactor: A vessel capable of supporting a cell culture in which a biological transformation takes place (also called a fermenter or reactor). Typically, the vessel contains microbes or other cells grown under controlled conditions of temperature, aeration, mixing, acidity, and sterility. Classically, it represented fermentation when bacteria were used; still differentiated forms from other cells are used.

Biosafety test: A class of tests that determine whether a chromatographic media or filter's material of construction can induce systemic toxicity, skin irritation, sensitization reaction, or other biological responses. These tests are often completed by labs in vivo or in vitro.

Biosensors: The powerful recognition systems of biological chemicals (enzymes, antibodies, DNA) are coupled to microelectronics to enable rapid, accurate low-level detection of such substances as sugars and proteins (such as hormones) in body fluids, pollutants in water, and gases in air.

Biosimilar medicine: Medicine that is approved by the regulatory authorities as being similar in terms of quality, efficacy, and safety to a reference biological medicine with which it has been compared.

Biosimilarity: Property of a medicine to show similarity and lack of significant differences in terms of quality, efficacy, and safety to a reference biological medicine to which it has been compared.

Biotechnology: Technology that manipulates living organisms so that they produce a certain specific protein including hormones or monoclonal antibodies.

BLA: Biologics license application

Blinded: When a filter is blinded, it means that particles have filled the pores and the flow through the filter from the feed side to the permeate side is reduced or stopped.

Blotting methods and applications: Following gel electrophoresis, probes are often used to detect specific molecules from the mixture. However, probes cannot be applied directly to the gel. The problem can be solved by three types of blotting methods: see Southern blotting, Northern blotting, and Western blotting.

BPCI Act: Biologics Price Competition and Innovation Act.

Broth: The contents of a microbial bioreactor: cells, nutrients, waste, and so on.

BSE: Bovine spongiform encephalopathy (mad cow disease)

Bubble point: The minimum pressure required to overcome the capillary forces and surface tension of a liquid in a fully wetted membrane filter. The bubble point value is determined by observing when bubbles first begin to emerge on the permeate side or downstream side of a fully wetted membrane filter when pressurized with a gas on the feed (upstream) side of the membrane filter.

Buffer: A mixture of an acid and its conjugate base; its pH changes slowly upon the addition of small amounts of acid or base.

Buffer exchange: Filtration process used for the removal of smaller ionic solutes, whereby the feed solution is washed, usually repeatedly, and one buffer is removed and replaced with an alternative buffer.

Calibrator: A term in clinical chemistry commonly referring to the standard used to "calibrate" an instrument or used in construction of a standard (calibrator) curve.

Capillary electrophoresis: Used as a complement to high-performance liquid chromatography (HPLC), particularly for peptide mapping. This technique is faster and will often separate peptides that coelute using HPLC. Separation is accomplished by relative mobility of the peptides in a buffer in response to an electrical current.

Capture step: The initial purification of the target molecule from crude or clarified source material. The objectives are the rapid isolation, stabilization, and concentration of the desired molecule. The initial chromatographic purification step following harvest with the purpose of removing cell.

Carbohydrate analysis: Used to determine the consistency of the composition of the covalently bound monosaccharides in glycoproteins. Unlike the polypeptide chain of the glycoprotein where production is controlled by the genetic code, the oligosaccharides are synthesized by posttranslational enzymes. Microheterogeneity of the carbohydrate chains is common. Determination can be accomplished on underivatized sugars after hydrolysis by high-performance liquid chromatography separation with pulsed amperometric detection or by gas chromatography after derivatization.

Cartridge or cartridge filter: A filtration or separation device having a membrane encapsulated within a housing. The housing normally has feed and permeate ports and, in the case of cross-flow filters, a retentate port. All of these ports may be used to control the flow parameters of fluid into and out of the housing and through the membrane.

Cartridge pressure drop: The differential pressure between cartridge inlet and outlet.

Cascade effects: A series of events that result from one initial cause.

Cassette: A device used for cross-flow filtration, typically in a rectangular form comprised of stacked flat sheets of membrane integrally bonded together. Most cassettes are typically designed to fit into a standard cassette holder where the feed, permeate, and any retentate ports mate with appropriate fittings on the cassette holders.

Catabolites: Waste products of catabolism, by which organisms convert substances into excreted compounds.

Cation-exchange chromatography: The form of ion-exchange chromatography that uses resins or packing with functional groups that can separate cations. A sulfonic acid would be an example of a strong cation-exchange group; a carboxylic acid would be a weak cation-exchange group.

CBE30: Changes being effected in 30 days

CBER: Food and Drug Administration Center for Biologics Evaluation and Research

CDC: Centers for Disease Control and Prevention

CDER: Food and Drug Administration Center for Drug Evaluation and Research

cDNA: Complementary DNA

CDRH: Center for Devices and Radiological Health

Cell bank: A cell bank is a collection of appropriate containers, whose contents are of uniform composition, stored under defined conditions. Each container represents an aliquot of a single pool of cells.

Cell culture: Cells taken from a living organism and grown under controlled conditions ("in culture"). Methods used to maintain cell lines or strains. The in vitro growth of cells isolated from multicellular organisms. These cells are usually of one type.

Cell differentiation: The process whereby descendants of a common parental cell achieve and maintain specialization of structure and function.

Cell fusion: The formation of a hybrid cell with nuclei and cytoplasm from different cells, produced by fusing two cells of the same or different species.

Cell harvesting: The process of concentrating (dewatering) the cell mass after fermentation. Cell slurries in excess of 70% wet cell weight are achievable. The cells may also be washed to prepare them for further processing, such as freezing or lysing. Unlike clarification processing, with cell harvesting, the cells are the target material.

Cell line: Type of cell population that originates by serial subculture of a primary cell population, which can be banked.

Cell substrate: Cells used to manufacture product.

Cells, type of: All cells are divided into two types: prokaryotic cells and eukaryotic cells. The basic component of living organisms is composed of cells with the ability to multiply. All cells contain cytoplasm, plasma membrane, and DNA.

Central dogma: The flow of genetic information is generally in the following direction: DNA > RNA > protein. This rule was dubbed the "central dogma," because it was thought that the same principle would apply to all organisms. However, we now know that for RNA viruses, the flow of genetic information starts from RNA.

cfu: Colony-forming unit

cGMP: The minimum requirements by law for the manufacture, processing, packaging, holding, or distribution of a material as established in Title 21 of the Code of Federal Regulations. Examples are Part 211 for Finished Pharmaceuticals, Part 606 for Blood and Blood Components, Part 820 for Medical Devices and Quality System Regulations.

Change management: A systematic approach to proposing, evaluating, approving, implementing, and reviewing changes (ICH Q10).

Channel height: The height of the path that the feed/retentate solution must pass through for a flat membrane cassette.

Channel length: The total length that the feed solution must travel along a flat cassette to reach the retentate outlet.

Chaperones: Proteins that help other proteins fold correctly, either by preventing them from folding incorrectly or by catalyzing their correct formation, used to maximize the usable protein produced by a variety of expression systems.

Characterization: Tests to determine the properties of a molecule or active substance, for example, molecular size/weight, chemical structure, and purity. These tests are also called physicochemical characterization.

Chemical compatibility or resistance: The ability of the components of a packed column or a filter to resist chemicals that can influence its performance. For example, some chemicals could cause the filter to shed particles, swell, or dissolve filter components. Repeatable performance requires that filters are resistant to all the chemicals that they are exposed to at a given concentration, temperature, and total exposure time.

Chemokines: Chemokines are the cytokines that may activate or chemoattract leukocytes. Each chemokine contains 65–120 amino acids, with molecular weight of 8–10 kDa. Their receptors belong to G-protein-coupled receptors. Since the entry of HIV into host cells requires chemokine receptors, their antagonists are being developed to treat AIDS.

Chemostat: A growth chamber that keeps a bacterial culture at a specific volume and rate of growth by limiting nutrient medium and removing spent culture.

Chemotaxis: Net-oriented movement in a concentration gradient of certain compounds. Various sugars and amino acids can serve as attractants, while some substances such as acid or alkali serve as repellants in microbial chemotaxis. White blood cells and macrophages demonstrate chemotactic movement in the presence of bacterial products, complement proteins, and antigen-activated T cells to contribute to the local inflammatory reaction and resistance to pathogens.

Chimeric transgene: A transgene that contains sequences derived from two different genes from two different species.

CHMP: Committee for Medicinal Products for Human Use (EMA).

Chromatogram: A plot of detector signal output versus time, elution volume, or column volume during the chromatographic process.

Chromatographic column: Vessel, typically in cylindrical shape including attached accessory parts like valves, for harboring the chromatography medium.

Chromosome: A long and complex DNA chain containing the genetic information (genes) of a cell. Prokaryotes contain only a single chromosome; eukaryotes have more than one, made up of a complex of DNA, RNA, and protein. The exact number of chromosomes is species-specific. Humans have 23 pairs.

Circular dichroism: With optical rotary dispersion, one of the optical spectrophotometric methods used to determine secondary structure and to quantitate the specific structure forms (a-helix, B-pleated sheet, and random coil) within a protein. The resultant spectra are compared to that of the natural protein form or to the reference standard for the recombinant.

Cistron: The smallest unit of genetic material that is responsible for the synthesis of a specific polypeptide.

Clean in place: A way to clean large vessels (tanks, piping, and associated equipment) without moving them or taking them apart, using a high-pressure rinsing treatment, sometimes followed by steam-in-place sanitization.

Clean room: A room in which the concentration of airborne particulate matter is controlled at specific limits to facilitate the manufacture of sterile and high-purity products. Clean rooms are classified according to the number of particles per volume of air.

Cleaning frequency: The number of chromatographic runs, after a cleaning cycle has to occur.

Clearance: Demonstrated removal according to specified parameters.

Clinical study or trial: Study with the objective of determining how a medicine is handled by, and affects, humans. Clinical studies or trials are conducted in healthy volunteers or in patients. Pivotal clinical studies involving a larger group of patients provide evidence on whether the medicine can be considered both safe and effective in a real clinical setting.

Clone: To duplicate exactly, whether a gene or a whole organism or an organism that is a genetically identical copy of another organism. A cell line stemming from a single ancestral cell and normally expressing all the same genes. If this is a B-lymphocyte clone, they will normally produce identical antibodies, that is, monoclonal antibodies.

Cloning vectors: Methods of transferring desired genes to organisms that will be used to express them. Cloning vectors are used to make recombinant organisms. Vector is an agent that can carry a DNA fragment into a host cell. If it is used for reproducing the DNA fragment, it is called a cloning vector. If it is used for expressing certain gene in the DNA fragment, it is called an "expression vector." Commonly used vectors include plasmid, lambda phage, cosmid, and yeast artificial chromosome. Plasmids are circular, double-stranded DNA molecules that exist in bacteria and in the nuclei of some eukaryotic cells. They can replicate independently of the host cell. The size of plasmids ranges from a few kb to near 100 kb. A typical plasmid vector contains a polylinker that can recognize several different restriction enzymes, an ampicillin-resistant gene (ampr) for selective amplification, and a replication origin for proliferation in the host cell. A plasmid vector is made from natural plasmids by removing unnecessary segments and adding essential sequences. To clone a DNA sample, the same restriction enzyme must be used to cut both the vector and the DNA sample. Therefore, a vector usually contains a sequence (polylinker) that can recognize several restriction enzymes so that the vector can be used for cloning a variety of DNA samples. A plasmid vector must also contain a drug-resistant gene for selective amplification. After the vector enters into a host cell, it may proliferate with the host cell. However, since the transformation efficiency of plasmids in *Escherichia coli* is very low, most *E. coli* cells that proliferate in the medium would not contain the plasmids. Therefore, we must find a way to allow only the transformed *E. coli* to proliferate. Typically, antibiotics are used to kill *E. coli* cells that do not contain the vectors. The transformed *E. coli* cells are protected by the ampr that can express the enzyme, □-lactamase, to inactivate the antibiotic ampicillin.

CMC: Chemistry, manufacturing, and controls

Codon: Group of three nucleotide bases in DNA or RNA that determines the composition of one amino acid in "building" a protein and also can code for chain termination.

Cohesion termini: DNA molecule with single-stranded ends with exposed (cohesive) complementary bases.

CoI: Circular of information

Column back pressure: The pressure of a chromatography system is measured right after the system pump. This pressure is called system back pressure.

The additional pressure if a column is attached to the system is called column back pressure.

Column dead volume (V_d): The volume outside of the column packing itself. The interstitial volume (intraparticle volume + interparticle volume) plus extra column volume (contributed by injector, detector, connecting tubing, and end fittings) all combine to create the dead volume. This volume can be determined by injecting an inert compound (i.e., a compound that does not interact with the column packing). Also abbreviated V_0 or V_m.

Column equilibration: To achieve a stable and equal distribution of a desired buffer, a column packed with a chromatography medium has to be run in the respective buffer to a point where pH, conductivity, and ultraviolet, measured at the column outlet, are identical to the respective values of the applied buffer.

Column performance: The performance of a column can be determined by checking the height equivalent to a theoretical plate and the asymmetry factor.

Comparability bridging study: A study performed to provide nonclinical or clinical data that allow extrapolation of the existing data from the drug product (DP) produced by the current process to the DP from the changed process.

Comparability exercise: The activities, including study design, conduct of studies, and evaluation of data, which are designed to investigate whether the products are comparable.

Comparable: A conclusion that products have highly similar quality attributes before and after manufacturing process changes and that no adverse impact on the safety or efficacy, including immunogenicity, of the drug product occurred. This conclusion can be based on an analysis of product quality attributes. In some cases, nonclinical or clinical data might contribute to the conclusion.

Complementary DNA (cDNA) library: The advantage of cDNA library is that it contains only the coding region of a genome. To prepare a cDNA library, the first step is to isolate the total mRNA from the cell type of interest. Because eukaryotic mRNAs consist of a poly-A tail, they can easily be separated. Then the enzyme reverse transcriptase is used to synthesize a DNA strand complementary to each mRNA molecule. After the single-stranded DNA molecules are converted into double-stranded DNA molecules by DNA polymerase, they are inserted into vectors and cloned.

Composite membrane: A membrane that is made up of two or more layers that are usually chemically or structurally different.

Concentrate: Also called retentate. The part of the process solution that does not pass through a cross-flow membrane filter.

Concentration: Cross-flow filtration process in which the components that do not pass through the membrane remain in the feed loop and increase in concentration.

Concentration factor: The concentration factor equals the ratio of the initial feed volume to retentate volume after separation. For example, if the initial flow volume is 1000 mL and the final retentate volume is 10 mL, the concentration factor is 10 times.

Concentration polarization: The buildup of molecules of dissolved substances (solutes) on the surface of the membrane filter during filtration. The concentration polarization layer increases resistance to filtrate flow and reduces the permeate flux, thus decreasing filtration efficiency.

Concurrent process validation: Establishing documented evidence that a process does what it purports to do based on information generated during actual implementation of the process.

Conductivity: Measurement of a substance's ability to conduct an electric current. Measured in siemens/cm. See also ionic strength.

Conjugated product: A conjugated product is made up of an active ingredient (e.g., peptide, carbohydrate), bound covalently or noncovalently to a carrier (e.g., protein, peptide, inorganic mineral), with the objective of improving the efficacy or stability of the product.

Conjugated proteins: Covalently bonded to prosthetic groups such as glycoprotein and metalloprotein.

Constitutive promoter: A DNA sequence that controls gene expression and is always available.

Contaminants: Any adventitiously introduced materials (e.g., chemical, biochemical, or microbial species) not intended to be part of the manufacturing process of the drug substance or drug product.

Continual improvement: Recurring activity to increase the ability to fulfill requirements (ISO 9000 2005).

Continuous cell line: A cell line having an infinite capacity for growth. Often referred to as "immortal" and previously referred to as "established."

Continuous fermentation mode: Continuous fermentation is an open system. Sterile nutrient solution is continuously introduced, while an equal amount of waste products are removed. Cell growth may or may not be adjusted: In a chemostat in a steady state, cell growth is controlled by adjusting the concentration of one substrate. In a turbidostat, cell growth is kept constant by using turbidity to monitor the biomass concentration, and the rate of feed of nutrient solution is appropriately adjusted. The nutrients involved can include carbohydrates, fatty acids, amino acids, and sources of nitrogen and sulfur. Sugars such as glucose, lactose, sucrose, and starch provide carbohydrates and nitrogen. Vitamins, minerals, or growth factors may be necessary for some microbe species. Stirring and mixing adds an air supply, removes carbon dioxide, and distributes nutrients, but an antifoam chemical agent is necessary to keep excess bubbles from forming. Fermentation is a multiphasic reaction in which gaseous components (N_2, O_2, and CO_2) must be mixed continuously with the liquid medium and solid cells. For optimal yields, the whole process must be carried out at a constant temperature.

Continuous mode: An open system of fermentation in which nutrient solution is continuously added and removed from the fermenter.

Continuous process verification: An alternative approach to process validation in which manufacturing process performance is continuously monitored and evaluated.

Control strategy: A planned set of controls, derived from current product and process understanding that ensures process performance and product quality. The controls can include parameters and attributes related to drug substance and drug product materials and components, facility and equipment operating conditions, in-process controls, finished product specifications, and the associated methods and frequency of monitoring and control (ICH Q10).

Corrective action: Action to eliminate the cause of a detected nonconformity or other undesirable situation (NOTE 2005). Corrective action is taken to prevent recurrence, whereas preventive action is taken to prevent occurrence (ISO 9000).

Cosmid (cosmid vector): An artificially constructed plasmid vector that contains a specific bacteriophage gene, which allows it to carry up to 45,000 base pairs of desired DNA. A vector that is similar to a plasmid, but it also contains the cohesive sites (cos site) of bacteriophage lambda to permit insertion of large fragments of DNA and in vitro packaging into a phage. The cosmid vector is a combination of the plasmid vector and the cos site which allows the target DNA to be inserted into the λ head.

Counterion: In an ion-exchange process, the ion in solution used to displace the ion of interest from the ionic site. In ion pairing, it is the ion of opposite charge added to the mobile phase to form a neutral ion pair in solution.

CP: Comparability protocol.

Creutzfeldt-Jakob disease: A disease affecting the human nervous system, believed to be caused by a prion that also causes bovine spongiform encephalopathy or "mad cow disease" in cattle.

Critical process parameter: A process parameter whose variability has an impact on a critical quality attribute and therefore should be monitored or controlled to ensure the process produces the desired quality.

Critical quality attribute: A physical, chemical, biological, or microbiological property or characteristic that should be within an appropriate limit, range, or distribution to ensure the desired product quality.

Cross-flow filtration (CFF): Also called tangential flow filtration. In CFF, the feed solution flows parallel to the surface of the membrane. Driven by pressure, some of the feed solution passes through the membrane filter. Most of the solution is circulated back to the feed tank. The movement of the feed solution across the membrane surface helps to remove the buildup of foulants on the surface.

Cross-flow rate: Also called retentate flow rate. The flow rate of solution that remains in the feed loop as measured in the retentate line.

Cross-linking: During the process of copolymerization of resins to form a 3D matrix, a difunctional monomer is added to form cross-linkages between adjacent polymer chains. The degree of cross-linking is determined by the amount of this monomer added to the reaction. For example, divinylbenzene is a typical cross-linking agent for polystyrene ion-exchange resins. The swelling and diffusion characteristics of a resin are governed by its degree of cross-linking.

Cross reaction: Antibodies against an antigen A can react with other antigens if the latter has one or more determinants in common with the determinants present on the antigen A or carry one or more determinants that are structurally very similar to the determinants present on antigen A.

Cryopreservation: Maintenance of frozen cells, usually in liquid nitrogen.

CSF: Colony-stimulating factor.

Cutoff: Nominally, the smallest entity that will pass through a separation device to become permeate (filtrate) larger particles (where retention is >90%) are thus "cut off" from the permeate. Actual cutoff values of any given device or lot of devices usually must be determined empirically. See also molecular weight cutoff and nominal molecular weight cutoff.

Cytokines: A protein that acts as a chemical messenger to stimulate cell migration, usually toward where the protein was released. Interleukins, lymphokines, and interferons are the most common. Small, nonimmunoglobulin proteins produced by monocytes and lymphocytes that serve as intercellular communicators after binding to specific receptors on the responding cells. Cytokines regulate a variety of biological activities. Cytokines regulate immunity, inflammation, apoptosis, and hematopoiesis.

Cytopathic: Damaging to cells, causing them to exhibit signs of disease.

Cytopathic effect: Morphological alterations of cell lines produced when cells are infected with a virus. Examples of cytopathic effects include cell rounding and clumping, fusion of cell membranes, enlargement or elongation of cells, or lysis of cells.

Cytoplasm: The protoplasm of a cell outside the nucleus (inside the nucleus is called nucleoplasm). Protoplasm is a semifluid, viscous, translucent mixture of water, proteins, lipids, carbohydrates, and inorganic salts found in all plant and animal cells.

Cytostat: Something that retards cellular activity and production. This can refer to cytostatic agents or to machinery, such as those that would freeze cells.

Cytotoxic: Damaging to cells.

Dalton: The unit of molecular weight, equal to the weight of a hydrogen atom.

Darcy's law: In 1856, a French hydraulic engineer named Henry Darcy published an equation for flow through a porous medium that today bears his name. $Q = KA(h_1 - h_2)/L$, where Q is the volumetric flow rate (m^3/s or ft^3/s), A is the flow area perpendicular to L (m^2 or ft^2), K is the hydraulic conductivity (m/s or ft/s), L is the flow path length (m or ft), h is the hydraulic head (m or ft), and $h_1 - h_2$ denotes the change in h over the path L.

Dead-ended filtration: Also called normal flow filtration. In dead-ended filtration, liquid flows perpendicular to the filtration media, and all of the feed passes through.

Decision maker(s): Person(s) with the competence and authority to make appropriate and timely quality risk management decisions.

Degassing: The process of removing dissolved gas from the mobile phase before or during use. Dissolved gas may come out of solution in the detector cell and cause baseline spikes and noise. Dissolved air can affect electrochemical detectors (by reaction) or fluorescence detectors (by quenching). Degassing is carried out by heating the solvent or by vacuum (in a vacuum flask) or online using evacuation of a tube made from a gas-permeable substance such as PTFE or by helium sparging.

Degradation products: Molecular variants resulting from changes in the desired product or product-related substances brought about over time and/or by the action of, for example, light, temperature, pH, and water, or by reaction with an excipient and/or the immediate container/closure system. Such changes may occur as a result of manufacture and/or storage (e.g., deamidation, oxidation, aggregation, and proteolysis). Degradation products may be either product-related substances or product-related impurities.

Denaturation: Unfolding of a protein molecule into a generally bioactive form. Also the *disruption* of DNA duplex into two separate strands.

Depth filter: A thick filter that captures contaminants within its structure. A membrane filter primarily captures contaminates on its surface.

Depyrogenate: The removal of pyrogens (lipopolysaccharides) from a process solution.

Desalting (see also diafiltration): Technique in which low-molecular-weight salts and other compounds are removed from nonionic and high-molecular-weight compounds. An example is the use of a reversed-phase packing to retain sample compounds by hydrophobic effects but to allow salts to pass through nonretained. Use of a size-exclusion chromatography column to exclude large molecules and retain lower-molecular-weight salts is another example.

Design space: The multidimensional combination and interaction of input variables (e.g., material attributes) and process parameters that have been demonstrated to provide assurance of quality. Working within the design space is not considered as a change. Movement out of the design space is considered to be a change and would normally initiate a regulatory postapproval change process. Design space is proposed by the applicant and is subject to regulatory assessment and approval.

Desired product: The protein that has the expected structure or the protein that is expected from the DNA sequence and anticipated posttranslational modification (including glycoforms) and from the intended downstream modification to produce an active biological molecule.

Desorption: The opposite of adsorption, that is, from the ion exchanger; essentially, two possibilities exist to desorb sample molecules: reducing the net charge by changing pH or adding a competing ion to "block" the charges on the ion exchanger.

Detectability: The ability to discover or determine the existence, presence, or fact of a hazard.

Developmentally regulated promoters: A DNA sequence that controls gene expression and is available only at certain times or stages.

Diafiltration: Diafiltration is a unit operation that incorporates ultrafiltration membranes to remove salts or other microsolutes from a solution. Small molecules are separated from a solution while retaining larger molecules in the retentate. Microsolutes are generally so easily washed through the membrane that for a fully permeated species about three volumes of diafiltration water will eliminate 95% of the microsolute.

Dialysis: Removal of small molecules from a solution of macromolecules by allowing them to diffuse through a semipermeable membrane into water or a buffer solution. This osmotic pressure separation method is controlled by the concentration gradient of salts across the membrane.

Differential pressure: In cross-flow filtration, the pressure drop along the cartridge between the feed (inlet) port and the retentate (outlet) port.

Diffusion: Movement of gas molecules caused by a concentration gradient.

Diploid cell line: A cell line having a finite in vitro life span in which the chromosomes are paired (euploid) and are structurally identical to those of the species from which they were derived.

Direct flow filtration: Filtration process where the entire feed stream flows through the filter's media. Also referred to as normal flow filtration and dead-end filtration.

DNA (deoxyribonucleic acid): The nucleic acid based on deoxyribose (a sugar) and the nucleotides G, A, T, and C. Occurring in a corkscrew-ladder shape, it is the primary component of chromosomes, which thus carry inheritable characteristics of life. The basic biochemical component of the chromosomes and the support of heredity. DNA contains the sugar deoxyribose and is the nucleic acid in which genetic information is stored (apart from some viruses). A DNA molecule has two strands, held together by the hydrogen bonding between their bases. Due to the specific base pairing, DNA's two strands are complementary to each other. If we know the sequence of one strand, we can deduce the sequence of another strand. For this reason, a DNA database needs to store only the sequence of one strand. By convention, the sequence in a DNA database refers to the sequence of the 5′–3′ strand (left to right). In a DNA molecule, the two strands are not parallel, but intertwined with each other. Each strand looks like a helix. The two strands form a "double helix" structure, which was first discovered by James D. Watson and Francis Crick in 1953. In this structure, also known as the B form, the helix makes a turn every 3.4 nm, and the distance between two neighboring base pairs is 0.34 nm. Hence, there are about 10 pairs per turn. The intertwined strands make two grooves of different widths, referred to as the major groove and the minor groove, which may facilitate binding with specific proteins. In a solution with higher salt concentrations or with alcohol added, the DNA structure may change to an A form, which is still right handed, but every

521

2.3 nm makes a turn, and there are 11 base pairs per turn. Another DNA structure is called the Z form, because its bases seem to zigzag. Z DNA is left handed. One turn spans 4.6 nm, comprising 12 base pairs. The DNA molecule with alternating G–C sequences in alcohol or high salt solution tends to have such structure.

DNA cloning: Production of many identical copies of a defined DNA fragment. DNA cloning is a technique for reproducing DNA fragments. It can be achieved by either cell-based or polymerase chain reaction–based technique. In the cell-based approach, a vector is required to carry the DNA fragment of interest into the host cell.

DNA fingerprinting: Sequences of nucleic acids in specific areas (loci) on a DNA molecule are polymorphic, meaning that the genes in those locations may differ from person to person. DNA fragments can be cut from those sequences using restriction enzymes. Fragments from various samples can be analyzed to determine whether they are from the same person. The technique of analyzing restriction fragment length polymorphism is called DNA typing or DNA fingerprinting.

DNA hybridization (dot blot) analysis: Detection of DNA to the nanogram level using hybridization of cellular DNA with specific DNA probes. Manifestation can be by 32P labeling, chemiluminescence, chromogenic or avidin–biotin assays.

DNA library: Set of cloned DNA fragments that together represent the entire genome or the transcription of a particular tissue.

DNA polymerase: An enzyme that catalyzes the synthesis of double-stranded DNA from single-stranded DNA.

DNA replication: DNA molecules are synthesized by DNA polymerases from deoxyribonucleoside triphosphate (dNTP). The chemical reaction is similar to the synthesis of RNA strands. Both DNA and RNA polymerases can extend nucleic acid strands only in the 5′–3′ direction. However, the two strands in a DNA molecule are antiparallel. Therefore, only one strand (leading strand) can be synthesized continuously by the DNA polymerase. The other strand (lagging strand) is synthesized segment by segment. DNA replication is triggered by the expression of all required proteins, such as DNA polymerase, DNA primase, and cyclin. In yeast, the transcription factor regulating the expression of these proteins is called master cell bank–binding factor. In mammals, the corresponding transcription factor is E2F.

DNA screening: Once a particular DNA fragment is identified, it can be isolated and amplified to determine its sequence. If we know the partial sequence of a gene and want to determine its entire sequence, the probe should contain the known sequence so that the detected DNA fragment may contain the gene of interest.

DNA sequencing: The Sanger method being used today was pioneered by Fred Sanger in the 1970s. It is also known as the dideoxy method, because a small amount of dideoxynucleotides are mixed with normal deoxynucleotides during sequencing. The dideoxynucleotides lack both 2′ and 3′ hydroxyl groups, while the deoxynucleotides lack only the 2′ hydroxyl group. The 3′ hydroxyl group is essential for forming the phosphodiester bond that connects two nucleotides. Therefore, the dideoxynucleotide will be the terminator of a polynucleotide chain since it lacks the essential 3′ hydroxyl group. The Sanger method may sequence a DNA fragment containing up to 500 nucleotides. For large-scale sequencing (such as the entire human genome), a strategy known as the shotgun sequencing is commonly used. In this approach, the DNA molecule of interest is randomly chopped into numerous small pieces. After these small pieces are sequenced, the whole sequence is assembled by their common overlaps.

DNA synthesis: The formation of DNA by the sequential addition of nucleotide bases.

DNA vaccine: A nucleic acid vaccine where genes coding for specific antigenic proteins are injected to produce those antigens and trigger an immune response.

DNase: An enzyme that produces single-stranded nicks in DNA. DNase is used in nick translation.

Downstream processing: Starting with a feed stream free of cells and cell debris, the purification sequences involving chromatography and membrane separations to achieve final product purity. Capture, intermediary purification or polishing comprising the purification part of the process.

Drain valve: Valve for draining off material in a filter housing usually at the lowest point.

Drug master file: A submission to Food and Drug Administration that can be used to provide detailed information about facilities or articles used in the manufacturing, processing, packaging, and storing of one or more human drugs.

Drug product: A pharmaceutical product type that contains a drug substance, generally in association with excipients.

Drug substance: The material that is subsequently formulated with excipients to produce the drug product. It can be composed of the desired product, product-related substances, and product- and process-related impurities. It may also contain excipients including other components, such as buffers.

Dynamic binding capacity: Dynamic capacity describes the amount of sample that will bind to a gel packed in a column run under defined conditions. The dynamic capacity for any media is highly dependent on running conditions, sample preparation, and even origin of the sample. In general, the lower the flow rates, the higher the dynamic capacity. As the flow rate approaches zero, the dynamic capacity approaches the available capacity. Dynamic binding capacities are determined by loading a sample containing a known concentration of the target molecule and monitoring for the molecule in the column flow through while applying sample.

Edman degradation: A type of protein sequencing from the amino-terminus.

EEA: European Economic Area.

Effective area: In a membrane separation device, the active area of the membrane exposed to flow.

Efficacy: The ability of a substance (such as a protein therapeutic) to produce a desired clinical effect: its strength and effectiveness.

Efficiency: Description of the peak width and shape. The efficiency of a column is usually expressed as a plate number and an asymmetry factor.

Effluent: The stream of fluid leaving the filter.

Electrophoresis: Methods in which molecules or molecular complexes are separated on the basis of their relative ability to migrate when placed in an electric field. An analyte is placed on an electrophoretic support and then separated by charge (isoelectric focusing) or by molecular weight (sodium dodecyl sulfate polyacrylamide gel electrophoresis). Visualization is accomplished by staining of the protein with nonselective (Coomassie blue) or selective (silver) staining techniques. The dye-binding method using Coomassie blue is a quantifiable technique when a laser densitometer is used to read the gels. The silver stain method is much more sensitive

523

and therefore used for detection of low levels of protein impurities, but due to variability of staining from protein to protein, it cannot be used for quantitation.

ELISA: Enzyme-linked immunosorbent assay, a test to measure the concentration of antigens or antibodies.

Elongation: A ribosome contains two major tRNA-binding sites: A site and P site. After the large subunit joins the initiation complex, the initial Met-tRNAiMet enters the P site and the newly arrived aminoacyl-tRNA is always placed at the A site ("A" for "aminoacyl"). Then, methionine is transferred to the new aminoacyl-tRNA, forming a "peptidyl-tRNA" where a peptide is attached to the tRNA. Subsequently, the empty tRNA at the P site is ejected from the ribosome and the peptidyl-tRNA jumps to the P site ("P" for "peptidyl"). During this translocation step, the ribosome also moves one codon down the mRNA chain. Similar steps are repeated in the next cycles of elongation.

Eluate: Combination of mobile phase and solute exiting column, also called effluent.

Elution: The removal of adsorbed material from an adsorbent such as the removal of a product from an enzyme bound on a column.

EMA: European Medicines Agency.

Enabler: A tool or process that provides the means to achieve an objective (ICH Q10).

Endogenous: Growing or developing from a cell or organism or arising from causes within the organism.

Endogenous pyrogen assay: An in vitro assay based on the release of endogenous pyrogen produced by endotoxin from human monocytes. This assay appears to be more sensitive than the USP rabbit pyrogen test, but is much less sensitive than the limulus amoebocyte lysate assay. It does have the advantage that it can detect all substances that cause a pyrogenic response from human monocytes.

Endogenous virus: Viral entity whose genome is part of the germ line of the species of origin of the cell line and is covalently integrated into the genome of animal from which the parental cell line was derived. For the purposes of this document, intentionally introduced, nonintegrated viruses such as EBV used to immortalize cell substrates or bovine papillomavirus fit in this category.

Endonuclease: A restriction enzyme that breaks up nucleic acid molecules at specific sites along their length. Such enzymes are naturally produced by microorganisms as a defense against foreign nucleic acids.

Endonucleases: Enzymes that cleave bonds within nucleic acid molecules.

Endoplasmic reticulum: A highly specialized and complex network of branching, interconnecting tubules (surrounded by membranes) found in the cytoplasm of most animal and plant cells. The rough endoplasmic reticulum (ER) is where ribosomes make proteins. It appears "rough" because it is covered with ribosomes. The smooth ER is the site for synthesis and metabolism of lipids, and it is involved in detoxifying chemicals such as drugs and pesticides.

Endotoxin: The outer cell wall of gram-negative bacteria. A poison in the form of a fat/sugar complex (lipopolysaccharide) that forms a part of the cell wall of some types of bacteria. It is released only when the cell is ruptured and can cause septic shock and tissue damage. Pharmaceuticals are tested routinely for endotoxins. A heat-stable lipopolysaccharide associated with the outer membrane of certain gram-negative bacteria. It is not secreted and is released only when the cells are disrupted.

Enhancers: Enhancers are the positive regulatory elements located either upstream or downstream of the transcriptional initiation site. However, most of them are located upstream. In prokaryotes, enhancers are quite close to the promoter, but eukaryotic enhancers could be far from the promoter. An enhancer region may contain one or more elements recognized by transcriptional activators.

Enzyme-linked immunosorbent assay: A multiantigen test for unknown residual (host) cellular protein and confirmation of desired protein. It may be used to determine the potency of a product. It is extremely specific and sensitive, basically simple, and inexpensive. It requires a reference standard preparation of host cell protein impurities to serve as an immunogen for preparation of polyclonal antibodies used for the assay.

Enzymes: Proteins that catalyze biochemical reactions by causing or speeding up reactions without being changed in the process themselves. Enzymes are the catalysts of biochemical reactions in the cell and include such examples as oxidoreducatases, transferases, hydrolases, proteases, nucleases, phosphatases, lyases, isomerases, and ligases.

Epithelium (epithelial): The layer(s) of cells between an organism or its tissues or organs and their environment (skin cells, inner linings of the lungs or digestive organs, outer linings of the kidneys, and so on).

Ethylene oxide (EtO) sterilization: A sterilization process still common for biomedical products, in which product is subjected to steam and highly toxic ethylene oxide gas. Because many materials change properties when exposed to moisture and EtO by-products, products destined for this process must be specially engineered for EtO sterilization.

Eudravigilance: Data processing network and management system for reporting and evaluating suspected adverse reactions during the development and following the marketing authorization of medicinal products in the European Economic Area.

Eukaryote: An organism whose cells contain a membrane-bound nucleus. Sometimes spelled eukaryote. The eukaryotic cell contains a nucleus. Eukaryotes include Protista, fungi, animals, and plants. The eukaryotic cell contains organelles, which are defined as membrane-bound structures such as nucleus, mitochondria, chloroplasts, endoplasmic reticulum, Golgi apparatus, lysosomes, vacuoles, and peroxisomes. For animal cells, the cell surface consists of the plasma membrane only, but plant cells have an additional layer called cell wall, which is made up of cellulose and other polymers. All biological membranes, including plasma membranes and all organelle membranes, contain lipids and proteins. The lipids found in biomembranes are mainly phospholipids and cholesterol. In the plasma membrane and some of organelle membranes, proteins and phospholipids are attached to carbohydrates, forming glycoproteins and glycolipids, respectively.

Excipient: An ingredient added intentionally to the drug substance that should not have pharmacological properties in the quantity used.

Exclusion limit: In size-exclusion chromatography (SEC), the upper limit of molecular weight (or size), beyond which molecules will elute at the same retention volume, called the exclusion volume. Many SEC packings are referred to by their exclusion limit.

Exclusion volume (V_e): The retention volume of a molecule on a size-exclusion chromatography packing; all molecules larger than the size of the largest pore are totally excluded and elute at the interstitial volume of the column.

Exogenous: Developing from outside, originating externally. Exogenous factors can be external factors such as food and light that affect an organism.

Exonucleases: Enzymes that catalyze the removal of nucleotides from the ends of a DNA molecule.

Expanded-bed adsorption (EBA) mode: Expanded-bed mode is a single-pass operation in which desired proteins are purified from crude, containing feedstock containing particles without the need for separate clarification, concentration, and initial purification. The expansion of the adsorbent bed creates a distance between the adsorbent particles, that is, increased void-age (void volume fraction) in the bed, which allows for unhindered passage of cells, cell debris, and other particles during application of crude feed to the column. EBA is a unit operation that uses adsorbents and columns for recovering proteins directly from crude feedstock. The system of EBA comprises several distinct steps. First, the settled bed is expanded by the upward liquid flow of equilibration buffer; the crude feed, a mixture of soluble proteins, contaminants, cells, and cell debris, passes upward through the expanded bed. Target proteins are captured on the adsorbent media, while particles and contaminants pass through the expanded bed. Loosely bound material is washed out with the upward flow of buffer; a change to elution buffer while maintaining the upward flow desorbs the target protein in expanded-bed mode. Alternatively, if flow is reversed, the adsorbent particles will quickly settle, and the proteins can be desorbed by an elution buffer as in conventional packed-bed chromatography. The mode used for elution, expanded bed, or settled bed depends on the characteristics of the feed. After elution, the adsorbent is cleaned by a predefined cleaning-in-place solution, followed by either regeneration with application buffer for further use or equilibration with storage solution.

Express: To translate a cell's genetic information, stored in its DNA (gene), into a specific protein.

Expression construct: The expression vector that contains the coding sequence of the recombinant protein and the elements necessary for its expression.

Expression system: Organisms chosen to manufacture (by expression) a protein of interest through recombinant DNA technology.

Expression vector: A way of delivering foreign genes to a host, creating a recombinant organism that will express the desired protein.

Extractables: Substances that may dissolve or leach from a membrane during filtration and contaminate the process solution. For example, the leachates might include wetting agents in the membrane, membrane-cleaning solutions, or substances from the materials used to encase the membrane.

FACS: Fluorescence-activated cell sorter.

FBS: Fetal bovine serum.

FDA: Food and Drug Administration.

Fed-batch mode: Fermentation in which substrate is added incrementally throughout the process.

Feed: Material or solution that you apply on chromatography column or introduce into a membrane separation system device.

Feed pressure: The pressure measured at the feed port of a separation device such as a chromatographic column, a cartridge, or cassette.

Feed stream flow rate: Volumetric flow rate of the feed. Measured in volume per unit time.

Feedback: The modification or control of a process or system by its results or effects.

Feedforward: The modification or control of a process using its anticipated results or effects (*Oxford Dictionary of English*, Oxford University Press, Oxford, U.K., 2003). Feedback/feedforward can be applied technically in process control strategies and conceptually in quality management (ICH Q10).

Fermentation: An anaerobic bioprocess, fermentation is used in various industrial processes for the manufacture of products such as alcohols, acids, and cheese by the action of yeasts, molds, and bacteria. The fermentation process is used also in the production of monoclonal antibodies. Whereas the use of fermentation goes back several millenniums, its use as an expression system, however, began in 1973 with the first genetic engineering experiment: a gene from the African clawed toad was inserted into laboratory *Escherichia coli* bacteria. Besides the familiar alcoholic beverages, products made by microbial fermentation (some recombinant, others not) are far ranging: from ethanol, dyes, and other chemicals to enzymes, foods and food additives, vitamins, soy products, vaccines for animals and people, antibiotics and antifungal agents, steroids, diagnostic agents, and enzyme inhibitors. Bacteria and yeast grow fast in low-cost media; offer high expression levels of the proteins they can make, which they sometimes secrete into their circulating medium; and both can withstand rough treatment compared with animal cells. Fermentation of recombinant cell lines begins with genetic engineering of microbes classified "generally recognized as safe," such as *E. coli*, *Bacillus subtilis*, *Streptomyces* species, and *Saccharomyces cerevisiae* and *Schizosaccharomyces pombe* yeasts.

Fermentation: Chemical reactions induced by living organisms (or enzymes derived from living organisms) to produce raw material for pharmaceutical products.

Fermenter: A bioreactor used to grow bacteria or yeasts in liquid culture. See fermentation or fermentation technology.

Fiber: More correctly called hollow-fiber membrane.

Filter area: The surface area of filter media inside a separation device.

Filter efficiency: Filter efficiency represents the percentage of particles that are removed from the fluid by the filter.

Filtrate: Also called permeate. The portion of the process fluid that passes through the membrane.

Filtrate flow rate: The instantaneous volume per unit time of filtrate produced by a system, typically measured on a filtrate flow meter.

Filtration: Removal of particles, normally solids, from a fluid. These can be contaminants or valuable products.

Filtration efficiency: A filter's ability to remove particles of the specified size, expressed as a percentage or as a beta ratio.

Flanking control regions: Noncoding nucleotide sequences that are adjacent to the 5′ and 3′ end of the coding sequence of the product that contains important elements that affect the transcription, translation, or stability of the coding sequence. These regions include promoter, enhancer, and splicing sequences and do not include origins of replication and antibiotic resistance genes.

Floc: A fluffy aggregate that resembles a woolly cloud.

Flow path length: The total length that a feed solution travels from inlet to outlet. Flow path length is an important parameter to consider when doing any process development, system design, or scale-up or scale-down experiments. The flow path length and other fluid channel geometries such as lumen diameter or channel height can impact the fluid dynamics of the system and will directly impact pump requirements and differential pressure of the filtration step.

Flow rate: The volumetric or linear rate of flow of mobile phase through a chromatography column.

Flux: Flux represents the volume of solution flowing through a given membrane area during a given time. Expressed as LMH (liters per square meter per hour).

Flux rate in LMH: Flux rate in LMH = {permeate flow (mL/min) ÷ cassette surface area (m2)} × 0.06

Form FDA 2567: Transmittal of Labels and Circulars.

Form FDA 356h: Application to Market a New Drug, Biologic or an Antibiotic Drug for Human Use.

Form FDA 3674: Certification of Compliance, under 42 U.S.C. § 282(j)(5)(B), with Requirements of ClinicalTrials.gov Data Bank (42 U.S.C. § 282(j)).

Formal experimental design: A structured, organized method for determining the relationship between factors affecting a process and the output of that process. Also known as "design of experiments."

Formulation: Conversion of drug substance (the API) to drug product.

Fouling: Accumulation of material on the surface of the chromatography media or on the membrane that can slow and alter the process.

Fractionation: Separation of molecules in a solution based on differences in the properties of the molecules.

Frit: The porous element at either end of a column that serves to contain the column packing. It is placed at the very ends of the column tube or, more commonly, in the end fitting. Frits are made from stainless steel or other inert metal or plastic, such as porous PTFE or polypropylene.

FSH: Follicle-stimulating hormone.

Fungal fermentation: Fungal fermentation can manufacture many recombinant therapeutic proteins. Yeasts offer certain advantages over bacteria as a recombinant expression system such as their ability as eukaryotic organisms to perform certain posttranslational modifications on the proteins they make. *Saccharomyces cerevisiae* (baker's or brewer's yeast) is the fungal species most commonly found in fermentation processes, recombinant or not. It is the one with which people have the most experience (thousands of years' worth) and thus the best understanding (the full genome was sequenced in 1996). *S. cerevisiae* has been genetically engineered to produce a wide range of proteins: antigens to hepatitis B, influenza, and polio, human growth hormone and insulin, antibodies and antibody fragments, human growth factors, interferon and interleukin, blood components such as human serum albumin,; and tissue plasminogen activator. Other species that have been studied include *Schizosaccharomyces pombe* and *Pichia pastoris*. In yeast, the expression of recombinant proteins to at least 150 mg/L in shake flasks. Coexpression of a necessary enzyme has helped some yeasts produce correctly folded collagen molecules. Some heterologous proteins can be lethal to the yeast cells that make them—but a methanol-induced promoter has been developed to meet that challenge. Once the yeast culture has reached a certain cell density, changing its growth medium to methanol induces expression of the therapeutic protein in large amounts (effectively ending the batch). Similar inducible promoter genes have been used with bacteria and *S. cerevisiae*, but *P. pastoris* only works with methanol, and it is the only one that does.

Fusion of protoplast: Fusion of two cells whose walls have been eliminated, making it possible to redistribute the genetic heritage of microorganisms.

Fusion partner: The gene for a protein that can be joined to the gene for a medically useful protein to optimize production in bacterial fermentation expression systems. When making a small protein or peptide in *Escherichia coli*, it is often necessary to produce the protein fused to a larger protein to get high levels of stable expression. The resulting fusion protein must be cleaved (chemically or enzymatically broken) to yield the desired protein or peptide. The nonproduct fusion partner is left over and usually thrown away.

Gamma sterilization: A type of sterilization process accomplished by bombarding the object to be sterilized with electron beam, x-ray, or ^{60}Co or ^{137}Cs irradiators. All generate forms of gamma rays, radiant energy at short wavelength (0.1 nm or less). The governing standard is ISO 11137-Sterilization of Healthcare Products—Requirements for Validation and Routine Control—Radiation Sterilization. Because some product materials can be adversely affected by gamma radiation, objects destined for gamma sterilization must be engineered specifically for this process.

Gaucher's disease: A rare inherited disorder of metabolism; people with this disease do not have enough of a specific enzyme called glucocerebrosidase and may be treated with enzyme replacement therapy.

Gel electrophoresis: Gel electrophoresis is a technique for separating charged molecules with different sizes. Two kinds of gels are commonly used: agarose and polyacrylamide. Agarose gels can be applied to a wider range of sizes than polyacrylamide gels. By using standard agarose electrophoresis, nucleic acids up to 50 kb may be separated. If pulsed field gel electrophoresis is used, the upper limit can be extended to 10 Mb. Polyacrylamide gels may separate nucleic acids that differ in length by only one nucleotide if their length is less than 500 bp. In a gel (either agarose or polyacrylamide), the negatively charged DNA fragments move toward the positive electrode at a rate inversely proportional to their length. After the electric field is applied for a certain period, DNA fragments with different lengths will be separated, which can be visualized by autoradiography or by treatment with a fluorescent dye (e.g., ethidium bromide). The relationship between the size of a DNA fragment and the distance it migrates in the gel is logarithmic. Therefore, from the band positions, the lengths of DNA fragments can be determined. Also used is a 2D gel electrophoresis.

Gel filtration: A separation method based on the molecular size or the hydrodynamic volume of the components being separated. This can be accomplished with the proteins in their natural state or denatured with detergents. Also called size-exclusion chromatography.

Gel layer: During the filtration process, the thin layer of particles or molecules that may build up at the membrane surface. It is also referred to as the concentration polarization layer. High *trans* membrane pressure can lead to an increase in the thickness of the gel layer and negatively impact the filtration process by reducing flux and inhibiting passage though the membrane.

Gene: The basic unit of heredity, which plays a part in the expression of a specific characteristic. The expression of a gene is the mechanism by which the genetic information that it contains is transcribed and translated to obtain a protein. A gene is a part of the DNA molecule that directs the synthesis of a specific polypeptide chain. It is composed of many codons. When the gene is considered as a unit of function in this way, the term cistron is often used. By definition, a gene includes the entire nucleic acid sequence necessary for the expression of its product (peptide or RNA). Such sequence may be divided into regulatory region and transcriptional region. The regulatory region could be near or far from the transcriptional region. The transcriptional region consists of exons and introns. Exons encode a peptide or functional RNA. Introns will be removed after transcription. "Gene family" refers to a set of genes with homologous sequences.

Gene expression: An organism may contain many types of somatic cells, each with distinct shape and function. However, they all have the same genome. The genes in a genome do not have any effect on cellular functions until they are "expressed." Different types of cells express different sets of genes, thereby exhibiting various shapes and functions. The essential steps involved in expression of protein genes are transcription where a DNA strand is used as the template to synthesize a RNA strand, which is called the primary transcript; RNA processing that modifies the primary transcript to generate a mature mRNA (for protein genes) or a functional tRNA or rRNA; for RNA genes (tRNA and rRNA), the expression is complete after a functional tRNA or rRNA is generated. However, protein genes require additional steps of nuclear transport, where mRNA is transported from the nucleus to the cytoplasm for protein synthesis and finally protein synthesis in the cytoplasm where mRNA binds to ribosomes, and synthesize a polypeptide based on the sequence of mRNA.

Gene regulatory elements: Transcriptional regulation is mediated by the interaction between transcription factors and their DNA binding sites that are the *cis*-acting elements, whereas the sequences encoding transcription factors are *trans*-acting elements. The *cis*-acting elements may be divided into the following four types, that is, promoters, enhancers, silencers, and response elements. The transcription region consists of exons and introns. The regulatory elements include promoter, response element, enhancer, and silencer. Downstream refers to the direction of transcription and upstream is opposite to the transcription direction. The numbering of base pairs in the promoter region is as follows. The number increases along the direction of transcription, with "+1" assigned for the initiation site. There is no "0" position. The base pair just upstream of +1 is numbered "−1", not "0."

Gene transfer: The use of genetic or physical manipulation to introduce foreign genes into a host cells to achieve desired characteristics in progeny.

Generic medicine: Medicine that has the same composition in active substance(s) and the same pharmaceutical form as the originator reference medicine, and whose bioequivalence with the originator reference medicine (i.e., the same behavior in the body) has been demonstrated by appropriate bioequivalence studies.

Genetic code: Protein synthesis is based on the sequence of mRNA, which is made up of nucleotides, while proteins are made up of amino acids. There must be a specific relationship between the nucleotide sequence and amino acid sequence. This relationship is the so-called genetic code, which was deciphered by Marshall Nirenberg and his colleagues in the early 1960s. One of their approaches is shown as follows. Three nucleotides (a codon) code for one amino acid. Synthesis of a peptide always starts from methionine (Met), coded by AUG. The stop codon (UAA, UAG, or UGA) signals the end of a peptide. This applies to mRNA sequences. For DNA, U (uracil) is replaced by T (thymine). In a DNA molecule, the sequence from an initiating codon (ATG) to a stop codon (TAA, TAG, or TGA) is called an open reading frame, which is likely (but not always) to encode a protein or polypeptide. The genetic code is not randomly assigned. If an amino acid is coded by several codons, they often share the same sequence in the first two positions and differ in the third position. Such assignment is accomplished by the design of wobble position, but "the evolutionary dynamic that shaped the code remains a mystery." The standard genetic code applies to most, but not all, cases. Exceptions have been found in the mitochondrial DNA of many organisms and in the nuclear DNA of a few lower organisms.

Genetic engineering: A technique used to modify the genetic information in a living cell, reprogramming it for a desired purpose (such as the production of a substance it would not naturally produce). Altering the genetic structure of an organism (adding foreign genes, removing native genes, or both) through technological means rather than traditional breeding.

Genome: All the genes carried by a cell. "Genome" is the total genetic information of an organism. For most organisms, it is the complete DNA sequence. For RNA viruses, the genome is the complete RNA sequence, since their genetic information is encoded in RNA. The genomes of prominent organisms are given as follows:

Genome Size	Organism	Genome Size (Mb): 1 Mb = 1 Million Base Pairs (for Double-Stranded DNA or RNA) or 1 Million Bases (for Single-Stranded DNA or RNA)	Gene Number
Hepatitis D virus	0.0017	1	
Hepatitis B virus	0.0032	4	
HIV-1	0.0092	9	
Bacteriophage λ	0.0485	80	
E. coli	4.6392	4,400	
S. cerevisiae (**yeast**)	12.155	6,300	
C. elegans (**nematode**)	97	19,000	
D. melanogaster (**fruit fly**)	137	13,600	
M. musculus (**mouse**)	3000	30,000–70,000	
Homo sapiens (**human**)	3000	30,000–70,000	

Genomic library: The genomic library is normally made by λ phage vectors, instead of plasmid vectors because the entire human genome is about 3×10^9 bp long, while a plasmid or λ phage vector may carry up to 20 kb fragment. This would require 1.5×10^5 recombinant plasmids or λ phages. When plating *Escherichia coli* colonies on a 3 in. petri dish, the maximum number to allow isolation of individual colonies is about 200 colonies per dish. Thus, at least 700 petri dishes are required to construct a human genomic library. By contrast, as many as 5×10^4 λ phage plagues can be screened on a typical petri dish. This requires only 30 petri dishes to construct a human genomic library. Another advantage of λ phage vector is that its transformation efficiency is about 1000 times higher than the plasmid vector. Preparation of the genomic library using λ phage vectors. It is basically the cloning of all DNA fragments representing the entire genome.

Genotype: The genetic composition of an organism (including expressed and nonexpressed genes), which may not be readily apparent.

Germ cell: The "sex cells" in higher animals and plants that carry only half of the organism's genetic material and can combine to develop into new living things.

GLP: Good laboratory practice, regulations issued by the Food and Drug Administration describing practices for conducting nonclinical laboratory studies.

531

Glycosylation: The type and length of any sugar or carbohydrate groups attached to a given molecule. The addition of one or more oligosaccharide groups to a protein. The covalent attachment of sugars to an amino acid in the protein portion of a glycoprotein. Adding one or more carbohydrate molecules onto a protein (a glycoprotein) after it has been built by the ribosome; a posttranslational modification.

GM-CSF: Granulocyte–macrophage colony-stimulating factor

GMPs: Good manufacturing practices required by Food and Drug Administration regulations. cGMP stands for current good manufacturing practices.

Golgi body: A cell organelle consisting of stacked membranes where posttranslational modifications of proteins are performed; also called Golgi apparatus.

Gradient elution: Technique for the separation of molecules by increasing mobile phase strength (i.e., conductivity) over time during the chromatographic separation. Gradients can be continuous or stepwise.

Growth hormone: A protein produced in the pituitary gland to control cell growth.

GVHD: Graft-versus-host disease

Hapten: A low-molecular-weight substance that alone can react with its corresponding antibody. In order to be immunogenic, haptens are bonded to molecules having molecular weights greater than 5000. An example would be the hapten digoxin covalently bonded to bovine serum albumin, forming the digoxin-bovine serum albumin (BSA) immunogen.

Harm: Damage to health, including the damage that can occur from loss of product quality or availability.

Harvesting: Separation of raw biological material from cell culture.

Hazard: The potential source of harm (ISO/IEC Guide 51).

Hemocytometer: A device for counting blood cells.

HETP: Height equivalent to a theoretical plate. A carryover from distillation theory: a measure of a column's efficiency. For a typical high-performance liquid chromatography column well packed with 5 µm particles, HETP (or H) values are usually between 0.01 and 0.03 mm. HETP = L/N, where L is the column length and N is the number of theoretical plates.

High-affinity antibody: Antibodies with a high affinity for antigen. These antibodies are predominantly IgG and produced during a secondary response to antigen. Cells producing a high affinity antibody can be triggered by low concentration of antigen.

High-performance liquid chromatography (HPLC): A separation technique that uses small particle size, narrow-bore columns, and high inlet pressures to achieve separation in short periods of time with high resolution. Any form of column chromatography that uses a liquid mobile phase can be extended to HPLC. An instrumental separation technique used to characterize or to determine the purity of a biological drug product by passing the product (or its component peptides or amino acids) in liquid form over a chromatographic column containing a solid support matrix. The mode of separation, that is, reversed phase, ion exchange, gel filtration, or hydrophobic interaction, is determined by the column matrix and the mobile phase. Detection is usually by ultraviolet absorbance or by electrochemical means.

HLA: Human leukocyte antigen

Holdup volume: Quantity of fluid remaining within the filtration media after draining the system.

Hollow fiber: The tubelike structure made from a membrane and sealed inside a cross-flow cartridge where the feed stream flows into the inner diameter of one end of the hollow fiber and the retentate (the material that does not permeate through the walls of the hollow fiber) flows out to the other end. The material that passes through the membrane (walls of the hollow fiber) is called permeate. In hollow-fiber bioreactor cell culture system, cells are separated from the medium using semipermeable membranes arranged into hollow fibers.

Hormone: A protein released by an endocrine gland to travel in the blood and act on tissues at another location in the body.

Host: A cell whose metabolism is used for the growth and reproduction of a virus, plasmid, or other form of foreign DNA.

Host-related impurities: Impurities related to the culturing of cells (e.g., cell debris, nucleic acids, host cell proteins, cell culture media components, and endotoxins) or transgenic milk components.

Housing: The mechanical structure that surrounds and supports the membrane or filter element. The housing normally has feed, retentate, and permeate ports that direct the flow of process fluids into and out of the filter assembly.

HPLC: High-performance liquid chromatography or high-pressure liquid chromatography, a commonly used method for separating liquid mixtures.

HRSA: Health Research Services Administration.

HSV: Herpes simplex virus.

Hybridoma: An immortalized cell line (usually derived by fusing B-lymphocyte cells with myeloma tumor cells) that secretes desirable antibodies.

Hybridoma technology: Fusion between an antibody forming cell (lymphocyte) and a malignant myeloma cell ("immortal"), which will result in a continuously growing cell clone (hybridoma), which can produce antibodies of a single specificity. Hybrid cells made by combining tumor cells and plasma cells; the combination of normal B lymphocytes and myeloma cells is commonly used in cell culture expressions systems to produce monoclonal antibodies.

Hydrophilic: "Water loving." Refers both to stationary phases that are compatible with water and to water-soluble molecules in general. Most chromatography media used to separate proteins are hydrophilic in nature and should not sorb or denature protein in the aqueous environment.

Hydrophobic: "Water hating." Refers both to stationary phases that are not compatible with water and to molecules in general that have little affinity for water. Hydrophobic molecules have few polar functional groups: most are hydrocarbons or have high hydrocarbon content.

Hydrophobic interaction chromatography (HIC): A technique in which reversed-phase packings are used to separate molecules by virtue of the interactions between their hydrophobic moieties and the hydrophobic sites on the surface. High salt concentrations are used in the mobile phase; separations are effected by changing the salt concentration. The technique is analogous to salting out molecules from solution. Gradients are run by decreasing the salt concentration over time. HIC is accomplished in high-salt medium by binding the hydrophobic portions of a protein to a slightly hydrophobic surface containing such entities as phenyl or short-chain hydrocarbons. The protein can be eluted in a decreasing salt gradient, with the most hydrophobic proteins eluting from the column last.

Immortalize: To alter cells (either chemically or genetically) so that they can reproduce indefinitely.

Immune response/reaction: Production of antibodies by the human body in reaction, for example, to viruses and substances recognized as foreign and possibly harmful.

Immunoassay: A qualitative or quantitative assay technique based on the measure of interaction of high affinity antibody with antigen used to identify and quantify proteins.

Immunoblotting: A technique for transferring antibody/antigen from a gel to a nitrocellulose filter on which they can be complexed with their complementary antigen/antibody.

Immunodiffusion (double, Ouchterlony technique): A technique in which an antigen and antibody are placed in two adjacent wells cut into a medium such as agar. As they diffuse through the medium, they form visible precipitation lines of antigen/antibody complexes at the point where the respective concentrations are at the optimum ratio for lattice formation.

Immunodiffusion (single): An identity diffusion technique whereby the product (antigen) is placed in a well cut into a medium such as agar containing its complementary antibody. The product diffuses into the medium forming a ring-shaped precipitate whose density is a function of antigen concentration.

Immunogenicity: Capability of a specific substance to induce the production of antibodies in the human body. The biological response to such a substance is termed an immune response or reaction.

Immunospecificity: A performance characteristic determined by conducting cross-reactivity studies with structurally similar substances that may be present in the analyte matrix. Specificity studies are determined with each new lot of polyclonal antibodies used in the immunoassay. For monoclonal antibody, each subsequent new lot is usually characterized by biochemical and biophysical techniques in lieu of comprehensive specificity studies.

Immunotoxin: Monoclonal antibodies coupled with toxins that are capable of delivering the toxin moiety to a target cell.

Impurity: Any component present in the drug substance or drug product that is not the desired product, a product-related substance, or an excipient including buffer components. It may be either process or product related.

In situ hybridization: Hybridization with an appropriate probe carried out directly on a chromosome preparation or histological section.

In vitro: An experiment performed in a test tube, petri dish, or other lab apparatus with parts of a living organism, such as testing a drug with tissue samples. From Latin, meaning "in glass." Performed in the laboratory rather than in a living organism (in vivo). Biological reactions taking place outside the body in an artificial system.

In vitro cell age: A measure of the period between thawing of the master cell bank vial(s) and harvest of the production vessel measured by elapsed chronological time in culture, population doubling level of the cells, or passage level of the cells when subcultivated by a defined procedure for dilution of the culture.

In vivo: An experiment performed using a living organism. From Latin, meaning "in live (subjects)." Biological reaction taking place inside a living cell or organism.

Inactivation: Reduction of virus infectivity caused by chemical or physical modification.

Inclusion bodies: Very high expression levels of heterologous proteins expressed in bacteria may lead to the formation of inclusion bodies. In such cases, the protein molecules clump together (aggregate) in the cytoplasm to create irregular organelle-like structures (about 1 μm in diameter). This presents a good-news–bad-news scenario: Dense inclusion bodies are easily separated from broken cells by centrifugation, thus facilitating product purification after the cells are homogenized.

But the aggregated, misfolded proteins are also insoluble, which can make further processing difficult. Organic solvents, detergents, or chaotropic substances can be used to denature those clumps and solubilize the proteins. But the next problem encountered will be that their 3D structure is almost always wrong by that point. To get correct, biologically active proteins, a renaturation step must follow: Inclusion bodies are dissolved using chaotropes and then diluted so the proteins can properly refold. A few bacterial expression options are available that avoid the inclusion body issue entirely. For example, some species secrete products rather than retaining them within cellular walls. Cultivation of *Escherichia coli* at lower temperatures (30°C rather than 37°C) sometimes prevents aggregation of heterologous proteins. Coexpression of chaperone proteins (or increased production of innate cofactors) that encourage proper folding of the biotherapeutic molecule may also help. And combining the gene for the protein of interest with one expressing a highly soluble native cytoplasmic protein (a fusion partner) may offer an answer. But the resulting fusion proteins must be chemically or enzymatically cleaved in downstream processing so the fusion partner can be purified away.

Inducer: A chemical or conditional change that activates the expression leading to the production of a desired product. A small molecule that interacts with a regulator protein and triggers gene transcription.

In-house working reference material: A material prepared similarly to the primary reference material that is established solely to assess and control subsequent lots for the individual attribute in question. It is always calibrated against the in-house primary reference material. Potency: the measure of the biological activity using a suitably quantitative biological assay (also called potency assay or bioassay), based on the attribute of the product that is linked to the relevant biological properties.

Inlet: The initial part of the column or a filtration device, where the solvent and sample enter. In case of a chromatographic column, there is usually an inlet frit that holds the packing in place and, in some cases, protects the packed bed.

Inlet pressure: The pressure driving a fluid into the feed port of a separation device.

Innovation: The introduction of new technologies or methodologies (ICH Q10).

Inoculate: To introduce cells into a culture medium.

Inoculum: Material (usually cells) introduced into a culture medium.

Installation qualification (IQ): The IQ provides a systematic method to check the system/equipment static attributes prior to normal operation. A detailed description of the system should be included as a part of IQ. This description includes all important major/minor components of the system. The availability of the applicable standard operating procedures must be verified. These include system/equipment operation, maintenance, cleaning, and/or sanitization.

Integration site: The site where one or more copies of the expression construct is integrated into the host cell genome.

Interchangeability: Refers to the medical/pharmaceutical practice of switching one medicine for another that is equivalent, in a given clinical setting. A product is considered to be interchangeable if it can be administered or dispensed instead of another clinically equivalent product.

Interferons: A cytokine that inhibits virus reproduction. Interferons (IFNs) also affect growth and development (differentiation) in certain normal and tumor cells. IFNs are the cytokines that can "interfere" with viral growth.

They also have the ability to inhibit proliferation and modulate immune responses.

Interleukins: Interleukins are the cytokines that act specifically as mediators between leucocytes. The following table shows the major source and effects of various types of interleukins.

Intermediary purification: Chromatographic purification step(s) following capture with the purpose of removing cell culture and process-related impurities.

Intermediate: For biotechnological/biological products, a material produced during a manufacturing process that is not the drug substance (DS) or the drug product (DP) but for which manufacture is critical to the successful production of the DS or the DP. Generally, an intermediate will be quantifiable, and specifications will be established to determine the successful completion of the manufacturing step before continuation of the manufacturing process. This includes material that may undergo further molecular modification or be held for an extended period before further processing.

Intermediate purification step: Further removal of bulk impurities with the main objectives on concentration and purification.

International nonproprietary name (INN): Scientific or generic name of an active substance. INNs for new active substances are allocated by the World Health Organization in Geneva. The INN is a unique and universally accessible name. For generic and biosimilar medicines cross-referring to originator products, it is the regulatory authority that decides whether the INN of the active substance as submitted for the generic or the biosimilar medicine is scientifically acceptable.

Interspersed repeats: Interspersed repeats are repeated DNA sequences located at dispersed regions in a genome. They are also known as mobile elements or transposable elements. A stretch of DNA sequence may be copied to a different location through DNA recombination. After many generations, such sequence (the repeat unit) could spread over various regions. In mammals, the most common mobile elements are LINEs and SINEs. LINEs stands for long interspersed nuclear elements. Its basic organization is shown as follows. All mobile elements contain direct repeats. The most common LINE in humans is the L1 family. A human genome contains about 60,000 to 100,000 L1 elements. "SINEs" stands for short interspersed nuclear elements. Its length is about 300 bp. In humans, the most abundant SINE is the Alu family. A human genome contains about 700,000–1,000,000 Alu sites. Although most LINEs and SINEs are located in extragenic regions, some of them are located in introns. For example, the human retinoblastoma gene is as long as 180 kb, consisting of 27 exons. Its introns contain many Alu and a few L1 elements.

Ion-exchange chromatography (IEC): A mode of chromatography in which ionic substances are separated on cationic or anionic sites of the packing. The sample ion (and usually a counterion) will exchange with ions already on the ionogenic group of the packing. Retention is based on the affinity of different ions for the site and on a number of other solution parameters (pH, ionic strength, counterion type, etc.). A gradient-driven separation based on the charge of the protein and its relative affinity for the chemical backbone of the column. Anion/cation exchange is commonly used for proteins. A mode of chromatography in which ionic substances are separated on cationic or anionic sites of the packing. Separation in IEC is based upon the selective, reversible adsorption of charged molecules to an immobilized ion-exchange group of the opposite charge. An ion exchanger consists of an insoluble porous matrix to which charged groups have been covalently bound.

Ionic strength: The weight concentration of ions in solution, computed by multiplying the concentration of each ion in solution (C) by the corresponding square of the charge on the ion (Z) summing this product for all ions in solution and dividing by 2.

Isocratic elution: Use of a constant-composition mobile phase in liquid chromatography.

Isoelectric focusing: An electrophoretic method that separates proteins by their pI (isoelectric point). They move through a pH gradient medium in an electric field until they are located at their isoelectric point where they carry no net charge. Prior to reaching their pI, protein mobility also depends upon size, conformation, steepness of pH gradient, and the voltage gradient. This method is used to detect incorrect or altered forms of a protein as well as protein impurities.

Isoelectric point: The isoelectric point is the pH of a solution or dispersion at which the net charge on the macromolecules or colloidal particles is zero.

kb: Kilobase.

Knowledge management: Systematic approach to acquiring, analyzing, storing, and disseminating information related to products, manufacturing processes, and components (ICH Q10).

lac Operon: The lac operon governs the production of enzymes for metabolizing lactose. In the absence of lactose, the repressor substance binds to the operator, inhibiting the production of three enzymes. Lactose, however, represses the repressor, allowing the enzymes to be produced. The lac operon of *Escherichia coli* consists of three genes: lacZ, lacY, and lacA, encoding β-galactosidase, lactose permease, and thiogalactoside transacetylase, respectively. Lactose permease is located on the cell membrane, capable of pumping lactose into the cell. β-galactosidase can convert lactose into glucose and galactose. Thiogalactoside transacetylase is responsible for degrading small molecules.

lac Repressor: In the absence of lactose, transcription of the lac operon is inhibited by the lac repressor. The lactose can bind to the lac repressor, preventing it from interacting with its DNA binding site. Hence, in a medium containing lactose, the lac operon is quickly transcribed, producing the enzymes to generate glucose, which is the major energy source for *Escherichia coli*.

Lambda (λ phages): Enterobacteria phage λ is a bacterial virus, or bacteriophage, which infects the bacterial species *Escherichia coli*. These phages are viruses that can infect bacteria. The major advantage of the λ phage vector is its high transformation efficiency, about 1000 times more efficient than the plasmid vector. The λ phages are commonly used in DNA cloning. They have two life cycles: lytic and lysogenic. In the lytic cycle, λ phages replicate rapidly and eventually cause lysis of the host cell. In the lysogenic cycle, the viral DNA circularizes and integrates into the host DNA. Then, λ phages may replicate with the host cell. Under certain conditions (e.g., ultraviolet irradiation of cells), the λ phages may transform from the lysogenic cycle to the lytic cycle. This transformation is mainly controlled by two proteins: cI (also known as λ repressor) and Cro. Increase in cI proteins promotes the lysogenic cycle, whereas increase in Cro proteins promotes the lytic cycle.

Leakage: Leakage occurs when resin-derived substance (ligands or other compounds) disintegrate from the matrix.

537

Library, genome or cDNA: Suppose you have known the partial sequence of a gene (e.g., from the sequence of a homologous gene) and want to determine its entire sequence, then you may use the technique described in this section. DNA library is a collection of cloned DNA fragments. There are two types of DNA libraries. The genomic library contains DNA fragments representing the entire genome of an organism. The cDNA library contains only complementary DNA molecules synthesized from mRNA molecules in a cell.

Life cycle: All phases in the life of a product from the initial development through marketing until the product's discontinuation (ICH Q8).

Ligase: Enzyme used to join DNA molecules. An enzyme that causes fragments of DNA or RNA to link together; used with restriction enzymes to create recombinant DNA.

Limulus amoebocyte lysate (LAL) test: A sensitive test for the presence of endotoxins using the ability of the endotoxin to cause a coagulation reaction in the blood of a horseshoe crab. The LAL test is easier, quicker, less costly, and much more sensitive than the rabbit test, but it can detect only endotoxins and not all types of pyrogens and must therefore be thoroughly validated before being used to replace the USP rabbit pyrogen test. Various forms of the LAL test include a gel clot test, a colorimetric test, a chromogenic test, and a turbidimetric test.

Linear velocity: The velocity of the mobile phase moving through the column, expressed in cm/h. Related to flow rate by the cross-sectional area of the column.

Locus: The site of a gene on a chromosome.

Lumen: The inner open space or cavity of a single hollow-fiber element that is used in the construction of hollow-fiber cartridges.

Lymphocytes: White blood cells that produce antibodies.

Lymphokines: Substances released predominantly from T lymphocytes after reaction with the specific antigen. Lymphokines are biologically highly active and will cause chemotaxis and activation of macrophages and other cell-mediated immune reactions. Gamma interferon is a lymphokine.

Lyophilization: Freeze-drying, used for the long-term preservation of microorganisms and some finished therapeutics.

Lysis: The process whereby a cell wall breakdown occurs releasing cellular content into the surrounding environment. Destruction of bacteria by infective phage.

Lysosomes: Cell organelles containing enzymes, responsible for degrading proteins and other materials ingested by the cell.

MAb: Monoclonal antibody; a highly specific, purified antibody that recognizes only a single antigen.

Macrokinetics: Movement of whole cells and their media within a bioreactor.

Macrovoid: A generally undesirable open space in a membrane filter that is appreciably larger than the average of the pore openings in a given filter. Macrovoids can lead to pinhole defects resulting in unwanted passage that directly impacts final product yield. Macrovoids can also impact the overall membrane strength and thus the device's ability to maintain integrity under pressure.

Manufacturing scale production: Manufacture at the scale typically encountered in a facility intended for product production for marketing.

Mass spectrometry: A technique useful in primary structure analysis by determining the molecular mass of peptides and small proteins. Often used with peptide mapping to identify variants in the peptide composition. Useful to locate disulfide bonds and to identify posttranslational modifications.

MCB (master cell bank): An aliquot of a single pool of cells that generally has been prepared from the selected cell clone under defined conditions, dispensed into multiple containers, and stored under defined conditions. The MCB is used to derive all working cell banks. The testing performed on a new MCB (from a previous initial cell clone, MCB or working cell bank) should be the same as for the MCB unless justified.

Media: A (usually sterile) preparation made for the growth, storage, maintenance, or transport of microorganisms or other cells.

Media exchange: A filtration step used to exchange one type of media for an alternative type of media during an aseptic cell culture separation.

Media migration: Media migration occurs when solid components of a filter (particles, adhesives, etc.) break free of the filter and enter the process solution.

Medium (media): The component of a separation device. For example, the chromatographic matrix in a chromatography column or the membrane in a membrane cassette.

Meiosis: Cell division in which the daughter cells have half the number of chromosomes as the parent cell.

Membrane: A thin layer of a highly engineered material with pores used to separate particles, biological matter, and molecules from a solution.

Membrane recovery: The degree to which the original performance of a membrane can be restored by cleaning.

Membrane test: A process, based on membrane bubble point characteristics, for testing the integrity of the membranes.

Messenger RNA (mRNA): RNA that serves as the template for protein synthesis; it carries the transcribed genetic code from the DNA to the protein synthesizing complex to direct protein synthesis.

Metabolites: Chemical byproducts of metabolism, the chemical process of life.

Metazoan: Organism of multicellular animal nature.

Microbiology: The study of microscopic life such as bacteria, viruses, and yeast.

Microcarrier: A microscopic particle (often, a 200 µm polymer bead) that supports cell attachment and growth in suspension culture.

Microencapsulated: Surrounded by a thin, protective layer of biodegradable substance referred to as a microsphere.

Microfiltration (MF): It usually refers to removing submicron-sized particles. MF is a pressure-driven membrane-based separations process in which particles and dissolved macromolecules larger than 0.1 µm are rejected (collected). The process removes particles, primarily from liquids, by passing the liquid sample through a microporous membrane.

Microheterogeneity: Slight differences in large, complex macromolecules that result in a population of closely related but not identical structures. Protein microheterogeneity can arise from many sources: genetic variants, proteolytic activity in cells, during translation into protein, during attachment of sugars, and during commercial production.

Microinjection: Manually using tiny needles to inject microscopic material (such as DNA) directly into cells or cell nuclei; computer screens provide a magnified view; a technique by which part of one cell is injected into another cell, as DNA into ova or other cells to create transgenic animals.

Microkinetics: Movement of chemicals into, out of, and within the cell.

Micron (micrometer, µm): One one-millionth of 1 m.

Microporous membrane: A thin, porous film or hollow fiber having pores ranging from 0.01 to 10 µm. Science and industry use microporous membranes to separate suspended matter from liquids.

Microtubules: Cellular organelles common in microorganisms: thin tubes that make structures involved in cellular movement.

Minimum exposure time: The shortest period for which a treatment step will be maintained.

Mitochondria (mitochondrion singular): Animal cell organelles that reproduce using their own DNA. They metabolize nutrients to provide the cell with energy and are believed to have once been symbiotic bacteria. Chloroplasts are their plant cell equivalents; cellular organelle responsible for oxidative metabolism and phosphorylation in eukaryotic cells, widely believed to have originated as a symbiotic bacterium.

Mobile phase: The solvent that moves the solute through the column.

Molecular weight cutoff (MWCO): The size designation in daltons for ultrafiltration membranes. The molecular weight of the globular protein that is 90% retained by the membrane. No industry standard exists; hence, the MWCO ratings of different manufacturers are not always comparable.

Monoclonal antibodies: Monospecific antibodies that are produced by a single clone of immune cells. They have become an important tool in molecular biology and medicine, and the basis of many biopharmaceuticals.

Monoclonal antibodies (MAb): Antibodies that are produced by a cellular clone and are all identical.

Motif: The motif is a characteristic domain structure consisting of two or more ☐ helices or ☐ strands. Common examples include coiled coil, helix–loop–helix, zinc finger, and leucine zipper. Many proteins also contain specific domains such as the Src homology 2 domain.

Multicellular organisms: Referring to organisms composed of more than one cell—often billions of them, arranged in various organs, tissues, and systems.

Multimerization of genes: Low yields often result due to the susceptibility of the peptides to proteolysis. Fusion peptide improves stability but often this is only a small portion of the fusion protein, still resulting in low yields of the target peptide. One way of increasing the molar ratio, and hence increasing the amount of peptide produced, is to produce a fusion protein with multiple copies of the target peptide. An additional beneficial effect is often obtained by this strategy, since the gene multimerization has also been shown to increase the proteolytic stability of the produced peptides. When the gene multimerization strategy is employed to increase the production yield, subsequent processing of the gene product to obtain the native peptide is needed. By flanking a peptide gene with codons encoding methionine, CNBr cleavage of the fusion protein, containing multiple repeats of the peptide, has successfully been used for obtaining native peptide at high yield. A pentapeptide multimerized to 3, 14, and 28 copies, fused to dihydrofolate reductase, is engineered to be separated by trypsin cleavage.

Mutagen: An agent (chemicals, radiation) that causes mutations in DNA.

Mutagenesis: The induction of genetic mutation by physical or chemical means to obtain a characteristic desired by researchers.

Mutation: A change in the genetic material, either of a single base pair (point mutation) or in the number or structure of the chromosomes. A permanent change in DNA sequence or chromosomal structure.

Mycoplasma: Parasitic microorganisms that infect mammals, possessing some characteristics of both bacteria and viruses.

Myeloma: Lymphocytic cancer; a malignancy normally found in bone marrow. Tumor cell line derived from a lymphocyte.

Nanofiltration: Separation processes targeted for solutes having molecular weights from 500 to 1000 Da.

Necrosis: Localized nonapoptotic death of cells and tissues.

NF-λB: This regulator of transcription consists of two subunits: p50 (green) and p65 (red). They belong to the Rel family.

Nick: A break in the sugar–phosphate backbone of a DNA or RNA strand.

Nick translation: In vitro method used to introduce radioactively labeled nucleotides into DNA.

NIH: National Institutes of Health.

NMDR: National Marrow Donor Registry.

Nominal filter rating: A rating that indicates the percentage of particles of a specific size or molecules of a specific molecular weight that will be removed by a filter. No industry standard exists; hence, the ratings from manufacturer to manufacturer are not always comparable.

Nominal molecular weight (MW) cutoff: In ultrafiltration, the MW size of a protein or other solute (in thousands of Daltons) that will be retained to 90% by the membrane.

Nonendogenous virus: Virus from external sources present in the master cell bank. See also virus.

Nonspecific model virus: A virus used for characterization of viral clearance of the process when the purpose is to characterize the capacity of the manufacturing process to remove and/or inactivate viruses in general, that is, to characterize the robustness of the purification process.

Normal flow filtration: Also called dead-ended filtration. In normal flow filtration, liquid flows perpendicular to the filter media, and all of the feed passes through.

Normalized water permeability: The water flux rate at 20°C. Flux (normalized to 20°C) = cassette flux measured temp. (°C) × 20°C ÷ measured temperature (°C).

Northern blot (Blotting): Technique for transferring RNA fragments from an agarose gel to a nitrocellulose filter on which they can be hybridized to a complementary DNA. Northern blotting is used for detecting RNA fragments, instead of DNA fragments. The technique is called "Northern" simply because it is similar to "Southern," not because it was invented by a person named "Northern." In the Southern blotting, DNA fragments are denatured with alkaline solution. In the Northern blotting, RNA fragments are treated with formaldehyde to ensure linear conformation.

Nuclear transfer: Moving a part or all of an organism's genetic information into an unfertilized egg (whose nucleus had previously been removed); can be used for cloning or to produce transgenic animals (if the genes put into the egg have been recombined with genes from others species).

Nuclear transport: After RNA molecules (mRNA, tRNA, and rRNA) are produced in the nucleus, they must be exported to the cytoplasm for protein synthesis. On the other hand, many proteins operating in the nucleus must be imported from the cytoplasm. The traffic through the nuclear envelope is mediated by a protein family that can be divided into exportins and importins. Binding of a molecule (a "cargo") to exportins facilitates its export to the cytoplasm. Importins facilitate import into the nucleus.

Nucleic acid chain: In a nucleic acid chain, two nucleotides are linked by a phosphodiester bond, which may be formed by the condensation reaction similar to the formation of the peptide bond.

Nucleic acids: DNA or RNA; long, chainlike molecules composed of nucleotides.

Nucleotides: Molecules composed of a nitrogen-rich base, phosphoric acid, and a sugar. The bases can be adenine (A), cytosine (C), guanine (G), thymine (T), or uracil (U). A nucleotide is composed of three parts: pentose, base, and phosphate group. In DNA or RNA, a pentose is associated with only one phosphate group, but a cellular free nucleotide (such as adenosine triphosphate [ATP]) may contain more than one phosphate group. If all phosphate groups are removed, a nucleotide becomes a nucleoside. In cells, a free nucleotide may contain one, two, or three phosphate groups. The energy carrier ATP has three phosphate groups; adenosine diphosphate has two; and adenosine monophosphate has one. If all phosphate groups are removed, a nucleotide becomes a nucleoside such as adenosine.

Nucleus: The largest organelle, a sphere that contains all the cell's genetic material and a nucleolus that builds ribosomes.

Oleophobic: Membranes that repel nonpolar fluids such as oil and lubricants.

Oligonucleotides: Short segments of DNA or RNA, that is, a chain of a few nucleotides.

Oncogene: A gene that, when expressed as a protein, can lead cells to become cancerous, usually by removing the normal constraints on its growth.

Operational qualification (OQ): The documented evidence that the system or equipment performs as intended throughout all anticipated operating ranges. The OQ protocol contains the procedures to verify specific dynamic attributes of a system or equipment throughout its operating range, which may include worst-case conditions. Applicable standard operating procedures and training procedures are documented in the appropriate protocol section. The executed OQ protocol verifies that the system or equipment performs as intended.

Operator gene: A gene that switches on adjacent structural gene(s).

Operon: A segment of DNA containing adjacent genes including structural genes and an operator gene and a regulatory gene. Complete unit of bacterial gene expression consisting of a regulator gene(s), control elements (promoter and operator), and adjacent structural gene(s).

Organelle: A structurally discrete component that performs a certain function inside a eukaryotic cell.

Organism: A single, autonomous living thing. Bacteria and yeasts are organisms; mammalian and insect cells used in culture are not.

Originator company: Company that was first to develop and produce a specific medicine (biopharmaceutical or pharmaceutical).

Originator reference medicinal product: Medicine that has been developed and produced by an originator company and that has been approved by the national regulatory authorities or the European Commission on the basis of a full registration dossier.

Outsourced activities: Activities conducted by a contract acceptor under a written agreement with a contract giver (ICH Q10).

Overload: In preparative chromatography, the overload condition is defined as the mass of sample injected onto the column at which efficiency and resolution begin to be affected if the sample size is further increased.

p53 (LocusLink): Involved in control of transcription, this is a tumor suppressor protein, also known as "guardian of the genome." It plays an important role in cell cycle control and apoptosis. Defective p53 could allow abnormal cells to proliferate, resulting in cancer. As many as 50% of all human tumors contain p53 mutants.

Packed-bed mode: A traditional chromatography mode, where the resin is confined between the bottom of the column and the flow adapter.

Parental cells: Cell to be manipulated to give rise to a cell substrate or an intermediate cell line. For microbial expression systems, it is typical to also describe the parental cells as the host cell. For hybridomas, it is typical to also describe the parental cells as the cells to be fused.

Particle size distribution: The distribution of particle sizes (number or weight fraction) in a fluid.

PAS: Prior approval supplement.

PASS: Postauthorization safety studies.

Patent: Set of exclusive rights granted to a company for a given period of time in exchange for the disclosure of its invention.

PBPC: Peripheral blood progenitor cell.

PCR: Polymerase chain reaction, a method of duplicating genes exponentially. See polymerase chain reaction.

Pd: Pharmacodynamic(s).

Peak shape: Describes the profile of a chromatographic peak. Theory assumes a Gaussian peak shape (perfectly symmetrical); peak asymmetry factor describes shape as a ratio.

Peptide bond: Chemical bond between the carboxyl (–COOH) group of one amino acid and the amino (–NH$_2$) group of another.

Peptide mapping: A powerful technique that involves the breakdown of proteins into peptides using highly specific enzymes. The enzymes cleave the proteins at predictable and reproducible amino acid sites, and the resultant peptides are separated via high-performance liquid chromatography or electrophoresis. A sample peptide map is compared to a map done on a reference sample as a confirmational step in the identity profiling of a product. It is also used for confirmation of disulfide bonds, location of carbohydrate attachment, sequence analysis, and identification of impurities and protein degradation.

Peptides: The peptide is a chain of amino acids linked together by peptide bonds. Polypeptides usually refer to long peptides, whereas oligopeptides are short peptides (<10 amino acids). Proteins are made up of one or more polypeptides with more than 50 amino acids. The primary structure of a protein refers to its amino acid sequence. The amino acid in a peptide is also called a residue. Proteins consist of fewer than 40 amino acids.

Performance indicators: Measurable values used to quantify quality objectives to reflect the performance of an organization, process, or system, also known as "performance metrics" in some regions (ICH Q10).

Performance qualification (PQ): The documented evidence that the system, equipment, or process is capable of consistently producing a safe product of high quality. The PQ protocol describes the procedures that verify the specific capabilities of a process equipment/system through the use of simulation material and/or actual product.

Permeate: Also called filtrate. The portion of a process fluid that passes through a membrane.

Pharmaceutical quality system: Management system to direct and control a pharmaceutical company with regard to quality (ICH Q10 based upon ISO 9000 2005).

Pharmaceuticals: Conventional or traditional chemical medicines.

Pharmacodynamic tests or studies: Study of the actions and effects of a medicine on living systems over a period of time.

Pharmacokinetic tests or studies: Studies to determine how medicines are absorbed, distributed, metabolized, and eliminated by the body.

Pharmacovigilance: Science and activities relating to the detection, assessment, understanding, and prevention of any adverse effects of medicinal products placed on the market.

Phase I clinical study or trial: Study with the objectives of determining how a medicine is handled by, and affects, humans and of helping to predict the initial dosage range for the medicine. Although such studies are often conducted in healthy volunteers, phase I studies in patients are also possible in some situations.

Phase II clinical study or trial: Study with the objectives of proving the efficacy concept of a medicine and of collecting data to establish the correct dose of that medicine. Phase II studies are not formally required for the development of biosimilar medicines as efficacy and the dose are already established for the reference product.

Phase III clinical study or trial: Study involving a larger group of patients, which aims to provide definitive evidence on whether the medicine can be considered both safe and effective in a real clinical setting. Comparability is the scientific evaluation of a comparison of two medicinal products to determine equivalence and any detectable differences at the level of quality, efficacy, and safety.

Phenotype: The part of an organism's genotype that is expressed and thus is generally apparent by observation.

Physicochemical characterization: Tests to determine the properties of a molecule or active substance, for example, molecular size/weight, chemical structure, and purity.

Pilot plant: A medium-scale bioprocessing facility used as an intermediate in scaling up processes from the laboratory to commercial production.

Pilot plant scale: The production of a recombinant protein by a procedure fully representative of and simulating that to be applied on a full commercial manufacturing scale. The methods of cell expansion, harvest, and product purification should be identical except for the scale of production.

Plaque: Clear area in a plated bacterial culture due to lysis by a phage.

Plasmid: A circular molecule of DNA that can replicate autonomously of other replicons and is commonly dispensable to the cell; used in genetic engineering. Hereditary material that is not part of a chromosome. Plasmids are extrachromosomal, circular, and self-replicating and found in the cytoplasm of cells (naturally in bacteria and some yeasts). They can be used as vectors (along with viruses) for introducing up to 10,000 base pairs of foreign DNA into recipient cells.

Plasmid cloning: Process by which a plasmid is used to import recombinant DNA into a host cell for cloning. In DNA cloning, a DNA fragment that contains a gene of interest is inserted into a cloning vector or plasmid.

Plasmid insertion: Plasmids are similar to viruses, but lack a protein coat and cannot move from cell to cell in the same fashion as a virus. Plasmid vectors are small circular molecules of double-stranded DNA derived from natural plasmids that occur in bacterial cells. A piece of DNA can be inserted into a plasmid if both the circular plasmid and the source of DNA have recognition sites for the same restriction endonuclease. The plasmid and the foreign DNA are cut by this restriction endonuclease (EcoRI in this example; see restriction enzymes also) producing intermediates with sticky and complementary ends. Those two intermediates recombine by base pairing and are linked by the action of DNA ligase. A new plasmid containing the foreign DNA as an insert is obtained. A few mismatches occur, producing an undesirable recombinant.

Plasmid plastid: Any of several types of cellular organelle found in plants and algae but not in animals or prokaryotes.

Plate number: Refers to theoretical plates in a packed column.

PLC: Programmable logic controller, a purpose-made device for industrial control. Microprocessors, now common in desktop computers, were originally devised in the 1970s for PLCs or for the types of operations common to PLCs (polling or checking sensors and activating/deactivating valves and switches compared against programmed presets or default levels).

Pleating: Folding filter media to increase the surface area that can be fitted into a given separation device

Point of breakthrough: The breakthrough volume is useful in determining the total sample capacity of the column for a particular solute.

Polishing step: Final removal of trace impurities to gain high level purity of end product.

Polyclonal: Derived from different types of cells.

Polyhedrin: Protein some viruses use to protect themselves from ultraviolet light.

Polymerase: An enzyme that catalyzes the production of nucleic acid molecules.

Polymerase chain reaction (PCR): In vitro technique for amplifying nucleic acid. The technique involves a series of repeated cycles of high-temperature denaturation, low-temperature oligonucleotide primer annealing, and intermediate-temperature chain extension. Nucleic acid can be amplified a millionfold after 25–30 cycles. PCR is a cell-free method of DNA cloning. It is much faster and more sensitive than cell-based cloning.

Polypeptides: Molecules made up of chains of amino acids, which may be pharmacologically active in the human body. They contain fewer amino acids and hence have lower molecular weights than proteins.

Pore: Small interconnecting passage through the membrane. The size and irregular path of a pore determines the removal rating of a membrane.

Pore size distribution: The range of pore sizes in a membrane. The tighter the pore size distribution, the better control one has over the filtration process.

Porosity: A measurement of the open space in a membrane. Also called open area or voids volume.

Postauthorization safety study: Any study with an authorized medicinal product conducted with the aim of identifying, characterizing, or quantifying a safety hazard, the safety profile of the medicinal product, or of measuring the effectiveness of risk management measures.

Posttranslational processing: Protein processing done by the Golgi bodies after proteins have been constructed by ribosomes. Enzymatic processing of a protein such as the addition of carbohydrate moieties or the removal of a signal sequence to direct a protein through a cell or organelle membrane. Endoplasmic reticula (ER) are either smooth or rough. These membrane-bound networks of branching, interconnected tubules are like little manufacturing plants inside the cells of most eukaryotes. The smooth ER synthesizes and metabolizes lipid molecules and helps detoxify cells. The rough ER is covered with ribosomes, which are the site of protein synthesis, where RNA from the cell nucleus is translated into amino acid sequences based on the genetic code. Bacteria have ribosomes, but they do not have ER. Yeasts have most of the same organelles as other eukaryotes (plants and animals), but they do not function quite the same way. In a eukaryotic cell, protein synthesis does not stop at the amino acid chain. Complex posttranslational modifications are performed by the Golgi apparatus in a way that has barely begun to be understood by cellular biologists. These stacked-membrane structures put the finishing touch on glycoproteins and other complex polypeptides. It is not well understood how they differentiate one from the other or how they recognize molecules.

Prefiltration: Removal of coarse particles/contaminants prior to final normally finer filtration.

Pressure: An increase in the force exerted on something above standard atmospheric conditions, measured as gauge pressure (psig, barg).

Pressure drop: The difference in pressure between two points.

Pressure, absolute: Gauge pressure plus 14.7 psi (1 bar).

Pressure–flow rate curve: To determine the optimal packing flow rate and pressure, a pressure versus flow rate curve for each lot of chromatography media is performed.

Pretreatment: The chemical or physical cleaning of a fluid prior to filtration or chromatography.

Preventive action: Action to eliminate the cause of a potential nonconformity or other undesirable potential situation. *Note*: Preventive action is taken to prevent occurrence, whereas corrective action is taken to prevent recurrence (ISO 9000 2005).

Prions: Resembling viruses, these pathogens are composed only of protein, with no detectable nucleic acid.

Probes: A probe is a piece of DNA or RNA used to detect specific nucleic acid sequences by hybridization (binding of two nucleic acid chains by base pairing). They are radioactively labeled so that the hybridized nucleic acid can be identified by autoradiography. The size of probes ranges from a few nucleotides to hundreds of kilobases. Long probes are usually made by cloning. Originally they may be double stranded, but the working probes must be single stranded. Short probes (oligonucleotide probes) can be made by chemical synthesis.

Process analytical technology: A system for designing, analyzing, and controlling manufacturing through timely measurements (i.e., during processing) of critical quality and performance attributes of raw and in-process materials and processes with the goal of ensuring final product quality.

Process characterization of viral clearance: Viral clearance studies in which nonspecific "model" viruses are used to assess the robustness of the manufacturing process to remove and/or inactivate viruses.

Process-related impurities: Impurities that are derived from the manufacturing process. They may be derived from cell substrates (e.g., host cell proteins, host cell DNA), cell culture (e.g., inducers, antibiotics, or media components), or downstream processing (e.g., processing reagents or column leachables).

Process-related impurities: Target protein derivatives (e.g., des-amido forms, oxidized forms, scrambled forms, di- and polymeric forms).

Process robustness: Ability of a process to tolerate variability of materials and changes of the process and equipment without negative impact on quality.

Process scale chromatography: The chromatographic procedure and equipment used in industrial production are referred to process scale.

Process validation: Establishing documented evidence that provides a high degree of assurance that a specific process will consistently produce a product meeting its predetermined specifications and quality attributes.

Product life cycle: All phases in the life of the product from the initial development through marketing until the product's discontinuation.

Product realization: Achievement of a product with the quality attributes appropriate to meet the needs of patients, health-care professionals, and regulatory authorities (including compliance with marketing authorization) and internal customers' requirements (ICH Q10).

Product-related impurities: Molecular variants of the desired product (e.g., precursors, certain degradation products arising during manufacture and/or storage) that do not have properties comparable to those of the desired product with respect to activity, efficacy, and safety.

Production cells: Cell substrate used to manufacture product.

Prokaryote: An organism (e.g., bacterium, virus, blue-green algae) whose DNA is not enclosed within a nuclear membrane or whose cell contains neither a membrane-bound nucleus nor other membrane-bound organelles such as mitochondria and plastids. Includes the "true bacteria" and the archaeans. Sometimes spelled procaryote.

Promoters: DNA sequence that initiates transcription of a gene to produce mRNA, used in genetic engineering to direct cells to manufacture a protein of interest. Promoter is the DNA region where the transcription initiation takes place. In prokaryotes, the sequence of a promoter is recognized by the sigma (□) factor of the RNA polymerase. In eukaryotes, it is recognized by specific transcription factors. *Escherichia coli* has five sigma factors: sigma 70 (regulates expression of most genes), sigma 32 (regulates expression of heat shock proteins), sigma 28 (regulates expression of flagellar operon [involved in cell motion]), sigma 38 (regulates gene expression against external stresses), and sigma 54 (regulates gene expression for nitrogen metabolism).

Prospective validation: Validation conducted prior to the distribution of either a new product, or product made under a revised manufacturing process, where the revisions may have affected the product's characteristics. It is also to ensure that the finished product meets all release requirements for functionality and safety.

Protease: Enzyme that speeds the breakdown of proteins into amino acids.

Protein: Macromolecules whose structures are coded in an organism's DNA. Each is a chain of more than 40 amino acids folded back upon itself in a particular way. A polypeptide consisting of amino acids. In their biologically active states, proteins function as catalysts in metabolism and, to some extent, as structural elements of cells and tissues. See enzymes, interleukin, interferon,

Protein passage: The passage of protein into the permeate stream.

Protein quantification: Quantitation of the total amount of protein can be done by a number of assays. There is no one method that is better than the rest; each has its own disadvantages ranging from the amount of protein required to do the test to a problem with variability between proteins. Some of the types include Lowry, bicinchoninic acid, Bradford, Biuret, Kjeldahl, ultraviolet spectroscopy.

Protein sorting: Proteins are synthesized on ribosomes that are located mainly in the cytosol. Only a small number of ribosomes are located in mitochondria and chloroplasts. Proteins synthesized on these ribosomes can be directly incorporated into the compartments within these organelles. However, most mitochondrial and chloroplast proteins are encoded by nuclear DNA and synthesized on cytosolic ribosomes. These and all other proteins synthesized in the cytosol must be transported to appropriate locations in the cell. This is made possible by the specific signal sequence in the newly synthesized peptide.

Protein synthesis: The process of protein synthesis goes through the following steps: Peptide synthesis always starts from methionine (Met). Therefore, the initial aminoacyl-tRNA is Met-tRNAiMet, where the subscript "i" specifies "initiation." In bacteria, the methionine of the initial aminoacyl-tRNA has been modified by the addition of a formyl group (HCO) to its amino group. The modified methionine is called formylmethionine (fMet), which is unique for bacteria. Thus, fMet is an obvious foreign substance in eukaryotes. It can elicit a strong immune response. In humans, the immune response elicited by the peptide "fMet-Leu-Phe" is about a thousand times greater than "Met-Leu-Phe."

Protein transport: A protein destined for the nucleus and/or cytoplasm contains a specific sequence that can be recognized directly by importin/exportin or through an adaptor protein.

Proteins: Large molecules made of amino acids arranged in chains, for example, erythropoietin.

Proteolytic: Capable of lysing (denaturing or breaking down) proteins.

Proteomics: Study of "proteome," which contains all proteins in a cell at a particular time. While all cells in an organism have the same genome, they usually have different proteomes.

Proven acceptable range: A characterized range of a process parameter for which operation within this range, while keeping other parameters constant, will result in producing a material meeting relevant quality criteria.

PSI: Pounds per square inch. A unit of pressure. 1 psi = 6.78 kPa.

PSUR: Periodic safety update reports.

Purification: Processes used to remove impurities (foreign or undesired materials) from a medicinal product.

Pyrogen: A substance (e.g., endotoxin) that produces a fever within a warm-blooded animal when injected into the bloodstream. Filtration materials of construction that come in contact with injectable liquids must meet pyrogenicity standards.

Pyrogenicity: The tendency for some bacterial cells or parts of cells to cause inflammatory reactions in the body, which may detract from their usefulness as pharmaceutical products.

QC–QA: Quality control–quality assurance.

Quality: The degree to which a set of inherent properties of a product, system, or process fulfills requirements (see ICH Q6A definition specifically for "quality" of drug substance and drug [medicinal] products).

Quality attribute: A molecular or product characteristic that is selected for its ability to help indicate the quality of the product. Collectively, the quality attributes define identity, purity, potency and stability of the product, and safety with respect to adventitious agents. Specifications measure a selected subset of the quality attributes.

Quality by design: A systematic approach to development that begins with predefined objectives and emphasizes product and process understanding and process control, based on sound science and quality risk management.

Quality manual: Document specifying the quality management system of an organization (ISO 9000 2005).

Quality objectives: A means to translate the quality policy and strategies into measurable activities (ICH Q10).

Quality planning: Part of quality management focused on setting quality objectives and specifying necessary operational processes and related resources to fulfill the quality objectives (ISO 9000 2005).

Quality policy: Overall intentions and direction of an organization related to quality as formally expressed by senior management (ISO 9000 2005).

Quality risk management: A systematic process for the assessment, control, communication, and review of risks to the quality of the drug (medicinal) product across the product life cycle.

Quality system: The sum of all aspects of a system that implements quality policy and ensures that quality objectives are met.

Quality target product profile: A prospective summary of the quality characteristics of a drug product (DP) that ideally will be achieved to ensure the desired quality, taking into account safety and efficacy of the DP.

Rabbit pyrogen test, USP: An assay for the presence of pyrogens (not restricted to endotoxins as is the limulus amoebocyte lysate test) involving the injection of the test material into rabbits that are well controlled and of known history. The rabbits are then monitored for a rise in temperature over a period of 3 h.

Radioimmunoassay (RIA): A generic term for immunoassays having a radioactive label (tag) on either the antigen or antibody. Common labels include I125 and H3 that are used for assay detection and quantitation. Classical RIAs are competitive binding assays where the antigen and tagged antigen compete for a limited fixed number of binding sites on the antibody. The antibody bound tagged complex is inversely proportional to the concentration of the antigen.

RCR: Replication-competent retrovirus.

RCV: Replication-competent virus.

rDNA: Recombinant DNA.

rDNA technology: The rDNA technology relates generally to the manipulation of genetic materials and, more particularly, to recombinant procedures making possible the production of polypeptides possessing part or all of the primary structural conformation and/or one or more of the biological properties of naturally occurring proteins such as erythropoietin.

Real-time release testing: The ability to evaluate and ensure the quality of in-process and/or final product based on process data, which typically include a valid combination of measured material attributes and process controls.

Recirculation rate: Same as retentate flow rate.

Recombinant: Containing genetic material from another organism. Genetically altered microorganisms are usually referred to as recombinant, whereas plants and animals so modified are called transgenic (see transgenics).

Recombinant DNA: DNA that contains genes from different sources that have been combined by methods of genetic engineering as opposed to traditional breeding experiments.

Recombinant protein production: Many proteins that may be used for medical treatment or for research are normally expressed at very low concentrations. Through recombinant DNA technology, a large quantity of proteins can be produced. This involves the cloning of the gene encoding the desired protein into an "expression vector" that must contain a promoter so that the protein can be expressed.

Recovery: The amount of solute (sample) that elutes from a chromatography column or can be collected in the retentate or permeate solution of a filtration device relative to the amount applied.

Reequilibration: The equilibration phase after a chromatographic run.

Reference standards: International or national standards.

Regeneration: Returning the packing in the column to its initial state after gradient elution. Mobile phase is passed through the column stepwise or in a gradient. The stationary phase is solvated to its original condition. In ion-exchange chromatography, regeneration involves replacing ions taken up in the exchange process with the original ions that occupied the exchange sites. Regeneration can also refer to bringing back any column to its original state (e.g., the removal of impurities with a strong solvent).

Relevant genotypic and phenotypic markers: Those markers permitting the identification of the strain of the cell line that should include the expression of the recombinant protein or presence of the expression construct.

Relevant virus: Virus used in process evaluation studies that is either the identified virus or of the same species as the virus that is known or likely to contaminate the cell substrate or any other reagents or materials used in the production process.

Requirements: The explicit or implicit needs or expectations of the patients or their surrogates (e.g., health-care professionals, regulators, and legislators). In this document, "requirements" refers not only to statutory, legislative, or regulatory requirements but also to such needs and expectations.

Residence time: The time required for an incremental unit of feed solution to pass through a separations device.

Resolution (R_s): The resolution is defined as the distance between chromatographic peak maxima compared with the average base width of the two peaks. The resolution is a measure of the relative separation between two peaks. Ability of a chromatography media to separate chromatographic peaks. It is usually expressed in terms of the separation of two peaks. Resolution can be calculated as follows: $R_s = 1.18 \times (t_{R2} - t_{R1})/(w_{h1} + w_{h2})$, where w_{h1} is the peak width at half height (in units of time) of the first peak, w_{h2} is the peak width at half height (in units of time) of the second peak, and t_{R1} and t_{R2} refer to the retention times of the first and the second peak, respectively. A value of 1 is considered to be the minimum for a measurable separation to occur and to allow good quantification. Values of 1.7 or larger are generally desirable for rugged methods. Resolution is more important in analytical techniques than in preparatory techniques.

Response elements: Response elements are the recognition sites of certain transcription factors. Most of them are located within 1 kb from the transcriptional start site.

Restriction enzyme: Bacterial enzyme that cuts DNA molecules at the location of particular sequences of base pairs. The role of these enzymes in bacteria is to "restrict" the invasion of foreign DNA by cutting it into pieces. Hence, these enzymes are known as restriction enzymes. Restriction enzymes, also called restriction nucleases (e.g., EcoRI from *Escherichia coli*), surround the DNA molecule at the point it seeks (sequence GAATTC). It cuts one strand of the DNA double helix at one point and the second strand at a different, complementary point (between the G and the A base). The separated pieces have single-stranded "sticky ends," which allow the complementary pieces to combine.

Restriction map: Linear arrangement of various restriction enzyme sites.

Restriction site: Base sequence recognized by an enzyme.

Retentate: The portion of the feed solution that does not pass through a cross-flow membrane filter.

Retention: The ability of a separation device to retain particles of a given size.

Retention time: The time between injection and the appearance of the peak maximum.

Retention volume: The volume of mobile phase required to elute a substance from the column. $V_R = V_m - KD\, V_s$, where V_m is the void volume, KD is the distribution coefficient, and V_s is the stationary phase volume.

Retrospective validation: Validation of a process for a product already in distribution based upon establishing documented evidence. The review and analysis of historical manufacturing and product testing data that verify a specific process can be consistently produced meeting its predetermined specifications and quality attributes.

Retrovirus: RNA virus that replicates via conversion into a DNA duplex.

Reverse osmosis: Type of cross-flow filtration used for removal of very small solutes (<1000 Da) and salts. It uses a semipermeable membrane under high pressure to separate water from ionic materials. High pressure is necessary to overcome the natural osmotic pressure created by the concentration gradient across the membrane.

Reverse transcriptase: An enzyme that catalyzes the synthesis of DNA from RNA.

Reversed-phase chromatography (RPC): A chromatographical separation method based on a column stationary phase coated to give nonpolar hydrophobic surface. Analyte retention is proportional to hydrophobic reactions between solute and surface. Retention is roughly proportional to the length of the bonded carbon chain. RPC is in theory closely related to hydrophobic interaction chromatography. This is the most common high-performance liquid chromatography mode. Mobile phase is usually water and a water-miscible organic solvent such as methanol or acetonitrile. There are many variations of RPC in which various mobile phase additives are used to gain a different selectivity.

Ribosome: Cell organelles that translate RNA to build proteins.

Risk: The combination of the probability of occurrence of harm and the severity of that harm (ISO/IEC Guide 51).

Risk acceptance: The decision to accept risk (ISO Guide 73).

Risk analysis: The estimation of the risk associated with the identified hazards.

Risk assessment: A systematic process of organizing information to support a risk decision to be made within a risk management process. It consists of the identification of hazards and the analysis and evaluation of risks associated with exposure to those hazards.

Risk communication: The sharing of information about risk and risk management between the decision maker and other stakeholders.

Risk control: Actions implementing risk management decisions (ISO Guide 73).

Risk evaluation: The comparison of the estimated risk to given risk criteria using a quantitative or qualitative scale to determine the significance of the risk.

Risk identification: The systematic use of information to identify potential sources of harm (hazards) referring to the risk question or problem description.

Risk management: The systematic application of quality management policies, procedures, and practices to the tasks of assessing, controlling, communicating, and reviewing risk.

Risk reduction: Actions taken to lessen the probability of occurrence of harm and the severity of that harm.

Risk review: Review or monitoring of output/results of the risk management process considering (if appropriate) new knowledge and experience about the risk.

RNA: Ribonucleic acid.

RNA polymerase: An enzyme that catalyzes the synthesis of RNA in transcription.

RNA processing: RNA processing is to generate a mature mRNA (for protein genes) or a functional tRNA or rRNA from the primary transcript.

Robustness: The robustness of a procedure is a measure of its capacity to remain unaffected by small, but deliberate variations in method parameters and provides an indication of its reliability during normal usage.

Roller bottle: A container with large growth surfaces in which cells can be grown in a confluent monolayer. The bottles are rotated or agitated to keep cells in suspension, but they require extensive handling, labor, and media. In large-scale vaccine production, roller bottles have been replaced by microcarrier culture systems that offer the advantage of scale-up.

Running buffer: The solution used to perform a chromatographic run.

Sanitization: A cleaning process that destroys most living (pathogenic) microorganisms.

Sanitizing agent: An agent introduced into a system to kill organisms and prevent the growth of organisms.

Scale-up: To take a biopharmaceutical manufacturing process from the laboratory scale to a scale at which it is commercially feasible.

Seed stock: The initial inoculum or the cells placed in growth medium from which other cells will grow.

Selectivity: Same as separation factor or relative retention ratio. The separation factor is a measure of the time or distance between the maxima of two peaks. If $a = 1$, then the peaks have the same retention and coelute. $a = k_2/k_1$, where k_1 is the retention factor of the first peak and k_2 is the retention factor of the second peak.

Senior management: Person(s) who direct and control a company or site at the highest levels with the authority and responsibility to mobilize resources within the company or site (ICH Q10 based in part on ISO 9000 2005).

Separation: During operation, the separation device divides a liquid or gas feed stream into separate components.

Sequence: The precise order of bases in a nucleic acid or amino acids in a protein.

Serum: The watery portion of an animal or plant fluid (such as blood) remaining after coagulation. When cheese is made, whey is the milk serum that is left.

Severity: A measure of the possible consequences of a hazard.

Shear rate: A ratio of velocity and distance expressed in units of s^{-1}. The shear rate for a hollow-fiber cartridge is based on the flow rate through the fiber lumen and can be calculated as follows: $g = 4q/\pi r^3$, where g is the shear rate, s^{-1}; q is the flow rate through the fiber lumen, cm^3/s; and r is the fiber radius, cm.

Sieving: Removal of particles from a feed stream as a result of entrapment within the depth of the membrane pore structure.

Silencers: Silencers are the DNA elements that interact with repressors (proteins) to inhibit transcription. In prokaryotes, silencers are known as operators, found in many genes such as lac operon and trp operon. In a few cases, a DNA element may act either as enhancer or silencer, depending on the binding protein. For example, certain genes contain an element called E box (consensus CACGTG) that can bind either Max/Myc dimer or Max/Mad dimer. The Max/Myc dimer activates transcription, whereas the Max/Mad dimer suppresses transcription of these genes.

SIP: Steam in place or sterilize in place.

Size-exclusion membrane separation: Mechanism for removing particles from a feed stream. Based strictly on the size of the particles versus the pore size that the feed stream is being filtered through. Retained particles are held back because they are larger than the pore opening.

Size-exclusion chromatography: A chromatographic technique in which analytes are excluded from the stationary phase, and thus separated, based on their size.

Slurry: A thick mixture of adsorbent and solvent used to pour columns. Excess solvent is drained out as the adsorbent settles and more slurry is added until the column is filled.

Sodium dodecyl sulfate polyacrylamide gel electrophoresis (SDS-PAGE): An electrophoretic separation of proteins based on their molecular weights. A uniform net negative charge is imposed on the molecules by the addition of SDS. Under these conditions, migration toward the anode through a gel matrix allows separation via size, not charge, with the smaller molecules migrating the longest distance. This technique is not reliable for sizes below a MW of ca. 8000. Proteins are observed via Coomassie blue or silver staining or can be further transferred to membranes for antigen/antibody specificity testing.

Solute: An ionic or organic compound dissolved in a solvent, for example, the sugar in a cup of coffee is a solute.

Somatic cell: In higher organisms, a cell that (unlike germ cells) carries the full genetic makeup of an organism.

SOP: Standard operating procedure.

Southern blot (blotting): Technique for transferring DNA fragments from an agarose gel to a nitrocellulose filter on which they can be hybridized to a complementary DNA. Southern blotting is a technique for detecting specific DNA fragments in a complex mixture. The technique was invented in the mid-1970s by Edward Southern. It has been applied to detect restriction fragment length polymorphism and variable number of tandem repeat polymorphism. The latter is the basis of DNA fingerprinting.

Sparge: To spray. A sparger is the component of a fermenter that sprays air into the broth.

Specific model virus: Virus that is closely related to the known or suspected virus (same genus or family), having similar physical and chemical properties to those of the observed or suspected virus.

Specification: A list of tests, references to analytical procedures, and appropriate acceptance criteria that are numerical limits, ranges, or other criteria for the tests described. It establishes the set of criteria to which a drug substance (DS), drug product (DP), or materials at other stages of its manufacture should conform to be considered acceptable for its intended use. Conformance to specification means that the DS and DP, when tested according to the listed analytical procedures, will meet the acceptance criteria. Specifications are critical quality standards that are proposed and justified by the manufacturer and approved by regulatory authorities as conditions of approval.

Spiking: Adding a known amount of substance being measured to a sample in order to determine the original sample concentration by the known addition technique or to determine the accuracy of a direct measurement technique.

Stakeholder: Any individual, group, or organization that can affect, be affected by, or perceive itself to be affected by a risk. Decision makers might also be stakeholders. For the purposes of this guideline, the primary stakeholders are the patient, health-care professional, regulatory authority, and industry.

Starling flow: A portion of filtrate (permeate) that is driven back through the membrane in the reverse direction near the outlet of the cartridge, due to the high permeability of these membranes in the presence of permeate pressure. This phenomenon is most often associated with the operation of microfiltration membranes using permeate flow control.

State of control: A condition in which the set of controls consistently provides assurance of continued process performance and product quality (ICH Q10).

Steam-in-place: The process of sterilizing a tank or process device, such as a hollow-fiber cartridge, with steam, without removing the device from the separation system.

Stepwise elution: Use of eluents of different compositions during the chromatographic run. These eluents are added in a stepwise manner.

Sterilization: A process that removes/destroys all (pathogenic) microorganisms from a solution or a solution processing system.

STN: Submission tracking number.

Strain: A population of cells all descended from a single cell. A group of organisms of the same species having distinctive characteristics, but not usually considered a separate breed or variety.

Substitution/substitutability: (i.e., the ability to substitute) Refers to the practice of dispensing one medicine instead of another equivalent and interchangeable medicine at pharmacy level and without requiring consultation with the prescriber. The term "automatic substitution" refers to the practice whereby the pharmacist is obliged to dispense one medicine instead of another equivalent and interchangeable medicine due to national or local requirements.

Substrate: Reactive material, the substance on which an enzyme acts.

Substratum: The solid surface of which a cell moves or on which cells grow.

Supernatant: Material floating on the surface of a liquid mixture (often the liquid component that has the lowest density).

Surface area: In an adsorbent, refers to the total area of the solid surface as determined by an accepted measurement technique such as the Brunaeur–Emmett–Teller (BET) method using nitrogen adsorption.

Surface filter: A filter in which particles larger that the pores are retained on the surface of the filter.

Surfactant: Any substance that changes the nature of a surface, such as lowering the surface tension of water.

Suspension: Particles floating in (not necessarily on) a liquid medium, or the mix of particles and liquid itself.

Swelling, shrinking: Process in which chromatographic resins increase or decrease their volume because of their solvent environment. Swelling is dependent upon the degree of cross-linking; low-cross-linking resins will swell and shrink more than highly cross-linked resins. If swelling occurs in a packed column blockage, increased back pressure can occur, and column efficiency can be affected.

Symbiotic: Living together for mutual benefit.

Synthesis: Creating products through chemical and enzymatic reactions.

T-helper cells: T lymphocytes with the specific capacity to help other cells, such as B lymphocytes, to make antibodies. T-helper cells are also required for the induction of other T-lymphocyte activities. Synonym is T inducer cell, T4 cell, or CD 4 lymphocyte.

T-suppressor cells: T lymphocytes with specific capacity to inhibit T-helper cell function.

Tailing: The phenomenon in which the normal Gaussian peak has an asymmetry factor >1. The peak will have skew in trailing edge. Tailing is caused by sites on the packing that have a stronger-than-normal retention for the solute.

Tandem repeats: Tandem repeats are an array of consecutive repeats of DNA sequence. They include three subclasses: satellites, minisatellites, and microsatellites. The name "satellites" comes from their optical spectra. The size of a satellite DNA ranges from 100 kb to over 1 Mb. In humans, a well-known example is the alphoid DNA located at the centromere of all chromosomes. Its repeat unit is 171 bp, and the repetitive region accounts for 3%–5% of the DNA in each chromosome. Other satellites have a shorter repeat unit. Most satellites in humans or in other organisms are located at the centromere.

Tangential flow filtration: Also called cross-flow filtration. In tangential flow filtration, the feed solution flows parallel to the surface of the membrane. Driven by pressure, some of the feed solution passes through the membrane filter. Most of the solution is circulated back to the feed tank. The movement of the feed solution across the face of the membrane surface helps to remove the buildup of foulants on the surface.

TCID50: Tissue culture infectious dose, 50%.

Telomerase: In eukaryotes, the chromosome ends are called telomeres that have at least two functions, to protect chromosomes from fusing with each other and to solve the end-replication problem. In the absence of telomerase, the telomere will become shorter after each cell division. When it reaches a certain length, the cell may cease to divide and die. Therefore, telomerase plays a critical role in the aging process.

Termination: Protein synthesis will terminate when the ribosome arrives at one of three stop codons. The termination process is assisted by special proteins called termination factors that recognize the stop codons. Their association stimulates the release of the peptidyl-tRNA from the ribosome. Subsequently, the released peptidyl-tRNA divides into tRNA and a newly synthesized peptide chain. The ribosome also divides into the large and small subunits, ready for synthesizing another peptide.

Theoretical plate (N): Column efficiency is expressed by the number of theoretical plates (N). The number of theoretical plates can be calculated using the following equation. Theoretical plate is a concept, and a column does not contain anything resembling physical distillation plates or any other similar feature. Theoretical plate numbers are an indirect measure of peak width for a peak at a specific retention time. Columns with high plate numbers are considered to be more efficient (i.e., higher column efficiency) than columns with lower plate numbers. A column with a high number of plates will have a narrower peak at a given retention time than a column with a lower number of plates. Length of column relating to this concept is called height equivalent to a theoretical plate. $N = 5.55 \times (t_R/w_{1/2})2$ or $N = 16 \times (t_R/w_b)2$, where t_R is the retention time and $w_{1/2}$ and w_b are the width of the peak at half the peak height ($h_p/2$).

Thermal stability: The ability of a membrane and filtering device to maintain its performance during and after exposure to elevated temperatures; for example, elevated temperature experienced during high-temperature processing or steam sterilization.

Three-dimensional structure: The 3D structure is also called the tertiary structure. If a protein molecule consists of more than one polypeptide, it also has the quaternary structure, which specifies the relative positions among the polypeptides (subunits) in a protein.

Throughput: The volume of solution that will pass through a separation device before the filtrate output drops to an unacceptable level. It is also the rate at which a separation system will generate filtrate.

Tissue culture: Growing plant or animal tissues outside of the body, as in a nutrient medium in a laboratory; similar to cell culture, but cells are maintained in their structured, tissue form.

Titer: A measured sample or to draw a measured, representative sample from a larger amount is to titrate.

Titer reduction: The measurement of a filter's ability to remove microbes or virus from a fluid.

Topoisomers: During replication, the unwinding of DNA may cause the formation of tangling structures, such as supercoils or catenanes. The major role of topoisomerases is to prevent DNA tangling.

Transcription: The first stage in the expression of a gene by means of genetic information being transmitted from the DNA in the chromosomes to messenger RNA. Transcription is a process in which one DNA strand is used as template to synthesize a complementary RNA. In eukaryotes, there are three classes of RNA polymerases: I, II, and III. In prokaryotes, binding of the polymerase's σ factor to promoter can catalyze unwinding of the DNA double helix. The most important σ factor is sigma 70, whose structure has been determined by x-ray crystallography, but its complex with DNA has not been solved. After the DNA strands have been separated at the promoter region, the core polymerase ($\alpha\alpha\beta\beta'$) can then start to synthesize RNA based on the sequence of the DNA template strand.

Transduction: Transfer of genes from one bacterium to another using a phage (virus).

Transfection: Permanently changing a cell using viral DNA.

Transformation: Permanent genetic change following incorporation of new DNA.

Transgene: A foreign gene incorporated by transformation into the germ line.

Transgenics: The alteration of plant or animal DNA so that it contains a gene from another organism. There are two types of cells in animals and plants, germ line cells (the sperm and egg in animals, pollen and ovule in plants) and somatic cells (all of the other cells). It is the germ line DNA that is altered in transgenic animals and plants, so those alterations are passed on to offspring. Transgenic animals are used to produce therapeutics, to study disease, or to improve livestock strains. Transgenic plants have been created for increased resistance to disease and insects as well as to make biopharmaceuticals.

Translation: The process by which information transferred from DNA by RNA specifies the sequence of amino acids in a polypeptide (protein) chain.

Transmembrane pressure (TMP): The force that drives liquid flow through a cross-flow membrane. During filtration, the feed side of the membrane is under higher pressure than the permeate side. The pressure difference forces liquid through the membrane. TMP = {(feed pressure + retentate pressure)/2} − permeate pressure.

Transplastomic: Transformation of plastids.

Trend: A statistical term referring to the direction or rate of change of a variable(s).

tRNA translation: Translation is a process by which the nucleotide sequence of mRNA is converted into the amino acid sequence of a peptide. It starts from the initiation codon and then follows the mRNA sequence in a strictly "three nucleotides for one amino acid" manner.

Trypsin: Trypsin allows the growth of cells as independent microorganisms distinct from tissue culture by causing cell disaggregation. Excised tissue is softened and treated with a proteolytic enzyme, normally trypsin, then washed, and suspended in a growth medium to produce a primary culture. Subculturing from the primary culture usually involves treatment with an antitrypsin (such as serum) to produce a secondary culture. Cell lines are established by repeated culture through cycles of growth, trypsinization, and subculture. Trypsin is also used to remove anchorage-dependent cells from their attached substratum.

Tryptic fragment analysis: Quantitating the resultant fragments caused by tryptic digestion.

Tubule: Tubelike structure (larger internal diameter [ID] fibers than hollow fibers) made from ultrafiltration or microfiltration membrane and sealed inside a cross-flow cartridge. When in use, the feed stream flows into one end of the tubule, and the retentate (the material that does not permeate through the walls of the tubule) flows out the other end. The material that does flow through the membrane (walls of the tubule) is called the permeate.

Tumor necrosis factors (TNFs): TNFs are the cytokines produced mainly by macrophages and T lymphocytes that help regulate the immune response and hematopoiesis (blood cell formation). There are two types of TNF:

TNFα (also called cachectin, produced by macrophages) and TNFβ (also called lymphotoxin, produced by activated CD4+ T cells).

Turbidity: The measure of relative sample clarity of a liquid. Measurements are based on the amount of light transmitted in straight lines through a sample. The more light that is scattered by fine solids or colloids, the less clear (and more turbid) the solution. Often reported in NTU (nephelometric turbidity unit).

Turbidostat: A variation on a chemostat. Whereas a chemostat is designed for constant input of medium, a turbidostat is designed to keep the organisms at a constant concentration. A turbidity sensor measures the concentration of organisms in the culture and adds additional medium when a preset value is exceeded.

Turbulent flow field: The state that results from mixing the contents of a fermenter or bioreactor to provide oxygen to the cells. That must be balanced against the shear that causes cell damage and death.

Two-dimensional gel electrophoresis: A type of electrophoresis in which proteins are separated first in one direction by charge followed by a size separation in the perpendicular direction.

Ultrafiltration (UF): The separation of macrosolutes based on their molecular weight or size. UF is a pressure-driven, convective process using semi-permeable membranes to separate macrosolutes based on their molecular weight or size. By removing solvent from solution, solute is concentrated or enriched. UF membranes may also be used for diafiltration to remove salts or other microspecies from solution via repeated or continuous dilution and reconcentration.

Unicellular: Composed of only a single cell.

Unprocessed bulk: One or multiple pooled harvests of cells and culture media. When cells are not readily accessible, the unprocessed bulk would constitute fluid harvested from the fermenter.

Upstream: The feed side of a separation process. Microbial fermentation, insect cell culture mammalian cell culture, animal care, or plant cultivation.

Upstream processing: Cellular separations including cell lysates, cell harvesting, clarification, and cell culture perfusion.

Ultraviolet spectroscopy: A quantitation technique for proteins using their distinctive absorption spectra due to the presence of side-chain chromophores (phenylalanine, tryptophan, and tyrosine). Since this absorbance is linear, highly purified proteins can be quantitated by calculations using their molar extinction coefficient.

Vaccines: Preparations of antigens from killed or modified organisms that elicit immune response (production of antibodies) to protect a person or animal from the disease-causing agent.

Vacuolation: In cell and tissue culture, excess fluid, debris (aggregates), or gas (from sparging) can form inside a cell vacuole. A vacuole is a cavity within the cell that can be relatively clear and fluid filled, gas filled (as in a number of blue-green algae), or food filled (as in protozoa).

Validation: Validation is establishing documented evidence that provides a high degree of assurance that a specific system (an interacting or interdependent group of items that function together to achieve a specific function), process, or facility will consistently produce a product meeting its predetermined specifications and quality attributes. Validation can be subdivided into three activities; installation, operational, and performance qualifications.

Validation change control: A formal monitoring system by which qualified representatives review proposed or actual changes that might affect validated status and take preventive or corrective action to ensure that the system retains its validated state of control.

Validation protocol: A validation protocol is a documented set of instructions designed to confirm specific static and/or dynamic attributes of the installation, operation, or performance of a utility/system, equipment, or process.

Vector: The plasmid, virus, or other vehicle used to carry a DNA sequence into the cell of another species.

Vessel jacket: A temperature control method consisting of a double wall outside the main vessel wall. Liquid or steam flows through the jacket to heat (or cool) the fluid in the vessel. Because biopharmaceutical products are so sensitive and vessel jackets can cause uneven heating (hot or cold spots), shell-and-tube or plate-and-frame heat exchangers are more common in biopharmaceutical production systems.

Viability: Life and health, ability to grow and reproduce; a measure of the proportion of live cells in a population.

Viral clearance: The removal of viral contamination using specialized membranes or chromatography.

Virus: The simplest form of life; RNA or DNA wrapped in a shell of protein, sometimes with a means of injecting that genetic material into a host organism (infection). Viruses cannot reproduce on their own, but require the aid of a host.

Virus removal: Physical separation of virus particles from the intended product.

Viruslike particles: Structures visible by electron microscopy that morphologically appear to be related to known viruses.

Viscosity: A measurement of a fluid's resistance to shear. A slow-flowing liquid such as gear oil has a higher viscosity than a free-flowing liquid such as mineral spirits. In a given separation process, higher-viscosity, Newtonian fluids have a lower flow rate through a cartridge than do lower-viscosity fluids.

Viscosity: Thickness of a liquid; determines its internal resistance to shear forces.

Void time: The time for elution of an unretained peak (t_m or t_0).

Void volume (V_i): The total volume of mobile phase in the column (the remainder of the column is taken up by packing material). It can be determined by injecting an unretained substance that measures void volume plus extra column volume. Also referred to as interstitial volume. Instead of V_i, V_0 or V_m are sometimes used as symbols. For example, for nonrigid gels like Superose™, Sephacryl™, and other gel filtration gels, one can estimate the void volume of a column to be approximately 30% of the total bed volume. Also, it is the amount of open space within membrane filter media.

Volumetric flow rate: Units for measuring quantities of a substance by the volume they take up.

Wall effect: The consequence of the looser packing density near the walls of the rigid column. Mobile phase has a tendency to flow slightly faster near the wall because of the decreased permeability. The solute molecules that happen to be near the wall are carried along faster than the average of the solute band and, consequently, band spreading results.

Water flux: Measurement of the amount of water that flows through a cartridge. Clean water flux refers to the flux measurement made under standardized conditions on a new (and cleaned) membrane cartridge.

Water for injection: Very pure water suitable for medical uses.

Western blot (blotting): This test is used to detect contaminating cell substrates and to evaluate recombinant polypeptides. After electrophoretic separation, the negatively charged proteins (the antigens) are electrophoretically transferred from the polyacrylamide gel onto a nitrocellulose membrane positioned on the anode side of the gel. Following incubation of the membrane with a specific antibody, they are labeled with another anti-antibody for detection. Western blotting is used to detect a particular protein in a mixture. The probe used is therefore not DNA or RNA, but antibodies. The technique is also called "immunoblotting."

Wetting: The process of filling pores of a hydrophobic membrane with water. Typical methods include use of alcohol as a wetting solution or high pressure to drive air out.

WHO: World Health Organization.

Wobble pairing: See tRNA translation.

Working cell bank (WCB): The WCB is prepared from aliquots of a homogeneous suspension of cells obtained from culturing the master cell bank under defined culture conditions. A working cell bank is created from the master cell bank by reviving the live cells: thawing and then culturing them on agar medium. From a frozen bank, it may take 2 days to grow enough new cells to begin fermentation. From a lyophilized bank, it can take longer. Refrigerated cultures need only a day or so.

Worst case: A set of conditions encompassing upper and lower processing limits and circumstances, including those within standard operating procedures, which pose the greatest chance of process or product failure, when compared to ideal conditions. Such conditions do not necessarily induce product or process failure.

Yeast: A single-celled fungus.

Yeast artificial chromosome (YAC): The YAC vector is capable of carrying a large DNA fragment (up to 2 Mb), but its transformation efficiency is very low. A vector constructed from the telomeric, centromeric, and replication origin sequences needed for replication in yeast cells used to clone pieces of DNA

Yield: The amount of target molecules that can be recovered from cross-flow filtration and/or chromatography. Also called recovery.

Index